The Encyclopedia of Evolution

HUMANITY'S SEARCH FOR ITS ORIGINS

Richard Milner

Foreword by Stephen Jay Gould

An Owl Book

Henry Holt and Company
New York

for

MRS. THELMA CARR MILNER

who gave me life, twice

Henry Holt and Company, Inc.
Publishers since 1866
115 West 18th Street
New York, New York 10011

Henry Holt® is a registered trademark
of Henry Holt and Company, Inc.

Library of Congress Cataloging-in-Publication Data
Milner, Richard.
    The encyclopedia of evolution : humanity's search for its origins
 / Richard Milner ; foreword by Stephen Jay Gould.—1st Owl Book
 ed.
      p.   cm.—(A Henry Holt reference book)
    Originally published: New York : Facts on File, c1990.
    Includes index.
    1. Human evolution.   I. Title.   II. Series.
    GN281.M53   1993
    573.2'03—dc20   93-2293
                                                              CIP

ISBN 0-8050-2717-3

Henry Holt books are available for special promotions and premiums.
For details contact: Director, Special Markets.

First published in hardcover
by Facts on File, Inc., in 1991.

First Owl Book Edition—1993

Printed in the United States of America
All first editions are printed on acid-free paper. ∞

10 9 8 7 6 5 4 3 2 1

CONTENTS

ACKNOWLEDGMENTS

This book owes its existence directly to three people: Mr. Carl Sifakis, who convinced me (and his many readers) that authoritative reference works can also be entertaining; Professor Stephen Jay Gould, who generously extended a helping hand to welcome me back to evolutionary studies; and my original publisher, Mr. Ed Knappman, who conceived, suggested, and edited this encyclopedia. For four years he prodded and encouraged me to expand its scope far beyond anything I had expected to write.

Major funding for the project was provided by the Miriam and Ira Wallach Foundation, New York City. Thanks also to Ms. Barbara Tate and the Henry Street Settlement's Arts for Living Center, where the author has been a writer-in-residence.

Various parts of the manuscript have benefitted from the comments, criticisms, and encouragement of Dr. Ralph Colp, Dr. Michael MacRoberts and Dr. Barbara MacRoberts, Dr. James R. Moore, Professor Stephen Jay Gould, Mr. Gene Kieffer, Dr. Malcolm Kottler, Ms. Jude Bruno, Dr. Steven Poser, Professor Elisabeth Vrba, Professor Donald Stone Sade, Mr. Dean Hannotte, Mr. Doug Murray, Mr. Melvin Van Peebles. Thanks also to Dr. Eric Delson, Dr. Richard Wrangham, Dr. Susan Parker, Dr. Leonard Horner, Dr. Richard Sayfarth, Dr. Donald Johanson, Dr. Antony J. Sutcliff, Dr. Alan Walker, Dr. Howard Topoff, Mr. Peter Gautrey, Mrs. Rhoda Knight Kalt, and Stephen and Sylvia Czerkas for responding to my queries, and to Ms. Margo Crabtree for permission to use her wonderful photographs of fossil hominids.

I am particularly indebted to Ms. Rebecca Araya, who was my chief assistant throughout the entire project; to Mr. Dave Bergman for contributing several entries on early geologists and substantial bibliographic assistance; to Dr. Dolores Bentham for material on "new Lamarckism"; to Dr. Ralph Colp for assistance with Darwiniana; to Ms. Delta Willis for helping sort out the East African hominids; to Ms. Jennifer Minichello; and to a longtime friend who wrote the genetics entries and prefers to remain anonymous. I thank the meticulous copyeditor Carol Campbell, and jacket photographer Eliot Goldfinger.

Ms. Nina Root's excellent Library Services staff at the American Museum of Natural History, particularly photo archivists Barbara Mathe and Carmen Collazo, and senior librarian Sarah Granato, did more than their share to help. Also at the AMNH, Mr. Martin Cassidy, director of the Museum Reproductions studio (where I have worked as a sculptor-caster) has been a valued mentor, friend, and critic.

I had the pleasure of reading original Darwin letters at the American Philosophical Society in Philadelphia, where then head librarian of the Darwin Library, Dr. William Montgomery, was most helpful. In England, thanks to Mr. Peter Gautrey, head of the Darwin Research Library at the Cambridge University Library, and his staff; thanks also to the staff of the Colindale Newspaper Library, the British National Archive, Maidstone County Archives, the British Museum Library, Miss Silverthorne at the Bromley Library, and Mr. Philip Titheradge of Down House.

At Facts On File, I wish to thank the dedicated people who saw the hardcover version through production, particularly copy editor Paul Scaramazza and editor-in-chief Jerry Helferich. For the help and encouragement they have given on this new paperback edition, I must acknowledge the astute readers who informed me of errors: Mr. L. Balasubramaniam, Mr. Joseph J. Brigham, Ms. Fabienne Smith, and Mr. James D. Williams. Thanks also to those professionals who lent their expertise: Professor Daniel Gasman of the City University of New York; Professor Ernst Mayr of Harvard; Dr. Ian Tattersall of the American Museum of Natural History; Professor K. L. Seshagiri Rao, Professor of Religious Studies at the University of Virginia. (They are, of course, not responsible for any errors that remain or for my failure to follow good advice.) Also, thanks to all my wonderful colleagues at *Natural History Magazine*, particularly Ms. Ellen Goldensohn, Mr. Vittorio Maestro, Mr. Bob Anderson, and Ms. Lisa Stillman. My daughter Ivory is a continuous source of inspiration. And special thanks to the enthusiastic, helpful editor of this edition at Henry Holt and Company, Inc., Ms. Mary Kay Linge.

Objects for the jacket photograph were generously lent by Maxilla & Mandible of New York City, Mr. Eliot Goldfinger, Mr. Gary Sawyer, Mr. Anibal Rodriguez, Mr. Martin Cassidy, Mr. Richard Mirissis, Mr. Dave Bergman, Ms. Melissa Kalt, and Mr. Roy Pinney. For helping me obtain rare Victorian books and Darwiniana, I thank booksellers extraordinaire: Mr. David Bergman (Dave's Books), Mr. John Chancellor (Kew Books), Mr. Donald Hahn (Natural History Books), and, in London, Dr. Eric Korn.

Among my many teachers, my special gratitude to Professor John T. Hitchcock for his courses at UCLA in history of anthropology, and also to Professor Jack H. Prost and Professor Earl W. Count. And I wish to honor the memories of: Professor Charles Seeger, Ms. C. Diane Camp, Mr. Irving Milner, Dr. James A. Oliver, and Mr. Peter Wolff.

FOREWORD
Stephen Jay Gould

We have, as a society, lost our bearings in so many ways. Perhaps other things are more important in a world of poverty and pollution, but I would rank our growing preference for automated sameness over a personal touch as one of the greatest ills of our age. Our corner stores are dying, and motorists will actually wait on line (yes, New Yorkers like me say "on line," not "in line") to throw their toll money into the automated bin, while booths staffed by real people draw no business at all.

But primates are social creatures, and (for all its tragedies) perhaps the one great legacy of our cultural history resides in our stated respect for individuality. This principle applies with special force to scholarship. With a misplaced definition of "objectivity," many people think that books of nonfiction, particularly reference works like encyclopedias, should be impersonal, and devoid of style or idiosyncrasy. No, and a thousand times no. The truly great books of reference have a personal stamp, as any work of passion worthy of our attention must. We still speak of Johnson's Dictionary, Roget's *Thesaurus* and Roberts' *Rules of Order*. The great textbooks may not be best-sellers in a bland, commercial world, but among the monuments in our intellectual history are Lyell's *Geology* and Marshall's *Economics*. Milner's *Encyclopedia of Evolution* now joins this noble tradition of reference books stamped with the vibrant idiosyncrasy of true scholarship. If camels are horses built by committees, then Milner's *Encyclopedia* has provided Darwin with a full-blooded thoroughbred.

Muscular and vibrant idiosyncrasy avoids crankiness and uses the personal touch both to impart a vision and to supply details that conventional accounts would never treat or even discover. This book finds just the right balance between the necessarily universal and the unique. It can be used as a work of reference in the most worthy and conventional manner—to find the names, the dates, the leading concepts clearly and crisply defined, the major arguments and the chief actors.

But the strength of Milner's *Encyclopedia* does not reside in this worthy material, for others have worked in this mode. Milner's different approach follows the variety of his own life. He has loved and studied evolution and paleontology since we were boys together at a junior high school in Queens. (We were the only two dinosaur nuts at a time when such an interest marked a kid as a geek or a weirdo, not a red-blooded normal child in the phase between policeman and fireman. He was called "dino" on the playground; I was "fossil face.") He completed five years of graduate work in anthropology, but then, through a series of interesting circumstances, spent 15 years in the midst of pop culture before returning, in mid-life, to his first (and longest) love for natural history.

From this vantage point both in and outside academia, Milner views evolution and its impact in a comprehensive way often ignored (through true ignorance, that is, not condescension) by more traditional scholars. Milner's uniqueness and idiosyncrasy (in the good sense) lies in his range of chosen topics— a full spectrum of influences on evolutionary thought, from the abstractions of high culture to the come-ons of sideshow hucksters. I see two special strengths in this ecumenical approach:

1. If we make an artificial division into high and vernacular culture, and consider just the former in a narrowly confined and misplaced concept of importance, then we will never understand the impact of science in society. Consider the role of garden clubs, of bird watchers, of car mechanics, of telescope makers, of horseplayers with sophisticated knowledge of probability. Academic knowledge of science may be abysmal in America, but scientific styles of thought are widespread, if only we could harness them to a context of more general understanding.

2. Science pursues an external truth, but only from a perspective inextricably embedded in social contexts. If scientists ignore the context, they not only act in an elitist way, they also preclude any real understanding of scientific change and utility. Social context is particularly important for evolutionary science, because this subject, more than any other, impacts the great myths and hopes of Western soci-

ety—ideas of progress, God, human origins, the meaning of life, the basis of ethics, to name just a few.

Richard Milner has given us so much more than a conventional encyclopedia. His book is a series of mini-essays on the highways and byways of this most socially contentious of all scientific ideas. Scores of essays on people, many unknown to almost all professionals, but as important to the diffusion of the idea of evolution as the wisest professor: Henry Ward Beecher, the great Victorian liberal clergyman who defended Darwin; Carl Akeley and the canonical vision of taxidermy in the romantic mode; E. Ray Lankester, the head of the British Museum, who exposed the medium Henry Slade; Chief Red Cloud, who befriended the great paleontologist O. C. Marsh while Custer advanced on the Black Hills; Darwin's aristocratic neighbor and premier naturalist, John Lubbock; Willis O'Brien, who developed stop-motion photography and made the great dinosaurs of early movies (including the King Kong *vs.* pterodactyl scene); Julia Pastrana, the sideshow "gorilla woman" who fascinated Ernst Haeckel, Alfred Wallace, and Darwin; E. L. Youmans, the intellectual impresario who founded *Popular Science Monthly* and who saw to the publishing of Darwin, Herbert Spencer and others under the Appleton imprint; Charles R. Knight, whose paintings established the image of *the* dinosaur for professionals and amateurs alike and who, thereby (despite his invisibility in scholarly publications), has had as much influence in paleontology as any scientist on earth.

But Milner does not stop with historical personnages. He treats the fictional (Tarzan) and the fossil, both real (Lucy, Zinjanthropus) and fake (Piltdown, Calavaras Skull). He also writes about individual nonhuman animals who have made a difference in evolutionary theory—all the leading personalities of apedom, from Lana to Nim Chimpsky, and even Lonesome George, the last surviving (and mateless) saddlebacked Galapagos tortoise.

Milner's events are as fascinating, and diverse, as his people. I never knew that the trial of the medium Henry Slade featured a battle between Darwin and Wallace, as the rationalist (and wealthy) Darwin contributed funds to the prosecution, while the spirtu-alistic (and impecunious) Wallace testified directly for the defense.

I never heard about T. H. Huxley's arrest in connection with agitation by the Sunday Leagues against pervasive blue laws that had made almost everything but churchgoing illegal on the Christian sabbath. (He preached a "sermon" to 2,000 people on Darwinism one Sunday, surrounded with enough music and recitation to court definition as a "secular service." The baffled police knew that this wasn't kosher, but couldn't devise an appropriate charge and finally ran Huxley in for keeping a disorderly house.)

It never occurred to me that the concocted language of the "primitives" in *Quest for Fire* was not just gibberish, but based on a scholarly best-guess about what proto-Indo-European (our supposed ancestral tongue) sounded like.

I had never encountered "The Happy Family," a popular side-show attraction of the 19th century, featuring animals calmly living together that would tear each other apart in the wild. Although started before Darwin, and meant (in part) as a commentary on Isaiah ("the wolf also shall dwell with the lamb"), these exhibits were later viewed as an argument against Darwin and the struggle for existence. (There was no particular secret in rearing such a group. Good trainers can recognize unusually docile animals, and most creatures raised together from birth and given adequate food on a regular schedule will grow to live in peace.)

Details in the hands of an anecdotalist or an antiquarian remain just that—interesting in themselves perhaps, but incoherent and without aim. Details in the hands of a scholar with a general purpose are the essence of understanding. Milner has given us details in the sublime and coordinated sense. He has also fused high with vernacular culture to show us the full impact of science's most seminal idea—evolution.

In so doing, he has produced a popular book worthy of Darwin's words to Huxley as he importuned his old friend and combative champion to write a work on evolution for the general public: "I sometimes think that general and popular treatises are almost as important for the progress of science as original work."

INTRODUCTION

Artist Paul Gauguin, dying in Tahiti, scrawled onto his final masterpiece: *"What are we? Where do we come from? Where will we go?"* Simple, childlike questions, they have tantalized creative geniuses for centuries, inspiring Michelangelo's Sistine ceiling, Emily Dickinson's poetry, Steven Hawking's cosmogony. When Victorian biologist Charles Darwin tackled those questions, he connected all three with one overarching concept: evolution. Although not the first to do so, he was by far the most persuasive, unleashing a tidal wave in Western thought.

"Nothing in biology makes sense," wrote geneticist Theodosius Dobzhansky, "except in the light of evolution." Every branch of the life sciences, from biochemistry and botany to zoology, has an evolutionary underpinning. Practical applications are so ubiquitous that we hardly notice their common yoke: DNA "fingerprinting" in the courtroom, bioengineering of livestock and plants in agriculture, medical strategies to combat resistant strains of bacteria, conservation programs for endangered species, growing awareness of planetary ecology. All are joined together by a basic, shared assumption—the idea of evolution, which only 150 years ago was considered unscientific and vaguely immoral.

Entire libraries are devoted to its scientific implications—histories of geology, journals of animal behavior and population genetics, ecological studies of jungle canopies and ocean floor vents, speculations about dinosaurs and "death stars." An encyclopedist of evolution could spend a lifetime exploring abundant riches without venturing beyond this scientific literature. But we are dealing with an impolite and unruly idea that simply refuses to stay put.

Vast and far-reaching in its effects, the theory of evolution by natural selection long ago leapt the boundaries of biology to influence an astounding variety of fields: science fiction novels, children's toys, Hollywood movies, even religious and political movements. Therefore, a reference work devoted to this single, unifying idea has to include not only diverse scientific developments, but also the major historical repercussions in literature, art, philosophy and "pop culture" as well. And people, lots of interesting and unusual people.

During the colorful history of this great cultural enterprise, contributions to evolutionary knowledge have been applied (and misapplied) by a variegated spectrum of remarkable individuals: the world's richest man and his strange evolutionary "religion" [ANDREW CARNEGIE], a college president and a school dropout, both obsessed with collecting slabs of fossilized footprints [REVEREND EDWARD HITCHCOCK and ROLAND T. BIRD], the dinosaur expert who tried to become king of Albania [BARON NOPCSA], the occupational therapist from Louisville, Kentucky whose love for African gorillas cost her her life [DIAN FOSSEY], the artist who went blind painting prehistoric animals [CHARLES R. KNIGHT], even the proverbial Doctor [GIDEON MANTELL], Lawyer [SIR CHARLES LYELL], and Indian Chief [RED CLOUD].

Even as evolutionists study change, their very perceptions of nature—and of change itself—are constantly evolving. Information about the web of life, animal species and their behavior, shifting theories and fads in science, all defy attempts to "freeze" them in encyclopedia entries. Upon close scrutiny, they seem to be perpetually transforming themselves.

Chimpanzees, for instance—a few years ago we thought of them as natural clowns, like Cheetah, the comic relief in Tarzan movies. In the 1970s they became newly respected as hand-signing symbol users; next—during the ensuing "ape language controversy"—they were viewed as clever, manipulative brats that only *pretended* to know language in order to yank the human investigator's chain. During Jane Goodall's ongoing African field studies, chimps went from peaceful, sweet-natured vegetarians to cannibalistic, infanticidal killer apes. Of course, the chimps remain chimps, but how drastically human perceptions of them changed over 30 years! [See APE LANGUAGE CONTROVERSY; CHIMPANZEES; INFANTICIDE.]

The status of fossil man is probably the most rapidly changeable area at present; it has gone from a search for "the missing link" to the "oldest man," from reconstructions of "noble savages" to cannibals, from "killer apes" to food sharers. Species names change, genera are split or lumped. Experts attack each other's credibility, even while wriggling out of their own extravagant pronouncements of a few years earlier. There are more fossil hominids known than ever, but paleoanthropologists cannot agree on just how many and which are ancestral to ourselves. [See AUSTRALOPITHECUS BOISEI; CANNIBALISM CONTROVERSY; FOSSIL MAN; "LUCY"; "MISSING LINK"; NOBLE SAVAGE; "OLDEST MAN"; TAUNG CHILD.]

Several fossil hominids, such as the Calavaras skull and Piltdown man, have been exposed as deliberate frauds. A few others (Nebraska man, Ramapithecus) were ludicrously oversold by the fame hungry and were later retired when they proved to be mirages based on fragments. Still, there is a core of solid evidence for a radiation of diverse species of small-brained near-men walking upright in Africa three or four million years ago. Most authors mention Darwin's "fossil free" hunch that man originated in Africa, but history has forgotten that his "junior partner" Alfred Russel Wallace had scooped him again; for it was Wallace who first suggested (to Darwin, no less) that Africa was the probable "cradle of mankind." [See CALAVARAS SKULL; "CRADLE OF MANKIND"; NEBRASKA MAN; PILTDOWN MAN (HOAX); RAMAPITHECUS.]

This encyclopedia attempts to rescue many such "unknown" incidents from oblivion. Did you know that Gregor Mendel never formulated Mendel's laws? Or that when Francis Crick, codiscoverer of the structure of DNA, established the famous "central dogma of genetics" he had no idea what "dogma" meant? That arch "materialist evolutionist" Thomas Huxley liked to astonish audiences with his "supernatural power" to summon spirits? Or that Darwin said he could not really prove that any species had evolved—but didn't think that his theory should be judged on that basis? [See "CENTRAL DOGMA" OF GENETICS; HUXLEY, THOMAS; MENDEL'S LAWS; THEORY, SCIENTIFIC.]

When Charles Darwin established evolutionary biology, he was standing on the shoulders of giants. Such great naturalists as the Comte de Buffon, the Chevalier de Lamarck and his own grandfather Erasmus Darwin richly deserve the credit they have received as his forerunners. A nonscientist, however, exerted a far greater influence in establishing the idea with a wider public. Fifteen years before *Origin of Species* (1859) appeared, a "pop" evolution book titled *The Vestiges of Creation* (1844)—based on inaccurate facts and published anonymously—became a best-seller that remained in print for seventy years. [See BUFFON, COMTE DE; DARWIN, ERASMUS; LAMARCK, JEAN-BAPTISTE; VESTIGES OF CREATION.]

At the time, scientists dismissed the *Vestiges* as sheer junk. Even the unorthodox zoologist Thomas Henry Huxley, soon to be a leading evolutionist, tore it to shreds in a review—which embarrassed him in his later years. Other leading natural scientists assured the public that no reputable professional believed in evolution (it was then known as "the development hypothesis" or "transmutation theory"). The most popular geologist of the day, Hugh Miller, denounced it as vulgar "sciolism" (pseudoscience).

Somehow, the public knew better than the scientists, and kept buying the *Vestiges* through 13 editions, although the name of its author, Robert Chambers, remained a secret until after his death. The possible identity of the "unknown evolutionist" was a common topic at dinner parties; Darwin jokingly referred to him as "Mr. Vestiges." [See CHAMBERS, ROBERT; MILLER, HUGH; "MISTER VESTIGES"; SCIOLISM; VESTIGES OF CREATION.]

The expedition of H.M.S. *Beagle* in 1831 had also captured the popular imagination, but not because of any anticipated connection between that historic voyage and the theory of evolution. A major impetus to Captain Robert FitzRoy's setting sail for South America was to return three young Indians to their homeland—part of an experiment dear to the captain's heart. The public neither knew nor cared that the young Charles Darwin was aboard as unofficial naturalist. But all English newspaper readers knew the outlandish names that sailors had given to the Fuegian Indian passengers: York Minster, Fuegia Basket, and Jemmy Button.

Gathered to England's bosom on the *Beagle*'s previous expedition, the Indians had been baptized and civilized at FitzRoy's expense, and he was now delivering them back to their "primitive heathen" kin, to spread the light of Britannia. Jemmy Button, who was purchased for a fancy button according to some accounts, was a particular favorite of the public. Dressed in top hat and kid gloves, this "civilized savage" was introduced to society and even presented at Court to the king and queen.

Sadly, the return to Tierra del Fuego was a disaster for Jemmy and his fellows. Darwin told the world in his best-selling *Journal* (1839; republished 1845) how the good-natured Indian became a pathetic victim of Captain FitzRoy's well-intentioned experiment with human lives. [See BUTTON, JEMMY; FITZROY, CAPTAIN ROBERT; VOYAGE OF THE H.M.S. BEAGLE.]

The historical interface of the evolutionary enterprise and popular culture has produced other episodes equally tragic and bizarre. For instance, around 1906, a favorite of New York City newspapers was the Congo Pygmy Ota Benga, who became famous as an "evolutionary exhibit" in the Bronx Zoo's Monkey House. Benga's stint at the New York Zoological Park, during which he shared a cage with an ape, triggered a storm of righteous outrage. The park's director insisted he had been "employed," not imprisoned, but Ota Benga was summarily retired. Deprived of the public attention he had come to enjoy, Benga found employment in a factory and, a few years later, committed suicide. [See BENGA, OTA.]

In recent decades, pop culture acclaimed a new "ape woman" in the sad, strange saga of field primatologist Dian Fossey, whose deeply felt kinship for wild gorillas precipitated her private war with African hunters. Like so many of the great apes to which she had dedicated her life, Fossey was mercilessly hacked to death. A modern martyr to evolutionary consciousness, her story became famous through books, documentaries, and—the apotheosis of our time—a major motion picture. [See "APE WOMEN," LEAKEY'S; FOSSEY, DIAN.]

Bitter conflicts over evolutionary ideas have frequently spilled out of scientific circles into the forums of popular culture, but creationism was often *not* the issue. Everyone has heard of the Scopes "monkey trial" in Tennessee (although most of the "facts" are usually wrong), but the Slade trial of Victorian England is now almost forgotten.

In 1876, British Spiritualists and evolutionists fought this emotional courtroom battle over the authenticity of "Doctor" Henry Slade, an internationally celebrated spirit-medium (or channeler in today's jargon). The issue was "materialism"—the idea that human personality is inseparable from the biological brain.

"Materialist" scientists insisted that Slade's communion with "departed spirits" had to be a criminal fraud to bilk the bereaved. While many among the general public may have privately agreed, for a scientist to publicly insist that a human "soul" cannot exist apart from a body was shocking—much worse than claiming apes for relatives. Evolutionists Charles Darwin and Alfred Russel Wallace (a prominent Spiritualist) supported opposing sides in the Slade affair, each convinced that his own vision of evolutionary science was at stake. [See CREATIONISM; MATERIALISM; SCOPES TRIAL; SLADE TRIAL; SPIRITUALISM.]

Two years earlier, distinguished physicist John Tyndall had shocked the British Association with the uncompromising "materialist evolutionism" of his "Belfast Address"—in which he openly declared the church's "monopoly" on questions of human origins, nature and destiny to be over and done with. And in America, in 1925, Carl Akeley's statue *Chrysalis* (displayed in a Unitarian church) became the center of another public furor, reviled in newspaper editorials as "pagan nature worship." Akeley had depicted man emergent from a cracked-open gorilla skin, offering a triumphant vision of Ascent from the Primates in place of a guilt-ridden Primal Fall. [See AKELEY, CARL; BELFAST ADDRESS; THE CHRYSALIS; MATERIALISM.]

Evolution as a subject of art has its own special and fascinating history, most recently exemplifed in the unprecedented explosion of dinosaur imagery in popular culture. They are absolutely inescapable, appearing on everything from U.S. postage stamps and cans of Chef Boyardee Dinosaur Pasta to whole emporiums like the DinoStore in Birmingham, Alabama ("Gifts of Extinction"). Only 200 years ago, science did not even suspect the existence of ancient reptilian giants, nor had British zoologist Sir Richard Owen yet coined the term "dinosaurus." [See DINOSAUR RESTORATIONS; OWEN, RICHARD.]

Owen and his artist Waterhouse Hawkins began the public's infatuation with dinosaur imagery when they constructed their life-size *Iguanodon* statues at the Crystal Palace in 1853—an instant crowd pleaser. Thirty years later, the definitive dinosaur paintings and murals of the American Charles R. Knight became the international standard, inspiring such "pop" derivatives as the Sinclair Oil "dino" logo, and the "carnivorous machines" of Walt Disney's *Fantasia*. (Disney also "evolved" Mickey Mouse from a nasty, angular rodent to the round-faced icon of an international megacorporation—an artistic transformation in tune with solid biological principles!) [See CUTENESS, EVOLUTION OF; FANTASIA; HAWKINS, BENJAMIN W.; IGUANODON DINNER; KNIGHT, CHARLES R.; SINCLAIR DINOSAUR.]

The mystery of mankind's origin, nature, and destiny obsessed and inspired not only the biologists Charles Darwin, Alfred Russel Wallace and Thomas Henry Huxley, but a diverse "who's who" of great authors whose influence has permeated the far corners of our culture: Playwright George Bernard Shaw constructed *Back to Methuselah* (1920) as a sugar-coated manifesto for Lamarckian evolution, devoting the play's preface to an attack on Darwinism. *War of the Worlds* (1898) creator H.G. Wells, a student of Thomas Huxley's, wrote the first "Darwinian" science fiction. (His extraterrestrials were products of an alien evolution terrorizing Earth in a cosmic battle for "survival of the fittest.") [See SHAW, GEORGE BERNARD; WELLS, H.G.]

"Father of psychiatry" Sigmund Freud developed fanciful scenarios of Ice Age "primal hordes" to explain the "childhood" of the human race. Freud's theories were based on the belief, common among 19th-century biologists, that human evolution mirrored each individual's developmental stages. [See BIOGENETIC LAW; FREUD, SIGMUND.]

In our own century, Hindu mystic Gopi Krishna taught that "the secret of Yoga" is the systematic attempt to accelerate the evolution of the human brain. Churchman Pierre Teilhard de Chardin, despite heavy censure by ecclesiastical authorities, strove mightily to reconcile Catholicism with Darwinian evolution. And filmmaker Stanley Kubrick produced the cinematic epic *2001*, tracing the human odyssey from tool-wielding man-apes to the exploration of outer (and inner) space. Creative thinkers, whether scientists or not, have shared in probing the meaning of evolution; their stories are included as well. [See GOPI KRISHNA; TEILHARD DE CHARDIN, FATHER PIERRE; 2001.]

Most adults who yearn to "unscrew the unscrutable" begin as kids hooked on some seemingly trivial enthusiasm or, in my case, by several of the following: a summer spent collecting beetles; a towering *Tyrannosaurus* skeleton in a museum; animals sketched at the zoo; *Tarzan* comics; pet snakes; "dinosaurabilia" collectibles; or Hollywood's King Kong climbing the Empire State building—doomed to extinction by a changed world that no longer had a place for him. (A remarkably apt metaphor for today's real gorillas, driven to ever-higher ground by creeping human cultivation of their mountain slopes.) Images, objects or tales that ignite youthful imaginations constitute the "romance of natural history"; that, too, is what this encyclopedia is about. [See BEETLES; DINOSAURABILIA; GORILLAS; GOULD, STEPHEN JAY; KING KONG; TARZAN OF THE APES.]

Twenty years before publication of the *Origin of Species* (1859), Charles Darwin's exciting *Narrative of the Voyage of the Beagle* (1839) instantly won a place in the popular culture, inspiring many younger naturalists to seek their own adventures. Alfred Russel Wallace remembered into old age how he and Henry Walter Bates dreamed of exploring the Amazon after reading Darwin's rhapsodic descriptions of its rain forests. (They also shared his passion for beetle collecting.) The same book, written for nonspecialists, entranced the great botanist Joseph Dalton Hooker on his own voyage to the Antarctic. He slept with pages of Darwin's *Journal* under his pillow. [See BATES, HENRY W.; HOOKER, SIR JOSEPH; ORIGIN OF SPECIES; VOYAGE OF THE H.M.S. BEAGLE; WALLACE, ALFRED RUSSEL.]

One of the most peculiar aspects of the worldwide spread and acceptance of evolution is the amazing range of perceptions and attitudes that have claimed it. Lamarckians equate evolution with progress, increasing complexity and perfection. Darwin himself saw the possibility of degeneration as well as progress, of neutral "sideways" changes as well as adaptation. Some evolutionists, such as the German

Ernst Haeckel and the Englishman Ray Lankester, became obsessed with degeneration, expressed in their rantings about nationalistic "racial hygiene." As playwright George Bernard Shaw put it, "Darwinism was fortunate enough to be adopted by anyone with an ax to grind." [See DEGENERATION THEORY; HAECKEL, ERNST; LAMARCKIAN INHERITANCE; LANKESTER, E. RAY.]

For Haeckel, evolution meant the moral justification of ruthless competition, while for the Russian Prince Kropotkin it was altruism and social cooperation that ensured the survival of species. Some rejoiced in identifying their roots in the animal world and establishing a blood connection to all other living creatures. Others were shocked, as if they had discovered the revered founder of their family was, in the phrase of novelist Thomas Carlyle, "nothing but frog spawn." [See KIN SELECTION; ALTRUISM; KROPOTKIN, PRINCE PETER; SOCIAL DARWINISM.]

Science views evolution as contingent history—a branching bush sending forth divergent twigs as circumstance and variation allow at each juncture. More mystical philosophers have insisted there must be an ultimate goal or unfolding of a preordained destiny: a ladder or escalator to the infinite. Evolutionary data have been used as evidence for the biological unity of mankind and also its converse, dividing so-called "races" into "higher or lower" depending on their supposed time of evolution. (A scientifically discredited idea that nevertheless lingers among apologists for racism.) [See BRANCHING BUSH; CONTINGENT HISTORY; ORTHOGENESIS; RACE; RACISM IN EVOLUTIONARY SCIENCE; TEILHARD DE CHARDIN, FATHER PIERRE.]

Certainly, Darwinism, natural section, "survival of the fittest," and other key concepts keep changing. Darwin tinkered with his theory from one edition of *Origin of Species* to the next, and no one since has left it alone. History has seen it evolve into "Neo-Darwinism," "The Synthetic Theory of Evolution," and, in the 1970s, a significant and wide-ranging shift in emphasis regarding rates and pattern of change, known as the theory of Punctuated Equilibrium (or "Punk Eek"). [See NATURAL SELECTION; NEO-DARWINISM; PUNCTUATED EQUILIBRIUM; "SURVIVAL OF THE FITTEST"; SYNTHETIC THEORY.]

Even that complex Victorian hero of science, Charles Darwin himself—or, more properly, our image and understanding of him—keeps changing, as historians of the new "Darwin Industry" study newly accessible records, letters and private papers. (Some 14,000 letters are presently in the process of being published.) [See DARWIN, CHARLES.]

Darwin knew he could not hope to effect a major revolution in thought by himself and constantly consulted with the world's leading botanists (Joseph Hooker and Asa Gray), the top geologist (Charles Lyell), his "bulldog" and knight champion, zoologist Thomas Henry Huxley, and others. Although they never coauthored papers (sharing credit for each morsel of knowledge is a 20th century fashion), they were true collaborators in the evolutionary enterprise.

Ironically, the one top naturalist in his circle with whom Darwin did *not* regularly collaborate was Alfred Russel Wallace, the very man he was forced to accept as coauthor of the theory of evolution by natural selection (which is properly known as the Darwin-Wallace theory). [See "DELICATE ARRANGEMENT"; GRAY, ASA; HENSLOW, JOHN; HOOKER, SIR JOSEPH D.; HUXLEY, THOMAS HENRY; LYELL, SIR CHARLES; WALLACE, ALFRED RUSSEL.]

These Victorian naturalists were remarkable human beings, whose personalities, friendships and rivalries left their stamp on an era; often their private correspondence was even more interesting than their public contributions to science. If this encyclopedia's biographical entries retain some lingering flavor of these wonderfully documented lives, you may want to know them better through their own letters and writings on natural history. (Bibliographies are given below the entries.)

Thomas Henry Huxley once remarked that when he was perplexed about a question, he liked to "talk it over" with men "of real intellectual grasp and power" of a previous age. Despite our unquestioned conviction that present scientific knowledge far surpasses anything known a century ago, we can still learn a great deal by "talking over" the great questions of evolution and science with Darwin and his friends. (And there is also the added delight of wandering about in what Wallace aptly called "the wonderful century.")

Contact with the personalities of Darwin, Wallace, Huxley, Hooker and the rest also imparts something else: an intimate insight into lives extraordinarily well lived. All shared a fierce dedication to finding truth; a resigned humor about their own human limitations; a deep sense of loving wonder about the natural world; and, above all, a cheerful willingness to commit a lifetime's unstinting labor to a worthwhile quest. And all at a time when information couldn't be "faxed" overnight and words weren't "processed"; their overseas letters had to be still worth reading three months after posting, and their durable books were written with scratchy pens dipped in ink.

During most of the 19th century, there was little money to be made in science. It was carried forward on the shoulders of a volunteer army, in which great physical courage was taken for granted. Wallace habitually risked his life in the jungles, often dragging himself through bouts of malaria to capture new, undescribed butterflies. Darwin never hesitated to board the *Beagle*, though that ship was of a design and manufacture sailors had nicknamed "floating coffins," so commonly did they capsize. (He also rode 400 miles on horseback through hostile Indian country in Argentina, where even the hardy gauchos feared to accompany him.)

Perhaps the most delightful example of a naturalist's courage and resourcefulness concerns Darwin's closest friend and confidant Joseph Dalton Hooker, an expert on trees and flowers. Convinced that Hooker was a spy for the British government, the Rajah of Sikkim sent 100 armed tribesmen to ambush and kill the adventurous botanist at a remote mountain pass high in the Himalayas, near the Indo-Tibetan border. Within a half hour, Hooker had turned them into a platoon of plant collectors. [See HOOKER, SIR JOSEPH.]

Recently, in the midst of a discussion about Darwin's closest "lieutenants" with psychiatrist-historian Dr. Ralph Colp, he suddenly paused and exclaimed, "Gee, weren't those *great guys?*" They were indeed. And anyone who, by a leap of imagination, is capable of making friends in another century, can get to know them amazingly well through their abundant letters and memoirs. Colp himself, for example, has assembled a more complete medical dossier for Charles Darwin than most physicians keep on their more ambulant patients. [See COLP, RALPH; DARWIN, CHARLES, ILLNESS OF.]

Looking back after a century's scientific "progress," you may be surprised to find the old naturalists fighting over many of the same evolutionary questions offered as new in our own day, if couched in slightly different terminology. Often they argued more passionately than their intellectual descendants and with superior powers of expression. Darwin and his circle never disappoint. And no matter what hour you'd like to visit, they are always home.

They were well aware that they were making history and breaking new ground using very fragmentary evidence. How they would have enjoyed recent developments in evolutionary biology: molecular clocks; biochemical distance between species; radioactive dating of rocks; the African hominid fossils; the discovery of DNA; field observers living peaceably and intimately among free-living chimpanzees and gorillas.

But rather than answering their questions, we have only deepened the mysteries. No longer do we wonder why paleontologists cannot produce a "missing link"; we've got almost more "early man" fossils than we know what to do with. The question now is, where do they fit in the increasingly complicated picture of hominid radiation? We no longer wonder whether the abrupt gaps in the fossil record are the result of "imperfect preservation," as Darwin supposed; now we believe that they reflect great and sweeping mass extinctions—but, despite promising research directions, we still really don't know why they occurred.

Scientists no longer debate the existence of "pangenes" or "blending inheritance." Today, they can construct gene maps and actually photograph the shape of DNA. But still they puzzle over just what it is that is naturally selected, how "random" are mutations and the relative importance of adaptation and dumb luck. (During conditions of global mass extinctions, for instance, a lot of extremely well adapted forms get wiped out.) [See GREAT DYING; NEUTRAL TRAITS; PAINTPOT PROBLEM; PANGENESIS.]

Darwin insisted that uncertainty is part of science. Does that mean evolutionary theory is worthless, as some religionists have argued, because its truths are not eternal? Thomas Huxley once asked his listeners to suppose they were lost in the countryside on a dark night, with no clue to the road. If someone came along and offered a flickering lantern, would you refuse it because it was only an imperfect and unsteady light? "I think not," Huxley answered himself, "I think not."

We have had a century and a half of undeniable advances in our knowledge, a virtual deluge of new evidence, and light-years of progress in technology since the days of the Victorian naturalists. Each year brings exciting discoveries: new hominid fossils, more geophysical clues to the reasons for mass extinctions, deeper understanding of how the biosphere works, insights into the neural network of the human brain and the miniature universe of living cells. [See BIOSPHERE; EUKARYOTES; GAIA HYPOTHESIS; NEURAL DARWINISM; PLATE TECTONICS.]

Is it time, then, as we look back, to congratulate ourselves on our present understanding of human origins, nature, and destiny—so far advanced beyond that of the science's founders? In Huxley's phrase, "I think not." Rather, it is past time to appreciate the magnificence of what they gave us and to teach intellectual history to our young. (You can find Jack the Ripper and the Beatles in London's waxworks museum, but not Darwin or Huxley.)

It is also time to end this Introduction, but without the customary Hallelujah chorus in praise of our exalted state of knowledge. Does anyone really believe, as they did a few years ago, that after another study or two of baboon behavior, or stone tools, or fossil skulls, all will come clear? Better by far to humbly admit how much we still don't know about those eternal questions with which we began: *What are we? Where do we come from? Where will we go?* For now, these great enigmas that fascinated Darwin and Huxley still beckon, and—as another Victorian mystery solver said to his friend Dr. Watson—"The game is afoot."

—————*RICHARD MILNER*

NOTE TO THE READER: Entry headings with asterisks (*) refer to related entries in the Update section on page 471.

ABANG
Orang Stone Toolmaker

Abang, a male orang-utan in the Bristol zoo, learned to do something once thought impossible for apes—to manufacture and use a stone knife.

Apes in the wild do not use stones to shape other stones (at least not while humans are watching), although they have been observed using rocks to smash hard fruit rinds. To an anthropologist, a real stone tool is made by chipping, shaping or flaking; prehistoric sites contain thousands of such implements made by early humans and perhaps by other hominids.

One common type is a long, sharp-edged flake struck off a larger flint core. Some scientists insisted that australopithecines could not have made such tools; only humans had the mental competence and manual dexterity required for toolmaking.

In 1971, R. V. S. Wright set out to disprove such pronouncements. Working with Abang, then five-and-a-half years old, Wright repeatedly demonstrated how to use a flint blade to cut a nylon rope tied round a food box. After about an hour of demonstrations, Abang mastered the use of the knife to get the food.

After three hours (eight short sessions), Abang learned to make his own knife by striking off a flint flake with a hammerstone. With practice, he produced thinner, sharper flakes by striking nearer the rim. And he added his individual touch of holding and steadying the core with his feet.

Now, when Abang is given a tied-up box of food, a flint nodule and a hammerstone, he quickly makes a stone knife and cuts the rope. Wright never claimed Abang gained "insight" above other apes; it appears to be a straightforward case of learning by imitation and improvement by trial and error. For his achievement Abang deserves a place in the pongid pantheon.

See also APES, TOOL USE OF; ORANG-UTAN.

For further information:

Desmond, Adrian. *The Ape's Reflexion*, London: Blond & Briggs, 1979.

ACTUALISM
Continuity of Causality

Actualism is the assumption that the Earth's past can be explained in terms of natural processes observable in the present. Many historians credit James Hutton (*Theory of the Earth*, 1788) with first applying it to geology, though he was an indifferent field worker who based theories on very casual observations. The works of Sir Charles Lyell, a meticulous field geologist, solidly established actualism. Lyell made it the keystone in his cluster of ideas later known as uniformitarianism—the foundation of modern geology.

Lyell's subtitle for his *Principles of Geology* (1830–1833) is *An Attempt to Explain the Former Changes of the Earth's Surface by Reference to Causes Now in Operation.* His systematic observations of erosion, sedimentation and volcanic formations enabled him to clarify many longstanding mysteries about the Earth's features. Unlike his predecessors, Lyell did not use miracles or divine intervention to explain ancient events. (He was stymied, however, by the origin of the Earth and its earliest composition, since there appeared to be no comparable processes observable now.)

Catastrophist geologists believed that most of the Earth's history was a series of cataclysms or upheavals unlike anything known today, involving drastically different processes. Though actualism eventually led to the downfall of catastrophist geologists, some of them initially embraced Lyell's approach. If more ordinary geologic features had been produced by known causes, they reasoned, then those that defied explanation could safely be assigned to forces outside the range of human knowledge. And, despite their very different conclusions, catastrophists admired Lyell's success in applying his actualist methods to a broad range of geological problems.

By 1840, Lyell's uniformitarian principles (encompassing actualism, progressionism, vast geological time scale, a steady-state Earth and more) had exerted a huge influence. Not only did they set the tone for the next century of geological research, they directly influenced Charles Darwin's view that species originated gradually through ordinary reproduction, rather than suddenly through supernatural agencies. Darwin had read Lyell's books during his voyage aboard H.M.S. *Beagle* and later claimed his theories came "half out of Lyell's brain."

But while Lyell had no problem in imagining that great canyons or mountain ranges had been shaped slowly by natural forces over eons, he balked at accepting a similar process for the human species. It was not until 1863, in his book on *The Antiquity of Man,* that he publicly supported Darwin's ideas about the continuity of life in the natural world, though he still skirted the issue of man. Perhaps, he grudgingly conceded, "community of descent is the hidden bond which naturalists have been unconsciously seeking while they often imagined that they were looking for some unknown plan of creation."

See also CATASTROPHISM; GRADUALISM; LYELL, CHARLES; PROGRESSIONISM; STEADY-STATE EARTH; UNIFORMITARIANISM.

For further information:

Bowler, Peter J. *Evolution: The History of an Idea.* Berkeley: University of California Press, 1984.

Gould, Stephen Jay. *Time's Arrow, Time's Cycle.* Cambridge: Harvard University Press, 1987.

Greene, John C. *The Death of Adam: Evolution and Its Impact on Western Thought.* Ames: Iowa State University Press, 1959.

ADAM AND EVE
Biblical Primal Couple

In the Old Testament, the Lord creates the first man out of clay, breathes life into him and calls him "Adam," which means "from the dust of the Earth." When Adam becomes lonely in his Garden paradise, the Creator fashions the woman Eve from the man's rib, as "an help meet [suitable] for him"—commonly misread as "a help-mate." (Despite common belief, men do *not* have one fewer pair of ribs than women.)

Until the first couple violates the Lord's instruction not to eat the fruit of the forbidden Tree of the Knowledge of Good and Evil, they live in total harmony with nature, over which they are given "dominion." Before the Expulsion, they spend their days enjoying the benevolent works of the Creator and inventing names for all the plants and animals.

Linnaeus, the great Swedish botanist who founded the system of biological classification, insisted Adam was "the first naturalist." To spend one's life studying, classifying and giving names to plants and animals, Linnaeus wrote, was therefore sanctified in tradition as a manner of worshiping God. (Not incidentally, an excellent response to bookish churchmen who scorned his practice of natural science as vulgar.)

Did Adam and Eve have navels? For painters and sculptors of religious subjects, this was a controversial question of great practical concern. During the Renaissance, the primal couple was sometimes represented without them, since neither was born of

PRIMAL PAIR was often depicted with navels by earlier church artists, but 19th-century illustrator John Tenniel avoided controversy by covering up the umbilical question.

woman. Other artists avoided the question with strategically placed shrubbery. Some theologians argued the first man and woman were "finished" creations; they sported belly buttons though they never actually had an umbilicus—just as Eden's tree trunks were created complete with growth rings.

A popular 19th-century natural history writer, Philip Gosse, seized on that idea in an ill-conceived effort to finally reconcile scripture and science. If Adam and Eve's hair, fingernails and navels were created complete in an instant, bypassing growth and development, he argued, then God must have created all the Earth's fossils and geologic strata as false remnants of a past that never actually happened.

Gosse's infamous 1857 book *Omphalos* (Greek for navel) was uncredible to scientists and clerics alike. The Reverend Charles Kingsley, for instance, sadly confessed that if such mental contortions were really necessary to reconcile geology with the Bible, it shook his faith in scripture.

Even today, Adam and Eve linger in the scientific imagination. When paleoanthropologist Louis Leakey wrote his review of fossil man discoveries, he called it *Adam's Ancestors* (1934). And when biochemists believed they had traced modern humans back to an African female of 200,000 years ago, they nicknamed her the "Mitochondrial Eve" (1987).

Mark Twain often complained that Adam and Eve, the "founders" of the human race, deserve more in the way of a memorial by their descendants. A century ago, Twain campaigned for all the world's peoples to take up a collection for a colossal statue of Adam and Eve—towering over all the rival religious shrines—to be erected in the Holy Land. For once he was not joking, but no one took his proposal seriously.

See also MITOCHONDRIAL EVE; ORIGIN MYTHS (IN SCIENCE).

ADAPTATION
Shaped for Survival

Fins and flukes evolved as swimming adaptations, wings for flight and camouflage for defense. Structures and behaviors useful to an organism in a particular environment are adaptations, called that long before Charles Darwin. Some creatures show striking combinations of adaptive traits, all used to exploit a specialized niche or habitat.

Woodpeckers (Darwin's favorite example) get their living by climbing tree trunks and extracting insects from bark. Adaptive features include a thick skull, "shock absorber" neck construction, chisel bill, long, barb-tipped tongue, claws like grappling hooks and stiff tail feathers for stability. Admiring such adap-

tations—as narrators of television wildlife films often do—can lead to using them to "explain" evolution.

In fact, the concept of adaptation is one of the most troubling and puzzling in natural history. Since an animal is the product of a long contingent history, its adaptation is relative at any given time. Feathers may now be adaptive for flight, yet they evolved before birds were good fliers. How could early wings have been adaptive? Or, as S. J. Gould put it, "What good is forty percent of a wing?"

Many biologists believe the answer lies in a change or shift in function. A structure's eventual use may be quite different from its origin, a process known as "exaptation." Early flightless "wings" may have been used to stabilize swift-running birds, as ostriches use them today, or they may have first functioned as heat regulators. Some living birds' wings are not adapted primarily for flying. Penguins use them to swim and some wading birds (African black herons) curl them into glare shields while fishing in shallow water. For many structures, we cannot determine how they originated, nor what may be their future.

The explanation of the origin of adaptations is one of the most controversial areas in evolutionary biology. One persistent question is whether—as seems likely—new behaviors usually appear first and new structures subsequently evolve. Darwin thought behavior changed first, but rejected Lamarckian "willing" and "striving" of organisms in new directions.

Biologists are still seeking a mechanism by which such seemingly "Lamarckian" results can be accomplished by the selection of random mutations or another explanation compatible with current knowledge of genetics. [See BALDWIN EFFECT.] Dissatisfaction with the traditional model has led some biochemists to seek new understandings of how variation is produced. For example, genetic theory is only beginning to come to terms with Barbara McClintock's "jumping genes," which suddenly change loci, or with discoveries of extra-cell DNA with still unknown functions. Newly discovered areas of ignorance are always a healthy sign for science.

Adaptation, which at first seems such an easy, common-sense concept, turns out to be slippery, sometimes even circular and paradoxical. A species is adapted if it survives in its environment, but how do we know it has not simply been dumb-lucky or neutral, while some calamity eliminated its competitors? (Mass extinctions have several times wiped out 97% of species on Earth.) If a species becomes extinct tomorrow, how well adapted was it? And how to explain seemingly maladaptive structures such as the peacock's glorious tail, which gets in the way of efficient flight or food-getting? (This last question led

Darwin to propose his theory of "sexual selection.")

To complicate the situation, many species we think are not adapted to certain environments have moved into them anyway. When Darwin visited the Galapagos, he was fascinated by the plentiful marine iguanas. When not basking on rocky beaches, these large lizards spend their days underwater, grazing on seaweed. Although excellent divers and swimmers, these reptiles have never (or, perhaps, not yet) evolved webbed feet, streamlined forms or other obvious aquatic adaptations. Their bodies give no obvious indications that they get their living on the ocean floor.

Biologists studying wildlife in a South American rain forest in 1987 found fish that have recently become fruit eaters. When human activity caused regular flooding of the river banks, the fish learned to swim among submerged treetops, feeding on the fruit. In Oregon, a local population of deer (considered vegetarians) now patrol a river's banks, eating beached fish that jump out of the water while spawning. And back in the Galapagos, members of one of the gannet species are usually found clumsily perched on tree branches—bizarre behavior for web-footed birds.

One of Darwin's enduring demonstrations was that adaptations are usually not marvels of perfection at all, but historical compromises. On closer examination, they usually turn out to be jerry-built contraptions—products of a unique, opportunistic history.

Less sophisticated naturalists, lacking any experimental or evidentiary foundation, have, over the years, created fanciful "just-so" stories, to explain the origins of particular adaptations. Some biologists have even suggested eliminating the concept of adaptation altogether. They find it too vague to be useful and historically abused as a substitute for thoughtful investigation.

See also EXAPTATION; "JUST-SO" STORIES; ORCHIDS, DARWIN'S STUDY OF; PANDA'S THUMB.

For further information:
Grant, Verne. *The Evolutionary Process.* New York: Columbia University Press, 1985.
Provine, William B. "Adaptation and Mechanisms of Evolution after Darwin: A Study in Persistent Controversies." In *The Darwinian Heritage,* edited by D. Kohn, pp. 825–866. Princeton: Princeton University Press, 1984.
William, G. C. *Adaptation and Natural Selection.* New Jersey: Princeton University Press, 1966.

ADAPTATION, ORIGIN OF

See EXAPTATION; MIVART, ST. GEORGE J.

ADAPTATIONIST PROGRAM
Controversy About Utility

Darwin thought selective forces shaped creatures like 1,000 wedges from all directions, molding them into the optimum adaptation for their way of life. Some biologists still assume all structures and behaviors of organisms that have survived must have an adaptive function. If no adaptive use could be found for a structure or behavior, some theorists got into the bad habit of inventing them.

Critics of such "just-so" stories argue that all traits don't have to be adaptive. Some might be neutral or adaptive only in particular environments, while others may be incidental by-products of adaptive structures. Some species are not really not well adapted at all, but have survived in a region where they were lucky enough to have no effective competition. Therefore, they argue that the adaptationist program blinds us to possible alternative explanations for evolution.

Adaptationists think that giving up the search for adaptive function is breaking with a basic principle of Darwinian evolution. Verne Grant, in *The Evolutionary Process* (1985), gives the example of red and yellow onions *vs.* white. "It would not be obvious at first glance," he says, that these color differences are adaptive. But the colored onions are resistant to a smudge fungus that commonly kills the white variety. Chemicals toxic to the fungus turn out to also produce the red and yellow pigments.

"The color of onions is not an isolated instance," Grant concludes. "Time and again, a character which has been considered to be of no adaptive significance has later been found to serve a definite useful function in the life of the organism."

See also ADAPTATION; "JUST-SO" STORIES.

ADAPTIVE RADIATION

See ADAPTATION; DARWIN'S FINCHES; HAWAIIAN RADIATION.

AFAR HOMINIDS
Ethiopian Fossils

Afar is a triangular area in Ethopia between the Blue Nile and the Red Sea, referred to as "Ophir" in the biblical story of Solomon and the Queen of Sheba. But recent discoveries of fossil hominids push its known history even farther back—perhaps to near-humans who inhabited the region three million years ago.

To geologists, the Afar Depression is one of the rare places where three of the Earth's gigantic plates

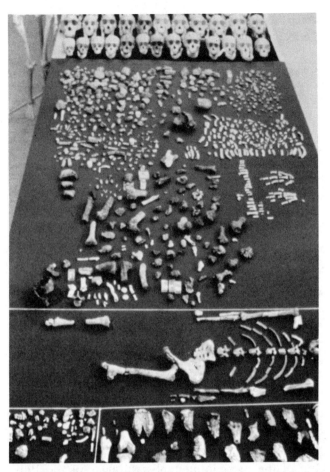

AFAR HOMINID FOSSILS from Donald Johanson's Ethiopian expedition are spread out with chimpanzee skulls from the Cleveland Museum's collection. "First Family" fragments lie near the famous "Lucy" skeleton.

pull and rub against one another—creating a junction of rifts. This geologic movement, and subsequent erosion of exposed sedimentary deposits, has caused scores of Plio-Pleistocene fossils to be exposed near the surface.

Maurice Taieb, a French geologist, began survey work in the region in 1971 with several colleagues, including Jon Kalb and Yves Coppens. The following year this international research team added an American graduate student, Donald Johanson, who had taken part in their 1967 survey of Ethiopia's Omo River Valley. It was a most fortunate collaboration, though not an entirely harmonious one.

While on this expedition to the Afar, Johanson discovered a fossil knee joint, which appeared to indicate an erect-walking hominid. Spurred by the hope of finding more, he returned to the same camp the following year (with Tom Gray) and recovered parts of a pelvis, ribs, arm and hand bones, and skull and teeth. With almost 25% of the original skeleton, it was the most complete fossil of an early hominid

(three million years old) ever found. Since the individual appeared to be female (judged from the pelvis), she was nicknamed "Lucy." (Some of the fossil hunters had been playing "Lucy in the Sky with Diamonds" on a portable tape recorder.)

Johanson concluded they had found evidence of the "oldest" human, a previously unknown species of *Homo*. His colleague Tim White, however, convinced him the partial skull was different enough to belong to a new species of near-men, which they named *Australopithecus afarensis*: "Southern ape of the Afar." Some experts agree with their assessment, but the Leakeys and others see Lucy as a species of *Homo*—or even a mistaken composite of *two* species, one *Homo* and one australopithecine.

When the Johanson–Taieb team returned the following season, they surpassed the previous year's success by unearthing the largest collection of hominid skeletal bones ever found at that time depth, about three million years. This extraordinary fossil trove has been nicknamed "The First Family" and is the first real glimpse of a population sample.

Thirteen individuals of various ages and both sexes are represented; the males appear to have been almost twice as large as the females. All were relatively short (four feet) with fully bipedal hips, legs and feet. From general conformation of the creature, brain size was conjectured to be about that of a chimpanzee, though not one good skull was among all these remains. Their arms were longer than any known human's, but shorter than those of modern apes. They were fully upright walkers, suggesting that human intelligence or language ability is not related to the earlier adaptation of bipedal locomotion.

Anatomist C. Owen Lovejoy expressed amazement at the richness of the Afar discoveries. When Johanson showed him the first knee joint, Lovejoy "told him to go back and find me a whole animal. He obliged with Lucy. So I told him to go back and get me some variety. The next year he found Mom and Pop and the kids."

According to some experts, *afarensis* fragments from East Africa are even more ancient than those from Afar. The oldest known hominid fossil is believed to be a five-million-year-old jawbone of *A. afarensis* found by Yale's Andrew Hill and Kiptalan Cheboi in 1984 at Lake Baringo, Kenya.

See also JOHANSON, DONALD; "LUCY"; OLDUVAI FEMALE.

For further information:
Delson, Eric, ed. *Ancestors: The Hard Evidence.* New York; Alan R. Liss, 1985.

Johanson, Donald, and Edey, Maitland. *Lucy: The Beginnings of Humankind.* New York: Simon & Schuster, 1981.

———, and Shreeve, James. *Lucy's Child*. New York: Wm. Morrow, 1989.

Leakey, Richard. *Origins*. New York: Dutton, 1977.

Lewin Roger. *Bones of Contention*. New York: Simon & Schuster, 1987.

Reader, John. *Missing Links: The Hunt for Early Man*. London: Collins, 1981.

Willis, Delta. *The Hominid Gang*. New York: Viking, 1989.

AFRICAN GENESIS (1961)
Evolutionary Bestseller

"Not in innocence, and not in Asia, was mankind born. [In Africa] we came about . . . on a sky-swept savannah glowing with menace . . . we held in our hand the weapon . . . the legacy bequeathed us by those killer apes, our immediate forbears." Thus begins Robert Ardrey's *African Genesis*, a 1961 bestseller that had enormous impact in shaping the popular image of early man.

Ardrey was a dramatist by profession. During the 1930s, he was one of the young playwrights who created the Theater of Social Protest; they included Sidney Kingsley (*The Dead End Kids*), Clifford Odets (*Golden Boy*) and Irwin Shaw. Ardrey's best known play, *Jeb*, was a controversial exploration of American race relations.

After many years of writing successful dramatic fantasies, Ardrey was drawn to exploring scientific fact, though his critics considered *African Genesis* to be a continuation of his first career. As far as Ardrey was concerned, he was "the accountant and interpreter of . . . a contemporary revolution in the natural sciences."

He traveled to Africa, interviewed fossil hunters Raymond Dart and Robert Broom, studied the literature of animal behavior and emerged with a view of early man as a "killer ape" who ruthlessly defended territory and lived in a society built on male dominance hierarchies. This hominid was a bloodthirsty hunter who turned his weapons against his own kind.

African Genesis was a success for many reasons. First, Ardrey bolstered his arguments with a dramatist's vivid imagery; his prose excites emotion and inspires belief. Second, he was publicizing revolutionary discoveries that had not yet spread beyond scientific circles. Among them were the newer studies of animal social behavior and territoriality and the sensational early "near-man" fossils, found by Dart and Broom, that pointed to an African "cradle of mankind."

Yet his conclusions—based largely on Raymond Dart's speculations—were really not proven by the vast amount of data he marshaled. Parallels were drawn from baboon behavior (as it was then perceived) to the supposed behavior of early man. Damage to fossil bones was attributed to human murder and cannibalism, where other explanations were equally or more plausible. Ardrey wrote a classic of pop sociobiology, followed by *The Territorial Imperative* (1966) and *The Social Contract* (1970).

"We are Cain's children," Ardrey concluded, "man is a predator whose natural instinct is to kill with a weapon. The sudden addition of an enlarged brain to the equipment of an armed already-successful predatory animal created . . . the human being." It was a chilling, pessimistic vision of our past and future, but certainly not the inevitable conclusion to be drawn from the "contemporary revolution in the natural sciences."

See also BABOONS; DART, RAYMOND ARTHUR; HUNTING HYPOTHESIS; KILLER APE THEORY; OSTEODONTOKERATIC CULTURE.

For further information:
Ardrey, Robert. *African Genesis*. London: Collins, 1961.

———. *The Hunting Hypothesis*. New York: Atheneum, 1976.

———. *The Social Contract*. New York: Atheneum, 1970.

———. *The Territorial Imperative*. New York: Atheneum, 1966.

AGASSIZ, LOUIS (1807–1873)
Geologist, Zoologist

One of the most influential naturalists of the 19th century, Louis Agassiz (pronounced AGG-uh-see) was a comparative anatomist of the old school, who had studied with Georges Cuvier in Paris. His comprehensive, meticulous volumes on fossil fish practically established the field, and his work on European glaciations provided the foundation for all future research on Ice Ages.

Raised in the Swiss Alps, he suspected their glaciers were remnants of vast continental ice sheets, which had left such telltale evidence as isolated boulders, deeply etched grooves in surface rock, and characteristic rubble heaped by the moving ice. Geological studies in Scotland and Ireland produced supporting evidence for his glacial hypothesis, which at first was rejected as fanciful by such leading geologists as Sir Charles Lyell. Nevertheless, within a few years his system of periodic ice ages was universally accepted.

After university stints in Germany, Austria, and France, Agassiz immigrated to America; in 1848, he was invited to join the Harvard faculty. (During his travels, his first wife remained in Switzerland, where she died.) In mid-life, Agassiz took an American wife, brought his children to Boston and became a major influence on the establishment of zoology and paleontology in America.

At Harvard, he founded the Agassiz Museum of Comparative Zoology, completed in 1860. It is preserved today as a Victorian-style natural history museum, just as he designed it—a three-dimensional textbook of the Plan of Creation as reflected in classification.

As an educator, he hoped he had "taught men to observe." He advised students to "read nature, not books . . . If you study nature in books, when you go out-of-doors you cannot find her." One of the standard ordeals he imposed on new students was to leave them alone with a preserved fish or bird, telling them to describe as much about it as possible. No books or instruments were allowed. Frequently, he would send them back for additional hours—or even days—of communing with the reeking specimen until he was satisfied with their observations.

Despite his many studies of fossil animals and ancient climatic changes, Agassiz was a staunch antievolutionist. A believer in divine plans and ideal forms, he saw no continuous lineages in the fossil record. Because a diversity of types existed in the earliest strata, he thought that later species must have had separate, successive creations.

His extensive reconstructions of glacial activity only confirmed his view that there could have been no continuous evolution of life through geologic time. When great glaciers covered the continents, he concluded that all living things must have perished. Each time the ice sheets retreated, the Earth was repopulated by divine creation of new species.

TOPPLED FROM HIS PEDESTAL as leading zoologist of the 19th century, Professor Louis Agassiz refused to recognize the Darwinian revolution in biology. After the California earthquake of 1906, a large statue of Agassiz was found upended, with its head stuck firmly in the ground.

In his *Methods of Study in Natural History* (1863), Agassiz compared the idea of continuous evolution—then known as the development hypothesis—to medieval alchemy. "The philosopher's stone is not more to be found in the organic than the inorganic world," he insisted, "and we shall seek as vainly to transform the lower animal types into the higher ones by any of our theories, as did the alchemists of old to change the baser metals into gold."

Standing on his reputation as the greatest naturalist in America, he ridiculed the Darwinian theory when it appeared in 1859 and refused to reconsider his position to the end of his life. Since Harvard botanist Asa Gray was Darwin's friend and ally, the powerful Agassiz missed no opportunity to undercut and discredit his colleague. With his charm, and prestige, he could be a formidable opponent.

Agassiz was stunned, however, when his best students, including his own son Alex, a marine biologist, abandoned his system of thought and adopted Darwinian theory. Because his teacher Cuvier had appeared to prevail over Lamarck's earlier evolutionism, Agassiz had convinced himself that he would "outlive this mania." Instead, his influence gradually eroded until he slipped off his pedestal as a leader of biological thought.

Years after his death, during a California earthquake, a huge marble statue of Agassiz actually did topple from the Zoology Building of Stanford University. Though unbroken, Agassiz was found upside down, head firmly planted in the ground. Some biologists, who recalled his intellectual posture in life, considered the memorial to now be strikingly appropriate. Stanford's first president, the zoologist David Starr Jordan, recorded in his memoirs that campus wits declared Agassiz was "better in the abstract than in the concrete."

See also ICE AGE; MILLER, HUGH; PROGRESSIONISM.

For further information:

Agassiz, Elizabeth. *Louis Agassiz: His Life and Correspondence.* 2 vols. Boston: Houghton-Mifflin, 1886.

Agassiz, Louis. *Geological Sketches.* 2 vols. Boston: Houghton-Mifflin, 1886.

———. *Methods of Study in Natural History.* Boston: Ticknor & Fields, 1863.

Barber, Lynn. *The Heyday of Natural History.* Garden City, N.Y.: Doubleday, 1980.

Lurie, E. *Louis Agassiz, A Life in Science.* Chicago: University of Chicago Press, 1960.

AGNOSTICISM
Seeking Evidence for Belief

Evolutionist Thomas Henry Huxley (1825–1895) may not have been the first agnostic, but he was the first to call himself one. Comparative physiologist, innovative educator, and "Darwin's bulldog," Huxley's interests ranged widely over science, religion and philosophy. He coined the term "agnostic" in 1869, when he joined London's Metaphysical Society, a group of outstanding theologians, scientists and other intellectuals who met to explore issues of truth and belief.

When asked whether he was an atheist, a Christian, a theist, a materialist, an idealist, a freethinker or a pantheist, Huxley was at a loss. Later, he recalled he hadn't "a rag of a label to cover [myself] with" and felt like the proverbial fox who had left his tail in a trap and thereafter was not recognized by his companions.

"The one thing in which most of these good people were agreed," he wrote, "was the one thing in which I differed from them. They were quite sure they had attained a certain 'gnosis'—that is, a revealed knowledge of the truth about existence."

> So I took thought, and invented what I conceived to be the appropriate title of "agnostic," (meaning without revealed knowledge). It came into my head as suggestively antithetic to the "gnostic" of Church history, who professed to know so much about the very things of which I was ignorant; and I took the earliest opportunity of parading it at our Society, to show that I, too, had a tail, like the other foxes. To my great satisfaction, the term took.

Agnosticism, Huxley took pains to point out, "is not a creed but a method," a skeptical, experimental approach to personal belief as well as to science. "In matters of the intellect," he advised, "follow your reason as far as it will take you, without regard to any other consideration," and "do not pretend that conclusions are certain which are not demonstrated or demonstrable." Then, a man "shall not be ashamed to look the universe in the face, whatever the future may have in store for him."

Nevertheless, he believed in "the sanctity of human nature," had "a deep sense of responsibility" for his actions, and could nurture a profound religious feeling in the entire absence of theology. When Huxley's young son died, the Reverend Charles Kingsley asked if he now regretted his lack of belief in the soul's immortality.

In an uncompromising and moving letter, Huxley replied:

> If a jeering devil asked me what profit it was to have stripped myself of the hopes and consolations of the mass of mankind . . . [I should answer] truth is better than much profit. I have searched over the grounds of my belief, and if wife and child and name and fame were all to be lost to me one after the other . . . still I will not lie . . . [I] refuse to put faith in

that which does not rest on sufficient evidence, I cannot believe that the great mysteries of existence will be laid open to me on other terms.

See also: HUXLEY, THOMAS HENRY; METAPHYSICAL SOCIETY; SECULAR HUMANISM.

For further information:

Budd, Susan. *Varieties of Unbelief: Atheists and Agnostics in English Society, 1850–1960.* London: Heinemann Educational Books, 1977.

Coley, Noel G., and Hall, V. *Darwin to Einstein: Primary Sources on Science and Belief.* New York: Longman, 1980.

Huxley, Leonard, ed. *Life and Letters of Thomas Henry Huxley.* London: Macmillan, 1900.

AGRICULTURE (AND EVOLUTION)

See ARTIFICIAL SELECTION; GENETIC VARIABILITY.

ARENS, W.

See CANNIBALISM CONTROVERSY.

AKELEY, CARL (1864–1926)
Artist Who Saved African Apes

Gorillas, one of our close evolutionary kin, have yielded important clues to the roots of human behavior. But without the impassioned concern of artist-taxidermist Carl Akeley, there might have been no gorillas left in the mountains of central Africa for anyone to study.

Akeley became interested in animals while still a boy in the farming town of Clarendon, New York. A self-taught taxidermist by age 13, he practiced on neighbors' pets that had died and then moved to Rochester to work at Ward's Natural Science Establishment. There he helped mount P. T. Barnum's famous elephant, Jumbo, who had been killed in a circus train accident—a big step up from the neighbors' canaries. (Later, Akeley would be the first to mount an entire herd of elephants.)

Unsatisfied with the traditional methods of stuffing animal skins, he invented an entirely new approach for Chicago's Field Museum. He took careful measurements of muscles and bones and then skillfully sculpted clay models in realistic action poses. These anatomically correct sculptures were cast in plaster and a hollow shell was made from the mold. Finally, the skin of the animal was carefully fitted over the completed form, producing a dramatic effect. So startlingly lifelike were his animal groups, especially when combined with meticulous background paintings, that the Akeley Method became the standard for world-class museums.

Akeley went to Africa in search of specimens for the Field Museum in 1896 and again in 1905. He fell in love with its wildlife—though the animals did not always return his affection. On his first trip, he was attacked by a female leopard he had wounded, which bit deeply into his left arm. Somehow, he managed to kill the enraged cat with his bare hands and narrowly escaped with his life.

After his African exhibits won fame and an invitation to dine at the White House with President Theodore Roosevelt, he was hired by the American Museum of Natural History in New York. In 1909, accompanied by his second wife, Delia, he joined Roosevelt's hunting safari in Uganda. Akeley asked the President to shoot an elephant for the museum; Roosevelt's specimen stands guard today in the mounted herd.

During the same field trip, Akeley had another close brush with death when a bull elephant attacked. While recuperating, he conceived the museum's great African Hall: a wide-ranging depiction of the continent's ecology and wildlife. "My fondest dream," he called it, "a great museum exhibition, artistic in form, permanent in construction, the unifying purpose of my work." Millions who would never visit Africa could experience the land and creatures he loved. Despite service in World War I, the breakup of his second marriage and the difficulty of raising funds, he persisted.

Inspired by the accounts of explorer Paul du Chaillu, Akeley became increasingly attracted by the elusive gorillas of the then-Belgian Congo (now Zaire). When he returned to Africa in 1921, he sought them in their remote forest home, became completely fascinated and was the first to photograph wild gorillas on motion picture film. Although their very existence was still new to science, the gorillas were already being ruthlessly hunted by both Europeans and Africans.

Despite intense feelings of affection and kinship for the great apes, Akeley shot five of various ages and sexes, took casts of their faces and hands and brought their skins back to New York for the African Hall. His mounted family group, frozen in time as they browse in their lush mountain forest, is a masterpiece that still excites millions of visitors to the museum. Even Dian Fossey, the most fanatic of gorilla conservationists, said when she saw them for the first time in 1983, that she didn't begrudge Akeley the taking of those gorillas—so respectful had he been in creating the classic exhibit.

Starting about 1922, Akeley became an insistent, lone voice calling out for conserving gorillas as a world treasure, and his campaign impressed the Belgian ambassador. Finally, he took his case directly to King Albert, who convinced the Belgian government to create the Parc National Albert in 1925.

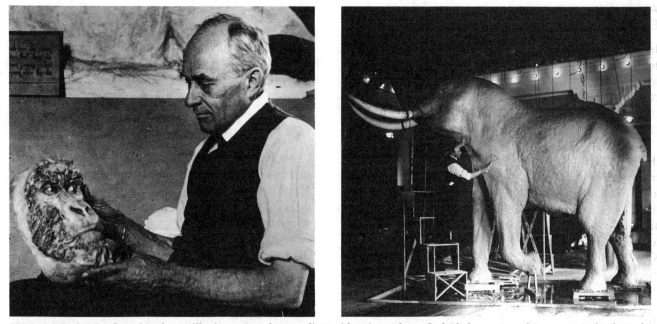

SEEING HIMSELF reflected in the gorillas he protected, naturalist-taxidermist-sculptor Carl Akeley contemplates an ape's death mask. In the American Museum of Natural History's African Hall, he mounted an entire herd of elephants using his "Akeley Method"—a new world-class standard of dramatic artistry and scientific accuracy.

In 1926, Akeley returned to the Virunga volcanoes to work with photographers, artists and botanists on the background for his gorilla habitat group. Although afflicted with dysentery, he led his party through the soaking, misty forests to the heart of the gorillas' homeland, which he considered "the most beautiful spot in the world." There Carl Akeley died and was buried by Mary, his third wife, and a small group of friends.

Concrete was brought to the remote site, and Akeley's forest grave was turned into a permanent memorial, later visited by such noted naturalists as George Schaller and Dian Fossey when they followed him in observing wild gorillas. However, in 1979, at the peak of hostilities between Fossey and local gorilla poachers, the grave was broken into and Carl Akeley's bones were stolen.

Journalists saw this as the Africans' ultimate expression of their resentment toward Europeans who had come into their country to demand they stop killing apes—while other interlopers offered ready cash for gorilla trophies and captured infants. Poachers, it was assumed, had desecrated the shrine of the gorilla defender's "tribal ancestor."

The truth appears to be a bit less dramatic. During the late 1970s, rumors circulated among the local population that the colonial Europeans had left sealed caches of buried treasure around the countryside. Virtually every monument or structure with a concrete base had been broken into in the search for hidden riches, and Akeley's tomb was no exception.

In 1990, some bones and a skull, presumed to be Akeley's, were returned to a naturalist's cabin in the vicinity of Kabara Meadow and were taken to the gravesite by Akeley's biographer, Penelope Bodry-Sanders, who repaired the memorial.

Although the Akeley African Hall is a monument to its creator as well as to the animal, "The greatest of all his achievements," says Bodry-Sanders, "his greatest memorial, is that the mountain gorillas still walk the earth today. They are endangered . . . and under constant pressure from poachers and human encroachment, but they still exist. If it had not been for Carl Akeley, they would probably have been extinct a long time ago." And the gorillas' evidentiary link to human beings might have been lost forever, within a century of their discovery.

See also CHRYSALIS; FOSSEY, DIAN; GORILLAS.

For further information:
Bodry-Sanders, Penelope. *Carl Akeley: Africa's Collector, Africa's Savior.* New York: Paragon House, 1990.

AKELEY METHOD (OF TAXIDERMY)

See AKELEY, CARL.

ALBERT NATIONAL PARK (former Belgian Congo)

See KARISOKE RESEARCH STATION.

"ALEX" (TALKING AFRICAN GRAY PARROT)

See APE LANGUAGE CONTROVERSY.

ALLELOCHEMICALS
Evolutionary Poisons

In the course of the evolutionary "struggle for existence," many animals and plants have evolved chemical defenses against predators. Such natural chemicals that affect the health, behavior, growth or population biology of members of other species are called allelochemicals.

Many plants produce substances that are toxic to the insects, bacteria, rodents and even viruses that try to feed on them. One of the most common allelochemicals, known as L-canavanine, has been identified in 1,200 legumes, including clover, wisteria, alfalfa and some trees. When hookworm larva were fed L-canavanine in experiments at the University of Kentucky, it interfered with their protein metabolism, thereby disrupting development and producing defective adults.

Some creatures utilize allelochemicals even though they cannot produce them. Certain caterpillars, for instance, have evolved an immunity to the poison leaves on which they feed. When attacked by ants, they repel them by regurgitating a chemical derived from these leaves at their foes.

Research into alleolochemicals is providing a new source of pesticides and other agricultural poisons; natural products without the harmful side effects of DDT and other artificial chemicals.

See also NATURAL PRODUCTS CHEMISTRY.

ALLOPATRIC SPECIATION
Evolution and Geographic Isolation

Evolution is profoundly affected by the mobility of populations as well as by restrictions and barriers.

We enjoy a freedom of movement today that—except for some birds and insects—is new in the Earth's history. A woman can fly from New York to Kenya in a day with her dog (and any insects or microbes she and it might be carrying). During most of the Earth's history, land creatures were often blocked in their travels by mountains, oceans, rivers, forests or deserts.

Animal species may consist of populations spread over an area with no natural barriers—no mountains or rivers that prevent them from intermingling. Yet, two related species may evolve from a single, diffused population, perhaps specializing in different kinds of foods in different parts of the range. Harvard's evolutionary theorist Ernst Mayr has called them "sympatric" ("same place") species.

Other species have no choice; population mating restrictions are thrust upon them. For instance, a river may divert and cut right across a population's range, leaving breeding groups on either side. When populations are genetically isolated over a long period, they may diverge, first into geographic subspecies and finally species. In Mayr's terminology, these are "allopatric" ("different places") species.

Because each subgroup has only a part of the total variability of the species' population, they take on different characteristics. By sampling error (see GENETIC DRIFT), one cut-off group may have a higher proportion of certain genes or the presence of particular alleles or mutations.

In small populations, such random differences can easily become established and pass on to a large number of descendants. Thus, the most common "origin of species" may occur on the very edge of species ranges, with very small subpopulations that become isolated from the main population.

Allopatric species, then, are closely related species that occupy geographical areas separated by some barrier. After a time apart, even if the two populations should remingle, they may have evolved differences in behavior, color or vocalization ("isolating mechanisms") that keep them from mating.

See also ISOLATING MECHANISMS; MAYR, ERNST.

ALMAGEST

See FLAT–EARTHERS.

ALTRUISM

See KIN SELECTION; KROPOTKIN, PRINCE PETER.

ALVAREZ, LUIS AND WALTER

See ALVAREZ THEORY.

ALVAREZ THEORY*
Evidence of Ancient Impact

A 1986 headline on *Time* magazine asked: DID COMETS KILL THE DINOSAURS? National attention centered on two geologists, the father and son team of Luis and Walter Alvarez. The Alvarezes were fascinated with a layer of iridium (a rare metal of the platinum family) that had been found at three sites sandwiched in with 65-million-year-old rock. It occurred near the Cretaceous boundary—the time when the great families of dinosaurs died out.

Iridium is a rare element on Earth, but is commonly found in meteors. From their early studies, the Alvarezes concluded that more iridium layers would be found all over the world at the same position in the ancient rock layers; their prediction has since proved true. But their subsequent pronouncement remains controversial. They argued that the iridium was laid down during a catastrophic event occurring at the end of the Cretaceous; this catastrophe also caused climates to cool suddenly and precipitated the great dying. Showers of giant meteors may have bombarded the Earth, raising massive dust clouds that screened the sun's rays. Paleontologists who disagreed claimed that this speculation did not explain the worldwide nature of extinctions, since meteors would not fall uniformly over the whole Earth; they also maintained that the theory failed to account for the survival of turtles, crocodilians, and other cold-blooded groups.

Critics also argued that this new catastrophism is contrary to the tenets of geology—the uniformitarian notions of slow, gradual change. Seeking extraterrestrial causes for great changes in life on Earth, they charged, was a step backwards for science. Even the *New York Times* got into the act, stating in an editorial that "terrestrial events . . . are the most immediate possible causes of mass extinctions. Astronomers should leave to astrologers the task of seeking the causes of earthly events in the stars." Stephen Jay Gould replied that if the *Times* was now advising scientists on what direction to take their theories, then he, as a paleontologist, should be qualified to set the price of newspapers.

During the 1980s, iridium was found in hundreds of localities, as the Alvarezes had predicted, and scientists accepted a worldwide iridium layer as established fact—although some suggested that the layer may have been formed by volcanic forces on earth. In 1987, however, Bruce F. Bohor and his collaborators at the U. S. Geological Survey announced a discovery that lent strong support to the Alvarez Theory. They had studied many samples of the iridium layers and found shocked quartz—an altered structure of quartz crystals caused by tremendous impact pressure—throughout. Volcanic activity does not produce shocked quartz. As more evidence of massive extraterrestrial impacts accumulates, the Alvarez Theory continues to gain adherents.

See also DINOSAURS, EXTINCTION OF; NEMESIS STAR.

For further information:

Muller, Richard. *Nemesis, the Death Star.* With an introduction by Luis Alvarez. New York: Wiedenfeld & Nicolson, 1988.

Raup, David. *The Nemesis Affair: A Story of the Death of Dinosaurs and the Ways of Science.* New York: Norton, 1986.

AMAZON RAIN FORESTS

See BATES, HENRY WALTER; BELT, THOMAS; RAIN FOREST CRISIS; WALLACE, ALFRED R.

AMERICAN PALEOZOIC MUSEUM
Victorian Dinosaur Disaster

After Benjamin Waterhouse Hawkins, the first great dinosaur artist, created a successful prehistoric park in England, Americans wanted a bigger, better one. Under anatomist Richard Owen's direction, Hawkins had sculpted life-size models of the giant saurians at London's Crystal Palace—perhaps inventing the first fantasy theme park. His monsters were so popular with the public that he was invited in 1869 "to undertake the resuscitation of a group of animals of the former periods of the American continents" in New York City's Central Park.

Working with the great paleontologist Edward Drinker Cope in Philadelphia, Hawkins absorbed his dramatic tooth-and-claw view of prehistoric life and planned suitably dramatic poses for the new Paleozoic Museum. He took molds of American fossil dinosaurs and a year later began work in a special studio in Central Park.

There were to be huge hadrosaurs being attacked by smaller dinosaurs, while others fought over carcasses. Giant aquatic reptiles were to be half-submerged in real pools. Visitors would enter a huge, domed cavern, where mammoths and giant sloths filled out this "Complete visual history of the American continent from the dawn of creation to the present time."

Unfortunately, the entire project ran afoul of the infamous Tweed Ring, a nest of corrupt, powerful politicians who controlled most municipal projects. In 1870, their leader, Boss (William M.) Tweed, managed to wrest control of all projects in Central Park from the Commission that had hired Hawkins. With the false claim that the project was costing the city too much money, the Tweed Ring ordered work on the project stopped. And Hawkins was ridiculed in the press as an impractical dreamer.

By the spring of 1871, Hawkins had finished the plaster casts of seven major figures and was battling to keep the project alive. In response, Tweed's hired vandals smashed molds with sledge hammers, destroyed sketches and small models and taunted Hawkins about his *"alleged* pre-Adamite animals." To this day, Tweed's motives are unclear. Was it pure greed, because he couldn't make a huge profit from the project? Or was there also a warped religious conviction to account for Tweed's deep implacable antagonism for the Paleozoic Museum?

Whatever Tweed's reasons, the result was a shattered studio and a Hawkins broken in spirit. He retreated to the Princeton Natural History Museum, where he spent a few years painting imaginative prehistoric landscapes, then returned to his native England in 1877. The smashed remnants of his work, his models, and his studio remain buried today, near 63rd Street, in the southwest corner of Central Park.

A few of his small models and sketches were rescued and give some inkling of what the magnificent exhibit might have looked like. His restorations were fanciful and inaccurate by present standards,

but embody the unique style and viewpoint of their time. Hawkins's first prehistoric garden built for England's Crystal Palace survives at Sydenham Park, near London, where his splendid Victorian dinosaurs remain to delight new generations.

See also DINOSAUR RESTORATIONS; IGUANODON DINNER; NEMESIS STAR.

ANAGENESIS
Single Line Evolution

When a species evolves into a new species, the process is known as anagenesis ("single line origin"). If a splitting or divergence of populations occurs and subsequent development runs along two or more lines, it is called cladogenesis ("branching origin"). Visualized as a family tree, anagenesis refers to the straight, sturdy boughs and cladogenesis to the smaller, forking branches.

Single-line versus branching interpretations of human evolution are endlessly debated by paleoanthropologists. Some are inclined to place most fossil hom-

VICTORIAN VISION OF MONSTERS appears in this drawing by Benjamin Waterhouse Hawkins of the Paleozoic Museum he hoped to build in New York's Central Park during the 1860s. Hawkins's American dream was smashed by corrupt city officials, but a few of his earlier dinosaur models (top, left) can still be seen in a small park in Sydenham, England.

inids directly on the human line, while others consider most of them side branches.

Donald Johanson, for instance, traces *Homo sapiens* back to *Australopithecus* on a single line, of which his discovery *A. afarensis* [see "LUCY"] is the most ancient representative. From the 1960s through the 1970s, it was fashionable to arrange most of the major fossils in a single line: from *Australopithecus* through *Homo erectus* to *Homo sapiens*.

Other experts, notably Richard Leakey, consider *Australopithecus* and other near-men to be side branches, and they seek for ever more ancient "human-looking" humans. Leakey believes the human lineage is distinct, going back through his *Homo habilis* to ancestors as yet unknown.

During the 1960s, Richard Leakey found remains of *Homo habilis* at Lake Turkana, at the same time depth as some of the australopithecines. It had previously been thought that australopithecines predated *H. habilis* and was therefore its ancestor, but Leakey's research seemed to indicate they coexisted at about the same time and place. More recent fossil finds have further complicated any attempts to reconstruct a simple evolutionary tree for hominids. The "Black Skull" discovered in 1986, for instance, provides evidence that a hyper-robust type of australopithecine existed concurrently with the other, more lightly-built hominids.

It seems increasingly likely there was a branching radiation (cladogenesis) about six million years ago, after which several types of apes, humans, man-apes, and near-humans all coexisted. Enough evidence has not yet accumulated to determine whether they shared their habitat amicably or were warring rivals (we may never know), nor do we know which were actually our ancestors and which our side-branching relatives.

See also HOMO HABILIS; JOHANSON, DONALD; LEAKEY, RICHARD; "OLDEST MAN."

For further information:

Delson, Eric, ed. *Ancestors: The Hard Evidence*. New York: Alan R. Liss, 1985.

Johanson, Donald, and Edey, Maitland. *Lucy: The Beginnings of Humankind*. New York: Simon & Schuster, 1981.

————, and Shreeve, James. *Lucy's Child*. New York: Wm. Morrow, 1989.

Lewin Roger, *Bones of Contention*. New York: Simon & Schuster, 1987.

Willis, Delta. *The Hominid Gang*. New York: Viking, 1989.

ANAXIMANDER (600 B.C.)
Ancient Greek Philosopher

One of the earliest recorded theorizers about human origins, Anaximander was a student of the Ionian philosophers at Miletus. His teacher, Thales, believed the primary substance of the universe was water, but Anaximander disagreed.

Water was already a derivative substance, he argued, and thought "the material cause and first element of things was the Infinite . . . neither water nor any other of what are now called the elements." He tried to define a concept of the "boundless" nature of matter, "one eternal, indestructible substance out of which everything arises, and into which everything once more returns"—anticipating the "materialist" view of matter and energy.

He was the first philosopher in the Western tradition to assert the origin of life from nonliving matter, and the first who taught man was "like another animal, namely, a fish, in the beginning." Anaximander also observed that "while other animals quickly find food for themselves, man alone requires a prolonged period of suckling," recognizing the profound significance of lengthy infancy on human society.

However, it is not always possible to understand the real meaning and context of the ancient philosophers. As historian Edward Clodd writes of Anaximander, "In dealing with speculations so remote, we have to guard against reading modern meanings into writings produced in ages whose limitations of knowledge were serious, and whose temper and standpoint are wholly alien to our own."

While there are certainly precursors to modern ideas in antiquity, one must beware of picking and choosing only those fragments that seem to fit in with our current perspective. [See WHIG HISTORY.] The scientific theory of evolution, less than 200 years old, appeared rather suddenly—it is not a system of ideas that was slowly elaborated over centuries.

See also THALES.

ANDERSSON, JOHANN G.

See PEKING MAN.

ANGIOSPERMS, EVOLUTION OF
Origin of Flowering Plants

"That abominable mystery" was Charles Darwin's description of the evolutionary history of angiosperms, or flowering plants. This group, which includes grasses and hardwood trees, seemed to appear suddenly during the Cretaceous period (about 100 million years ago), evolved and diverged rapidly and became the Earth's dominant form of vegetation. But how they originated and why their evolution exploded during the Cretaceous has puzzled botanists since Darwin's day. Recently, paleontologist Robert Bakker proposed an exciting new theory that flowers were "invented" by the dinosaurs!

Angiosperms (meaning enclosed seeded) are the most successful group of plants in Earth's history, with 250,000 living species, compared with 50,000 species of all other green plants combined. They spread widely over the planet, completely altering a landscape that had been dominated by the gymnosperm (naked seeded) group, which includes ginkgo trees and such conifers as pines and firs. Botanist Anthony Huxley (great-grandson of Thomas H. Huxley) describes angiosperms as developing "all kinds of costumes; their sex lives are incredibly varied, with partners of every imaginable kind; and they dabble in drugs, strange perfumes and oils."

Rapid adaptive radiation of flowering plants in the Cretaceous has recently been documented by core samples from the Atlantic coastal plain; these samples show increasingly complex leaf forms and pollen grains developing during a relatively short period. Early kinds of pollen are simple, with a single germination furrow, while later grains show three furrows and a rougher texture, which would make them stickier and more easily transported by insects. Although some flowers are wind-pollinated, most living species rely on animals, usually insects, which implies a long coevolutionary relationship. According to Huxley, insects "sought food and at first went after pollen. To distract them from this plants developed nectar as a substitute, while their flowers became increasingly insect attractive, with scents, colors, and guide patterns."

But coevolution must also have involved the larger animals that fed on plants. Dinosaurs, Robert Bakker points out in *The Dinosaur Heresies* (1986) held the roles of large land herbivores for longer than any other vertebrate group, so there must have been a rich history of adaptive attack and counterattack between planteater and plant."

Whatever their still-obscure origins from gymnosperms (magnolia and buttercup-like flowers seem to resemble the earliest blooms), the early angiosperms certainly evolved for 40 million years in a landscape dominated by plant-eating dinosaurs. While many botanical theories have been advanced to explain their rapid rise, including coevolution with insects or mammals, dinosaurs have been consistently overlooked as a major factor. "That is an extraordinary oversight," argues Bakker, "considering the dinosaurs were the only herbivores large enough to gobble an entire flowering shrub in one gulp." Today's rhinos and elephants can level trees and convert African bushland into open woodland in a few days; thus it seems likely that dinosaurs exerted a major selection pressure on vegetation.

As Bakker reads the fossil record, plant-eating dinosaurs evolved more quickly than plants, with a dinosaur species lasting two or three million years before being replaced by a new one, while plants were changing at a much slower rate, the average species lasting eight million years. While Cretaceous meat eaters like tyrannosaurs were essentially unchanged from their Triassic ancestors, herbivores were developing all kinds of new feeding adaptations—they were the fastest evolving dinosaurs of all.

Herbivorous dinosaurs went through three major periods of development. First there was the Age of Archisaurs, early dinosaurs of the Triassic and Jurassic with comparatively simple teeth and digestive systems. Next, in the late Jurassic, came the Age of High Feeders, like brontosaurs (apatosaurs) and stegosaurs. (In Bakker's reconstructions, stegosaurus customarily stands on his hind legs and balances on his tail, like a kangaroo, to browse on high vegetation.) And finally, the Cretaceous, Age of the Low Feeders, is characterized by many kinds of beaked and wide-mouthed dinosaurs, which fed close to the ground.

Bakker believes the key to the history of this coevolution of dinosaurs and flowering plants is the timing of extinctions. The first flowering plants appeared in the early Cretaceous, after the extinctions of stegosaurs and apatosaurs, just as the Age of the Low Feeders was beginning. He thinks this dramatic shift from Jurassic-type to Cretaceous-type dinosaurs opened the way for flowering plants. Instead of the tall browsers of the Jurassic, herbivores of the Cretaceous were low feeders, grazing close to the ground. This gave an advantage to fast-growing, fast-spreading and fast-reproducing plants—all characteristics of the new angiosperms.

So herbivorous dinosaurs may have given the angiosperms their first break, enabling them to gain a foothold in environments that had previously been dominated by the older types of plants. Of course, stegosaurus didn't purposely die out to let his low-cropping cousins take over. And ankylosaurs ate what tasted good—they didn't plan to munch down the flowers' competition. "Nonetheless, because of the way they suffered extinction and then rebuilt their herbivorous groups," Bakker concludes, "the dinosaurs played a central role in one of the grandest dramas of the flora. In their way, dinosaurs invented flowers."

See also COEVOLUTION; DINOSAUR HERESIES; DIVINE BENEFICENCE, IDEA OF.

For further information:

Bakker, Robert T. *The Dinosaur Heresies.* New York: Wm. Morrow, 1986.

Huxley, Anthony. *Plant and Planet.* Bath, England: Readers Union/Pitman, 1974.

Thomas, Barry. *The Evolution of Plants and Flowers*. New York: St. Martin, 1981.

ANIMAL RIGHTS
Inter–Species Morality

Man's absolute right to use animals as he sees fit—to hunt them, eat them, wear their skins, experiment on them—was for many years justified by the passage in Genesis where God gives man "dominion over all the beasts." Indeed, throughout the 19th century, many scientists maintained that the purpose of animals was to be used by man. Horses were made for human transport, foxes for fur, and sheep and cows for meat. Anatomist Richard Owen of the British Museum even noted that there was a convenient gap in the row of horse's teeth, where the bit was "meant" to be inserted.

When Darwin established an evolutionary kinship between species, he saw at once it might imply a moral obligation not to abuse them. In one of his early private notebooks, he had jotted that "animals our fellow brethren in pain, disease, death, suffering and famine, our slaves in the most laborious works, our companions in our amusements, they may partake from our origin in one common ancestor, we may be all netted together."

His disciple in America, the Harvard botanist Asa Gray, pursued the point: "It seems to me that there is a sort of meanness in the wish to ignore that tie. I fancy that human beings may be more humane when they realize that, as their dependent associates live a life in which man has a share, so they have rights which man is bound to respect." The philosopher Bertrand Russell snorted that "such a philosophy could logically end with the demand of Votes for Oysters."

Yet many religions—including the Jewish and Christian—have a tradition that grants animals the status of independent beings, created without reference to human wants. Those who chose to look beyond the famous passage in Genesis, would also have found this rarely quoted passage:

> For the fate of the sons of men and the fate of the beasts is the same; as one dies, so the other. They all have the same breath, and man has no advantage over the beast for all his vanity. All go to one place; all are from the dust, and all turn to dust again. Who knows whether the spirit of man goes upward and the spirit of the beast goes down to earth? (Ecclesiastes 3:18)

Naturalist Henry Beston put it another way. "Animals," he wrote "are not brethren. They are not underlings. They are other nations—caught with ourselves in the net of life and time, fellow prisoners of the splendor and travail of the earth."

See SPECIESISM.

For further information:

Salt, Henry S. *Animal Rights*. London: Bell & Sons, 1915.
Singer, Peter. *Animal Liberation*. New York: Avon, 1975.
Vyvyan, John. *The Dark Face of Science*. London: Michael Joseph, 1971.

ANNAUD, JEAN-JACQUES

See QUEST FOR FIRE.

ANT–ACACIAS
A Marvel of Coevolution

Acacia trees of Central America, unprotected by poisonous leaves or sap, may seem hopelessly vulnerable to the region's thousands of leaf-devouring insects. But each tree has its own resident ant colony, acting as its immune system by attacking and repelling all intruders. It is a marvel of coevolution between a species of tree and social insect.

Ant-acacias provide the ants with food, partly through special structures at the bases of the leaves that excrete proteins. They are also the only plants that can produce the animal starch glycogen. (Carbohydrate-producing growths in some species are called Beltian bodies, after Thomas Belt, the 19th-century naturalist who discovered this amazing relationship between the acacias and the ants.) Ants keep the tree clean, repelling not only animal or insect intruders, but also destroying any climbing vines.

One acacia species has swollen thorns in which the ants hollow out nests. When an ant queen discovers an unoccupied tree, she burrows into a green thorn, and lays her eggs. The larvae are fed with carbohydrates from the leaf tips. Nine months later a new generation patrols its tree day and night, attacking any other insects they encounter and killing any nearby tree seedlings that might compete for sunlight. Eventually, the colony may number 30,000; when all the thorns on the tree are occupied, the colony may split in two, with one group migrating to another tree.

Ecologist Daniel Janzen has done a classic study of the relationship between ant and acacia and is impressed with their close mutual interdependence. "The experimental work with unoccupied swollen-thorn acacias," Janzen wrote, "indicates that if the acacia-ants were abruptly exterminated, the acacia tree population would be drastically reduced to the point of extinction."

See also BELT, THOMAS; BELTIAN BODIES; GUANACOSTE PROJECT; RAIN FOREST CRISIS.

For further information:

Belt, Thomas. *The Naturalist in Nicaragua.* London: John Murray, 1874.

Holldobler, Bert, and J. Wilson, Edward. *The Ants.* Cambridge, Mass.: Harvard University Press, 1990.

Janzen, Daniel. "Coevolution of Mutualism Between Ants and Acacias in Central America." *Evolution* 20:3, (1966): 243–275.

APE LANGUAGE CONTROVERSY
Capacity for Symbolic Communication

Can animals learn human language? Dyak tribesman told anthropologists they thought wild orang-utans can speak, but pretend to be dumb when humans are around. (The apes, they explained, are afraid men would put them to work if they found out.) Charles Darwin's neighbor, Sir John Lubbock, tried in 1882 to teach his dogs the Sign Language for the Deaf, prompting Samuel Butler's remark: "If I was his dog and he taught me, the first thing I should tell him would be that he is a damned fool." A century later—with Lubbock's dogs long forgotten—attempts to teach chimps human sign language became fashionable in research.

Robert Yerkes, the pioneering psychologist who began to study captive chimps around 1900, noticed they were more apt to mimic movements and facial expressions than the sounds of human speech. Early on, he suggested trying a visual language, but no one did until the 1960s, when a few psychologists began teaching chimpanzees American Sign Language (Amslan).

Soon after, one researcher developed a system of arbitrary printed symbols that was free of inadvertent cues from humans. In honor of Yerkes, he dubbed this language "Yerkish." Washoe and Sarah, the first of the signing chimps, created excitement among scientists and in the popular press. By 1971, psychologists Beatrice and R. Allen Gardner had taught Washoe, a young female chimp, 150 hand gestures in simplified Amslan. And Herbert Terrace's Nim Chimpsky mastered a vocabulary of 132 signs.

Sarah, one of the first famous language-learning chimps, was taught by psychologist David Premack to communicate with plastic chips on a magnetic board. He had hoped to eliminate some of the vaguaries of body language by using standardized chips. Soon Duane Rumbaugh taught another chimp, Lana, to use a system of geometric symbols on a specially designed computer keyboard.

There were two basic questions. Could chimps learn to associate a given sign, symbol or word with its referent? And were they capable of combining them into "sentences" according to some kind of rules (grammar)?

Lana, for instance, learned that a blue triangle meant apple. Each trainer wore a different symbol, so she could identify them as individuals. Later, she was taught a sign for "give." To give an apple, Sarah had to line up four signs: "trainer (name)" / "give" / "apple" / "Sarah."

Eventually, she learned to do this. But just what had she learned? Was she really using language with understanding or was she going through rote actions no different from circus tricks?

Herbert Terrace, one of the founders of ape language research, became deeply troubled by this question and reexamined his data. He knew psychologists had been impressed before by "remarkable animals" in flawed experiments. (Last century, Clever Hans, a horse, amazed scientists when he tapped out answers to mathematical problems, until his apparently neutral trainer was removed from view.) In 1979, with rare courage, Professor Terrace publicly challenged his own methods and accomplishments.

Reviewing the videotapes and records, Terrace could not honestly establish that his chimp had really used the symbols conceptually or mastered any of the grammatical rules that structure human language. Yes, Nim had learned scores of gestures, but he was often haphazard about their sequence. (He didn't seem to know the difference between "Nim eat banana" and "Banana eat Nim.") And he often produced a lot of irrelevant signs, which the experimenters had ignored.

Apes were playing a game all right, Terrace concluded, but not the language game; they were just running off various signs until they got what they wanted. No one had really demonstrated that chimps understood that signs carry definite meanings.

Those who wanted to preserve human uniqueness in the natural world were delighted, echoing Samuel Butler's sentiments: Researchers were deluded fools. Grant funding dried up and no new chimp language programs were begun. But the story was far from over; in fact, some surprises lay just ahead.

Duane Rumbaugh and Sue Savage-Rumbaugh (at the Language Research Center of Yerkes and Georgia State University) never doubted their apes had learned to use symbols. The challenge was to convince everyone else, so they redesigned their experiments. The chimps were now given tasks that were impossible without real symbol using.

Their chimps, Sherman and Austin, would see a symbol flashed on a screen. The chimps then left the room, entered another room containing many objects and were to bring back only the object named (symbolized) on the screen. To do this, they must associate and remember what the symbol stands for. Not only did the chimps retrieve the correct objects, but they came back empty-handed if the named object was not in the other room.

Going even further, Sherman and Austin began to communicate with each other using the signs. After being taught they must share food, the chimps used the symbols to ask each other for specific items. The Rambaughs' work finally convinced many scientists that apes can use signs to convey meaning.

Sue Savage-Rumbaugh and Rose Sevik together made some breakthroughs with pygmy chimps, or bonobos. While trying with little success to teach language behavior to a wild-born female, they inadvertently discovered that her captive-born son "had learned everything we'd been trying to teach the mother."

Still, no one has yet shown that apes can master syntax—the abstract rules for combining symbols. But whether apes have the capacity for symbolizing and language is no longer an all-or-nothing question. They have some of it, and so do other, nonprimate animals. The question now is how much do they have?

During the 1980s, a growing body of research established that language-like abilities (conceptualizing and symbolizing) belong to some very diverse creatures: sea lions, dolphins, pigeons, parrots.

At the University of Hawaii, Lou Herman's dolphins not only associate hand signs with particular objects ("ball," "disk"), but respond correctly to sentence-like instructions. Without having seen the sequence before, they can tell the difference between "Take the disk to the ball" and "Take the ball to the disk."

But perhaps the most astounding language learner of all is an African grey parrot named Alex. Irene Pepperberg, at Northwestern University, has worked with Alex for 13 years. According to their published reports on experiments, Alex can identify seven colors, five shapes, and quantities of up to six. And he needs no plastic chips, computers, or signing gestures—he does it in English!

That parrots mimic human speech sounds is no surprise, but Alex seems to know what he is saying. Pepperberg can present Alex with two wooden triangles, one green and one blue. When he is asked "What's the same?" about them, he replies "Shape." If asked "What's different?" Alex says "Color." He has correctly analyzed hundreds of shape and color combinations he had never seen before. Until the publication of Pepperberg's work in 1987, scientists considered such conceptual use of symbols unique to primates.

The parrot asks for things by name ("Give Alex cork." "Give Alex key."), then demonstrates he really wanted the items by cleaning his beak with the cork

or scratching his legs with the key. Alex knows 80 words and uses them appropriately in simple sentences. He says "No!" when he doesn't want to do something.

Although Pepperberg's research has been published in reputable scientific journals—and even reported in *Newsweek* magazine—Alex has yet to make anything like the popular impact of Jane Goodall's chimps poking termite mounds with twigs. Yet, if confirmed, the bird's accomplishments may well revolutionize our estimation of conceptual capacities among nonhuman beings.

See also CLEVER HANS PHENOMENON; KOKO; NIM CHIMPSKY; WASHOE.

For further information:
Desmond, Adrian. *The Ape's Reflexion.* London: Blond & Briggs, 1979.
Linden, Eugene. *Apes, Men and Language.* New York: Dutton, 1975.
———. *Silent Partners; The Legacy of the Ape Language Experiments.* New York: Times Books, 1986.
Sebeok, Thomas, and Umiken, Jean, eds. *Speaking of Apes: A Critical Anthology of Two-way Communication with Man.* New York: Plenum Press, 1979.

APES
Closest Relatives of Man

Mankind's closest living evolutionary relatives are the four great apes: the gibbons and orang-utans of Asia and the chimpanzees and gorillas of Africa. Also known as anthropoid (meaning man-like) apes, they are tailless with fully flexible shoulder joints and relatively big brains. Although the gibbon is a small creature with proportionally longer arms, most of the apes are large primates; male gorillas often weigh more than 400 pounds.

Apes often make night nests in the crotches of trees, show complex social behavior and subsist mostly on fruits and leafy plants. All live in the deep forest, finding safety among the trees, though gorillas and chimps spend most of their days on the ground. Their common manner of locomotion is by shifting weight from the feet to the underside of the folded fingers—often miscalled "knucklewalking." Gibbons are smaller arboreal specialists that rarely come to the ground; among the apes, they are the most distantly related to man.

Chimps occasionally hunt monkeys or small young antelopes for meat; they are also the only apes that have been observed to regularly make and use simple tools. Chimps use twigs to poke for termites in their mounds and rocks to smash open nuts. They also make "sponges" of dried leaves, with which they gather water from tree hollows.

To Darwin's delight, his friend Thomas Henry

OLD QUESTION about men and apes is expressed in this *B.C.* cartoon by Johnny Hart. Actually, humans did not "come from" any existing ape. Chimps, gorillas, and mankind each represent twigs that share common ancestors.

Huxley proved that, muscle for muscle and bone for bone, apes resemble humans much more than they do monkeys. In *Man's Place In Nature* (1861), his classic of comparative anatomy, Huxley demonstrated that apes are our closest kin in the natural world—a conclusion that biochemistry would confirm a century later.

In 1871, Charles Darwin suggested in his *Descent of Man* that humans probably originated in Africa, because the chimp and gorilla still live there. Closely related species, he reasoned, came into existence in the same general area. He would have been excited, but not surprised, at the dozens of hominid fossils discovered there since 1924, evidence for an African hominid radiation.

The 1970s brought new techniques for studying chromosomes and molecular chemistry that showed an even closer kinship than anyone had suspected. Chimpanzees were found to share 98% of their genetic material with man and gorillas about 95%.

While monkeys have lived next to humans for thousands of years (they appear in the ancient art of Assyria and India), only a handful of tribal people had seen the great apes until the last century. In the late 1600s, a few European anatomists dissected and described the rare chimp and orang carcasses that had been sent them, but confusion abounded about the number of species and their behavior. One pioneering Dutch anatomist, Peter Camper, marred his otherwise brilliant monograph on the chimpanzee by misnaming it a human "Pygmie." Buffon, France's top naturalist, portrayed the chimp standing fully upright and carrying a walking stick.

Gorillas were scarcely known until the 1860s, when explorer Paul du Chaillu created a sensation on his European lecture tour with the stuffed gorillas he had shot. His observations were more those of an excited hunter than a scientist; he recounted how ferocious gorillas bit his gun barrels closed or twisted them into pretzel shapes—stories that have not stood the test of time. At the beginning of the 20th century, naturalists still believed fanciful tales of chimps carrying lighted torches at night and gorillas kidnapping African women.

Alfred Russel Wallace, codiscoverer of the principle of natural selection, was the first European to study the natural behavior of apes in the wild. During his exploration of the Malay Archipelago, the great naturalist observed orang-utans, shot a few for museums and raised an orphaned youngster in his camp.

African apes have been studied in the wild only since the late 1950s, when George Schaller made his pioneering study of the mountain gorilla. Some years later, Dian Fossey was able to gain these gorillas' confidence to the point where she became an accepted member of their troop; Jane Goodall was doing similar work with chimpanzees. Despite their great strength and exaggerated reputation for fierceness, most apes proved to be gentle, approachable and curious about the human observers. On occasion, they can also be extremely aggressive–especially toward rival males from neighboring groups.

All apes are now endangered, as humans invade and destroy their forest environments. Yet apes continue to be in great demand for medical and psychological research, zoos and circuses and they continue to be trapped and hunted by poachers.

Several individual apes have achieved worldwide fame as subjects of experiments on language and cognition (thought). Among those specially educated in the use of sign language are Nim Chimpsky and a gorilla named Koko. Other ape celebrities are Ham, the chimpanzee astronaut, Gargantua, the circus gorilla, and Flo, the wild chimp who befriended Jane Goodall. So many readers of Goodall's books came to know Flo that on her death she became the first nonhuman to receive an obituary in the staid *London Times*.

Of fictional apes, Hollywood's monstrous but vulnerable King Kong—though completely fantastic in size and behavior—is an enduring and tragic character. While his power and menace inspires fear of real gorillas, his helplessness in the face of man's lethal civilization also evokes sympathy.

Another Hollywood legend was created in the 1960s from the French novel *Planet of the Apes* (1963), which was originally intended as a satire on anthropology. Orangs represented the stupid but official scientific establishment, chimps are the irreverent intellectuals, and gorillas the dim-witted, brutal militarists.

Recently, there has been increasing pressure to treat apes more humanely, stop unnecessary or repetitive experiments and intensify efforts to preserve their natural habitats. Today in Zaire there is a stiffer penalty—ten years in jail—for killing a gorilla than for murdering a human being.

Such laws cannot measure the value of an ape against a human life; but they can address the real possibility that gorillas could easily be wiped out within a century of their discovery. As philosopher Walter Gernert has put it, "Since only 5,000 gorillas are left and there are five billion humans, killing one gorilla is proportional to killing several million people, as far as the loss to the gene pool is concerned."

See also "APE WOMEN," LEAKEY'S; CHIMPANZEES; GORILLAS; KOKO; NIM CHIMPSKY; ORANG-UTAN.

For further information:

Kevles, Bettyann. *Watching the Wild Apes: The Primate Studies of Goodall, Fossey & Galdikas.* New York: E. P. Dutton, 1976.

Morris, Ramona and Desmond. *Men and Apes.* New York: McGraw-Hill, 1966.

Reynolds, Vernon. *The Apes.* New York: Dutton, 1967.

APES, TOOL USE OF
A Puzzling Pattern

Great apes are often considered to be of similar intelligence, but their tool-using behavior is not at all alike. Scots primatologist William McGrew tried to compare them and wound up "genuinely puzzled."

As is well known, chimpanzees strip the leaves from twigs to prepare them for poking into termite mounds to capture insects. They wet a stick with saliva so the thirsty insects grab onto it, then withdraw the probe and eat the termites. Chimpanzees chew leaves and use them as sponges for gathering water from tree hollows in the dry season, and have also been observed using two stones to smash open nuts.

In a famous series of experiments conducted by the German psychologist Wolfgang Kohler, a captive chimp piled up several boxes, climbed atop this construction, and used a stick to knock down a banana dangling from the ceiling. Yet these feats pale beside a captive orang's use of a stone tool to make another tool—which captive chimps have been unable to learn. [See ABANG.]

Orang-utans have never been seen making or using tools in the wild. In captivity, though, they are extremely persistent and adept at picking all kinds of locks and latches. Abang, an orang in a Netherlands zoo, startled anthropologists by learning to make flake-stone tools with a hammerstone—something only humans were supposed to be able to do. Although he was taught the skill by a trainer, the orang often chipped stone tools spontaneously, particularly if he needed one immediately.

Gorillas, though closely related to chimps and humans, have never been observed to use tools in the wild and only very rarely in captivity. They seem to have little interest in or ability for dextrous activity, though they do build night-nests out of sticks and leafy branches.

McGrew wondered whether evolutionary closeness to humans is correlated with degree of tool use but found none. Chimps and gorillas are most closely related genetically to humans, but their tool-using behavior seems poles apart. The chimp's spontaneous tool use is great, but it is the more distantly related orang that runs him a close second in the wild and surpasses him in captivity.

Gorillas' tool use is practically nil, on a par with that of gibbons—the apes most distantly related to humans. Although both are vegetarians, gibbons are totally tree-living, while gorillas spend their days on the ground. Results of McGrew's survey for evolutionary patterns in ape tool use, he concluded, ranged ". . . from the puzzling to the perverse."

"APE WOMEN," LEAKEY'S
Primate Field Observers

Dr. Louis Leakey, famous for his discoveries of early hominid fossils in East Africa, wanted more than dry bones to help reconstruct early man—he wanted behavior. During the 1960s, he sought a few exceptional individuals to study our closest living relatives, the African apes.

His recruits would face dangerous, difficult years in the wilderness, trying to approach unpredictable animals capable of tearing a human apart—and they would not carry guns. From the first, Leakey believed that women were better suited for the job than men. Women, he thought, were more perceptive about social bonds and maternal behavior, more patient and capable of long-term dedication and would perhaps appear less of a threat to the dominant male leaders of the ape groups.

Of course, there are men with those qualities who have studied wild animals (George Schaller among the gorillas, George Adamson among the lions), but Leakey was an opinionated man, and his hunches often bore fruit. Nicknamed Leakey's "ape women," his protegees would persevere and develop extraordinary empathy with their subjects.

Jane Goodall, a young Englishwoman, was the first. She came to visit Dr. Leakey when he was curator of what is now the National Museum in Nairobi, and was hired as assistant secretary on the spot. "He must have sensed," she later wrote, "that my interest in animals was not just a passing phase." She accompanied Louis and Mary Leakey on their next paleontological expedition to Olduvai Gorge, where she was exposed to Leakey's enthusiasm for understanding the roots of human behavior. He began to interest her in studying a group of wild chimpanzees he had seen near Lake Tanganyika.

Leakey was particularly interested in the lake shore

JANE GOODALL'S CURIOSITY led her to seek out fossil-hunter Louis Leakey in Kenya, never dreaming he would steer her toward studies of wild chimpanzees that would eventually bring her international acclaim. (Photo by Kenneth Love.)

habitat because early hominid remains showed evidence of lake-side living. When Goodall protested that she was inadequately trained to undertake such a study, "Louis [told me he felt] a university training was unnecessary, [and] even that in some ways it might have been disadvantageous," she later recalled. "He wanted someone with a mind uncluttered and unbiased by theory who would make the study for no other reason than a real desire for knowledge; and, in addition, someone with a sympathetic understanding of animals."

Once she agreed to take on the project, Leakey set about to raise funds for "a young and unqualified girl," and succeeded in persuading the Wilkie Foundation in Illinois to fund her first six months in the field.

Next, Leakey wanted to find a "gorilla girl." Like Goodall, Dian Fossey had no university background in animal behavior, but had loved animals since childhood. She was working as an occupational therapist in Louisville, Kentucky, and had traveled to Africa on vacation to see wildlife; she met Leakey—who spoke of his search to find a woman to study wild gorillas.

In 1966, when Leakey gave a lecture in her home city, Fossey spoke with him about her continuing interest in studying gorillas. "After a brief interview," she wrote in *Gorillas in the Mist* (1983), "he suggested I become the 'gorilla girl' he had been seeking . . . Our conversation ended with his assertion that it was mandatory I should have my appendix removed before venturing into the remote wilderness" of the Zaire volcanoes. She promised she would have the operation, and Leakey wrote to a friend that he had found "another Jane," who would "completely dedicate herself to the study of the Mountain Gorilla."

Six weeks later, on returning home from the hospital minus her appendix, Fossey found a letter from Leakey. It began, "Actually, there really isn't any dire need for you to have your appendix removed. That is only my way of testing applicant's determination!" "This was my first introduction," she wrote, "to Dr. Leakey's unique sense of humor."

A few years later, inspired by the efforts of Goodall and Fossey, another young woman, Birute Galdikas, undertook a prolonged field study of the orang-utan in Indonesia. Three great ape species had acquired dedicated female observers, just as Leakey had envisioned.

The "ape women's" work has become famous far beyond scientific circles. Less well known, however, is that their remarkable and historic studies would not have come about without the prodding and determination of Louis Leakey. Not only had he found a way to flesh out the dusty fossil bones of Olduvai, but he did it with unqualified amateurs, working outside of—and conspicuously outperforming—every established university department of zoology and anthropology in the world!

See also APES; FOSSEY, DIAN; LEAKEY, LOUIS.

For further information:

Fossey, Dian. *Gorillas in the Mist*. Boston: Houghton Mifflin, 1983.

Goodall, Jane van Lawick. *In the Shadow of Man*. Boston: Houghton Mifflin, 1971.

AQUATIC THEORY
Water-Dwelling Ape Ancestry?

Since most primates are covered with hair, why are humans relatively hairless, with a well-developed layer of subcutaneous fat? And why do human infants show an amazing aptitude for holding their breath and swimming underwater almost from birth? These and other questions have led a few unorthodox biologists to the startling theory that in the not-too-distant past, the human species went through a semiaquatic phase.

Originally proposed by German biologist Max Estenhofer in 1912 and more recently by British marine biologist Sir Alister Hardy, the notion of an aquatic phase in human evolution has never been respectable in anthropology. But it has been ably championed by Welsh writer Elaine Morgan in two books, *The Aquatic Ape* (1982) and *The Descent of Woman* (1972)—her response to Darwin's *Descent of Man*, published almost exactly a hundred years earlier.

In addition to smooth, hairless skin, the curiously low placement of the larynx in humans allows the airway to be closed while swallowing food underwater. Morgan argues that these features are rare among land mammals, but are found in seals and dolphins, which began as land-dwelling mammals.

When human babies are brought to deep water for short stays, they adapt happily and quickly. Among the Tupi–Guarani Indians of Brazil and in parts of the Soviet Union, babies are delivered underwater and begin to swim immediately after the umbilical cord is severed. Pearl divers in Japan and Korea and the Varua tribe of New Guinea are in the water constantly from childhood, routinely making dives of 40 feet.

Morgan believes our ancestors moved to tropical seashores in search of food and to escape predators. Gradually, they ventured deeper than the shallow tide pools and began to dive for food. During this time they lost their body fur, became more streamlined in the water and kept only the hair on the head, which was useful in protecting the braincase from the sun and—particularly in females—for providing a handhold for the babies swimming with their mothers. (Men could grow bald without harm to the species, since children, as in all primates, stayed with the mother when foraging.)

Even if the aquatic theory is too far fetched (some have shamelessly called it "all wet" or "out to sea"), Morgan has stirred up some useful discussion about the currently orthodox explanations for body fat and hair loss. "The clumsy explanation currently on offer," she writes, "is that we lost our hair because we got too hot, and then we needed fat to keep us warm." But experiments show that:

> dipilating (removing hair from) a mammal in hot climates sends its temperature up, not down. [If it is argued that sweating won't work without hairlessness, consider that] few mammals sweat as freely as people, but of those that do not a single one—from the patas monkey to the desert donkey—has ever

needed to lose its hair for the process to become effective. [As for needing fat to keep warm] in most mammals, even in cold climates, the subcutaneous fat is the first to be withdrawn. It is often poorly developed in [land-dwelling] neonates, even those like sheep which are born in the coldest season of the year.

Morgan would like to see a search for early man (and woman) conducted in the Danakil Alps of Ethiopa, not far from the supposed cradle of mankind but isolated as a swampy marshland by advancing seas some two-to-five million years ago. If local groups were separated from their savannah food sources by encroaching seas and marshes, they could have adapted by spending more and more time in aquatic environments.

A maverick thinker, Morgan knows her theory is ridiculed by most anthropologists and will probably remain so until someone finds some three-million-year-old australopithecine fossils with piles of cracked seashells in the Danakil Alps. Even if such an aquatic adaptation occurred, it seems to have been very limited. At depths of greater than 48 feet, nitrogen bubbles form in the human bloodstream, producing the disorienting "rapture of the deep," while surfacing too quickly from such depths can cause the fatal "bends."

Nevertheless, she thinks it valuable to keep on asking the question "Why is it adaptive for a savannah primate to become naked and bipedal? When such a question remains unanswered for more than 100 years," she writes, "it may be that it is the wrong question. The alternative 'In what environment would an anthropoid be most likely to become naked and bipedal?' is more productive of new ideas."

For further information:
Morgan, Elaine. *The Aquatic Ape.* New York: Stein & Day, 1982.

ARCHAEOPTERYX
Transitional Fossil

At Solenhofen, Germany, a particularly fine-grained stone has been quarried for centuries because it is smooth and porous, thus particularly suitable for lithography. (Inks are rolled onto the treated stone, which is then used to make prints as in the famous posters of Toulouse Lautrec.) In these Jurassic stones, at least 150 million years old, a beautifully preserved imprint of a feather was discovered in 1861.

That discovery created a sensation, since no birds had been known from that remote period. German paleontologist Hermann von Meyer began to keep watch on the quarry, and soon a complete skeleton of the remarkable creature emerged; von Meyer named it *Archaeopteryx lithographica*—ancient winged creature from the printing stone.

Coming just two years after the publication of *Origin of Species*, the discovery was hailed as a "missing link" between reptiles and birds, a proof of the theory of evolution written in stone. Archaeopteryx was a truly transitional creature, combining attributes of two classes of vertebrate animal. Its feathers were entirely like those of modern birds, but the skeleton was very reptilian.

Some scientists didn't know what to make of archaeopteryx. Was it a bird-like reptile or a reptile-like bird? Anatomist Richard Owen, a bitter enemy of the evolutionists, described it as an aberrant bird. But a sharp-eyed paleontologist, John Evans, noticed that Owen had overlooked a set of fine, perfectly formed teeth within the beak. Congratulating Evans on that crucial observation, Hugh Falconer, a friend of Darwin, joked that perhaps next he would find the creature's "fossil song."

Archaeopteryx is one of the most interesting and important fossils ever found (eventually, five specimens turned up). Its feet are suitable for perching, its pelvis is bird like, yet it does not have the keeled breastbone to which the powerful flight muscles attach in modern birds. Its bones are heavy, not hollow as in later flying birds; it has a lizard-like tail, claws on its wings, and teeth.

Attempts have been made over the years, including one as recently as the 1980s by astronomer Fred Hoyle, to prove that the fossil is a fake cooked up by evolutionists to prove their theories. But the microscopic structure of the feathers, preserved in the fine stone, is too intricate to be copied; the fossils have withstood the most intense scrutiny and have passed every conceivable test for authenticity. They remain an important corroboration of the kind of transitional forms that illuminate the history of life, but are rarely preserved.

See also: BRANCHING BUSH; "MISSING LINK"; TRANSITIONAL FORMS.

ARCHEAN (EON)
Earliest Earth History

Geophysicists and paleontologists are showing increasing interest in the very earliest period of Earth's history, known as the Archean, meaning "oldest." Until recently, this period was usually neglected; attention focused on the fossil-rich records that came later.

Comprising 43% of our planet's history, the eon encompasses Earth's first 2,000 million years. Archean continents were about as thick as they are today, but were devoid of any life—which appears to have originated in the seas some 3,600 million years ago. For most of the Archean, evidence of life consists of microfossils, geochemistry, and stromatolites (structures identical to those produced today by mattes of blue-green algae). Geologists reckon the end of the Archean at 2,500 million years ago, when oxygen first came to dominate the chemistry of the atmosphere.

The bacterial life forms of the Archean never became extinct, despite the rise of their oxygen-breathing descendants. As James Lovelock wrote, "They live on, wherever the environment is free of oxygen. They run the vital and extensive ecosystems of the anoxic zones beneath the sea floor, in the wetlands and marshes, and in the guts of nearly all consumers including ourselves."

ARDREY, ROBERT

See AFRICAN GENESIS.

ARGUMENT FROM DESIGN

See BRIDGEWATER TREATISES; PALEY'S WATCHMAKER; PANDA'S THUMB.

ARISTOGENESIS
Evolution by the "Superior Few"

Paleontologist Henry Fairfield Osborn, the influential director of the American Museum of Natural History in its formative years, was an aristocrat. Upper crust in manners, education and social standing, he believed evolutionary progress was driven by the "superior few"—his theory of aristogenesis.

Osborn extended his notion of superior and inferior to ethnic groups, never doubting he and his circle were the highest products of evolution. To the lasting embarrassment of his admirers, he wrote enthusiastic introductions for his friend Madison Grant's notorious books *The Passing of the Great Race* (1916) and *The Conquest of a Continent* (1933). Grant warned of "Nordic debasement" by a flood of "alien" Italians, Jews, Orientals and Africans, unless America maintained strict immigration quotas and laws against "racial intermarriage."

His theory of aristogenesis also led Osborn to disavow the "apish ancestry" of man—a strange position for America's leading evolutionist. Instead of Darwin's ape-man, he imagined a separate, superior line of "dawn-man," who never lived in forests (*Man Rises to Parnassus*, 1927). "We have all borne with the monkey and ape hypothesis long enough," he told an audience, "and are glad to welcome this new idea of the aristocracy of man back to . . . a remote period." Anatomist William K. Gregory claimed Os-

AUTOCRATIC president of the American Museum of Natural History and a leading paleontologist, Henry Fairfield Osborn startled fellow evolutionists by disavowing an ape ancestry for humans. In this 1930 newspaper cartoon, a chimpanzee is greatly relieved at the news.

born was "afflicted with pithecophobia—the dread of apes as relatives or ancestors."

See also EUGENICS; OSBORN, HENRY FAIRFIELD.

For further information:
Hellman, Geoffrey. *Bankers, Bones and Beetles: The First Century of the American Museum of Natural History.* Garden City, N.Y.: The Natural History Press, 1968.
Osborn, Henry F. *Man Rises to Parnassus.* Princeton: Princeton University Press, 1927.

ARTIFICIAL SELECTION
Breeding Domestic Varieties

Natural selection is a familiar concept from Charles Darwin's *Origin of Species* (1859), but he also wrote several volumes on what he called artificial selection—the creation of domestic varieties by breeders.

In *Origin*, Darwin contrasts natural and artificial selection at some length. He makes it clear that natural selection does not create variability—it merely acts on "the individual differences given by nature." Man selects the variations he can see; Nature "can act on every internal organ . . . on the whole machinery of life."

Man breeds animals for what he finds beneficial to himself, while Nature selects for the benefit of the plant or animal species. Since human breeders have produced the fastest horses, the most succulent fruits and the most ornate pigeons, "What may not natural selection effect?"

Characteristically, Darwin was not content to draw the comparison and rest; he methodically found out everything he could about plant and animal breeding. He built a dovecote behind his house, bred pigeons and attended meetings of the local pigeon enthusiasts. He bombarded horse breeders, gardeners and agricultural experts with his questions. Thousands of botanical experiments were performed in his garden and greenhouse.

In 1868, he published a two-volume work on the subject—*The Variation of Animals and Plants Under Domestication.* Here Darwin covers virtually everything that was then known about the breeding of dogs and cats, horses and asses, pigs, cattle and other barnyard animals, pigeons, canaries, fowl, bees, fruit, flowers and vegetables. Though little read today, its wealth of detail was valuable to contemporary gardeners and breeders.

In fact, Darwin had gathered so much material on pigeons that the bird figured prominently in several of his books. When his publisher submitted *Origin of Species* to an independent evaluator for an opinion, the critic wrote back suggesting he cut the theoretical parts and turn it entirely into a book on pigeon breeding! "Everybody is interested in pigeons," he enthused. "The book would be reviewed in every journal in the kingdom, and would soon be on every library table."

Every major theoretical point was translated by Darwin into a practical research program. Whatever subject he tackled, he did so with an energy and thoroughness that is exhausting even to contemplate. To understand and appreciate his achievement, it is necessary to look beyond the *Origin*, at his lifelong body of work. More than just a "keen observer" or "naturalist"—he was a tireless practical experimenter. When he decided artificial selection could prove the reality of evolution, he became an expert among experts on breeding domestic plants and animals.

See also NATURAL SELECTION.

For further information:
Darwin, Charles. *Origin of Species.* London: John Murray, 1859.
———. *The Variation of Animals and Plants Under Domestication.* 2 vols. London: John Murray, 1868.

ARYAN "RACE," MYTH OF
Racial Supremacist Theory

One of the most infamous and disastrous attempts to trace the racial ancestry of Europe was born as a minor issue in comparative linguistics, developed into a pseudo-Darwinian theory of history and ended by almost destroying civilization.

Originally, the term "Aryan" was applied to a

language group also known as Indo-European. In the 1780s, Sir William Jones, an English Orientalist, concluded that ancient Indian Sanskrit was related to Persian, Greek, Latin, Celtic and the Germanic languages. Detailed comparisons of vocabularies and grammars enabled him to demonstrate that they all must have branched off from a lost mother tongue. He called the language "Aryan" after the "Aryas," an ancient people who had invaded India and Persia. By the mid-19th century, linguists, including the Brothers Grimm and Franz Bopp in Germany, developed "Aryan" studies into an important branch of inquiry, sifting evidence from linguistics, folklore, religious traditions and archeology.

It was but a short, illogical step from the notion of a single mother tongue to that of a single original race that civilized Europe. The period's romanticism fostered the conjecture that an Aryan migration had happened long ago, though no one agreed on when or where. Instead of asking about the reality of the Aryan race, scholars concentrated on determining the Aryan's racial characteristics and land of origin.

Count Arthur de Gobineau, a French journalist, Orientalist, diplomat and historian, identified himself with the French nobility, though in fact he came from a prosperous bourgeois family. Appalled by the Revolution of 1848 and the "democratization" of French politics, he developed an elitist theory that mankind was divided into three races, differing in degrees of superiority: black at the bottom, yellow in the middle, and white at the top. In his *Essays on the Inequality of the Races*, published in the 1850s, he developed his notion that within the white race, the Aryan branch was the highest of the high. Aryans originated in central Asia, he believed, and were tall, blond, alert, honorable and powerful.

Gobineau wrote that he was "sure that everything great, noble and fruitful that man has created on earth . . . issues from a single root, results from a single idea, and belongs to a single family—the Aryan race." In England, Max Muller, a professor of comparative philology at Oxford, was the most eloquent champion of the Aryan source of European civilization.

But as Muller gathered more evidence his belief that European culture was founded by a pure Aryan race evaporated. In 1888, he about-faced, arguing that language has nothing to do with race and that a person of any race can learn to speak any language. "An ethnologist who speaks of Aryan race, Aryan blood, Aryan eyes and hair," he wrote, "is as great a sinner as a linguist who speaks of a dolichocephalic (long-headed) dictionary or a brachycephalic (broad-headed) grammar." But Muller's earlier teachings had been enormously influential.

With the spread of European colonial empires and the inequities of economic domination, came the rise of Social Darwinism as a convenient justification for conquest. If evolutionists had taught that it is "natural selection" for the "fittest" to survive, then it was only right that the "superior" white race should dominate and subjugate people with yellow or brown skin. And blond, blue-eyed people should rule over brown-eyed people, Germans over Jews and so on. Darwin would have been appalled. Many times he had emphasized that he was not a Social Darwinist, detested slavery and had complained that his theories about the natural world were being misapplied to commerce and politics.

In America, the most famous advocate of "Aryan" supremacy were Madison Grant who, in 1916, wrote *The Passing of the Great Race* and Lothrop Stoddard whose *Rising Tide of Color Against White World Supremacy* appeared in 1920. American advocates of the myth of Aryan supremacy propagandized mostly against "mixing" with people of color, although they also tried to bar immigration of "inferior" European types such as gypsies and Jews. European writers such as the Englishman Houston Stewart Chamberlain (*The Foundations of the Nineteenth Century*, 1899) or the German composer Richard Wagner (who published his anti-Semitic diatribe *Judaism in Music* in 1850) directed their venom at European Jews. In their popular works, everything that was good, true and pure was Aryan; everything that was low and degraded was Jewish.

Any serious examination of language stocks or ethnic histories was now completely overwhelmed by polemics, hatred and politics, as the Aryan hysteria continued to mount. Chamberlain wrote this prophetic statement in his *Foundations*: "Though it were proved that there never was an Aryan race in the past, yet we desire that in the future there may be one. That is the decisive standpoint for men of action." When asked to define an Aryan during the height of the Nazi madness, Josef Goebbels proclaimed, "*I decide who is Jewish and who is Aryan!*"

During the German Third Reich (1933–1945), the ideal of Aryan purity and supremacy became that nation's official policy. Adolf Hitler's program of herding "inferior" races into concentration camps and gas chambers was rationalized as making way for the new order of superior humanity. Meanwhile, S.S. officers were encouraged to impregnate selected women under government sponsorship to produce a new "master race"—an experiment that produced a generation of ordinary, confused orphans.

Hitler was furious when the black American Jesse Owens outraced "Aryan" athletes at the 1936 Berlin Olympics, contradicting his theories of racial supremacy. And when the "Brown Bomber" Joe Louis knocked out boxer Max Schmeling, German propaganda became even more vehement that white superiority would be vindicated. However, when Hitler needed the Japanese as allies in World War II, he promptly redefined the Asians as Aryans.

Historian Michael Biddis has commented that "the history of the Aryan myth demonstrates the power of belief over the power of knowledge . . . We may now hear more often of Caucasians than of Aryans, but the substance and errors of the belief in white supremacy linger."

See also LEBENSBORN MOVEMENT; RACISM (IN EVOLUTIONARY SCIENCE).

ATAVISM

See VEBLEN, THORSTEIN.

ATRAHASIS
Babylonian Flood Myth

Clay tablets from ancient Sumer preserve the oldest known account of a great flood or deluge, written thousands of years before the story of Noah appeared in Hebrew. Later, the myth was elaborated by neighboring Babylonians, who conquered Sumer about 3000 B.C., led by King Sargon I of Akkad. In the 19th century B.C., when the powerful Hammurabi extended his empire over both Akkad and Sumer, the old Babylonian (Akkadian) language replaced the Sumerian—but much of the old Sumerian religion and literature was retained.

In the Babylonian flood story, the survivor-hero is called Atrahasis, which means "the exceeding wise"—the man who was saved because of his wisdom. (In the original Sumerian, the hero's name was Ziusidra.) Two versions of the story were translated from tablets excavated by the British at Nineveh during the 19th century: Atrahasis and the Gilgamesh epic.

In the Atrahasis version, at the beginning the gods—sky-god (Anu), Earth diety (Enlil), and god of the deep waters (Enki)—were alone in the universe. These Great Ones put many lesser gods to work digging canals to irrigate food crops, but after 40 years' hard labor they rebelled—so the Great Ones decided to create the human race to provide workers. They proceeded to slaughter a lesser god, mix his flesh and blood with clay, and feed it to 14 birth goddesses, who then bore the original 14 humans: seven males and seven females.

As the human race multiplied and spread over the Earth, they made so much noise that the Earth-god Enlil could not sleep. He sent a plague to kill some, but the wise king Atrahasis appealed to Enki, and the plague ceased. But again the humans multiplied

and made so much noise, that Enlil could not sleep, so he sent a drought to Earth. Again Atrahasis prayed until Enki intervened and, behind Enlil's back, sent rain through the storm god Adad. Once again the human race multiplied and became noisy; Enlil sent a second drought; but Enki again saved mankind—defying his fellow gods by sending fish in a whirlwind for humans to eat.

Finally, the major gods had a meeting and decided that never again should a god save mankind, but Enki refused to join in and laughed at the other gods' decision. Enlil then decided to take matters into his own hands and wipe out this troublesome mankind once and for all with a terrible flood.

Atrahasis foresaw the disaster in a dream and asked "his lord" Enki to show him its meaning. Enki instructed him to destroy his reed house, build a boat with the reeds and "spurn property and save life." The boat was to be roofed, waterproofed with pitch, and loaded with clean animals, fat animals from the farm and also birds and wild creatures. (In the earlier Sumerian story, only farm animals were taken aboard, but in the Babylonian versions birds and wild creatures were added. In the Gilgamesh epic "the seed of all living things" is included for the first time.)

After loading the animals, Atrahasis invited his people to a banquet, but sent his family to board the boat. He was heartbroken and ill as he anticipated the fate of the majority. Soon the storm god Adad thundered, and Atrahasis boarded his ship, sealed the door with pitch and cast off as the flood came and destroyed everything living. As the earth became dark, the mother goddess, Mami, repented agreeing to the destruction and denounced Enlil for killing her offspring. It stormed for seven days and nights, while some of the gods wept.

Fragments of the text are missing but, when the flood subsides, Atrahasis made an offering and the gods smelled its smoke and gathered round. Nintu, the birth goddess, berated the others for trying to destroy humanity and vowed she would remember the disaster always. As they squabbled, Enlil saw the boat and asked angrily how mankind could have possibly survived. Enki admitted he had warned Atrahasis and had instructed him in the building of the ship.

With some additions and changes in emphasis, the Atrahasis story appears in the long epic of Gilgamesh, king of the city of Uruk in southern Babylonia. In about the 10th century B.C., a revised version was first written down by the Hebrews; most scholars believe they received it from their neighbors in Mesopotamia.

See also CATASTROPHISM; "NOAH'S ARK PROBLEM"; ORIGIN MYTHS IN SCIENCE.

AUSTRALOPITHECINES
Man—Apes of Africa

Difficult to pronounce and misleading in meaning, *Australopithecus* refers to the genus of the early hominids, who are our closest evolutionary relatives and possible ancestors. Coined by South African anatomist Raymond Dart in 1924 to describe his famous "Taung baby" skull, it means "Southern ape." But subsequent discoveries showed they were neither southern nor apes. They were upright, bipedal walking hominids that ranged over the length of the African continent, from south and east to the northern areas of Ethiopia.

Four species are currently recognized. Two are of lighter build (gracile), the *Australopithecus africanus*, from South and East Africa, and the *Australopithecus afarensis*, from East Africa and Ethiopia. They were about four feet tall with arms longer than humans but shorter than modern apes; their feet and pelvises indicate bipedalism. Their faces were flatter than modern apes, but jut out farther than *Homo*. *Afarensis* in the oldest, dated at three-to-five million years, and *africanus* is somewhat younger at two-to-three million. Many experts, though by no means all, believe *afarensis* may be ancestral to *Homo* and all other australopithecines. Since they are neither apes nor humans, it was proposed to call them Dartians (after their discoverer, Raymond Dart) to emphasize their uniqueness, but the name never caught on.

Two larger, heavy-boned australopithecines are recognized: *Australopithecus robustus* and *Australopithecus boisei*, the hyperrobust form represented by the spectacular "Black Skull." Formerly known as *Zinjanthropus* or "nutcracker man," for its heavy grinding molar teeth and bony crests that supported strong jaw muscles, the *boisei* were thought to be a later side branch, but now it appears they coexisted in the same times and places with some of the smaller, more lightly built australopithecine species.

See also: AFAR HOMINIDS; AUSTRALOPTIHECUS BOISEI; DART, RAYMOND ARTHUR; DARTIANS.

AUSTRALOPITHECUS AFRICANUS

See AUSTRALOPITHECINES; TAUNG CHILD.

AUSTRALOPITHECUS BOISEI
"Hyperrobust" Hominid

Australopithecines, or man-apes, are generally split into two groups: gracile and robust. The more lightly built, or gracile, species include "Lucy" (*A. afarensis*) from Hadar, Ethiopia and the South African (*A. africanus*) fossils, such as the Taung child. The robust, or heavy, forms have larger teeth, massive bones,

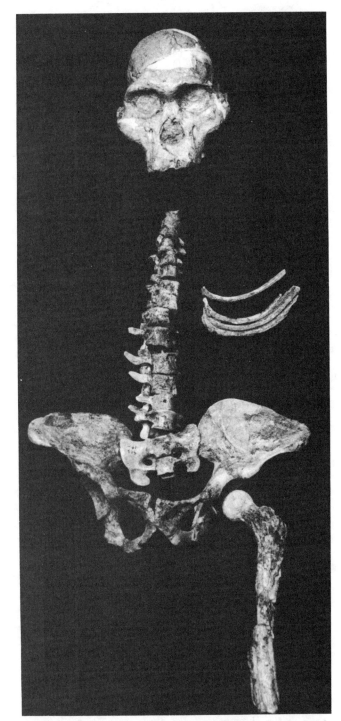

"MRS. PLES" was Dr. Robert Broom's nickname for the South African fossils he named *Plesianthropus*, but lumped with *Australopithecus africanus*. By the 1950s, Broom and his colleagues had collected not only skulls, but also ribs, spinal vertebrae, pelvises and leg bones that clearly indicated an erect, bipedal walker. (Photo by Margot Crabtree.)

and, sometimes, bony crests atop their skulls, which supported strong jaw muscles. Anthropologists have divided the robust types into two species: *A. robustus* and the "hyperrobust" *A. boisei*. Famous among the *boisei* is Mary Leakey's 1959 discovery "Nutcracker Man," nicknamed for his powerful molars.

The more recently discovered "Black Skull," code-labeled KNM-WT 17000, is a very heavy-boned, small-brained australopithecine. Detailed geological work by Frank Brown of the University of Utah on the beds in which the skull was found dated it at between 2.5 and 2.6 million years old.

It was discovered in 1985 by Alan Walker of the Johns Hopkins Medical School, who was working closely with Richard Leakey's group west of Kenya's Lake Turkana. Walker's site, a gully west of the lake, is an hour's drive south of Nariokotome, where the extraordinary skeleton of a *Homo erectus* adolescent had been unearthed the previous year. [See TURKANA BOY.]

The blackened *boisei* fossil hominid had one of the smallest hominid brains on record (410 cubic centimeters), smaller than the much less massive "Lucy," which lived a million years earlier. Also, the "Black Skull" sports the most prominent bony ridges or crests of any known hominid; huge jaw muscles attached to them may have been used for chewing tough plant stalks or crushing pits or seeds. Its head tilted farther forward than those of the much more ancient australopithecines, and the face protrudes more. Its overall structure is immense; two teeth found with the skull are four or five times as large as modern human molars.

Because the new "Black Skull" is three-quarters of a million years older than "Nutcracker Man," it suggests that robust australopithecines were a long-lived, successful branch of the hominid family—not a late, "degenerate" dead end as many researchers had supposed.

Several species of early hominids appear to have been well established in Africa two or more million years ago. Presumably, only one was ancestral to *Homo sapiens*, yet all diverged and coexisted for millions of years during a very complex history that is still unknown.

Walker, Leakey, and their colleagues consider the skull to be an early member of the *A. boisei* lineage, but suggest it might be a separate species, ancestral to "Nutcracker man." Its early date, they believe, makes it unlikely that *A. africanus* was ancestral to *boisei*, but leaves open the question of what its direct ancestry might be.

Whatever the precise origin of humans, the discovery of this ancient, hyperrobust skull has helped destroy both the notion of a single australopithecine lineage and the idea that australopithecines grew increasingly robust through time. While the puzzle appears to grow more complex, the discovery of each new hominid fossil enlarges the overall evidence for

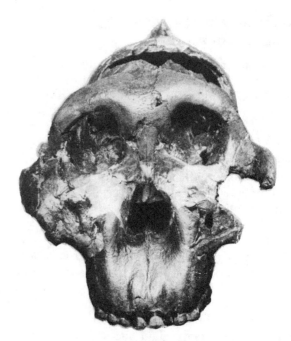

"ZINJ" OR "NUTCRACKER MAN" are two nicknames for this heavy-boned fossil, the *Australopithecus boisei* found by Mary Leakey in East Africa's Olduvai Gorge. This hyperrobust hominid had huge grinding molars and powerful jaw muscles anchored in a special bony crest atop the skull. (Photo by Margot Crabtree.)

the emergence of the human species from a diversified group of ground-dwelling bipedal primate species in ancient Africa.

See also: AFAR HOMINIDS; AUSTRALOPITHECINES; JOHANSON, DONALD; LEAKEY, RICHARD; "LUCY"; TAUNG CHILD.

AVATARS (HINDU ORIGIN MYTH)
Progression of God—Man

Five-thousand-year-old Hindu myths contain striking evolutionary images, such as the rebirths of an avatar (god-man) in successively "higher" forms on Earth. The Sanskrit word means "descent," in the sense of a god descending to the world of men.

Since Hindus believe in cyclic rather than linear time, each descent marks the end of a cycle of destruction and a new creation. Nine avatars, or incarnations, of God seem to represent an evolutionary progression.

As the first avatar, the god Vishnu appeared in the form of a giant fish called Matsya. A prior creation of humanity had ended in a deluge because the people were wicked, and all had been drowned except the king Meanu. Matsya arose from the watery depths to rescue and protect the four Vedas, which contained all knowledge about creation of the inner and outer worlds.

Vishnu's second embodiment was as an immense tortoise that carried the sacred mountain on his back, while other gods sought divine ambrosia in the seas around him. His third incarnation was as a boar; he saved the Earth from a second flood by diving into the sea and carrying the Earth up on his tusks. When humanity again faced destruction, this time from a tyrant king, the avatar returned as Narasingha, the man-lion, and killed the king. In subsequent incarnations, Vishnu took the form of a wild man of the woods; a dwarf-man; and then of the epic hero Rama, whose faithful follower was Hanuman, the monkey-king. Next, he returned as the blue goatherd Krishna, and finally as Buddha.

Evolutionists cannot fail to recognize the implicit progressionism in the sequence of animal incarnations of the avatar—fish, tortoise, mammal, primate, hominid, man, god-man.

Hindus believe that the sequence is cyclic in each age, or Yuga, and that humanity has been created and destroyed many times. In the tenth incarnation, Hindus believe the avatar will appear armed with a scimitar, riding a white horse, to signal the destruction of the present creation. Then he will sleep on the waters and begin a new world when he wakes.

See also: ATRAHASIS; NOAH'S FLOOD; ORIGIN MYTHS (IN SCIENCE); PROGRESSIONISM.

AVEBURY, LORD

See LUBBOCK, SIR JOHN.

AVISE, JOHN C., geneticist

See MITOCHONDRIAL DNA.

AXOLOTL

See NEOTENY.

BABBAGE, CHARLES
See NINTH BRIDGEWATER TREATISE.

BABOONS
Evolved for Fighting or Finesse?

Baboons originally attracted the interest of anthropologists because these large primates had left the forest for the open plains of Africa, the presumed cradle of mankind. Some thought the baboons' social organization and behavior might give clues about the influence of a similar ecology on our ancestors' way of life.

Reconstructions of early human evolution on the African savannah were strongly influenced for more than 20 years (1950s through the 1970s) by the famous study of Kenyan baboons of S. Washburn and I. DeVore. Observing from jeeps, they had focused on big males with dagger-like canines, whose aggressive dominance battles seemed to shape baboon social organization. This older view holds that because males guard the group against leopards and other predators, the biggest, fiercest individuals have their pick of food and consort females. Females were seen as subordinate to the dominant males; the rank order of females was not studied. Dramatic, tooth-and-nail conflicts for patriarchal dominance, reminiscent of Freud's imagined "primal horde," became embedded in scientific literature as the central fact of baboon behavior. This was the image that underlay the "killer ape" theory of human origins popularized by Robert Ardrey in *African Genesis* (1961) and endlessly repeated in anthropology textbooks.

But baboons, like chimpanzees, have "changed" drastically over the past few years, depending on the watcher and length of observation. Shirley Strum of the University of California, San Diego, has been observing a group of Kenyan baboons for more than 17 years. She long ago abandoned her van to follow them on foot and gradually acclimated them to her presence in their midst. She avoids making friends (or enemies) and claims the baboons pay her no more attention than they do a rock or a tree.

Shirley Strum's years among the baboons led her to construct an entirely different picture of baboon social life. She believes their complex behavior is based on social finesse, female rank hierarchies, alliances, friendship and cooperation.

Olive baboons, which Strum studied in a high-altitude savannah in the Kenya Rift, live in groups of from 45–150 individuals. Of the 45 baboons in her "Pumphouse Gang," only seven or eight are big males; the rest are females and juveniles. All spend a good part of their days foraging for plant food; insects, and anything else edible. Occasionally, a male will kill and eat a young antelope or small mammal.

To her surprise, Strum could find no set dominance hierarchies among males, who might bluff-fight several times in a day, reversing position as winner and loser each time. Among the females, however, she discerned a very complex and long-lasting rank system, maintained by groups of close relatives. A low-ranking female, for instance, will not allow her infant to play with another whose mother is of much higher rank. In many cases, the rank is hereditary and maintained by tradition as much as by defense.

Females make friends and alliances with each other and with males. Males can befriend females or juveniles; adult males do not make friends with other adult males. A male new to the group does what Shirley Strum did—he sits on the edge, observes behavior and tries to figure out relationships. He does not, as previous observers thought, attempt to move into the center of the group and take over from the dominant male.

Although the big males are certainly fierce fighters, what appears most interesting is the way other members of the troop get around their strong-arm behavior. A dominant male can compel a female to accompany him; if she doesn't like him, however, she will show it by, for instance, moving away when he tries to mount her and in general being difficult until he tires and loses interest.

When fights did occur between the big males, they frequently had unexpected outcomes. Several times, while large males fought over a female, a much-less-powerful male quietly walked up to the female, groomed her and led her away to copulate while his more powerful rivals were still fighting. Strum describes this as a triumph of "wit and understanding" over brute force.

Sometimes, males attacked by stronger ones will grab a female or juvenile to use as a living shield between him and the aggressor. But this tactic only works if the baby or female is a friend who shrieks at the attacker. If the "shield" has no relationship with the defending male, it may shriek at him instead, causing double trouble—for then he will be mobbed by the whole troop.

Male baboons may hunt but always alone—never cooperatively. When a male makes a kill he does not offer to share it with the group, although he may give some to a favored female.

Strum observed one of the big males kill a small Thomson's gazelle; he was soon approached by a female who liked meat. He didn't offer to share, so she walked behind him and began to groom his back until he was in an ecstatic "grooming stupor." Then she grabbed part of the carcass and ate it.

But this male would not be suckered again. Some days later, he obtained more meat, and she tried the same tactic of lulling him into a pleasure trance. He allowed her to groom him, but every time she made

"So then Sheila says to Betty that Arnold told her what Harry was up to, but Betty told me she already heard it from Blanche, don't you know..."

BABOON TROOPS, once thought to be ruled by their large males, are now seen as a complex web of alliances controlled largely by females, as parodied in this Gary Larson cartoon. Unarmed and on foot, anthropologist Shirley Strum observed groups closely for years on Kenyan plains.

for the meat, he slapped his hand down on it to block her. Finally, she walked away and attacked his favorite female.

This strategy put the male in a real quandary. He looked at the carcass, then at the female he should defend, then at the meat, then at the female. ("Fortunately for us," Strum says, "baboons think with their faces.") Finally, he went to the side of his favorite female; as he did so, the wily one bounded over and grabbed the meat.

According to Shirley Strum, the baboon social organization is one of reciprocity, understanding, friendship and complex behavioral trade-offs. At a public lecture, a questioner suggested that perhaps the old view might have reflected the investigators' own male-dominated hierarchies in their university. "How do we know," he asked, "that your new view does not reflect the rise of feminism in the intervening years?"

"They were in the field for only a couple of months," Strum replied, "and we have been at this for many years now—not just myself but qualified male observers. Washburn was my teacher, and considers me a 'black sheep' for so completely contradicting his conclusions. But we believe we are describing what is actually there." Nevertheless, a London literary critic, reviewing her book *Almost Human* (1987), described Strum's preoccupation with the network of baboon relationships as "very California."

See also: CHIMPANZEES; HUNTING HYPOTHESIS; KILLER APE THEORY.

For further information:
Ardrey, Robert. *African Genesis*. London: Collins, 1961.
Haraway, Donna. *Primate Visions: Gender, Race, and Nature in the World of Modern Science*. New York and London: Routledge, 1989.
Strum, Shirley, *Almost Human: A Journey into the World of Baboons*. New York: Random House, 1987.
Washburn, S., and DeVore, I. "The Social Behavior of Baboons and Early Man." In *Social Life of Early Man*, edited by S. Washburn, 91–105. New York: Wenner-Gren Foundation, 1963.

BACON, SIR FRANCIS (1561–1626)

See INDUCTION; "TWO BOOKS," DOCTRINE OF.

BAKKER, ROBERT

See ANGIOSPERMS, EVOLUTION OF; DINOSAUR HERESIES.

BALDWIN EFFECT
Genetic Assimilation

If changes in offspring can only come about through genes, with no input from behavior or experience,

how can a species respond to its environment? How can an anteater develop a long tongue and snout for preying on anthills or an ostrich develop knee pads for kneeling in the sand?

One answer that was proposed around the turn of the century was genetic assimilation or the "Baldwin Effect" (after American psychologist James Mark Baldwin, one of several who came up with it).

First, said Baldwin, there is an environmental stimulus favoring certain behaviors and their effects—such as ostriches getting calloused knees from kneeling in sand. Next, it becomes common and widespread in the population. Finally, those individuals who have a genetic predisposition to develop larger, thicker pads have a selective advantage and the trait becomes genetically assimilated—mimicking the inheritance of acquired characters.

Random mutations might produce a tendency to form callousities all the time in populations. The point is, if the ostriches were not all engaged in the kneeling behavior, it would not make any impact as a response to experience with environment.

Imagine ancestors of anteaters that find themselves living where anthills are the most abundant source of food. Whatever they were eating before, a change in behavior has to come first. Everybody in the population has to be licking at anthills in order for those individuals with longer snouts and tongues to "take hold."

In his book *The Neck of the Giraffe: Where Darwin Went Wrong* (1987), Francis Hitching complains that genetic assimilation "is so close to what Lamarck himself suggested as to be virtually indistinguishable" from his theory. But there is a real distinction between transmission from parental experience directly to offspring and species "experience" mediated by Mendelian inheritance. The irony is that neither Darwin nor Lamarck made that distinction. Therefore the "Baldwin Effect" can equally claim to be Lamarckian and Darwinian.

See also ADAPTATION; LAMARCKIAN INHERITANCE.

For further information:
Hitching, Francis. *The Neck of the Giraffe: Where Darwin Went Wrong*. New Haven, Conn.: Ticknor & Fields, 1987.

BARLOW, LADY NORA (1885–1989)
Darwin's Devoted Granddaughter

Lady Nora Barlow, granddaughter of Charles Darwin, cheerfully took on many tasks connected with preserving her grandfather's literary legacy. Born Emma North Darwin, she was the daughter of Horace Darwin, an innovative manufacturer of scientific measuring devices at his Cambridge Instrument Company.

Lady Barlow was a good-humored link between her grandfather and 20th-century Darwin scholars; she spent years gathering and preparing many family papers for several books. She edited republications of Darwin's *Diary of the Voyage of the Beagle* (1932), a collection titled *Charles Darwin and the Voyage of the Beagle* (1945), Darwin's notes on birds (1963) and the correspondence between Darwin and his Cambridge mentor, the botanist Joseph Henslow (*Darwin and Henslow*, 1967).

In 1934, while working on the manuscript of Darwin's *Beagle* diary, Lady Barlow visited Admiral Robert FitzRoy's daughter, Laura, who was then in her nineties.

"Charles Darwin," said Miss Laura FitzRoy, "was a great man—a genius—raised up for a special purpose. But he overstepped the mark. Yes, he overstepped the mark for which he was intended." Lady Barlow dutifully recorded her words—a final summation of the fateful relationship between the commander of the *Beagle* and his ship's naturalist so long ago.

See also (H.M.S.) BEAGLE; FITZROY, CAPTAIN ROBERT; GAUTREY, PETER.

For further information:

Barlow, Lady Nora, ed. *Charles Darwin and the Voyage of the Beagle.* New York: Philosophical Library, 1945.

———. *Charles Darwin's Dairy of the Voyage of H.M.S. Beagle.* Cambridge: Cambridge University Press, 1932.

———. *Darwin and Henslow: The Growth of an Idea.* London: John Murray, 1967.

BARNACLES
Darwin's Most Tedious Monograph

In July 1837, a year after returning from his world travels as ship's naturalist, Charles Darwin began a series of notebooks on "the species question." He knew his theories would not be taken seriously unless he first proved himself as a sober, serious scientist capable of detailed work in biology. So for eight years (1846–1854), he devoted himself to the study of barnacles.

Known to zoologists as *Cirripedia* ("curl-footed"), these marine invertebrates first attracted Darwin's attention when he discovered a new parasitic Chilean species that burrows into shellfish. He found no place for it in the established classification, then realized the zoology of the entire group was in complete disorder.

Darwin's four highly detailed volumes on *Cirripedia* (1850, 1851, 1854, 1858) brought accuracy and order to the study of all fossil and living barnacle species. More than a century later, it is still a basic reference for barnacle specialists.

In the course of the long, tedious work, he became a master taxonomist (classifier) and sharpened his understanding of species and their variability—though he wondered whether it was really worth eight years of his life. About halfway through the work he wrote to a friend, "I hate a Barnacle as no man ever did before, not even a sailor in a slow-sailing ship." After Darwin's death, his son, Francis, asked Thomas Henry Huxley if he thought the long barnacle project had been worthwhile and received this reply:

> Your sagacious father never did a wiser thing . . . Like the rest of us, he had no proper training in biological science, and it has always struck me as [a] remarkable . . . insight, that he saw the necessity of giving himself such training, and [courageous], that he did not shirk the labor of it. . . . It was a piece of critical self-discipline, the effect of which manifested itself in everything your father wrote afterwards . . .

During his *Cirripedes* work, Darwin wrote Captain FitzRoy that he was "for the last half-month daily hard at work in dissecting a little animal about the size of a pin's head . . . and I could spend another month, and daily see more beautiful structure."

After several years of this routine, the Darwin children accepted it as a normal part of life. One of his young sons was overheard asking a neighbor's child, "Where does your father work on *his* barnacles?"

For further information:

Darwin, Charles. *A Monograph of the Sub-Class Cirrepidia.* 4 vols. London: Ray Society, 1850–1858.

Darwin, Francis, ed. *Life and Letters of Charles Darwin.* London: John Murray, 1887.

BARNUM, PHINEAS T. (1810–1891)
Entrepreneur of Natural Hoaxtery

P. T. Barnum, America's great showman and shameless hoaxer, knew exactly what the 19th-century public would pay to see: "fossilized" men, "primitive" savages, gorillas and orang-utans, "unnatural" animals and such extremes of human variation as dwarfs and giants. This same wellspring of curiosity also motivated serious students of evolution and natural science.

While he publicized the "educational" value of his exhibits, Barnum was often accused of "nature faking" by scientific experts. Thus he invented the classic ploy of urging the public to come see the disputed exhibit and "judge for yourself."

In 1869, a huge "fossilized man" was found buried on a farm in upstate New York. Known as the Cardiff giant, it became a popular sensation. Thousands of people paid a dollar apiece to file past the pit where

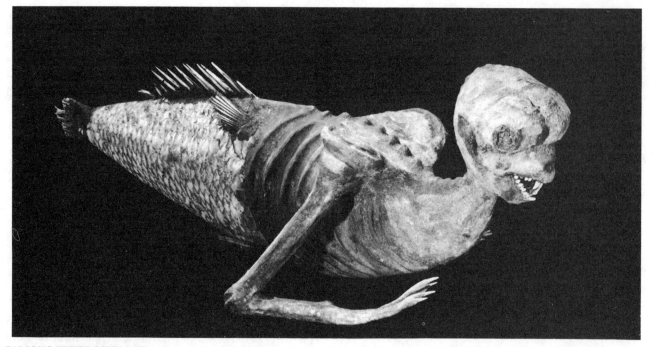

FAMOUS FEEJEE MERMAID was an outrageous fake made of various animal and fish bones skillfully pasted together by Japanese craftsmen. Showman P. T. Barnum challenged curiosity-seekers to "judge for themselves"—by paying the price of admission.

the "remarkable find" lay exposed. Actually, it was a figure carved in stone that had been artificially aged and buried a year before its "discovery" by a local hoaxer.

Recognizing a gold mine when he saw one, Barnum offered to lease the "petrified giant" for $60,000, but the owners were doing so well by exhibiting it that they turned him down. Undaunted, Barnum commissioned a sculptor to carve a copy, which proved even more lucrative than the original fraud after Barnum's promotion campaign. He publicly admitted both giants were frauds, but insisted his was the *real* fake—the other was only a copy.

One of the classic early exhibits at his American Museum in New York City was the celebrated "Feejee Mermaid," which appeared to be a small human skull, torso and arms with the lower body of a fish. Actually, it was a monkey's partial skeleton skillfully joined to that of a large fish by Japanese craftsmen. Although it was denounced by zoologists, thousands paid to "judge for themselves."

Another of Barnum's exhibits, the "Happy Family" was a large cage containing many types of animals and birds, including predators and prey, all living together in peace. Natural enemies like owls and mice, hawks and squirrels all dwelt amicably "like the animals in Noah's Ark."

Barnum claimed he had discovered a great secret of how the laws of nature could be suspended and "the lion lay down with the lamb." He also claimed it was the "one and only original" exhibit of its kind;

in fact, he had copied animal trainers' tricks that had been performed for decades in the streets of London.

Once, when a ship docked in New York harbor fresh from an African voyage and carrying a large primate aboard, newspapers described it as a rare gorilla, which stirred public interest. When Barnum went to see the animal, he recognized it at once as a baboon, but bought the animal anyway for his show.

Billed as a "The Most Remarkable Gorilla in Captivity," Barnum's new star attraction packed in the customers. A zoologist from Yale traveled to New York to see it and was outraged. "Mr. Barnum," he said, "it has a tail. Gorillas don't have tails." "Yes, but this one does," replied the master of humbug. "That's what makes him so remarkable."

See also CARDIFF GIANT; THE HAPPY FAMILY.

For further information:

Barnum, Phineas T. *Struggles & Triumphs, or Forty Years' Recollections.* Buffalo: Worven Johnson, 1872.

Harris, Neil. *Humbug: The Art of P. T. Barnum.* Boston: Little, Brown, 1973.

Saxon, A. H. *P. T. Barnum: the Legend and the Man.* New York: Columbia University Press, 1989.

BATES, HENRY WALTER (1825–1892)
Naturalist on the Amazon

Henry Walter Bates, one of the great pioneers in tropical biology, came from a Leicester family of hosiery makers, but refused to spend his life dyeing socks. While working 13 hours a day, he educated himself by night, reading classics and natural history. Young Bates developed a special passion for collecting beetles—the same enthusiasm that started Charles Darwin and Alfred Russel Wallace on careers as naturalists and evolutionists. [See BEETLES.]

Beetle-mania brought Bates together with Wallace, who was then a young schoolmaster. On weekend bug hunts, the two friends talked of famed German naturalist and traveler Baron von Humboldt and of the *Beagle*'s Charles Darwin, whose writings had inspired them. They, too, wanted to see the South American rain forests and explore the Amazon. They decided to pursue that dream together, financing the journey by selling the specimens they would collect. And they had another, remarkable purpose—to gather data "toward solving the problem of the origin of species." (Darwin, in contrast, had no such plan when he set off for the tropics aboard H.M.S. *Beagle*.)

Wallace and Bates reached Para, at the mouth of the Amazon in May 1848, where they stayed for a year and a half before venturing up a great branch of the river; eventually, they parted company to continue independent explorations.

Wallace collected along the Rio Negro, about which he wrote the classic *A Narrative of Travels on the Amazon and Rio Negro* (1853). He had to write it from unaided memory. On his way back to England, Wallace's ship caught fire, and all his notes and specimens, the product of four years' labor and hardship, were lost. Undaunted, the unsinkable Wallace set out for Malaysia, where he did much collecting and scientific work, including the papers on natural selection that would link his name with Darwin's as cofounder of the new evolutionary biology.

Wallace sent Bates his 1855 paper (the "Sarawak Law") from Malaysia. "The theory I quite assent to," Bates replied from Amazonia, "and, you know, was conceived by me also, but I profess that I could not have propounded it with so much force and completeness."

Four years later, when Bates returned to England (after 11 years in the rain forests), he read Darwin's *Origin of Species* (1859) and became an instant convert. Bates had managed to return to England with almost 15,000 specimens, mostly insects, including 8,000 species that were new to science. Darwin's theories gave him a unifying framework for interpreting his collections and observations.

In 1862, he published a paper ("Insect Fauna of the Amazon Valley") on his Amazonian butterflies in which he showed that good tasting species mimicked the colors and forms of poisonous, foul tasting ones. He explained this "mimicry" as "a most beautiful proof of natural selection," and Darwin praised it as "one of the most remarkable and admirable

MOBBED BY CURL-CRESTED TOUCANS, Amazon explorer-naturalist Henry Walter Bates is surrounded by screaming flock after shooting one of the birds. He and Alfred Russel Wallace had set out for the rain forests to gather evidence on evolution and the origin of species.

papers I ever read in my life." [See MIMICRY, BATESIAN.]

At Darwin's urging, Bates wrote his memoir *The Naturalist on the River Amazons* (1863). Although its composition was a long and painful process, Bates produced one of the great books about the South American rain forest and its peoples. Still in print, it remains indispensable to ecologists and travelers to the tropics. Sadly, it has become a baseline against which to measure the often devastating changes of the past century. It is a vivid personal chronicle, scientifically accurate yet full of metaphor and poetic imagery. One contemporary naturalist enthused, "Bates, I have read your book—I've seen the Amazons." Bates replied that he would rather spend another 11 years in the jungle than ever go through the ordeal of writing another book. He never did either.

See also MIMICRY, MULLERIAN; RAIN FOREST CRISIS; SARAWAK LAW; WALLACE, ALFRED RUSSEL.

For further information:

Bates, Henry W. *The Naturalist on the River Amazons*. London: John Murray, 1863.

Bedall, B. G., ed. *Wallace and Bates in the Tropics: An Introduction to the Theory of Natural Selection*. New York: Macmillan, 1969.

BATESON, WILLIAM

See BIOMETRICIAN–MENDELISM CONTROVERSY; HARDY–WEINBERG EQUILIBRIUM.

(H.M.S.) *BEAGLE*
Vessel of Discovery

In later years, Charles Darwin recalled that he acquired his "real education" during his five-year voyage of discovery aboard H.M.S. *Beagle*—when he was not being seasick.

Built in a shipyard at Woolwich-on-Thames in 1820, the *Beagle* was a 235-ton sloop brig with ten cannons. In strange contrast to this formidable artillery, her unimposing carved figurehead depicted the head and paws of a beagle. It was not an isolated example of the shipbuilder's sense of humor; other vessels in the fleet were H.M.S. *Porcupine* and H.M.S. *Opposum*.

The *Beagle* was of a design sailors had nicknamed "floating coffins" or "half-tide rocks," because they went down so easily in foul weather. During her first surveying expedition along the rough, stormy coastline of Patagonia (1826–1830), she seemed in constant danger of living up to these nicknames. By 1828, she was in poor repair and out of provisions, her crew debilitated by scurvy and disease, and Captain Pringle Stokes, in despair, retired to his cabin and shot himself.

Lieutenant Robert FitzRoy, then a young officer stationed on H.M.S. *Ganges* at Montevideo, was sent to replace the unfortunate Stokes. Shortly after assuming command, FitzRoy almost lost the ship in a violent storm, but he soon mastered the art of sailing her through difficult and jagged channels on the roughest seas. He managed to put the expedition back together and to accomplish its difficult charting, exploring and mapmaking mission, even to establishing the exact longitude of Rio de Janeiro.

For the *Beagle*'s second survey of South America (1831–1836), FitzRoy, then age 23, again commanded her and was promoted to captain; he selected 21-year-old Charles Darwin as his ship's naturalist.

With his assistant, Syms Covington, Darwin worked in a tiny forward poop cabin, preparing literally tons of natural history specimens for shipment to England. He attended mess with the officers and at night shared the Captain's cabin.

"FLOATING COFFINS" or "Half-Tide Rocks" were sailor's nicknames for the sister ships of H.M.S. *Beagle*, the surveying vessel on which young Charles Darwin sailed round the world. (Sketch and reconstruction by Lois Darling.)

Newly refitted, the *Beagle*'s mission was to chart the coast of Patagonia and Tierra del Fuego, then circumnavigate the globe. She sailed from Devonport, England, on December 27, 1831, and returned to Falmouth almost five years later, on October 2, 1836.

Barely 90 feet long, the *Beagle* accommodated 74 men, including eight marines. Also aboard were three lieutenants, a surgeon, and artists Augustus Earle and, later, Conrad Martens. Passengers included three Fuegian Indians known as York Minister, Jemmy Button, and Fuegia Basket, who were returning to Tierra del Fuego, their homeland. Richard Matthews, a missionary, disembarked in New Zealand.

Neither Darwin nor FitzRoy ever sailed on the *Beagle* again, but she made a third surveying voyage. In 1837–1843, she sailed to New Zealand and Australia under the command of Captain Wickham, who had served as an officer under FitzRoy.

Upon her return, the ship was retired from ocean duty, stripped of masts and gear and used as a Coast guard watch vessel (1845–1870) on the River Roach in Essex. After 25 years on the river, she was sold for scrap and towed into the Thames estuary.

A century later (1964), the Charles Darwin Research Station in the Galapagos Islands equipped a modern research vessel and christened her *Beagle II*. Naturalists and scientific specialists use her to learn more about the unique life forms of the Galapagos, which the original *Beagle*'s naturalist said brought him "nearer to that mystery of mysteries, the origin of species."

See also BUTTON, JEMMY; DARWIN, CHARLES; FITZROY, CAPTAIN ROBERT; VOYAGE OF THE H.M.S. BEAGLE.

"BEANBAG GENETICS"
Original Mendelian Theory

"Beanbag genetics" is a nickname for the original Mendelian theory of particulate inheritance current around 1900. A gene was assumed to be a discrete unit, analogous to an individual bean. Even when jumbled together in chromosomes, each remains separate and independent—like beans in a beanbag.

Each gene was believed to carry instructions for a particular trait. Father Gregor Mendel's breeding experiments in his monastery garden isolated one-gene traits (unit characters) such as smooth or unwrinkled peas. Early Mendelists thought there must also be one gene for blue eyes, one for blond hair, and so on. Some fancifully imagined single genes would be discovered for such traits as musical genius or even criminality.

With its emphasis on large mutations, beanbag genetics seemed incompatible with Darwin's idea of "continuous variations." An advantageous or adaptive trait would be swamped before it could be established in a population. [See PAINTPOT PROBLEM.]

After the rise of population genetics (1920–1930), the discovery of recombination, polygenic traits and other sources of variation, beanbag genetics became only a starting point for understanding the subtle and complex mechanisms of heredity.

See MENDEL, ABBOT GREGOR; SYNTHETIC THEORY; WRIGHT, SEWALL.

BEECHER, HENRY WARD (1813–1887)
Influential American Preacher

Reverend Henry Ward Beecher, the superstar of liberal American preachers during the 1870s, reconciled science and religion without antagonizing his congregation of tradesmen. It was inefficient for God to design each species separately, he explained, so He created laws and forces that generated everything else. Belief in evolution was not disrespectful to God because, in Beecher's words, "design by wholesale is grander than design by retail."

Once called "the most admired man in America after Abraham Lincoln," Henry Ward Beecher was on a pedestal of his own making. Crowds from three states ferried to Brooklyn Heights every Sunday to hear his sermons at the Plymouth Church.

His biographers Lyman Abbott and S. B. Halliday claim Beecher's public advocacy of evolution required courage, since:

. . . evolution had come to be identified in the public mind with infidelity. It does undoubtedly involve a recasting of the philosophic statements of creation, sin, revelation, and redemption [which appears] equivalent to an entire abandonment of these truths.

Beecher's series of sermons on "Evolution and Religion," originally delivered to capacity crowds at his church, gained wide popularity. They were printed in newspapers across the country and republished as a book in 1885.

In Beecher's evolutionary view, a personal, conscious God gave successive revelations to a changing human race as it was prepared to receive them—an idea revived a century later in Stanley Kubrick's *2001: A Space Odyssey.*

Beecher taught his congregation that the Book of Genesis "is a poem, not a treatise in cosmogony." Adam and Eve's tragedy in Eden is a poetic parable, not a scientific account of the origin of humanity. Man himself is a composite creature—animal at the bottom, spiritual at the top—gradually being lifted up to a higher moral and spiritual condition ("regeneration").

Those who resist all divine elevating influences, "go down steadily lower and lower until they lose

MOST INFLUENTIAL EVANGELIST in 19th-century America, the Reverend Henry Ward Beecher fused Darwinism and religion into a new gospel of spiritual evolution. At the peak of his popularity, Beecher's reputation was destroyed by a sex scandal.

the susceptibility, the possibility of moral evolution, moral development; let them keep on, and in the great abyss of nothingness there is no groan, no sorrow, no pain, and no memory." Heaven and hell are replaced by the choice between progressive evolution versus degradation. Human moral degeneration is described in terms a naturalist might apply to a parasitic worm. [See DEGENERATION THEORY.]

Its opposite is "Regeneration," brought about by a willingness to live according to divine principles, which raise man to spiritual heights. Design in nature is not disproved by evolution; on the contrary, a grander design—the evolution of souls beyond the physical realm and into higher realms of the spiritual—is illustrated and exemplified. Religion will not be weakened by the idea of evolution, but "vivified and enlarged."

In Beecher's cheerful church, the concept of Original Sin was considered a libel on a kindly, caring God. Sin was now to be considered a relapse into animality. Beecher himself was guilty of just such a relapse when he had an adulterous affair with a woman in his congregation: Mrs. Tilton, who was also his best friend's wife. The resulting front page scandal and uproar was known as "The Great Sensation" and preoccupied the nation for weeks. The elders of his church considered barring Beecher from preaching, but they realized his personality and charisma had brought millions of dollars to Plymouth Church.

During a civil suit brought by Mr. Tilton, Beecher vacillated and was evasive; his testimony was inconclusive. Although he escaped conviction and retained his pulpit, he lost his moral authority and never recovered it.

For further information:

Beecher, Henry Ward. *Evolution and Religion: Part I. Part II.* New York: Fords, Howard, & Hulbert, 1885.

Lyman, Abbot. *Henry Ward Beecher: A Sketch of His Career.* Hartford: American Publishing Co., 1887.

BEE-EATERS

See SOCIAL BEHAVIOR, EVOLUTION OF.

BEETLES
Evolutionist's Inspiration?

According to Alfred Russel Wallace, he and Charles Darwin shared a "child-like" passion for collecting beetles. That was one key similarity between them, Wallace suggested, that may have eventually led them independently to a theory of evolution by natural selection.

Wallace and Darwin recalled into old age their early love for their beetle collections; both could instantly recall the exact place and circumstance of each acquisition. So dedicated a collector was Darwin that he sought beetles between classes at Cambridge and even put one in his mouth so he could hold one in each hand. To his great disgust, the one between his lips gave off a noxious spray, causing him to lose all three prizes.

Both served as professional collectors of natural history specimens. Darwin spent a major part of his time on the *Beagle* collecting fish, insects, rocks, fossils, reptiles and plants for shipment back to England. He later wrote a manual on collecting methods, including such practical matters as labeling, cataloging and packing. When his voluminous, high-quality collections reached the British Museum, his reputation as a naturalist was assured.

For years, Wallace earned his entire living from collecting. He sold thousands of insects to the museum and to private collectors for two cents apiece—his sole means of financing his expeditions to the tropics.

Wallace thought beetle collectors were more likely to have developed an evolutionary perspective than anatomists or physiologists. Thomas Huxley, for instance, focused on living things as "mechanisms." If you dissected one crayfish, following Huxley's lab manual, you pretty much knew them all. Variations or anomalies were seen as departures from the normal type—curios, not sought-after prizes.

But collectors reveled in variations. They liked beetles because there are so many different species, subspecies and seemingly endless varieties. Even a boy, as young Darwin discovered, could find, name and publish a new beetle. (He remembered the thrill for the rest of his life.) When arranged in a collection, beetles formed series that ranged from tiny to the size of a large potato, with hundreds of different colors, patterns, wing shapes.

Eventually, a philosophical collector had to wonder how many good species he had and how many among them were varieties, or hybrids, and why there should be so many differences in the group. A very complete series looked like a subtly blended spectrum.

One might easily visualize a Creator forming several species of big cats, bears, or eagles, but separate creations for 150,000 kinds of beetles gave one pause. As Compte de Buffon wrote years earlier, it was hard to believe God "made a different kind of fold for each beetles's wing." Secondary causes appealed more to reason. The Reverend Henry Ward Beecher would tell his congregation that belief in natural processes is no insult to God—"because design by wholesale is much grander than design by retail."

In Argentina, Darwin was bitten by a very dangerous beetle, the Benchuca, known as the "Great Black Bug of the Pampas." Some biographers attribute his chronic illness to a blood parasite they commonly carry. However, other historians believe the naturalist's mysterious malady was lifelong "psychomatic" stress and neurosis.

During the 1930s, the great British geneticist and evolutionist J. B. S. Haldane was asked by a priest what his long study of nature had taught him about the attributes of God. "For one thing," Haldane replied, "He has an inordinate fondness for beetles."

See also "DELICATE ARRANGEMENT"; SPECIES, CONCEPT OF; TRANSITIONAL FORMS; WALLACE, ALFRED RUSSEL.

BEIJING MAN
Old Fossil's Name Updated

Spectacular finds of early-man fossils in limestone caves near Beijing (formally Peking), China during the 1920s became world famous as Peking man. For years textbooks carried its original scientific name *Sinanthropus pekinensis,* meaning Chinese man from Peking.

Transliteration of Chinese characters into the Roman alphabet has always been difficult and somewhat subjective. For almost two centuries, most Western writers used the Wade–Giles system of transliteration, devised by 19th-century Englishmen. However, in 1979, the Chinese Information Service officially adopted the Pin yin system, which more accurately reproduces Chinese sounds in English spelling. Thus, Mao Tse-Tung became Mao Zedong and Peking became Beijing.

Since the species name *pekinensis* was dropped in 1950, when these early-man fossils were "lumped" into *Homo erectus,* no spelling change in the older scientific name was necessary. For convenience, this encyclopedia will retain the old spelling of the popular name, Peking man, until English speakers switch from "Peking duck" to "Beijing duck" or refer to Pekinese dogs as "Beijinese."

See also CHINA, DARWINISM IN; HOMO ERECTUS; PEKING MAN.

BELFAST ADDRESS (1874)
Tyndall's Shocking Materialism

Among the leading lights of Victorian science, physicist John Tyndall (1820–1893) was an outspoken proponent of evolution who did not shrink from attacking religion's orthodoxies.

As president of the British Association, the nation's leading scientific organization, Tyndall delivered a speech on August 19, 1874 at the Belfast (Ireland) meetings that caused shock waves on both sides of

HOMO ERECTUS skulls were excavated from deep layers in this "Hill of the Dragons" near Beijing. In the 1920s, when the discoveries were made, the fossil hominid was called *Sinanthropus* and the nearby city spelled "Peking."

the Atlantic. It summarized the advances of science, particularly evolutionary biology, and contained an explicit manifesto that scientific truth would replace religious revelation.

"We claim, and we shall wrest from theology," he trumpeted, "the entire domain of cosmological theory." His address began with a sweeping history of science from the Greek philosophers onward, and concluded with a detailed exposition of natural selection. Its major thrust was that the church's long monopoly on questions of human origins and nature was at an end and that all biology was now founded on evolutionary theory.

But what raised the most violent outcry was his assertion that one could discern in matter "the promise and potency of every form and quality of life." For focusing on the creative properties of matter, rather than the Creator, critics branded him with the severe epithet of "Materialist Atheist."

Tyndall was impervious to the insult. Materialist science, he concluded, is the best hope for understanding nature. While "you are not urged to erect it into an idol" he told his audience, "science claims unrestricted right of search."

Most of Tyndall's ideas in "the notorious Belfast Address" were increasingly acceptable to many in the scientific community, but an English gentleman just didn't say such things in public. The Victorian attitude on public expression of unscriptural sentiments is epitomized by a genteel lady's famous remark about *Origin of Species:* "Descended from the apes? Well, let us hope that it is not true—but if it is, let us hope it won't become generally known."

See also AGNOSTICISM; MATERIALISM; MILITARY METAPHOR.

BELLAMY, EDWARD (1850–1898)
Utopian Novelist

Edward Bellamy created the publishing sensation of 1887 with *Looking Backward 2000–1887,* an evolutionary novel about time travel to the end of the 20th century. A futuristic tale about a gentleman who falls asleep in 1887 and awakens in the year 2000, it gave Victorian readers an imagined glimpse of the glorious future just ahead and quickly became a perennial bestseller on both sides of the Atlantic.

"Although in form a fanciful romance," Bellamy explained to one of his critics, *Looking Backward 2000–1887* "is intended, in all seriousness, as a forecast, in accordance with the principles of evolution, of the next stage in the industrial and social development of humanity."

Darwin himself avoided detailed social predictions, but guessed "man in the distant future will be a far more perfect creature than he now is." Future men, he supposed might consider him and his friends "lowly savages." Other evolutionists were not so reticent about pontificating about humanity's next stage of development. English philosopher Herbert Spencer and German anthropologist Ernst Haeckel penned volumes about social evolution.

Former British Prime Minister A. J. Balfour recalled that in his youth everyone, "even my barber rattled on about evolution, Darwin and Huxley and the lot of them—hashed up somehow with the good time coming and the universal brotherhood."

Born in Massachusetts, Bellamy had been trained as a lawyer, but never practiced, preferring to pursue a career in journalism. During a year of study abroad, he was exposed to Haeckel's ideas on social evolution, which had become a political force in Germany. Outraged by the institutionalized injustices of his day, Bellamy's keen interest in social problems and reform led to his own vision of social evolution.

At the heart of *Looking Backward* is a complicated scheme for transforming the existing industrial, political, economic and social system into a more rational and humane one. A superefficient socialistic government would keep track of everyone's needs and provide for them in fair exchange for services. While history has not yet fulfilled Bellamy's all-encompassing vision of an ideal state, many of his incidental predictions—such as credit cards and great music cheaply available for the masses—have proved remarkably accurate.

Music would be piped into private homes ("as many as 150,000 at once") on telephone lines, which greatly impresses the book's time-traveling hero. In his day, he explains, most women played and sang current songs for their families, which caused many to detest music:

> . . . if we should have devised an arrangement for providing everybody with music in their homes, perfect in quality, unlimited in quantity, suited to every mood, and beginning and ceasing at will, we should have considered the limit of human felicity already attained, and ceased to strive for further improvements.

Money had been abolished in Bellamy's world of the year 2000. Everyone is expected to work a certain number of years in "Industrial Service," unless they could qualify for special positions in the arts or professions. Their work earned them certain credits that were put on their credit card accounts. Merit, talent and achievement would replace hereditary class or property as the measure of social worth.

Citizens would receive guaranteed medical coverage, free university education, voluntary occupational retraining and other benefits. Criminals would be treated as "atavistic throwbacks"—reversions to the savage or barbaric stage of evolution—and treated firmly, but with compassion for their misfortune of being less highly evolved than the majority.

Hundreds of thousands of copies of *Looking Backward* were sold in the United States and England; it even became the basis of an influential philosophical-political cult. By the turn of the century, active Bellamy Societies all over America and Europe were optimistically promoting evolutionary socialism as the road to Utopia.

See also MONIST LEAGUE; SHAW, GEORGE BERNARD; SOCIAL DARWINISM.

For further information:
Bellamy, Edward, *Looking Backward, 2000–1887.* Boston: Houghton Mifflin, 1890.

BELT, THOMAS (1832–1878)
Naturalist, Geologist, Engineer

Charles Darwin thought "the best of all natural history journals which have ever been published" was *A Naturalist in Nicaragua* (1874), written by the self-taught ecologist Thomas Belt. Belt's work as a mine geologist and engineer took him to wilderness areas all over the world; during his off hours he explored and recorded the natural wonders he found.

Tropical biologist Daniel H. Janzen has lived for years in the Parque Nacional Santa Rosa in Costa Rica, where he is succeeding in bringing back some of the dry forest plants and animals from the brink of extinction.

In the late 20th century, Janzen often rereads Belt for new insights and precise observations of what the forest was like 100 years ago. But he laments, "Belt was working amidst an ocean of nature, while all that remains to us are small and rapidly shrinking ponds . . . [For modern workers] the consuming tragedy is that . . . Belt's book is a litany of habitat destruction."

On almost every page, Belt introduces a bird, plant or insect that holds his attention, then records his musings about its adaptation or evolution. Often, the engineer's precision is evident, as in his descriptions of the angle at which an insect enters a flower.

Among the adaptations he describes are those of insects that resemble leaves in shape, color, even veins. Some are faded and blotched as if dying; others appear brown and withered with "a transparent hole through both wings that looks like a piece taken out of the leaf."

Even as he marvels at the realistic mimicry, he recalls anti-Darwinian objections that such detailed fidelity "could not have been produced by natural selection, because a much less degree of resemblance would have protected" the insect. What could account for such details as false holes?

Belt answers that natural selection not only protects mimetic (mimicking) prey animals, but also sharpens the perception of the predators, "a progressive improvement in means of defense and attack." On one side the disguise gets better, and on the other so does the ability to penetrate it, until the camouflage becomes very fine-tuned indeed.

This progressive refinement, he goes on, is not different from the way natural selection works on carnivores and their prey on the plains. Faster hares will survive, which will in turn select for faster hounds. From generation to generation, each species pushes up the other's running speed until both become more streamlined and fleet.

See also ANT-ACACIAS; BELTIAN BODIES; GUANA-COSTE PROJECT.

For further information:
Belt, Thomas, *The Naturalist in Nicaragua.* London: John Murray, 1874.

BELTIAN BODIES
Ant Food Factories

Beltian bodies are bright orange growths that appear at the tips of new leaves in the ant-acacia trees of Latin America. These structures have developed to feed the trees' resident ant colonies, which protect and defend them.

Ants get their fats and protein from the Beltian bodies, named for 19th-century naturalist Thomas Belt. He discovered and was first to describe the remarkable coevolutionary relationship between acacias and ant colonies. Belt's classic, *The Naturalist in Nicaragua* (1874), was Charles Darwin's favorite work of natural history.

See also ANT-ACACIAS; BELT, THOMAS; COEVOLUTION; GUANACOSTE PROJECT.

BENGA, OTA (1881–1916)
Man in the Monkey House

Visitors to the New York Zoological Park, better known as the Bronx Zoo, flocked to the monkey house in 1906. There they could see a living spectrum of their "lower" evolutionary cousins: monkeys, chimpanzees, America's first gorilla (Dinah) and an African pygmy tribesman. His name was Ota Benga, and he shared a cage with a parrot and an orangutan named Dohong.

In 1904, Benga was brought to America from the Belgian Congo by the noted African explorer Samuel Verner. According to Verner, he found Benga held captive by another tribe, and paid a ransom to free

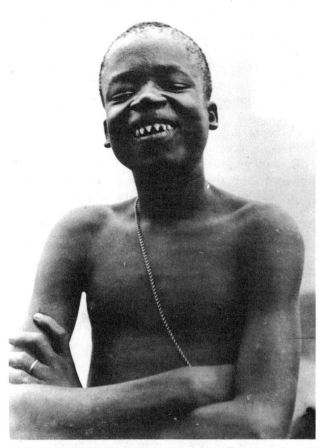

MAN IN THE ZOO, the African "Pygmy" Ota Benga was the only human ever exhibited in an American zoo. Although the New York Zoological Park's director insisted Benga was an employee, free to leave his cage whenever he wished, the incident created a public furor.

him. Then the little fellow helped Verner find other pygmies to accompany him to St. Louis, Missouri—to appear at the St. Louis Exposition. But he couldn't have had the vaguest idea about what he was letting himself in for.

In 1904, the entire "Pygmy village" was exhibited at the St. Louis Exposition along with exotic African wildlife. Afterwards, Verner handed Benga over to eccentric zoologist Dr. William T. Hornaday, the director of the Bronx Zoo during its early years.

Hornaday believed he could read the thoughts of zoo animals and even published a book of interviews with some of his charges. His writings about beasts attribute to them very nearly human thoughts and personalities. But while he considered apes almost human, he saw nothing wrong with keeping a little black man on view in the monkey house. The bizarre display was the only time a caged human being was ever exhibited at an American zoo.

Hornaday did not realize how controversial his monkey house had become. New York's black community was outraged, and church organizations insisted that Benga be freed from his captivity. White fundamentalist clergymen also expressed strong disapproval, but not because the pygmy was a human being caged like an animal. What these churchmen feared was that the exhibition in the monkey house would convince the masses of the truth of Darwin's theory of evolution.

Hornaday explained that he was merely offering an interesting exhibit and that Benga was happy and knew he was absolutely free to leave his cage at any time. This statement could not be confirmed, as Benga spoke no English. However, Hornaday finally bowed to the threat of legal action and insisted the pygmy leave his cage. Dapper in a white suit, Benga paraded around the zoo with huge crowds trailing him, but at night he still returned to the monkey house to sleep.

After a while, Ota Benga began to hate being mobbed by curious tourists and mean children. He fashioned a little bow and a set of arrows and began shooting at zoo visitors he found particularly obnoxious. After he wounded a few gawkers, he had to leave the Zoological Park for good.

For a few years he was employed in a tobacco factory in Lynchburg, Virginia. Several individuals and institutions tried to care for him, but the displaced African grew hostile, depressed and irrational. His malaise could probably have been cured by a return to his native forests, but he had no resources to get back to his homeland. Bitter, lonely and tormented, Ota Benga got hold of a gun and, in 1916, committed suicide.

For further information:

Bridges, William. *A Gathering of Animals: An Unconventional History of the N.Y. Zoological Park.* New York: Harper & Row, 1974.

Sifakis, Carl. *American Eccentrics.* New York: Facts On File, 1984.

BERINGER, JOHANNES (c. 1667–1738)
Victim of Fossil Fraud

There have been several famous attempts to test the credulity of learned men with forged fossils. One of the earliest of these geological hoaxes occurred in 18th century Germany. The victim was Dr. Johannes Beringer, dean of the medical school at the University of Wurzburg, court physician and respected amateur collector of antiquities. It was not a harmless student prank, as some books still claim, but a deliberate attempt to destroy a man's reputation. The villains were two jealous academics: a professor of geography and a librarian.

The plotters forged a number of intriguing stones,

"LYING STONES," including three-dimensional images of strange creatures and bizarre shapes, were dutifully described in Professor Johannes Beringer's book *Lithographia Wirceburgensis*. All were hand-carved hoaxes deliberately planted by malicious rivals to destroy his reputation.

buried them and hired three young students to dupe Dr. Beringer into "finding" them. First, they dug up what appeared to be petrified impressions of small animals and skeletons, but the "fossils" grew more and more improbable. All were carved with considerable skill and buried with great care in local fossil quarries. They included butterflies sipping nectar at flowers, spiders with their webs and even letters of the Hebrew alphabet "fossilized" in stone!

Each successive find delighted the gullible doctor, and he wrote a learned treatise on the wonders of his stone collection and his "search for the truth" about how they might have been formed. His book, the *Lithographia Wirceburgensis*, was published in 1726. Shortly after it came off the press, he found the stone that finally gave the game away—it had his own name on it.

Beringer furiously defended his honor by hauling the perpetrators into the municipal court. They were exposed and duly disgraced, while Beringer recovered his reputation and position. The plotters confessed they had sought revenge on Herr Doktor Beringer because he had treated them with insufferable arrogance and condescension. One historian calls the episode "a devastating comment on the effects of professional competition in science," which required a "tremendous amount of work, expense and time" by the malicious pranksters.

BIGFOOT
Surviving Man-Ape?

Strictly speaking, "Bigfoot" originally referred to the legendary ape-like creatures of northern California, but the name has spread to Oregon and Washington as a synonym for the local Indian term "Sasquatch."

Bigfoot was a nickname invented by a California newspaper in the early 1920s, reporting that huge man-like tracks had been discovered in the woods.

The earliest record of large mysterious man-like footprints in North America dates from 1811 in Alberta, Canada. In British Columbia, Salish Indians claimed the existence of a large hairy "man of the woods," which they called Sasquatch. Other tribes in the area also believed there was such a creature, variously called "Omah" and "Seeahtik."

Reports of sightings and tracks have appeared sporadically since 1900—at least a couple in each decade. Several plaster casts of these footprints exist, but most experts have dismissed them as either fakes or the weathered prints of grizzly bears. A very few anthropologists believe the prints are genuine, among them Professor Grover Krantz of Washington State University, who has spent years searching for the creatures.

Naturalist Ivan Sanderson believed there could exist a relict population of extinct hominids, possibly *Paranthropus* or *Gigantopithecus*. He thought small pockets of them might exist in other isolated parts of the world, such as the Himalayas, where local Sherpa tribesmen call them "Yetis," the "Abominable Snowman" of the Western press.

One lone newspaper account of the capture of a Sasquatch appeared in the *Daily Colonist* published in Victoria, British Columbia on July 4, 1884. According to the article, a creature (nicknamed Jacko) was captured by a train crew along the Fraser River near the town of Yale.

Jacko was said to be less than five feet tall, exceptionally strong and covered with glossy hair all over his body. However, no historian has any further scrap of evidence of what became of him, and there are no photos or other backup to the story.

In December 1971, an amateur photographer named Roger Patterson claimed to have made a home movie film of a Sasquatch at Bluff Creek in northern California. The film was studied and minutely analyzed by experts. Frame by frame, they studied the biped's stride, timing of gait, stature and anything else they could to determine whether it was truly a Sasquatch or just a blurry sequence of a man in a monkey suit.

One expert in biomechanics, Professor D. W. Grieve, wrote:

> My subjective impressions have oscillated between total acceptance of the Sasquatch on the grounds that the film would be difficult to fake, to one of irrational rejection based on an emotional response to the possibility that the Sasquatch actually exists. That seems worth stating because others have reacted similarly to the film.

In other words, he didn't know what to make of it.

See also CRYPTOZOOLOGY; GIGANTOPITHECUS.

BINOMIAL NOMENCLATURE
Linnaean Scientific Names

Crowds gather whenever a zoo is fortunate enough to exhibit a certain roly-poly Chinese mammal. They read the large print on the sign, GIANT PANDA, but usually ignore the scientific name *Ailuropoda melanoleuca* just below.

These two-part Latin names (binomials or binominals) are understood by scientists worldwide. The first names a Genus (a group of related species), while the second indicates the particular species. Sometimes the name of the discoverer or namer is added in third place.

Binomials locate an animal in the scheme of classification, just as a telephone number locates a home phone. The first part is like the area code plus prefix, which identifies the state and locality within it. Genus places the creature in some large class (mammals, birds, reptiles, etc.) and also narrows it down to a few related species within that class.

Just as the last digit of a telephone number identifies only one particular phone, the second name pinpoints the exact species.

Binomial nomenclature was invented by the Swedish naturalist Carl Linnaeus in the 1750s. He thought similarities between groups expressed a divine plan of organization in which species were fixed and unchanging. After Charles Darwin published his *Origin of Species* in 1859, the static classification of Linnaeus was given a dynamic, evolutionary dimension. Similar groups are now thought to be genetic relatives descended from common ancestors.

See also LINNAEUS, CARL.

BIOGENETIC LAW
Ontogeny Recapitulates Phylogeny

As far back as the ancient Greeks, curious observers noticed that an individual's development resembles the larger history of its kind. Knowing that humans are born from a womb filled with liquid, the Ionian philosopher Anaximander supposed humankind itself originated as fishy creatures that emerged from the ocean.

During the intellectual ferment of the 19th century, a German embryologist named Karl von Baer made detailed studies of the stages that individuals pass through in fetal development, or ontogeny. Zoologist Ernst Haeckel, the most influential evolutionist in Germany, took the embryological concepts and wedded them to his notion of Darwinian evolution. The result became known as Haeckel's "Biogenetic Law": Ontogeny recapitulates phylogeny.

That famous phrase, memorized by generations of uncomprehending schoolchildren, means that the fetal development of an individual (ontogeny) is a speeded-up replay of millions of years of species evolution (phylogeny). In other words, a human embryo passes through various stages during its nine months in the womb: invertebrate; fish; amphibian; reptile; mammal; primate; ape; man. A fascinating concept, but the "law" is untrue and was rejected by biologists around 1900. Nevertheless, it has become embedded in many school courses and textbooks and continues to be taught.

Although there is a very general resemblance between the embryo's development and man's evolutionary history, there are too many differences in the particulars. Haeckel's major fallacy was in equating the growth pattern of an embryo with the *adult* forms of various kinds of creatures. If there is any recapitulation, it is of previous types of *embryo* development, which is not at all what Haeckel had in mind.

During the late 19th century, Haeckel's Biogenetic Law was considered one of the proofs of evolution. As detailed research showed it to be a sweeping and superficial generalization, untenable in most particulars, science abandoned it.

For 50 years no biologist tackled an overview of the problem until Stephen Jay Gould's *Ontogeny and Phylogeny* appeared in 1977 and revived interest in the subject. Current attempts to understand relationships between ontogeny (individual development) and phylogeny (evolutionary history) are difficult and technical, and the issue remains unresolved.

See also ANAXIMANDER; FREUD, SIGMUND; HAECKEL, ERNST.

BIOGRAM

See COUNT, EARL W.

BIOLOGICAL DETERMINISM
Possibilities and Constraints

Biological determinism is a hot potato in science and society. Broadly defined, it means that certain behaviors, potentials, even destinies are shaped by the genetic "cards" an individual is dealt at conception. Certainly all scientists can agree that our legs and feet are shaped to make us walkers, our brains process language, our sex determines whether we can give birth.

But many apparent biological differences between individuals cannot be used to predict a human being's potential. Skin color, height, weight, or brain size—all are easily noticed features that have nothing whatever to do with an individual's character or capacity. There have been great short basketball players and brilliant scientists with small brains. The "genetically defective" bodies of Charles Proteus Steinmetz, the electrical genius, or the painter Toulouse-Lautrec did not keep them from lives of brilliant achievement.

Nineteenth-century criminologist Cesare Lombroso, among others, pushed theories of biological determinism far beyond what was justified by the facts. Lombroso, for instance, claimed he could tell a prostitute from a "normal moral woman" just by the shape of her skull, nose, arms, thighs and facial wrinkles. According to Lombroso, the "normal moral woman" was characterized by "passivity, docility, and apathy towards sex." Male criminals, he wrote, could be easily identified by their "feeble cranial capacity, heavy and developed jaws, projecting ears, and crooked or flat noses." For decades, such theories were taken very seriously by university professors, government policymakers, and police departments.

So much human suffering was caused by such ill-founded theories that later generations of biologists, reacting against facile determinism, refused to accept *any* conclusions about biological patterns and constraints of human behavior. Since the 1930s, attempts to measure intelligence of various ethnic groups have provoked bitter controversies; opponents argue that social, rather than genetic, factors are the key determinants of performance.

Still, an evolutionist must try to understand biological bases of behavior, and genetic or sexual differences are legitimate areas for investigation. A major challenge to future researchers is whether such studies can be designed with full awareness of the scientist's or researcher's ingrained cultural biases and expectations.

See also PHRENOLOGY; RACISM (IN SCIENCE).

For further information:
Caplan, Arthur, ed. *The Sociobiology Debate.* New York: Harper & Row, 1978.
Gould, Stephen Jay. *The Mismeasure of Man.* New York: Norton, 1981.
Lewontin, R.; Rose, St.; and Kamin, L., eds. *Not in Our Genes: Biology, Ideology, and Human Nature.* New York: Random House, 1984.
Singer, Peter. *The Expanding Circle: Ethnic and Sociobiology.* New York: New American Library, 1981.

BIOLOGY
Origins of the Discipline

Pioneer French evolutionist Jean-Baptiste Lamarck (1744–1829) coined the word *biologie* to embrace the study of all living things in his *Philosophie Zoologique* (1804). The term was rapidly adopted in Europe, but did not become popular in England and the United States until the 1850s. Few modern biologists realize that their discipline is still a fairly recent field of study.

Until the mid-19th century, students of plants and animals were usually called "naturalists," and their subject "natural history"—which included not only botany and zoology, but geology and sometimes ethnology as well. Darwin came aboard the H.M.S. *Beagle* as "ship's naturalist," equally adept with a microscope, skinning knife, insect net or geologist's hammer. (The fossils and rocks he collected on his travels were as influential as the living species in forming his ideas.)

By the late 1880s, biology was in vogue in England; natural history began to sound vague and old-fashioned. Geology was banished to a separate department, and a caste division arose between the professional biologist and the amateur naturalist, who was now relegated to Sunday butterfly-chaser and bird-watcher. However, some modern, unorthodox zoologists honor the old tradition by calling themselves naturalists, meaning they are curious about everything in nature.

The teaching of life sciences in schools and universities only gained a foothold in the late 19th century; one of the major forces for its establishment was Thomas Henry Huxley, Darwin's friend and colleague. Previously, natural history was something one pursued outside of school. Classics, Greek and Latin dominated academic studies, and it was considered vulgar to dissect frogs in a classroom. After all, how could tearing apart a plant or animal compare with the intellectual virtues to be gained by studying Homer? Professor Huxley, as always, had an eloquent answer:

> Let those who doubt the efficacy of science as moral discipline make the experiment of trying to come to a comprehension of the meanest worm or weed—of its structure, its habits, its relation to the great scheme of nature. It will be a most exceptional case, if the mere endeavour to give a correct outline of its form, or to describe its appearance with accuracy, do not call into exercise far more patience, perseverance, and self-denial than they have easily at their command . . . There is not one person in fifty whose habits of mind are sufficiently accurate to enable him to give a truthful description of the exterior of a rose!

See also HUXLEY, THOMAS HENRY; LAMARCK, JEAN-BAPTISTE.

BIOMETRICIAN–MENDELISM CONTROVERSY (c. 1890–1920)
Darwinism's "Thirty Years' War"

Many textbooks still tell the story of how, when Mendelian genetics was independently rediscovered by three botanists around 1900, it was rapidly embraced as the key to evolution. In fact, those most involved in investigating heredity and evolution—

the mathematical biologists called biometricians—swore eternal hostility to the new ideas. Mendelism and Darwinism, they insisted, were conflicting, incompatable theories; one or the other must triumph, but science could not contain both.

For their part, the early Mendelians correctly assessed the lasting value of their founder's contribution, but completely misunderstood its implications for evolution. According to geneticist R. A. Fisher, "They thought of Mendelism as having dealt a death blow to [natural] selection theory." Debate raged for decades, often growing acrimonious and personal. Whichever faction won, it seemed the Darwinian theory of evolution would be the major casualty.

Charles Darwin and Alfred Wallace had based their theory of natural selection on the idea of continuous variation "by slow and insensible degrees," allowing living populations to diverge "indefinitely from the original type." To the early Mendelians, if genes were really nonblendable units [see "BEANBAG GENETICS"], then evolution must proceed by hops and skips rather than smooth skating.

Following Darwin, biometricans had insisted that evolution must proceed gradually by natural selection of small heritable variations; his dictum, *"Natura non facit saltum"* ("Nature makes no jumps"), had become their battle cry. In England, Raphael Weldon and Karl Pearson led this school, which reached its zenith during the closing years of the 19th century.

Biometricians were so convinced they were discovering how inheritance worked, they scorned any help from Mendel's long-forgotten pea plants. Karl Pearson had built upon his mentor Francis Galton's "Law of Ancestral Heredity" (1897), an attempt to find the "formula" for "blending inheritance." Working with the occurrence of two colors in a thousand pedigreed basset hounds, Galton worked up statistics based on the carefully kept stud books of dog breeders.

Galton's "Law of Ancestral Heredity," in his own words, states "that the *average* contribution of each parent is one-fourth, of each grandparent one-sixteenth and so on. Or [the contribution] of the two parents taken together is one-half, of the four grandparents together one-fourth, and so on. The hero-worshipping Pearson believed Galton's useless formula "must prove almost as epoch-making as the law of gravitation to the astronomers."

Opposing the biometricians were the arch-Mendelian William Bateson and mutationist Hugo deVries, who maintained evolution must proceed by a series of jerky leaps. Continuous variations were too small, they argued, to offer any real selective advantage. New species must have appeared instantaneously, since selection could not create new species, although it did sharpen and refine them by whittling away at the extremes of variation in each generation.

Arguments grew increasingly bitter, then vicious. Some of Mendel's critics published outrageously offensive articles, laced with personal insults. Incensed by one particularly arrogant, condescending attack on Mendel, William Bateson wrote a crackling, impassioned reply, which grew into the first genetics textbook, *Mendel's Principles of Heredity: A Defense* (1902). Now the battleground shifted to the research front.

Up to about 1910, results of experiments seemed to favor the Mendelians, which won over most general biologists (but not the biometricians). Genetic ratios were confirmed in many animals, including man, genes were actually seen and located on chromosomes, multiple gene effects and linkages were discovered and "pure line research" showed Darwinian gradualism was bounded by internal constraints, beyond which selection was useless. [See PURE LINE RESEARCH.]

However, by the 1920s, the pendulum swung again in favor of the Darwinians. New research had proven many mutations result in very small differences, that pure line limits can be pushed by multifactor inheritance, that "modifier genes" can change the expression of other genes or combinations of them. These modifiers, which at first seemed unpredictable, were found to act in a Mendelian (discontinuous) manner, but produce a Darwinian (gradualist) result.

Between the 1920s and 1930s, the *students* of Bateson, Weldon and Pearson realized their conflicting views could be synthesized to produce a comprehensive theory of evolution. R. A. Fisher, Sewall Wright and J. B. S. Haldane (among others) laboriously worked out mathematical models that reconciled the statistical study of populations with classical recombinant Mendelian heredity.

Although students of the biometricians and Mendelists at last managed to reconcile the long-standing controversy, the principals never accepted their ingenious synthesis. When R. A. Fisher submitted his first paper pointing the way to the "genetical theory of natural selection"—now considered a classic—it was sent to Bateson and Pearson by the Royal Society. For once the sworn enemies were in agreement: Both recommended rejection! (It was subsequently published in 1918 in Scotland.)

Neither Bateson, Weldon, nor Pearson ever gave an inch. Weldon died in 1906, still immersed in baiting Bateson over the inheritance of coat color in horses. Pearson, as late as 1930, was an unrepentant anti-Mendelian. And Bateson, with no feeling for mathematics or statistics, never could see how Mendelism could be fitted into Darwinian evolution.

Historians still disagree over the causes of the

disagreement. Some have argued that the depth of the prolonged animosity was irrational—though the factions understood each other's work, they were driven by irreconcilable sociopolitical influences.

Karl Pearson and the biometricians belonged to a rising professional middle class fired up about eugenics and other "progressive" solutions to social problems. Their commitment to inexorable gradualism was in tune with their deep belief that the world could be improved by evolutionary science without waiting for "hopeful monsters" to arise. The Mendelians (so the argument goes) were largely conservative fellows from complacent social classes, who favored the status quo, with the occasional spontaneous appearance of great or gifted individuals. They wanted no part of social experiments to flatten out the class structure and make "Jack as good as his master."

Opposing historians offer another, "rational" explanation. Hidden agendas aside, the two sides simply did not understand one another on the scientific level. Each was locked into a different research tradition, concerned with different kinds of evidence. "When building a house," goes an old epigram, "one must first be sure that the bricks are, in fact bricks." Early Mendelians and biometricians simply could not agree on what was a brick.

See also HOPEFUL MONSTERS; MENDEL, ABBOT GREGOR; MENDELIAN RATIOS CONTROVERSY; MENDEL'S LAWS.

For further information:
Bateson, William. *Materials for the Study of Variation: Treated with Especial Regard to Discontinuity in the Origin of Species.* London: Macmillan, 1894.
Box, J. F. *R. A. Fisher: The Life of a Scientist.* New York: Wiley, 1978.

BIOPHILIA
Mutual Attraction of Life Forms

Sociobiologist Edward O. Wilson wrote the book *Biophilia* (1984) about the deep need of living things to reach out for each other. In coining the term "biophilia," Wilson suggested the phenomenon of "life liking other life" may be rooted in the basic evolutionary kinship of all creatures.

Disparate species, despite differences in size or lineage, often seem to seek out each other's company. Capuchin monkeys and black tamarins, for instance, traverse the treetops in mixed troops. Although widely different in behavior and diet, they are habitual boon companions.

In the Galapagos Islands, seals sometimes tease the large marine iguana lizards, which are not noted for their sense of humor. Seals poke them, gently pull their tails, or nudge them underwater—and the iguanas neither bite nor become fearful.

Some reactions of captive animals to other species are quite remarkable. An enormous killer whale in a San Diego sea park often deliberately throws bits of fish to the wild gulls who wait at the corners of his tank. He enjoys feeding them just as people toss crumbs to pigeons. And chimps and gorillas in captivity have been known to relish the company of a pet kitten, tenderly holding and petting it just as a human would.

On the other hand, friendliness for other species sometimes ends abruptly where hunger begins. Chimpanzees often socialize with young baboons, play with them, groom them; then may suddenly kill and eat them. Although fieldworkers have been shocked by such seemingly "schizophrenic" behavior, the situation is reminiscent of humans who affectionately raise a "pet" calf or turkey for slaughter. Emotional conflict in such situations is a common theme in the folk tales of peoples who raise domestic animals for food.

Charles Darwin was amazed at the tameness of all kinds of birds and animals in the Galapagos. Hawks would fly down and perch on his gun barrel, and birds refused to budge from bushes unless knocked off. Later, he realized these creatures evolved in the absence of any large land predators. The normal flightiness of continental species was a response to thousands of years of persecution by man and other predators.

Darwin himself was a biophiliac when he wrote that animals, "our companions in amusement . . . work and suffering . . . may all be one—we may all be netted together." Perhaps all naturalists are attracted to study the Earth's creatures as a celebration of their own aliveness.

For further information:
Wilson, Edward O. *Biophilia*. Cambridge: Harvard University Press, 1984.

BIOSPHERE, EVOLUTION OF
The World of Life

Eduard Suess first used the word "biosphere" in 1875, in a geological treatise on the Alps, and the concept was developed around 1911 by the Russian scientist Vladimir Vernadsky: "The biosphere is the envelope of life . . . the area of the Earth's crust occupied by transformers which convert cosmic radiations into effective terrestrial energy." In other words, the biosphere is that part of the Earth where living things normally exist. Recently, it has become clear that the biosphere itself evolves. According to the Gaia hypothesis, life may play an important role in creating, maintaining and expanding its own favorable environment.

Analysis of early microfossils and geochemistry show that the Earth's original biosphere contained no oxygen. But about two billion years ago, purple and green microbes began to transform water and sunlight into food. They used the hydrogen and expelled oxygen as waste into the atmosphere. To these anaerobic bacteria, oxygen was a deadly gas.

According to biochemist Lynne Margulis, "This toxic-waste crisis turned out to be a blessing, though, for it inspired further innovation among the microbes. Some bacteria . . . came up with means to detoxify and eventually exploit it; they invented oxygen breathing."

These new oxygen-breathing (aerobic) bacteria survived in local niches for hundreds of millions of years. As levels of oxygen in the atmosphere rose higher and higher, some aerobic and anaerobic bacteria joined together into a new form of life: the cell with a nucleus.

"This new piece of microbial technology," says Margulis, "was as different from the basic bacterium as the space shuttle is from a paper airplane." Known as the eukaryotic, or nucleated, cell, these composite creatures have dividing membranes, including one that cordons off a nucleus from the rest of the cell, or cytoplasm. Self-reproducing parts known as organelles perform specialized functions, such as respiration (breathing) and photosynthesis (food making). And their nuclei contain long twisted protein chains of DNA, libraries of information that enable them to produce copies of themselves.

Today the biosphere is relentlessly assaulted by pollutants introduced by automobiles and industry and the unprecedented destruction of habitat. Oil spills, contaminating pristine Arctic waters, holes in the ozone layer caused by freon and other gases, reduction in oxygen caused by burning of vast rain forests and the consequent "greenhouse" effect—all pose grave dangers to the biosphere. In ecologist Paul Ehrlich's metaphor, destruction of parts of the biosphere is like removing hundreds of the thousands of rivets from an airplane's wing. One never knows at which point the wing will fall off.

See also EXTINCTION; GAIA HYPOTHESIS; GREAT DYING; GREENHOUSE EFFECT; RAIN FOREST CRISIS.

BIPEDALISM
Upright Walking Adaptation

Walking upright on two legs is a locomotor pattern called bipedalism. It is rare among mammals as a habitual mode, although chimpanzees and gorillas can manage it for short distances, especially when they want to check for danger over tall grass or brush. (So can meerkats, woodchucks, squirrels, monkeys and kangaroos, but most of these have tails to help keep balance. Meerkats, mongoose-like social mammals of the African deserts, take turns standing upright during sentry duty, scanning the sky and horizon for predators, while their fellows are feeding head down.

Ancient fossil hominid skeletons indicate that several species of near-man walked bipedally some three million years ago. Bipedalism in hominids is associated with a basin-shaped pelvis with short upper blade (ilium), long straight femurs, and flat (plantigrade) feet with a double arch.

Upright posture is also indicated by the large hole, or foramen magnum, at the base of the skull. In humans it lies directly beneath the cranium, in ape skulls it is set farther back. Chimps and gorillas also have long narrow pelvises and shorter, bowed leg bones.

For many years, upright posture was considered a most important distinguishing factor between men and apes and was assumed to have appeared late in human evolution, well after the expansion of the brain and development of tool use abetted by hands freed from locomotion functions.

As African fossil hominids came to light over the past 60 years, one of the very surprising discoveries was that several species were fully erect bipedal walkers—even those with brains the size of a chimp's. Australopithecines had human-like feet, fully erect posture and were fully bipedal, though many doubt they yet had what we could call language or human cultural capacities. Bipedal walking was the original adaptation of our branch of ground-dwelling primates; apparently, it had little to do with advanced intelligence.

The long-standing tradition about the uniqueness of bipedalism in humans goes back to the ancient Greek philosophers. They knew that birds walked on two legs, but observed that man seemed to be the only other creature that did. Consequently, when one of Plato's students defined man as a "featherless biped," no one was able to offer a logical refutation. The next day, however, another student showed up waving a plucked chicken.

BIRD, ROLAND THAXTER (1899–1978)
Dinosaur Tracker

Junius and Roland Bird, brothers raised in New York's Catskill Mountains, were a most remarkable pair. Each spent a separate, individualistic career trekking through wilderness in search of clues to America's prehistoric past.

Roland (known as R. T.) was one of the last of the old-fashioned dinosaur hunters, best known for dis-

STANDING ERECT for short periods is not uncommon among mammals, especially when scanning distant landscape for danger. Grizzly bears, prairie dogs, and meerkats (a South African mongoose) are all sometimes bipedal.

covering and preserving the spectacular dinosaur trackways from the Paluxy River, near Glen Rose, Texas.

Junius, who became an archeologist at the American Museum of Natural History, sought traces of ancient man in the New World, from the Arctic to the tip of South America. (He found the famous Inca fabrics in Peru and, in Tierra del Fuego, proof that humans had lived there 11,000 years ago, while giant ground sloths still roamed.)

R. T's true calling came after some aimless years of dabbling in cattle farming, land speculation and roaming the American wilderness in a motorcycle-camper he invented. (Recreational vehicles were still in the future.) His chance discovery of a fossil amphibian jaw sparked a lifelong passion: excavating dinosaur bones and footprints preserved in stone.

As a junior-high-school dropout turned drifter, R. T. did not appear a likely candidate for making important scientific discoveries. But with his father's encouragement, he sent his fossil amphibian skull to Barnum Brown, then America's greatest dinosaur

digger, who decided to take him on as an assistant.

Throughout the 1930s, Bird energetically excavated, prepared, shipped and catalogued uncounted tons of dinosaur bones for the American Museum of Natural History, often so absorbed he let his paychecks pile up for months uncashed.

R. T. felt especially close to the ancient behemoths when following the trackways they had left in the mud, long since hardened into stone. R. T. cleared river banks and built dams to enable him to follow the tracks, stalking them even under rock formations by dynamiting away hills. (One local workman insisted R. T. was trying to make a fool of him, because "There's no way that big fella could have walked under there.")

In water, R. T. discovered, the creatures propelled themselves as modern hippos do, by pushing only their front feet against the bottom. Rarely, there was a single imprint of a hind foot, where the animal kicked into a sharp turn.

At a time when scientists believed the huge sauropods were slow-moving and had to stay half sub-

FOSSILIZED BEHAVIOR intrigued R. T. Bird, who collected these trackways at the Paluxy River in Texas and mounted them behind an apatosaur (brontosaur) skeleton years later. The remarkable sequence clearly shows a meat-eater (the three-toed prints) following and perhaps attacking a large plant-eating dinosaur.

merged to support their bulk, R. T. showed they could move quickly on land and moved in herds. He uncovered one dramatic sequence where sauropods were being pursued by a hungry two-legged meat-eater. Forty years later, Robert Bakker noticed the tracks showed a social structure, with young and juveniles protected in the middle of the herd—which delighted R. T.

For years, he had argued that the trackways were really fossilized behavior, but museums (and sponsor Harry Sinclair, the oil magnate) were more interested in skeletons. Nevertheless, R. T. pestered Brown to let him work on the trackways until he got his way. Over several years, he meticulously chose, partitioned and quarried stone blocks of petrified footprints and shipped them back to the Museum of Natural History in New York.

But while Brown's dinosaur skeletons were catalogued, prepared and mounted, R. T.'s crates of fossil footprints languished in the museum yard. Stricken by illness, R. T. went into forced retirement and obscurity.

During the 1950s, Edwin Colbert, the museum's curator of vertebrate paleontology, was constructing a new hall of Jurassic dinosaurs and remembered the tracks. If they could be restored behind the giant dinosaur skeletons, he decided, it would make a marvelous exhibit.

By this time they had lain neglected outdoors through 14 New York winters, a "weathering mess." Crates were rotted, stones broken, labels washed away. More than 100 heavy blocks were heaped in jumbled disarray; a few more years of neglect and they would be lost to science. Colbert sent for R. T.—the only man in the world who could rescue the achievement of his youth.

Despite very frail health, Bird accepted the challenge, although he was so weak he was only able to work for two sessions a day of 20 minutes each, separated by a two-hour rest. Nevertheless, he doggedly persevered. In six months, with help from assistants, he had reconstructed the massive jigsaw puzzle; it was cemented in place behind Barnum Brown's dinosaurs.

R. T. died in 1978, but not before he had the satisfaction of seeing a permanent on-site museum built in Texas to preserve and display some of his beloved trackways. At his funeral, relatives lovingly placed in his pocket one of his prized, polished gizzard stones, which had aided a dinosaur's digestion 60 million years earlier. His unique tombstone in the Grahamsville, New York cemetery is carved with the likeness of a brontosaur and this epitaph: R. T. BIRD, DISCOVERER OF SAUROPOD DINOSAUR FOOTPRINTS.

See also BROWN, BARNUM; FOSSIL FOOTPRINTS, PALUXY RIVER; "NOAH'S RAVENS"; SINCLAIR DINOSAUR.

For further information:

Bird, Roland T. *Bones for Barnum Brown: Adventures of a Dinosaur Hunter.* Fort Worth: Texas Christian University Press, 1985.

BIRDS, EVOLUTION OF

See ARCHAEOPTERYX.

BLACK, DAVIDSON

See PEKING MAN.

"BLACK SKULL"

See AUSTRALOPITHECUS BOISEI.

BLENDING INHERITANCE

See PAINTPOT PROBLEM.

"BLIND CHANCE"
Objection to Evolution Theory

Many critics of evolutionary theory point to the intricate mechanisms of organisms, such as the vertebrate eye or the eagle's wing, as being explicable only by some sort of conscious design. Before Charles Darwin's day, natural theologians like the Reverend William Paley believed the naturalist's mission was to demonstrate everywhere in nature the evidence of divine design or purpose.

Paley argued that if he saw a stone in a field, he could accept its being there as a result of "blind" natural processes or chance. But if he came upon a watch, with all its intricate parts shaped to work together for a definite purpose, he would have to assume the existence of a watchmaker. [See PALEY'S WATCHMAKER.]

The argument does not do justice either to the creative potential of "chance" nor to the stone in the field, which one could equally well argue has regularities of design and composition. It is made of grains of certain minerals bound together in specific ways and may contain crystals of geometric shape or stratified layers of various colors. Indeed, Darwin started out with a strong interest in geology, and his understanding of the creation of rocks through regular, natural processes encouraged him to explore the possibility that life forms also developed over vast periods.

Most people misunderstand the mathematical concept of chance. It does not mean that a few thousand cells got together by accident to form a frog, that everything is random in nature, or that a roomful of monkeys typing for a thousand years will come up with the works of Shakespeare.

What it does mean is that natural forces and probabilities, given enough time, can account for the seeming designs without the direct intervention of a deity. If it appears highly unlikely that such a thing as a human being could be created in this way, no mathematician would disagree. On the other hand, she might point out that the chances against almost anything in history happening just the way it did are almost infinite.

From a statistical perspective, any individual human being is the highly improbable result of an unlikely sequence of events. In order to exist just as he is, his father and mother had to each have had the parents they had on both sides, going back thousands of generations. Some of his forebears may have had to travel to far distant lands to meet their eventual mates; some may have met under very unlikely circumstances.

Once his parents did meet and procreate, the chances of the one particular egg being fertilized by the one particular sperm that formed his genetic combination are exceedingly slim. If one were to calculate the probability for the existence of that unique individual, the chances would be almost infinitesimal. Each one of us is the very improbable, highly unlikely result of a unique history—and yet here we all are.

It has been calculated that a bridge hand containing all the spades in the deck comes up about once in 800 billion deals. But, the amazing thing is, so does any other hand. It may happen to be a winning hand, but the cards don't know that. Except from the point of view of the winner, there's really nothing incredible about a long-shot bet coming in or most of them losing. That rare big winner is as unremarkable as the thousands of losers; in a casino, both kinds of events happen all the time.

This paradox of the extremely unlikely being not only possible, but inevitable, is one aspect of chance in evolution. Another key concept is how natural selection keeps changing the odds. It is a two-step process. In the first step, pure chance prevails; while the second step is directed and anti-chance. In effect, it keeps throwing out the "losing" cards and returning spades to the deck until the chance of drawing all spades becomes more and more likely. As geneticist Sir R. A. Fisher put it, "Natural selection is a mechanism for generating an exceedingly high degree of improbability."

To many—even to mathematicians and physicists—it is distasteful to accept any element of blind chance as a creative force, perhaps shaping human consciousness as well as anatomy. Sir John Herschel disdainfully referred to natural selection as "The law of higgelty-piggelty." The position is epitomized in the oft-quoted statement by Albert Einstein, "I refuse to believe that God plays at dice with the universe." His colleague Neils Bohr once challenged the great physicist on the point, remarking, "Albert, please don't tell God what to do."

See also CONTINGENT HISTORY; NATURAL SELECTION.

BLYTH, EDWARD (1810–1873)
Zoologist

Edward Blyth devoted his life to natural history, although his passion to learn about plants and animals left him perpetually penniless. A pharmacist who neglected his business in England, he finally found a zoological post at a museum in Calcutta, India. While there, he corresponded with Charles Darwin and sent specimens and information for *Variation in Plants and Animals under Domestication* (1868).

Blyth, along with Lyell, drew Darwin's attention to the importance of Alfred Wallace's species paper published in 1855. Darwin thought him an excellent naturalist, and "a very clever, odd, wild fellow, who will never do what he could do, from not sticking to any one subject."

Blyth would be all but forgotten today, except for a book by respected evolutionist Loren Eiseley, *Darwin and the Mysterious Mr. X* (1979). Eiseley advanced the astonishing notion that Darwin had lifted a good part of his theory from Blyth—Eiseley's "Mysterious Mr. X." Because of Eiseley's excellent reputation, several scholars took the trouble to investigate his conclusions about Blyth, but not one found the case or the evidence convincing.

See also EISELEY, LOREN.

For further information:
Eiseley, Loren. *Darwin and the Mysterious Mr. X: New Light on the Evolutionists.* New York: E. P. Dutton, 1979.

BONE CABIN QUARRY
Dinosaur Digger's Delight

Bone Cabin Quarry, located at Como Bluff in eastern Wyoming, is one of the most important sites for the discovery of Mesozoic reptiles, especially dinosaurs. It was mined first by the great 19th century paleontologist O. C. Marsh, and later by Barnum Brown, who garnered many of the great dinosaur skeletons now on display at New York's American Museum of Natural History.

When early paleontologists first explored the site, only a lonely sheep-herder lived at the remote place. Fossils were so plentiful that he had built his cabin out of the long bones of dinosaurs. Major finds were hacked out of the rocks by hand and thousands of tons were transported out of the area by horse and cart.

See also BROWN, BARNUM; COPE-MARSH FEUD.

BONZO

See REAGAN, RONALD.

BOUCHER DE PERTHES (1788–1868)
Founder of Prehistoric Archeology

Scientists of the 1830s were not prepared to believe in the existence of 30,000-year-old man-made tools and artworks. [See FOUR THOUSAND AND FOUR B.C.] That tons of such artifacts were sealed into rock strata, along with the bones of extinct European rhinoceroses and mammoths, seemed even more far-fetched. So no one paid the slightest attention to their discovery by an eccentric French amateur in the rural village of Abbeville. Even his name sounded like an impossible joke—Monsieur Boucher de Crevecoeur de Perthes, literally "Mr. Butcher Broken-Heart of Lost-Town."

Born to wealth, Boucher de Perthes also inherited his position as head of the Abbeville customshouse, which allowed him the leisure to attend local social functions and to dabble in literature and politics. He wrote novels that didn't sell, plays that were never performed and ran for parliament, but was never elected. Even to his friends, Boucher de Perthes seemed a most unlikely individual to revolutionize archaeology and reshape science's understanding of the human past.

In 1837, Boucher became intrigued by some unusual man-made objects turned up by workmen dredging the Somme Canal, and he purchased an ancient polished stone ax hafted in staghorn from them. During visits to several gravel pits in the Somme River terrace, just outside Abbeville, he obtained a number of bifacially chipped stone "hand-axes." Within a few weeks, he had become an enthusiastic collector.

Some years later, after accumulating thousands of worked flints and studying their contexts in excavations, he decided to publish an illustrated description of his discoveries. Since any pre-Roman artifact was then considered "Celtic," he entitled his book *Celtic and Diluvian Antiquities* (1847). Boucher asserted that the older tools must have been buried during the biblical Flood, which put off geologists, while churchmen rejected his insistence on the great antiquity of man. In addition, many of the prizes in his collection appeared to be crude forgeries chipped by quarrymen, who thought it great fun to con a gullible gentleman out of a few francs.

Boucher was often the victim of such hoaxes, which was a major reason most scientists refused to take him seriously. Deception of the "toffs" (gentry) for profit by clever tricksters from the underclass was an international sport, for which science was just another arena. Perhaps Lewis Carroll had such real-life nonsense in mind when he wrote one of his verses in *Through the Looking-Glass* (1872):

I sometimes dig for buttered rolls,
Or set limed twigs for crabs;
I sometimes search the grassy knolls
For wheels of Hansom-cabs.
And that's the way (he gave a wink)
'By which I get my wealth—
And very gladly will I drink
Your Honour's noble health.

In his classic *Antiquity of Man* (1863), Sir Charles Lyell tells how each geologist, fearing all his colleagues had been duped, went into Boucher's excavation determined to "not quit the pits till he had seen one of the hatchets extracted" from a sealed deposit with his own eyes.

"The great point" in making valid discoveries in prehistoric archeology, geologist Alphonse Gaudrey wrote in 1859, "was not to leave the workmen for a single instant, and to satisfy oneself by actual inspection whether the hatchets were found *in situ* [sealed in a closed deposit]" with the bones and teeth of extinct horses and other mammals. Like detectives on stakeout, a scientist would have to watch closely for hours, or several would take shifts, never taking their eyes off the quarrymen. Exactly as at spiritualistic seances and carnival shell games, it was a contest of alertness and acuity between those who would fool the gentry and those who would catch the fraud.

One after another came the British geologists to Abbeville: the paleontologist Hugh Falconer, the clergyman-geologist Sir Joseph Prestwich, the evolutionist Sir John Lubbock and England's leading geologist, Sir Charles Lyell. Acting separately, each man personally witnessed (or performed) the extraction of a flint tool from sealed deposits. (Prestwich even took a photo of one half uncovered.) Once they knew what to look for, the geologists returned to England and promptly found similar artifacts in the gravels of the Thames Valley.

In later years, Charles Darwin was somewhat chagrined to recall how heartily he and his colleagues once scoffed at the reports of Boucher de Perthes. After publication of his *Origin of Species* in 1859, however, scientists were positively eager to find the "prehistoric man" themselves. Riding the wave of his success, Boucher decided to offer 200 francs to any workman who discovered human remains in the stone-tool bearing deposits. Sure enough, on March 26, 1863, he was shown a human jaw stuck in the gravels at a site called Moulin-Quignon.

After prolonged examination by an impressive committee of French and British scientists, the Moulin–Quignon jaw was accepted as authentic. Theories were spun about its resemblence to such "primitive peoples as the Lapps and Basques" and it received

an honored niche among relics of the prehistoric past. Unfortunately, the jaw later proved to be a hoax and not at all ancient. Boucher's workmen, impressed with his newly acquired fame and credibility, apparently had felt entitled to share in his success.

Boucher turned his huge ancestral mansion into a private museum, where he displayed his thousands of prehistoric artifacts and more than 1,600 Flemish, Dutch and French oil paintings by old masters. After his death, Boucher's heirs had the picture collection appraised by art historians and received some bad news: Most of them were fakes.

See also LUBBOCK, SIR JOHN; PRESTWICH, SIR JOSEPH.

For further information:
Brodrick, Alan. *Father of Prehistory: The Abbe Henri Breuil.* New York: Wm. Morrow, 1963.

BRACHIATION
Arm-Swinging Locomotion?

Man shares with the anthropoid apes a remarkable shoulder joint capable of rotation through a full 360 degrees. Known as a "brachiation complex," it presumably developed when human ancestors climbed in trees or hung from branches. Monkeys' shoulders are stiff, with very limited flexibility, but the rotating shoulder is common to man, chimps, orangs, gorillas and gibbons.

Early anthropologists assumed this anatomy was originally adapted to swinging from branch to branch as gibbons do. But recent studies indicate the "brachiating" gibbon is an aerial specialist, and that human ancestors may never have habitually swung through the trees.

Experts in locomotion anatomy believe early ancestral hominoids may have evolved for climbing vertical tree trunks, grasping branches and feeding with the arms and hands—a more generalized adaptation than acrobatic arm-swinging.

BRACHIOSAURUS
The Biggest Dinosaur

The largest land animal that ever lived, a species of brachiosaurus, probably weighed 80 tons, had a 40-foot neck, was about 80 feet long, and was about 60 feet tall—tall enough to be able to look into the top window of a five- or six-story building.

In 1979, a fossil skeleton of the creature was excavated from an ancient, dry riverbed in a Colorado mesa on the western slope of the Rocky Mountains above Escalante Valley. The discoverer, Dr. James A. ("Dinosaur Jim") Jensen of the Earth Sciences Museum at Brigham Young University in Utah, believes the animal lived in the late Jurassic period—some 150 million years ago—and was a plant eater. Either it fed from the high limbs of trees or kept its body submerged in water and reached that long neck out to browse along river banks.

An interesting question about the brachiosaurus is what kind of heart could pump blood up its long neck. "A giraffe has a four-chambered heart and we think most dinosaurs had two-chambered hearts," said Dr. Jensen, who considers the brachiosaur heart an unknown marvel. "Their heads were the highest of any known animal, and tremendous pressure must have been needed."

If the complete skeleton were assembled, it would dwarf the tyrannosaurs and apatosaurs on display in most of the great museum dinosaur halls.

BRANCHING BUSH
Model of Evolution

Charles Darwin pictured evolution as a "Tree of Life." From a common trunk, branches diverge in several directions, each in turn sprouting numerous twigs. Paleontologist Stephen Jay Gould speaks of Darwin's tree as "actually more like a branching bush." The trunk and heavy boughs of a tree are not very flexible; an intricate bush, with delicate twigs burgeoning in all directions, is closer to our current idea of evolutionary history.

Prior to Darwin, natural history was influenced for centuries by the medieval idea of the Great Chain of Being, which arranged living things in a hierarchy. The model was a ladder or staircase. Simple, lowly (base) creatures occupied the lower rungs; higher on the ladder were creatures successively closer to man, the pinnacle of all forms of life; only angels or other spiritual entities were higher. God was on the top rung.

The Chain of Being was a hard mental habit to break. Early evolutionists, such as Lamarck and Erasmus Darwin, transformed the ladder into an escalator. Animals were thought to aspire to the next higher rung, a constant striving upward. Even Charles Darwin realized he sometimes imagined development from "less perfect to more perfect," and reminded himself "never to say higher or lower."

Determining if a clam is "higher" than a mussel or a hamster higher than a field mouse, is impossible. Each species is the product of a unique history, influenced by origin, habitat, competition, predators, climate, opportunities, and luck of the draw. Galapagos island finches, for instance, diverged from mainland ancestors to become seedeaters, insect eaters, woodpecker-like species, etc. None can be said to be "higher" or "lower"; they simply adapted to various niches in the new environment.

Lamarck measured evolutionary "progress" in terms

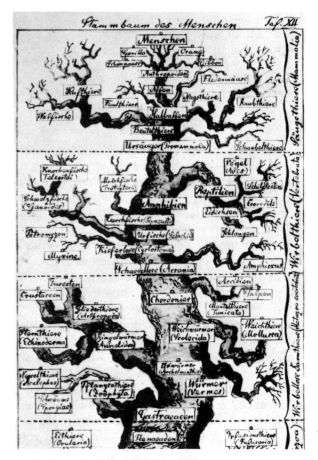

DARWIN'S GERMAN DISCIPLE, Professor Ernst Haeckel, had a special fondness for drawing evolutionary "trees of life," or phylogenies. Evolutionists today have modified the tree image to that of a spreading bush, lacking the rigid, heavy "trunk."

of closeness to man, an appealing idea from the human perspective but hardly fair to the rest of living things. (Old habits die hard; many texts still refer to the "higher primates" or the "man-like apes.")

This most common misunderstanding of evolution—a fallacy that distorts man's true place in nature—can be corrected by picturing the branching bush. Where it is very bushy, representing a cluster of many related species, we have a successful group that has radiated into many niches.

If most of the cluster became extinct, leaving only one surviving species and a few fossils, the usual procedure has been to bring in the old evolutionary "ladder." The fossils are arranged in a direct line, making "progress" or "leading up" to the single surviving twig. This is the illusion of "finalism"—that species evolve toward a final goal, be it the modern horse or man.

If many closely related species continue to survive, no one would dream of arranging them in a hierarchy leading to the "highest" one. For instance, there are many species of rodents on Earth today and many different antelopes. No one wonders which is the highest antelope or rodent. But because only a single twig remains in the case of *Homo sapiens*, we commonly see ourselves as the goal or culmination of all hominid evolution.

There were in fact, many species of man-like creatures, most of which died out. We are the only surviving twig of a once successful cluster that became nearly extinct. Yet we imagine the history of our line as if all its branches were striving to become ourselves, the end and summit of evolutionary development. That, says Stephen Jay Gould, is "life's little joke."

See also GREAT CHAIN OF BEING; DIVERGENCE, PRINCIPLE OF; HORSE, EVOLUTION OF; ORTHOGENESIS.

BRANCHING EVOLUTION (CLADOGENESIS)

See ANAGENSIS.

BRIDGEWATER TREATISES (1830s)
Expositions of Natural Theology

When the Earl of Bridgewater died in 1828, he left £8,000 to be spent furthering the good work of the Reverend William Paley in natural theology. Bridgewater's bequest sponsored a series of books illustrating the "Power, Wisdom and Goodness of God as Manifested in the Works of Creation."

Seventeenth-century naturalist John Ray had begun the tradition of using natural history to discover God's designs—later elaborated and popularized by Reverend Paley's classic *Natural Theology* (1802) which remained popular for 60 years. Its cornerstone was the "argument from design," built on compilations of examples from the natural world. Whether analyzing delicate structures of the human eye or the eagle's wing, natural theologians believed exquisite adaptations were conceivable only as products of divine workmanship. The universal tailoring of structure to function, they believed, illustrated the benevolence and wisdom of a caring God. (They ignored the clumsy structures and "made-over contraptions" that also abound in the natural world.)

"Just as the intricate structure of a watch implies a watchmaker," writes science historian Peter Bowler, so natural theologians argued that "The incredible complexity of living things proclaims the power of their Designer. In this tradition, each case of adaptation is considered individually, and the argument gains strength from the sheer number of examples cited." Several eminent scientists contributed volumes to the Bridgewater series, but, in Bowler's words, "this endless catalogue of adaptations produced not a sense of divine benevolence, but sheer boredom in the reader."

The president of the Royal Society, the archbishop of Canterbury, and the bishop of London were appointed to find suitable authors. Among the first chosen was William Buckland, dean of Westminster Abbey and an accomplished geologist, whose work became a much-imitated classic of scriptural geology.

Buckland interpreted the history of the Earth on catastrophic principles. Evidence of earlier life forms was assigned to earlier creations, while "higher" species were introduced by the Creator when conditions on Earth had sufficiently improved. Although the explanations were not simpleminded, they nevertheless relied on miracles—phenomena different in kind from those explainable by natural laws.

In 1838, the brilliant Cambridge mathematician Charles Babbage published an unofficial *Ninth Bridgewater Treatise* to offer a natural explanation for the seemingly miraculous. The result pleased neither churchmen nor scientists.

See also NATURAL THEOLOGY; NINTH BRIDGEWATER TREATISE; PALEY'S WATCHMAKER.

BROOM, ROBERT (1866–1951)
Paleontologist Extraordinary

Dr. Robert Broom was an amazing man, a nonconforming individualist who made enemies and admirers throughout a long, productive and adventurous life. Although almost 40 years were required, he and Raymond Dart changed the course of paleoanthropology by finding the first australopithecine fossils in South Africa and convincing the scientific world they were the earliest hominids.

Born in Scotland in 1866, Broom trained as a surgeon, but retained a strong interest in paleontology and comparative anatomy. He considered himself the "scientific son" of the British Museum director Richard Owen, who thought highly of his talents. Had Owen lived to see how those talents were used, he would probably have been appalled that his protege's discoveries supported the theories of Charles Darwin, his hated rival.

After a stint in Australia, Broom moved to South Africa, where he established a successful medical practice around 1900. After a few years, however, he became restless with city life and took off to explore the Karroo Desert region, where he prospected for fossils. Had he done nothing else in science, he would be remembered for his brilliant work in finding, excavating, classifying and making sense of the mammal-like reptiles he found in the Karroo. Broom documented the transition from reptiles to mammals and was awarded the Medal of the Royal Society in 1928 for this achievement.

When Raymond Dart found the first australopithe-

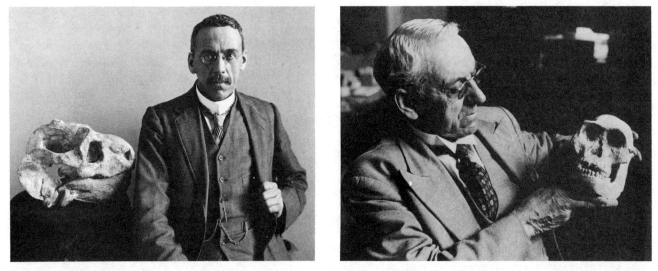

BRILLIANT, ECCENTRIC physician Robert Broom devoted years to finding and describing fossils of the remarkable, mammal-like reptiles of South Africa. Late in life, he capped his career by unearthing a series of spectacular australopithecine remains, helping shift science's search for early man to Africa.

cine fossil, the "Taung Baby," Broom, as a fellow South African fossil hunter, was overjoyed. Dart never forgot how Broom "burst into my laboratory unannounced. Ignoring me and my staff, he strode over to the bench where the skull reposed and dropped to his knees 'in adoration of our ancestor.'"

At the age of 69, Broom decided to find "an adult Taung ape" and searched in a limestone quarry near his home. This site, called Sterkfontein, near Krugersdorp in the Transvaal, yielded the first adult *Australoptithecus* (originally named *Plesianthropus* by Broom).

A few years later, he heard that some teeth had been found by a schoolboy at a farm at Kromdraai, near his first site. He invaded the local classroom, bought the teeth for a few shillings and five chocolate bars, and persuaded the boy to take him to the cave where he had found them.

The result was Broom's discovery of the large or robust form of australopithecine, to which he originally gave the name *Paranthropus* in 1938. But British scientists, notably Sir Arthur Keith, refused to accept the accumulating evidence of these South African renegades, Dart and Broom. Their near-men simply did not fit the image expected for early man.

Australopithecines had small, ape-sized braincases and man-like jaws. "Piltdown man" the forged fossil Keith believed to be the authentic "first Englishman," had a large man-like braincase and an ape-like jaw—exactly the reverse.

But Broom triumphed, lived to see Piltdown exposed as a fraud and his own work established as a landmark in the search for human origins. As historian of science John Reader sums up:

By 1948 when Broom was 81 years old, he had assembled enough fossil teeth, jaws, skulls and skeletal remains to convince most people that *Australopithecus* had existed in gracile (slender) and robust form, had possessed teeth more man-like than ape-like, had walked upright and was a good candidate for the ancestry of man—just as Dart had said.

See also AUSTRALOPITHECINES; DART, RAYMOND ARTHUR; KARROO BASIN; KILLER APE THEORY.

For further information:

Broom, Robert. *Finding the Missing Link*. London: Watts, 1950.

Reader, John. *Missing Links: The Hunt for Earliest Man*. London: Collins, 1981.

BROWN, BARNUM (1873–1963)
Dinosaur Fossil Collector

Named after the great showman P. T. Barnum, indefatigable dinosaur digger Barnum Brown assembled his own version of "The Greatest Show on Earth": A parade of giant dinosaur fossils wrenched from the cliffs and arroyos of the American West. Brown's lasting contribution—hundreds of tons of dinosaur fossils—formed the nucleus of the American Museum of Natural History's world-famous collection. During the 1960s, Brown, near ninety, could still be seen leading visitors around the crammed dinosaur halls, announcing "Here's another one of my children," as he pointed out the bones of a reptilian giant. But when he began his career in 1897, the museum had not a single dinosaur.

As a child, Brown, born in Carbondale, Kansas, collected fossils from freshly ploughed fields. He attended the University of Kansas, then moved to New York City, where he studied paleontology at Columbia University and worked at the American Museum of Natural History. Henry Fairfield Osborn, the museum's dynamic director, sent Brown on fossil hunts while he was still a graduate student.

His first assignment took him to Como Bluff, Wyoming, to prospect the rich Jurassic deposits that had been discovered and excavated by Yale's Othniel C. Marsh. Brown and his colleagues discovered new beds containing enormous quantities of fossils, including the *apatosaurus* (then called brontosaurus) that still dominates the museum's Jurassic Hall. There were so many fossils in the area a sheepherder had used them to build a little house, so Brown named the site "Bone Cabin Quarry."

But dinosaur hunters were—and still are—very proprietary about "their" sites, even if there are more than enough fossils to be carried away in several lifetimes. Marsh was furious about his former sites being worked and began a bitter feud with Osborn that lasted to the end of his life.

During the early years of the 20th century, Brown dug up fossils all over the West. One of his greatest discoveries, a nearly complete skeleton of *Tyrannosaurus rex*, was blasted out of tons of sandstone near Hell Creek, Montana in 1902. Encased in huge blocks of matrix (surrounding sandstone), the fossils were taken by horse-drawn wagon to the nearest railroad 130 miles away.

As his exploits became known, Brown became nationally famous. Nicknamed "Mr. Bones" by the newspapers, his dinosaur-finding ability became legendary. Crowds would meet his train and offer to help him find ancient monsters near their town. Brown dressed in expensive, fashionable outfits even while exploring remote, dust-blown sites, as befitted his celebrity status.

In 1909, a chance remark by a Canadian visitor to the museum about the richness of fossil beds along the Red Deer River in Alberta, Canada, prompted Brown to lead an expedition there. The party navigated downriver on a large raft and found fossil deposits near the town of Content. Brown was ecstatic. "Box after box," he wrote, "was added to the collection till scarcely a cubit's space remained unoccupied on board our fossil ark."

Over the next decades, he traveled widely, searching for fossils and prospecting for oil in India, South America, Ethiopia, and the Greek islands. Brown's second wife, Lilian, chronicled her adventures accompanying him on field trips in such books as *Bring 'em Back Petrified* (1956) and *I Married a Dinosaur* (1950). When she first decided to join her bonehunting husband in the field, the family maid expressed grave concern. "After all," she warned, "who knows what the beasts died of?"

One of Brown's most famous discoveries was the "great dinosaur graveyard" at the Howe Ranch, near the base of the Bighorn Mountains in Montana. After some preliminary work in 1933, he convinced the Sinclair Oil Company to put up the money for major excavations at the site. The team's efforts soon paid off when they uncovered a vast bone deposit; in Brown's words "a veritable herd of dinosaurs." More than 4,000 bones (about 20 dinosaurs) packed in 144 crates weighing 69,000 pounds were shipped to New York.

Sinclair Oil, which used a *Diplodocus* as its company logo, garnered a windfall of publicity from public interest in Brown's digs. During the 1930s and 1940s, they gave free dinosaur stamps and booklets at their service stations, a promotion created and supervised by "Mr. Bones" himself.

But what accounted for the incredible concentration of dinosaurs at the Howe Ranch "graveyard"?

BONE-HUNTER Barnum Brown, when almost ninety, supervised construction of Sinclair Oil's lifesize dinosaur models, shown being barged down the Hudson River to be exhibited at the 1964 New York World's Fair.

Brown concluded from the surrounding silt that the animals were crowded together in standing pools. Geological changes that raised the mountains had also elevated the lowlands. "The large lakes were drained and the swamps vanished. The dinosaurs became more and more concentrated in the remaining pools as they were pushed together in huge herds."

Although Brown was describing a local phenomenon, not a worldwide upheaval, his picturesque descriptions contributed to a popular image of dinosaurs "perishing in a parched, drying landscape." Such was the power of this vision of dinosaur extinction that Walt Disney chose to depict it in his animated epic *Fantasia* (1940); and, half a century later (though we now know the dinosaur's world became cooler, not hotter) Don Bluth's *Land Before Time* (1989) carried on this vision.

In 1956, when he was 83, Brown got funding to explore a site at Lewiston, Montana, where he discovered and excavated a plesiosaur skeleton. Two years later he raised funds to use a helicopter to prospect the Isle of Wight, where fossils abounded in the steep sea cliffs. After spotting skeletons from the air, he planned to strap himself into a bosun's chair and be lowered down the cliff, where he would excavate while dangling above the English Channel.

While planning this expedition, he was approached by his old sponsor, the Sinclair Refining Company, to supervise the construction of life-size dinosaur models for the 1964 New York World's Fair. They were to be built in the town of Hudson, north of New York City, and transported to the World's Fair via the Hudson River. Delighted at being offered a "new job" at the age of 89, Brown looked forward to startling Manhattanites with the bizarre sight of a bargeful of dinosaurs floating down the Hudson River.

Brown supervised the dinosaurs' construction, but never did witness their journey to the Fair. He died in February, 1963, just a week short of his 90th birthday, and was buried beside his first wife, Marion. When his second wife, Lilian, died some years later, according to his daughter's memoirs, she "was buried on the other side of Barnum, who undoubtedly would have had a good chuckle over being sandwiched between his two wives."

See also BIRD, ROLAND T.; FANTASIA; OSBORN, HENRY FAIRFIELD; SINCLAIR DINOSAUR.

For further information:

Brown, Frances R. *Let's Call Him Barnum.* New York: Vantage, 1988.
Brown, Lilian. *Bring 'em Back Petrified.* New York: Dodd, Mead, 1956.
Preston, Douglas. *Dinosaurs in the Attic: An Excursion into the American Museum of Natural History.* New York: St. Martins, 1986.

BROWN, ROLAND (1893–1961)
Paleobotanist

Roland Brown spent his life studying the remains of ancient trees and flowers, though his professor in the subject strongly warned him against it. E. W. Berry, Brown's teacher at Johns Hopkins University, was an outstanding paleobotanist with an uncanny ability to find fossil plants. But there were so few paying positions available, he told his students they would starve to death if they followed in his footsteps. Brown was the only one of Berry's many students to ignore the advice.

He did not go jobless, but enjoyed a long career with the U.S. Geological Survey (1928–1961). During many field trips to the western states, he became an expert on the geology of the Cretaceous–Tertiary boundary, which marked the end of the Age of Dinosaurs about 70 million years ago.

Brown discovered hundreds of fossil plant species—fan-palms and other tropical vegetation that grew all over ancient Wyoming, in country that is sagebrush semidesert today. He found similar fossils of the lush vegetation that once thrived in the Rocky Mountains and Great Plains.

When he was not studying the origin and evolution of plants, Brown was studying the origin and evolution of words (etymology). Shortly after four o'clock each afternoon, he would put away all books and notes on fossil plants and concentrate on his dictionary of scientific terms.

In 1954, he published his classic *The Composition of Scientific Words*, designed to help create new names for genera and species. Latin and Greek roots are defined, so modern biologists can understand the old names and devise new ones.

Brown was an unassuming individual, who did the work at hand without standing on ceremony. Once, during the 1940s, the Geological Survey sent out a young geologist to lead a field party to survey for coal in the Montana wilderness. Word reached them that Dr. Roland Brown would be coming out from Washington, D.C. to lend his expertise. Special efforts were made to get everything set up before the arrival of the distinguished visitor; a laborer was sent for from a distant town to help prepare the campsite.

A few days later a shabbily dressed, meek-looking individual showed up in camp and was immediately handed a pick and shovel to dig drainage ditches. He set right to work and labored steadily for several hours, but progress was not fast enough for the young man in charge. Everything had to be done quickly, he ordered, because Dr. Brown might arrive at any time. The laborer laid down his tools and admitted, "I am Dr. Brown."

For further information:
Andrews, Henry. *The Fossil Hunters: In Search of Fossil Plants.* Ithaca, N.Y.: Cornell University Press, 1980.

BRULLER, JEAN

See "VERCORS" (JEAN BRULLER).

BRYAN, WILLIAM JENNINGS (1860–1925)
Antievolutionist Crusader

During his lifetime, politician and great orator William Jennings Bryan won fame as a progressive reformer with a strong social conscience. Secretary of state under Woodrow Wilson, he had been the Democratic nominee for president three times. Bryan campaigned vigorously for women's suffrage, justice for the working poor and curbs on corporate greed. He was also the architect of legislation prohibiting teaching evolution in the schools, thus leaving a reactionary legacy of bitter legal battles 60 years after his death.

Bryan has been vilified as an ignoramus and a demagogue who pandered to uneducated bigots in the backwaters of the United States. Movies and plays portray him during his last stand, as the adversary of Clarence Darrow in the celebrated Scopes "Monkey Trial" of 1925, which was not his finest hour. Bryan understood neither scientific issues nor the meaning of academic freedom; Darrow maneuvered him into offering ridiculous testimony in the courtroom. Fatigued and on the verge of collapse, he died a few days after the trial.

He is often portrayed as his vitriolic enemy H. L. Mencken described him: a religious fanatic, obstructing intellectual progress with a mulishly stubborn belief in the literal interpretation of the Bible. In fact, Bryan had not always opposed evolutionary ideas, and had arrived at his seemingly reactionary position with the best of intentions for America's welfare.

He feared that the Darwinian theory, as many at the time understood it, was "a merciless law by which the strong crowd out and kill off the weak." Bryan preferred to believe "that love rather than hatred is the law of development." He also thought that "class pride and the power of wealth" were using Darwinism to justify exploiting the poor, just as European kings had once used the doctrine of Divine Right.

And his fears were justified. Industrial giants like John D. Rockefeller and Andrew Carnegie did indeed adopt Social Darwinist views about being "the fittest," their ruthlessness justified as part of a great law of nature. That this was a misreading of evolutionary theory occurred neither to Bryan nor the industrialists, since it was also taught by many biology professors of their day.

In addition, the Darwinian banner was being carried by militarists and, in Bryan's words, "was at the basis of that damnable doctrine that might makes right that had spread over Germany." He knew that during World War I, German intellectuals believed natural selection was irresistably all-powerful (*Allmacht*), a law of nature impelling them to bloody struggle for domination. Their political and military textbooks promoted Darwin's theories as the "scientific" basis of a quest for world conquest, with the full backing of German scientists and professors of biology.

Bryan also perceived another evil resulting from the interpretation of Darwinism by the intellectuals of his day: an ill-conceived faith in eugenics as the wave of the future. It would paralyze the hope of social reform, Bryan railed, and "its only program for man is scientific breeding, a system under which a few supposedly superior intellects, self-appointed, would direct the mating and the movements of the mass of mankind—an impossible system!"

For these compelling reasons, as Stephen Jay Gould has pointed out, Bryan saw Darwinism as a many-faceted evil, quite apart from its conflict with biblical accounts of creation. Science had too easily lent respectability to political and social programs that went far beyond its proper sphere. Bryan "had the wrong solution," Gould says, "but he had correctly identified a problem!"

See also BUTLER ACT; MENCKEN, H. L.; SCOPES TRIAL.

For further information:
Koenig, Louis. *Bryan: A Political Biography of William Jennings Bryan.* New York: Putnam, 1971.

BUFFON, COMTE GEORGES-LOUIS LECLERC DE (1707–1788)
French Naturalist

"This orang-outang," wrote the Comte de Buffon in the mid-18th century, "is a very singular brute, which man cannot look upon, without contemplating himself, and being convinced that his external form is not the most essential part of his nature."

Such tantalizing passages in Buffon's *Histoire Naturelle* (1749–1767) clearly hint at a common ancestry for man and apes. But although he appears to have glimpsed it, Buffon never could entirely break with the "chain of being" ideas sanctioned by the church.

Although Georges-Louis Leclerc, Comte de Buffon was born to the wealth and power of the French aristocracy, his curiosity about nature drew him to a life of scientific pursuits. As a youth, he sought to please his family first by studying law and then medicine. But neither could hold his interest so much as exotic species of plants or animals.

While still quite young, he was made head of the King's Botanic Gardens (*Jardin du Roi*, later the *Jardin des Plantes*) in Paris; there he had ample opportunity to pursue his diverse interests. Soon after, he began his great work of natural history, the *Histoire Naturelle*, a compendium of everything then known about the natural world. But Buffon was not only a cataloger; he was also a philosopher who looked for causes and explanations. Eventually, the published work which occupied him for 50 years came to 44 volumes.

Nearly a century before Charles Darwin, Buffon can be seen groping for an evolutionary interpretation to tie together his varied observations. He was struck time and again by such phenomena as offspring varying from their parents, the "fit" between organisms and environment, competition for resources and development of races and varieties. His observations were astute, but a satisfying general theory eluded him.

Nevertheless, by 1766, Buffon became convinced that related species could arise from a common ancestor. But, in his view, various species could only diverge from a particular plan or type as they adjusted to climate and environment. He called it "degeneration."

Although he proposed a principle of common descent, lineages came only from a family that seemed fixed by permanent constraints. For instance, Buffon thought that an ancestral cat could "degenerate" into tigers, lions, and leopards, but the original cat family was a "given." Buffon uses "degeneration" and "degradation" in a very general sense, similar to what later naturalists would refer to as "varieties"— a departure from an earlier type or species.

The species he describes are not diminished in complexity, as later degeneration theorists would say of snakes being "degraded" lizards, "losing" their limbs. Buffon simply believed that any change from a "pure" ancestral type was by definition a "degenerate" form.

Between species he often sought "intermediate gradations," for he thought nature allowed no gaps. These linking species, he thought, came into being by gradual, regular processes, which he did not specify. But he did assert that God did not occupy Himself "with the particular fold in a beetle's wing" but worked through natural "Second Causes."

Buffon also tackled geology and was the first European Christian to state openly that he thought the world much older than the 6,000 year limit imposed by church authorities. [See FOUR THOUSAND AND FOUR B.C.] In 1788, he published *Les Epoques de la Nature*, in which he described the formation of the Earth's features by normal, currently observable processes. Some of his passages are strikingly similar to Sir Charles Lyell's statements of "uniformitarianism" that came 40 years later.

Buffon's musings on the relationship of apes and man seem very much ahead of his time. If we only study its body, he says, "we might look on that animal as the one in which the ape species begins, or . . . the human species ends . . . The mind, thought, and speech, therefore do not depend on the form or organization of the body. Those are gifts bestowed on man alone . . . Forced to judge by external appearance alone, the ape might be taken for a variety of the human species . . . [but] The Creator has . . . infused this animal body [of ours] with a divine spirit." It could have been Alfred Russel Wallace writing a century later.

Buffon also sounds surprisingly modern in his suggestion that distinctly human behavior is directly correlated with prolonged education of the young. Man is the only creature, he points out, where mother and infant continue an intense, intimate association well past the age of three years. It is a time of intense socialization and the learning of language, after which the child will still not be capable of surviving on its own for another 10 years.

Therefore, Buffon concludes that a "state of pure nature, wherein we suppose man to be without thought and speech, is imaginary, and never had existence."

> This needful and long intercourse of parents with their children produces society in the midst of a desert . . . the parents communicate to it not only what they possess from Nature but also what they have received from their ancestors, and from the society of which they form a part.

Buffon's *Histoire Naturelle* was extremely influential; it was translated into many languages, and updated editions continued to be printed and read for 100 years after his death.

See also DEGENERATION THEORY; FOUR THOUSAND AND FOUR B.C.; GREAT CHAIN OF BEING; WALLACE, ALFRED RUSSEL.

BUMPUS'S SPARROWS
Study of Survivors

Since Charles Darwin's day, biologists have published thousands of books and articles about the "struggle for existence," but remarkably few experiments or statistical studies have been done on natural selection. The earliest attempt to investigate differential mortality—the differences between surviving and nonsurviving individuals in a natural population—was made by Professor Hermon Carey Bumpus (1862–1942), a zoologist at Brown University.

During the winter of 1898, while out strolling after a severe ice storm in Providence, Rhode Island, Bumpus found 136 stunned house sparrows lying on the ground. He brought them to his laboratory, where he tried to care for them; 72 eventually revived and 64 died.

Instead of simply disposing of the dead birds, Bumpus weighed them and made careful measurements of length, wingspan, beak, head, humerus, femur, skull and so on. Then he did the same for the survivors. Comparing the statistics of the two groups, he found the measurements of the birds that survived were closer to the mean of the group than were those of the birds that died.

This type of mortality, where extremes are eliminated, is referred to as balanced phenotype, or stabilizing selection.

Since "Bumpus's Sparrows" was not just the first but also one of the only studies done until the mid-20th century, it was frequently cited by selectionists from Alfred Russel Wallace onward. For almost 40 years, it was the sole experimental demonstration of natural selection given in textbooks, causing geneticist Raymond Pearl to lament in 1934, "If ever an idea cried and begged for experimental [research programs] surely it is this one . . . but there have been so very, very few of them." By the 1950s, the famous sparrows were at last displaced as the classic textbook example of natural selection by H. B. D. Kettlewell's much more elaborate study of peppered moths in England's polluted woods.

Even today "Bumpus's Sparrows" continues to be quoted in about five published scientific articles every year, a very respectable figure considering that the average paper cited receives only 1.7 references a year. Besides, very few papers that were published 90 years ago now receive any citations at all. Pearl's complaint of a half-century ago holds good: The number of experimental tests of natural selection is pitiful; the few that have been conducted still do heavy duty as exemplars.

See also NATURAL SELECTION; PEPPERED MOTH; STABILIZING SELECTION.

BURGESS SHALE
Earliest Known Animals

For years fossil hunters were stymied in their search for very early types of animals, since most remains were of creatures with hard parts. Fossils of the earlier animals without shells or skeletons were extremely rare. However, in 1907, Charles D. Walcott discovered an assemblage of previously unknown worms, jellyfish-like animals and strange creatures high in the Rocky Mountains of British Columbia, near the Burgess Pass.

Estimated to be more than 550 million years old, the Burgess shale contains many soft animals preserved in rapidly deposited muds, which were the result of repeated underwater landslides. There are Annelid worms (related to the living earthworms), Priapulidae (an archaic type of sea worm) and early Chordates (animals with notocords, or primitive backbones.)

But the Burgess shale is most interesting because it contains several animals belonging to phyla totally unknown on the Earth today. These lineages arose from unknown ancestors in Precambrian times, and all became extinct.

One segmented swimmer, *Opabinia*, has short stalks containing five compound eyes on its head and spiked pincers at the end of its flexible trunk. Another of these strange creatures has seven tentacles on its back and seven pairs of stilt-like legs. It was named *Hallucigenia* by its discoverer, who could scarcely

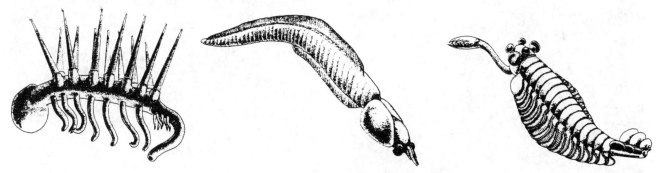

STRANGEST CREATURES known to science were recently re-studied fossil remains in Canada's Burgess Shale. Shown here are *Hallucigenia* (left), so called because its discoverer could scarcely believe his eyes, the enigmatic *Nectocaris* (middle) and *Opabinia*, with its clawed nozzle and five eyes—all unlike any known living phyla.

believe his eyes. Fifteen of them were found clustered around a large worm, possibly preparing to dine when all were buried in the ancient mud.

Still another Burgess creature (*Pikaia*) is a slim, tentacled, cigar-shaped animal with no head, no eyes, no fins or limbs—but a rudimentary backbone. It may well be the ancestral vertebrate.

These early oceanic creatures make us wonder what turn of evolutionary history eliminated the extinct phyla, which seem so bizarre, and favored *Pikaia*, the early chordate from which mammals and birds may have eventually developed. Biologists cannot even begin to speculate on what kind of creatures might have colonized the land if *Hallucigenia* had represented the more successful phylum in Cambrian seas.

See also CONTINGENT HISTORY.

For further information:
Gould, Stephen Jay. *Wonderful Life: The Burgess Shale and the Nature of History.* New York: Norton, 1989.
Whittington, Harry. *The Burgess Shale.* New Haven: Yale University Press, 1985.

BURIAN, ZDENEK (1905–1981)
Czech Dinosaur Artist

One of the greatest—and fastest—of all painters of prehistoric life, Zdenek Burian (BOOR-i-yan) left a magnificent legacy of more than 15,000 artworks and an indefinite number of sketches and preparatory studies. In a few days, without sacrificing quality,

Burian was able to turn out a finished painting that would take another artist weeks or months.

Born in Moravia, in what is now Czechoslovakia, Burian was graduated from the Academy of Fine Arts in Prague and started as an illustrator in publishing. Fascinated with natural history since his boyhood and inspired by the work of Charles R. Knight, Burian turned his remarkable abilities to depicting the lost world of prehistory. Almost immediately, his ability to imbue extinct creatures with vitality, as well as the dramatic tension of his compositions, began to win recognition.

In 1932, paleontologist Dr. Josef Augusta saw some of his illustrations in an early book, *The Hunters of Mammoth and Reindeer* and sought out the painter. Together, they produced the most beautiful series of popular literature on prehistoric subjects ever created, including *Prehistoric Animals* (1960), *Prehistoric Man* (1960), *The Age of Monsters* (1966), *Prehistoric Sea Monsters* (1964), and *The Book of Mammoths* (1962). With another scientific collaborator, he produced three more profusely illustrated volumes on the life of prehistoric man. These books are now scarce and coveted by collectors; inexplicably, there have been no recent reprints of them.

In Czechoslovakia, Burian's paintings have been declared national treasures, which by law cannot be sold or exported from the country. The government built a small Burian Museum to house a permanent exhibition of his works. The museum, located in

PAINTER OF PREHISTORY Zdenek Burian was noted for his skillful brushwork, dramatic moodiness, and scientific accuracy. The government of his native Czechoslovakia has declared his unique body of work a National Treasure.

Dvur Kralove nad Labem in the northwest part of the country, adjoins the city zoo and exhibits sixty of his works, a tiny fraction of his prodigious output.

A few years after his death in 1981, Burian's long-time student, Vladimir Krb, immigrated to Canada, where he creates exciting dinosaur murals for the Tyrell Museum of Paleontology at Drumheller, Alberta in the tradition of his teacher.

See also DINOSAURABILIA; DINOSAUR RESTORATIONS; KNIGHT, CHARLES R.

For further information:
Augusta, Josef, and Burian, Zdenek. *Prehistoric Animals.* London: Paul Hamlyn, 1960.
————. *Prehistoric Man.* London: Paul Hamlyn, 1960.
————. *The Age of Monsters.* London: Paul Hamlyn, 1966.

BURNET, REVEREND THOMAS (c.1635–1715)
Founder of Scriptural Geology

Many of the early geologists and naturalists were churchmen who wanted to reconcile their observations of nature with the word of God. During the 17th century, Reverend Thomas Burnet founded a style of scriptural geology that flourished in England until the 19th century. His popular book *Sacred Theory of the Earth* (1691) was a pioneering attempt to explain geologic features by mechanical forces and principles and yet be true to scriptural teachings.

Following the lead of Newton in physics and Descartes in cosmogeny and geology, Burnet tried to explain the role of natural forces in shaping the Earth. His book's full title is *Sacred Theory of the Earth: Containing an Account of Its Original Creation, and of All the General Changes Which It Hath Undergone, Or Is to Undergo, until the Consummation of All Things.* Burnet's premise was that God had a plan for the Earth and for man, which was translated into natural principles that could be known. If we can discover the natural patterns God created, we would know how to fit in and so further God's plan. That notion, in different forms, was later shared by Christian fundamentalists, Social Darwinists, eugenicists, environmentalists, and—minus the deity—Marxist socialists. All assume that we can understand the forces of history in terms of simple laws with which we can harmonize. Those who are out of tune with natural laws are opposing the inevitable.

Burnet also believed the Bible itself "providentially conserved . . . the Memory of Things and Times so remote, as could not be retrieved." From it he took his notions of the origin of man, the primal paradise, destruction of the ancient world by a "Universal Deluge or flood," and the "peopling of the second Earth." All would end, he believed, with "the fire next time."

In *Sacred Theory*, the world was created as a perfect sphere, containing a fluid mass. But, as the crust dried, it cracked, allowing the inner waters to flow out and flood the land. Thus were created mountains, rivers, earthquakes and all the untidy features of this "dirty little planet," which was pristine when first formed by God.

Burnet wrote clearly of how ordinary processes of erosion, deposition and volcanic activity had shaped the surface of the Earth. But he saw them mostly as destructive forces, a decline from original perfection.

His book was considered heretical, because Burnet described Noah's flood more as a natural event than a punishment for human sin. Trying to have it both ways, he added that Earth crises were sychronized with human events—that it was "the great Art of Divine Providence to adjust the two Worlds, human and natural." Nevertheless, he was denounced by fellow churchmen for taking liberties with sacred texts in order to mesh them with his theories.

In a later work, *Archaeologiae Philosophicae* (1692), Burnet tried to reconcile his account of earth history more closely with Genesis, but never succeeded in winning over his critics. If God was acting to punish man's sins, the critics preferred to read of direct divine intervention rather than such mundane secondary causes as erosion. In addition, they could not sanction Burnet's idea of a degraded Earth, ruined by the flood, at a time when most scholars sought evidence of "Divine Beneficence" in nature.

See also DIVINE BENEFICENCE, IDEA OF; "TWO BOOKS," DOCTRINE OF.

BURROUGHS, EDGAR RICE

See TARZAN OF THE APES.

BUTLER, SAMUEL (1835–1902)
Novelist, Polemicist

Novelist Samuel Butler created two Victorian classics that continue to enjoy popularity almost a century after his death—the utopian satire *Erewhon* (1872) (an anagram for "Nowhere") and his prickly family saga *The Way of All Flesh* (1903). Practically forgotten now is the incredibly bitter and personal vendetta Butler launched against Charles Darwin in books such as *Life and Habit* (1877), *Unconscious Memory* (1880), *Evolution Old and New* (1879) and *Luck or Cunning?* (1887).

Once an admirer and correspondent of Darwin's (he had even visited Down House several times), Butler gradually turned against him. Darwin's theory of evolution wasn't even original, Butler argued. Buffon, Lamarck and Erasmus Darwin had invented it years before, and natural selection was a gross error, which added nothing. If habits and strivings of an individual's lifetime could not be passed on to offspring, there could be no progress, only a "nightmare of waste and death."

But Darwin's star was riding high in the last quarter of the 19th century, and Butler's attacks antagonized a public newly enamored of science. Since Butler had also criticized and attacked the follies of religion, he stood in a no-man's land. George Bernard Shaw recalled "how completely even a man of genius could isolate himself by antagonizing Darwin on the one hand and the Church on the other."

It was Shaw who recognized Butler's bile was not stirred by a mere argument about biology. Samuel Butler had declared "with penetrating accuracy that Darwin had 'banished mind from the universe'." He extended the attack to Darwin's character, the playwright believed, because Butler was "unable to bear the fact that the author of so abhorrent a doctrine was an amiable and upright man."

In the preface to his evolutionary drama *Back to Methusalah* (1923), Shaw marveled that scientists who casually dismissed Butler's anti-Darwinian arguments were blind to the real source of "the provocation under which he was raging."

> They actually regarded the banishment of mind from the universe as a glorious enlightenment and emancipation for which he was ignorantly ungrateful . . . [He was] a prophet who tried to head us back when we were gaily dancing to our damnation across the rainbow bridge which Darwinism had thrown over the gulf which separates life and hope from death and despair. We were intellectually intoxicated with the idea that the world could make itself without design, purpose, skill, or intelligence . . .

The feud became intensely bitter and personal shortly after the publication of Butler's *Evolution Old and New*, in which he examined Erasmus Darwin's evolutionary theory and concluded it was far superior to his grandson's. At the time, Charles Darwin was publishing a translation of a German article, by a Dr. Krause, about Erasmus's life and work.

Darwin sent a copy of Butler's book to Krause, who then added a few caveats to his admiring biography of Darwin's grandfather. Erasmus's theory was great in its time, Krause concluded, but "to wish to revive it at the present day . . . shows a weakness of thought and mental anachronism no one can envy."

Butler was furious and insisted Darwin publicly admit this passage (and others) had not appeared in the original German version but had been inserted as a devious attack on his book. Darwin claimed innocence, though his son and biographer Francis later admitted that Butler "had some cause of complaint." Charges and countercharges flew by letter, though Darwin made no public response to Butler's published accusations. Huxley told Darwin to grin and bear it, for "every great whale has its louse."

The Darwin–Butler feud became convoluted, blown out of proportion and was still the subject of argument by champions of both sides long after the principals were dead. Butler was certainly thinking of Darwin when he wrote this prophetic poem about immortality:

> We shall not argue, saying " 'Twas thus" or "Thus,"
> Our argument's whole drift we shall forget
> Who's right, who's wrong, 'twill be all one to us;
> We shall not even know that we have met,
> Yet meet we shall and part and meet again
> Where dead men meet, on lips of living men

See also NEO-LAMARCKIAN; SHAW, G. BERNARD.

For further information:
Butler, Samuel. *Evolution, Old and New; or the Theories of Buffon, Dr. Erasmus Darwin, and Lamarck, as Compared with Those of Mr. Charles Darwin.* London: Hardwick and Bogue, 1879.
———. *Luck, or Cunning, as the Main Means of Organic Modification?* London: Trubner, 1887.
Willey, B. *Darwin and Butler: Two Versions of Evolution.* London: Chatto and Windus, 1960.

BUTLER ACT
Tennessee Antievolution Law

On March 21, 1925, the Tennessee legislature passed the Butler Act, making it illegal to teach human evolution in the state's public schools. Section 1 of the Act provided:

> That it shall be unlawful for any teacher in any of the universities, normals and all other public schools of the State . . . to teach any theory that denies the story of the Divine Creation of man as taught in the Bible, and to teach instead that man has descended from a lower order of animal.

Two months after the law was adopted, John T. Scopes, a Dayton, Tennessee high school teacher, was charged with teaching evolution and made front-page news across the country. The celebrated legal confrontation, which became famous as the "Monkey Trial," began on July 10, 1925, with Clarence Darrow defending Scopes and William Jennings Bryan prosecuting.

Scopes was convicted, but the verdict was later overturned on a technicality. As a result, it was not possible to appeal and thereby challenge the constitutionality of the law. Forty years later, antievolutionary laws were struck down in Tennessee, Arkansas and Mississippi.

But antievolution laws were soon replaced in Arkansas and some other states by balanced treatment laws, which required "creation science" and evolutionary biology to be given equal time in classrooms. On January 5, 1982, the day that Judge William Overton ruled that the Arkansas law was incompatible with the U.S. Constitution, the Mississippi Senate passed its own balanced treatment law by an overwhelming majority.

See also BRYAN, WILLIAM JENNINGS; CREATIONISM; INHERIT THE WIND; SCOPES TRIAL; SCOPES II.

BUTTON, JEMMY (c.1816–c.1870)
A Victorian "Experiment"

Had it not been for Captain Robert FitzRoy's promise to return Jemmy Button and two other Yahgan Indians to their homeland at the tip of South America, Charles Darwin's epochal voyage of discovery might never have taken place.

"Jemmy Button" was the English name given to Orundellico, a 14-year-old Indian boy who was taken from Tierra del Fuego to England in 1830 aboard H.M.S. *Beagle*. According to Captain Robert FitzRoy, he had "paid" an adult Indian a large mother-of-pearl button for the lad; hence his nickname.

Imperious Captain FitzRoy "collected" three other Fuegians at the same time, not by force, but by describing the wonders of England to the naive tribesmen. On his own responsibility, the Captain was attempting an experiment with human lives. He planned to take these "savages" from "brute creation," and expose them to the light of British civilization. They would be taught to speak and read English, given Bible studies, shown how to dress and eat with knife and fork and then returned to their homeland. Jemmy's companions were given the names Boat Memory, York Minister, and (a young girl) Fuegia Basket.

In FitzRoy's view, it was a philanthropic project to benefit both the British and the Fuegians. When Jemmy and his friends returned to their tribe, they would bring superior knowledge and prosperity to their benighted relatives. They could spread their new knowledge of clothing, English, Christianity, higher morals and how to cultivate food plants. When the next British sailors arrived at this remote, wild spot, FitzRoy believed, they would be greeted by friendly, English-speaking natives who would cheerfully supply food, wood, water and other provisions.

Jemmy Button and his companions behaved remarkably well for people who were wrenched and dislocated from their culture and treated as specimens. They were bright, curious and eager to learn the white man's ways. Most of the time they lived with missionaries in England, who took them on occasional excursions around London, and even had

JEMMY BUTTON (top), one of the "civilized" Fuegian Indians brought to England as an experiment, was sketched by his would-be benefactor, Captain Robert FitzRoy of H.M.S. *Beagle*. His compatriots Fuegia Basket and York Minister, all in European dress, are also shown in FitzRoy's drawing.

a long audience with the King and Queen. Jemmy enjoyed dressing as a dandy in highly polished boots, short oiled hair and white kid gloves. However, Boat Memory died soon after arriving in England.

FitzRoy's Fuegian experiment was the original impetus for the *Beagle's* second voyage to South America. When the Admiralty hesitated to finance another long expedition, he appealed to his wealthy relatives. Finally, fearing the surviving Fuegians would all grow discontented and die in England, he decided to finance their return himself.

After the captain had put up some of his own money, the Admiralty came through with funds for a new surveying mission to chart the coast of South America. It was then that he decided to use the opportunity to advance another of his pet theories: He would gather evidence on the voyage that would prove the truth of Genesis.

Both of FitzRoy's major projects became complete personal disasters. His experimental subject Jemmy Button ended by turning his back on British civilization and reverting to the "wild" ways of his naked tribe. And the naturalist he chose to help him prove the creation theory turned out to be Charles Darwin!

(FitzRoy commited suicide years later in despair about the complete failure of his grand aims in life.)

Jemmy, 16 years old on the *Beagle's* second surveying voyage of 1832, became a special favorite of the sailors. The Fuegian's eyesight was very keen, and he was able to spot distant objects before anyone else on board. When Jemmy couldn't get his way with the officer on watch he would pout and say, "Me see ship, me no tell."

Many gifts had been given to the Fuegians with which to start their new civilized life in the wild country at the tip of South America. If the project's ludicrous Victorian smugness was unrecognized by the captain, it did not escape Darwin. He wrote in his diary:

> The choice of articles [by the Missionary Society] showed the most culpable folly and negligency. Wine glasses, butter-bolts, tea trays, soup tureens, mahogany dressing case, fine white linen, beaver hats and an endless variety of similar things, show how little was thought about the country where they [the Fuegians] were going to. The means absolutely wasted on such things would have purchased an immense stock of really useful articles.

A young missionary, Richard Matthews, had come along on the voyage to help Christianize the Fuegians and establish a settlement. Jemmy Button, Fuegia Basket and York Minister were ferried in small boats up the Beagle Channel to Ponsonby Sound, along with supplies and a crew of sailors. They were met by a fleet of canoes manned by grease-covered Indians, naked in the cold except for skimpy capes of guanaco and otter skins. The sailors put up tents to house the cargo and constructed three wigwams: one for the missionary, one for Jemmy, and one for York and Fuegia Basket.

All hands began digging a vegetable garden, while 100 or so native tribesmen stood around, staring in wonder. Finally, Jemmy's mother, two sisters and four brothers arrived. Darwin wrote that "it was laughable, but almost pitieable, to hear him speak to his wild brother in English, and then ask him in Spanish whether he did not understand him." Jemmy had forgotten his own language, although he was soon to regain it. Matthews stayed with the Indians, while FitzRoy and his men departed the camp.

When the *Beagle* crew returned 10 days later, everything was in shambles. Matthews reported that as soon as the crew left, the natives had started to take everything in sight; when he tried to stop them, he was beaten and almost killed. The vegetable garden was trampled. York Minister had sided with the natives and was let alone, but Jemmy had been in several fights for defending the missionary and had been beaten as well. FitzRoy was shocked and disappointed with the results of his good intentions. He took Matthews back on board, but left the Fuegians with their kinsmen and departed.

A year later, when the *Beagle* returned after further explorations, the camp was no more. York Minister and Fuegia Basket had taken all Jemmy's goods and departed with the other Fuegians. Jemmy had remained, but had replaced his European clothes with a loincloth, taken a native woman for his wife, and was hunting and fishing for a living. Although he visited his old friends aboard *Beagle,* he told them he was finished with civilization forever. Darwin poignantly recorded his last view of Jemmy, standing on shore near his campfire, waving "a long farewell."

See also FITZROY, CAPTAIN ROBERT; VOYAGE OF H.M.S. BEAGLE.

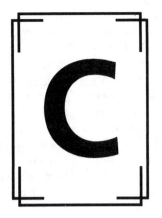

CADUCEUS
Symbol for Secret of Life

A powerful symbol, both ancient and modern, the caduceus (Ka-DOO-shus), depicts two snakes coiled around a vertical staff. Medical schools emboss it on diplomas, and medical associations use it as a logo. To physicians, it represents the secret of life as embodied in the healing arts; thus, it also stands for knowledge and wisdom.

A spiral staircase, or corkscrew, is a single helix. If one intertwines another helix going in the opposite direction, the form becomes a double helix, like the two snakes. An ancient Hindu tradition interprets the caduceus as the symbol of evolutionary energy or life-force (kundalini), which lies coiled like a serpent at the base of the spine. Yogis attempt to awaken this energy and encourage it to ascend the spinal column, passing through the various body centers (chakras) until it reaches the top of the skull—the seat of highest consciousness.

In Hindu paintings thousands of years old, left and right spinal nerves (ida and pingala) are depicted as wrapped around the vertebral column in a double

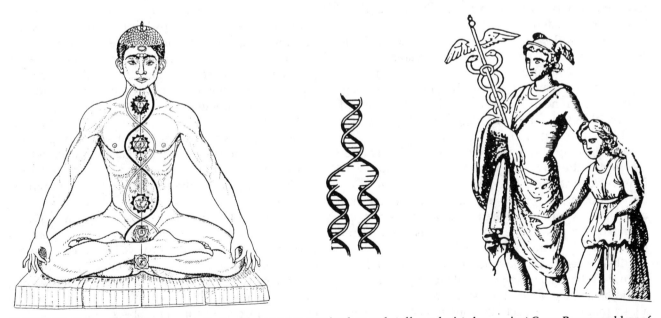

SECRET OF LIFE and art of healing, symbolized by serpents entwined around staff, are depicted on ancient Greco-Roman emblem of Hermes (Mercury). Associated with sexual/evolutionary energy in Hindu tradition, the double helix also represents a Western scientific "secret of life"—the structure of a DNA molecule.

helix pattern, represented by the caduceus. In the Indo-Tibetan tradition, the same symbol is also used to connote wisdom, knowledge and the secret of life.

In 1952, James Watson and Francis Crick discovered the biochemical structure of DNA, the creative blueprint for all organisms. Knowledge of DNA's special properties has led to an entirely new understanding of genetics, including undreamed-of methods for mapping chromosomes and artificially altering life-forms. The structure: a double helix, the ancient symbol for "evolutionary energy" and knowledge of the secret of life.

See also DNA; KUNDALINI.

CALAVARAS SKULL (HOAX)
Californian Paradise Lost

Twenty-five years before the famous Piltdown hoax, the fraudulent "missing link" that bamboozled anthropology for 40 years, a similar fake caused great excitement in the United States. The skull of what was thought to be a remote human ancestor turned up in Calavaras County, California in 1876, the same Calavaras County made famous for tall tales by Mark Twain in his 1865 story about the jumping frog contest.

J. D. Whitney, chief geologist of the California Geological Survey, was convinced the skull was that of a genuine fossil man, and proved that humans had inhabited central California during Tertiary times (two million years ago). That would have made "Calavaras Man" much earlier than any other human fossils known from anywhere in the world. And Calavaras County would be the cradle of mankind, a New World Eden.

Not until 1911, after Whitney had died, did the truth come out. Gravel beds at Calavaras actually contained some ancient fossils, mainly stumps of extinct palm trees. At about the time these were discovered, a nearby stream flooded and washed several skeletons of recent Digger Indians out of a local Indian burial ground.

With the help of some buddies, a young local miner convinced the town physician that the skull was found with the palm tree fossils in the gravels. The duped doctor, who was a collector of fossils and curiosities, went farther than the pranksters had anticipated. He passed the skull on to his good friend J. D. Whitney at the Geological Survey.

Public interest in the find soon reached a fever pitch, but the miners remained silent—probably fearing to bring disgrace on themselves and the town for a joke that had unexpectedly gone too far. However, the truth eventually came out when the skull was reexamined and the earth matrix found inside the skull did not at all match the gravels of the mining pit.

There have been other attempts to find the earliest man in the world in the North Americas, in contradiction to the established theory that the human species originated in the Old World, most probably in Africa. Such other presumed fossils as "Nebraska man" and another "California man" have turned out to be false leads, not necessarily deliberate. And about the time the Calavaras hoax was finally exposed, another joker was putting a human skull together with an ape's jaw and planting it in an old gravel pit in Sussex, England at the site known as Piltdown.

See also NEBRASKA MAN; PILTDOWN MAN (HOAX).

CAMBRIAN–SILURIAN CONTROVERSY
Rival Geological Periods

The Old Red Sandstone of England was the oldest system of fossil-bearing rock known until the 1830s. Packed with remains of an "age of fishes," it was called the Devonian, after Devon, where it was found. But what came before the Devonian? Adam Sedgwick, a former theology student turned geologist, and an imperious aristocrat named Roderick Impy Murchison traveled all over Europe attacking rock formations with their geologist's hammers to find the answer.

Once good friends, the two became bitter rivals in the quest to discover, describe and name the oldest geological systems (and corresponding periods) in Earth's history.

Murchison was a wealthy gentleman who spent years doing nothing but fox hunting six days a week, retaining eight professional hunters and a pack of hounds. A man of restless temperament and energy, he tired of that life and became obsessed with making a name for himself in geology, then a wide-open field for a wealthy amateur.

After publishing many papers on the geology of Scotland, the Alps and other parts of Europe—and gaining the presidency of the Geological Society in 1831—he set out in search of the peculiar "greywacke" formations, then unstudied but thought to be very old.

His search led him to Wales, where in 1834 he found a formation of fossil-bearing rock that lay underneath the Devonian and was therefore more ancient. With romantic flair, he named it for the Silures, an ancient Welsh tribe that had resisted the invading Romans. Thus the Silurian came into being; it was to become Murchison's lifelong obsession.

After almost a decade of work, he brought out his massive two-volume treatise *The Silurian System* (1854), which all geologists bought, though few could plow through its dull compilation of detail. But even while he was working to establish the Silurian as the oldest rock system, his friend Sedgwick discovered an even more ancient one underneath it. Sedgwick gave it the archaic name for Wales, which is Cambria, and thus established the Cambrian period in geology.

Now began the great rivalry, with Murchison and Sedgwick trekking to rock formations all over Europe, each attempting to enlarge his geological domain. Murchison extended the Silurian downward, while Sedgwick pushed his original Cambrian upward. The two men were completely unable to agree on where the natural boundaries occurred. Murchison, however, found a way to resolve the dispute. He got himself appointed director of the National Geological Survey and simply ordered that the name "Cambrian" be deleted from all government books and geological maps.

Science is not always fair—at least not while its contributors are still alive. But it also has a way of correcting itself in time, and after both men were dead, Sedgwick was vindicated. The Cambrian has grown in importance over the years, while the Silurian has diminished.

Since Murchison's time, his beloved rock kingdom has been shown to encompass several systems, one of which has been split off and renamed the Ordovician. Thousands of fossil finds of extinct early phyla, including those of the remarkable Burgess shale, have literally put the Cambrian back on the maps for good. Sir Roderick Murchison, sometimes known as "The King of Siluria," would not have been pleased.

See also BURGESS SHALE; MURCHISON, SIR RODERICK IMPY; SEDGWICK, REVEREND ADAM.

For further information:
Rudwick, M. J. S. *The Great Devonian Controversy. The Shaping of Scientific Knowledge among Gentlemanly Specialists.* Chicago: University of Chicago Press, 1985.
Secord, J. A. *Controversy in Victorian Geology: The Cambrian–Silurian Dispute.* Princeton: Princeton University Press, 1986.

CANINE DIASTEMA
Dental Gap

Many male primates have large canine teeth, which are used in fighting and defense. Where the upper canines meet, or occlude, with the lower jaw, there are spaces, or gaps, between the opposing teeth.

Canine diastemas are characteristic of the jaws of baboons, gorillas and monkeys. They are used as a diagnostic feature in studying fossils because they are absent in hominids (men or near-men). A primate jaw with canine diastemas is considered probably related to apes or monkeys, not close to the human family.

CANN, REBECCA (AND COLLEAGUES)

See MITOCHONDRIAL "EVE."

CANNIBALISM CONTROVERSY
Are Humans Man-Eaters?

Amazing though it seems, discoverers of almost every fossil man-like creature have rushed to announce they have found associated "evidence of cannibalism." Later, in most cases, their colleagues declare the evidence insufficient, and it is eventually dropped from discussion.

African man-apes (australopithecines), Peking Man (*Homo erectus*), Neandertal and Cro-Magnon were all at first believed by their discoverers to have had a taste for their fellows' flesh. Debate has gone on for years about whether our ancestors habitually dined on each other, or if the usual horrific interpretation reveals more about the minds of anthropologists than it does about prehistoric cannibalism.

Robert Broom and Raymond Dart, the South African paleontologists who discovered many australopithecine fossils, thought battered bones and perforated skulls proved we descended from a predatory ape who didn't stop at his own species.

Some years later, Professor Franz Weidenreich helped excavate remains of *Homo erectus* (Peking man) in a Chinese cave and noticed many skulls were battered at the base. He concluded they ate each other's brains, but later changed his mind. At various times, other experts have tarred both Neandertal man and early *Homo sapiens* with the same brush.

Deep scratches on fossil hominid bones could have had other causes than so-called "gourmet" cannibalism; they may have been gnawed and dragged by hyenas or other scavengers. *Homo erectus* might have been the prey, not the butcher. At the Peking man site, only skulls were found, which could also mean that heads were carried into the cave for some ritual. (Many contemporary tribal peoples use the skulls of dead relatives in ancestor worship.)

Anthropologists are still debating the nature and extent of cannibalism among tribal peoples in recent times. Professor W. Arens created an uproar among experts when he challenged the whole idea in his book *The Man-Eating Myth* (1979). Like most anthropologists, Arens had always taken for granted that 19th-century explorers had visited cannibalistic tribes in Africa, New Guinea and South America. But when he sifted the massive literature on the subject, he could not find one satisfactory first hand account of cannibalism as a socially approved custom in any part of the world.

When Arens went to Africa to collect stories about tribal cannibalism, he got the surprise of his life. Azande villagers he was studying had decided that *he* was a "blood-sucker," a kind of vampire. Although he lived there for a year and a half, Professor Arens never succeeded in convincing them that he was not secretly feeding on human blood at night.

While European colonials had nurtured their fear of African cannibals, they never realized that Africans had the same suspicions about Europeans. And the Africans had evidence. Some years before, during a war, Europeans had tried to persuade the locals to give blood for their wounded soldiers. Villagers were still afraid of being summoned to the hospital, where their blood would be taken. Their memories of the urgent pleas for their blood had become a belief that Europeans need to drink African blood to live.

At first, Arens was condescending about this belief but later kicked himself for not grasping the underlying political metaphor. Of course it made sense for Africans to feel that Europeans were robbing their vitality and draining them of their life's blood.

He realized that charges of cannibalism have been made throughout history to unite the accusing group as moral people and to place the enemy group outside humanity. Colonial Europeans justified subjugation of tribal people since the 1600s on the grounds they were uncivilized cannibals. Koreans thought Chinese were cannibals and the Chinese thought the same of the Koreans. Arens began to suspect that accusations and beliefs about cannibalism are much more widespread than the actual practice had ever been.

Arens concluded that there never had been reliable reports of any widespread practice of "gourmet" cannibalism. It was a myth among the anthropologists, he said, and challenged his colleagues to prove otherwise. Soon after his book appeared, several field-workers came forward to offer their evidence.

George Morren of Rutgers University had done fieldwork in New Guinea in the late 1960s, where older Miyanmin tribesmen who had participated in cannibalism gave him detailed accounts. He cross-checked complex descriptions with several informants. He also studied court records of a 1959 trial involving more than 30 Miyanmin who were accused of killing and eating 16 people from a neighboring tribe.

When he particularly pressed the Miyanmin about the possible religious or symbolic significance of the incident, they insisted there was none. "No, we just went after the meat." It appears to be as close to an authentic, documented account of culturally sanctioned cannibalism as we are likely to get.

Meanwhile, other anthropologists have not been persuaded by Arens to throw out most of the "cannibal" literature as biased and second-hand. And a growing body of evidence from recently deciphered writings and artworks shows the ancient Mayans and Aztecs practiced bloody sacrifices and cannibalistic rites on a large scale. The debate over a cannibal past for the human species continues.

Observers of primate behavior have also contributed to this fascinating debate. After a decade of watching chimpanzees in the forests of Zaire, Jane Goodall had concluded that they were gentle, peaceful, social, sometimes clownish vegetarians. But she has since watched them hunt and kill other animals for meat, deliberately kill infant chimps from neighboring groups and even kill and eat babies from within their own community.

Two chimpanzees, a mother and daughter team, suddenly began a series of cannibalistic infanticides. One would distract a new mother, another would snatch the baby, then they'd both kill and eat it! Jane Goodall was sad and disillusioned. She admitted that she had thought chimps were "better" than humans, but now realized that a chimp's heart contains dark secrets as well.

See also CHIMPANZEES; HOMO ERECTUS; KILLER APE THEORY.

For further information:
Arens, W. *The Man-Eating Myth*. New York: Oxford University Press, 1979.

CARDIFF GIANT
Petrified Man Hoax

Fossil evidence for human evolution was still so scant in the 1870s that a bewildered public paid a fortune to see the most brazen scientific hoax in history. The Cardiff Giant, a crude stone statue, was successfully promoted as the petrified remains of a huge, extinct species of man that once inhabited upper New York state.

George Hull, a former cigar maker from Binghamton, New York, conceived the plot to create the giant and, in 1868, obtained a five-ton block of gypsum in Iowa and had it fashioned into the shape of an immense man by a Chicago stonecutter. He then shipped the statue to his cousin, William Newell, near Cardiff, New York, who supposedly discovered it while digging a well behind his barn a year later. It is not clear whether the hoax was originally planned as a swindle or if, as Hull later claimed, he had the giant built to ridicule clergymen who insisted on the literal truth of every word in Genesis including "there were giants in the earth in those days."

A Syracuse newspaper headlined the find as A WONDERFUL DISCOVERY, and the pair pitched a tent

GIANT FRAUD being laid to rest at the Farmer's Museum in Cooperstown, New York in 1948. Thousands had paid to see the fake "petrified man" that had supposedly been plowed up on a Cardiff farm; irreverent workmen pay their last respects with appropriately phony sentiments. (Courtesy of New York State Historical Society, Cooperstown, N.Y.)

on the farm and began exhibiting the giant in a ditch, charging a nickel a look. News of the find flashed around the world. Thousands swarmed to see it, and admission was raised to a dollar.

Meanwhile, experts argued about the fossil's authenticity. The director of the New York State Museum thought the giant was really a statue, but was indeed most ancient and "the most remarkable object yet brought to light in this country." Others, including Oliver Wendell Holmes and Ralph Waldo Emerson, concurred. Cornell's president pronounced the giant a gypsum forgery, and Yale paleontologist O. C. Marsh muttered it was "remarkable—a remarkable fake."

But the crowds, now arriving by special trains,

continued to grow, and great showman P. T. Barnum offered $60,000 to lease the giant from Newell for three months. The farmer refused. Undeterred, Barnum hired a sculptor, Professor Carl C. F. Otto, to make an exact copy of the giant.

When Hull and Newell brought their giant to New York City in 1871 for exhibit, they discovered Barnum was already displaying his version in Brooklyn. While they hauled Barnum into court, newspapermen were investigating Hull's activities and uncovered his purchase of gypsum in Iowa. They located the stonecutter in Chicago, one Edward Salle, who admitted to carving the giant, aging it with sand, ink and sulfuric acid and punching pores into it with darning needles. Faced with the growing evidence of fraud, Hull confessed. Barnum now was able to avoid prosecution by claiming all he had done was show a fake of a fake—which could not be considered a forgery. His giant, after all, was guaranteed to be an authentic fake.

Their fraud netted Hull and Newall about $44,000 after expenses of $2,200. Barnum, who continued showing his version for years, made more than $150,000. Today, the Cardiff Giant—Hull's original phony, not Barnum's authentic fake copy—is displayed in an earthen pit at the Farmer's Museum in Cooperstown, New York.

See also BARNUM, PHINEAS T.; BERINGER, J.; CALA-VARAS SKULL (HOAX); PILTDOWN MAN (HOAX); TASADAY TRIBE (HOAX).

CARNEGIE, ANDREW (1835–1919)
Darwinian Industrialist

When steel was king, Andrew Carnegie, "the richest man in the world," was king of steel. A Scots immigrant from a poor working-class family, Carnegie rose in business to become a powerful, ruthless tycoon who exploited man and Earth, crushed competition, and justified his actions by a philosophy of Social Darwinism.

Entrepreneurial competition, he believed, does a service to society by eliminating the weaker elements. Those who survive in business are "fit," and therefore deserve their positions and rewards. Carnegie elevated the capitalist ethic to a law of nature.

Although he proclaimed himself a "Darwinist," Carnegie drew his inspiration from the English philosopher Herbert Spencer (1820–1903). Unlike Darwin, Spencer had sought to apply evolutionary thinking across a broad spectrum of political and social questions. "Before Spencer," Carnegie said repeatedly, "all for me had been darkness, after him, all had become light—and right." It was Spencer, after all, and not Darwin, who was the author of the phrase "survival of the fittest."

Carnegie believed that Spencer had revealed to man his own destiny: social evolution toward a peaceful industrial world, whose mass-produced products and technologies would be available to all. Man's striving had a purpose: progressive evolution to higher levels of efficiency and happiness. One biographer wrote "Carnegie's happy little slogan, 'All is well, since all grows better' was for him the satisfying distillation of thirty volumes of Spencer's philosophy."

Yet Carnegie seemed blithely unaware that, in many crucial areas, his beliefs and actions were very un-Spencerian. For instance, he did not really favor *laissez-faire* capitalism, though he paid lip service to Spencer's enthusiasm for open competition. In fact, he fought for high protective tariffs, engaged in price fixing, patent monopolies and other artificial reductions of free competition.

Unlike Spencer, Carnegie rejected the idea of eugenics and expected men of "genius" to rise from the poverty-stricken families Spencer thought were the "unfit." The immigrant American rejected the idea of hereditary social classes and believed the people who held vast wealth should give most of it away to foster the public good rather than hoard it. "The man who dies thus rich, dies disgraced," he wrote. Carnegie's desire to help and protect the weak, which he thought was the moral duty of the strong, was another very un-Spencerian notion.

Despite his loose reading of Spencer's teachings, Carnegie revered him as the greatest thinker alive. Apparently, he believed Spencer had "proved" that evolution leads to progress and that ruthlessness can enable a man to acquire the means to promote the higher good. Historians disagree on how important Darwin's or Spencer's ideas became in the business community as a justification for exploitive entrepreneurs. Most businessmen had probably never heard of Spencer; the old Ben Franklin tradition of individualism, hard work and Yankee ingenuity was all the rationale they needed. Carnegie was a notable exception in actively promoting his faith in Social Darwinism.

Carnegie courted Spencer's friendship for several years and invited him to visit North America. In 1882, Spencer made the voyage with Carnegie aboard the ship. After a brief tour of Canada, he visited Pittsburgh, which Carnegie extolled as the model of the Spencerian future, the well-regulated industrial beehive. Historian Joseph Frazer Wall relates that Spencer:

> . . . did not recognize utopia when it was shown to him. [He complained that] in this smoky, polluted air a man would be fortunate if he could recognize his own hand held close to his face. [At the Bessemer steel plant] the heat and noise of the mills reduced Spencer to a state of near collapse. When the tour was over, he could only gasp out to Carnegie, "Six months' residence here would justify suicide."

Near the end of Spencer's visit, on November 9, 1882, Carnegie held a great testimonial dinner for him in New York. Newspaper and railroad magnates, politicians and such national celebrities as the Reverend Henry Ward Beecher gathered to honor Spencer at the peak of his popularity. The chairman welcomed them by stating that the glittering assemblage had been brought together "by natural selection."

Spencer's speech turned out to be not a clarion call

ANDREW CARNEGIE, the richest man of his time and a self-proclaimed Social Darwinist, revered the teachings of Herbert Spencer's evolutionary philosophy. Proud of having sponsored discovery of the huge *Diplodocus carnegiei* dinosaur, he sent full-size skeletal casts to museums all over the world.

in this graphic proof of evolutionary development . . . an organism so successful in its physical development that it outgrew the brain and nervous system necessary to control it." Although biologists no longer characterize dinosaurs in this way, the metaphor has been applied to Carnegie's own Pittsburgh Steel corporation, which became so vast it could no longer be controlled by its head.

See also SOCIAL DARWINISM; SPENCER, HERBERT.

For further information:
Wall, Joseph F. *Andrew Carnegie.* New York: Oxford University Press, 1970.

CARPENTER, CLARENCE RAY (b. 1905)
Pioneer of Primate Field Studies

Charles Darwin was fascinated by living primates and was convinced their behavior must hold important clues to the evolution of man. While his friend and champion Thomas Huxley studied the kinship between humans and apes by dissecting muscles and bones (*Evidence as to Man's Place in Nature*, 1863), Darwin spent weeks at the London Zoo watching live monkeys. One result was his treatise on *Expression of the Emotions in Man and Animals* (1872), the ground-breaking work on the evolution of behavior.

For a half century after Darwin's death, few field naturalists attempted to follow his lead by studying the behavior of monkeys and apes in the wild. One rare exception, the American zoologist R. L. Garner, built a cage for *himself* in the midst of an African forest for protection from the "ferocious" chimps and gorillas he hoped to see. Most primatologists followed Huxley's lead, sticking close to their laboratories. Then, during the 1930s, a young American psychologist, C. Ray Carpenter, had what seemed to him an obvious idea: study the behavior of free-living primate societies to gain insights into the origin and evolution of the human way of life.

Despite volumes of speculation on the emergence

to competitive battle, but a reclusive philosopher's reaction to the feverish activity of American industry. His message to the assembled capitalists was to slow down, learn to relax and enjoy leisure time.

In 1895, Carnegie endowed a museum in Pittsburgh; its most famous attraction was the fossil skeleton of a gigantic new species of sauropod dinosaur, measuring more than 87 feet long, discovered in Colorado and Wyoming. Of all the scientific projects Carnegie sponsored, the long-necked behemoth—named *Diplodocus carnegiei* in his honor—was his special favorite.

Full-size copies of *diplodocus* were cast, mounted and sent all over the world at Carnegie's expense. His friend King Edward VII had admired it, so one was sent to the British Museum. Others went to the great museums in Germany, France, Austria, Mexico and Argentina.

According to his biographer, Carnegie "delighted

of man from a primate base, practically nothing was known about how monkeys and apes lived in the wild. What kind of communications systems did they have? How large were their groups and what was the normal proportion of males to females? Do they establish territories? How much aggression is normal between individuals or groups? Carpenter began to establish a program for asking questions that would allow comparisons. Was there a way of life that set apes apart from monkeys? Do any nonhuman primates use tools? Carpenter urged evolutionists to go out to remote forests and savannahs and find out firsthand how monkeys and apes actually live.

In the 1930s, Carpenter began pioneering field studies of spider monkeys in South America and the howler monkeys of Barro Colorado, Panama. Both are arboreal and difficult to approach and observe from the ground. Field conditions were rough, and little money was available for staff or equipment. Nevertheless, Carpenter's patience, perserverance and dedication went a long way and he produced new data and insights on each field trip.

Howlers are large, reddish-haired monkeys whose incredible sounds had long puzzled naturalists. They don't really howl, but produce a roaring sound, building in volume and intensity until it resembles the din of a passing train. After observing the movements of groups of 20 or 30 animals, Carpenter concluded the sounds were "spacing mechanisms" that kept various groups from coming too close to each other's established feeding territories. Monkeys had substituted making noises at each other for physical fighting over territory—a behavior perhaps analogous to human diplomacy (and perhaps not).

Carpenter founded the first research colony for studying the naturalistic behavior of monkey populations under semiwild conditions. During the 1930s, he brought 409 rhesus monkeys from India and liberated them on the tiny island of Cayo Santiago, off the coast of Puerto Rico. Fifty years later, their thousands of descendants have enabled scientists to learn much about their kin groups, communications, hierarchies and social behavior. (All the monkeys on Cayo Santiago have been tattooed with numbers, so individuals can be identified and families traced through generations.)

In the spring of 1937, Carpenter and a few colleagues trekked through the jungles of northern Thailand (then known as Siam), where he conducted the first systematic field studies of gibbon apes, analyzing their communication, locomotion, foraging and social behavior.

He recorded their early morning vocalizations, made shortly after dawn, when all groups of gibbons in the area call back and forth and establish the locations of their fellows. As with the howlers, Carpenter suggested that gibbons' calls establish the positions of groups in relation to one another; sometimes there are vocal battles to determine which group gets first crack at a favored fig tree.

As a trailblazing elder among monkey watchers, Ray Carpenter was able to watch anthropology and zoology catch up with his vision of completing Charles Darwin's research program. From a small trickle in the 1950s, by the 1960s scores of scientists were watching monkeys and apes all over the world. Journals were founded, departments expanded and primate field studies became well established in universities. It seemed only a matter of time before the comparative method would answer many of the questions about the continuity of human behavior with that of monkeys and apes.

Since the 1960s, popularized by European ethologists, British and Japanese zoologists and American anthropologists, field studies of monkeys and apes practically became an academic industry. Yet, three decades later, the thousands of papers on primate social behavior have thrown disappointingly little light on the origin and evolution of the human species—though they continually open up new and puzzling areas for further investigation of "man's poor relations."

See also "APE WOMEN," LEAKEY'S; CAYO SANTIAGO; GARNER, RICHARD LYNCH.

For further information:
Bourne, Geoffrey. *Primate Odyssey*. New York: Putnam, 1974.
Carpenter, C. Ray. *Naturalistic Behavior of Nonhuman Primates*. University Park: Pennsylvania State University Press, 1964.

CARTESIAN DUALITY
Does A Frog Have A Soul?

Before Charles Darwin and Alfred Wallace established an evolutionary perspective in Western thought, it was widely believed that man and animals were two unconnected, utterly different orders of beings. The rationale for this duality was proposed by Rene Descartes, the French genius of the 17th century.

Descartes was an influential founder of modern philosophy and the father of analytical geometry. As a scientist, he adopted the mechanistic viewpoint: phenomena of nature were viewed as machines, whose parts and workings could be analyzed according to mechanical principles. [See MATERIALISM; MECHANISM.]

Man's body also was a machine made of physical matter, like those of the animals, but Descartes had a problem. His traditional Christian religious beliefs maintained that man was more than matter, more than a machine—he had an immortal soul. How could Descartes continue to advocate scientific materialism, yet escape heresy?

His solution was to postulate a major difference in kind between man and the animals, which became known to philosophers as the Cartesian duality. Humans, he decided, must be the only conscious beings, because we alone are endowed with a soul by God. Animals are automata; robots; clockwork droids without consciousness or feeling.

Descartes believed his duality between man and beast, spirit and machine, soul and soulless provided a reason for believing in life after death. If "the souls of animals are of the same nature as our own," he wrote, we would "have no more to fear or hope for after this life than flies or ants." Such conclusion would lead to immoral conduct and social chaos.

Duality also resolved another disturbing problem. If animals can expect no justice in an afterlife and they are untainted by Adam's original sin, why would a just God allow them to suffer? Descartes' very unfortunate answer was that animals never suffer because, as mere machines, they are incapable of feeling pain.

Others were quick to embrace the Cartesian duality. Belief in a profound difference in kind between humans and animals relieved all guilt about killing or eating animals and enabled scientists to vivisect live creatures without paying any attention to their cries.

Descartes' ideas persisted for years. They were finally displaced by Darwin's demonstration that all life is related and connected. At last there was scientific support for what many nonscientists with common sense had known all along—that animals share many of our feelings, certainly including physical pain.

See also ANIMAL RIGHTS.

CATARRHINES
Old World Primates

Catarrhines are the large division of primates that include Old World monkeys, apes and man. The name comes from differences between the noses of Old World monkeys, which project downward, and those of New World monkeys, or Platyrhines, which are flatter.

Old World monkeys, or cercopiths, tend to walk on all fours, even on high tree branches; they lack flexible shoulder joints needed to swing or hang. None have prehensile (grasping) tails, which are

BARBARY "APES" of Gibraltar are misnamed; actually, they are the last "wild" monkeys in Europe, cousins of the rhesus macaques and baboons. Old World monkeys, true apes, and humans comprise the Catarrhine group, only distantly related to New World primates. (Photo by MacRoberts.)

found only in New World monkeys. Most live in social groups and eat a wide variety of plants, fruits, insects and roots.

Macaques are a major group of catarrhines, which include the rhesus monkeys of India, Japanese snow monkeys, Barbary "apes," and high-tree-dwelling colobus and masked monkeys of Africa. Closely related to the macaques are the larger baboons, including mandrills, hamadryads and olive baboons of North and East Africa. Baboons live on the ground—some on rocky cliffs, some on the open grassland—but often sleep in tree branches at night.

Hominoids, the other great division of Catarrhines, include the apes (Pongids) and the single living species of man. The two pongids most closely related to man are the African chimpanzee and gorilla. Gorillas and chimps are "knuckle-walkers," spending their days foraging among the vegetation of the forest floor. The Asian forest apes rarely come to the ground. Gibbons are aerial specialists (brachiators or arm-swingers), and orangs spend most of their time foraging in high branches.

All living humans belong to one species, *Homo sapiens*. Differences among geographic populations in stature, skin color, blood types, etc. are zoologically insignificant; many other widely distributed species show similar variations throughout their range. Evidence from fossils, anatomy and biochemistry indicates an Old World origin for the human species, probably, as Darwin originally suggested, in Africa.

See also APES; BABOONS; CHIMPANZEES; GORILLAS; ORANG-UTAN; PLATYRRHINES.

CATASTROPHISM
Upheavals and Cataclysms

Catastrophism (as a label) has been applied to many different theories about the history of life on earth all with the core idea of dramatic, fairly rapid and discontinuous change.

Georges Cuvier, the influential 18th-century anatomist, read a history of great "catastrophes" and "debacles" in the geologic record: widespread earthquakes, floods and inundations, volcanic upheavals. In the classic version, there was a succession of extinctions, which wiped all life off the face of the Earth, followed each time by a new creation of a different set of plants and animals. Enlightenment Frenchman that he was, Cuvier called these violent convulsions "revolutions," sometimes (without conscious irony) translated by Englishmen as "catastrophes."

Until well into the 19th century, geologists thought the Earth too young to have acquired many of its features (including fossil-bearing rock layers) by a slow and gradual buildup. Clear and abrupt boundaries between strata, fossils of seashells on mountains and successive extinctions suggested a picture of rapid and disastrous events, unlike anything observable in the contemporary world.

One common view was that the layers of fossil-bearing rock, with their different worlds of plants and animals, represented separate, "successive creations by the Author of nature." Some geologists thought this view blasphemous, as it implied the Creator was an inept craftsman, smashing his works again and again in an attempt to create a more perfect world. Others balked at accepting so many instances of life coming out of nonlife, since no such process could be seen currently operating on Earth.

Generally, catastrophism fit in well with such accepted biblical miracles as Noah's Flood, which lent authority to the concept. But Cuvier and others (notably the discoverer of Ice Ages, Swiss-American Louis Agassiz) were excellent scientists; to characterize them as religious zealots who ignored the evidence when it contradicted scripture is unfair. At the time, available evidence was still ambiguous, thus allowing room for many competing scientific interpretations.

Catastrophism is usually contrasted with Sir Charles Lyell's doctrine of uniformitarianism, which transformed the study of geology and had a great influence on young Charles Darwin.

This battle between catastrophists (as champions of scriptural miracles) and uniformitarians (as knights of rational science) has been vulgarly oversimplified in hundreds of books. Lyell's methods of geology, particularly his insistence on understanding present processes as the key to interpreting the past, were admired by many geologists who have been labeled catastrophists; some invoked catastrophes only where no known (observable) processes could account for the geological facts.

As Thomas Huxley described the catastrophists of his youth, they went further than imagining great events like poisoned atmospheres or worldwide floods; they insisted these events operated according to different processes than those we consider natural today. In the past, they thought, nature operated according to different rules. As the mathematician Charles Babbage, inventor of the mechanical computer, insisted, if God could set up natural laws, he could also reprogram the universe at will.

See also ACTUALISM; LYELL, SIR CHARLES; NEMESIS STAR; "NEW CATASTROPHISM"; NOAH'S FLOOD; STEADY STATE EARTH; UNIFORMITARIANISM.

For further information:
Eisely, Loren. *Darwin's Century.* New York: Doubleday, 1958.
Gould, Stephen Jay. *Time's Arrow, Time's Cycle.* Cambridge: Harvard University Press, 1987.
Greene, J. C. *The Death of Adam: Evolution and Its Impact on Western Thought.* New York: Mentor Books, 1959.

CAYO SANTIAGO
Scientist's Monkey Island

The world's largest and most studied captive colony of Indian rhesus monkeys inhabits the tiny tropical island of Cayo Santiago, off the coast of Puerto Rico. Professor C. Ray Carpenter, who pioneered field studies of monkeys and apes, released a shipload of them on the uninhabited island during the 1930s. Almost alone during that period, Carpenter insisted knowledge of free-ranging primates was a necessary background for understanding human behavioral evolution.

Originally stocked with 409 monkeys, the colony has survived for nearly 60 years, and now contains thousands of their descendants. Many studies have been made of their dominance behavior, aggression, social organization, communication and reproductive biology. Individuals are tattooed with identifying numbers while young, so their life histories can be traced. Some families have been observed through several generations.

One important study by anthropologist Donald Stone Sade (1960s) revealed the importance of kin relationships throughout a monkey's life. During a dispute between two adult males, he observed a third monkey burst through the foliage to drive off the attacker. Identification records showed it was the persecuted male's mother who had come to his rescue—evidence for unexpectedly long-term bonds of kinship.

In later years, the research program was thrown into turmoil by a fight for control between medical researchers and behavioral scientists. There were charges of plagiarism, falsification of data, misuse of grants and disputes over power and dominance among the island's human primates. So far, the monkeys have kept their observations of the scientists' behavior to themselves.

See also CARPENTER, CLARENCE RAY.

"CENTRAL DOGMA" OF GENETICS
From DNA Outward

A dogma is a doctrine or belief one accepts as true without question—an article of faith. How peculiar, then, that the idea that genetic information only flows outward from DNA, and never into it, is called the "Central Dogma of molecular genetics." Francis Crick, the codiscoverer of the structure of DNA, had stated the "Dogma" in these words:

> The transfer of information from nucleic acid to nucleic acid, or from nucleic acid to protein may be possible, but transfer from protein to protein, or from protein to nucleic acid is impossible.

It has proved a fruitful principle, ever since James Watson and Crick discovered the double-helix structure of DNA in the 1950s. DNA is the blueprint; it gives instructions to RNA and to proteins about how to arrange themselves. However, there are some mechanisms being investigated that may run backwards and influence the DNA, perhaps in the presence of certain enzymes. And, for DNA itself to evolve, there must have been a time in the past when organisms did without it.

Robert Shapiro, author of *Origins: A Skeptic's Guide to the Creation of Life on Earth* (1986), put the question about the famous Central Dogma to Francis Crick himself.

Astoundingly, the great scientist cheerfully admitted that, at the time, he had not known what "dogma" meant. Later, a friend explained to him that a dogma must be accepted and believed on faith, without question.

"I didn't know it meant that," said Crick. "I thought it meant a hypothesis, some arbitrary thing which was laid down for no particularly good reason. Otherwise it would have been called the 'Central Hypothesis,' and then nobody would have made all this fuss."

See also "NEW" LAMARCKISM.

For further information:
Crick, Francis. *What Mad Pursuit: A Personal View of Scientific Discovery.* New York: Basic Books, 1988.
Shapiro, Robert. *Origins: A Skeptic's Guide to the Creation of Life on Earth.* New York: Summit/Simon & Schuster, 1986.

CHAMBERS, ROBERT (1802–1883)
Evolution's Pioneer Popularizer

Fifteen years before Charles Darwin's *Origin of Species* (1859) appeared, a popular Scots author named Robert Chambers published his own highly controversial treatise on evolution. Chambers's *Vestiges of the Natural History of Creation* (1844) argued that all species had gradually developed according to natural laws, without direct intervention of a Creator. It became an instant bestseller, though scientists condemned it as unworthy of serious attention.

Despite its continuing success (11 editions were published), Chambers never put his name on the book nor would even admit to writing it. With his brother William as lifelong partner, Robert, the self-taught son of a poor weaver, worked his way up from street bookseller, to publisher, to popular author and man of letters. During their early years, the Chambers brothers had struggled and scrabbled through poverty, facing obstacles and heartbreak straight out of a Charles Dickens tale.

Most of Chambers's many publications were very popular and thoroughly noncontroversial. Such titles as *Walks in Edinburgh, Biographical Dictionary of Eminent Scotsmen, Life and Works of Robert Burns, Chambers's Cyclopedia of English Literature* and *Chambers's Educational Course* give an idea of what kind of books put bread and butter on the table. And they were among the first publishers to offer cheap editions, affordable to the working classes.

By the time they had achieved solid success as Edinburgh publishers, both men were models of Scots diligence and enterprise. Certainly, they could not allow any hint of scandal or public controversy to threaten all they had built. Yet Robert Chambers was a man of real curiosity and intellectual boldness, who did not hesitate to tackle two of the great questions of his age: evolution and spiritualism. Both interests stemmed from his willingness to explore the possibility of phenomena "invisible" to most people.

Chambers himself was highly visible when, in 1848, he was a candidate for the post of Lord Provost of Edinburgh. As a highly popular local son, who had glorified the city in his writings, he seemed a shoo-in on the eve of the election. However, his unscrupulous opponent issued an ultimatum: Either face a public challenge that he was the anonymous "Mr. Vestiges," author of that "blasphemous, materialist

MYSTERIOUS AUTHOR of a controversial best-seller about organic evolution before Darwin's *Origin of Species*, Robert Chambers kept his identity secret. Though not a scientist, his *Vestiges of Creation* (1844) was a milestone in the history of evolutionary biology.

book," or withdraw from the race. Thus ended his political career.

The nature of the mixed reception given *Vestiges* was exactly the opposite of what Chambers had expected. He believed his ideas would interest scientists, but that the mass of general readers would find it unpalatable. To his amazement, it was the scientists who attacked and reviled the book, while the wider public eagerly snapped it up.

Although scorned by most established naturalists (including Darwin and Thomas Huxley), *Vestiges* did have a profound influence on some younger men. For instance, it inspired Alfred Russel Wallace and Henry W. Bates to search for (and find) evidence of evolution in the Amazon rain forests. In later years, Wallace and Chambers became fast friends, attracted by their mutual interests in evolution and Spiritualism.

Like Wallace, Chambers was impressed by the enthusiastic reports of friends who believed they had seen and heard spirit manifestations at seances. Appalled that most scientists simply dismissed such accounts (just as they had at first rejected evolution), Chambers wrote a pamphlet *Testimony: Its Posture in the Scientific World*, which examined the scientific criteria for accepting evidence of supernatural phenomena. He published it in 1859, the same year Charles Darwin's *Origin of Species* appeared.

Chambers challenged the physicist Michael Faraday, who had urged that "any extraordinary natural facts" reported by nonscientists should be routinely distrusted. The unaided senses alone were unreliable, Faraday argued, and most untrained observers prone to self-deception, bias and delusion.

"We can all apprehend a fact or event and describe it," Chambers replied. A habitually negative, overly skeptical view of the ordinary person's testimony would make it impossible to function in everyday life:

> Could a merchant believe in a market-report? Could the politician believe in the genealogy of his monarch? Could any jury not present at the crime convict a criminal? Would each geologist distrust his neighbor about what was found and in which deposit or strata? Could anyone believe a single fact of history or geography on the testimony of authors?

Too much reliance on "the skeptical method" leads to a "vicious circle," in which the search for knowledge cannot thrive, said Chambers. We cannot credit a fact until it accords with accepted laws of nature, but we cannot determine the laws of nature without accumulating facts. If we reject every novel fact that doesn't fit in with our conception of nature's regularities, "We can't know what is possible till we've learned everything."

In 1861, Darwin wrote Chambers, "You fulminate against the scepticism of scientific men. You would not fulminate quite so much if you had had so many wild-goose chases after facts stated by men not trained to scientific accuracy. I often vow to . . . utterly disregard every statement made by anyone who has not shown the world he can observe accurately."

But Chambers shared Wallace's view that the thousands of eyewitness reports of supernatural phenomena, stretching back through history to biblical times, could not *all* have been the babblings of fools. Even if most were rightly discarded as superstitious delusion, still there must be a few "golden grains" amidst such an enormous mass of earnest testimony. This precious residue, small but incontrovertible, points to "an immaterial and immortal part within us, and a world of relation beyond that now pressing on our senses."

For Chambers and Wallace, it was this skepticism toward skepticism that allowed them to embrace both

Spiritualism and evolution. Both phenomena were invisible to most scientists of the day. Yet both men shared an open-minded, risky willingness to seek evidence of the unseen. As Chambers wrote Wallace, "We have only to enlarge our conception of the natural, and all will be right."

Chambers invited Wallace to write the article on Spiritualism in *Chambers's Encyclopedia*, and he himself wrote the introduction to the autobiography of Daniel D. Home, the prince of Victorian spirit-mediums. In 1867, he wrote to Wallace: "I have, for many years, known that these phenomena are real, as distinguished from impostures; and it is not of yesterday that I concluded they [could] explain much that has been doubtful in the past; and, when fully accepted, revolutionise the whole frame of human opinion on many important matters." But Chambers's major book on Spiritualism was never published and—fearing for his reputation even at a century's distance—his descendants still nervously keep the manuscript under lock and key.

See also "MR. VESTIGES"; SPIRITUALISM; VESTIGES OF CREATION; WALLACE, ALFRED RUSSEL.

For further information:

Chambers, Robert. *Vestiges of the Natural History of Creation.* New York: Wiley & Putnam, 1844.

Milhauser, Milton. *Just Before Darwin: Robert Chambers and "Vestiges."* Middletown, Conn.: Wesleyen University Press, 1959.

CHIMPANZEE LANGUAGE

See APE LANGUAGE CONTROVERSY; NIM CHIMPSKY; WASHOE.

CHIMPANZEES
Fascinating Forest Ape

Human perceptions of the African forest apes has changed drastically over the past few decades and keeps on changing. Chimpanzees were first described as "wild" tribes of "Pygmie" men. Nineteenth-century explorers believed apes roamed the jungle at night carrying blazing torches and occasionally carrying off an African woman.

Victorian zoo-goers watched chimps drink from teacups, then make a shambles of the tea table: a naughty burlesque of the audience. Thirty years later they became costumed circus clowns who rode motor bikes. To medical researchers, they were sullen and unpredictable stand-ins for humans. Psychologists viewed them as retarded surburban children who couldn't learn to speak.

But maybe they could learn language without speech, through hand signs. By the 1970s, we saw them either as symbol-learning prodigies or uncom-prehending brats who deliberately gave false data for a few grapes. In movies they were Tarzan's antic jungle sidekick or Ronald Reagan's educated Bonzo. Will the real chimpanzee please stand up on your knuckles?

Our greatest field observer of the natural behavior of chimpanzees is Jane Goodall, and even she finds her image of the chimp constantly changing. After her first 10 years in the forest, she thought them peaceful, gentle vegetarians who did little but socialize and pull up plants all day. A few years later, she redefined them as toolmakers and users of tools. After another decade of observation, she saw them as warring "communities," cooperative hunters of meat and sometimes even cannibalistic baby killers.

Three communities of about 50 individuals each inhabit the Gombe Stream Reserve where Goodall has watched chimpanzees for 30 years. Six to ten are males; twice that number are females. Each group ranges over about 30 square miles, and its relations with neighboring communities are hostile. Males sometimes silently patrol the "border" and are now known to attack their neighbors, killing all males and infants. Females are taken into the marauding group, and all usually mate with the victors and produce new infants.

This kind of violently destructive behavior, which Goodall finds so sadly like human warfare, is not rare among animals. Some species of social birds, such as acorn woodpeckers, will often smash eggs within their own kin group. Swallows sometimes carry their eggs to other's nests, pushing out the original eggs. Cuckoos have long been known for nest parasitism as their major adaptation. Evolution favors reproductive strategies that produce the most offspring, without regard for human values of justice or fair play. (As Thomas Huxley maintained, if we want a more nurturing and compassionate society, we must build one without looking to a natural ethic based on ape behavior or evolution, for the "fittest" in the struggle for existence are often "the ethically worst.")

Although chimps feed mostly on wild fruits, vegetables and palm fibers, they frequently hunt small monkeys, bushbuck, piglets and sometimes kill and eat young baboons with whom they have regularly played and socialized. Only males hunt and never alone. Hunting is a social activity, preceded by several males working themselves up to a pitch of excitement. Prey are stalked, surrounded and deliberately ambushed by several cooperating apes. They catch and kill with their bare hands, then simply tear up the carcasses.

Meat is distributed among the group in a leisurely food-sharing ritual, which is still not understood by

OUR CLOSEST KIN, chimpanzees are genetically very similar to humans. These African forest apes have been variously perceived by naturalists as vegetarians and meat-eaters, clowns and killers. Over the past thirty years, Jane Goodall and her colleagues have made extensive observations of free-living chimp communities at the Gombe Stream Reserve. (Photo by Kenneth Love.)

anyone but the chimps. Some males are more generous than others, but all respond to the gesture of upturned palms or fingers pressed gently to the hunter's lips. Unlike their behavior in most other situations, the more dominant individuals patiently wait their turn.

Chimps show regional cultures or traditions, especially in food getting. They learn by observing, then imitate and practice. "Termiting" has become their most famous use of tools, where they strip leaves off twigs, chew them to the proper shape, and insert them in termites' burrows in the clay mounds. A human fieldworker who tried it was impressed with the skill required to shape the twigs properly, find the right places to insert them and capture termites without being painfully bitten. He eventually gave up.

Males compete to establish social dominance, but Goodall observed there are ways other than strength or fighting skill to attain top position. In 1960, a middle-ranking male hit on the trick of banging a large empty kerosene can while he charged about, terrifying others into submission without a fight. All had access to the cans, but he maintained his position for several years. When humans took the cans away, he brandished a chair and palm fronds. Some of the strongest males who could have easily beaten him showed no motivation to rise in rank.

Genetically, chimps are more like humans than any other creature; we have 98% of our genetic ma-

terial in common. Their behavior includes embracing, back patting, open-mouth "kissing" and hand-holding. In addition to hunting, tool using and food sharing, many researchers agree they demonstrate reasoned thought, memory, directed communication and ability to plan for the immediate future. In captive experiments, chimps were able to learn at least 300 signs in computer language or sign language at an early age. Goodall believes they have more social complexity and intellect than gorillas.

Yet, they are still hunted for food and cash by Africans, and many infants continue to be taken for biomedical research. Labs commonly lock them up in small, isolated cages until they literally go insane. With what we now know about them, says Goodall, it's time to treat chimpanzees with more kindness and respect, like the relatives they are.

See also APE LANGUAGE CONTROVERSY; "APE WOMEN," LEAKEY'S; HUNTING HYPOTHESIS.

For further information:
De Waal, Frans. *Chimpanzee Politics: Power and Sex Among Apes.* London: Unwin, 1982.
Goodall, Jane. *The Chimpanzees of Gombe: Patterns of Behavior.* Cambridge, Mass.: Belknap Press, 1986.
———. *In the Shadow of Man.* Boston: Houghton Mifflin, 1971.

CHINA, DARWINISM IN

During the 19th century, the West regarded China as a "Sleeping Giant," isolated and mired in ancient traditions. Few Europeans realized how avidly Chinese intellectuals seized on Darwinian evolutionary ideas and saw in them a hopeful impetus for progress and change.

According to the Chinese writer Hu Shih (*Living Philosophies*, 1931), when Thomas Huxley's *Evolution and Ethics* was published in 1898, it was immediately acclaimed and accepted by Chinese intellectuals.

Rich men sponsored cheap Chinese editions so they could be widely distributed to the masses "because it was thought that the Darwinian hypothesis, especially in its social and political application, was a welcome stimulus to a nation suffering from age-long inertia and stagnation."

Within a few years, evolutionary phrases and slogans became accepted Chinese proverbs. Thousands named themselves and their children after them to "remind themselves of the perils of elimination in the struggle for existence, national as well as individual."

A famous general called Chen Chiung-ming renamed himself "Ching-tsun" or "Struggling for Existence." Author Shih himself adopted the name "Fitness" (Shih), from the phrase "survival of the fittest." He recalled that because of "the great vogue of evolutionism in China . . . two of my schoolmates bore the names 'Natural Selection Yang' and 'Struggle for Existence Sun.' "

Although disrupted for a decade during the 1960s by the antiforeign, anti-intellectual Red Guard, China now boasts a fine Paleontological Institute in Beijing and a cadre of paleontologists who have recently made wonderful discoveries of dinosaurs and mammal-like reptiles in their country.

CHRONOMETRY
Methods of Dating

When the ancient stone tools or fossil bones of early man are found, the obvious question is: How old is it? Where does it fit into the sequence of evolutionary history? During the last two centuries, the applied science of chronometry has developed ingenious techniques for finding the answers.

Relative dates tell if something is older or younger than other objects, whether it seems to belong to a certain rock layer, or whether it can be correlated with time sequences established elsewhere. Absolute dates give an age in number of years, usually by taking advantage of some kind of built-in geophysical "clock" in the material.

Usually, relative dates are deduced by relating the object to a context that is already known. For instance, if a stone hand ax comes from a sediment layer whose date is already established at 100,000

years, that is the relative date. If it has been found washed out of a hillside, but closely resembles the characteristic tools or fossils of that assemblage, it can also be assigned that date. If it is not known precisely where it fits, sometimes all that can be said is that it is older or younger than others.

The fluorine test, which exposed the Piltdown hoax, is a method of determining whether several bones were buried at the same time or at different times by measuring the level of fluorine they have absorbed from ground water. It cannot give age in years but is a sophisticated technique for establishing ages of bones relative to one another. As one expert put it in describing the fluorine descrepancy in Piltdown, it was "unlikely that after the owner of the skull had died, the jaw lingered on for another thousand years."

Absolute dating techniques were undreamed of a century ago; all take advantage of built-in "clocks" in the material. The most famous is carbon-14 (C^{14}), but other important methods based on radioactive isotopes are potassium-argon and fission-track. Another method, geomagnetic dating, is based on patterns of past changes in the Earth's magnetic field.

Carbon-14 is used to date organic materials, such as shells, bone, or wood. C^{14}, an isotope of carbon, enters the cells of all living things from their food and accumulates throughout their lifetimes. When organisms die, their C^{14} begins to change to an isotope of nitrogen (N^{14}). Since the rate of this radioactive decay is a known constant—a half-life of 5,720 years—the proportion of C^{14} left gives a measure of how long ago the plant or animal died. But C^{14} technique has an effective range of only 50,000 years. Beyond that, so little C^{14} is left that dating is impossible. Therefore, it covers the relatively recent period of prehistory: from 10,000 to 50,000 years ago, including the recent Ice Age. (Newer enrichment processes can extend the technique back to 70,000 years.)

Potassium-argon uses the same basic techniques, but is based on the rate of change of an isotope of potassium into argon, especially in volcanic ash. Since potassium (K^4) changes much more slowly than C^{14}, it is suitable for dating much older material, extending into the range of human evolution. The *Zinjanthropus* skull was the first fossil hominid dated with this method, astounding anthropologists with an age of 3.75 million years.

Fission-track dating is based on the microscopic trails left in natural volcanic glass by the decay of U^{238}, an isotope of uranium. The number of tracks in a specimen are compared with the total produced by placing it in a nuclear reactor. When the total amount of U^{238} is known, it is possible to determine what proportion of the original amount has decayed, again at a known constant rate, which yields a date. This method has been used to cross-check potassium-argon dates. (Whenever possible, dates are correlated with other methods and sequences, since converging results increase the probability of accuracy.)

When rocks do not contain enough potassium or uranium, it is sometimes possible to learn something about their position in earth history by examining fossilized magnetic fields, which can be deduced by the way tiny particles of metal line up in the rocks. Since the Earth's magnetic field is known to have changed direction many times—and the rocks record those reversals—it may be possible to fit the pattern into a known sequence.

Each of these methods has its limitations and appropriate applications. Care must be taken to use uncontaminated samples, somewhere in the effective range of the techniques, which fall off near upper and lower limits. Moreover, there are "gaps" where no technique works. For instance, between C^{14} and K-Ar, there is a space of about 50 thousand years where no "absolute" dating technique yet exists.

See also FOUR THOUSAND AND FOUR B.C.; FLUORINE ANALYSIS; RADIOCARBON DATING; SMITH, WILLIAM; STRATIGRAPHIC DATING.

For further information:
Oakley, Kenneth P. *Frameworks for Dating Fossil Man*. Chicago: University of Chicago Press, 1964.

(The) CHRYSALIS
Akeley's Evolutionary Sculpture

Two years before his final, fatal trip to his beloved East African wilderness, Carl Akeley (1864–1926) created an evolutionary sculpture for a church that caused a public sensation. The bronze depicted a handsome "modern" man emerging from a cracked-open gorilla skin; he titled it *The Chrysalis*.

Akeley is best remembered as the genius behind the magnificent African Hall at New York's American Museum of Natural History. He was also instrumental in saving the mountain gorillas when they were being slaughtered without limit, 70 years before Dian Fossey's famous battles with poachers. But though an accomplished taxidermist, anatomist and naturalist, Akeley was first and foremost a sculptor, renowned for his wildlife bronzes.

The Chrysalis distilled in one sculpture Akeley's strong feelings of evolutionary kinship with the animal kingdom. The title refers to the cocoon in which caterpillars transform into butterflies; humans, he implied, emerged from apes. He hastened to add in a public lecture he knew full well that humans had not literally sprung from the gorilla. "They undoubtedly had a common ancestor. Science is on the trail of this ancestor and will locate it."

CONTROVERSIAL CHURCH SCULPTURE entitled *Chrysalis* was considered very daring in 1925. Carl Akeley's man emerging from a cracked-open gorilla skin was meant to symbolize an evolutionary "Ascent of Man," rather than the more traditional Fall from Grace.

The piece was commissioned for New York's West Side Unitarian Church, where it was on display for many years (the church no longer exists). Creationists were outraged and publicly criticized the Unitarians for placing it in their house of worship. *The Chrysalis* became the focus of a spirited public controversy.

"Could anything be more degrading," asked one newspaper editorialist, "than for a church to contain Akeley's statue and for the pastor of that church to say of it: 'I know of no concrete symbol which so well expresses the religious message which I am trying to preach every Sunday!'"

When asked about his church affiliation, Carl Akeley replied, "Most of my worshipping has been done in the cathedral forests of the African jungles, with the voices of birds and animals as music."

The Unitarian pastor, Reverend Charles Francis Potter, was unperturbed by the fundamentalist tempest. "The point of the statue," he explained, "is not the gorilla, but the man, who has risen above his animal ancestry. This statue shows the rise of man as opposed to the fall of man."

See also AKELEY, CARL; GORILLAS; MILITARY META-PHOR.

CLADOGENESIS ("BRANCHING EVOLUTION")
See ANAGENESIS.

CLASSIFICATION*
Identifying "Natural" Groups

Classification of living things (also known as taxonomy or systematics) started out, even before Linnaeus (1707–1778), as an exercise in revealing God's plan or "unmasking nature." Creatures or plants were thought to form "natural" groups, and the scientist's task was to find out what these were.

Sometimes early classifiers relied on resemblances that now seem naive: for instance, some grouped birds and bats together because both have wings. Mid-nineteenth-century zoologist Richard Owen, among others, looked beyond locomotion or function to the underlying anatomy. Birds' and bats' wings, for instance, are only superficially similar. While avian wings are formed of wrist bones, a bat's are made of elongated fingerbones. Owen coined the term *homologues* to mean similar underlying anatomical structures. Despite different outward appearances, whales' flippers, bats' wings, and human hands are all variations on the basic five-fingered bony hand.

Charles Darwin gave the story an evolutionary dimension. Homologues became the key to classifying related species, because they *were* the same structures, modified through descent from a common ancestor. Darwin's original test of a biological species was whether its members habitually interbred in the wild. But years later, as he worked on classifying long-dead barnacle specimens, he had to rethink the problem. Finally, he concluded that classification was an art that one learned through years of experience in biology, and that any "objective" definition of a species must fail. In his view species were always changing and therefore had no natural boundaries. [See SPECIES, CONCEPT OF.]

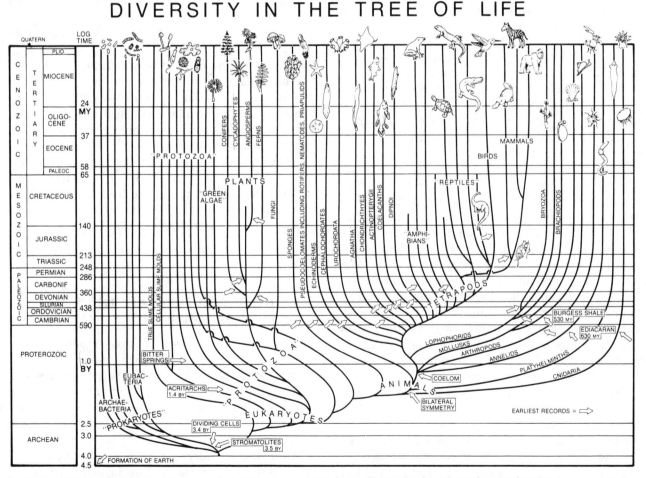

DIVERSITY IN THE TREE OF LIFE

TREE OF LIFE compiled by Professor Elisabeth Vrba incorporates latest discoveries in paleontology and molecular comparisons: a cladistic hypothesis of evolutionary history. An arrow at basal stem of a lineage points to earliest known fossil record for that group. Designed for her students at Yale University, Vrba's phylogeny is published here for the first time. (Graphic artist: Susan Hochkgraf.)

For a century after the *Origin* appeared, each systematist more or less followed his own intuitive method of comparison and contrast. During the 1940s and 1950s, authorities George G. Simpson and Ernst Mayr dominated the field. In the 1960s, however, German zoologist Willi Hennig, frustrated by traditional taxonomy's lack of rigor, designed a new approach that he claimed had consistent, teachable methods: cladistics. Taking their name from the Greek *clados*, meaning a branch, cladists construct hierarchies of species clusters ("sister groups") that attempt to show the patterns of species derivation. Their object is to identify the earlier anatomical characters and distinguish them from those that were derived later. Usually working with special computer programs, they draw up branching trees or cladograms that represent the results of their detailed comparisons.

Cladists have produced such new insights as placing birds and crocodiles in sister groups of equal rank, since both derive from a common ancestor. But many cladistic classifications that are supposed to be repeatable by other cladists have turned out to be controversial. When results don't jibe, cladists accuse each other of not following proper procedure or of selecting the wrong characters for comparison—a problem older than Linnaeus.

Sometimes a valid character for classifying separate species may be unexpected or invisible. It was recently discovered, for instance, that two different species of identical-appearing African elephant-nosed fish, *Mormyrides*, give off patterns of electrical signals as distinct as any colors or markings—but they could be seen only with an oscilloscope.

Classification studies have been profoundly affected in recent years by another formerly "invisible" characteristic of a species: its genetic material. Molecular biochemistry as a tool for classification dates back only to 1962, when Linus Pauling and Emil Zuckerkandl of the California Institute of Technology recognized that protein sequences from living species can be compared for relative genetic closeness or distance. They proposed that such molecules can serve as an evolutionary "clock," since mutations in a protein sequence can indicate evolutionary age. By 1967, Allan Wilson and Vincent Sarich at the University of California at Berkeley had refined the technique and used it to calculate that humans, chimps, and gorillas shared a common ancestor 4 to 6 million years ago—much more recently than the 20 to 30 million years that "bone" paleontologists had believed.

In the 1970s, Sarich's molecular comparisons scored a telling triumph. When the fossil primate *Ramapithecus* was declared a hominid ancestor on the basis of jaw fragments, he argued that protein comparisons showed hominids split off no longer than eight million years ago, and insisted, "Anything older than that cannot be hominid, no matter what it *looks* like."

Paleontologists howled in outrage, but their own subsequent fossil finds proved Sarich right.

"CLAWS"
British Amateur's Carnosaur

Some of the largest fossil claws ever seen came from a clay pit in Surrey, England in January, 1983, long after most experts had concluded that all the largest meat-eating dinosaurs were already known. In fact, it was the single huge claw sticking out of the clay that first attracted attention to the site. The monster was immediately nicknamed "Claws," a take-off on the giant movie shark "Jaws."

Found by Bill Walker, a plumber and amateur fossil hunter, the claw was 310mm long: about one-and-a-half times the size of the largest tyrannosaur claw known to that time. Three truckloads of fossil bones and rock were excavated and carted off to the British Museum (Natural History), where pieces of the giant puzzle were examined and assembled: a 70% complete skeleton.

"Claws" turned out to be somewhat smaller than the size of its talons at first indicated, but it was still the largest carnosaur yet discovered. British Museum scientists named it *Baryonyx walkeri*, meaning "Walker's heavy claw." The creature was bipedal (two-legged), had dagger-like teeth in a long, crocodile-like jaw, stood about four meters high and lived about 125 million years ago.

Large carnosaurs are very rare anywhere, but "Claws" was described as "the first major British find for over a century" in the land that coined the term "dinosaur."

See also DINOSAUR; MANTELL, GIDEON A..

CLAY THEORY (of Origin of Life)
Shaping Organic Molecules

An extraordinary number of religious traditions among diverse peoples—Jews, Christians, Moslems, Native Americans, Polynesians, Australian aborigines—describe living things as having been originally shaped from clay. Some scientists, too, are exploring the notion that clay may have played a special part in the origin of life.

In the 19th century, scientists first proposed that life might have arisen that from tarry masses of organic chemicals, in a primordial soup. Looking back on a century of unsuccessful experimental attempts to produce life from "carbon chemistry in water," during the early 1980s, Professor Graham

Cairns-Smith at the University of Glasgow sought an alternative line of research: the properties of clay.

Cairns-Smith sees in the microstructure of clays the possibility of a selection mechanism for the creation of certain densities and structures of organic molecules, and their replication. Water action, for instance, gathers materials of compatible density together in lumps, eliminating both lighter and heavier ones. Clay minerals have a minutely layered structure caused by the buildup of thin plates, the arrangement and order of which is often copied by the new particles that adhere.

If clay has the ability to store and express information, arranging new particles in complex patterns, Cairns-Smith asks, could this be a possible cradle for life? Suppose, for example, that such structures served as molecular templates for biochemicals, arranging them in new patterns. If they gained new properties and were able to reproduce themselves, genetics might "take over" the established structure and reproduce without it. Cairns-Smith compares it to the "takeover" of the original television set by transistors and microelectronics once the basic relationship between components had been established using large glass tubes and other devices.

How life began is still a wide open question in science. Writer Robert Shapiro has commented that if Graham Cairns-Smith's experiments prove to be more productive than those using gases, sparks, and tarry water, the clay theory could emerge as a "peculiarly satisfying" explanation for the origin of life.

See also LIFE, ORIGIN OF.

For further information:
Cairns-Smith, A. G. *Genetic Takeover and the Mineral Origins of Life.* New York: Cambridge University Press, 1982.
Shapiro, Robert. *Origins: A Skeptic's Guide to the Creation of Life on Earth.* New York: Bantam, 1986.

CLEVELAND, LEMUEL ROSCOE

See SEX, ORIGIN OF.

CLEVER HANS PHENOMENON
Mystery of the "Talking" Horse

Evolution of the capacity for thought and speech has long fascinated anthropologists, but recent "ape language" experiments sparked heated controversy. Can Koko the gorilla really communicate in sign language? Why did Nim Chimpsky's longtime trainer decide he never really "spoke"? In these debates, scientists often cite the case of a famous "talking" horse who lived 80 years ago. His name was Clever Hans.

Billed as the smartest animal in history, the stallion was the star of a German circus. It was claimed Clever Hans could read, spell, do arithmetic and work out musical harmonies. His trainer, Herr von Osten, posed dozens of mathematical and verbal questions, and the horse, with amazing accuracy, tapped out answers with his hooves.

Herr von Osten really believed in Hans. He was a man of integrity who swore he did not cheat by giving Hans the answers, and his sincerity was believable. To prove his point, he let strangers question the horse, and Hans still gave correct answers. Audiences were fascinated, and scientists baffled until the mystery was unraveled by a psychologist named Oscar Pfungst.

In a series of systematic experiments, Pfungst tried to isolate the factors that were present in successful performances. He rearranged elements of the question and answer proceedings and received his first clue when he discovered that if the human didn't know the answer to the question, the horse was stumped. Next, he searched for deliberate sound or hand signals by the trainer, but found none. Some thought telepathy might be the answer, but Pfungst sought a more conventional explanation.

His answer came after he discovered that the horse was baffled when the questioner was hidden from view. Eventually, Pfungst concluded the animal responded to very minute cues the questioner wasn't even aware he was giving.

Hans performed best with men who began the session by leaning forward slightly in tense expectation, and then relaxed with barely perceptible movements when the horse had completed the correct number of taps—at which point Hans would stop. He was simply responding to human approval, not to the content of the questions.

Many of the "ape language" programs of the 1970s were greeted with initial enthusiasm but have since been shown to be tainted by the Clever Hans phenomenon. Involuntary human shaping of the animal's responses proved to be a major flaw and embarassment. In the future, the only experiments that will have credibility must be carefully designed to eliminate human expectations from creating the result. The "talking horse" of long ago is still telling us something.

See also APE LANGUAGE CONTROVERSY.

CLINES

See RACE.

COEVOLUTION
Species Evolving Together

When Charles Darwin wrote of a "struggle for existence," he did not simply mean that an individual was pitted against all others in a contest for "fitness." From his studies of plants and insects carried out in his own greenhouse and garden, Darwin discovered that evolution also includes mutually beneficial partnerships between species. For example, he found that sexual parts of orchids mimic the sexual colors and smells of the wasps that pollinate them. Structures of all organisms, he wrote in *Origin of Species* (1859), are "related in the most essential yet often hidden manner to that of all other organic beings with which it comes into competition for food or residence, or from which it has to escape, or on which it preys . . ."

In the 1960s, population biologists Paul R. Ehrlich and Peter H. Raven coined the term "coevolution" to describe the reciprocal evolutionary changes that occur when unrelated species influence changes in each other. Their concept arose out of a combination of the older concerns of evolutionary biology with the newer ones of ecology and population genetics.

In the 1920s, two brilliant mathematical biologists—the American Alfred Lotka and the Italian Vito Volterra—worked out equations to describe natural cycles of ecological relations. Volterra devised formulas that would predict annual abundance and scarcity cycles for marine fish, and Lotka showed how the populations of foxes and rabbits were affecting the other's numbers. Later, scientists were able to demonstrate that the rabbits and foxes were actually influencing not only each other's population size, but their physical evolution as well and that of the vegetation in their habitat.

Coevolution may involve animals that prey on others (predation), competition for food, shelter or other resources, or symbiosis, a close association between two species, which may be mutually beneficial or (as in parasitism) beneficial only to one.

Predation relationships act in nature like the human arms race. Each new weapon inspires a defense or a counterweapon. As carnivorous dinosaurs developed more powerful teeth and jaws, their herbivorous prey developed armored plates and spikes. Predatory cats developed greater ability to chase down their prey, even as antelopes evolved greater speed. High-speed chases between cheetahs and impalas at 70 miles per hour or better across the African plains approach the upper speed limits possible for four-legged animals. Similarly, hooves and legs of horses evolved in a feedback interaction with the speed of their carnivorous foes, just as the horses' teeth and digestive system coevolved with the grasses on which they fed.

In northern Alaska, snowshoe hares attain great peaks of population every 11 years or so, when thousands of them devastate the bark and twigs of young birch and willow trees. But the trees fight back by secreting a resin that stymies digestive bacteria in the hare's gut. If a hare eats enough resinous bark and twigs, it will die of starvation. Each year, the plants respond to the browsing hares by putting out higher and higher concentrations of resins until the hare population crashes and the cycle begins again. Another part of the complex coevolution relationship concerns the foxes, lynxes and other predators, which also increase during years of the hares' abundance, when they feast, and decline as the hare population diminishes. During the low ebb of the hare population cycle, most female lynxes become infertile.

A classic example of coevolution is the monarch butterfly and the milkweed plant. Milkweed long ago evolved a defense against most birds, insects and mammals—a poison milky latex so deadly that South American Indians tip their arrows with it. But the monarch butterfly has evolved a defense to the poison: Females lay their eggs on the milkweed where their larvae have adapted to feeding on the leaves while packing away the deadly, active ingredient in special sealed-off body cells. While the poison does the caterpillar no harm, it makes the insect distasteful to predators.

Later, when the larvae turn to butterflies, they use the poison to repel birds that try to eat them. Experiments have shown that blue jays promptly regurgitate monarchs they eat and thereafter shun them as food. But the story isn't over. Another butterfly, the viceroy, has coevolved to mimic the monarch in pattern and color, so that—even though it's good to eat—birds confuse it with the poisonous monarch and leave it alone.

See also GUANACOSTE PROJECT; MIMICRY, BATESIAN; MIMICRY, MULLERIAN; ORCHIDS, DARWIN'S STUDY OF.

For further information:
Futuyma, Douglas, and Slatkin, M., eds. *Coevolution*. Sunderland, Mass.: Sinauer, 1983.

COLBERT, EDWIN H.

See BIRD, ROLAND T.; KARROO BASIN; LYSTROSAURUS FAUNA.

COLP, RALPH JR. (b. 1924)
Historian, "Darwin's Psychiatrist"

Charles Darwin must be the only man in history to have his own devoted physician and psychiatrist a century after his death. A distinguished New York City practitioner, Dr. Ralph Colp Jr., considers Darwin one of his patients and has spent years combing the evolutionist's personal papers and medical records for clues to his long, mysterious illness.

Like Darwin, Colp is the son of a prominent physician. As a boy, he was fascinated by the portraits of Darwin, Huxley, Pasteur, Lister and Freud that hung in his father's surgical office. After reading an idealized biography of Darwin, he thought these heroes of science "nobler and more perfect than physicians like my father who were influenced by practicing for money."

Nevertheless, Colp followed his father into medicine, did a stint as a military surgeon and completed a psychiatric residency. Soon after, a debilitating identity crisis interrupted his life, and Colp entered a long period of analysis.

At the 1959 centennial of the publication of the *Origin of Species*, a new flurry of interest in Darwin led Colp back to viewing his boyhood hero through the eyes of a psychiatrist. He began to wonder if Darwin's "long and protracted illness—mysterious both in its symptoms and causation—was a manifestation of his identity crisis." Colp suggested in a medical journal that there might be a correlation between the onset of Darwin's illness and the beginning of work on his theory of evolution.

Little by little, Colp became a serious, self-taught Darwinian scholar and made the pilgrimage to the Darwin Manuscript Library at Cambridge, England. There he was befriended by archivist Peter Gautrey, himself a self-taught scholar with a complete command of Darwiniana.

Colp has a photographic memory for Darwin's life and work and can identify thousands of quotations and sources from Darwin's books, papers and correspondence without apparent effort.

His major book on the subject has been *To Be An Invalid: The Illness of Charles Darwin* (1977). Colp's thesis is that Darwin's bouts of weakness, nausea, inability to work, depression, insomnia and other symptoms were part of a complex psychosomatic condition brought on by deep conflicts associated with his life's work.

For instance, Darwin had recurrent nightmares about being hanged. At first he kept his ideas on evolution secret for many years; to openly defy biblical authority by writing his real views, he once said, was "like confessing a murder." Colp believes this guilt and ambivalence was what kept Darwin from writing *The Origin of Species* for decades, until he was stung into action by the fear of losing priority to Alfred Russel Wallace. [See the "DELICATE ARRANGEMENT."]

Continuing his research beyond Darwin's medical problems, Colp's published journal articles have elucidated Darwin's metaphors, his attitudes towards slavery and the American Civil War, his relationship (or lack of one) with Karl Marx, scandals concerning his favorite doctors, the tragic episode of his young daughter's death, the development of his theories and many others.

Colp does not view his Darwinian studies as an escape into a comfortable gaslit past. "Charles Darwin," he has written, "is an intellectual time-bomb that is still going off."

See also DARWIN, CHARLES, ILLNESS OF.

For further information:
Colp, Ralph Jr. *To Be an Invalid: The Illness of Charles Darwin.* Chicago: University of Chicago Press, 1977.

COMPARATIVE METHOD
Projecting from Present to Past

Some scholars still speak as if there were a reliable procedure for reconstructing the past from the present: the comparative method. Once considered a foundation of anthropology and ethology (animal behavior), it is not really a single method, but encompasses several kinds of comparisons.

The idea was first developed in anatomy and early paleontology, when adaptations to special conditions were viewed as deviations—not always progressive advances—from the older form. Early systematists looked for a divine plan in animals, seeing specializations as distortions of an original ideal type.

Georges Cuvier became famous for being able to predict what kind of fossil animal was entombed in a rock by examining a protruding bone or tooth. Today, we would consider his feat a judgment of diagnostic features or homologies based on comparison, rather than a prediction. For instance, if he saw a very large tooth with the cusp pattern of a modern pig, it was a safe bet the block contained a giant fossil pig.

In the late 18th century, James Hutton's study of the Earth (which was continued and refined by Sir Charles Lyell) compared ancient rock formations with processes presently going on. This uniformitarian view projected the work of observable natural forces into the past; rock formations made sense as the result of depositions, erosion, river or volcanic action.

During the mid-19th century, students of language (philologists) began to compare grammars and vocabularies of various existing languages, to group them in families and even to reconstruct words of a presumed "ancestral" language (Indo-European). In his *Antiquity of Man* (1862), Lyell devotes a whole chapter to the similarity of reconstructing ancient languages with deducing the Earth's history by the comparative method.

Comparative anatomists used to reason backwards in time to what ancestors of living forms might look like. A transitional type between birds and reptiles was predicted and, sure enough, the fossil *Archaeop-*

teryx, a bird with teeth, claws and a lizard-like tail, was found in 1861. But the method proved wholly unreliable, since the history of life is unique and ancestors cannot often be reconstructed from their descendants, which may have specialized in very different directions. Best to have a fossil in hand before talking about an ancestral form. The Piltdown hoax was so readily accepted because a "missing link" halfway between man and ape was expected. With its combination of a human skull and ape jaw, Piltdown fit these expectations.

In anthropology, a comparative method was established by Edward Tylor and Sir John Lubbock to reconstruct cultural evolution. The idea was that if one compared "primitive" tribes on Earth today, one could separate the commonalities from the special developments, or rank the stages of development in ascending order. Thus, it was believed science could identify those living tribes that were closest to our ancestors in art, economy or religion. After more than a century of trying to make the scheme work, it became clear that modern Kalahari Bushmen or other hunting-gathering cultures do not necessarily represent our ancestors; they are not primitive any more than the Eskimo or other peoples who have adapted to various environments.

Darwin's ambitious book *The Expression of the Emotions in Man and Animals* (1872) compares infants, monkeys and other mammals to try to understand the evolution of facial expressions, gestures and vocalizations. Seventy-five years later, the ethologists Niki Tinbergen and Konrad Lorenz picked up where Darwin left off, attempting to reconstruct the evolution of behavior. Lorenz, for instance, compared courtship behavior in many related species of ducks and postulated what ancestral courtship rituals might have been like and how they became elaborated. However, such studies remain conjectural despite continuing attempts by sociobiologists to refine the method.

Molecular biologists now compare the "distances" of serum proteins, seeking to establish evolutionary relationships and times of divergence by a biochemical version of the comparative method. Critics question its validity as a reliable evolutionary "clock," claiming it has not yet been demonstrated that the rates of change are uniform.

Darwin himself kept a sense of humor about the comparative method. In 1878, in a letter to George J. Romanes, who was then writing his book on *Mental Evolution* (1883), Darwin slyly suggested "you ought to keep an idiot, a deaf mute, a monkey and a baby in your house!" Romanes replied that this "idea of a 'happy family' is a very good one," but he was afraid his wife would object that the baby "would stand a poor chance of showing itself the fittest in the struggle for existence."

Two years later, Romanes brought home a monkey from the London Zoo, "a very intelligent, affectionate little animal." He wanted to raise it in the same room as the infant "for purposes of comparison," he wrote his mentor Darwin, "but the proposal met with [maternal] opposition . . . I am afraid to suggest the idiot, lest I should be told to occupy the nursery myself."

See also CUVIER, BARON GEORGES; LORENZ, KONRAD; LUBBOCK, SIR JOHN; UNIFORMITARIANISM.

For further information:
Darwin, Charles. *The Expression of the Emotions in Man and Animals.* London: John Murray, 1872.
Lorenz, Konrad. *Evolution and Modification of Behavior.* Chicago: University of Chicago Press, 1965.

COMPOSITE PHOTOS
A Comparative Image

Charles Darwin's clever cousin, Sir Francis Galton, sought an objective way to picture a "typical" individual of any group. In his day, each learned professor worked out his own supposed criminal types or "racial" classifications, which led to endless, insoluble arguments. Galton's interest in heredity also led him to ask what traits were shared in common among close relatives and which were variable or unique to the individual. In 1877, in response to a request from the director of prisons, he created the first composite photographs.

In *Inquiries into Human Faculty and Development* (1878), Galton explained his method. First, he collected frontal portraits of the faces to be compared, similar in attitudes and size. Next, he centered each picture on an easel, using pinholes through the eye pupils as fixed points. When other pins were punched through the photos and anchored in the board, each face could be exactly superimposed in all the others.

To find the "typical" features of a family group, Galton combined eight different faces: parents, sons and daughters, and close relatives. Using a slow photographic plate that took 80 seconds for full exposure, he shot each face for 10 seconds. (The lens was covered after each shot, while the next photo was put in place.)

Only features common to all eight faces appear on the finished composite, because they receive the longest exposure. Individual variations are not exposed long enough to leave an image, and are eliminated.

American artist Nancy Burson (b. 1948) updated Galton's technique for the electronic age, using television cameras and digital computors. After an MIT

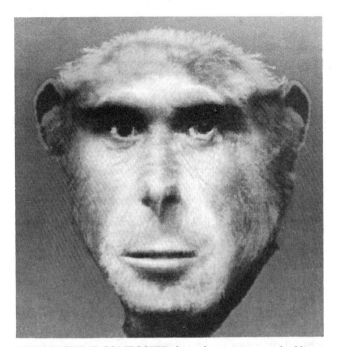

HUMAN-CHIMP COMPOSITE photo, by computer artist Nancy Burson, bears a striking resemblance to images of early hominids produced by reconstructing anatomy on fossil skulls. Basic technique of making such composites was pioneered by Sir Francis Galton, cousin of Charles Darwin.

engineer developed the basic system for her, special software devised by Richard Carling and David Kramlich permitted her to wed technology to art.

During the early 1980s, Burston made composites of the "typical" movie star, businessman, U. S. president and "Big Brother" dictator. She sometimes got surprising results. For instance, when she combined *Three Assassins* (1982), "I had expected a portrait of evil and ended up with the boy next door."

For *Evolution II* (1984), she combined the faces of a man and a chimpanzee. The man face was already a composite made up of a dozen popular film actors. Burson's intent, she explained, "was to see if by combining ape and man a credible image of early man would result."

Burson's "early man" image appears strikingly similar to recent reconstructions of *Homo habilis* made by painters and sculptors using traditional anatomical methods.

For further information:
Burson, Nancy. with Carling, R., and Kramlich, D. *Composites: Computer-generated Portraits.* New York: Morrow, 1986.
Galton, Sir Francis. *Inquiries into Human Faculty and Its Development.* London, 1878.

COMTE, AUGUSTE

See POSITIVISM.

CONGO
Chimpanzee Painter

As director of the London Zoo, Desmond Morris became intrigued with a chimp called Congo who loved to draw and paint with colors. In a book on the *Biology of Art* (1962), Morris reproduced dozens of her productions, arguing they show "a recognizable, personal style."

He reported that Congo was in no way trained; as soon as she was given a pencil, she began producing her characteristic designs, described as radiating fan shapes. She became famous in England for creating her artworks on "Zootime," a popular 1950s television show.

An exhibit of Congo's paintings at a London art gallery in 1957 spread her fame worldwide. Many were sold for fancy prices. Salvador Dali declared that Congo was a better abstract painter than Jackson Pollack; Miro and Picasso hung original "Congos" in their studios.

For further information:
Morris, Desmond. *The Biology of Art.* London: Methuen, 1962.

CONSPICUOUS COLORATION
Warning to Predators

Among Alfred Russel Wallace's many remarkable discoveries was his demonstration that some of the most gaily colored animals in the world are also the most poisonous. Brilliantly banded coral snakes or boldly striped wasps seem to advertise their presence in bright warning colors. Indeed, the most beautiful frogs in the world—the so-called poison-dart frogs of South and Central America—produce toxins so potent that forest Indians use them to tip their blowgun darts. Yet these lethal amphibians are jewel-like in gorgeous reds, yellow and purples. A Victorian naturalist, Thomas Belt, wrote charmingly of a Nicaraguan species:

> The little frog hops about in the daytime, dressed in a bright livery of red and blue. He cannot be mistaken for any other, and his flaming vest and blue stockings show that he does not court concealment . . . I was convinced he was uneatable so soon as I made his acquaintance and saw the happy sense of security with which he hopped about.

One recently discovered species, a shimmering irridescent yellow frog, is so venomous that if a man holds it in his hand for more than 60 seconds, he may die from contact with the neurotoxic poison.

When Charles Darwin first became interested in animals with bright colors, he thought the colors might be part of sexual display. But then he realized the evolution of gaudy caterpillars could not be connected with courtship because they are immature organisms. Why advertise themselves to predators? He appealed to Wallace for a solution.

As he did so often (to Darwin's delight and sometimes consternation), Wallace came up with an unexpected answer. Bright colors had evolved in many different types of poisonous or bad-tasting animals, he noted, so there must be some advantage to being easily noticed. Perhaps, he thought, the colors served as warning to predators. Maybe a hungry bird would learn to avoid certain insects or frogs more easily if it received a shock (a jolt of poison or disgusting taste) at the same time it saw a burst of bright colors. In the days before the development of behavioral psychology, this was not at all an obvious conclusion. "You are the man to apply to in a difficulty," Darwin wrote Wallace, "I never heard anything more ingenious than your suggestion." Wallace's idea has held up very well, and is still inspiring research today.

Bright color in prey can affect a predator in two ways: It can facilitate the association of a bad experience with a bold color, or it might invite more frequent attack at first, resulting in repeated negative lessons over a short period.

During the early 1980s, Paul Harvey at the University of Sussex set up domestic chicks (as predators) and bad-tasting crumbs (as prey) in a drab enclosure. Half the crumbs were dyed gray to match the floor and the other half were brightly colored. At first, chicks pecked more often at the colored crumbs, but learned to avoid them after a short time. By contrast, they never entirely learned to leave the equally bad-tasting, dull crumbs alone.

Building on this experiment, Harvey's colleague Tim Roper put in fewer bright crumbs and found chicks remembered to avoid them much more easily than the dull ones. Thus, writes Roper, "Wallace was right; coloration can affect rapidity of learning *and* ease of remembering."

Those unfamiliar with the older naturalists assume they simply were observers of nature, in contrast to the behavioral experimenters of our own day. Actually, Darwin and many of his contemporaries were tireless empiricists whose so-called observations were achieved through informal experimental designs. Darwin gave the following evidence on the function of conspicuous coloration in 1871:

> [Mr. Wallace's] hypothesis appears at first sight very bold; but . . . Mr. J. Jenner Weir, who keeps a large number of birds in an aviary, has made, as he informs me, numerous trials, and finds no exception to the rule that [smooth, dull] caterpillars are greedily devoured by his birds [while hairy or gaudy ones] are invariably rejected . . . When the birds rejected a

caterpillar, they plainly shewed, by shaking their heads and cleansing their beaks, that they were disgusted by the taste.

Yet, puzzles still remain a century later. Researchers at Gottingen University in Germany gave starlings and domestic chicks a choice between mealworms painted dull green or painted boldly in black and yellow stripes. Inexperienced birds showed an innate (unlearned) tendency to avoid the boldly striped worms. Why these birds should avoid stripes without having learned they taste bad is still unknown.

Bold stripes, Roper theorizes, may perform a double function for prey animals. The stripes may act both as a camouflage *and* a warning. From a distance, stripes break up the outline of a creature (as in zebras), and, up close, they may serve as a warning of poison, as in the coral snake. "Striping may, therefore, allow the prey to have the best of both worlds; it may make detection by a predator less likely in the first place, while offering protection should detection nevertheless occur." The poisonous black-and-yellow striped cinnabar caterpillar is an example; it is highly visible from up close, but almost invisible from a distance against the bright flowers of its normal food plant.

Of course, all of nature's defenses can backfire when used against a creature that is immune. Bee-eater birds of Africa, for instance, can counter a bee's toxicity by removing the stingers before eating the insects. In such case, the warning coloration becomes an invitation to a meal.

For further information:
Cott, H. B. *Adaptive Coloration in Animals*. London: Methuen, 1940.
Poulton, E. B. *The Colors of Animals: Their Meaning and Use, Especially Considered in the Case of Insects*. London: K. Paul, Trench, Trubner, 1890.

CONTINENTAL DRIFT
The New Geology

Alfred Wegener, a German meterologist, had studied masses of geological evidence from all over the world and come to a startling conclusion: All the Earth's continents had been joined together at the beginning of the Dinosaur Age and had been slowly drifting apart for 225 million years. Geologists ridiculed his idea, but the evidence was compelling.

What else explains why Cambrian rocks from Canada and Scotland are identical or why formations on Africa's west coast are same as those in Brazil? Africa and South America, he pointed out, look like jigsaw puzzle pieces that fit into each other. (Brazil would nestle snugly into the curve of West Africa.) All had once congealed, he thought, in an ancient supercon-

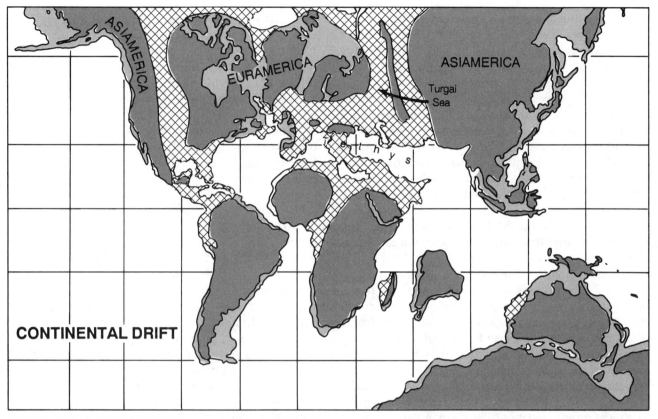

CONTINENTAL DRIFT

JIGSAW PUZZLE OF CONTINENTS led meteorologist Alfred Wegener to propose that continents had once been joined, then somehow drifted apart. Accumulated evidence a century later proved him correct. Map shows position of the continents in Late Cretaceous, 80 million years ago.

tinent he named Pangea, which means "all lands."

Although some geologists were intrigued by the evidence, most ridiculed the idea, as there was no known mechanism for how the continents could move around. The issue went unresolved until the 1960s, when geologist Harry Hess founded the modern study of plate tectonics, which is based on geophysical data that had accumulated since Wegener's day.

That the continents merged and drifted apart at least twice in the past and are on their way to doing it again in another 200 million years has now been established.

See also GONDWANALAND; KARROO BASIN; LAURASIA; PANGEA; PLATE TECTONICS.

For further information:
Wegener, Alfred. *The Origin of Continents and Oceans.* New York: Dover, 1966; first published 1915.

CONTINGENT HISTORY
Quirky Pathway of Evolution

Science has traditionally sought to discover the fixed laws underlying nature, the reliable regularities that allow prediction of particulars. Evolutionary biologists, following the example of physicists or chemists, sought laws to explain the past and perhaps predict future trends. They thought if there were no such laws to be found, then the study of evolution might become just history—incapable of prediction, and therefore not a science at all.

For a century, many paleontologists adhered to the basic assumption that life followed an inevitable course of gradual progress from simpler to more complex forms, ultimately leading to man as its highest production. Since the 1970s, however, there has been a quiet revolution in thought, stimulated by very technical and unglamorous studies of the strange little creatures, preserved in Canadian shale, that inhabited the oceans some 550 million years ago.

Their discoverer, Charles Walcott, head of the U.S. Geological Survey, described these Burgess shale creatures as primitive arthropods (trilobites, insects), annelids (worms) and relatives of other known groups of sea creatures. He "made out" body segments and mouth parts, legs and eyes in a manner consistent with the idea that they were early versions of modern phyla, prior to the "improvements" that came later.

With the pioneering work of invertebrate paleontologist Harry Whittington in the 1960s, carried forward by Conway Morris and others a decade later, the Burgess fossils were reexamined and dissected for the first time with startling results. These creatures were unlike anything known in zoology. Some had five eyes, body plans radically different from arthropods or mouth parts and spikes structurally unrelated to anything that came after them. They might as well have come from Mars.

Attuned to expect similarities with known phyla, Walcott mistook heads for legs and broken parts for whole organisms. What he thought were dry, objective descriptions turned out to be as wildly fanciful as anything in the emotional literature on fossil man.

Of the approximately 15 Burgess shale phyla, primitive vertebrates and arthropods represented only a tiny, inconspicuous minority. There is no obvious reason why they should have survived while the majority of the flourishing Burgess fauna disappeared forever, leaving no descendants. Had conditions been slightly different and a tiny Burgess "worm," which possessed a central notocord (spine), become extinct, animals with backbones (vertebrates) might never have developed—no birds, fish, frogs, or mammals. Instead, life on Earth might have sprung entirely from some of the other 14 body plans that thrived in those early seas.

A picture emerges of evolution as a unique, quirky history, whose details are chancy and contingent. The most minute twist or turn at any given point can have a very great effect later on, and the path is unpredictable. In evolutionary studies, "just history" can be a legitimate method and concern of science,

one that carries its own special fascination. Each development is contingent on the last and on the vagaries of changing conditions.

To appreciate contingent history, paleontologist Stephen Jay Gould suggests a mental game of "what if." Play back the tape of events and alter just one little detail, as in the old nursery rhyme: "For want of a nail, the shoe was lost, for want of a shoe, the horse was lost," next the rider, then the message, then the battle, finally the kingdom. It is a theme more familiar from literature and movies than science.

Gould's favorite example is Frank Capra's classic motion picture *It's A Wonderful Life* (1946), in which Jimmy Stewart's character is allowed to see what the world would have been like if he had never been born. Rescued from suicide by Clarence the Angel, the depressed hero believes his existence doesn't matter to anyone. But Clarence proves him wrong.

If he hadn't saved his brother from drowning one summer, that brother wouldn't have later saved a shipful of sailors during the war. His wife might have become a lonely, aged spinster. Even the town's name would have been different. Gould applies the same principle to understanding evolution in his book *Wonderful Life: The Burgess Shale and the Nature of History* (1989).

"Contingency is both the watchword and lesson of the new interpretation of the Burgess Shale,"

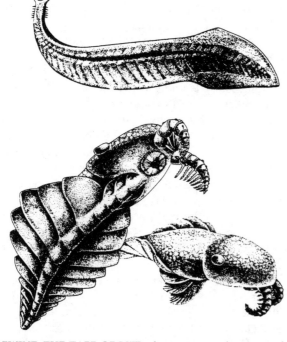

REWIND THE TAPE OF LIFE, change one contingency and the long-term effects could have been very different. In the film classic *It's a Wonderful Life,* the Jimmy Stewart character gets to see how much his own life has affected the history of his town. Paleontologist Stephen Jay Gould applies the same concept to the ancient Burgess shale creatures that affected the history of all life.

writes Gould. "The fascination and transforming power of the Burgess message—a fantastic explosion of early disparity followed by decimation, perhaps largely by lottery—lies in its affirmation of history as the chief determinant of life's directions."

If the modern order was not guaranteed by basic laws such as natural selection or mechanical superiority in anatomical design, but is largely a product of contingency, no particular outcome is inevitable. "It fills us with a new kind of amazement," says Gould, "at the fact that humans ever evolved at all. Replay the tape a million times from a Burgess beginning, and I doubt that anything like *Homo sapiens* (a tiny twig on an improbable branch of a contingent limb on a fortunate tree) would ever evolve again. It is, indeed, a wonderful life."

See also BURGESS SHALE; TELEOLOGY.

For further information:
Gould, Stephen Jay. *Wonderful Life: The Burgess Shale and the Nature of History.* New York: Norton, 1989.

CONVERGENT EVOLUTION
Caused by Similar Pressures

Convergent evolution is the process whereby differing species independently evolve similar traits as a result of similar environments or selection pressures. For example, the streamlined torpedo shape for cutting through water has been evolved by sharks and bony fish, by whales and dolphins (mammals), and by the extinct sea-going reptiles (ichthyosaurs).

Apparently, there is an optimum streamlined shape for gliding through water. Animals from widely divergent groups, with very different anatomies, have evolved amazing convergencies, even to the shape of tail flukes and position of dorsal fin. Similarly, the wings of bats, birds, and insects, though constructed very differently, show the effects of convergent evolution for sustaining flight.

CONVERGENT INTEGRATION

See TEILHARD DE CHARDIN, FATHER PIERRE.

COOKE, HAROLD

See NEBRASKA MAN.

COON, CARLETON

See RACISM (IN EVOLUTIONARY SCIENCE).

COOPER, MERIAN C.

See KING KONG.

COPE, EDWARD DRINKER (1840–1897)
Dinosaur Discoverer

For 30 years, Edward Drinker Cope was a major, driving force in American paleontology, possessed of an inexhaustable energy for research, discovery and excavation of new dinosaur fossils. He was also jealous, combative and downright underhanded. But without his raw egotism and ambition, the dinosaur halls of the great museums would not have become so quickly packed with Mesozoic saurians. His legacy also includes 1,400 books and articles describing and interpreting his fossil discoveries.

By the age of 21, the Philadelphia-born Cope had already published 31 papers on reptilian classification. This before he began his zoological education at the University of Pennsylvania under Joseph Leidy, the effective founder of American paleontology. He quickly made a name for himself, and was a museum curator and professor of zoology at Haverford College at age 24.

But a sedentary professor's life was not for the adventurous Cope. He made frequent trips to marl pits in New Jersey, paying quarrymen to notify him of interesting fossils. In the summer of 1866, a hadrosaur bone turned up, and Cope began to spend all his time at the pits. Soon after, he wrote his father that he had found "a totally new gigantic carnivorous Dinosaurian . . . the devourer and destroyer of Leidy's [his teacher's] Hadrosaurus, and of all else it could lay claws on." He named the beast "Laelaps," gave up his teaching post and bought a house near the quarries.

Although he came from a respectable Philadelphia Quaker family, Cope was anything but quiet and peaceable. Almost a century after his death, his reputation still lingers as one of the most aggressive, belligerent scientists of all time. He was known to settle scientific arguments with his fists and did not shrink at outright thievery from his competitors.

Once he stole the entire skeleton of a whale. When the huge carcass washed up at Cape Cod, the Agassiz Museum at nearby Harvard sent out a team to collect its bones. They labored all day under a hot sun, hacking away tons of blubber and decomposing flesh and finally loaded the bones onto a railroad flatcar. Cope, watching from a distance, waited until the workers left, then bribed the stationmaster to change the shipment's destination from Harvard's Museum to his own in Philadelphia. For several years, the Harvard curator had no idea what had become of his whale skeleton.

In 1868, Cope began a series of digs in the American West that would lead to discovery after discovery. He held rebellious field crews together and charmed the hostile Cheyenne, who were warring on white intruders in the region.

His most acrimonious battles were with his chief rival, another wealthy and unscrupulously competitive paleontologist, Othniel C. Marsh, of Yale University. Marsh was equally adept at gathering dinosaur skeletons in hostile Indian country, bribing railway employees to reroute shipments and claiming first discovery of species that Cope had already written up.

In later years, the feud died down, and the fortunes of both men declined. Cope had a bitter dispute with the government over a fossil collection he had made on the Hayden expedition. The National Museum claimed total ownership, even though Cope had spent $75,000 of his own money to insure the project's success. He had also made some bad investments and finally had to sell most of his vast lifelong dinosaur collection to the American Museum of Natural History for only $32,000.

He ended his days in a Philadelphia brownstone that was packed floor to ceiling with fossils and papers, but had virtually no furniture. Here the great dinosaur artist Charles R. Knight spent several weeks with him shortly before he died. Their intense discussions were fruitful, and Knight produced a remarkable series of paintings based on Cope's insights about the extinct monsters he loved.

See also COPE–MARSH FEUD; KNIGHT, CHARLES R.; LEIDY, JOSEPH.

For further information:

Cope, Edward D. *Origin of the Fittest.* New York: Appleton, 1887.

———. *Tertiary Vetebrata.* Washington, D.C.: U. S. Geological Survey, 1884.

Osborn, Henry F. *Cope: Master Naturalist.* Princeton: Princeton University Press, 1931.

COPE–MARSH FEUD
Dueling Dinosaur Diggers

One hundred years ago, the world's greatest collections of fossils were made by two rival paleontologists who loved dinosaurs and hated each other. Edward Drinker Cope of Philadelphia and Othniel C. Marsh of Yale's Peabody Museum would try almost any underhanded trick to beat the other to a new discovery.

Because both rushed to publish and never checked with each other, they often wrote up the same species at about the same time. Many dinosaurs therefore entered the scientific literature with two names: a Cope name and a Marsh name. It took museum paleontologists 30 years to straighten out the resulting confusion, determine who had named a species first, and unify those that were named twice.

Once Cope even managed to insult his rival by naming a mammal fossil *Anisconchus cophater,* the "jagged-toothed Cope-hater." His joke was not spotted; the species name was officially accepted and still stands.

Both found the American West a treasure trove of ancient giants. Cope spent several years in the badlands of South Dakota and Como Bluff, Wyoming, when the West was still rugged and wild. Born a Quaker, he carried no pistol but was able to befriend hostile Indians; his workers were never bothered by unfriendly Indians as they removed tons of fossils in horse-drawn wagons across open country.

Often, when Cope had gotten what he'd wanted from a site, he would order his workers to dynamite the remaining exposed fossil fields in order to prevent others from working the area after him—a scientific outrage. When he found that Marsh did not destroy fossil beds, Cope moved into his chief rival's abandoned excavations, which Marsh considered his own private domain.

One early incident, which caused years of enmity between the two former friends, occurred when an amateur fossil hunter offered some important bones for sale. Cope promised to buy them, wrote them up for publication, but didn't get around to sending a check. Marsh promptly wired the man a phony message cancelling the deal and then bought the fossils himself.

Another clash came about when Cope had finished mounting his huge *Elasmosaurus* at Philadelphia. The creature had an enormous neck as well as a very long, tapering tail. When Cope proudly unveiled the skeleton, Marsh pointed out that he had put two tails on the creature, one at either end. The head was actually mounted on the end of one of the tails. Cope could not bear the thought that his rival had demonstrated that he didn't know one end of a dinosaur from the other, and tried to buy up all the copies of his monograph. He succeeded in finding and destroying all but two of the published volumes, which Marsh carefully tucked away in his own personal library.

Both men came from wealthy families, but Marsh had the advantage: His uncle was the fabulously rich George Peabody. When he sought a professorship at Yale, Marsh announced to the trustees that he had persuaded his uncle to contribute $150,000 to establish a major museum at the institution. He was promptly appointed professor, as well as director of the new Yale Peabody Museum.

In 1890, the Cope–Marsh feud went public, with the two acrimonious dinosaur diggers attacking each other in the newspapers. Cope said Marsh had plagiarized his work, and Marsh ridiculed some of the

DINOSAUR FINDER and evolutionist Othniel C. Marsh (right) of Yale discovered thousands of new species, often rushing to beat his chief rival Edward Cope (left) to the sites in their legendary "Great Bone War." Marsh's tough-looking gang of Western desperados (middle) was really a field team of Yale-trained geologists and graduate students.

egregious errors in Cope's monographs. But their battle of the bones had been a boon to the new science of paleontology: The warring pair had discovered and named 1,718 new genera and species of fossil animals and packed several major museums with dinosaur skeletons.

See also COPE, EDWARD DRINKER; (CHIEF) RED CLOUD.

For further information:

Howard, Robert W. *The Dawnseekers: The First History of American Paleontology.* New York: Harcourt, Brace, 1975.

Plate, Robert. *The Dinosaur Hunters: Marsh and Cope.* New York: McKay, 1964.

CORAL REEFS
Darwin's First Book

When the young naturalist Charles Darwin set sail on his five-year voyage of discovery aboard H.M.S. *Beagle* in 1831, he intended to concentrate his researches on geology and marine invertebrates, including corals. At the time, sailors and explorers were quite familiar with coral reefs and coral islands, but no one knew how they got there.

Young Darwin was a skilled geologist well before he had a comparable knowledge of botany or zoology. (Geology was then considered part of natural history.) While exploring South America, he devoured the just-published *Principles of Geology* (1830–1833) by Sir Charles Lyell. Its uniformitarian approach inspired him to look for still active "small causes" that might have created major geological features over time.

He later recalled he began thinking about coral reefs before he had ever seen one, while still on the west coast of South America. For two years he had observed the shorelines, which seemed to show contradictory evidence of having been repeatedly built up and worn down. He began to form a theory, based on the effects of subsidence and uplift, of how coral reefs were formed.

At Tahiti he made his first field studies of reefs; other evidence gathered at Cocos–Keeling and Mauritius convinced him he was on the right track. Upon his return to England, he communicated his results to Lyell, who helped him get a hearing at the Geological Society of London in 1837. After 20 months' additional work on details, maps and charts, he published the study as his first scientific book: *The Structure and Distribution of Coral Reefs* (1842).

Darwin described three types of coral reefs: atolls, which are circular and enclose a lagoon; barrier reefs, which are long walls near a coastline, separated from the land by a channel; and fringing reefs, which stretch along a shoreline. All are built by tiny colonial invertebrates, thousands of soft-bodied polyps, that secrete protective limestone cells around themselves.

He studied water temperature, reef plants and animals and analyzed the natural community: an approach far ahead of its time. Darwin's early grasp of the web of interrelationships in nature foreshadowed his founding of the science of ecology in the *Origin of Species* (1859) years later.

Darwin's theory of coral reefs starts with the fact that live corals grow only in shallow water. He combines that with the observation that reefs seem to be associated only with areas of subsidence (land that is settling or sinking). Where there is uplift, as in the vicinity of volcanoes, atolls and barrier reefs do not occur.

Basic reef formation, he deduced, is caused by the

coral animals building on the accumulated limestone base of former colonies in places where the shore is subsiding. Since the "live area" thrives only in shallow water, the colony keeps pushing upwards to maintain itself at the same depth, while the ground beneath keeps sinking. It was a perfect demonstration of Lyell's uniformitarianism; given enough time, small, persistent natural forces create major geological features. Some reefs are more than 1,000 feet thick.

Professor John W. Judd, a well-known geologist who knew Darwin in his later years, has left the following delightful record of what happened when he returned to England:

> I well recollect a remarkable conversation I had with Darwin, shortly after the death of Lyell. With characteristic modesty, he told me that he never fully realised the importance of his theory of coral-reefs till he had had an opportunity of discussing it with Lyell, shortly after the return of the *Beagle.* Lyell, on receiving from the lips of its author a sketch of the new theory, was so overcome with delight that he danced about and threw himself into the wildest contortions, as was his manner when excessively pleased.
> He wrote shortly afterwards to Darwin: 'I could think of nothing for days after your lesson on coral-reefs, but of the tops of submerged continents. It is all true, but do not flatter yourself that you will be believed till you are growing bald like me, with hard work and vexation at the incredulity of the world.'

Despite Lyell's fears, Darwin's explanation of coral reefs, backed up by complete maps and data, won the admiration and approval of geologists and secured him instant recognition as an up-and-coming naturalist.

When geologist Alexander Agassiz (Louis's son) challenged some of his views on reefs in 1881, the now aged and ailing Darwin wrote that he wished "that some doubly rich millionaire would take it into his head to have borings made in some of the Pacific and Indian atolls, and bring home cores for slicing."

It was not until 1951, almost 70 years after Darwin's death, that American geophysicists drilled through 1,411 meters of reef limestone before hitting volcanic rock. These cores—and the new study of plate tectonics that revolutionized geology—only confirmed the basic correctness of Darwin's explanations.

In his last book, *Formation of Vegetable Mould through the Action of Earthworms* (1881), he referred back to his first on coral reefs to emphasize a major theme of his life's work: How great effects are produced by seemingly insignificant forces acting over immense periods of time.

See also EARTHWORMS AND THE FORMATION OF VEG-ETABLE MOULD; LYELL, SIR CHARLES; VOYAGE OF THE H.M.S. BEAGLE.

For further information:
Darwin, Charles. *The Structure and Distribution of Coral Reefs.* London: Smith, Elder, 1842.

CORN BLIGHT

See GENETIC VARIABILITY.

COUNT, EARL WENDEL (b. 1899)
Evolutionary Anthropologist

Born just inside the 19th century, anthropologist Earl W. Count has pursued its great evolutionary questions throughout the 20th. In the tradition of the Victorian generalists who dedicated themselves to probing the evolutionary background of the human species. Count's scholarly expertise encompasses cultural anthropology; mythology; philosophy of science; comparative anatomy; brain physiology; animal behavior; fossil man; and zoogeography. For more than 20 years, he was a one-man anthropology department at Hamilton College in upstate New York.

An ordained priest (Episcopal), he considers himself "a scientist in orders," a participant-observer doing extended fieldwork among the anthropology tribe. Count has written and published on human evolution in 12 languages, including Bulgarian, German, Swedish, Russian, French, Dutch and Greek. Sometimes he seems genuinely puzzled as to why his American colleagues have not read most of them.

Born to British-American parents, Count spent his early years in Bulgaria, where his father headed the Methodist missions. Returning to the United States during World War I, he worked as government translater, ranger-naturalist at the Grand Canyon and professor of anatomy at a medical college. Count did fieldwork among the Yurok Indians of Oregon in 1934, for which he received his doctorate in anthropology under A. L. Kroeber at Berkeley. His books include *Brain and Body Weight in Man* (1947), *4,000 Years of Christmas* (1950), *This Is Race* (1950) and *Being and Becoming Human: Studies of the Biogram* (1973).

One of Count's major contributions is his sustained attempt to follow Charles Darwin's lead in anthropology, to integrate the physical or biological study of man with the cultural. His systematic approach to tracing the evolution of human social behavior in the animal world preceded by 20 years the sociobiological movement founded by E. O. Wilson.

In a landmark paper ("The Biological Basis of Human Sociality," 1958), Count introduced the "biogram"—an approach for comparing the lifeways of various species and more inclusive groups of organisms, taking into account patterns of food-getting,

LAST VICTORIAN ANTHROPOLOGIST in an age of specialists, Earl W. Count has contributed to animal behavior, mythology, ethnology, evolutionary biology and "metascience." An
Episcopal clergyman, he writes and publishes in a dozen different languages.

reproduction, anatomy-physiology and biosocial behavior. Count has analyzed the discoverable outlines
of a "vertebrate way of life," "mammalian way of
life," "primate way of life," and "hominoid way of
life."

Married for a second time when in his mid-eighties,
Count continues to pursue what he considers the
major quest of the next century: to integrate scientific
knowledge with a new basis for human values.

See also MILITARY METAPHOR.

For further information:
Count, Earl W. *Being and Becoming Human*. New York: Van
Nostrand, 1973.
———. "The Biological Basis of Human Sociality." *Ameri-
can Anthropologist* 60:6(1958):1049–1085.
——— ed. *This Is Race*. New York: Schuman, 1950.

"CRADLE OF MANKIND"
Location of Human Origins

By the late 19th century, scientists no longer believed
humans originated in a Garden of Eden somewhere

in the Middle East. But what was the location of the
actual "cradle of mankind"? And where were the
fossils that could prove man had evolved?

Charles Darwin, in his *Descent of Man* (1871), was
an early advocate of an African origin. His reasoning
was that man's closest living relatives, gorillas and
chimpanzees, are found in Africa, perhaps still in the
forests where they evolved. Therefore, although no
early hominid fossils had yet been found anywhere
in the world, he suggested Africa was the place to
look.

Many recent books celebrate Darwin's remarkable
early conclusion but fail to mention that he had been
scooped (again!) by his junior partner Alfred Russel
Wallace. Seven years earlier, Wallace had attended a
Geological Society meeting at which Sir Roderick
Impy Murchison lectured on the subject of Africa's
antiquity. Not only was it the oldest continent, Murchison thought, but it was never submerged during
the Tertiary epoch.

Enthusiastically, Wallace related Murchison's views
in a letter to Darwin. "Here then is evidently the
place for finding early man," Wallace wrote in 1864.
"I hope something good may [also] be found in
Borneo and that the means may be found to explore
the still more promising regions of tropical Africa,
for we can expect nothing of man very early in
Europe." Twenty-five years later, the "Java Ape-
Man" was discovered not far from Borneo. In one
sentence, Wallace had predicted *two* major sites of
early man discoveries. But the brilliant hunches of
Darwin and Wallace seemed to die with them.

Some of the men who dominated evolutionary
theory during the late 19th and early 20th centuries,
imbued with visions of "Nordic" or "Aryan" racial
superiority, wanted no part of an African origin for
mankind. Black Africans were "retrogressive," said
Henry Fairfield Osborn, president of the American
Museum of Natural History, so he looked for humanity's roots to the plains of Asia. Once an ardent
Darwinian, Osborn later insisted there were no tropical forest apes in mankind's family tree. [See ARIS-
TOGENESIS.]

After the discoveries of *Homo erectus* in Java, Osborn was convinced more than ever that Asia was
the cradle of mankind. To prove it, he helped organize and fund one of the greatest paleontological
expeditions ever mounted: The East Asiatic Expedition through the Gobi Desert, led by Roy Chapman
Andrews. Its major objective was to find early hominid fossils.

The expedition found many fossils—dinosaur skeletons, previously unknown early mammals, the first
known clutch of dinosaur eggs—but no Asiatic "Dawn-
Man."

After 1924, African anatomists Raymond Dart and Robert Broom started finding near-men (the australopithecines) in South Africa. No European anatomist or paleontologist was ready to accept them as anything but "aberrant chimpanzees," so sure were they that Africa could not have been the "cradle."

Some British experts, with patriotic zeal, even accepted the blatant hoax of an "earliest man" (Piltdown) perpetrated in the English countryside, within 30 miles of Darwin's home. When Kenya-born Louis Leakey told his British professors in the 1930s he was going back to Africa to find early man, they told him not to waste his time.

During the next half century, the evidence began piling up. Robert Broom, Louis Leakey, Mary Leakey, Richard Leakey, Don Johanson turned up fossil after fossil. There were more new hominids than anyone could keep track of—"nutcracker man," "Lucy" (*Australopithecus afarensis*), *Homo habilis, Homo erectus*. No other continent has yielded so many hominid skulls and bones or such early ones (three to four million years old).

Darwin and Wallace, two Victorians who were not averse to having apes as ancestors, had been right about an African cradle of mankind. And they had made a lot more sense out of a lot less evidence than many who followed them.

See also AFAR HOMINIDS; "DELICATE ARRANGEMENT"; PILTDOWN MAN (HOAX).

For further information:
Leakey, Richard, and Lewin, Roger. *Origins*. New York: E. P. Dutton, 1977.
Reader, John. *Missing Links: The Hunt for Earliest Man*. London: Collins, 1981.

CRANIAL CAPACITY
Volume of Braincase

When describing fossil men or near-men, paleoanthropologists are always interested in the cranial capacity—the volume of the braincase.

Since there are variations in tissues and fluids, the cranial capacity is never exactly equal to brain size, but can give an approximation. A skull's capacity is determined by pouring seeds or buckshot into the large hole at the base of the skull (foramen magnum), then emptying the pellets into a measuring jar. The volume is usually given in cubic centimeters (ccs).

Living humans have a cranial capacity ranging from about 950cc to 1,800cc, with the average about 1,400cc. Brain size does not directly correlate with intelligence; one of the larger known human brains (1,600cc) belonged to an idiot, and the brain of the brilliant writer Anatole France was far below average in volume.

Among fossil hominids, australopithecine brain cases hover around 450cc (within the range of living chimps), *Homo erectus* about 1,000cc and *Homo habilis*, 750cc. While there seems to be a trend in hominid evolution toward increase in brain size, some of the smallest brained near-men were already fully upright, bipedal walkers. There are no significant differences in cranial capacity between the living races of mankind.

Several scientists have studied casts of the inside of fossil skulls, attempting to compare evidence for gross details of the brain. The shape of major lobes and grooves (sulci), and even major veins often leave impressions on the inner cranium. Ralph Holloway of Columbia University has amassed a roomful of such endocasts and believes studying organization of the brain is a better indicator than size alone. For instance, on some australopithecine skulls, Holloway has observed (as Raymond Dart did before him) that its brain form shares more similarities with the human pattern than with those of gorillas or chimps. Such an observation suggests (but cannot prove) that the australopithecines may have used language.

CREATION (FILM)

See O'BRIEN, WILLIS.

CREATIONISM
History of a Belief

Creationism is the Judeo–Christian version of the origin of the world and its inhabitants: God willed everything into existence. In the general sense, a creationist is one who believes in a divine Creator; it does not imply anything about *how* everything was brought into existence. Many evangelical Christians have considered themselves creationists while believing that God has worked through the process of evolution. Others, particularly Protestant fundamentalists and Orthodox Jews, define creationism more narrowly: a belief in fixity of species, six-day creation, young Earth, global Flood and the impossibility of understanding the mechanisms of Creation.

Duane Gish, of California's Institute for Creation Research, defines the fundamentalist doctrine as a supernatural Creator "bringing into being the basic kinds of plants and animals by the process of sudden, or fiat, creation." In his textbook proposed for public schools, Gish argued that "we cannot discover by scientific investigation anything about the creative processes used by the Creator [because they] are not now operating anywhere in the natural universe."

Henry Morris, Gish's colleague and founder of the Institute for Creation Research, disavows evolutionary biology because "we are completely limited to what God has seen fit to tell us, and this information is His written Word. [The Bible] is our textbook on the science of Creationism."

Historically, however, the manner in which creationists have interpreted the Bible has always looked beyond Genesis to the existing state of natural knowledge. These extra-biblical views then became so entwined with scripture in the beliefs of adherents, they came to be equated with the Bible itself. As scientific understanding changes, many religionists cling to its older views, which they have come to associate now with the preservation of their faith and morality. Such was the case, for instance, with the Flat-Earth movement, which opposed acceptance of global Earth theories as Satanic, anti-scriptural and destructive of the social order. [See FLAT-EARTHERS.]

Prior to the 18th century, church authorities did not insist on the fixity of type or permanence of species. "From Augustine through Thomas Aquinas to Francis Bacon and beyond," writes historian James Moore, "different 'kinds' of plants and animals could be naturally generated; one kind could give rise spontaneously to another, and these kinds were by no means necessarily identical with those of the Book of Genesis."

It was not until the doctrine of fixity or permanence of biological species was asserted by the Cambridge botanist John Ray (1686) and reinforced by the great taxonomist Carl Linnaeus (1740s), that it became the prevailing view, synonymous with creationism. According to James R. Moore's theological-historical studies:

> It was the massive dual alliance of creationism with the fixity of species, and of species with the "kinds" in Genesis, that Charles Darwin confronted in 1859 . . . Creationists were forced to decide whether their fundamental belief in God as the Creator of all things had become too closely tied to a particular scientific statement of the manner in which living species originated. Within . . . thirty years . . . many creationists . . . especially liberal Protestants, embraced theories of creative evolution.

When creative evolution theories began to fall from scientific favor in the 1920s with the rise of Mendelian genetics and its mechanisms, Protestant fundamentalists broke from the majority of creationists to proclaim a return to the "fundamentals" of Christian doctrine. But, in fact, the elements of their creed [see FUNDAMENTALISM] were distinctly modern innovations, unknown two centuries earlier.

However, during the 20th century, charismatic and dedicated leaders like Henry Morris, Duane Gish, the Reverend Jerry Falwell, and others have brought this once-minority movement to prominence in the United States, from whence it has spread worldwide.

According to Moore, possibly as much as a quarter of the population of the United States now:

> live in a universe created miraculously only a few thousand years ago, and on an earth tenanted only by those fixed organic kinds that survived a global Flood. The Book of Genesis and its [fundamentalist] interpreters now command an audience unknown since before the time of Darwin . . . The creationist cosmos of Protestant Fundamentalism has acquired an authority rivaling that of the established sciences.

The present creationist belief system owes much to a self-taught, armchair geologist named George McCready Price, who taught in Seventh-Day Adventist colleges. In 1923, he published a 700-page textbook, *The New Geology*, which explained all rock strata as dating from a worldwide catastrophic deluge, claimed the Earth was young, and that it had literally been created in six solar days. In the half-century preceding Price's attacks on evolution, many evangelicals had comfortably accepted the idea that God had performed his miracle of creation through evolution. A succession of different kinds of creatures through the ages did not trouble them, nor did the prospect that the "days" of Genesis might be interpreted as millions of years in duration.

Fundamentalist creationism, particularly after the national ridicule it received at the Scopes trial in 1925, remained for many years a quietly held, minority view. Then, in 1961, it received a tremendous impetus from a book called *The Genesis Flood*, coauthored by Henry M. Morris, a hydraulic engineer and disciple of George McCready Price's works, and John C. Whitcomb Jr., an Indiana professor of Old Testament theology. Spurred by the self-congratulatory conference of evolutionary biologists at the centennial celebration of *Origin of Species* at the University of Chicago in 1959, Morris and Whitcomb mounted their attack. Julian Huxley, the guest of honor at the Chicago celebration, had infuriated them by telling the assembled media "there is no longer need or room for the supernatural . . . the Earth was not created; it evolved."

In 1963, Morris and a group of fellow creationists, many with legitimate scientific credentials, organized the Creation Research Society, and, in 1972, Morris set up the Institute for Creation Research. Despite its claim to be a nonpolitical, nonreligious scientific organization, the Institute requires a statement of faith from all members on the fixity of created species, the universality of the Flood, and the historical reality of the Genesis creation. The group has mounted legal challenges to the teaching of evolution in public schools unless equal time is given to creation science, though the courts have generally held that their "science" is in fact a sectarian religion.

They are also agreed in condemning "evolution science" (which they consider a religion) as a major source of social problems. In one of the books Morris has coauthored, *The Bible Has the Answer*, evolution is described as "not only anti-Biblical and anti-Christian, but utterly unscientific and impossible as well. But it has served effectively as the pseudoscientific basis of atheism, agnosticism, socialism, fascism, and numerous other false and dangerous philosophies over the past century."

See also CREATION RESEARCH, INSTITUTE FOR; CREATIONISM, POPULAR SUPPORT FOR; SCOPES II.

CREATIONISM, POPULAR SUPPORT FOR

According to a 1982 Gallup Poll, the American public is almost evenly divided between those who believed that God created man in his present form in a single act of creation in the last 10,000 years and those who believe in evolution or an evolutionary process involving God. Americans had not been previously polled on their beliefs in creationism versus evolutionary theory.

Researchers from the Gallup organization conducted interviews with 1,518 adults over 18 years old in more than 300 regions throughout the United States. Forty-four percent of the participants—nearly a quarter of them college graduates—said they accepted the statement that "God created man pretty much in his present form at one time within the last 10,000 years." Nine percent agreed with the statement: "Man has developed over millions of years from less advanced forms of life. God had no part in this process." And 38% agreed with "Man has developed over millions of years from less advanced forms of life, but God guided this process, including man's creation." Nine percent said they did not know.

Protestants in the sample were more likely to believe in the biblical account of creation, while Catholics tended to favor the view of evolution guided by God. Fundamentalist beliefs were not concentrated in a particular area, although the South and Midwest had a slightly higher percentage of people holding creationist views.

Prominent mainline religious leaders interviewed in the *New York Times* were surprised and dismayed. Episcopal Bishop John S. Spong said he "did not know of a single reputable biblical scholar who would say that God created man in the last 10,000 years, since there is an enormous amount of evidence to the contrary."

Although the Gallup organization did not conduct a follow-up study, a more recent survey of beliefs among a collegiate population was made in 1986 by social scientists at the University of Texas. Nearly 1,000 students at five colleges were asked whether they accepted certain propositions as true, including the story of Adam and Eve. A surprisingly large majority, 60 percent, of the students in the survey said they do believe that "Adam and Eve were created by God as the first two people." If the study is accurate, a higher proportion of college-educated Americans believe the Adam and Eve story than the general population polled by Gallup four years earlier.

CREATION RESEARCH, INSTITUTE FOR
Grad School of Bible Biology

At a state-approved institute in southern California, a student can obtain a master's degree in geophysics, studying such things as geology and the origin of the cosmos. But this graduate school is different from most others in the country—its students learn that mountain ranges can be thrust up in a day, that the universe was created within the last 10,000 years and that fossils represent animals killed in a worldwide flood.

The graduate school is part of the Institute for Creation Research, a private organization that advocates a theory of divine creation and actively opposes the theory of evolution. It is approved to award master of science degrees by the California State Board of Education and is the only place in the United States where an advanced degree in science taught from a creationist viewpoint is granted.

Henry Morris, a former university professor with a doctorate in hydraulics, founded the nonprofit institute in 1972. His original goal was to publish creationist literature and to campaign for scriptural interpretations of human origins in public schools. In 1981, he obtained state approval for the graduate school, which offers degrees in science education, geology, astrophysics, geophysics and biology. By 1986 he was able to move from the campus of Christian Heritage College in El Cajon, California to his own campus.

As well as spacious class and seminar rooms, the institute also has a creationist museum with exhibits that question generally accepted evolutionary biology and present arguments in favor of supernatural creation.

In its 1986 catalogue, the Institute's philosophy of scientific creationism is spelled out:

Each of the major kinds of plants and animals was created functionally complete from the beginning and did not evolve from some other organism . . . The first human beings did not evolve from an animal ancestry, but were specially created in fully human form from the start.

Since the institute is not accredited by the Western Association of Schools and Colleges, most accredited schools will not recognize its degrees or accept its class credits for transfer. Professor Morris said he has no intention of applying for accreditation by the Western Association, which he describes as a "secular organization, pretty firmly committed to evolutionary theory." Nevertheless, under California law the institute's master's degree "is supposed to be the same as any other master's degree," according to state education consultant Morris Kear.

One student, interviewed in a California newspaper, summarized his reasons for attending the institute as "reconciling my scientific beliefs with the Bible, but that doesn't prevent me from being a scientist . . . I can't prove that the Earth is 6,000 years old, but evolutionists can't prove it's 4.5 billion years old. Sure, I'm biased, but don't tell me scientists on the other side of the fence are not biased."

See also CREATIONISM; "EQUAL TIME" DOCTRINE; FUNDAMENTALISM; "SCIENTIFIC CREATIONISM"; SCOPES II.

CREATIVE EVOLUTION

See ELAN VITAL; SHAW, GEORGE BERNARD.

CROCODILIANS
Evolutionary Enigma

When seeking patterns in nature, Charles Darwin always looked carefully at the exceptions, the facts that "didn't fit." Crocodilians (alligators, crocodiles, caimans, gharials) are an evolutionary "exception."

Known from at least 200 million years ago, their fossil bones look much the same as modern species, only larger. While most reptiles disappeared from the Earth in the "Great Dying" at the end of the Cretaceous era (75 million years ago), they managed to survive almost unchanged to the present day.

The persistence of crocodilians (and turtles, the other ancient surviving reptile family) poses a puzzle to theorists about Cretaceous mass extinctions. If asteroidal impacts raised dust that cooled the atmosphere, caused climates to vary and sea levels to fall or made other profound environmental changes that killed off most reptilian families, why were these two groups immune?

There is no definite answer yet, but some suggestive clues have recently emerged from studies of their reproductive biology and behavior.

Crocodilians and turtles share a special reproductive trait that is not found in any other living group of reptiles. In all other vertebrate species, the sex of offspring is determined by genetics; in crocodilians and turtles, it is determined by environment. Amaz-ingly, whether an egg will develop into a male or female depends on the temperature at which it was incubated!

Hotter conditions produce females in most turtles, and males in crocodilians. Hatched under lower temperatures, turtle eggs yield mostly males and crocodile eggs females.

Some researchers speculate that this apparently opposite effect may be related to body size; in both cases, high temperatures produce larger individuals. Female turtles are larger than males, which equips them to carry more eggs. Male crocodilians are the larger sex; their reproductive success, unlike turtles, depends on male-male competition for females and territory.

Because recently evolved (postdinosaur) reptiles like snakes have genetically determined sexes, while the more ancient lineages—crocodilians and turtles—do not, some scientists believe environmentally determined sex is the older system. Therefore, they speculate most reptiles on Earth during the Mesozoic period, including dinosaurs, had environmentally determined sexes.

If that were true, changes in climate could have produced a preponderance of one or the other sex, causing genetic bottlenecks and sharp curtailment of breeding. Dinosaurs may have become extinct, then, because their eggs produced too many individuals of one sex.

How would turtles and crocodilians have escaped the same fate? Perhaps by making compensations in their nesting behavior to adjust the incubation temperature. If climate became cooler, for instance, crocodilians may have been unique in mastering techniques for creating vegetation heaps of just the right temperature to go on producing enough individuals of both sexes. Using bacterial chemistry (decay) as a heat source, they may have adapted by placing some eggs higher in the heap and some lower, keeping the sex ratio viable. Such behavior would be favored by natural selection.

In fact, recent studies by Graham Webb in Australia, shows that sex ratios *are* maintained by distribution of eggs in a single nest. The top layer of eggs all developed into males, the middle layers produced a 50-50 ratio of sexes, and the bottom layers all hatched into females.

Rom Whitaker's observations at his Indian crocodile colony, near the Bay of Bengal, support the hypothesis that adult alligators do regulate the temperature at which they incubate eggs. For instance, he has often observed that the females dig several trial nests in different places before settling on one.

Previously, naturalists were puzzled by all the wasted motion, but Whitaker thinks the behavior is consistent with a search for just the right temperature of decaying plant material. He notes also that these partial nests are dug at night, when solar heat cannot confuse the choice. Also, the winter nest is always placed in a sunny spot, while the summer nest is always in a shaded area.

When Whitaker artificially incubates eggs, he keeps their temperature low, so as to hatch three females for every male. He has found this ratio produces the maximum number of offspring in his colony. In the wild, perhaps minor changes in nesting behavior can establish optimum breeding ratios for a given environment and population density.

Many seagoing reptiles disappeared at the same time as the dinosaurs, but grandfather turtle has persevered. Turtles may have survived by choosing different latitudes for their breeding beaches, varying depth of eggs in sand, height on beach, or by making some other unknown adjustments affecting incubation temperatures. But such conjectures become "just-so" stories without experimental evidence. This hypothesis can be tested against the behavior of living animals.

When more is known about the flexibility of their nesting behavior under a variety of conditions, it may well be discovered that a sort of behavioral "thermostat" maintains or adjusts sex ratios in these venerable lineages. Such studies could help construct a plausible explanation of why turtles and crocodilians are still with us and perhaps why the dinosaurs are not.

See also ALVAREZ THEORY; DINOSAURS, EXTINCTION OF; EXTINCTION.

For further information:
Bull, J. J. *Evolution of Sex Determining Mechanisms.* Menlo Park, N.J.: Benjamin/Cummings, 1983.

CROOKES, SIR WILLIAM

See SPIRITUALISM.

CRYPTOZOOLOGY
Searching for "Hidden" Animals

Every few decades, a lucky explorer, zoologist or fisherman finds a live creature that had been believed by experts to be legendary or long extinct. Sometimes, like the premature reports of Mark Twain's death, the extinction turns out to be "exaggerated."

In the 19th century, for example, hunters reported tales among Congo tribesmen of a large, cloven-hoofed animal with a giraffe-like head and zebra stripes on its hindquarters and legs. Most zoologists dismissed it as a local legend, but Sir Harry H. Johnston was fascinated when he read about this unknown beast of the deep forest. Years later, he launched an expedition in search of the creature, which the natives called okapi (o-CAP-ee).

After a nearly disastrous series of misadventures, he finally captured an okapi in 1906. One of the few large mammals discovered in the 20th century, the okapi turned out to be a living representative of a genus (*Palaeotragus*) known from fossils and believed by zoologists to have been extinct for 30 million years. A short-necked relative of the giraffe, okapis have since been bred in captivity and can be seen today in any large zoological garden.

A creature of even more ancient lineage is the lobe-finned fish called the coelacanth (SEE-la-canth), which had supposedly been extinct for 60-to-70 million years. It was well known from ancient rocks, but no living examples had ever been seen. On December 24, 1938, fishermen off the coast of East Africa, near Madagascar, hauled one up in their nets, and it reached ichthyologists J. L. B. Smith and Marjorie Courtenay-Latimer. However, the fish was dead and partially decomposed, and other scientists refused to believe their reports. Since then, eight more have been found alive, silencing skeptics forever.

Other examples of "living fossils" include the tuatara of New Zealand, a lizard-like remnant of a great family of dinosaurs, the duckbill platypus of Australia (a very archaic type of mammal that lays eggs) and giant isopods, huge crustaceans from the Gulf of Mexico.

In the 1980s, marine biologists discovered an entire unsuspected ecological system flourishing around thermal vents on the floor of the Pacific Ocean. To their astonishment, they found a hidden animal world that had been believed to be impossible—species that depend for their energy only on volcanic heat and minerals. These pale crabs, clams, giant tube worms, sea spiders and other creatures are revising scientist's views about the necessary conditions for life. Dependent on special bacteria that metabolize sulfides from the hot mineral water, they survive in pitch darkness, 8,000 feet below the surface—the only animals on Earth who rely on an energy source other than the sun.

Such sporadic discoveries, at a time when Earth has supposedly been thoroughly explored, encourage hopeful investigators to believe there may be still more surprises ahead. University of Michigan zoologist Roy Mackal has trekked through African swamps in search of a possible surviving dinosaur, called by fearful locals mokele-mbembe. Dr. Harold Edgerton of MIT and Robert Rhine have sought living plesiosaurs in Loch Ness, using specially developed underwater cameras and sonic devices. Anthropologist Grover Krantz of Washington State University continues to comb China and the northwest United

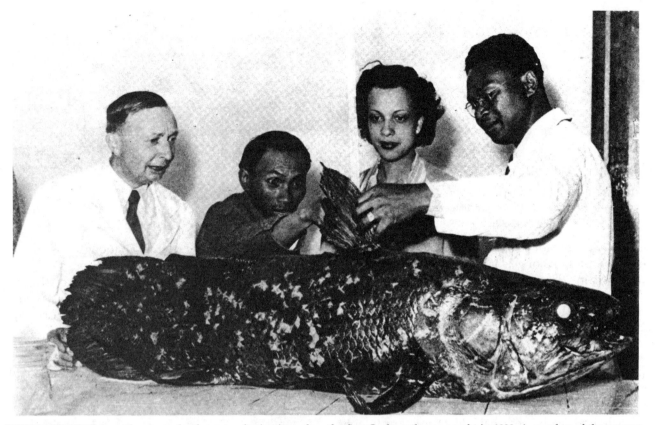

"LIVING FOSSIL" from the ocean depths amazed scientists when the first Coelacanth was caught in 1938. A member of the crossopterygian group, closely related to the first vertebrates to walk on land, it had been known only from fossils and was believed to have been extinct for several hundred million years.

States for the ape-like creatures known as yeti and bigfoot.

In 1978, these scientists and others gathered to form a new organization: The International Society of Cryptozoology, which is dedicated to the quest for "hidden" animals. Through the society, they share information, clues and techniques on locating animals that may still be undiscovered. As we grow sadly accustomed to more species becoming extinct each year, it is cheering to find a few that somehow survive in secret, defying human destructiveness and expert opinion.

See also BIGFOOT; LOCH NESS MONSTER.

CUTENESS, EVOLUTION OF
Biology of Infant Appeal

There seems to be a strong evolutionary basis for why we find the face of a baby cute and, by extension, the similar faces of baby animals, dolls and cartoon characters. A certain kind of face seems to release protective, parental responses; we want to cuddle and care for helpless infants.

In the 1950s, the famous Austrian ethologist Konrad Lorenz noticed adults find baby animals "cute"

because they share infantile features with newborn humans: "a short face in relation to a large forehead, protruding cheeks, maladjusted limb movements." He illustrated his point by comparing juvenile and adult faces of a duck, a dog and a human. The adult faces have relatively flat and shortened foreheads, medium-sized eyes, long or protruding snouts (or noses or beaks) and an overall angularity. In contrast, the young have relatively larger eyes, smaller noses and mouths, and high, curved foreheads.

These baby-like features elicit affection; toys and pets that have the same infantile proportions are also perceived as appealing. Stephen Jay Gould wrote an essay on Mickey Mouse in 1979, in which he traced the evolution of Walt Disney's nasty, angular rodent of the late 1920s through progressively more lovable versions. Mickey followed the Lorenzian formula exactly. His snout shrunk, eyes got larger and head proportions changed through the years to conform increasingly to the mass popular image of "cuteness" accepted widely throughout the world. (For example, Mickey is now immensely popular in Japan, where all cartoon characters sport the large-eyed, large-forehead look.)

Indeed, an infantalized version of feminine beauty

| 1928 | 1930's | 1941 | 1950's | 1960's | 1980's |

MICKEY MOUSE'S EVOLUTION to a "cuter" look shows increasing approximation to juvenile features. Earlier versions of Mickey had smaller forehead and eyes and longer snout. Ethologist Konrad Lorenz pointed out that we find baby animals' faces appealing because their proportions resemble those of human infants. (Copyright Walt Disney Inc., used by permission.)

is also widespread, as epitomized in the exaggerated facial proportions of geisha make-up and Betty Boop cartoons. In 1986, ethologists Robert Hinde and Les Barden wrote a paper on an unusual kind of "natural selection" in the only bear that reproduces asexually: *Ursus theodorus*, the Teddy bear. Named for President Theodore Roosevelt, the original toy bear looked like a long-snouted adult when it was created around 1900. But over the past 80 years, teddy bears show "a trend towards a larger forehead and a shorter snout," say the scientists.

As teddys with more baby-like faces are bought each year, the ones that remain unsold are "selected out" of the "evolving" population by human taste. Teddy bear faces, like those of cartoon mice and kewpie dolls, may be shaped by recognition patterns deeply imbedded in the human brain, a form of parental behavior that has helped our species survive.

See also ETHOLOGY; LORENZ, KONRAD; NEOTENY; ROOSEVELT, THEODORE.

CUVIER, BARON GEORGES (1769–1832)
French Comparative Anatomist, Paleontologist

A large block of sandstone was brought to Baron Georges Cuvier's laboratory at the Museum of Natural History in Paris. He examined the exterior of the stone with great care, then announced to the crowd of students and curiosity seekers that he would predict what kind of animal would be found inside the rock. It was, he said, a very large kind of pig.

Technicians went to work with hammers and chisels, chipping away carefully at the block. Most of the audience was still there hours later, when at last a fossil skeleton began to emerge. It was indeed a large pig, which prompted prolonged applause for Cuvier. Once again he had lived up to his reputation as the "Magician of the Charnel House"—the man who could reconstruct an entire animal from a single bone.

But it was no magic trick, although to some it appeared a marvelous feat. Cuvier, considered the founder of comparative anatomy, had studied thousands of skeletons and dissected carcasses of all kinds of animals; his studies had shown him that there was an order to animal structures.

Hooved mammals always had a particular type of molar teeth and never large canines. Retractile claws

"MAGICIAN OF THE CHARNAL HOUSE," as Cuvier was known, was a masterful teacher, zoologist, and comparative anatomist. One of his favorite tricks was to predict, from a protruding fragment, what kind of animal fossil would be found within a block of stone.

belonged only to members of the cat family. Though he hoped one day to work out "laws of form," Cuvier was simply demonstrating diagnostic features. He wasn't "predicting" what would be found; he knew from the protruding foot bones exactly what kind of creature was inside.

Cuvier's lifelong interest in animals began in boyhood; by his teens he was already a well-read zoologist. He studied for government service, but financial problems forced him in 1795 to accept a position at the museum, where he developed an interest in anatomy. Soon his expertise enabled him to extend and correct Linnaeus's system for classifying species.

The first to include fossils in his classification of animals, Cuvier noticed they were structurally unlike any living species, yet closely enough related to be grouped in one of the four phyla he had established. Still, Cuvier could not accept that discovery as sufficient proof for the evolutionary ideas proposed by his colleague Lamarck. He was, in fact, a militant antievolutionist who argued that fossil sequences were the result of periodic global floods, after which new forms of life appeared. Fossils were therefore always remnants of the most recent previous creation.

Although Cuvier refused to seek a naturalistic origin for man, he began to believe extinction of some creatures was a possibility, a daring thought for the time. The church had always taught that each animal was a link in God's chain of creation, which could not be broken. When mastodon bones were found in America, for instance, it was believed that some of the animals must still exist in the remote wilderness of Canada. But Cuvier put his foot down when it came to man-like creatures; he predicted no human fossils could ever be found. *"L'homme fossile n'existe pas!"* he proclaimed "Fossil man does not exist!"

For further information:
Cuvier, Baron Georges. *The Animal Kingdom.* (English translation.) London: W. H. Allen, 1886.
———. *Essay on the Theory of the Earth.* [1817] English trans., Edinburgh: Wm. Blackwood, 1827.

CUVIER—GEOFFREY DEBATE

See GEOFFREY ST. HILAIRE, ETIENNE.

DARROW, CLARENCE

See SCOPES TRIAL.

DART, RAYMOND ARTHUR (1893–1988)
Pioneering Paleoanthropologist

When South African anatomist Raymond Dart patiently chipped the first known skull of an australopithecine child from a slab of limestone sent to him in 1924, he knew he had made the "early man find" of the century. But for decades he was alone in that knowledge. England's greatest anatomist, Sir Arthur Keith, insisted Dart had found nothing but an aberrant chimpanzee.

With patience and tenacity, Dart defended his observations that the braincase was, in size and shape, unlike any known living or fossil ape. At a time when most anthropologists believed (on very little evidence) that humans had first appeared in Asia, Dart insisted Africa must be the cradle of mankind. His fossil near-men, he believed, were unlike anything found before. Certain details of the shape of the brain, as preserved in a natural cast of the skull's interior, appeared hominid, not ape-like. And the teeth, too, resembled the human pattern, with no large canines.

Pressing on alone, snubbed by European science, Dart continued to seek evidence of early man in Africa. He studied the compacted bone accumulations in rock cavities; to him, they indicated early man was a hunter so aggressive that he often murdered his own kind. Dart analyzed nearby accumulations of antelope bones and horns, which he believed hominids had used as tools and weapons—

his original "osteodontokeratic (bone-teeth-horn) culture." Recent reconsideration of these deposits by C. K. Brain, however, have led to the conclusion that early hominids most likely had nothing to do with the selection of these antelope and baboon bones.

Robert Broom, another South African (originally Scots) doctor and paleontologist, later joined the search and added many important discoveries of australopithecines, including another, larger species (Broom's *Plesianthropus*, now renamed *Australopithecus robustus*). European scientists continued to scoff and for

FIRST TO RECOGNIZE an australopithecine fossil, anatomist Raymond Dart chipped the skull of his famous "Taung baby" out of a block of limestone sent to him from a South African quarry. It took 30 years for science to accept his initial judgment that it represented a previously unknown hominid.

years Dart and Broom were, as Robert Ardrey described them in *African Genesis* (1961), "two wild men crying alone in the wilderness."

Raymond Dart was fortunate enough to live to be 95. By that time, science agreed that his "Taung Child" was indeed an epoch-making discovery. Even his old adversary, Sir Arthur Keith, admitted he had been wrong, and proposed the near-men be called "Dartians" (rhymes with Martians) in his honor, though the name never caught on.

Louis and Mary Leakey began to build on the work of Dart and Broom until, by the 1950s, they had established Africa as the probable birthplace of the human species. In the 1970s, Donald Johanson found much older and more complete australopithecine remains in Ethiopia. Dart enjoyed complete vindication and became a familiar figure at international scientific meetings, where the near-octogenaerian, now affectionately dubbed "Uncle Ray" by anthropologists, received standing ovations.

See also AUSTRALOPITHECINES; BROOM, ROBERT; KEITH'S RUBICON; KILLER APE THEORY; TAUNG CHILD.

For further information:
Dart, Raymond. *Adventures with the Missing Link.* New York: Viking, 1959.

DARTIANS
Man's Non-Ape Ancestry

Did humans come from apes? Most books on evolution describe our descent from "ape-like" ancestors and include anatomical diagrams comparing humans, chimps and gorillas. Yet, most also state man is not derived from any living ape, but from a "more generalized ape-like ancestor."

To Finnish paleontologist Bjorn Kurten, that is contradictory double-talk. We are not evolved from apes at all, he insists, but from an ancient line distinct before the apes branched off from it. Writers leave apes in our family tree because of sloppy thinking and historical misnomers of anthropology, he argues in *Not From the Apes: A History of Man's Origin and Evolution* (1972).

Before we had the wealth of hominid fossils accumulated since Darwin, there was some justification for imagining our ancestors as some kind of ape.

Scientific thought was still heavily influenced by the concept of the "great chain of being," a leftover from medieval tradition. Well before 1892, when Dr. Eugene Dubois dug up the Java skullcap, scientists were focused on finding the "missing link" between man and apes. As a result, Dubois called his discovery *Pithecanthropus*, meaning ape-man. Later, when many more remains of these creatures were found,

it became clear they were not apes, but an early kind of human, whose name was changed in 1950 to *Homo erectus*, meaning upright man.

Again, seeking that "link" between men and apes, in 1924 Raymond Dart named his famous Taung fossil "Southern ape" or *Australopithecus*—another unfortunate label, according to Kurten. Subsequent discoveries have shown that in their teeth and jaws, in their pelvic structure and upright posture as well as in their use of stone tools, these creatures do not resemble apes, but represent a distinctly human line, older than *Homo erectus*.

In 1947, the British anthropologist Sir Arthur Keith first proposed the colloquial name "Dartians" for *Australopithecus africanus* and its relatives, after the South African anatomist Raymond Dart. It's as good a name as any, Kurten argues. Sir Arthur pointed out that Dart not only discovered the creatures "but also so rightly perceived their true nature." It was a complete turnaround for Keith; for many years, he had insisted the skulls were merely aberrant apes and discounted Dart's view of their antiquity and ancestral place in the human family.

When compared to modern apes, the distinctiveness of the Dartians becomes apparent. Where apes have huge canines and gaps for them in the opposing tooth rows, the Dartians have small teeth and no canine gaps. Form and shape of the molars and premolars is similar in humans and Dartians, showing marked differences from ape tooth forms. And although Dartians had heavy jaws and brow ridges—superficially ape characteristics—the attachments of jaw muscles and overall skull shape are more like humans' than apes'. Dartian feet are human in form, without an opposable big toe, and their legs and pelvises indicate bipedal posture.

In contrast, the modern apes are not customarily bipedal; they lean forward and walk on folded fingers of their hands to aid their relatively short legs. Despite their demonstrated affinities to man, modern apes are specialized cousins, which branched off the Dartian line and went their own way several million years ago.

Unfortunately, the name *Australopithecus* is so well entrenched, it is doubtful it will ever be replaced by "Dartians." But Kurten's attempt to sweep away the cobwebs and place our ancestors in proper perspective is like a refreshing breeze through a musty museum of fossilized words.

See also AUSTRALOPITHECINES; DART, RAYMOND ARTHUR; DUBOIS, EUGENE; MAN'S PLACE IN NATURE.

For further information:
Kurten, Bjorn. *Not from the Apes: A History of Man's Origin and Evolution.* New York: Random House, 1972.

DA VINCI, LEONARDO (1452–1519)
Renaissance Genius

Five hundred years after his death, Leonardo da Vinci is still famous for such artistic masterpieces as the *Last Supper* and *Mona Lisa*. But he also earned a footnote to the history of biology as the only Renaissance man of record who took a careful look at fossils and concluded they were not anomalies of nature or relics of the biblical flood as many people then supposed.

In his notebooks, da Vinci wrote that fossil shells found in northern Italy were surely the remains of once-living creatures, since they were similar (down to minute details) to the mollusks he knew from the Mediterranean. He noted that even the same kinds of smaller organisms that attached to the living shells, or bored into them, were evidenced in the fossils.

Although da Vinci understood silting, sedimentation and the meaning of strata, he thought a constant exchange of land and sea was part of a steady-state Earth. His rejection of the idea that the shells were left over from a great flood was not so much an attack on religious orthodoxy as a confirmation of the rational causality of natural events.

Impressed above all by the "ordered layers" in which the fossil shells were found, he even urged a physician friend to study them for what light they might shed on Earth's past. But no one made the serious investigation da Vinci had suggested until the 17th century, when the work of the Dane Nicolaus Steno and the Englishman Robert Hooke established the underpinnings of paleontology, the scientific study of fossils.

See also BERINGER, JOHANNES; FOSSILS; NOAH'S FLOOD; SMITH, WILLIAM.

For further information:

Rudwick, M. J. S. *The Meaning of Fossils: Episodes in the History of Paleontology.* New York and London: McDonald and American Elsevier, 1972.

darwin
Unit of Evolutionary Change

In the International System of Units, the names of scientists have been given to various measures as a way of honoring their discoveries. A newton (after Sir Isaac Newton) is a measure of force, an ampere (after Andre M. Ampere) a measure of electrical current and a Kelvin (after Sir William Thomsen, Lord Kelvin) a measure of temperature.

British biologist J. B. S. Haldane, in 1949, proposed a unit of evolutionary change to be called the "darwin." As part of the efforts of biometricians to apply mathematics to evolutionary biology, Haldane thought such a unit might make comparison of rates of evolutionary change in various organisms easier.

According to his definition, one darwin equals a rate of change (in size, for example) of 0.1% per thousand years. Slow change in the fossil record would be measured in "millidarwins," while very rapid change, as in artificial selection, could be measured in "kilodarwins."

Unfortunately, no agreement exists on exactly *what* should be measured to compute rates of change. Size of organisms, for instance, can vary rapidly between generations with fluctuations in nutrition or hormonal levels, while the genetic make-up may not have changed much at all. On the other hand, certain important modifications in organisms (color, soft parts, biochemistry, behavior) may not leave any fossil evidence; it may be changing yet appear to remain the same.

So despite the brilliant Haldane's hopes and good intentions, the "darwin" is a standard unit in name only—it has so far been used very little to compare rates of evolution.

DARWIN, CHARLES ROBERT (1809–1882)
Naturalist, Evolutionary Theorist

For 40 years, in England's Kentish countryside, Charles Darwin, his wife Emma, and seven of their children lived quietly in a spacious old house, complete with gardens, fields, a patch of woods, greenhouse, later a clay tennis court and about fifteen servants—the estate of a modest millionaire.

A reclusive semi-invalid who shunned social functions, Darwin spent his days writing, reading, strolling his grounds, dissecting barnacles or orchids, talking with local pigeon and hog breeders, checking botanical experiments in his greenhouse or observing the activities of bees in his garden. A casual visitor would never have guessed that from this sleepy, idyllic retreat he was shaking the world.

Conservative in lifestyle, Darwin dutifully contributed to the village church, helped organize local philanthropies and served on the local magistrate's bench as a justice of the peace. (He adjudicated such cases as rabbit poaching, "furious driving" of carriage horses, disputes over livestock road crossings and vandalism to a farmer's fence.) So retired was his existence from the hubbub of London that when, around 1877, telephones were the latest rage among the well-to-do he refused to have one installed.

Despite this penchant for privacy, Darwin was world famous as a prolific (and popular) author, naturalist, philosopher, botanist, geologist, explorer and zoologist. Yet he had no professional training in biology, never passed a doctoral exam, accepted no formal students and became nauseated at the thought of delivering a public lecture.

Darwin is still best known for the theory of natural selection, an achievement for which he had to share credit with Alfred Russel Wallace, who formulated it independently. But a much more impressive achievement was Darwin's application of the theory to thousands of specific problems in natural history, years of labor that created the new science of evolutionary biology.

Although the Darwin–Wallace theory can be summarized in a few pages, how to use it to unravel nature's mysteries fills Darwin's 17 scientific books and more than 150 articles—an output that founded the modern research tradition. This incredibly productive life's work revolutionized every field he touched: botany; paleontology; physiology; taxonomy; comparative psychology; zoology; what we now call ecology; primatology; genetics; paleoanthropology; sociobiology; and all the life sciences.

These 40 years at Down House may have seemed outwardly uneventful and tranquil, but Darwin's inner life was one continuous adventure. He had stored up more memories, impressions and raw scientific data by age 27 than most people gather in a lifetime. On his epochal voyage of discovery aboard H.M.S. *Beagle*, he explored the wilds of South America, Australia, Tahiti, South Africa; observed and collected thousands of plant and animal specimens; chipped fossil skeletons of giant sloths out of cliffs; and discovered the secrets of coral islands and reefs.

It was an endless parade of wonders. For young Darwin it was a combination of adventure, hardship, scientific discovery and unremitting hard work, all packed into five years. He had galloped on horseback alongside Argentine gauchos, ridden rough seas, survived killer storms and earthquakes, and wandered awestruck through creeper-laced rain forests, teeming with gaudy birds and exquisite orchids. To the ordinary traveler, these events and experiences might register as striking but unrelated impressions. What was remarkable about the *Beagle*'s young naturalist was that he habitually sought underlying connections and regularities. Some years after returning home, he realized the key was the shared history of life-forms adapting to a changing Earth.

Gradually, he worked the notebooks and journals of his travels into a larger picture. His mind focused on a history measured in hundreds of millions of years, a puzzle that fascinated his brilliant grandfather, that "mystery of mysteries": How do species come into being? What is man's place in nature? Is the "tangled bank" of plants and creatures an impenetrable jumble or is there an order accessible to the human mind?

Exploring, blazing new trails—whether literally or intellectually—were the joy and spice of Darwin's life. However, his father, Dr. Robert Darwin, and grandfather Erasmus were both successful country physicians, and he was expected to follow in their footsteps. He agreed to pursue a safe, prosperous medical career, while inwardly yearning for travel and adventure.

Except for the informal tutelage of the Professor Reverend John Henslow, a botanist at Cambridge, young Charles considered his college education a complete waste of time. Medical school at Edinburgh was an even worse experience. General anesthetics had not yet been invented, and the practices of the day were often brutal. Once, unable to bear the screams of a strapped-down child during major surgery, he ran out of the operating theater and gave up all thoughts of pursuing a medical career.

To his father's disgust, Charles's only real interests seemed to be beetle collecting, shooting birds, and poking at rocks and plants—activities suitable only for an idle squire or a country parson. Doctor Darwin urged his son to become a clergyman: Calling on parishioners was at least a respectable excuse to go tramping about the countryside. (Professional naturalists barely existed at all; many dedicated amateurs came from the ranks of rural churchmen.)

Charles agreed; he had read Paley's *Natural Theology* (1816) (parts of which he knew by heart) and was attracted to the idea of studying God's designs in nature. While he was waiting for the term to begin at divinity school, he went off for a few months of "geologising" with Adam Sedgwick in Wales. But shortly after returning home, a fateful letter from Reverend John Henslow, his favorite Cambridge professor, was to change the course of his life forever.

Professor Henslow had been asked to suggest a ship's naturalist for a voyage round the world. Young Captain Robert FitzRoy (then 24 years old, only three years Darwin's senior) was taking the surveying ship *Beagle* to chart the coast of South America, and then to explore the Pacific islands, Tasmania, Australia and South Africa. Henslow offered to recommend Charles, expecting he would "snap at the opportunity." However, Dr. Darwin immediately put a damper on the young naturalist's enthusiasm, proclaiming it a dangerous, hare-brained scheme no "man of sense" would approve.

Charles appealed to his uncle Josiah Wedgwood, the famed ceramics manufacturer, a certified "man of sense" whose opinions carried considerable weight in family matters. "Uncle Jos" intervened, Darwin's father relented, and Charles went off to meet FitzRoy ("My beau ideal of a Captain.") Though some historians have painted Robert Darwin as a tyrannical

CHARLES DARWIN, shown in two very different seasons of his life. As a young father, holding son William, his warmth and humor are apparent (left), but most texts present him as the black-caped wintry sage of this melancholy Romantic portrait (right).

ogre for opposing the voyage, he paid his son's expenses, bought equipment and even provided him with a servant-assistant (Syms Covington) for the five-year trip.

On December 27, 1831 ("my real birthday"), H.M.S. *Beagle* sailed from Portsmouth harbor, with 21-year-old Charles Darwin aboard as unofficial naturalist and "messmate" to the Captain. His father was to take great pride in Charles's letters and accomplishments during the voyage. In retrospect, his opposition may have been simply a parent's legitimate fears for his son's life. Ships of the *Beagle*'s design had been nicknamed by sailors "floating coffins"; the ship was bound for the malarial tropics they called "the white man's graveyard."

Aboard ship Darwin kept careful notebooks, journals, and diaries, and read widely in natural history including Sir Charles Lyell's *Principles of Geology* (1828–1832), which had just come out ("My ideas come half out of Lyell's brain"). Although he gathered, packed and shipped home tons of natural history specimens, he began to believe "mere collecting" for its own sake was a vice of "the mob of naturalists without souls." His boyhood love of collecting insects had kindled his interest (as it had also for Alfred Russel Wallace), but "enlarged curiousity" eventually led both of them, independently, to join the ranks of

philosophical naturalists—combining observation with a search for general explanations or laws of nature.

The Charles Darwin who returned five years later (1836) was a very different person: seasoned, self-reliant and with a mission to devote his life to natural science. His thousands of specimen rocks, fossils, birds, mammals, plants and fish were to occupy naturalists at the British Museum for several years; hundreds of the species were new to science. And his first two publications (*Journal of Researches Aboard the H.M.S. Beagle*, 1839, and *The Structure and Distribution of Coral Reefs*, 1842) established him as a rising young scientist of talent and a popular author.

In 1839, Darwin married his first cousin Emma Wedgwood, whose traditional religious beliefs were opposed to his unorthodox inquiries into the origin of species. Soon after their marriage, she wrote him a letter, begging him to reconsider challenging the Bible's account of creation, lest they be separated for eternity in the hereafter. All his life he cherished her touching letter ("Many times have I kissed and cryed over this"), but remained committed to his scientific career.

After a couple of years spent in London organizing his collections at the British Museum and placing them in the hands of experts, Darwin and his wife decided to buy a country home. In 1842 they settled in the quiet little village of Downe, only 16 miles from London, but—even today—a place that seems remote and timeless. ("I am now fixed on the spot where I will live my life to the end.") When they first moved in, the village still had a feudal flavor; country folk doffed their hats or curtsied when they encountered such "gentry" as the Darwins.

As he settled into the life of a country gentleman and father of 10 children (three would be lost to disease and genetic defects), he purposely turned his attention to a project as dull as the voyage had been exciting—describing and classifying the thousands of preserved barnacles he had gathered. For eight years he labored on these tiny, drab creatures, although he found the work numbing and tedious. ("I hate a barnacle more than a sailor on a slow-moving ship.")

His purpose was to develop and establish his scientific judgment and credibility to deal with questions of species. Without paying his dues as a laborer ("hod carrier of science"), Darwin knew his theories on the "species question" would carry no weight with established zoologists.

And theories he had, notebooks full of questions, speculations, observations. Like his grandfather Erasmus, he had come to believe that all of living nature is related, that present species had evolved from ancient lineages. (He used the terms "transmutation" or "the development hypothesis." The term "evolution" was coined by philosopher Herbert Spencer and did not appear in Darwin's *Origin of Species* until the fifth edition in 1869.)

When he first met Thomas Huxley, later to become his great friend and champion, Darwin was examining some of his specimens at a laboratory table in the British Museum. "Isn't it striking," young Huxley remarked, "what clear boundaries there are between natural groups, with no transitional forms?" Glancing up from the tray of preserved specimens, Darwin quietly replied, "Such is not altogether my view." Huxley later recalled that "the humorous smile which accompanied his gentle answer . . . long haunted and puzzled me."

To the sights and sounds of exotic wildernesses, Darwin added more commonplace observations: the unexpectedly wide range of variation in individuals, which he learned from the thousands of barnacles he had dissected and classified. The production of domestic varieties of horses, hogs and pigeons bred by "artificial selection" in the quiet Kentish countryside, too, were grist for his mill of evolutionary theory. ("If these varieties can by produced by man's hand, what might Nature not achieve?")

Darwin had only confided his ideas on "descent with modification" to a few people: the botanist Joseph Hooker, his mentor in geology Charles Lyell, and the American botanist Asa Gray. Although reticent about publication for years—as if testing the waters before taking the plunge—his private correspondence reveals total dedication to revolutionizing the life sciences. Besides his tireless research, experimentation and the writing of books and papers, he actively campaigned behind the scenes to convince influential scientists there were new paths to knowledge with an evolutionary perspective and even "grandeur in this view of life."

Despite having written a sketch of his theory in 1842 and expressing it in a letter to British botanist Sir Joseph Hooker in 1844, the "great book" on evolution by natural selection was still unwritten in 1857. In the meantime, Wallace had published his paper, "On the law which has regulated the introduction of new species" (1855), causing Lyell to warn Darwin that he must publish soon or risk losing his chance to be first with his theory. But despite Lyell's prodding, Darwin continued to work at his own careful pace.

Just as Lyell had feared, Wallace, on an extended field trip in the Moluccas (now Indonesia), had formulated the identical theory of evolution by natural selection. It was an independent discovery, which Wallace promptly wrote up ("On the Tendency of Varieties to Depart Indefinitely from the Original Type," 1858) and sent to the one person he thought might appreciate its merit, Charles Darwin. Perhaps, wrote Wallace, if Darwin thought it worthy, he would pass it on to Sir Charles Lyell for publication. (Darwin had never told Wallace that he was working on a similar theory; he had only mentioned vaguely that he was working on a "long-term study" of the relationship of "varieties" to species.)

On receipt of Wallace's paper, Darwin went into a desperate panic, soon to be aggravated by a houseful of children quarantined with scarlet fever. "Your words have come true with a vengeance," he wrote Lyell, "that I should be forestalled." Wallace's section headings "could stand as the titles of my chapters . . . Never have I seen such a remarkable coincidence."

Unable to deal with the possibility of being preempted by Wallace, he appealed for help to his friends the geologist Sir Charles Lyell and botanist Sir Joseph Hooker. He was well aware some might accuse him of stealing from Wallace. ("I would as soon burn my whole book as any man believe I behaved in a paltry spirit.") He only hoped Wallace himself would not and was relieved, on Wallace's return, to discover the younger man had a "generous and noble disposition."

Although Darwin was a careful and meticulous scientist, he was imaginative and empathetic as well.

Living things were real beings to him, and he reacted to them personally. He wrote of trees with their "noble heads raised high" and complained that climbing plants were "little rascals who never do what I want them to." He saw "life running riot" in rain forests, "antedeluvian monsters" in harmless iguana lizards and, on all creatures, "the indelible stamp" of their evolutionary origins.

In the popular perception, Darwin is remembered best as a solitary observer and field naturalist. Actually, he excelled at two other modes of research for which he is not generally credited: collaboration and experimentation.

He experimented constantly—on the movements of plants, the habits of earthworms, the relationship of pollinating insects to flowers, digestion by carnivorous plants, cross-breeding of plant varieties, seed germination, tens of thousands of experiments upon which his later books were based. To test whether certain plants could have reached distant islands, he soaked seeds for months in barrels of brine, then planted them to see which could survive long immersion in salt water. He measured the activity of earthworms in his garden by calibrating the rate at which a heavy stone sank into the turf. He tested the reactions of the sundew plant to hundreds of substances, including dead flies, cobra venom, paper, atropine, nicotine and human hair.

Yet, he remained focused on profound riddles of existence, such as the origin and nature of man, while immersed in detailed observation and experiment, which enabled him to demonstrate mysteries that lurk within even the most commonplace and familiar forms of life.

Darwin knew he could not hope to effect a major revolution in thought by himself and frequently consulted and collaborated at various times with Reverend John Henslow, Thomas Huxley, Sir Charles Lyell, Joseph Hooker, Asa Gray and his own son Francis. Although they almost never coauthored papers (sharing credit for each morsel of knowledge is a 20th century fashion), they acknowledged intellectual debts by grandly dedicating their "great works" to one another.

Ironically, the one man with whom Darwin's name is forever linked as coauthor is Alfred Russel Wallace, whose opinions and social standing differed greatly from his own. If he had had to choose a collaborator, surely he would have much preferred Sir Joseph Hooker or Professor Thomas Huxley, but Wallace was thrust upon him by his independently arrived at theory of natural selection. (At first, Darwin was not very gracious about it, barely acknowledging Wallace in the first edition of *Origin of Species* (1859). However, in later life, Darwin became one of the controversial Wallace's staunchest allies and secured a government pension for him.)

Over the years, Darwin's books tackled the implications of his theory for human origins (*The Descent of Man*, 1871), behavioral evolution (*Expression of the Emotions in Man and Animals*, 1872), coevolution of insects and plants (*Orchids*, 1862), domestic breeding (*Variation in Domesticated Plants and Animals*, 1868), and botany and plant physiology (*Movements of Climbing Plants*, 1865, *Insectivorous Plants*, 1875, *Different Forms of Same Flower*, 1877). His first (*Coral Reefs*, 1842) and last books (*Earthworms*, 1881) were demonstrations of how great geological features may result from small, slow causes, acting regularly, over immense periods of time.

Darwin's scientific friends spoke of his intense honesty, the "central fire" of his character that never allowed him to cut corners in his work. He kept his vivid imagination disciplined by testing his ideas in thousands of careful experiments. Compulsively, he searched the scientific literature for facts that didn't fit into or contradicted his theories. Acknowledging all criticisms and suggestions with almost comically excessive gratitude, he carefully weighed every possible difficulty. In *Origin of Species*, he seems to be reluctantly driven to his conclusions, despite every possible objection, which he raises before his readers or critics can.

When he died of cardiac disease in 1882 ("I am not in the least afraid to die"), he expected to be buried in the old churchyard at Downe Village, but his powerful scientific friends petitioned for burial in Westminster Abbey, England's shrine of highest honor. Darwin's pallbearers included his old friends Huxley, Hooker, Wallace, Lubbock, and the presidents of the Royal and Linnaen Societies. His final resting place is a few paces away from that of Sir Isaac Newton, another scientific immortal.

Eulogizing his old friend, the scrappy, combative Professor Thomas Henry Huxley praised Darwin's lifelong restraint and diplomacy: "He delivered a thought-reversing doctrine to mankind with as little disturbance as possible to the deeply-rooted sentiments of the age." Yet Huxley well appreciated that, notwithstanding Darwin's gentility and conventional lifestyle, he had managed to reshape and revolutionize Western thought.

"None have fought better, and none have been more fortunate than Charles Darwin," wrote Huxley. "He found a great truth, trodden underfoot, reviled by bigots, and ridiculed by all the world; he lived long enough to see it, chiefly by his own efforts, irrefragably established in science, inseparably incorporated with the common thoughts of men . . . What shall a man desire more than this?"

See also CORAL REEFS; DARWIN, CHARLES, ILLNESS OF; DARWIN, CHARLES, NICKNAMES OF; DARWIN, CHARLES, WORKS OF; DARWINISM; "DELICATE ARRANGEMENT"; "DESCENT OF MAN"; EARTHWORMS AND THE FORMATION OF VEGETABLE MOULD; ORCHIDS, DARWIN'S STUDY OF; ORIGIN OF SPECIES; VOYAGE OF THE H.M.S. BEAGLE.

For further information:

Brent, Peter. *Charles Darwin: A Man of Enlarged Curiosity.* New York: Harper & Row, 1981.

Chancellor, John. *Charles Darwin.* New York: Taplinger, 1971.

Clark, Ronald. *The Survival of Charles Darwin.* New York: Random House, 1984.

Darwin, Charles. *Autobiography.* (Cuts restored). Edited by Nora Barlow. New York: Harcourt, Brace, 1958.

Darwin, Francis. *Life and Letters of Charles Darwin.* London: Murray, 1887.

Huxley, Julian, and Kettlewell, H. *Charles Darwin and His World.* London: Thames & Hudson, 1965.

Irvine, William. *Apes, Angels and Victorians: A Joint Biography of Darwin and Huxley.* London: Weidenfeld & Nicholson, 1956.

Ward, Henshaw. *Charles Darwin: The Man and His Warfare.* New York: Bobbs-Merrill, 1927.

DARWIN, CHARLES, ILLNESS OF
His Mysterious Malady

A few years after returning to England from his five-year voyage of exploration, Charles Darwin became a semi-invalid who suffered daily for the rest of his life. Doctors were baffled; they could find neither cause nor cure.

As a young man Darwin had uncommon strength and endurance. During the *Beagle* expedition, he endured rough seas, primitive conditions on overland treks and rode spirited horses with the roughest gauchos in Argentina. Whenever he encountered a mountain on his inland treks, he usually climbed it. Yet a few years later, he was afflicted with almost daily weakness, vomiting and chronic fatigue.

One theory, proposed by Israeli parasitologist Saul Adler in 1959, identified the illness as Chagas' disease, which is acquired through the bite of the Benchucha beetle, also known as the "Great Black Bug of the Pampas." In fact, Darwin wrote of having been bitten by the insect in 1835 in Argentina, where the illness is common. The disease is caused by a blood parasite that became known to science only some years after Darwin's death.

Other historians of biology suggest the illness was not organic but psychomatic in origin, exhibiting classic symptoms of neurotic anxiety. Dr. Ralph Colp Jr., a New York psychiatrist, has analyzed every scrap of evidence about Darwin's symptoms and the history of his condition and has made a meticulous case that the "mysterious malady" had a strong psychosomatic component (*To Be An Invalid: The Mysterious Illness of Charles Darwin*, 1977).

Darwin suffered from extreme anxieties as he developed his theories. Colp traces the beginning of Darwin's illness to his first work on evolutionary theory. From the first, his wife Emma worried whether his scientific investigations were going to cost him his eternal soul. Darwin dreamt of being beheaded or hanged; he thought a belief that went so contrary to biblical authority was "like confessing a murder."

Emma nursed Charles, carefully scheduling his work and visitors so as not to tire him. He admitted he was glad his illness saved him from a lot of frivolous social occasions. When his doctors could offer no help, he pursued quacks and water cures, sometimes going away for weeks to the mineral baths and spas that were then fashionable.

Corroboration of Colp's diagnosis comes from Gwen Raverat, Darwin's granddaughter, in her charming book of reminiscences *Period Piece: A Cambridge Childhood* (1952). "The attitude of the whole Darwin family to sickness was most unwholesome," she wrote. "At Down [House], ill health was considered normal."

> The trouble was that in my grandparents' house it was a distinction and a mournful pleasure to be ill. This was partly because my grandfather was always ill, and his children adored him and were inclined to imitate him; and partly because it was so delightful to be pitied and nursed by my grandmother.

Her father George, Aunt Etty and Uncle Horace—all children of Charles and Emma Darwin—were "most affected by the cult of ill health . . . [For Aunt Etty] ill health became her profession and absorbing interest . . . She was always going away to rest, in case she might be tired later on in the day, or even next day."

However, in 1989, Dr. Jared Goldstein of Randleman, North Carolina, published a new analysis of the Chagas' disease theory, in which he showed that a "sub-acute" form of the infection fit Darwin's symptoms very well. The pattern of the incubation period and a later stabilization during which the patient improves, all jibe with Darwin's medical history. Apparently, his immune system may have arrested the disease, for it never progressed to gross organ impairment. Colp, who had been unaware that Chagas' could take that form, agreed with Goldstein's findings, while Goldstein believes Colp's analysis of the psychomatic component is also correct.

It is remarkable that despite "never knowing a day of robust health" for 40 years, Charles Darwin managed to write his 17 scientific books and 155 articles—a lifetime output of more than 10,000 published pages—working no more than two or three hours a day. "I have always maintained," he said, "it is dogged as does it."

See also BEETLES; COLP, RALPH, JR.

DARWIN, CHARLES, NAMESAKES OF

Darwin scholar Richard Freeman collected long lists of places, plants and animals named for Charles Darwin to honor his achievements as explorer, geologist, botanist and zoologist. (There are probably close to 200.)

Among them are Darwin Bay, off Chile, and the Darwin Channel, which leads to Port Aysen there. Two major mountains in South America are called Mount Darwin, one in Tierra del Fuego and one in Peru. The most northern of the Galapagos Islands (formerly Culpepper Island) is now officially Darwin Island. There is a town called Darwin in Australia and a Darwin district and mountain in South Africa. (There are also mountains named for Captain FitzRoy and his officer Stokes in Chile and Argentina and a Beagle Channel.)

About 100 living things bear Darwin's name as part of their official scientific label. Plants include a fungi, algae, orchid, potato, tulip, cactus and tree. Animals include a ground beetle, a legless lizard, a jumping spider, an iguana, a fossil oyster, a fossil giant sloth, a tree frog, finches and the ostrich-like South American bird known as Darwin's rhea.

DARWIN, CHARLES, NICKNAMES OF

Charles Darwin seems to have had more than his share of nicknames, reflecting the many facets of his complex career and personality.

When an old man, respectful journalists dubbed him "The Sage of Down," while ironic ones sometimes referred to him as "The Saint of Science." But his irreverent friend, Thomas Henry Huxley, satirized these overblown compliments by privately calling him "The Czar of Down" and the "Pope of Science."

As a small boy, Darwin and his brother Erasmus used to conduct amateur chemical experiments, and Charles once caused an explosion. For his part in this misadventure, his brother called him "Gas" for years.

At Cambridge, he took regular strolls in the countryside with his mentor, Professor Henslow, the botanist. He considered their rambling talks more valuable than any of his courses and became known to the other students as "the man who walks with Henslow."

On board the H.M.S. *Beagle* as a young naturalist,

"THE POPE OF SCIENCE" was one of Thomas Huxley's more acerbic nicknames for Charles Darwin. In a letter requesting a recommendation for a friend, Huxley lampooned the great naturalist as head of the "Church Scientific," with his favor-seeking friend as a supplicant on bended knee.

he had two nicknames. One was "philosopher," shortened to "Philos" by the captain and officers with whom he liked to discuss intellectual questions. The other was "Flycatcher," when his shipmates got tired of seeing the deck full of his natural history collections.

His own favorite nickname for himself was "Stultis the Fool." To Darwin nothing was obvious, and he would try experiments most people would prejudge to be fruitless. When he was investigating pollination, he put a female flower in a bell jar together with some pollen from a male plant to see if there was any way the two could get together without the help of insects.

"I love a fool's experiment," he once told a friend, "and I am always making them." When describing his "fool's experiments" or new hypotheses in letters to scientific friends, he would wryly sign himself STULTIS, a Latin name meaning one who is absurd, futile, ridiculous, insane.

DARWIN, CHARLES, OFFSPRING OF

Emma and Charles Darwin had 10 children, of whom seven survived. Two, a son and daughter, were lost in infancy. Daughter Annie, a bright and beautiful youngster, died of an unknown fever at the age of ten. She had been the apple of Darwin's eye, and her loss was one of the Darwins' great sorrows in life.

Francis (Frank) became a botanist and edited the wonderful account of his father's life, *Life and Letters of Charles Darwin* (1887) and *More Letters of Charles Darwin* (1903), two of the greatest and more enduring volumes of annotated correspondence produced by the 19th century.

Son George became a noted astronomer and Plumian Professor of Astronomy at Oxford. His theories on the moon's influence on tides and earth motions are still respected. (George and Frank were elected members of the Royal Society, the fourth generation of Darwins in that prestigious scientific organization. The tradition extended into the sixth consecutive generation.)

Still another son, Leonard, had a career in the military; William became a banker; and Horace was a maker of scientific instruments. It was Horace who constructed the simple "wormstone" device, which still stands in the garden at Down House, to measure the rate at which earthworms bury large objects with the soil that passes through their bodies. Horace went on to found the Cambridge Instrument Company, still in existence today, known for its innovative, high-quality scientific equipment. For years, the company was known to customers simply as "Horace's Shop."

Gwen Raverat, a noteworthy granddaughter, wrote a charming, sparkling recollection about her Cambridge childhood and the Darwin family (*Period Piece*, 1952). Another granddaughter, Lady Nora Barlow, edited the diaries of Darwin's *Beagle* voyage and a volume of his correspondence with Henslow, the botanist. Grandson Bernard veered entirely away from science and became a champion golfer.

Descendants of the Darwins still favor first names like Etta, George, Emma and Charles, and have continued to intermarry not only with the Wedgwoods but also with the Huxleys, descendants of Darwin's friend Thomas Huxley.

DARWIN, CHARLES, WORKS OF

Charles Darwin's output was astounding, especially as he was a semi-invalid for most of his life after returning from his five-year voyage aboard H.M.S. *Beagle*. Generally, he worked only three hours a day, writing out all his manuscripts by hand.

All told, he published 17 books in 21 volumes; these works contain more than 9,000 printed pages. Another 1,000 pages appeared in scientific journals. (There were almost 10,000 additional pages of revisions added to editions of various books over 43 years.) Also, several thousand of his letters have been published, many of them for the first time during the 1980s.

Many of Darwin's superficial critics focus on a few passages in the *Origin of Species* (1859), unaware of the major body of his life's work. His wide-ranging contributions of new knowledge to geology, botany, animal behavior, reproductive biology and dozens of other fields has never been equaled. And he established research programs in these fields that are still being profitably pursued. Following is a list of his books.

1839   *Narrative of the Surveying Voyages of His Majesty's Ships Adventure and Beagle (1826–1836).* 3 vols. London: Colburn. Volume III written by Darwin and reprinted as a separate work with the following title.

1839   *Journal of Researches into the Natural History of Geology of the Countries Visited by H.M.S. Beagle from 1832–36.* London: Colburn.

1839–1843   *Zoology of the Voyage of the H.M.S. Beagle.* London: Smith, Elder. Edited by Charles Darwin, these were seven monographs by experts published in five parts; Darwin contributed notes and other material to each piece.

1842   *The Structure and Distribution of Coral Reefs.* London: Smith, Elder. First part of *Geology of Voyage of the Beagle.*

1844   *Geological Observations of the Volcanic Islands, Visited During the Voyage of H.M.S. Beagle.* London: Smith, Elder. Second part of *Geology of the Voyage of the Beagle.*

1846   *Geological Observations on South America.* London: Smith, Elder. Third part of *Geology of Voyage of the Beagle.*

1851–1855   Several volumes of monographs on fossil and modern barnacles (*Cirripedes*)

1859   *On the Origin of Species by Means of Natural Selection, or the Preservation of Favoured Races in the Struggle for Life.* London: Murray. 2nd edition, 1860; 3rd edition, 1861; 4th edition, 1866; 5th edition, 1869; 6th and last edition, 1872.

1862   *On the Various Contrivances by which Orchids are Fertilised by Insects.* London: Murray.

1865   *The Movements and Habits of Climbing Plants.* London: Linnean Society.

1868	*The Variation of Animals and Plants under Domestication.* 2 vols. London: Murray.
1871	*The Descent of Man, and Selection in Relation to Sex.* 2 vols. London: Murray.
1872	*The Expression of Emotions in Man and Animals.* London: Murray.
1875	*Insectivorous Plants.* London: Murray.
1876	*The Effects of Cross and Self Fertilisation in the Vegetable Kingdom.* London: Murray.
1877	*The Different Forms of Flowers on Plants of the Same Species.* London: Murray.
1879	*Erasmus Darwin.* London: Murray. Text by Ernst Krause, with preliminary essay by Charles Darwin. Essay is longer than Krause's text.
1880	*The Power of Movement in Plants.* London: Murray. Written with the assistance of his son, Francis.
1881	*The Formation of Vegetable Mould through the Action of Worms.* London: Murray.

For further information:

Barrett, Paul H., ed. *The Collected Papers of Charles Darwin.* Chicago: University of Chicago Press, 1977.

Jastrow, Robert, and Korey K., eds. *The Essential Darwin.* Boston: Little, Brown, 1984.

DARWIN, EMMA (1808–1896)
Wife of Charles

Soon after Charles Darwin's return to England, he married his first cousin Emma Wedgwood, who devoted her life to making his work possible even though she feared it might result in his damnation. Born into the famous family of prosperous potters, as Josiah Wedgwood's daughter, she brought considerable means to the marriage. Wedgwoods had been specially close to the Darwins for two generations, since Erasmus Darwin and the first Josiah Wedgwood founded their intertwined dynasties.

After their marriage, Emma wrote a deeply felt letter asking Charles to reconsider his defiance of biblical authority, so they might not be eternally separated in the afterlife if he was wrong. Her new husband was touched when he received it. Years later, after his death, it was found much handled with his jotted comment: "Many times have I kissed and cryed over this."

Emma soon grew used to having Charles's experiments about the house and was protective of his time and energies. She cautioned visitors to be brief and often cut them short by announcing it was time for his nap or his walk. She was nurse, companion, mother of 10 children (seven of whom survived) and supervised a large household staff.

She was tolerant of his friends but had little interest in their discussions. At one gathering, Charles saw her yawn and asked if she was bored by all the scientific talk. "Oh, no," she drily replied, "no more so than usual."

Emma regularly attended the village church on Sundays, and once heard a new pastor rail against Darwin's theories in a sermon. Although she held grave misgivings about her husband's philosophy, she promptly gathered her family, marched out of the church and did not return until the tactless cleric no longer occupied the local pulpit.

DARWIN, ERASMUS (1731–1802)
Grandfather of Evolution

Erasmus Darwin, grandfather of Charles, was much more than a prosperous physician with a good bedside manner. He was also a famous poet, philosopher, botanist and the first naturalist to publish a detailed theory of evolution—much of it in verse. His pioneering treatise on evolutionary theory, *Zoonomia; or, the Laws of Organic Life* (1794–1796), anticipated by more than 60 years the intellectual revolution his grandson would lead.

Born in the county of Nottingham in 1731, Erasmus showed a precocious intellect. Like his studious father, more interested in fossils than in sport, he later became the first in a line of six generations elected to membership in the Royal Society. After graduating from Cambridge, he studied medicine at Edinburgh, then moved back to England to Lichfield, where he established a medical practice. Some of the most brilliant and inventive men of his day lived in the area and enjoyed visiting together. Among these were inventor James Watt, metallurgist Matthew Boulton, chemist Joseph Priestley, and potter Josiah Wedgwood—all of them prime movers in creating the technology that would take England into the industrial era.

Dr. Darwin was the founder and ringleader of the Lunar Society, which brought these notables together regularly. The "Lunatics," as they called themselves, met only during the full moons, so that they might find their way home by bright moonlight. Their gatherings lasted from early afternoon until night, and their talk was probably the most interesting in England at that time—of Watt's steam engine or Wedgwood's new ceramic methods, of Priestley's oxygen experiments or Darwin's evolution theories.

Although he was a great and expansive talker, with a wry sense of humor, Dr. Darwin stammered.

ERASMUS DARWIN, physician, botanist, naturalist and poet wrote a series of books in which he developed a theory of evolution or descent with modification of all life from common ancestors. Although Charles Darwin denied his grandfather's influence, he clearly followed up many of Erasmus' original ideas.

When a rude young man asked whether his halting speech was inconvenient, he replied, "N-n-no Sir . . . it gives me . . . time for reflection . . . and . . . saves me . . . from asking . . . impertinent questions."

According to his biographer Hesketh Pearson, "there is hardly an idea [or] invention in the world of today that Dr. Erasmus Darwin did not father or foresee, from . . . eugenics and evolution to aeroplanes and submarines, from psychoanalysis to antiseptics."

In the late 18th century, Erasmus addressed two major evolutionary questions: whether all living things arose from a single common ancestor and how one species could develop into another. He assembled observations from embryology, comparative anatomy, systematics, geographical distribution and fossils. Overwhelming evidence, he thought, pointed to the development of all life from a single source, "one living filament."

Although he rejected the scriptural idea of separate creations for each species, that did not lessen his respect for "the Author of all things." It was just as wonderful, he argued, that "the Cause of causes" has set the whole evolving web of life in motion.

As for the means of evolution, Erasmus's writings touch on almost all the important topics except natural selection. He suggested that overpopulation sharpened competition, that competition and selection were possible agents of change, that man was closely related to monkeys and apes, that plants should not be left out of evolutionary studies and that sexual selection could play a role in shaping species. ("The final course of this contest among males seems to be, that the strongest and most active animal should propagate the species which should thus be improved.")

In addition, he wrote on related problems in natural history, which remarkably parallel the subjects taken up by his grandson for whole books: twining and movement in plants, theory of descent, cross-fertilization in plants, adaptive and protective coloration, heredity, domestication of animals. Behind these investigations was Dr. Darwin's strong belief in progress towards "ever greater perfection in all the productions of nature."

See also DARWIN, CHARLES; TEMPLE OF NATURE; ZOONOMIA.

For further information:

Darwin, Erasmus. *The Botanic Garden*. Dublin: J. Moore, 1790.

———. *The Temple of Nature or, the Origin of Society*. Baltimore: Bonsall & Niles, 1804.

King-Hele, Desmond. *Erasmus Darwin*. New York: Scribner's, 1963.

Pearson, Hesketh. *Doctor Darwin*. London: Dent, 1930.

DARWIN COLLEGE
Cambridge Graduate School

Charles Darwin had attended Christ's College at Cambridge during the 1820s, although he considered his stint there "an almost total waste of time." Nevertheless, one of the Cambridge mentors, the botanist Reverend John Henslow, later recommended him as naturalist on the surveying ship H.M.S. *Beagle;* Darwin always referred to this voyage as his "real education."

Today Cambridge's ancient cluster of schools has a recent addition: Darwin College, founded for postgraduate and postdoctoral students in 1964. Among its renowned alumnae is Dian Fossey, who interrupted her famous study of African mountain gorillas to take a degree in zoology there in the 1970s.

Its first building was the "Old Granary," which had been converted into a private residence years before by Professor George Darwin, the distinguished astronomer and son of Charles. After his death, it was donated to the University by the Darwin family as the nucleus of the new college.

DARWIN COLLEGE at Cambridge, England, founded in 1964. This building had originally been a granary, and for many years was the home of Charles Darwin's son George, a distinguished professor of astronomy at the university.

GROUCHO MARX takes over as dean of "Huxley College" in the classic comedy *Horse Feathers* (1932). In the film, the school's football rival was "Darwin College," which at the time was equally fictitious.

About 40 years before its actual founding, a wholly fictional version of "Darwin College" was featured in the classic Marx Brothers comedy *Horse Feathers* (1932). The film's plot revolves around a football game between two rival colleges: "Darwin" and "Huxley."

DARWIN CORRESPONDENCE

See SMITH, SYDNEY.

DARWINISM
Complex Cluster of Ideas

Darwin often spoke of "my theory" as if it were a single idea. Most historians of science agree that the core of Darwinism consists of two major ideas: the fact of evolution and the major mechanism of evolution he proposed, natural selection.

Other features implied in Darwin's thought are often neglected. As Harvard evolutionist Ernst Mayr points out, Darwinism also includes the concepts of common descent, gradualism, and multiplication of species.

Common descent is the idea that several related species trace back to a common ancestor from which they branched off. This idea cleared up puzzling similarities between various creatures, resolved the vague problem of archetypes or "common plans" and gave a new dimension to the classification of organisms into families, genera and species. Formerly puzzling similarities of structure in families and genera "made sense" if they shared a common ancestor.

Gradualism is the idea that evolution proceeds slowly and continuously, rather than by leaps or jumps. Some of Darwin's supporters, particularly Huxley, did not think it necessary to burden the

theory by insisting on gradualism. After all, no one had actually seen a species formed, and the fossil record looked more jumpy than continuous.

Mayr believes Darwin's insistence on small changes working over a long period were necessary to counter prevailing ideas of special creation or supernatural

POP DARWINISM inspired such popular kitsch as this pensive chimp seated on *Origin of Species*, once common on knick-knack shelves.

intervention. In addition, Darwin's geological research and admiration for Lyell's uniformitarianism came into play. If a great coral reef 1,000 feet thick could be produced over the centuries by tiny polyp animals, it was possible that species could be gradually transformed by steady forces acting over immense periods of time.

Darwin was impressed with the "creative force" of species radiation in the Galapagos Islands, producing so many different but closely related animals and plants in a relatively short time. He was among the first to reject the accepted view that "all places in the polity of nature are filled" and that new species only arose to replace those that became extinct. But although he helped establish that diversification of closely related species is a fact, the mechanism that caused it remained a puzzle to him. Fifty years after his death, population geneticists came up with plausable models of speciation, or how the multiplication of species occurs.

See also DIVERGENCE, PRINCIPLE OF; EVOLUTION; GRADUALISM; NATURAL SELECTION.

For further information:
Kohn, David, ed. *The Darwinian Heritage.* Princeton, N.J.: Princeton University Press, 1985.
Wallace, Alfred Russel. *Darwinism: An Exposition of the Theory of Natural Selection with Some of Its Applications.* London: Macmillan, 1889.

DARWINISM, NATIONAL DIFFERENCES IN

Charles Darwin's theory of evolution, as G. B. Shaw once observed, "had the good fortune to be adopted by anyone with an axe to grind." Interpretations of Darwinism were called on to support a bewildering variety of political beliefs, social theories and conflicting ideologies, differing from nation to nation.

England was of two minds about Darwinism. On the one hand, natural selection justified Britain's subjugation of other peoples, but Huxley argued that a civilized people should rebel against the evolutionary past and seek a more humane and compassionate vision than "survival of the fittest." Natural selection itself was eclipsed for awhile by the neo-Lamarckians but revived with renewed force after 1900 by neo-Darwinians, who were discovering it was compatible with the new population genetics.

German Darwinism was shaped by Ernst Haeckel, who combined it with anticlericalism, militaristic patriotism and visions of German racial purity. He encouraged the destruction of the established church in Germany, with its sermons about "the meek shall inherit the earth" and compassion for unfortunates. Such a "superstitious" doctrine would lead to "racial

suicide." During the 1930s, Adolf Hitler believed he was carrying Darwinism forward with his doctrine that undesirable individuals (and inferior races) must be eliminated in the creation of the New Order dominated by Germany's "Master Race."

In the United States, Social Darwinism was advocated by *laissez-faire* economists like Yale's Professor William G. Sumner, who found it compatible with free competition and rugged individualism. Philosopher Herbert Spencer, the prophet of progress, became more popular in America than Darwin. His conviction that unregulated competition is the key to progressive social evolution appealed to industrial capitalists.

America also had a strong resistance to wholly "mechanistic" explanations of the origin of man. Even Darwin's Harvard champion, Professor Asa Gray, could not imagine evolution without goals, without being directed by God. Many American biologists who retained their religious beliefs followed Alfred Russel Wallace in assigning to mankind a special spiritual origin not shared with animals.

French science resisted Darwinism for decades, perpetuating theories of Lamarckian evolution; national pride in the first evolutionary biologist being a Frenchman contributed to this stance. Influenced by the tradition of ideal types and classification, many scientists failed to see why Darwinism was so revolutionary outside France. Besides, how valid could it be if it had been become part of the ideology of resurgent German militarism?

Church influence remained strong in French education for much longer than in England. Yet, France was a treasure trove of human and animal fossils, prehistoric stone tools and the stunning Paleolithic cave paintings. Before long, a few outstanding prehistorians and paleontologists arose from the ranks of churchmen, particularly from the Jesuit order.

Darwinism was welcomed in Communist countries since Karl Marx and Friedrich Engels had considered *The Origin of Species* (1859) a scientific justification for their revolutionary ideology. As far as Socialist theorists were concerned, Darwinism had proved that change and progress result only from bitter struggle. They also emphasized its materialist basis of knowledge, which challenged the divine right of the czars.

An opposing branch of Russian Darwinism, led by Prince Peter Kropotkin, argued that "mutual aid" (and therefore socialism) was also a natural principle of evolution. The ideal society was based not on competition, but on cooperation. Individual needs were less important than those of the larger social entity.

Long after Lamarckian inheritance had been abandoned elsewhere, Russia stubbornly retained this

19th-century belief in the inheritance of acquired characteristics. Party theorists refused to accept that each generation must be educated anew, believing socialism would create permanent genetic transformations in the population.

Under Trofim Lysenko's dominance of Soviet science, "Mendelist" genetics was a forbidden doctrine, a bourgeois heresy. Lysenkoism was finally abandoned in the 1960s, but only after Lysenko's fraudulent research brought on agricultural disaster, which threatened the country with starvation.

Depending on a nation's culture, economic system and political history, Darwinian ideas shared the same fate as many religions and ideologies. They were interpreted so broadly that they could encompass anything. Identified with progressive change toward a final social good, Darwinism took on the qualities of each nation's history, self-image and aspirations.

See also CHINA, DARWINISM IN; ENGLAND, DARWINISM IN; HAECKEL, ERNST; KROPTKIN, PRINCE PETER; LAMARCKIAN INHERITANCE; LEBENSBORN MOVEMENT; MARXIAN "ORIGIN OF MAN"; MONISM; TEILHARD DE CHARDIN, FATHER PIERRE.

For further information:
Glick, T. F., ed. *The Comparative Reception of Darwinism.* Austin: University of Texas Press, 1974.

DARWIN LIBRARY

See GAUTREY, PETER J.; SMITH, SYDNEY.

DARWIN MUSEUM

See DOWN HOUSE.

(CHARLES) DARWIN RESEARCH STATION
Conservation of the Galapagos

The Galapagos Islands, 600 miles west of Equador, have become famous not only as an area of great natural wonders, but also for their pivotal role in inspiring Charles Darwin's theory of evolution. In the century and a half since his famous visit to the islands, now officially called the Archipelago of Colon, there have been many destructive intrusions on the unique and fragile ecosystem. Even in Darwin's day, giant tortoises were taken from the islands by the thousands, causing a great decrease in their numbers and the extermination of several species. Of the 15 subspecies that once existed, four have become extinct, and one variety is represented by only one surviving individual. [See LONESOME GEORGE.]

The islands now have 5,000 inhabitants, who live mainly on four islands (San Cristobal, Santa Cruz, Isabela, Floreana); some 25,000 tourists visit an-nually. Most of the 3,000 square miles of land has been designated as a national park by Equador, and the government has been making strong efforts at conservation and at stabilizing the number of residents.

However, the biggest threat to the islands' plants and animals are not the people, but the animals they have introduced, including goats, dogs, cats, rats, pigs and donkeys, which run wild. Cats eat the young of birds and lizards, while goats, rats and pigs attack the eggs of birds and tortoises. Donkeys and pigs graze on rare plants, causing erosion and loss of habitat.

The Charles Darwin Research Station, situated on Santa Cruz Island, was founded in 1964 to promote research and conservation. Personnel carry out basic investigations of the islands' ecosystems and conduct programs to hunt and eradicate imported pests, which thrive at the expense of the rare Galapagos creatures. In addition, they run a nursery where they breed the endangered giant tortoises with spectacular success. Naturalist-guides are trained at the Research Station to educate tourists about what Darwin called this "eminently curious little world within itself." The Darwin Research Station is the only place where visitors may conveniently get close to, stroke—but not ride on—the most famous species for which the islands are named. Galapagos means "horse saddle" in Spanish and refers to the saddle-shaped carapace (upper shell) of the centenarian reptiles.

See also DARWIN'S FINCHES; GALAPAGOS ARCHIPELAGO.

"DARWIN'S BULLDOG"
Thomas Huxley Unleashed

Charles Darwin hated the thought of publicly defending his controversial theory of evolution. But his pugnacious friend, the zoologist Thomas Henry Huxley, loved nothing better than a war of wits. Temperamentally the opposite of Darwin, Huxley relished public confrontations with critics, earning him the nickname "Darwin's bulldog."

It was Huxley who first reviewed *The Origin of Species* (1859) in the *London Times*, declaring it a "solid bridge of facts . . . [which] will carry us safely over many a chasm in our knowledge." His impromptu debate with Bishop Samuel Wilberforce at Oxford over the *Origin* has come down in history and folklore (somewhat inaccurately) as a milestone battle between "science and religion."

Certainly Huxley enjoyed deflating church authorities when they made pronouncements on scientific issues. He used to tell his anatomy students they could remember the mitral valve, shaped like a bishop's triangular hat, is on the left side of the heart "because a bishop is never known to be on the right."

Darwin was well pleased with Huxley's aggressive campaign to win over public opinion. In 1860, just after Huxley had bested Wilberforce, Darwin stressed the "enormous importance of showing the world that a few first-rate men are not afraid of expressing their opinion . . . I see daily more and more plainly that my unaided book would have done absolutely nothing."

One student in Huxley's anatomy class at London College was the American Henry Fairfield Osborn, later to become the president of the American Museum of Natural History. One memorable day in 1879, Osborn recalled, the reclusive Darwin paid a rare visit to the classroom.

The 22-year-old Osborn was thrilled when he was singled out to meet Darwin, who was then 70. After they exchanged a few words, Huxley hurried the great naturalist into the next room, saying "I must not let you talk too much." Years afterwards, Osborn recalled his sometimes formidable professor's touching solicitude toward his older friend. "You know, I have to take care of him," Huxley explained, "in fact, I have always been Darwin's bulldog."

See also HUXLEY, THOMAS HENRY; OXFORD DEBATE.

DARWIN'S FINCHES
Classic Case of Speciation

When Charles Darwin explored the Galapagos Islands as a ship's naturalist in 1835, he was struck by the peculiar species he found there and made notes on their behavior and distribution. One group of birds, the Galapagos finches, is often cited as a striking illustration of the evidence for speciation he discovered; they have become famous as "Darwin's finches."

Adapted to exploit different niches and habitats in the islands, the Galapagos finches diverged from one mainland species into many. For years they have been considered a great stimulus in forming Darwin's ideas, and no textbook on evolution fails to give them

a place of honor. But recent scholarship suggests that Darwin actually missed the lesson of the finches and that only years later he realized their significance.

What is remarkable about the little gray or black birds is how minimal—yet how striking—are the differences between them. All are rather undistinguished sparrow-like birds with thick pointed bills, but the various populations have evolved very different feeding adaptations.

Some of the modes of life involve changes in beak structure, while some are simply behavioral. On Hood Island, a large ground finch with a stout beak and sturdy legs rolls over stones to find the food underneath. On Wenman Island, a sharp-beaked finch creeps up on incubating seabirds, pricks their skin near the tailfeathers, and drinks their blood like a vampire. Ferdinanda Island contains a local population of small ground finches that leap onto the backs of basking marine iguanas and clean them of ticks.

Perhaps the most remarkable of Darwin's finches is one of the only birds to use tools: the woodpecker finch. Since it lacks a woodpecker-like beak but feeds on bark insects, it holds a twig or thorn in its bill, which it uses to poke grubs out of their holes. Then it holds its hunting tool in one foot while it eats its catch.

Ornithologist David Lack's study of Darwin's finches in the 1940s established the view that the birds were a crucial stimulus to Darwin, convincing him that divergent evolution occurs as a result of geographical isolation. Darwin, the well-known legend goes, realized these birds must have descended from a South American finch species that had somehow reached the islands. Adaptations to various niches in the new environment split the expanding population into varieties, then into species.

But, during the early 1980s, Frank Sulloway reexamined the evidence and showed that even though the birds remain a classic example of island speciation, Darwin did not, at first, recognize them as such.

When he collected them, the beak structures of the birds appeared so different that Darwin did not recognize how closely related they were. After the *Bea-*

CLASSIC EXAMPLE of divergence of closely related species, some finches of the Galapagos Islands are adapted for feeding on large seeds, others on small seeds and some for digging insects out of bark. Darwin did not realize their evolutionary significance until some years after his voyage.

gle's return, it was ornithologist John Gould of the Zoological Society who identified the finches as a closely related group.

Darwin did not really form his views on speciation until well after the *Beagle* voyage. Because he didn't recognize the significance of the finches, he neglected to label his specimens to show from which islands each bird had come. (He later kicked himself for the omission and, similarly, for not bothering to note the islands from which the various giant tortoise shells in his collection were gathered. It was only upon leaving the Galapagos that he was struck by the remark of the vice-governor, who claimed he could tell from which island any tortoise shell was taken by its pattern and shape.)

After returning to England and conferring with John Gould, Darwin belatedly realized the problem and tried to reconstruct the patterns of finch distribution from other collections. What made the situation even more difficult was the lack of a candidate for a hypothetically ancestral finch among the living birds of South America.

There was, however, a group of birds in the islands that *did* give Darwin a clue to evolution: the Galapagos mockingbirds. Although less dramatic in their adaptations than the finches, Darwin was able to identify several species that resembled their American cousins.

Once he recognized that the islands harbored distinct varieties allied to a mainland form, he realized he had evidence in favor of speciation from a common ancestor—rather than separate divine creations for each small island.

See also ADAPTIVE RADIATION; GALAPAGOS ARCHIPELAGO.

For further information:

Lack, D. *Darwin's Finches: An Essay on the General Biological Theory of Evolution.* Cambridge, England: Cambridge University Press, 1947.

Sulloway, F. J. "Darwin and His Finches: The Evolution of a Legend." *Journal of the History of Biology* 15: (1982): 1–53.

DARWIN'S POINT
Vestigial Ear Tip

A sculptor named Thomas Woolner called Charles Darwin's attention to a "little peculiarity in the external ear." He had first noticed it when working on a figure of Puck, to whom he had given pointed ears.

Commonly appearing among both men and women, it is "a little blunt point, projecting from the inwardly folded margin" of the ear, sometimes protruding both inwards and outwards. Darwin realized that thickenings of cartilage are variable, but believed "the

points are vestiges of the tips of formerly erect and pointed ears." The curious little structure has become known as "Darwin's point."

DATING METHODS

See CHRONOMETRY.

DAVID GRAYBEARD (Chimpanzee)
First to Contact Humans

A chimpanzee with gray chin-hair, David Graybeard was the first wild ape to fully accept the presence of a human scientist with tolerance and friendly curiosity. In 1960, Jane Goodall had already spent months tracking chimps at the Gombe Stream Reserve in Tanzania, but they only let her get close enough to observe them through binoculars.

David Graybeard changed that by deciding to trust her. He was the first to voluntarily visit her camp and to let her touch him. Soon after, others followed suit until she was accepted by the group.

Carrying on with his normal activities despite Goodall's presence, he allowed her the first observations of meat eating by apes. From David Graybeard she also got her first look at the now famous "termiting" behavior. He stripped leaves from a twig, wet it in his mouth, then poked it into the mound of hard clay to capture the thirsty insects on his "termite popsicle."

Inspired by Goodall's experiences with chimps, Dian Fossey sought out mountain gorillas on their forested peaks and, in time, became similarly accepted. But it was David Graybeard, no less than Jane Goodall, who pioneered changing the entire relationship between humans and free-living apes from fear to trust.

See also "APEWOMEN," LEAKEY'S; CHIMPANZEES.

DAWKINS, RICHARD

See "SELFISH GENE."

DECEPTION, EVOLUTION OF
Origin of Lying

Because humans have the unique ability for spoken language, it is commonly thought our species is the only one capable of lying. But deception is not confined to *speaking* falsehoods; it is a strategy that has evolved in many diverse organisms, including plants, insects, birds, dogs and apes. New observations of baboons, chimps, and gorillas especially seem to show deliberate intentions to deceive.

Misleading colors and structures are common in nature, though they have evolved by selection, not as conscious deception. Viceroy butterflies have come to resemble poisonous monarchs as a defense against predators, just as harmless king snakes mimic venomous coral snakes. [See MIMICRY, BATESIAN.] Some orchids fool male wasps into "mating" with them by imitating female wasp odor, which insures the pollination of the next flower he visits. Other plants attract flies by giving off a false scent of rotting meat.

Many creatures raise appendages, feathers or puff themselves up with air to appear larger and more fearsome when attacked. When shrimps are molting—and defenseless without their shells—they rush straight at predators as if to attack, though they are incapable of biting. Bluff charges are also common in rhinos and gorillas, and they are perfectly capable of following through. But they would rather not expend more energy or risk possible injury, when a bluff is usually enough to scare away an intruder.

A long-legged wading bird, the stilt, acts wounded to lure predators away from her nest and chicks. She flaps around near the ground, as if her wing is broken, always just out of range of the foe. When she has led the enemy a safe distance away from the nest, she suddenly takes off and flies away.

Just how conscious the bird is of this deception is unknown. Some naturalists call it an automatic "fixed-action pattern," while others give more credit to the animal for a deliberate and skillful performance.

Robert Mitchell, a researcher at Clark University, is particularly intrigued by animals' misleading behavior that appears to be deliberate. After five years of studying such "deceitful behavior," he believes many creatures plan a course of action, which he considers similar in motivation to human lying.

Mitchell found, for instance, that animals in contact with people sometimes fake injuries to get attention, just as a child might. A female zoo gorilla pretended to have her arm hopelessly stuck between the bars of her cage—a ploy to attract the keeper over for a hug. He also observed a dog that had broken its leg and became used to extra sympathy from its master, which ended when the leg healed. Thereafter, it faked a pitiful, convincing limp when it wanted attention. While Mitchell describes that behavior as "deception"; others might see in it only a "conditioned response." Limping behavior was learned and reinforced by petting, like the association Pavlov's dogs made between food and the dinner bell.

One chimpanzee in a free-ranging group was observed to give a "food call," indicating that there were bananas nearby. When the others ran off in the indicated direction, the "lying" chimp went the opposite way, where it knew the food was really hidden. "To my amazement," wrote Mitchell, "I learned that in some animal species deception may be more common than truth-telling."

These examples of deception operate on different levels and may be produced by different mechanisms. Some behaviors have evolved as innate patterns, some are learned or conditioned, while others—like the chimp's—appear to be deliberate attempts of an individual to get its own way with members of the same social group.

Richard Byrne and Andrew Whiten, psychologists at the University of St. Andrews in Scotland, have edited a compilation of primate deception (*Machiavellian Intelligence: Social Expertise and the Evolution of Intellect in Monkeys, Apes and Humans,* 1988). After finding what they believed to be deliberate deceptive behavior in their own observations of baboons, they asked other primatologists for examples from other species.

Among baboons, Byrne and Whiten observed a youngster who waited for an older animal to dig up a juicy root, then screamed as if it were being attacked. The juvenile's mother came running over, drove the industrious animal away and the juvenile went over and ate the root.

"Both of us" they add, "saw this young baboon go through the same routine with different 'victims' on different days . . . the behavior was not a coincidence but a tactic. [He] was not genuinely scared [but his] deception of his mother gave food that he could not have gotten in any other way."

Another of their observations of "tactical deception" among baboons concerned a hefty adolescent they named Melton, who had bothered a juvenile until it screamed. Several adults, including the young one's mother came tearing over, heading straight for Melton.

Instead of fleeing or showing submission, he immediately stood on his hind legs and scanned the distant hillside—in exactly the way that these baboons behave when they have seen a predator or a group of foreign baboons. The pursuers skidded to a halt and looked intently in the direction Melton was staring; they never resumed the chase . . . [there was] no genuine cause for alarm; the "outside threat" was a fiction.

Jane Goodall observed a chimpanzee move out of sight of hidden food, so that a second chimp couldn't even see him *look* at where the food might be hidden. When the second chimp went away, the first immediately went over and retrieved his banana.

Goodall's Gombe chimps produced an even more striking incident, observed by Frans Plooij. A male was zeroing in on some bananas he had hidden, when a second showed up. The first walked away, sat down and looked all around, everywhere but where the food was hidden. Then the second pretended to leave, but hid behind a tree to watch the deceiver. When he uncovered the stashed food, the hiding male showed himself and grabbed it away.

Byrne and Whiten suggest that if "fixed action patterns" or "conditioning" cannot explain such an incident, we may have to conclude apes are capable of thinking such devious thoughts as: "X thinks I think he doesn't know where the banana is, but I think he really does know, so I'll just wait and see where he goes."

Ability to deceive or mislead is therefore neither new nor unique to our species, but has evolved many times in a variety of creatures. Because it appears to have a "natural" origin doesn't make lying justifiable, for nature provides no moral guide to human behavior. Lying is not always effective, either. Abraham Lincoln once observed that "No man has a good enough memory to be a successful liar all the time."

See also EXPRESSION OF THE EMOTIONS.

For further information:

Byrne, Richard, and Whiten, Andrew, eds. *Machiavellian Intelligence: Social Expertise and the Evolution of Intellect in Monkeys, Apes and Humans.* Oxford: Oxford University Press, 1988.

DEDUCTION
A Method of Logic

To deduce means to reach a logical conclusion by comparing several facts or premises. The closely related word "deduct" means to subtract or exclude. As Sherlock Holmes put it: "After you eliminate the impossible, what remains—however improbable— must be the truth."

Charles Darwin once deduced the existence of an unknown moth with a 14-inch tongue from the co-evolved structure of a certain orchid. Much more recently, a young marine biologist deduced there must be a light source at the ocean floor after dissecting an eyeless shrimp. Both conclusions seemed incredible to experts—yet both were later proved correct.

While deductive logic often suggests experiments or prompts discoveries, it seldom produces revolutionary scientific insights or theories. Evolution by natural selection, for example, cannot be proved by deduction. It is built up of many strands of evidence, all contributing to the overall picture. Great theories often arise from mind-pictures, analogies or metaphors rather than deductions.

A true conclusion also requires true premises. If we say "All mermaids have scaly bellies; this woman has a scaly belly; therefore this woman is a mermaid," the first premise is based on fantasy, which makes the conclusion invalid.

Error can also arise from a shift in meaning during the reasoning, such as: The American Indian is disappearing; Chief Sitting Bull is an American Indian; therefore, Chief Sitting Bull is disappearing. Here, the deduction is wrong because of a shift in the meaning of "disappearing" between premise and conclusion.

Another fallacy is tautology. Where all the premises mean the same thing, the deductive conclusion contains no new information. Evolutionary theory is often accused of falling into this error, as in: Unfit are selected out; only the fittest survive; therefore, evolution promotes survival of the fittest. Stated in this way, nothing is deduced except that "Only the survivors survive." But this syllogism (chain of reasoning) is not really evolutionary theory, but a caricature of it.

Sir Arthur Conan Doyle popularized deduction as a method of reasoning backward from evidence to causes. For instance, if Sherlock Holmes saw a certain red clay on a suspect's boot, he might deduce that the person had walked along a certain road where the clay was found. But that kind of deduction works only where the possibilities are very narrow; if almost every street contains the red clay, there could be no inference about the suspect's route.

Darwin's faith in the logical "fit" of coadapted organisms led him to certain deductions. While studying orchids, he found one with a 14-inch nectary. He deduced: This flower's nectar is at the bottom of a long tube; all orchids with nectaries are pollinated by insects; therefore, an insect must exist in the vicinity with a 14-inch tongue. Experts on insects scoffed, but 40 years later they discovered the moth that fit Darwin's deduction. (It was given the scientific name *praedicta:* the moth that was predicted.)

More recently, in 1988, a very similar kind of deduction was made by a young marine biologist, Cindy Van Dover of the Woods Hole Oceanographic Institute, who was studying blind shrimps from the deep ocean floor.

Although they have no eyes, Van Dover noticed the shrimps have light receptors with reflective backings in their shell and primitive optic nerves. Rhodopsin, the same chemical found in retinas of eyes, was present. "These shrimp live at depths far too great for any sunlight to penetrate," Van Dover said, "but I reasoned that if they are capable of perceiving light, there should be some kind of light down there for them to perceive. So I suggested to John Delany [head of a University of Washington submarine expedition] that he look for the light when he visited a similar kind of habitat in the Pacific."

Using the research submarine *Alvin* and the special

camera that had photographed the wrecked *Titanic*, Delany's team discovered a mysterious glow in a scalding jet of water more than one mile beneath the surface of the Pacific Ocean. There was indeed a dim light source in the ocean's "darkest" depths. But without Cindy Van Dover's deduction from the shrimp's primitive light receptors, says Delany, they would not have known "what to look for, and we would not have found it."

See also INDUCTION; ORCHIDS, DARWIN'S STUDY OF; THEORY, SCIENTIFIC.

DEGENERATION THEORY
Evolution To "Lower" State

A peculiar byway of evolutionary theory during the 1870s held that life forms could evolve in three directions. Species could remain relatively unchanged over time; they could become more elaborate and complex; or they could become simpler and "degenerate."

German zoologist Anton Dohrn was a strong champion of the idea; his English friend E. Ray Lankester, director of the British Museum of Natural History, dedicated a little book on the subject to him in 1880. Their prime examples of degeneration were microorganisms and parasites, which they thought had "lost" the structures to move about or capture their own food. Others applied the concept to more complex animals perceived to have become simpler. Whales and snakes, for example, were described by one authority as "degenerate quadrupeds."

Charles Darwin's protege Sir John Lubbock, who made pioneering studies of social insects, was intrigued by ants that kept aphids as "slaves." The ants had "lost" some of their mouth parts and were now totally dependent on the aphids for the sugary food they produced. A similar fate, he warned, was in store for degenerate human slaveowners, who would soon be unable to feed or care for themselves.

Lankester did not hesitate to apply the degeneration principle across the board, from pond organisms to human history. "At one time," he wrote, "it was a favourite doctrine that [all] the savage races of mankind were degenerate descendants of the higher and civilised races." Although that idea "has been justly discarded . . . it yet appears that degeneration has a very large share in the explanation of the condition of the most barbarous races, such as the Fuegians, the Bushmen, and even the Australians."

There was a lesson here, Lankester thought, for "the white races of Europe." Because of "unreasoning optimism," Europeans believed themselves destined for continued progress. But, he warned:

It is well to remember that we are subject to the general laws of evolution, and are as likely to degenerate as to progress . . . It is possible for us—just as the Ascidian throws away its tail and its eye and sinks into a quiescent state of inferiority—to reject the good gift of reason with which every child is born, and to degenerate into a contented life of material enjoyment accompanied by ignorance and superstition.

Lankester's antidote to human degeneration was to gain knowledge of man's place in nature, so that "We shall be able by the light of the past to guide ourselves in the future . . . [to know] that which makes for, and that which makes against, the progress of the race. The full and earnest cultivation of Science—the Knowledge of Causes—[is the best] protection of our race—even of this English branch of it—from relapse and degeneration."

From a careful zoological study of microscopic invertebrates and parasites, Lankester leapt to saving the "English race." Well-meaning though he was, his hysterical, quasi-religious faith in science as salvation, and his fear of human degeneration was to have many sinister echoes. Restrictive immigration laws and sterilization of the "unfit" in America was founded on a similar belief; it was also Nazi Germany's rationale for the extermination of "inferior races."

See also BUFFON, COMPTE DE.

For further information:
Lankester, E. Ray. *Degeneration: A Chapter in Darwinism*. London: Macmillan, 1880.

"DELICATE ARRANGEMENT"
The Darwin–Wallace "Joint" Publication

Although Charles Darwin is usually credited with originating the theory of organic evolution by natural selection, he shares the discovery with a man few educated people could name.

Alfred Russel Wallace, a naturalist 14 years Darwin's junior, was ready to publish a crisp summary of the theory in 1857, before *Origin of Species* (1859) was written. How they came to publish together—without Wallace's knowledge or prior consent—has become known as "the delicate arrangement."

Darwin had been working in secret for years, confiding his theory to only a few close friends. "Outsiders" (including his sometime correspondent Wallace) knew only that he was interested in the relationship between varieties and species.

As a field naturalist in the Malay archipelago, Wallace studied the distribution of plants and animals and was struck by competition for resources among the native tribal populations. Having read the same book (Malthus, *On Population*, 1798) that inspired Darwin, he conceived the idea of evolution by natural selection during a malarial fever. Later, he wrote a well-thought-out essay on the subject (the "Sarawak Law," 1855) and mailed it to Darwin. If Darwin thought it worthy of publication, Wallace requested he pass it on to the influential geologist Sir Charles Lyell.

Lyell knew that Wallace was close to an evolutionary theory and had been warning Darwin to publish soon or lose the chance to be first. Receipt of Wallace's paper threw Darwin into a panic. He wrote his mentor Lyell, "Your words have come true with a vengeance . . . that I should be forestalled . . . So all my originality . . . [is] smashed . . . I never saw a more striking coincidence. If Wallace had my MS sketch written out in 1842, he could not have made a better short abstract! Even his terms now stand as heads of my chapters."

It could not have come at a worse time. One of his children had recently died and another was sick with scarlet fever. Darwin appealed to his friends Lyell and the botanist Sir Joseph Hooker to advise him—or rescue him—and they did.

They never considered consulting Wallace. Mail service between England and the Moluccas took several months each way; besides, Darwin, Lyell and Hooker were prosperous gentlemen-scientists, and Wallace a penurious beetle collector with no social connections. So Lyell and Hooker "arranged" the matter to protect their friend, whom they knew had labored for more than 20 years on the same theory Wallace had just proposed.

Darwin, who was usually generous and fair, admitted he was filled with "trumpery feelings." He could not bear the idea of being scooped by another naturalist publishing "his" theory first. Yet he would rather burn his book, he said, than for anyone to believe he stole his ideas from Wallace or behaved in a "paltry spirit." Seeking moral absolution, he asked his friends to handle the problem and agreed to accept whatever resolution they thought proper.

Since Sir Joseph Hooker was England's leading botanist and Sir Charles Lyell its most eminent geologist, the pair had great influence in scientific circles. In 1858, they persuaded the Linnean Society to publish extracts from two letters by Darwin (1842, 1844), in which he had sketched his views to friends, together with Wallace's essay. Darwin and Wallace were to share credit as codiscoverers of the theory of evolution by natural selection. Although Wallace's was the only real paper submitted, the fragmentary Darwin letters were published in first position by Hooker and Lyell.

In their cover letter "communicating" the manuscripts to the Society, they implied that Wallace and Darwin had agreed to joint publication. In fact, Wallace knew nothing of the "arrangement" until a year later; he had not even known that Darwin was working on an evolutionary theory.

The phrase "delicate arrangement" comes from a passage by Thomas Huxley's son Leonard in his *Life and Letters of Sir Joseph Hooker* (1918):

> Wallace's paper had come like a bolt from the blue . . . Yet . . . when this delicate situation had been arranged [Darwin wrote Hooker], "You must let me once again tell you how deeply I feel your generous kindness and Lyell's on this occasion, but in truth it shames me."

Darwin then went furiously to work on *The Origin of Species* and completed it in 13 months, after almost two decades of putting off the writing. Years later, in his *Autobiography* (1876), Darwin claimed he had "cared very little whether men attributed most originality to me or Wallace"—a statement that still jars even Darwin's staunchest admirers.

Darwin was apprehensive about Wallace's reaction on his return from Malaysia and was greatly relieved to discover his "noble and generous" disposition. (Some historians still think he acquiesced too willingly to becoming "a footnote to history.")

Wallace modestly stated his work on the problem was short compared with Darwin's decades of painstaking groundwork and never questioned his priority. ("I shall always maintain it is actually yours and yours only.") However, after Darwin's death, Wallace admitted he had "no idea" his paper had thrown the senior naturalist into such a panic.

To the end of his long life, Wallace insisted being first to publish is meaningless if an idea makes no impact, and that "my paper would never have convinced anybody." In *The Wonderful Century* (1904), he recalled:

> The whole literary and scientific worlds were violently opposed to all such theories . . . [but] the greatness and completeness of Darwin's work [caused a] vast change in educated public opinion . . . Probably so complete a [reversal] on a question of such vast difficulty and complexity, was never before effected in so short a time. [Establishing acceptance of evolution by natural selection] . . . places the name of Darwin on a level with that of Newton . . .

Journalist Arnold C. Brackman published a book titled *A Delicate Arrangement* (1980), which infuriated scholars by claiming Darwin actually stole a good part of his theory from Wallace—an accusation that went far beyond the evidence. Brackman's scholarship was attacked as sloppy, his argument inconclusive, and his strident charges of conspiracy and cover-up unconvincing.

Nevertheless, his attempt to restore some glory to Wallace was long overdue. And, however one interprets the machinations of Darwin's friends, the "delicate arrangement"—glorified in dozens of books as a classic example of unselfish collaboration—was certainly not one of the brighter episodes in the history of science.

See also "CRADLE OF MANKIND"; LINNEAN SOCIETY; SARAWAK LAW; WALLACE, ALFRED RUSSEL.

For further information:

Bedall, Barbara. "Wallace, Darwin and the Theory of Natural Selection." *Journal of the History of Biology* 1 (1968): 261–323.

Brackman, Arnold C. *A Delicate Arrangement: The Strange Case of Charles Darwin and Alfred Russel Wallace.* New York: Times Books, 1980.

DELUGE

See NOAH'S FLOOD.

DE MAILLET, BENOIT (1656–1738)
Cosmological Theorist

Benoit de Maillet thought planet Earth had developed by natural processes, still observable today, and considered fossils "libraries" of ancient plants and creatures that "exist no more." He even had a rough idea about the succession of strata, the kinship of man and apes and the origin of life from tiny organic atoms like those seen under the new microscopes. However, his fumbling attempts were also mixed in with stories of mermaids and other fantastic myths. And though he proposed the "development" of species, his notions of evolution were crude. He saw flying fish, for example, as being well on their way to becoming birds.

Yet he insisted that if God were a watchmaker (in the favorite metaphor of the period), he did not have to continually intervene to keep his machines running. Instead, de Maillet wrote, the Creator "had skill enough to make a clock so curiously, that by the Disorder which Time should produce in her Parts and Movements; there should be new Wheels and Springs formed out of Pieces, which had been worn and broken."

His book was translated from the French and widely read in English. It was entitled *Telliamed: Or Discourses Between an Indian Philosopher and a French Missionary on the Diminution of the Sea, the Formation of the Earth, the Origin of Men and Animals, etc.* (1750). Of course, there was no Telliamed. Philosophers whose theories challenged church teachings often attributed them to fictional Oriental sages. "Telliamed" is "de Maillet" spelled backwards.

DESCARTES, RENE

See CARTESIAN DUALITY.

(THE) DESCENT OF MAN (1871)
Darwin On Human Origins

Only a tiny hint about human evolution appeared in Charles Darwin's *Origin of Species* (1859): "Light will be thrown on the origin of man and his history." Although Darwin had often thought about man, for 30 years his published writing stuck to animals and plants. With the *Descent of Man*, he finally let the other shoe fall.

By this time, evolution had won over the majority of botanists and zoologists from the older idea that each species appeared instantly and separately. Others, more bold than he, had published pioneering works on human evolution in the wake of the *Origin*. Thomas Huxley's *Man's Place in Nature* (1863) and the German Ernst Haeckel's *Natural History of Creation* (1866) were among the most influential.

But the public was eager for Darwin's own version of human origins from ape-like creatures, though he fully expected the book would "be denounced by some as highly irreligious." But why, he wondered, is human evolution from "some lower form" any more shocking or incredible than the development of individuals from sperm and eggs?

The fact that man has "risen to the very summit of the organic scale" instead of having been placed there from the beginning, Darwin wrote, "may give him hope for a still higher destiny in the distant future." But hopes and fears aside, Darwin concluded that the time had come to acknowledge that "with all his noble qualities . . . god-like intellect [and] exalted powers—Man still bears in his bodily frame the indelible stamp of his lowly origin."

Three great "groups of facts" could no longer be denied: similarities in structure (and, we now know, biochemistry) among members of the same groups, patterns of geographical distribution and the worldwide succession of one group by another in the fossil record.

> It is incredible that all these facts should speak falsely. He who is not content to look, like a savage, at the phenomena of nature as disconnected, cannot any longer believe that man is the work of a separate act of creation.

Darwin noted that man is not descended from any existing monkey or ape but from ancestors that would be recognizably apish. They were hairy, social, had pointy ears and the males had large canine teeth.

Darwin thought human ancestors had dagger-like canines—similar to those found today in male gorillas and baboons—which became reduced in size as stone tools and weapons were mastered. That idea, found in *Descent of Man*, inspired bitter controversies a century later. A distorted view of baboon social behavior, based on the primacy of males with large canines, was promoted and widely accepted as the model for early man. [See BABOONS; HUNTING HYPOTHESIS.]

The Descent of Man and Selection in Relation to Sex (its full title) was actually two books in one. The first argues humans evolved from ape-like ancestors. Aside from a Neandertal skullcap, fossils of early hominids were not yet known to science. Darwin's "fossil free" conclusions about human origins were drawn entirely from embryology, comparative anatomy, animal behavior and anthropology.

Darwin also published his famous guess—correct, it now appears—that Africa was the cradle of mankind. He reasoned that it was a center of radiation, because our closest surviving relatives, chimps and gorillas, still live there. Over the following century, fossil hunters scoured Europe, Asia and North America for evidence of the earliest hominids in vain. Accumulated fossils point to Africa, as Darwin had thought.

The *Descent*'s second part deals entirely with Darwin's theory of "sexual selection": competition for mates as an important factor in evolution. In the final chapter, he draws together both themes in discussing the role of sexual selection in shaping the local variations ("races") of mankind.

Why did Darwin combine these two seemingly disparate topics in one work? In his attempt to identify uniquely human adaptations, he became baffled. Certain traits that humans do not share with apes, such as relative hairlessness (more pronounced in women) and musical ability, appear to have no adaptive advantage whatsoever. Then there was the puzzle of human racial differences. Could adaptation explain the evolution of so many different hair textures, nose shapes, skin colors, and bodily proportions among the world's peoples? He had to admit such traits might be unnecessary or neutral in making humans better fitted to survive.

Therefore, he took a long look at animal species that evolved traits that were "unnecessary" or even detrimental in relationship to their environments. Peacock's tails, for instance, only get in the way of feeding or escaping enemies. But they help the cock compete in a different arena, attracting the opposite sex. Similarly, Darwin thought, smoother skin in human females might have evolved to enhance attractiveness to males. Many other disparate characteristics, including musical ability, hair texture, or skin pigmentation, might be due to sexual selection.

Darwin sent questionnaires to missionaries and travelers all over the world, requesting information about ideals of beauty among various tribes. He was pleased to find that local standards of attractiveness vary greatly and often are exaggerations of the tribe's special characteristics. To complete his case, Darwin presented these anthropological inquiries alongside a huge compilation of evidence for sexual selection in antelopes, monkeys, birds and other animals.

The Descent of Man has inspired an astonishing spectrum of responses over the years, from reasoned to outlandish. In Darwin's day, one cartoonist depicted man "descending" an evolutionary staircase toward the worms!

Subsequent authors not only explored and dissented from the book's ideas, but played endlessly with the wording of its title. Theologian Henry Drummond wrote a popular work, *The Ascent of Man* (1894), in which "upward" biological progress is only the prelude to spiritual evolution. Much more recently, Elaine Morgan's *The Descent of Woman* (1972) challenged the male-oriented view of evolution and offered a feminist corrective.

See also AFRICAN GENESIS; MAN'S PLACE IN NATURE; SEXUAL SELECTION.

For further information:
Campbell, Bernard, ed. *Sexual Selection and the Descent of Man, 1871–1971*. Chicago: Aldine, 1972.
Darwin, Charles. *The Descent of Man and Selection in Relation to Sex*. London: John Murray, 1871.

DEVONIAN PERIOD
Age of Fishes

The Devonian period, named for Devon, England where it was first defined in thick formations of sedimentary rock, dates from about 306-to-408 million years ago.

It is in the Upper Paleozoic Era and often nicknamed "The Age of Fishes" because of its abundant species of jawless and bony fish, including some of the largest predatory fish that ever lived. But it is also notable as the time when colonization of the land began.

The climate was cool, and the first forests appeared along with the first winged insects. Toward the end of the Devonian, lungfish scrambled about the shoreline on stiffened fins, and some became the ancestors of amphibians.

Biologist Alfred Romer thought the reason these creatures first ventured across mud flats was not to "colonize the land," but simply to look for more water, just as lungfish and catfish do nowadays during droughts.

DE VRIES, HUGO (1848–1935)
Instantaneous Speciation

Hugo de Vries is remembered in the history of biology as being the first of three rediscoverers of Gregor Mendel's long-neglected paper on plant hybrids in the spring of 1900. With hindsight, we see de Vries as a midwife to the birth of genetic science, helping to usher in the modern synthesis of Darwinian and Mendelian thought, the dual foundation of evolutionary biology.

De Vries himself would have been appalled and outraged, for he considered his recognition of Mendel's work only a minor incident in an illustrious career. In the first two decades of the 20th century, he gained fame for reshaping science's view of the origin of species: tossing Darwin in the dustheap while giving a gracious nod to Mendel's ghost. It was his mutationism—an anti-gradualist theory of evolution by large discontinuous jumps—that he thought would ensure his enduring fame.

De Vries's story begins in 1886 when he found, in a field near Amsterdam, large numbers of the evening primrose growing wild. Among the plants were two new varieties that differed distinctly from the normal form of the species: "Both [these new species] come perfectly true from seeds. They differ from [the parental stock] in numerous characters, and are therefore to be considered as new elementary species."

There it was, in a field in the Netherlands, what no one steeped in the Darwinian tradition expected—brand new species! And unlike what the Darwinian gradualists had envisioned, they did not grade away imperceptibly from their progenitors, nor was any selection necessary to produce them. In a single step, in one generation, the new species had come into existence and, when self-fertilized, they remained absolutely constant.

During the 1890s de Vries conducted a prodigious number of plant crosses, not only between species but between varieties of the same species. Sometimes an apparently new species was created as he had hoped, but it was a rare occurrence. He kept careful records of his crosses and noticed that certain ratios of traits (such as the classic three-to-one) in hybrid generations came up again and again, but the results made no sense to him. Then, in early 1900, Professor Martinus Willem Beijerinck (1851–1931) of Delft sent de Vries a copy of Mendel's paper with the note: "I know that you are studying hybrids, so perhaps the enclosed reprint by a certain Mendel which I happen to possess, is still of some interest to you."

In a flash, all of the notebooks full of numerical ratios that de Vries had recorded over the past decade of crossbreeding experiments fell into place, and he immediately set about writing not one but three Mendelian papers.

But de Vries's enthusiasm soon evaporated when he saw that the simple, classic Mendelian ratios only appeared in special cases. Since the complexities of Mendelian theory had not yet been worked out, de Vries concluded that Mendel's laws had nothing to do with the origin of species. "It becomes more and more clear to me," he wrote Mendel's champion in England William Bateson in 1901, "that Mendelism is an exception to the general rules of crossing. It is in no way *the* rule!"

Between 1901 and 1903, de Vries published the two volumes of his magnum opus *Mutationstheorie* (*Mutation Theory*) in which he spelled out his conclusions. "The new species originates suddenly," he wrote, "it is produced by the existing one without any visible preparation and without transition."

For approximately two decades de Vries's theory eclipsed Darwin's and, during this period, most biologists were confident that de Vries had laid Darwin to rest. Erik Nordenskiold in his monumental *History of Biology*, written in the early 1920s, was able to say that Darwin's theory "has long been rejected in its most vital points." And, as late as 1932, Clarence Ayres, in his biography of T. H. Huxley, said: "All of Darwin's 'particular views' have gone down wind: variation, survival of the fittest, natural selection, sexual selection, and all the rest. Darwin is very nearly, if not quite, as outmoded as Lamarck."

De Vries devoted the remainder of his life to proving his theory of speciation by mutation, and it was through his experiments on plant hybridization—experiments designed to elucidate the formation of new species—that he was led to Mendel's paper.

De Vries's hostility to Mendel grew in later years. In his 1907 book *Plant Breeding*, he does not even mention Mendel, and he refused to sign a petition for the erection of a Mendel memorial in Brunn in 1908. In 1922 de Vries declined an invitation to attend the celebration of the centennial of Mendel's birth: "To my regret I cannot accede to your request. I just don't understand why the academy would be so interested in the Mendel celebrations. The honoring of Mendel is a matter of fashion which everyone, also those without much understanding, can share; the fashion is bound to disappear."

After recently restudying the de Vries–Mendel relationship, Dutch historian of science Onne Jeijer commented, "I cannot free myself from the impression that de Vries went a little out of control somewhere at the end of February and the beginning of March 1900 when he received Beijerinck's copy of Mendel's paper. Exactly what happened is not clear, but de Vries's temporary infatuation with Mendel's work does not read true to de Vries's character."

De Vries's theory of speciation by sudden jumps or mutations was finally eclipsed as Mendelian principles proved to be the soundest basis for understanding evolutionary change. By the 1930s, most of the early difficulties in applying them to more complex results had been ironed out. By then, it was clear that de Vries's "new species" were not new at all but only freaks created by abnormal chromosomal pairings. (Darwin himself had once warned that nature often presented ambiguities to the scientist, and that "She will lie to you if she can.")

Today, scarcely anyone remembers either that Darwin's theory fell on hard times around the end of the 19th century or that de Vries's mutation theory dominated evolutionary thinking during the first two decades of the 20th century. Instead, de Vries is enshrined in textbooks as one of the three rediscoverers of Mendel's work. Few recall that, after some momentary enthusiasm for Mendel, de Vries became convinced that the Austrian monk's work would take a definite back seat to his own monumental discovery of those two "new species" that had sprung up overnight in that field near Amsterdam in 1886.

For further information:

Dunn, L. C. *A Short History of Genetics.* New York: McGraw-Hill, 1965.

Mayr, Ernst. *The Growth of Biological Thought.* Cambridge: Harvard University Press, 1982.

Meijer, O. G. "Hugo de Vries no Mendelian?" *Annals of Science* 42 (1985): 189–232.

Olby, R. C. *Origin of Mendelism.* Chicago, University of Chicago Press, 1985.

Provine, W. B. *The Origins of Theoretical Population Genetics.* Chicago: University of Chicago Press, 1971.

DEWEY, JOHN (1859–1952)
American Philosopher, Educator

When the Darwinian revolution shook up Western thought during the latter half of the 19th century, many perceived the uproar as a conflict between science and religion. (Some still do.) Shortly after 1900, American philosopher John Dewey realized that the most profound rumblings were coming from within science itself. Long before the physicist Werner Heisenberg had made it respectable in science, Charles Darwin had introduced a principle of uncertainty.

In a famous lecture at Columbia University in 1906,

Dewey recalled that the history of science had been a search for certainties and fixed laws that represent "the truth" about nature. Species were thought of as both fixed forms and the divinely ordained patterns that caused them to exist. That conception of species, said Dewey, "was the central principle of knowledge as well as of nature. Upon it rested the logic of science."

> . . . for two thousand years . . . the familiar furniture of the mind, rested on the assumption of the superiority of the fixed and final . . . upon treating change and origin as signs of defect and unreality. In laying hands upon the sacred ark of absolute permanency, in treating the forms [species] that had been regarded as types of fixity and perfection as originating and passing away [becoming extinct], the "Origin of Species" introduced a mode of thinking that in the end was bound to transform the logic of knowledge, and hence the treatment of morals, politics, and religion.

It was Darwin who prepared the way for the 20th century's acceptance of Albert Einstein's relativity. Years later, agreeing with Dewey's assessment of the Darwinian revolution, English philosopher Bertrand Russell wrote: "Darwin threw down a challenge to the old rigidities . . . It seemed that everything, instead of being so or not so, as in the logic books, was only more so or less so. And in this mush of compromise all the old splendid certainties dissolved."

John Dewey had also pointed out that Darwin's way of looking at nature was something new. It was pluralistic, seeking different explanations with various kinds of questions, including an historical approach, which considered nature over long periods of time. (History previously had not been part of the scientific enterprise.)

Darwin was interested in function, how specific things *worked* in the living world: whether seeds could sprout after soaking for weeks in salt water, or how orchids attracted pollinating insects. He asked nature thousands of limited questions with his experiments. Darwin was not concerned with who made the world or why, said Dewey, but with what kind of a world is it: How does it work? Darwin was unconcerned with proving any "absolute truth"; he thought the best theory was the one that could link up the most facts. He was delighted with an obscure reviewer (April, 1861) who was:

> one of the very few who see that the change of species cannot be directly proved, and that the doctrine must sink or swim according as it groups and explains phenomena. It is really curious how few judge it in this way, which is clearly the right way.

Even today, the deepest argument between religious fundamentalists and scientists is not about conflicting accounts of how the world (or species) began, but over the nature of certainty. Can a creationist be sure of his answers? Certainly, since he's also convinced there is only one "right" interpretation of scripture (his own). When he asks a scientist directly if Darwin's theory could be disproved tomorrow by a new theory, the scientist must answer yes. To the fundamentalist, this admission that Darwin's truth is relative and less than eternal "proves" its worthlessness.

Darwin also broke down the old idea of a dualism between "nature" and the "artificial" world of man and his cultural traditions. Following up on Dewey's ideas, philosophers pointed out that Darwin sketched "a nature with man in it." Human beings had "a place in nature" (as in the title of Thomas Huxley's book) and were not alien to it. Culture, language, even philosphy and religion are as natural as trees and stones. Understanding that, even our own awareness becomes fodder for the evolutionary process.

Many supposed Darwinian evolution established a new law that must be followed: That we must try to conform to nature's "goals" of evolution for "improvement of the species." John Dewey thought Darwinism implied the opposite. Because there are no predetermined goals or necessary progress in Darwinian evolution, Dewey saw new possibilities for human freedom. "Conformity with laws," no matter how noble and perfecting, was still a belief in determinism—a new way of saying man had no real influence over his destiny. Part of the Darwinian revolution, Dewey believed, was the acknowledgement that we do have a newfound freedom, and consequent responsiblity, to understand and meet challenges that have direct bearing on the future of our species.

Dewey realized that "old ideas give way slowly; for they are more than abstract logical forms . . . [they are] deeply engrained attitudes." Questions in science cannot be answered in the terms they are posed:

> In fact intellectual progress usually occurs through sheer abandonment of questions . . . We do not solve them: we get over them . . . Old questions are solved by disappearing, evaporating, while new questions . . . take their place. The greatest dissolvant in contemporary thought of old questions, the greatest precipitant of new methods, new intentions, new problems, is the one effected by the scientific revolution that found its climax in the *Origin of Species.*

See also ESSENTIALISM; SPECIES, CONCEPT OF; THEORY, SCIENTIFIC.

For further information:

Appleman, Philip, ed. *Darwin: A Norton Critical Edition.* New York: Norton, 1970.

Dewey, John. *The Influence of Darwin on Philosophy and Other Essays on Contemporary Thought.* New York: Holt, 1910.

DIFFERENTIAL REPRODUCTION
Populational Concept of Fitness

Science redefined natural selection after the rise of population genetics in the 1930s. No longer was competition imagined as a struggle between the "fit" (strongest, best) and "unfit" (weakest, worst), but simply as success in leaving progeny. Darwin had that concept in mind, but rigorous populational thinking had to await the rediscovery of Mendelian genetics 20 years after Darwin's death.

Natural selection was redefined in terms of "differential reproduction." The "fittest" individuals are simply those that perpetuate the highest frequency of their genes in descendant populations. Nevertheless, though gene frequencies are amenable to mathematical models, it must not be forgotten, as Ernst Mayr puts it, that "it is the potentially reproducing individual, and not the gene, that is the target of selection."

Individuals may maximize passing on their genes in a number of ways besides simply being progenitors. Complex social behavior in many kinds of creatures may involve close kin who are "helpers" with eggs; "altruistic" behavior in defending a group of close relatives (bees, mole rats); killing neighbor's infants (chimpanzees); or destroying competitor's eggs (swallows, cuckoos).

But even success in breeding and reproducing the genes of individuals and kin groups cannot assure ultimate survival. Most of the new generation may be wiped out in a natural disaster or succumb to disease at a vulnerable stage of the life cycle.

See also FITNESS; "SURVIVAL OF THE FITTEST."

DIGIT
Gorilla Martyr

Dian Fossey first encountered the young male gorilla she called Digit (because of a twisted middle finger) in 1967, when he was about five years old. He was always the first member of his group to come forward to investigate human visitors and liked meeting newcomers brought by Fossey. She enjoyed his company regularly for a decade, during which she watched him mature into the leader of his own family group.

When the Rwandan government requested a gorilla picture to promote tourism, Fossey provided a shot of the gentle and curious Digit. His portrait appeared on large color posters, distributed throughout the world, with the caption: COME AND SEE ME IN RWANDA! But Fossey "could not help feeling that our privacy was on the verge of being invaded by the public."

It was not tourists, but African poachers who destroyed Fossey's beloved friend on New Year's Eve day in 1977. His head and hands had been hacked off, sending Fossey into a state of shock, anger and "withdrawal into an insulated part of myself" from which she never recovered. "Digit gave his life," Fossey wrote, "so his family group might survive for the perpetuation of his kind" though he never lived to see the only infant he sired.

As a tribute to Digit's memory, an organization was founded to raise money for antipoaching and conservation efforts for the magnificent apes: The Digit Fund.

See also FOSSEY, DIAN; GORILLAS.

For further information:
Fossey, Dian. *Gorillas in the Mist*. Boston: Houghton Mifflin, 1983.

"DIMA"*
Celebrated Infant Mammoth

A complete frozen carcass of a small baby mammoth was found by Soviet miners on June 23, 1977, as they thawed patches of frozen ground (permafrost) in Siberia, searching for gold. Mammoth remains had been found in this frigid region for hundreds of years, but this one, a six- or seven-month-old female was complete and exceptionally well preserved.

Found near a tributary of the River Kolyma, the ancient infant was immediately flown to Leningrad for study by Professor Nikolai Vereshchagin. Little "Dima" proved to be very similar to a modern African or Indian elephant, except that she had long reddish-brown hair and very small ears—just as Paleolithic cave artists had recorded.

A special traveling case was designed and built for

MAMMOTH-SHAPED TRAVELING CASE was built to transport "Dima," the mummified baby mammoth found in the Siberian tundra in 1977. Two years later, Soviet scientist Dr. Andrei Kapitsa accompanied her to a special exhibition in London, where she was kept under British police guard.

Dima's remains, which were sent round the world during the late 1970s. From Moscow to London, she proved a great attraction; museum-goers formed long lines and waited patiently to see her.

Dima was one of the first frozen mammoths to be investigated with modern biochemical techniques. Electron microscopes revealed her white and red blood cells, in perfect shape, and her albumen was still genetically active after thousands of years in the natural deep-freeze. When tested, her blood serum showed close kinship to that of modern Indian and African elephants.

While these results came as no surprise, it was a dramatic confirmation that biochemical techniques told the same story of age and relationship to modern elephants that had been determined from bones, anatomy and Ice Age artists' drawings. Lab scientists were ecstatic; it isn't every day they get to test the blood of an animal that disappeared from the face of the Earth 10,000 years ago.

DINOSAUR
Ruling Reptiles of the Mesozoic

Dinosaurs were a successful and varied group of ancient reptiles, ranging in size from 80 tons to creatures no larger than a chicken. Thousands of different species filled all the available niches or habitats: browsers, swamp and river dwellers, herbivores, two-legged runners and the largest, most fearsome meat-eaters ever to walk the Earth.

All told, they ruled the planet for about 140 million years before their comparatively sudden worldwide extinction. Their heyday, comprising the Triassic, Jurassic and Cretaceous periods, is known as the Mesozoic (middle life forms) Era. Despite such Hollywood fantasies as *One Million B.C.*, dinosaurs and men never shared the Earth. By the time early humans had appeared, the great reptiles had been gone for many millions of years.

Although descended from common ancestors—the small, swift, bipedal Pseudosuchins—dinosaurs formed two great groups (orders), which were anatomically distinct. Based on different structures of the pelvis, the first are the *Saurischia* (lizard-hipped) dinosaurs, while the second are *Ornithiscia* (bird-hipped). Two famous saurischian species are the long-necked plant-eater *Apatosaurus* (formerly called *Brontosaurus*) and the bipedal meat-eater *Tyrannosaurus*. Among the ornithiscians were horned and armored dinosaurs (including the familiar *Stegosaurus* and *Triceratops*) and the *Iguanodon*, which was the first dinosaur known to science.

Certain prehistoric reptiles that are popularly included with dinosaurs are not dinosaurs at all, but are classified with other groups. For instance, sail-backed reptiles (pelycosaurs), dolphin-like reptiles (ichthyosaurs), flying reptiles (pterosaurs), and paddle-flippered plesiosaurs are all *not* members of the dinosaur group.

Everything we know about dinosaurs has been discovered only since about 1830, as great masses of fossils were collected during the 19th and early 20th century. Since the 1960s, there have been exciting new interpretations of the material; some scientists have argued that the dinosaurs were quick-moving, warm-blooded, intelligent and social animals. [See DINOSAUR HERESIES.] And since the 1970s, so many new fossils and new species have come to light that knowledge of dinosaurs has more than doubled in two decades.

For further information:

Colbert, Edwin. *Dinosaurs: An Illustrated History.* Maplewood, N.J.: Hammond, 1983.

Glut, Donald. *The New Dinosaur Dictionary.* Secaucus, N.J.: Citadel Press, 1982.

Norman, David. *The Illustrated Encyclopedia of Dinosaurs.* New York: Crescent Books, 1985.

Paul, Gregory. *Predatory Dinosaurs of the World.* New York: Simon & Schuster, 1988.

Wallace, Joseph. *The Rise and Fall of the Dinosaurs.* New York: W. N. Smith, 1987.

DINOSAURABILIA
Saurians in Pop Culture

Two-hundred years ago no one had even heard of a dinosaur. Today children play with dinosaur toys, visit Dinosaur National Monument, even eat dinosaur-shaped breakfast cereals. During the 1980s a veritable explosion of dinosaur merchandising swept America, replacing Teddy bears and Disney characters as the most ubiquitous image in popular culture.

The term "Dinosaurabilia" was coined in 1980 by Dean Hannotte, a Manhattan computer expert whose passion is collecting anything connected with dinosaurs. Hannotte has assembled thousands of rare items, ranging from original paintings by Charles R. Knight (the father of dinosaur art) to early paleontological books, cereal premiums, toys, night lights, stereopticon slides, inflatables, jewelry, cards, model kits, neckties, pulp magazines, soaps and comic books.

Among the quality collectibles are the line of Sinclair Oil Company premiums, glassware and stamp books issued at gas stations in the 1930s and 1940s with the company's "Dino" logo. Rare illustrated books by artists such as Knight and the Czech painter Zdenek Burian are also coveted, as are promotional materials from early dinosaur movies, including the original *King Kong* (1933), with its classic dinosaur scenes based on Knight's illustrations. The most authoritative account of the origin and evolution of dinosaurabilia is *The Dinosaur Scrapbook* (1980), written by Donald F. Glut, one of the world's champion collectors of Dinosaurabilia.

See also DINOSAURS, RESTORATION; KNIGHT, CHARLES R.

For further information:
Glut, Donald F. *The Dinosaur Scrapbook.* Secaucus, N.J.: Citadel, 1980.

DINOSAUR EGGS
Clues to Fossilized Behavior

Dinosaur eggs were first discovered by Roy Chapman Andrews's expedition to Mongolia's Gobi Desert in 1923. When George Olson, a paleontology assistant, reported he had found fossil eggs weathering out of a cliff face, Andrews thought they must be sandstone concretions. Dinosaur eggs did not at first occur to him, as none had ever before been known.

Eggs they were, and very well preserved—13 of them, arranged in concentric circles, as they had been deposited in the ancient reptile's nest. They were eventually identified as those of *Protoceratops* and are still on exhibit at the American Museum of Natural History.

Later, the expedition gathered many more eggs from several species. Back in America, they caused a popular sensation. The following year, Andrews decided to auction off an egg at a fund-raising event for the expedition. One wealthy bidder paid $5,000, then donated the fossil to Colgate University's museum.

When the news reached the Chinese, they concluded each and every dinosaur egg must be worth $5,000—and the expedition had collected dozens. Their worst suspicions were confirmed: Andrews was no seeker of knowledge, but a greedy treasure hunter plundering the country's riches. Eventually, the flap over the high-priced dinosaur egg and other misunderstandings led to the American Museum of Natural History being barred from further fieldwork.

More than a half century later, dinosaur eggs made news again. Paleontologists Jack Horner and Robert Makela, working in Teton County, Montana, found so many dinosaur eggs that they named their sites Egg Island and Egg Mountain. Among the litter of bone fragments and eggshells, they found complete nests. The density of eggs suggested these creatures—hypsilophodonts—bred in colonies.

Each nest was separated by about five feet on each side, the length of an adult hypsilophodont. And there were several layers of egg remains, suggesting the tendency of a group to nest for many years in the same area.

In 1988, Horner and David Weishampel of the Johns Hopkins medical school used a CAT scanner to examine some of the unhatched dinosaur eggs. They photographed a fully formed fetus of a previously unknown species, which they named *Orodromeus makelai.* The development of its bones indicated the little saurian would have been able to function immediately after hatching.

By contrast, other embryonic dinosaurs (*Maiasaura*) Horner had found would have been born helpless, needing lots of parental care. (*Maiasaura* means good mother lizard.) Bones of young and juveniles were found all over the nest sites, unlike those of *Orodromeus,* which might have promptly taken off without a period of prolonged parenting.

Horner's interpretations of the eggs and patterns of nests suggests great behavioral and developmental differences between dinosaur species right from birth. In the case of maiasaurs, he has inferred a level of social behavior and prolonged juvenile dependency previously unsuspected in dinosaurs. These are exciting developments for paleontologists, as they open up new windows into dinosaur development and behavior. Dinosaur eggs have already hatched some intriguing new theories.

See also GOBI DESERT EXPEDITION.

DINOSAUR FOOTPRINTS

See BIRD, ROLAND THAXTER; FOSSIL FOOTPRINTS, PALUXY RIVER.

DINOSAUR HERESIES
New View of Ancient Creatures

During the past few years, a revolution led by younger paleontologists, including Jack Horner, Robert Bakker and his teacher John Ostrum, has sparked new interpretations of the dinosaurs.

In the older, traditional view, the great reptiles were seen as cold-blooded, solitary, sluggish creatures that could barely carry their great bulk, let alone engage in complex or intelligent behaviors. But the newer evidence and interpretations are sweeping these orthodoxies aside. Dinosaurs, it now appears, were warm-blooded, swift-moving social animals, with a much more complex range of behavior and intelligence than had been believed possible.

Much of the old stereotype was based on the mistaken belief that large dinosaurs needed two brains to control their vast bodies: a small one in the skull and another at the base of the tail. This was the view satirized by *Chicago Tribune* writer Bert Leston Taylor in 1912:

> You will observe by these remains
> The creature had two sets of brains . . .

BABY MAIASAUR showed up in X ray of fossil egg found by Jack Horner at Egg Mountain, Montana in 1987, the basis of this model from Museum of the Rockies. Unequipped to fend for themselves, hatchlings must have required a prolonged period of parental care.

Thus he could reason "A priori"
   As well as "A posteriori" . . .
If something slipped his forward mind
   'Twas rescued by the one behind.
And if in error he was caught
   He had a saving afterthought.

It has since been established that the sacral enlargement was not another brain at all, but a junction in the nervous system where many nerves are joined: an enlarged ganglia, common to many reptiles.

Dinosaurs could not really have been all that slow moving and inefficient. As Jack Horner puts it, "They dominated the earth for 140 million years. Let's stop asking why they failed and try to figure out why they *succeeded* so well." (Man-like creatures, in comparison, have been around for only three-to-four million years.)

Horner has excavated a hill he calls "Egg Mountain"—a veritable dinosaur rookery—in the Teton River country of Montana. One nest contains the bones of 11 baby dinosaurs; nearby were several other nests containing more than 400 eggs. This spectacular find is part of a new chain of evidence that dinosaurs were sociable, nested in colonies and showed parental care. These are traits that were thought to be nonreptilian—limited to mammals and birds. Horner suggests that dinosaurs were quite different from modern reptiles, that they were warm-blooded and perhaps as intelligent as many birds and mammals.

In 1969, John Ostrum announced at a paleontologist's convention his conclusion "that erect posture and locomotion probably are not possible without high metabolism and high uniform temperature," Parisian bone specialist Armand de Ricqles noticed at about the same time that the internal structure of dinosaur bones looked more like that of mammals

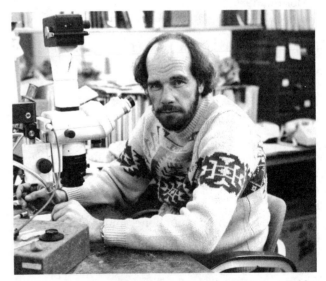

DINOSAUR "HERETIC" JACK HORNER, with Robert Bakker and other paleontologists, attacked old stereotypes of dim-witted, slow-moving "unsuccessful" dinosaurs. Their discoveries during the 1970s and 1980s point to complex social behavior, devoted parental care, possible warm-bloodedness, quick locomotion, bird-like courting rituals and remarkable adaptability.

than like lizards'. Dinosaur bone tissue is more porous, indicating a high rate of blood flow, like mammals, and bone growth patterns are more mammal-like.

Robert Bakker put together more evidence that pointed in the same direction during the early 1970s. Dinosaur behavior was unlizard-like, he argued, and their physiology was warm-blooded. (In this belief, he goes further than his teacher Ostrum.) His evidence is marshaled from comparative anatomy, geographic distribution and fossil ecology. For instance, the number of fossil dinosaur meat-eaters compared to herbivores is in the same ratio as that of mammal predators to prey in African field studies and totally different from the patterns in living reptile populations. Moreover, he maintains, the warm-blooded dinosaurs "never died out completely. One group still lives. We call them birds."

Now a curator at the University of Colorado Museum, Bakker believes orthodox paleontology restricted its own possibilities for learning about dinosaurs by classifying them as reptiles and viewing them as overgrown lizards. They should be grouped with birds, he believes, and their warm-bloodedness gave them the agility and stamina to dominate the Earth for an enormously long time. Debate still continues, but the number of "heretics" is growing among professional paleontologists.

See also ANGIOSPERMS, EVOLUTION OF; DINOSAUR EGGS.

For further information:
Bakker, Robert. *The Dinosaur Heresies.* New York: Wm. Morrow, 1986.
Desmond, Adrian. *The Hot-Blooded Dinosaurs.* New York: Dial Press, 1977.
Horner, Jack, and Gorman, James. *Digging Dinosaurs.* New York: Workman, 1988.

DINOSAUR NATIONAL MONUMENT
America's Prehistoric Preserve

As the world's only national park devoted to dinosaurs, Dinosaur National Monument in northeastern Utah, is unique—a protected natural wilderness area where thousands of the great creatures were buried by river sands about 150 million years ago. The incredible concentration of bones and skeletons, once buried a mile beneath the earth, were upthrust, then exposed to the surface by erosion.

During the late 19th century, when geologist Earl Douglass came to search the Morrison Formation for fossils, it was a dinosaur-hunter's paradise. He had been hired by steel magnate Andrew Carnegie to gather specimens for the new Carnegie Museum in Pittsburgh.

This same rock formation had already yielded rich treasures to other paleontologists in Colorado and at Como Bluff, Wyoming. On August 19, 1909, Douglass found eight tail bones of *Apatosaurus* (formerly *Brontosaurus*) in the Uinta Basin sandstone, which was to become one of the richest quarry sites ever discovered.

During the next 14 years, museums mined more than 350 tons of dinosaur skeletons, collecting numerous individuals of 13 species. Most came from the "Carnegie Quarry," but the site remained as rich as ever. It was, in fact, a national treasure, deserving of protection after the museums had carted off their prizes.

In 1915, an 80-acre area around the site was declared a national monument by President Woodrow Wilson. At the time, Douglass urged the government to enclose the remaining fossil quarry in a building, protected against the weather and vandals. Thousands of visitors could see the natural formation without damaging it—"one of the most astounding and instructive sights imaginable."

Douglass's vision was realized at last in 1958, when a large modern structure was built on the fossil-rich sandstone ledge he had discovered 50 years before. Within this visitor's center, preparators continue to work on the sandstone "island," exposing new fossil bones and sometimes making new discoveries while the public looks on.

See also CARNEGIE, ANDREW; DINOSAUR.

DINOSAUR RESTORATIONS
The Art of Science

Pictures of dinosaurs have become so ubiquitous in our culture it seems strange that less than 200 years ago no one had ever seen one. There were images of imaginary fire-breathing dragons in European folklore, but actual bones of brontosaurs and tyrannosaurs had yet to be discovered, let alone reconstructed by artists.

Dinosaurs were first discovered in 1822, when Dr. Gideon Mantell and his wife Mary Ann were searching for fossils in the English countryside. Mrs. Mantell found a large fossil tooth in a Kentish field, and soon the couple unearthed more teeth and bones embedded in rocks that predated ancient mammals. Since these peglike teeth reminded him of the living iguana lizard, Mantell named the animal *Iguanodon*—the first dinosaur to be described in a scientific paper.

British anatomist and paleontologist Richard Owen of the British Museum, was impressed with the fossils and coined the term "dinosaur" from the Greek for "terrible lizard." Owen soon joined forces with British artist and anatomist Benjamin Waterhouse

DINOSAUR ROBOTS are the latest wrinkle in reconstructions of the prehistoric reptiles. During the mid-1980s, several companies began manufacturing computerized animatrons that move, roll their eyes, open their jaws and roar. Some natural history museums now exhibit dinorobots alongside the more traditional mounted skeletons.

Hawkins to produce life-size reconstructions of the beasts. Under Owen's direction, Hawkins built huge dinosaur statues of concrete, stone and iron to be shown at the great Crystal Palace exhibition at Sydenham in 1853.

Grand opening of the dinosaur display was heralded by a dinner for the leading scientists held *inside* the iguanodon on New Years Eve, 1853. It was a triumph for Hawkins, according to historian Lynn Barber, "Even though Owen, with his usual charm, made a speech attacking Hawkins for getting the iguanodon wrong." [See IGUANODON DINNER.]

The squat rhino-like appearance of iguanodon was challenged by no less an authority than Thomas Henry Huxley, who was too polite to have brought it up at the dinner. Nevertheless he had studied birdlike fossil footprints and anatomical evidence, and thought that iguanodon should be shown as an upright creature that strode or hopped on two legs. Later, it was also found that the horn on the snout of the reconstructed iguanodon didn't belong there at all. It was, in fact, a thumb bone. A century later, paleontologist George Gaylord Simpson wrote that "the animal thus thumbed its nose at its first reconstruction." Hawkins's sculptures can still be seen at Sydenham, although the Crystal Palace is long gone.

Problems with reconstructions have been a constant worry to paleontologists. Now and again, new evidence comes to light that shows that a long-held view of the way a stegosaur or tyrannosaur looked has to be rethought and reinterpreted. As recently as 1986, a curator at the American Museum of Natural History admitted to his chagrin that the famous brontosaur (really, *Apatosaur*) specimen there had worn "the wrong head" for years.

The longstanding feud between Edward D. Cope and O. C. Marsh, which influenced the course of paleontology for decades, began when Marsh pointed out that Cope had placed a skull on the wrong end of his *Elasmosaurus*. He had mistaken the tail for the neck, and the error was caught just as Hawkins was to begin sculpting his reconstruction. Cope never forgave Marsh for proving to the world that he didn't know which end of a dinosaur was up.

By the 1870s, thousands of tons of dinosaur bones were being excavated in the American West, and many were intact skeletons with the bones in their proper order. Cope and Marsh, by then bitter rivals, each rushed their skeletons back to their respective museums, commissioning artists to create reconstructions as fast as they could assemble the skeletons.

One of the greatest of the anatomist-sculptor-painters was Charles R. Knight. For years, he was associated with the American Museum of Natural History in New York, where he worked closely with the museum's director, Henry Fairfield Osborn. Knight brought a new accuracy and vigor to the field of dinosaur reconstruction. His dinosaurs didn't simply stand and pose—they looked like real animals and often were shown in relationship to landscapes, habits, predators and prey. Knight captured the imagination of the scientific community as well as the general public; for years, imitators and Hollywood monster-makers drew on his work for the standard images of dinosaurs.

For several decades after Knight's pioneering work, there was a dearth of fine artists working closely with scientists. A notable exception was Rudolph Zallinger, whose Pulitzer-Prize-winning dinosaur mural at

SKIN TEXTURE AND MUSCLES are carefully researched and built up by California dinosaur artist Steve Czerkas, shown here working on face of Deinonychus for group in Los Angeles County Museum.

the Yale Peabody Museum of Natural History was reprinted in *Life* in 1955. The mural inspired some of the current generation of paleontologists, including Robert Bakker, to enter the field. Since the 1960s, a real renaissance has occurred, with excellent work by Bakker himself, Douglas Henderson, Stephen Czerkas, Ron Sequin, Gregory Paul, John Gurche, Mark Hallet, William Stout, Pete von Scholly and Vladimir Krb. An impressive collection of some of their work, along with that of past masters, toured the country under the title "Dinosaurs Past and Present."

Over the decades, various changes and trends can be seen in artists' reconstructions. In earlier works, such as Hawkins's iguanodons, the creatures are depicted as "violent, clumsy, and slovenly beasts, dressed in drab greys, browns and dark greens and standing by themselves." More recent works portray them as more lively, sleeker and living in complex groups, keeping pace with recent theories that dinosaurs were warm-blooded, quick-moving, social animals.

Posture, too, has changed in the paintings. Some quadrupedal animals, including iguanodon and stegosaurus, were later thought to stand upright. Huge sauropods were, until recently, drawn with their front legs splayed out and bent at the elbows, like modern lizards. But the newer artists reconstruct the elbows as straight, with the front legs pulled in close to the center of gravity.

One prominent present-day sculptor of dinosaurs, Stephen Czerkas, builds his models on resin casts of the skeleton, adding clay where muscles attached to the bone. He has even been able to add realistic skin texture, for fossil impressions of the scales of dinosaur skin are now well known. Contemporary painters also use more vivid colors, but they can only guess, since colors don't fossilize. "This makes a lot of sense," says paleontologist Robert Bakker, "because dinosaurs are closely related to birds. Colors were undoubtedly used, especially in the mating season."

See also COPE–MARSH FEUD; HAWKINS, BENJAMIN W.; IGUANODON DINNER; KNIGHT, CHARLES R.; PTEROSAUR, MACREADY'S.

For further information:
Burian, Z., and Spinar, Z. V. *Life Before Man.* New York: American Heritage Press, 1972.
Czerkas, Sylvia, ed. *Dinosaurs: Past and Present.* 2 vols. Seattle: University of Washington Press, 1986.
———, and Glut, D. *Dinosaurs, Mammoths and Cavemen: The Art of Charles R. Knight.* New York: Dutton, 1982.
Stout, William. *The Dinosaurs.* New York: Bantam, 1981.

DINOSAURS, EXTINCTION OF
Why the "Great Dying?"

Because humans are the species writing evolutionary history, we inflate the importance of the last two or three million years, which produced mankind. But from a larger perspective, the conquest of the land by hundreds of kinds of dinosaurs, lasting 140 million years, is the major triumph of land vertebrates. In their variety of species and long dominance of the planet, they were far more successful than we have yet to prove to be.

Yet, at the close of the Cretaceous period, about 70 million years ago, something devastating happened. What it was must still be considered unknown, although there is no lack of clever theories. Every large land animal then in existence was wiped out, as well as the flying pterosaurs and sea-going plesiosaurs and mosasaurs. Why? To some, the question may well be related to another: If such a successful and powerful lineage could disappear from the face of the Earth, are there lessons to be learned? Can our own species similarly disappear?

Theories about the extinction of the dinosaurs have a long and convoluted history. Here are a few of those that have enjoyed some prominence and popularity over the past century and a half.

1. *Pre-Evolutionary Theories.* Dinosaurs were antediluvian or "pre-Adamite" monsters that were wiped out in a great flood. Some early naturalists speculated they were part of a previous creation. Others thought that there may have been no room for them on Noah's Ark, and so they were doomed to drown.

2. *"Racial Senility" Theory.* Some paleontologists, like W. E. Swinton in the 1930s, thought that species, like individuals, may become old and senile. Extinction was therefore inevitable, like the death of an individual. Evidence of species "senescence" was found in "overossified" dinosaur skulls with bony frills and grotesque horns—supposedly produced by growth hormones run wild. Now these "monstrous" appendages are seen as gradually developed features, adaptive for defense and reproduction. Dinosaurs were a progressive group, constantly developing new niches and specializations. "Senility of species" is a false analogy with the life cycle of individuals.

3. *"Too Stupid and Slow"* vs. *Mammals.* Recent evidence and new interpretations show that many dinosaurs were fast-moving, perhaps warm-blooded and, in some cases, quite intelligent. Late Cretaceous wide-eyed "ostrich" dinosaurs (coelurosaurs) and dromaeosaurids like *Deinonychus* and *Sauronithoides* are examples of "bright" dinosaurs, which were de-

veloping just before the whole group became extinct. Some of these had such man-like characteristics as stereo-vision, correlated with opposable thumbs and large brains. Adrian Desmond writes they were "separated from other dinosaurs by a gulf comparable to that dividing men from cows." But these agile, alert creatures did not survive their comparatively dim-witted giant relatives. However, they were quite spectacular in contrast to the small and not very bright or prepossessing mammals of the time.

4. *Mammals Ate the Dinosaurs' Eggs.* No doubt they did, and so did other dinosaurs. The first dinosaur eggs found, which captured worldwide attention, were collected by Roy Champman Andrews on the Gobi Desert (Central Asiastic) Expedition during the 1930s. Rare and important skulls of small early mammals were found nearby in the same fossil beds, leading to the popular view that the mammals ate the dinosaurs' eggs. But Cretaceous mammals were small, and most dinosaur eggs were big and tough-shelled.

5. *Dinosaurs Were Poisoned.* Various scientists have suggested that the development of poisonous defenses by land plants did the dinosaurs in. Alkaloids in the newly developed flowering plants [see ANGIO-SPERMS, EVOLUTION OF] have been suggested, and one scientist even fed poisons to tortoises to prove they could not taste them. In fact, flowering plants were around for quite a few million years before the dinosaurs' demise and may even have been beneficial to them. Such explanations still would not account for the disappearance of sea-going plesiosaurs and for the mass extinction of many invertebrates and other nondinosaurian species during the same period.

6. *Failure to Adapt.* Perhaps the most common belief about dinosaur extinction is that when an animal becomes overspecialized and fails to adapt to changing circumstances it becomes extinct. First, this is a tautology—like saying that those that do not survive, do not survive. Second, it doesn't specify *what* changes. (No creature is adaptable enough to survive all changes, except within a certain range.) Finally, it has the ring of morality rather than science. It teaches the lesson of flexibility, like the Zen story of the bamboo that can bend in the wind, while the sturdy oak is felled by a storm. But nature's lessons depend on the attitude of the observer. Some of the most archaic and least flexible creatures of the Cretaceous—tortoises and crocodiles—managed to survive and are still with us.

7. *Cosmic Cataclysms.* The current favorite of geologists and paleontologists is the idea that some extraterrestrial event caused sweeping changes on Earth.

While earlier theories postulated an exploding supernova, which altered Earth's climate, giant meteors are now in vogue. Geophysicists Walter and Luis Alvarez have found layers of iridium at the Cretaceous–Tertiary boundaries—a substance not normally found in the Earth's crust except at sites of meteor impact. Giant meteor showers could have raised immense clouds of dust, which would block solar rays and cool the planet. Even a small change in worldwide temperatures could account not only for the extinction of the dinosaurs, but for the disappearance of many other species as well, including the sea-going reptiles, shellfish, ammonites and plankton that vanished at the same time. (Many theories address themselves only to dinosaurs, ignoring the other important extinctions that simultaneously occurred.)

Great meteor showers would explain many puzzles in the fossil record. But why would thousands of giant meteors crash to Earth in violent storms? Some astronomers postulate the existence of a stellar twin to our sun—the Nemesis Star [see NEMESIS STAR]—that makes its periodic swings every 28 million years or so and disrupts meteors in the Van Allen belt. These astronomical theories, the iridium evidence and the interest in "punctuated equilibrium" have fostered a new respectability for catastrophic explanations. [See "NEW" CATASTROPHISM; PUNCTUATED EQUILIBRIUM.] Ultimate acceptance will depend on the actual discovery of the Nemesis star by astronomers, who are searching the heavens for it. Within the decade, we will know if the "predicted" star exists, or whether we need yet another theory to understand the "great dying" of the Cretaceous dinosaurs.

See also ALVAREZ THEORY; CROCODILIANS; EXTINCTION.

For further information:
Hsu, Kenneth J. *The Great Dying: Cosmic Catastrophe, Dinosaurs, and the Theory of Evolution.* New York: Ballantine/Random House, 1986.
Wilford, John Noble. *The Riddle of the Dinosaur.* New York: Random House, 1985.

DIRECTIONAL SELECTION

See STABILIZING SELECTION.

DISNEY, WALT

See CUTENESS, EVOLUTION OF; FANTASIA.

DIVERGENCE, PRINCIPLE OF

Evolution is often pictured as a family tree or branching bush, bristling with divergent forks. Each lineage repeatedly splits and differentiates, lines splay out, in Alfred Russel Wallace's image, "like the twigs of a gnarled oak or the vascular system of the human body." Some of the early evolutionists, such as Ernst Haeckel, spent years working out detailed "trees of life," showing the divergence of families, genera and species over time.

But though they seem inseparable today, evolution and divergence have not always been associated. Erasmus Darwin drew no family trees nor did Jean-Baptiste Lamarck. Although Charles Darwin sketched such a tree in an early notebook, the principle of divergence occurred to him much later, about 15 years after he had developed his basic theory of natural selection. Divergence was a crucial missing piece even during the writing of *Origin of Species* (1859), and he called its last-minute inclusion "the keystone" of his book.

In the later views of both Darwin and Wallace, divergence serves a double function in evolution. First, it enables a given species under selection pressure to survive in modified form by exploiting new niches in the ecology. And second, the gain in diversity boosts the habitat's carrying capacity, enabling it to support a greater total amount of life. Typically, small isolated habitats (like Darwin's beloved Galapagos) exhibit a startling diversity of closely related species, adapted for exploiting different foods or parts of the habitat. (Long before he formulated his theories, Darwin was impressed by the "abundance of great creative force" in those tiny, isolated islands.)

There has been much recent debate among scholars about why Darwin came to understand divergence so late in the day, or even whether he might have lifted the idea from Wallace. Despite the extraordinary documentation of the Darwin correspondence (some 14,000 letters), it is disturbing that certain crucial documents are missing, such as Wallace's very first letter to Darwin, written in October, 1856 from Malaysia, and the "lost letters of 1858," which immediately preceded the composition of *Origin of Species*.

Wallace's letter reached Darwin in April, 1857, five months before Darwin sent Asa Gray, the American botanist, an updated summary of his evolution theory, the product of 20 years' thought and work. He had written out sketches of his ideas before (1842, 1844), but this latest version contained something significantly new—the "principle of divergence."

There is no doubt Darwin had worked out his theory of natural selection long before Wallace had independently formulated it, but historians are still puzzled by how and when Darwin arrived at the divergence principle. Some have recently attempted to reconstruct the Darwin–Wallace connection using ingenious, even outlandish, methods.

John L. Brooks, for instance, managed to locate the old Dutch records of mail boats between England and Malaysia and tried to document exactly when Darwin could have received Wallace's letter. If it was earlier than the date given in Darwin's official family biography, as Brooks believes, he would have had plenty of time to incorporate Wallace's ideas on divergence into his own writings. But Brooks's tenacious detective work has produced results that, to most specialists, remain ambiguous.

Barbara Bedell's minute examination of Darwin's notebooks and papers shows he was stimulated by Wallace's preliminary publication on evolution, but actually worked out divergence independently, as he later claimed. Wallace's 1855 paper (the "Sarawak Law"), began with the question: If one examines the numbers of closely related species within genera, geographic distribution of natural groups, and kinds of differences between species in a local area, what overall pattern (he called it a "law") would emerge? The answer was that the largest number of species seemed to be produced from those genera confined to a small area (such as islands) and their differences were related to feeding adaptations (sharp beaks, blunt beaks, long beaks, etc.)

When Darwin read this paper, he scrawled on it "Why should this law hold?" The answer, he later realized, was that under selective pressures, organisms evolve to fill "vacant places in the natural economy." He compared it to a division of labor, with efficient specialists exploiting the various food sources in a limited area. In one of his garden experiments, Darwin counted 20 species of wild plants (belonging to eight orders and 18 genera) that had sprouted on a cleared piece of turf measuring three by four feet.

What is now taken for granted was then a startling insight: contrary to "common sense" expectations, the fierce struggle for existence does not reduce the overall number of related species in an area of limited resources. Instead, it has the paradoxical effect of allowing many more species and individuals to thrive there, constantly evolving "endless forms most beautiful and most wonderful."

See also BRANCHING BUSH; DARWIN'S FINCHES; "DELICATE ARRANGEMENT"; HAWAIIAN RADIATION.

DIVINE BENEFICENCE, THE IDEA OF

Sherlock Holmes, the world's first consulting detective, found in a red rose evidence of God's goodness to man. In Sir Arthur Conan Doyle's story "The Adventure of the Naval Treaty" (1893), Holmes remarks that "Our highest assurance of the goodness of Providence seems to me to rest in the flowers. All other things, our powers, our desires, our food, are really necessary for our existence in the first instance. But this rose is an extra. Its smell and colour are an embellishment of life, not a condition of it. It is only goodness which gives extras, and so I say again we have much to hope from the flowers."

Holmes's deduction from the flowers was *not* one of his unique conclusions; in fact, it was a commonplace notion widely held by the Victorians. God's goodness expressed in nature had been a major theme of Reverend William Paley's *Natural Theology* (1802), studied by every aspiring naturalist including the young Charles Darwin, who memorized long passages. Divine beneficence, it was believed, had provided for all human needs—not only the physical requirements of food and water, but even humanity's need for uplift and beauty. Flowers were a prime example. They had been created with vivid colors and pleasant scents for the sole purpose of causing human delight and enjoyment. As late as the 1920s, the window of Walsh's flower shop in New York City proclaimed BEAUTIFUL FLOWERS—SMILES OF GOD'S GOODNESS.

Divine beneficence was a blanket explanation for natural phenomena in Victorian natural history. A delightfully dramatic example is Dr. David Livingstone's description of his near-fatal encounter with a lion, in *Missionary Travels and Researches in South Africa* (1858):

> Starting, and looking half round, I saw the lion just in the act of springing upon me. I was upon a little height; he caught my shoulder as he sprang, and we both came to the ground below together. Growling horribly close to my ear, he shook me as a terrier dog does a rat. The shock produced a stupor similar to that which seems to be felt by a mouse after the first shake of the cat. It caused a sort of dreaminess, in which there was no sense of pain nor feeling of terror . . . like chloroform . . . the shake annihilated fear, and allowed no sense of horror in looking round at the beast. This peculiar state is probably produced in all animals killed by the carnivora; and if so, is a merciful provision by our benevolent Creator for lessening the pain of death.

Of course, his explanation charmingly sidesteps the question of why a benevolent deity would have allowed Dr. Livingstone (or other prey) to be mauled by the lion in the first place.

Linnaeus believed that contemplation of God's gifts to man was the naturalist's original mandate. Adam, he wrote, was "the first and most intelligent Botanist," who spent most of his hours in the Garden of Eden examining "the admirable works of the Crea-

tor." Before the "primal curse," it was believed, there were no thorns on roses and no briars or weeds.

Every plant was believed to have a benefit for humanity. If no practical purpose could be discerned, the flower must be useful for aesthetic pleasure or moral enlightenment. The number of petals or leaves on certain species were reminders of the number of apostles or gospels or deadly sins. Despite the best efforts of an 18th-century experimental botanist named Christian Konrad Sprengel, the idea that a flower's beauty, scent and color were created solely as a gift to humans persisted well into the 19th century. Sprengel's evidence that blossoms and fragrances aid sexual reproduction by attracting insects to spread pollen was ignored as distasteful and vulgar. His triumphantly titled book *The Secret of Nature Revealed* (1783) was forgotten until Charles Darwin rescued it from obscurity some 60 years later.

Indeed, heated debates had been going on for centuries about the function of pollen and were finally settled only after Darwin resurrected Sprengel's ideas and conducted new botanical experiments in his own greenhouse and garden. Publication of the *Origin of Species* in 1859 stirred new controversies about the beauty of flowers. Critics declared it impossible that such gorgeous colors and bizarre structures could really be produced by such a mundane process as natural selection.

Darwin responded by conducting years of experimentation on the mechanisms of pollination, culminating in his books *On the Various Contrivances by which Orchids are Fertilised by Insects* (1862) and *Effects of Cross and Self Fertilisation in the Vegetable Kingdom* (1878). Also, in his *Descent of Man* (1871), he developed his theory of sexual selection, by which he explained the origin of such natural "artworks" as the peacock's tailfeathers. (Male birds with more impressive displays would be preferred by the hens for mating.) Finally, Darwin discussed the evolution of the human aesthetic sense itself and its variation among different nations.

But it remained for his protege and neighbor, Sir John Lubbock (Lord Avebury), to help resolve the problem from the other side, that of insect senses. Banker, anthropologist, botanist and entomologist, Lubbock had devoted himself to solving the puzzles posed by Darwin's work. In the 1870s, he did the first tests on color vision in bees. His classic *Ants, Bees, and Wasps* was published in 1882, the year Darwin died.

Before Lubbock's experiments, it was widely assumed that insects could see only in black and white. His promising start was followed up in 1913 by the Austrian Karl von Frisch, who proved in a long series of elegant experiments that bees really do respond to color; thus nature's glorious flower show is directed to insects, not to humans.

We read in the later reports of Dr. Watson that Sherlock Holmes eventually retired from solving London's crimes and devoted himself to bee-keeping on the Sussex Downs. As all Sherlockians know, he even produced a technical monograph, the *Practical Handbook of Bee Culture, with Some Observations upon the Segregation of the Queen*. In "His Last Bow" (1917), Holmes describes this treatise on apiculture as "the fruit of my leisured ease, the *magnum-opus* of my latter years! . . . pensive nights and laborious days when I watched the little working gangs as once I watched the criminal world of London."

For such an authoritative work, the meticulous Holmes must surely have consulted Lubbock's work on color vision in bees. However, the good doctor has left us no clue as to whether the great detective ever changed his views on the reasons for roses.

See also LUBBOCK, SIR JOHN; ORCHIDS, DARWIN'S STUDY OF; SEXUAL SELECTION.

DNA (DEOXYRIBONUCLEIC ACID)
Watson and Crick's Discovery

DNA, the biochemical substance that encodes the instructions for creating new life, appears in every chromosome-containing cell of all living organisms. The two complementary strands that entwine to form the molecule's double helix are composed of phosphates and sugars bonded together at intervals across the molecule's center.

When the American James Watson and British scientists Francis Crick and Maurice Wilkins shared a Nobel Prize in 1962 for their work in establishing this structure as the physical basis of heredity, they became famous as the discoverers of DNA. Watson and Crick's publication of their celebrated paper on DNA in 1953 marked a turning point in 20th-century biology: It led within a few decades to an unprecedented expansion of molecular research, as well as to genetic engineering, a new biotechnology based on artificial alteration of DNA.

Watson and Crick's contribution was immense, but they did not discover DNA. It had first been extracted from human cells in 1869 by the German scientist Friedrich Miescher. He called the substance "nuclein," thought its function was phosphorus storage and never suspected its role in heredity.

In 1938, others working with X-ray crystallography of DNA had discovered its characteristic patterns on photo plates, while many biologists had defined its role as a transmitter of heredity by 1944. Identification of the chemical units at the molecule's center was achieved by Austrian emigre Erwin Chargaffe,

working in the United States in 1950—the same year two British chemists published important new information on the outer strands.

According to historian of genetics Edward Yoxen, of the University of Manchester, the Watson–Crick breakthrough did not arise out of any startling new laboratory discoveries of their own, nor out of any original experiments. Rather, Watson and Crick took it upon themselves to review all the known experimental work with the aim of extracting new significance from it. Their hypothetical model of the three-dimensional structure of DNA was achieved by reasoning and imagination alone.

It was a model that immediately appealed to those working in the field, but it left much data unexplained and could have been wrong, since it was a hypothesis rather than a proof. Decades of experimental research have since provided abundant confirmation of the Watson–Crick model, which has proved a fruitful basis for much biological research and understanding.

What Watson and Crick discovered was a new way of looking at an already known substance, many of whose properties had been determined by others. Their plausible model of its mechanism, worked out under the pressure of an international race with other scientists who were seeking the same goal, explained DNA's vital role in genetic storage, transmission and evolution.

See also CADUCEUS; GENE MAPPING PROJECT; GENETIC ENGINEERING.

For further information:
Judson, H. F. *The Eighth Day of Creation: The Makers of the Revolution in Biology*. London: Cape, 1979.
Watson, James. *The Double Helix: a personal account of the discovery of the structure of DNA*. London: Weidenfeld, 1981.

DNA IDENTIFICATION
Biochemical Tool for Criminal Law

Despite the shared genetic chemistry of species, each individual has a uniqueness extending right down to the DNA. In fact, the structure of the DNA has proven so identifiably distinct in each individual that it may soon replace fingerprints in criminal identification.

In the past, prosecutors have nailed criminals by analyzing bits of hair or body fluids left at the crime site. But the outward form, color or thickness of hair, type of blood or semen may leave room for doubt or be considered merely circumstantial evidence. With analysis of the DNA from protein molecules within the blood, hair or semen, biochemists claim "absolute identification." (Identical twins, who developed from the same egg and share the same DNA, would be a possible exception.) A perfect match involves comparing thousands of more specific variable features than exist in fingerprints, which now seem crude by comparison.

Among the first to be convicted with the DNA test was Timothy Spencer—for a double murder in Virginia (1988). His attorney tried to discredit the new DNA test, but could not find one biochemist who would challenge its validity. Shortly thereafter, the Seattle police department began a project to catalogue the DNA types of all known violent criminals in its files, as a step toward replacing fingerprint evidence with the new technology.

DNA MAPPING

See GENE MAPPING.

DOBZHANSKY, THEODOSIUS (1900–1975)
Evolutionary Geneticist

Harvard's Ernst Mayr, one of the architects of the modern Synthetic Theory of evolution and a historian of biology, points to one particular book that heralded the beginning of the new understanding and was more responsible for it than any other: Theodosius Dobzhansky's *Genetics and the Origin of Species*, published in 1937.

It was Dobzhansky's first book, but he probably would never have found the time to write it if he hadn't been forced to spend weeks in bed after a horseback riding accident. A Russian immigrant who came to work at T. H. Morgan's "fly room" at Columbia University in 1927, Dobzhansky brought to America the innovative techniques developed by Russian geneticists before their science was crushed by the tragic madness of Lysenkoism. [See LYSENKOISM.] Versed in the research problems of both field naturalists and lab men, he was able to make connections between the two approaches.

During the first 20 years of the 20th century, Darwin's theory of natural selection had fallen out of favor among scientists. Many thought it insufficient to explain the origin of adaptations, while new discoveries of gene mutations seemed to them to be incompatible with Darwinian models of change. But in *Genetics and the Origin of Species*, as historian Bentley Glass put it, "for the first time, the profound significance of the work done in population genetics in Russia and Germany was combined with an exposition of the new neo-Darwinism stemming from R. A. Fisher, Sewall Wright, and J. B. S. Haldane, to produce what has been called the modern synthetic theory of evolution."

Dobzhansky's book was the first systematic overview encompassing organic diversity, variation in natural populations, selection, isolating mechanisms (a term he coined) and species as natural units. Later, working with Sewall Wright, he went on to demonstrate how evolution can produce stability and equilibrium in populations rather than constant directional change.

His studies of isolating mechanisms identified nongenetic barriers to reproduction, such as behavior or vocalizations, which may keep populations distinct even after geographical barriers to interbreeding are removed. Dobzhansky also helped demonstrate that a population arbitrarily divided into two subpopulations can diverge into two species even in the absence of any selection pressure. Among other projects, he studied the mechanisms of genetic variability within natural populations and showed how detrimental genes can spread in certain combinations with beneficial ones (heterozygote fitness.)

He was particularly fascinated with unraveling the multiple effects of a single genetic change (pleiotropy) and with the complex role of gene arrangement and chromosomal structure in producing evolutionary change. Never content with laboratory studies alone, he repeatedly stalked wild populations of fruit flies in the mountains of Arizona, New Mexico, California and even the rain forests of Brazil. Dobzhansky's intimate familiarity with the processes of variation and evolution in these fast-breeding insects also enabled him to apply his methods to understanding variation and change in human populations.

See also ALLOPATRIC SPECIATION; FLY ROOM; GENETIC DRIFT; ISOLATING MECHANISMS; MAYR, ERNST; SYNTHETIC THEORY.

For further information
Dobzhansky, Theodosius. *Genetics and the Origin of Species*. New York: Columbia University Press, 1937.
———. *Mankind Evolving*. New Haven: Yale University Press, 1962.
Dobzhansky, Theodosius, Ayala, F., et al. *Evolution*. San Francisco: W. H. Freeman, 1977.

DOLLINGER, I.

See NATURPHILOSOPHIE.

DOLLO'S PRINCIPLE
Irreversability of Evolution

A Belgian biologist named Louis Dollo attempted to formulate laws or principles for the evolutionary process. One of them—still known as Dollo's Principle—holds that evolution is irreversible. At a time when "degeneration" theories were in vogue, some assumed that man could return again to being an ape through a "reversal" in evolution. Various science fiction tales, like Jules Verne's *Island of Dr. Moreau* (1896) dealt with the theme. But Dollo tried to demonstrate that once evolution had taken a particular path, there was no retracing it.

Although his principle is enshrined in most textbooks, it is not strictly accurate. Dinosaurs, for instance, had a remote ancestor that was quadrupedal, walking on all fours. But the immediate ancestors of all dinosaurs were upright bipeds: the small, swift-running thecodonts. Some dinosaurs continued to evolve as upright bipeds, like tyrannosaurus, while another lineage—including giant herbivores like brontosaurus—returned to being quadrupeds.

In pointing out this error in Dollo's irreversability principle, paleontologist George Gaylord Simpson noted it is more accurate to say that evolution is irrevocable. When a series of complex changes has been encoded in the DNA, it is not likely that these will be undone and the original plan restored. However, there may be modifications that appear to "reverse" earlier traits.

The brontosaurs' "return" to quadrupedalism is one example; another is the teeth of whales. Ancestral whales had simple, peg-like teeth, but the *Zeuglodon* of 60 million years ago had cusped, flattened teeth. Yet its descendants went back to the simple conical teeth without cusps, as in modern whales and dolphins.

While some anatomical features may evolve to resemble an earlier version in their development, whole organisms do not. Besides, the notion of "forward" and "reverse" implies stages and irresistable straight-line evolution in a particular direction (orthogenesis). "Evolution does not work in straight lines," George Simpson once commented, "but the minds of some scientists do."

DOLPHINS, LANGUAGE ABILITY OF

See APE LANGUAGE CONTROVERSY.

DOMESTICATED VARIETIES

See ARTIFICIAL SELECTION.

DOUBLE HELIX

See CADUCEUS; DNA.

DOUGLASS, EARL

See DINOSAUR NATIONAL MONUMENT.

DOWN HOUSE
Home of the Darwins

Soon after Charles Darwin returned to England from his long voyage around the world, he married his cousin Emma Wedgwood and settled down to domestic life. After a few years in London, the young Darwins decided to move permanently to the English countryside. In 1842, they bought a Georgian home in the tiny village of Down, 16 miles from London.

Despite its closeness to the city, it is an isolated area, nestled in the gently sloping chalk downs from which it takes its name. The Darwin home was (and is) called Down House. (In 1842, during recurrent paranoia about Irish rebels, the village became "Downe," to distinguish it from County Down. Darwin thought the spelling change was ridiculous and continued to give his address as Down House.)

Down House sits on 18 acres, including fields, a garden and a clump of woods Darwin planted himself. A sand-covered path was laid out winding through the shady woods and then returning towards the house along a sunny straightaway bordered by hedges. Darwin called it the Sandwalk, "my thinking path." His work on his 17 books and innumerable scientific papers was punctuated daily by several brisk turns around the Sandwalk, often with his little dog Polly trailing along.

Sometimes, when deep in thought, he would stack up a few flints around the turn in the walk. He might have a "three flint problem," just as Sherlock Holmes had "three pipe problems." Darwin would walk around the loop, knocking away a flint with his walking stick each time he passed. When all the stones were gone, it was time to head back home.

Here Darwin lived and worked for 40 years. At various times he added to the house, put up a dovecote out back for his breeding experiments, added a greenhouse (which still stands), a laboratory (now in ruins) and, later, a clay tennis court for his children. Charles and Emma raised their seven surviving children (three were lost to disease) at Down House.

HOME OF THE DARWINS still stands in the village of Downe, near Bromley, about 16 miles south of London. Open to the public as the Darwin Museum, several rooms, including the study where *Origin of Species* was written, have been restored with their original furnishings and decor.

Downe is located at the junction of several rail lines and can be approached either through Orpington or Bromley. A mile up the road lived the banker-astronomer Sir John Lubbock, at a glorious estate called High Elms, which had 3,000 acres. His son, also called John Lubbock, grew up with Darwin as his mentor and became a noted evolutionary scientist (paleontologist, animal behaviorist, prehistorian) as well as a banker and member of Parliament. Unfortunately, the magnificent house at High Elms burned to the ground in 1977.

Down House was sold after Emma Darwin's death and was used as a private school for girls until the 1930s, when Sir Buxton Browne arranged to purchase it for the Royal College of Surgeons.

Relatives and friends of the Darwin family contributed original furnishings, which had been scattered, and the house was restored to its appearance when Charles and Emma lived there. Today it is a museum, open to the public, where one can walk along Darwin's Sandwalk and meditate in his study; his old microscope, books and writing table occupy their former places.

Another room has been turned into a display of Darwin memorabilia, and the Victorian drawing room where Emma played the piano as Charles relaxed on the chaise lounge takes the visitor right back in time. In the garden is the "wormstone" device that their son Horace made to measure the rate at which earthworms turn over the soil for his father's book on the subject.

See also: SANDWALK; WORMSTONE.

For further information:
Atkins, Sir Hedley. *Down: Home of the Darwins.* London: Phillmore, 1976.

DOYLE, SIR ARTHUR CONAN

See DIVINE BENEFICENCE, IDEA OF; (THE) LOST WORLD.

DRAGONS
Myth That Came True

Dragon legends have persisted for centuries in Norse epics, medieval English ballads, Wagnerian operas, Japanese art and Chinese folktales. To modern scientists, gigantic reptiles were pure myth until about 1825. About that time, marveling European geologists began turning up huge fossil teeth and bones. Huge reptilian monsters, it turned out, had once lived and breathed, though long before there were pure-hearted knights to battle them.

Why had this ancient tradition of dragons persisted in so many of the world's cultures, long before science knew of dinosaurs? Perhaps it was more than a coincidence of fact and fancy; fossils were found and collected among many peoples long before the rise of paleontology.

In Asia, dragons have been prominent in art and legend for thousands of years; they were considered wise, beneficent and bringers of good luck. Over the years, they took on attributes of other animals (elongated as snakes, antlered like reindeer) and symbolized the whole Chinese nation. Even today, huge dragon puppets manned by dozens of people are trotted out in American Chinese communities as the highlight of New Year celebrations.

In European folklore, dragons were evil monsters that guarded great treasures or imprisoned damsels. Ordinary folk were not sufficiently pure of heart to find them, but great warriors like England's St. George sought and slew them. (Perhaps some ambitious knights actually brought back dinosaur claws or teeth from their quests. Is it possible Round Table dragon-killers journeyed to purchase impressive "battle trophies" at a far distant fossil quarry?)

Fossils (not only of dinosaurs, but any large animal) were regarded in many parts of the world as "dragon bones" or "dragon's teeth." In China, country folk collected them for sale to apothecaries, who displayed bins full of them, ready to sell and be ground up into traditional medicines.

In 1899, K. A. Heberer, a German naturalist, brought back a large collection of these Chinese "dragon bones" purchased from druggists and had them examined by museum experts. Ninety fossil species were identified, including extinct bears, giraffes, rhinos and lions—none of them reptiles. During the 1920s, anatomists G. H. R. von Koenigswald and Franz Weidenreich, realizing these sellers of folk medicine might be enlisted in the cause of science,

LINDWURM FOUNTAIN at Klaagenfurt, Austria, depicts traditional European dragon. Some scientists think statue's peculiar head, sculpted around 1590, was inspired by fossil rhino skulls found in nearby countryside. It is perhaps the first known attempt to reconstruct a prehistoric beast from fossil remains.

began a systematic search of the shops for unusual fossils.

It was in such a druggist's bin that Weidenreich recognized a few enormous teeth of hominid pattern. Thus was discovered the giant extinct ape he named *Gigantopithecus*. Later, Weidenreich's conclusion was confirmed after more teeth and fragments were found, though paleontologists are still seeking a skull or skeleton.

A particularly rich fossil site near Beijing was long known as Chou-kou-tien, "Hills of the Dragons." There, in quarries and limestone caves, the famous Peking Man skulls were found by Davidson Black, Franz Weidenreich and W. C. Pei during the 1930s. Originally named *Sinanthropous pekinensis*, they have since been included in *Homo erectus*, a species of ancient man whose remains have also come to light in East Africa, Java and Europe.

In Beijing, there is a famous "Wall of the Dragons," an artistic treasure of old China carved with magnificent Oriental dragons. In 1987, modern Chinese paleontologists from Beijing's Institute of Paleontology quarried and exhibited a rock slab showing six fossil synapsids. They nicknamed their spectacular prize "The *Scientific* Wall of the Dragons."

Another famous dragon sculpture—a large stone fountain—dominates the town square of Klagenfurt, Austria. Carved in 1590 by the sculptor Ulrich Vogelsang, its body is a traditional representation of a European dragon. But several paleontologists have been intrigued with its head proportions. The facial shape, with its pointed, angular jaws seems extraordinarily similar to the extinct European rhinoceros.

According to local tradition, the artist was indeed inspired by fossil rhino skulls found in the nearby countryside. (No one has ever compared measurements of the statue's head with those of woolly rhino skulls, but the exercise might confirm the story.) If true, the earliest known artistic reconstruction of an extinct animal from a fossil skull would be Klagenfurt's famous Dragon Fountain.

See also DINOSAUR; DINOSAUR RESTORATIONS; KNIGHT, CHARLES R.; MANTELL, GIDEON A.

DRYOPITHECINES
Miocene Apes

During the early Miocene period (about 9-to-24 million years ago) a group of ancestral medium-sized apes was spread widely over Africa, Asia and Europe. In 1856, the first fossils were discovered in southern France by E. Lartet, who named them dryopithecines, meaning oak forest apes. Poorly known for a century, the fossil representation has been substantially filled in during the past 20 years. One site

AHEAD OF HIS TIME, brilliant French zoologist Edouard Lartet identified and named the fossil apes of Europe; he called them Dryopithecines or "oak forest apes" in 1856.

in Kenya alone yielded 13 skeletons, and collections of remains have been found recently in Spain and Hungary.

The best known of the dryopiths is *Proconsul*, which was discovered by Mary Leakey in East Africa and which she recognized as an early representative of the hominoid lineage. It was tailless, had large canines (as do modern apes) and appears not to have been an upright bipedal walker. A later, related genus, *Sivapithecus*, known from Java and the Near East, may be ancestral to the orang-utan.

The relationship of the dryopithecines to the human line is not yet known; a "gap" in the fossil record seems to have occurred about four-to-nine million years ago, thus no fossils showing connections between these widespread ancient apes and the earliest near-men, or australopithecines, exist.

See also AUSTRALOPITHECINES; "HOMINOID GAP"; PROCONSUL.

DUBOIS, EUGENE (1858–1940)
Discoverer of Java Man

With a stubborn insistence that appeared entirely unreasonable to his associates, a Dutchman named Eugene Dubois set out to find a needle in an immense haystack when hardly anyone thought there was a needle there at all! His discovery of Java man remains one of the most incredible stories of self-confidence and determination in the face of seemingly impossible odds in the history of science.

Since his boyhood in Roermond, Holland, Dubois had always been an amateur naturalist whose pockets were bulging with stones and bits of animal bone. In 1877, at the age of 19, he entered medical school. There he was exposed to the exciting ideas of Darwinian biologists, who were revolutionizing every field from embryology to paleontology. In particular, he was enthralled by a visiting German lecturer, the famous Ernst Haeckel, who was trying to fill in the "missing links" in the chain of evolution.

Compelling as Darwin's theory was, there was still only a scrap or two of fossil evidence for human ancestors, notably the Neandertal skullcap from Germany. Anatomists agreed it was certainly human and, therefore, no "missing link." Expert paleontologists in France and Germany declared "There is no such thing as fossil man."

Professor Haeckel insisted a creature intermediate between men and apes must have existed and gave it the Latin name *Pithecanthropus alalus*, meaning "ape-man without speech." In Haeckel's enthusiasm for his subject, he ignored a cardinal rule of scientific classification, or taxonomy: You cannot name a creature before you know whether it exists! In the case of Haeckel's *Pithecanthropus*, however, this topsy-turvy method served science well, because young Dubois made up his mind he was going to find the creature Haeckel had named. It would be the final "proof" that Darwin was correct.

Dubois completed his medical studies and became a teacher of anatomy at the University of Amsterdam, but was restless with academic life. He was increasingly consumed by the idea of discovering the "missing link," but in what part of the world would he find it? Haeckel had thought "Lemuria," a supposed continent that had sunk beneath the Indian Ocean might have been the cradle of mankind, from which early hominids could have spread westward to Africa, northward to Asia and eastward via Java. Dubois eagerly read Alfred Russel Wallace's book of explorations *Malay Archipelago, The Land of the Orang-utan and the Bird of Paradise* (1869) that sketched Sumatra as a fascinating, prehistoric-looking landscape; conveniently, it was under Dutch colonial rule. Wallace had written:

> It is very remarkable that an animal, so large, so peculiar, and of such a high type of form as the orang-utan, should be confined to so limited a district—to two islands . . . With that interest must every naturalist look forward to the time when the caves of the tropics be thoroughly examined, and the past history and earliest appearance of the great man-like apes be at length made known.

Dubois wanted to explore those caves for himself and announced to his colleagues that he was leaving Amsterdam for Sumatra to find the missing link. They thought he was throwing away a brilliant career on an impossible, crazy scheme, but his new, young wife believed in him. In 1887, he quit his teaching post and spent many frustrating months trying unsuccessfully to find backing for an expedition.

When he realized that neither private nor governmental sponsors were to be found, Dubois enlisted as a doctor in the Dutch Indian Army. He, his wife and small children made the seven-week voyage aboard a mail boat, then found their way to a small hospital in the interior of Sumatra. His duties were light, as he had hoped, and he spent his time and money investigating many limestone caves and deposits. Once, he crawled headfirst into a tiger's den and got stuck—luckily, the resident was out hunting. Fortunately, his wandering helpers returned in time to free him before Dubois himself became a skeleton in a Sumatran cave.

After an attack of malaria in 1890, he was transferred to neighboring Java, placed on inactive duty and finally given assistance in his search by the colonial government, which supplied him with a crew of convict laborers. Among many other difficulties, the workers were secretly selling fossils as fast as they dug them up to Chinese merchants, who ground them to powder for use as "dragon's bones" in popular medicines. Nevertheless, crates of all sorts of fossils started to pile up at the house in Tulungagung, and, by 1884, he had shipped 400 cases of interesting new fossils back to Holland, but no early men.

Dubois had a hunch that the most promising site was an exposed, stratified 45-foot embankment along the Solo River, near Trinil village. One of the first prehistorians to conduct a systematic search, he squared the bank off in grids. Workmen combed it inch by inch and came up with a molar tooth in August, 1891, which he thought belonged to an extinct ape. Months later, three feet away, his diggers found a brown, round object resembling a turtle shell. When cleaned, it proved to be a fossil skullcap with thick brow ridges, and Dubois wrote "that both specimens come from a great manlike ape."

Rains came, the river rose and Dubois pondered his skull for a year before he could dig again. When he did, a left femur, or thighbone, was unearthed at the same site. Although thicker and heavier than that of modern man's, it certainly came from a bipedal primate. Then another tooth turned up, and, in 1893, Dubois formally announced discovery of the creature Haeckel had named only seven years earlier. Because he found a leg bone rather than a jaw, Dubois dubbed his find *Pithecanthropus erectus*—ape-man who walked erect.

His fossils were one of the most important discoveries ever made in the quest for human origins, although they embroiled their discoverer in bitter controversies for the rest of his life. Today, Java man has been assigned to the species *Homo erectus*, along with similar fossils from Africa, Asia and Europe. Dated at 250,000 to half a million years old, it is now considered an early human species, not a "missing link" between ape and man.

Pithecanthropus became Dubois's constant companion, to which he became greatly attached. Returning from Java in 1895, a storm at sea sent him rushing to the ship's hold to embrace his precious cargo. "If something happens," he told his wife, "you see to the children. I've got to look after this."

Dubois carried the bones to scientific meetings all over Europe, presenting his case. Most scientists pronounced it a primitive man and no "missing link." Many anthropologists visited Dubois in Holland, and he traveled widely with the bones, carrying them in a battered suitcase. One day, in Paris, he had met the anthropologist Leonce Manouvrier in his laboratory, after which they dined in a small cafe. Dubois took his old suitcase along, but forgot it in the restaurant. As the two scientists walked the Paris streets deep in conversation, Dubois suddenly clutched his companion's arm. "Pithecanthropus!" he shouted, and suddenly dashed back through the traffic. "Where is Pithecanthropus?" he screamed at the startled proprietor, who was locking up. To his great relief, the suitcase had been stashed behind the bar by a waiter, who had no idea it contained such a distinguished ancestor.

For several years, Dubois made his fossils accessible and produced numerous photos, casts and reconstructions, but the scientific world would not accept his interpretation of their antiquity or their pivotal role in human evolution. Deeply hurt, he withdrew them from further study and became a recluse.

At last, after three decades of refusing to see any scientific visitors, Dubois relented and invited several prominent anthropologists to his home in 1927. But by then, the search for early man had entered a different phase. Dubois had spent most of his life trying to press a wrong conclusion, based on fragmentary evidence, and now new fossils were filling out the picture. More Java remains had been found, and the discovery of similar skulls at Peking, China was imminent. *Pithecanthropus* was becoming well defined as a widespread population of an early type of *Homo*, not a "missing link" to the apes.

By 1935, only one voice was raised claiming that *Pithecanthropus* was not a man at all, but a very large kind of gibbon-like hominid. Sadly, the voice was that of the aged Dubois.

See also HAECKEL, ERNST; HOMO ERECTUS; WALLACE, ALFRED RUSSEL.

DUGDALE, RICHARD

See JUKES FAMILY.

EARTHWORMS AND THE FORMATION OF VEGETABLE MOULD (1881)
Darwin's Last Book

One of the secrets of Charles Darwin's greatness was his pleasure in finding wonder in the small, commonplace features of the Earth most of us ignore. At the age of 72, the year before his death, he published his very last book *On the Formation of Vegetable Mould by Earthworms* (1881). This simple farewell treatise, researched in his own backyard, tells us much about the man and his methods as about worms.

While his wealthy neighbors in the English countryside amused themselves with social gatherings, bird shooting and cricket matches, Darwin loved strolling in his woods and fields, lifting rocks and poking logs, exercising his "enlarged curiosity" as he had done since childhood. To the end of his life, he never tired of watching the bees in his kitchen garden, the climbing plants in his greenhouse and the earthworms in his fields.

Today we take for granted the industriousness of worms and their important place in the ecology. In Darwin's day, these facts had still to be proved. Indeed, it was startling to some when he claimed for worms an important part in world history. "In many parts of England," he wrote, "a weight of more than ten tons of dry earth annually passes through their bodies and is brought to the surface on each acre of land; so that the whole superficial bed of vegetable mould [topsoil] passes through their bodies in the course of every few years." One critic, writing in the *Gardener's Chronicle* in 1869, indignantly rejected Darwin's conclusion. (He had first expressed it in a short paper in 1837, more than 20 years before the *Origin*

of Species, 1859, appeared.) "Considering their weakness and size," the critic wrote, "the work that they are represented to have accomplished is stupendous."

To Darwin, it was exactly that kind of "stupendous" accomplishment through the small but steady work of natural forces that was so wonderful, which is why he chose to return to the worms so many years later. The progress of science has often been retarded, he wrote, by "the inability to sum up the effects of a continually recurrent cause . . . as formerly in the case of geology, and more recently in that of the principle of evolution."

Worms had originally been brought to Darwin's attention by his uncle Josiah Wedgwood, the same benefactor who had convinced his skeptical father to let him go on the *Beagle* voyage. Darwin had noticed that stones scattered in fields tended to sink and become buried with time, and Wedgwood suggested that they were being covered by worm castings.

Darwin then began adding to his thousands of other observations the condition of several fields near his home. One in particular had been ploughed in 1841 and was called by his sons "the stony field," as it was thickly covered with small and large flints. In the *Earthworms* book, Darwin noted:

> When [my sons] ran down the slope the stones clattered together. I remember doubting whether I should live to see these larger flints covered with vegetable mould and turf . . . [but] after thirty years (1871) a horse could gallop over the compact turf . . . and not strike a single stone . . . The transformation was wonderful [and was] certainly the work of the worms.

PUNCH, OR THE LONDON CHARIVARI.

PUNCH'S FANCY PORTRAITS.—NO. 54.

CHARLES ROBERT DARWIN, LL.D., F.R.S.

IN HIS *DESCENT OF MAN* HE BROUGHT HIS OWN SPECIES DOWN AS LOW AS POSSIBLE—*I.E.*, TO "A HAIRY QUADRUPED FURNISHED WITH A TAIL AND POINTED EARS, AND PROBABLY *ARBOREAL* IN ITS HABITS"—WHICH IS A REASON FOR THE VERY GENERAL INTEREST IN A "FAMILY TREE." HE HAS LATELY BEEN TURNING HIS ATTENTION TO THE "POLITIC WORM."

CARICATURE OF DARWIN and his earthworms appeared in *Punch* magazine in 1881, the year before his death. His last book, the study of worms, stressed a major theme of his life's work— that small causes produce major effects in shaping the planet.

Once Darwin got hold of a subject, he pursued it for all it was worth. He had his son Horace design a round, heavy stone, known as the "Wormstone," which had a gauge in its center. He placed it in his garden, so he could measure precisely how far it would sink in a given amount of time. One of Horace's wormstones can still be seen in the garden at Down House.

Darwin, then an old man, even took the long rail journey to Stonehenge to observe how far worms might have buried the ancient "Druidical stones." His sons helped dig test holes near the monolith's bases, and Darwin duly recorded how far the great stones had sunk—though he was somewhat disap-pointed in the worm's industriousness in that region.

In his search for nature's truths, Darwin was a man who literally left no stone unturned. He even brought pots full of worms into the house to measure their activity at various temperatures and had his son play notes on the bassoon and the piano to them in the drawing room, to see whether they reacted to higher or lower-pitched notes. (How his wife Emma reacted to worms on her piano is not recorded.)

Punch magazine greeted the earthworm book with a caricature of Darwin pondering a giant worm, coiled like a question mark, next to this verse:

> I've despised you, old Worm, for I think you'll admit
>   That you never were beautiful, even in youth;
> I've impaled you on hooks, and not felt it a bit;
>   But all's changed now that DARWIN has told us the
>   truth
> Of your diligent life, and endowed you with fame—
>   You begin to inspire me with kindly regard;
> I have friends of my own, clever Worm, I could
>   name,
>   Who have ne'er in their lives been at work half so
>   hard.

In returning to earthworms in his last years, Stephen Jay Gould has pointed out, Darwin was not concentrating on some obscure, quirky topic as some have thought—"a harmless work of little importance by a great naturalist in his dotage." He was actually returning to one of the great themes of his life's work. In the last paragraph of the book, Darwin refers to another animal of "lowly organization," the coral. One of Darwin's very first treatises was *On the Formation of Coral Reefs* (1842), in which he suggested that the world's great ocean reefs were the work of millions of tiny creatures. "Clever old man," wrote Gould in a centenary appreciation of the worm book, "he knew full well. In his last words, he looked back to his beginning, compared those worms with his first corals, and completed his life's work in both the large and small."

See also CORAL REEFS; DARWIN, CHARLES; ECOLOGY; UNIFORMITARIANISM; WORMSTONE.

For further information:
Darwin, Charles. *The Formation of Vegetable Mould, through the Action of Worms.* London: John Murray, 1881.

ECOLOGY
Organism–Environment Relationship

Darwin had stressed the importance of adaptation, the evolutionary adjustment of an organism to its environment. It was his German disciple Ernst Haeckel who suggested the creation of a whole new field: the study of the "homes" or niches animals and plants occupy. In 1866, he coined the term "oekology," from the Greek word *oikos*, "a home."

Ecology has since become a scientific specialty, analyzing the relationship between an organism and its internal and external environment. The ecology of a particular bird, for instance, would include: the kind of trees it inhabits; the food it eats; the air it breathes; the parasites in its blood; the bacteria in its digestive system; its predators; its nesting materials; its relationships with other species; the extent of its range and territory; its population density; social dynamics; and so on.

Often, the word "ecology" is misused in a narrow political sense to mean environmental or conservation issues.

See also ADAPTATION; HAECKEL, ERNST.

For further information:

Krebs, John R., and Davies, N. B. *An Introduction to Behavioral Ecology.* Oxford: Blackwell Scientific Publications, 1987.

May, Robert M. *Theoretical Ecology.* Philadelphia: W. B. Saunders, 1976.

Odum, E. P. *Ecology.* New York: Holt, Rinehart and Winston, 1963.

EDELMAN, GERALD

See NEURAL DARWINISM.

EISELEY, LOREN (1907–1977)
Evolutionary Essayist

Evolutionary essayist Loren Eiseley was at his best when playing Hamlet to some sort of Yorick's skull. An Eiseley essay often begins with the writer musing over a stone tool or fossil, then takes off on a flight of informed evolutionary speculation that whisks readers through time and space, leading them to an extraordinary view of the ordinary.

Raised amidst a troubled family on the bleak plains of Nebraska, Eiseley himself resembled a severe landscape, his granite face grim and brooding. In his pictures, wrote Kenneth Brower, "he is always gazing off, like the poet at the picnic, except with Eiseley there is no picnic." His early days had been plagued with hardship and poverty, and he spent the Depression years riding the rails, submerged among the homeless and jobless moving aimlessly around the country. Eventually, he was to gain international recognition as a naturalist, essayist and poet.

His works sang, always in a minor key, of Pleistocene hunters, lost worlds and the mysteries of kinship among all life forms. By profession he was an anthropologist at the University of Pennsylvania and influential historian of science. His lively account of the rise of evolutionary theory, *Darwin's Century,*

(1958), is a perennial on college campuses. And Eiseley's collections of essays, *The Immense Journey* (1957) and *The Star-Thrower* (1978), are among the most literate, imaginative introductions to evolution ever written.

But what Eiseley wanted to do above all else was to find a fossil man—to wrest from the Earth a major discovery in the tradition of Eugene Dubois or Louis Leakey.

Unfortunately, his years of bone-hunting were a source of frustration, which turned to bitter humor. In an article titled "Obituary of a Bone Hunter" (1947) he wrote of wriggling through a remote cave passageway only to find it covered by "living blankets" of spiders and containing no fossils. Entering another rocky crevice, he disturbed a brooding owl, which flew in his face as he crawled through dust and excrement, but all to no avail.

Once, an eccentric old man showed up at Eiseley's office, unwrapped a strange fossil human jawbone and offered to take him to the site, if he would just "admit" to the newspapers it came from "the Golden Age of the Oligocene." Refusing to humor the stranger's offbeat theories, Eiseley allowed him to walk away, taking the tantalizing fossil with him, which he always regretted. Eventually, he decided to leave the bone-finding to others; his gift was finding words. In 1971, Eiseley was elected to the National Institute of Arts and Letters, a rare achievement for a scientist.

Eiseley wrote in *The Firmament of Time* (1960) of the dark, evolutionary consciousness he carried everywhere, even into a dull classroom. He was fumbling among his papers, about to begin a lecture, when a student asked, "Doctor, do you believe there is a direction to evolution?"

> Instead of the words, I hear a faint piping, and see an eager scholar's face squeezed and dissolving on the body of a chest-thumping ape . . . I see it then— the trunk that stretches monstrously behind him. It winds out the door, down dark and obscure corridors to the cellar, and vanishes into the floor. It writhes, it crawls, it barks and snuffles and roars, and the odor of the swamp exhales from it. That pale young scholar's face is the last bloom on a curious animal extrusion through time . . . I too . . . am a many-visaged thing that has climbed upward out of the dark of endless leaffalls, and has slunk, furred, through the glitter of blue glacial nights . . . I have no refuge in time, as others do who troop homeward at nightfall. As a result, I am one of those who . . . haunt the all-night restaurants. Nevertheless . . . there are hazards in all professions.

Looking down at a Pleistocene mammal fossil in an excavation, Eiseley once observed the skull seemed to be tilted up, staring "sightless, up at me as though I, too were already caught a few feet above him in the strata . . . The creature had never lived to see a man, and I, what was it I was never going to live to see?"

See also BLYTH, EDWARD.

ELAN VITAL
Vitalist Principle

French philosopher and metaphysician Henri Bergson had a rich literary style, clothing his arguments in emotionally affecting language. His influential book *Creative Evolution* (1911) was a treatise on evolution that purported to refute Darwinism on the basis of Bergson's intuitive feeling for a self-organizing vital principle he called the *elan vital*.

Scientists complained they had no way to work if Bergson denied them the possibility of finding causal explanations for life processes. Paleontologist George Gaylord Simpson said such theories "do not explain evolution, but claim it is inexplicable and then give a name to its inexplicability: *elan vital*, omega, aristogenesis, cellular consciousness, holism . . . As Huxley has remarked, ascribing evolution to an *elan vital* no more explained the history of life than would ascribing its motion to an *elan locomotif* explain the operations of a steam engine."

See also BUTLER, SAMUEL; DE CHARDIN, TEILHARD; SHAW, GEORGE BERNARD.

ELDREDGE, NILES

See GOULD, STEPHEN JAY; PUNCTUATED EQUILIBRIUM.

ELEPHANTS

See KEYSTONE SPECIES; SOCIAL BEHAVIOR, EVOLUTION OF.

ENDOSYMBIOSIS
Partners Within the Cell

Symbiotic organisms—animals or plants that have a mutually beneficial relationship—are a well-known marvel of evolution. Clown fish hide safely in sea anemones, scavenging crumbs in the midst of its poisonous tentacles. The fish cleans the invertebrate; the anemone's stingers protect the fish from predators. Ant colonies that live in acacia trees are another striking example; the insects repel all invaders in exchange for food and shelter.

Over the last few decades, however, researchers have found evidence that many types of cells (once thought to be the smallest unit of life) are actually combinations of originally different creatures living together as a cooperative entity. It is even possible that all complex cells evolved in this manner.

A well-studied case in point is the one-celled animal (protozoan) *Paramecium*. Tiny "hair-like" filaments (cilia) by which it moves with coordinated whip-like motions are actually separate creatures, moving in unison. Tiny termites have incorporated thousands of even smaller microorganisms, which have evolved in their guts to break down wood; they are, in effect, the termite's digestive system. Component organisms of this kind are called endosymbiants, helpful creatures that live inside (endo). The phenomenon is endosymbiosis. Since they reproduce asexually along with their host (or fellows), they are sometimes called nonchromosomal genes.

Before it was understood that some of these tiny parts of one-celled creatures were life-forms themselves, the various functioning blobs within a single cell were called organelles: little organs. But since we usually think of organs (stomachs, hearts, brains) as made up of millions of cells, how can we distinguish between an organelle and an endosymbiant?

If the association goes very far back in evolution, and the symbiant and host have become utterly interdependent, it can be called an organelle. But we really cannot make a hard and fast distinction; there is none in nature.

See also EUKARYOTES; MITOCHONDRIAL DNA; PHYLA; PROTOPLASM.

ENGLAND, DARWINISM IN

Within two decades after publication of the *Origin of Species* (1859), Darwinian evolution was completely accepted by British science, though his theories of natural selection, sexual selection and others were not pursued by experimental biologists.

Yet England was deeply conflicted about integrating Darwinism into its national belief system. On the one hand, natural selection could justify British colonial rule over "backward" nations and peoples. But a traditional sense of fair play and Christian values both urged compassion and social justice. Thomas Huxley concluded if survival of the fittest is the law of nature, civilized people must "turn their backs on it" and seek a higher ethic based on compassion and generosity.

England's poet laureate Alfred, Lord Tennyson had infused such popular works as *In Memoriam* (1850) (a favorite of Queen Victoria's) with evolutionary ideas even before Darwin's *Origin of Species* had appeared. One of Huxley's students, H. G. Wells, was the first to write popular fantasy novels on time travel, lost worlds and other scientific themes. By the end of the 19th century, unlike the situation in the

MIXED FEELINGS ABOUT DARWIN caused English officialdom to withhold state honors from the scientist while he was alive, yet he received a national hero's funeral. The British Museum's Darwin statue was originally installed on its grand stairway facing the main entrance, but was later moved to an upstairs gallery.

United States, there was no uproar over teaching Darwin in the schools. (Ironically, when Thomas Huxley became a minister of education he insisted the Bible also be retained, so students would not be deprived of its "great literature.")

But without a workable theory of heredity and no actual research on natural selection, Darwinism was fading from the attention of British science after the 1890s. However, after the rediscovery of Mendelian genetics around 1900, England again took the lead. At Oxford, Darwin's devoted disciple E. Ray Lankester promoted the work of August Weismann and encouraged research in population genetics, which was to revitalize Darwinian theory.

Lankester also served as director of the British Museum (Natural History), formerly the stronghold of the antievolutionist Richard Owen. To leave no doubt that the museum's orientation had changed, a large marble statue of Darwin had been installed on the stairway of the Great Hall in 1885, facing every visitor who entered. (Darwin has since been moved

to a small upstairs gallery, which he shares with a marble figure of Thomas Henry Huxley.)

Darwinism has also had a resurgence in 20th century British science. One of its leading lights was biologist Julian Huxley, grandson of Thomas. He helped develop ethology, the evolutionary study of animal behavior, which became a major research tradition in England. In addition, he was one of the founders of the modern Synthetic Theory (which he named), combining Darwinism with population genetics, paleontology and other fields.

Julian Huxley also churned out popular essays on his grandfather's favorite themes: evolution, ethics, science, religion, belief. His brother Aldous's well-known novels explored such kindred issues as eugenics and how science, religion and society impinge on the individual.

Despite this vigorous scientific and literary tradition, there still lurks some ambiguity in the national attitude towards Darwin. His statue is still in the museum, but somewhat tucked away. No likeness of the founder of evolutionary biology is to be found at Madame Tussaud's, although they have room for General William Booth, founder of the Salvation Army.

Down House, Darwin's home, has been made into a charming museum (through private efforts), but it is not listed in guides that steer tourists to the homes of Samuel Johnson or Robert Browning. And though it freely bestows knighthoods on actors and bankers, Britain never publicly honored Charles Darwin during his lifetime.

See also DARWINISM, NATIONAL DIFFERENCES IN; HUXLEY, SIR JULIAN S.; LANKESTER, E. RAY.

EOS
Greek Goddess of Dawn

Eos, the ancient Greek goddess of dawn or morning, has long been favored by poets and scientists. From 19th-century paleontologist O. C. Marsh to contemporary musician Michael Jackson, her name has symbolized new horizons and optimistic beginnings.

A daughter of the Titans Theia and Hyperion (at one time god of the sun), Eos appears in Greek art riding through the dawn sky in her chariot or sometimes astride the winged horse Pegasus. She is associated with new winds, the "rosy fingers" of first light and all fresh, young life. In Roman times, she was renamed "Aurora."

During Charles Darwin's century, the Greek prefix *eo* ("first" or "dawn") became very popular with geologists seeking origins. Sir Charles Lyell, Darwin's mentor in geology, named the second main division of Teritary rocks, containing many early mammals, the Eocene (dawn of the recent time) and

EO, MEANING "DAWN," became so popular among paleontologists that Thomas Henry Huxley sent his friend Charles Darwin this humorous sketch of "Eohomo" and "Eohippus"—dawn man riding the dawn horse.

the term has stuck. Another geologist named John W. Dawson coined the term Eozooan (first animal) for certain fossil invertebrates that were later discredited as far from the first animals.

Other coinages have since fallen out of favor, although for a time they were household words. For years every biology student knew *Eohippus* (the "dawn horse"), which stood about three feet high. For technical reasons, its name was later changed to *Hyracotherium*.

The most spectacular misnomer was the genus named *Eoanthropus*—the "dawn man," applied to a supposed "ape-man" dug up at Piltdown, England in 1912. Once hailed as "the first Englishman," *Eoanthropus* is now known as the Piltdown Hoax. The fossils turned out to be deliberate forgeries. Crude tools supposedly used by the "dawn man" were known as eoliths (dawn stones), but this vague, all-inclusive category has also been dropped from the vocabulary of prehistorians.

Eo has recently made a comeback in popular culture—not in science at all—but in rock star Michael Jackson's $40 million 3-D fantasy *Captain Eo* (1986), viewed by millions at the Walt Disney parks. It depicts a futuristic hero, Captain Eo (Captain Dawn), a mythologized dream of the "next step" in human evolution.

"EQUAL TIME" DOCTRINE
Creationists' Legal Strategy

After creationists lost their battle to prohibit public schools from teaching evolution, which they per-

ceived as threatening to their children's religious beliefs, they focused new legal efforts on establishing "equal time" for "scientific creationism" as an equally plausible theory. [See SCOPES II.]

It was a strategic attempt to turn the tables on those who accused them of stifling free expression and inquiry. Science was the bigoted belief system, they maintained, if it could not tolerate exposure to competing ideas. Why not, they argued, teach creationism and evolution side by side, and let the students freely choose?

"Equal time" is customarily granted to competing political candidates, so conflicting opinions may be heard. On its surface, a balanced presentation in classrooms seems reasonable. But creationists assume there are only two theories of human origins: Genesis and evolution. Actually, beliefs abound about how the first people appeared on Earth, including hundreds of African, Native American, Polynesian, Chinese and Hindu versions, none of which the "equal time" advocates want to include. [See AVATARS; NAPI.]

Nevertheless, students must face the reality that technical achievements depend on commonly accepted science. Oil companies will not hire geologists who believe the Earth is only 4,000 years old; lawyers and prosecutors will lose cases if they fail to use the new technology of DNA identification. And despite creationist assertions that no one has actually seen evolution in action, geneticists, agricultural botanists, ecologists, medical immunologists and public health officials deal with it every day on a very practical level.

At the same time, it is vital that students do not uncritically swallow evolutionary theory, or any other scientific conclusion of the moment, without understanding the limits of its methods and explanations. Thomas Huxley fought hard and long for science to become the dominant belief system, but his greatest fear was that future generations might come to accept it for the wrong reasons—as an official truth to be taken on faith and authority.

Creationist attacks on the teaching of evolution have done science a service. Many teachers have been forced to develop a more systematic presentation of evolutionary theory, marshal their evidence, understand the objections and value systems of religionists, and foster active, critical discussions of scientific issues. Whether in politics or philosophy, religion or biology, rote-learned dogmas banish the stress of uncertainty while straightjacketing the inquiring mind.

See also FUNDAMENTALISM; SCIENTIFIC CREATIONISM.

For further information:
Futuyma, Douglas. *Science on Trial: The Case for Evolution.* New York: Pantheon, 1983.

Gilkey, Langdon. *Creationism On Trial: Evolution and God at Little Rock*. Minneapolis: Winston, 1985.

Godfrey, Laurie, ed. *Scientists Confront Creationism*. New York: Norton, 1983.

ESSENTIALISM
Belief in Ideal Types

One of the great shifts in conceptualizing the natural world has been the change from essentialism to population thinking. Charles Darwin and Alfred Russel Wallace helped create this modern outlook, which departed from 2,000 years of intellectual tradition. In the century and a half since their initial insights on natural populations, confusion has lingered as the older focus on individual organisms was replaced. Indeed, Darwin himself sometimes appeared confused as to whether it was individuals or populations that should be the units of study.

Essentialism is a way of looking at nature that goes back to the ancient Greek philosophers. Plato and his student Aristotle believed an "ideal" reality or "essence" lies behind everything we perceive in the world. Plato's famous allegory of the cave was meant to point up the illusory nature of what we think we see. Viewers in his cave, like the audience in a movie theater, saw only projected shadows on the wall and mistook them for real people, rather than flat and imperfect copies.

When speaking of human art, essentialism can be a valid description. A craftsman makes a chair, for instance, which we know is only an imperfect representation of the ideal "Chair," a type he has in his mind. But it does not therefore follow, as Plato believed, that plants and animals in nature are also "imperfect" models of ideal types. Accepting that conclusion implies they began as distinct ideas in the mind of God, which is something we cannot know by deduction, inference or observation. (We can believe it on faith, but then we leave the realm of natural science.)

Linnaeus and other early naturalists were typologists: they ignored variations of individuals in their constant search for the perfect "type specimen" of each species. Despite their efforts to accept the evidence of observation, their view of nature was really a continuation of the old essentialist outlook. (Even today, in such trivial matters as so-called beauty contests, judges still seek Miss World or Miss America—the individual who comes closest in their opinion to the "essential" woman. Fortunately, such searches for the "ideal type" are no longer taken seriously.)

Ernst Mayr has traced the idea of essentialism in biology, and concludes that the whole science was transformed by what he calls "population thinking,"

as a consequence of the Darwinian revolution. Mayr summarizes the change as follows:

> [The Essentialist believes there are] a limited number of fixed, unchangeable "ideas" underlying the observed variability [in nature], with the *eidos* (idea) being the only thing that is fixed and real, while the observed variability has no more reality than the shadows of an object on a cave wall . . . [In contrast] the populationist stresses the uniqueness of everything in the organic world . . . For the typologist the type *(eidos)* is real and the variation is an illusion, while for the populationist, the type (average) is an abstraction and only the variation is real. No two ways of looking at nature could be more different.

So long as naturalists accepted the premise of essentialism, the idea of the fixity of species did not seem at odds with nature. Today's creationists still hold on to a notion of "kinds," which is the same as "ideal types." Divine "ideas" they believe, shape each species, which reproduce "after their kind." When variations are shown to them that fall between "types," they insist on assigning them to one "kind" or another and thus never confront the phenomena of blurred boundaries, which more accurately describes nature.

Darwin and Wallace, after systematically acquainting themselves with the enormous range of variability in organisms, dared to say that nature was not so tidy. Species were not easy to "typologize" once a naturalist got away from catalogued museum collections and out into the forests and islands. There, boundaries between natural populations are often problematical; species, varieties and races are often difficult to define.

During the first few decades of the 20th century, mathematicians developed new tools for analyzing population phenomena and describing variability. Population geneticists and biometricians made variability—not ideal types—the focus of their studies. In the 1930s, these newer methods were combined with the older disciplines of paleontology, comparative anatomy and systematics (taxonomy) to shape the modern approach to evolutionary biology. Because it fused together several diverse sciences, it is known as the Synthetic Theory.

See also NATURPHILOSOPHIE; POPULATION THINKING; SYNTHETIC THEORY.

For further information:

Desmond, Adrian. *Archetypes and Ancestors: Paleontology in Victorian London*. Chicago: University of Chicago. Press, 1982.

Mayr, Ernst. *The Growth of Biological Thought*. Cambridge: Harvard University Press, 1982.

Oken, L. *Elements of Physiophilosophy*. (Translation of *Lehrbuch der Naturphilosophie*.) London: Ray Society, 1847.

ETHOLOGY
Study of Animal Behavior

Charles Darwin took a new approach to understanding animal behavior in his book *The Expression of the Emotions in Man and Animals* (1872). At a time when most zoologists were mainly interested in the anatomy of preserved specimens, he studied the behavior of the living animal and how its behavior might have evolved.

Curiously, his exciting lead was not followed for 50 years, until the Austrian Konrad Lorenz and later the Dutchman Niko Tinbergen made a new attempt to study behavior from an evolutionary perspective.

Psychologists had dominated the field with their rats, mazes and puzzle boxes; the ethologists conducted experiments in the field, basing them on observations of the animal's natural behavior. Their work with the courtship displays of ducks and geese, territorial behavior of stickleback fish, imprinting of water fowl on their mothers and social behavior of herring gulls gave new impetus to the science of animal behavior.

They were especially successful studying the interaction between innate, species-specific behavior and learning. Also, they penetrated animal signal systems to an unprecedented extent, gaining new understanding of animal communication.

But after several decades of excellent work, the ethologists got tangled up in their own convoluted theories of "drive," "motivation," "releasing mechanisms," and built an intellectual structure that collapsed under its own weight.

In addition, they became discredited by a spate of facile books applying their theories of fish and bird behavior directly to humans. (Lorenz's *On Aggression* (1966); Robert Ardrey's *The Territorial Imperative* (1966); Desmond Morris's *The Naked Ape* (1967) used ethological studies to support a view of humanity as aggressive, incorrigible, territorial killers.)

They began by trying to understand the evolution of behavior—an enormous scientific undertaking—and ended by attempting quick and easy shortcuts. Although "pop ethology" enjoyed a tremendous vogue during the late 1960s to mid-1970s, it became scientifically discredited for its superficiality.

Recently, the concepts of ethology have been all but abandoned as investigators of animal behavior relabel themselves "behavioral ecologists." Another successor to ethology is the more comprehensive discipline sociobiology. Scientific fashions come and go, but the endlessly fascinating behavior of animals remains.

See also EXPRESSION OF THE EMOTIONS; LORENZ, KONRAD; TINBERGEN, NIKOLAAS.

For further information:
Lorenz, Konrad. *Evolution and Modification of Behavior*. Chicago: University of Chicago Press, 1965.
Thorpe, W. *The Origins and Rise of Ethology*. London: Hinnemann, 1979.
Tinbergen, Nickolaas. *The Study of Instinct*. London: Oxford University Press, 1969.

EUGENICS
Breeding Better Humans

In 1883, Charles Darwin's cousin, Sir Francis Galton, coined the word "eugenics." He had first advanced his theories in 1865 in a series of magazine articles, which were later expanded into his book *Hereditary Genius* (1869), so his eugenic ideas were advanced long before he coined the word itself.

Eugenics comes from a Greek root meaning "noble in heredity" or "good in birth." Pursuing the social implications of Darwinism, Galton wanted to create a science that could improve humanity by giving "the more suitable races or strains of blood a better chance of prevailing speedily over the less suitable." He believed natural selection should be given an artificial assist, much as the domestic stockbreeder improves horses or cattle. According to historian Daniel J. Kevles, Galton:

> . . . suggested that the state sponsor competitive examinations in hereditary merit, celebrate the blushing winners . . . foster wedded unions among them at Westminster Abbey, and encourage by postnatal grants the spawning of numerous eugenically golden offspring . . . The unworthy, Galton hoped, would be comfortably segregated in monasteries and convents, where they would be unable to propagate their kind.

Once almost obligatory in all biology textbooks, the promotion of eugenic programs was set back by the disastrous, barbarous attempts to create a "master race" in Nazi Germany. However, the notion lingered on for several decades in America, bolstered by miscegenation laws in Southern states, immigration quotas, and legislation to sterilize criminals and the "feeble-minded." Eventually, it became recognized that the presumed scientific underpinnings of these social policies did not, in fact, exist.

But though Galton's term for the genetic betterment of humans is tarnished, his ideas continue to figure heavily in several important areas of social and scientific debate. There are still those such as Professor Arthur Jensen who argue for the validity of racial-genetic intelligence testing, although most scientists are convinced that cultural factors invalidate the results. Some sociobiologists are now concerned with establishing genetic bases, determinants and limitations for what is to be considered the "natural" behavior of humans and societies. And there is the new field of genetic testing, parental selection, manipulation of embryos and human genetic engineering, which is presently groping for a guiding ethic. Over all these, Galtonian eugenics casts a long shadow.

See also GALTON, SIR FRANCIS; JUKES FAMILY; LEBENSBORN MOVEMENT.

For further information:

Galton, Sir Francis. *Hereditary Genius.* London: Macmillan, 1869.

Jones, Greta. *Social Darwinism and English Thought.* Atlantic Highlands, N.J.: Humanities Press, 1980.

Kevles, Daniel. *In the Name of Eugenics.* Berkeley: University of California Press, 1985.

EUKARYOTES
Cells with Nuclei

Eukaryotes are living cells that have a central nucleus suspended in cytoplasm: the whole wrapped in a cell wall, like an egg yolk surrounded by protein, enclosed in a shell. Nucleated cells make up trees, starfish, fish, insects, birds, mammals; what we usually think of as our fellow creatures. But for millions of years before cells with nuclei appeared, living prokaryotes (cells without nuclei) dominated the Earth. Their close relatives are still with us in the form of blue-green algae, or cyanobacteria.

According to biochemist Lynn Margulis, "the sudden appearence of eukaryotes in the fossil record, and the absence of any intermediate forms, suggest that they were not the gradual result of genetic mutation but a technical innovation forged by communities of symbiotic bacteria." In her view (built on the work of L. Cleveland and others), nucleated cells originated when non-nucleated bacteria engulfed one another, forming compound organisms.

Cleveland studied the tiny universe in a termite's gut: millions of microscopic creatures swallowing and merging with others in remarkable ways. Some of those that are eaten are not digested and destroyed, but live cooperatively within other microbes, producing parasitic offspring to live within their host's offspring.

In many cases the relationship is not really parasitic, but mutually beneficial. Cells within cells become interdependent (endosymbiotic) and form stable organisms—new wholes greater than the sum of their parts.

All eukaryotes, including ourselves, may really be "microbial concoctions." Biology has long taught that we are complex, coordinated colonies of cells, but in the new view each of our cells may in turn be colonies made up of cooperating organisms.

See also ENDOSYMBIOSIS; GAIA HYPOTHESIS; PROTOPLASM.

EVOLUTION
Species Change Through Time

By "evolution" biologists mean that the change in gene frequencies of populations over the generations in time produces new species. Darwin called it "descent with modification": a slow process, usually operating over hundreds of thousands, and even millions, of years.

There are four commonly confused meanings of evolution, which should be kept separate and distinct: (1) the general process of populational and species change, which is considered an established scientific fact; (2) inevitable "progress" from lower to higher life forms, a discredited notion; (3) the particular history of the "branching bush" of life and the origin of various groups, or phylogenies, which are interpreted from the fossil record and biochemical studies; and (4) the mechanism, or "engine," of evolution, which Darwin and Wallace proposed as "natural selection," but which is currently being investigated and modified by research.

Here are some of the major arguments and objections that opponents of evolution never tire of raising—and some answers from the perspective of evolutionary biology:

1. *Fact or Theory?* Evolution became established as fact, not because it won debates among armchair philosophers or logicians, but because it unified thousands of disparate observations by comparative anatomists, field naturalists, geologists, paleontologists, botanists and (later) geneticists and biochemists. Without the overarching concept of a changing world in process over eons of time, what we consider modern science would not exist.

    That species are related through common ancestry is supported not by one argument or chain of reasoning, but by scores of interlocking research fields, each of which feeds into and supports the rest. Evolution is as well established a fact as gravitation. To paraphrase a leading paleontologist, apples are not going to stop falling in midair while scientists debate whether Newton's law of gravitation has been superceded by Einstein's theories. And species keep on changing over time, while we continue to search for the why and how of evolution.

    If one insists evolution is merely one interpretation of nature, what is the alternative? That the thousands of dinosaurs and the species that came before and after them were *not* related to each other, appeared full-blown and had no common connections? Such a model, whether called religion or "creation science," cannot lead to fruitful inquiry. It is an answer that stops all further questions.

2. *"General" Evolution* vs. *Speciation.* While some critics concede that new species (of fruit flies, for instance) have been produced in laboratories, they claim that general evolution has never been experimentally demonstrated. By this, they mean breeding a succession of progressively higher or more complex species. But there is no such theory of general evolution; Victorian notions of inevitable progress in biology are outmoded.

3. *Transitional Forms.* The oft-repeated claim that there are no transitional forms is demonstrably false. The Karroo region of South Africa, for instance, is a vast graveyard of the remains of mammal-like reptiles, a whole array of species whose anatomy was intermediate between reptiles and mammals. There is the famous *Archeopteryx,* with its feathers, teeth, claws and lizard-like skeleton, a transition between reptiles and birds. And the African hominid fossils respresent creatures with human-like dental patterns, small brains, arms longer than humans but shorter than modern apes, with pelvis, feet and legs for upright walking.

Transitional fossils are notably rare because, according to current theory, most species remain stable for long periods. When change occurs it is fairly rapid (in geologic terms) and often among small, isolated populations. The fossil record has been compared to freezing a multilevel parking garage in time. Most cars would be found on the various floors, with very few on the ramps. The amount of time each car spends on the ramp is short compared to the length of time it remains parked, yet each must have traveled the ramp.

Another evidence of transition is found in geographical distribution of living species. On Pacific island chains, for instance, biologists have tracked populational species across thousands of miles, discovering intermediate forms from one end of the island chains to the other.

Darwin himself was so impressed with a series of such geographic variations in Amazonian butterflies over a vast area of rain forest, he was moved to remark, "We feel to be as near witnesses, as we can ever hope to be, of the creation of a new species on this earth." Among living creatures, there is a series of gradual, intermediate species between lizards and snakes, thrushes and wrens, sharks and skates.

4. *Evidence and "Proof."* There is a common misconception that Darwin thought he had "proved" by logic that species had evolved. He was, in fact, a much more subtle thinker and philosopher of science. "The change of species cannot be directly proved," he wrote a friend, "and . . . the doctrine must sink or swim according as it *groups and explains [disparate] phenomena.* It is really curious how few people judge it in this way, which is clearly the right way." (A few years later he wrote that he was "weary of trying to explain" this point, since most could not grasp it.)

5. *"Holes" and Questions.* That there are "holes" and unanswered questions in evolutionary theory (just as there are in particle physics) is undeniable, which is normal for a healthy science. Thomas Henry Huxley once asked his students to imagine being lost in the countryside on a dark night, with no clue to the road. If someone offered a dim, flickering lantern, would you reject it on the grounds that it gave an imperfect light? "I think not," said Huxley, "I think not."

6. *Tautology of "Survival of Fittest."* This old chestnut that evolutionary theory is built on the circular reasoning that "the survivors survive" was laid to rest long ago. Critics argue that "only the fittest survive" is an untestable proposition without a uniform definition of fitness and, therefore, meaningless as an explanation. Fair enough, and it is true that some fuzzy-thinking scientists have concocted fanciful "just-so" stories about the origins of particular adaptations.

But whatever the fate of those archaic catch phrases "natural selection" and "survival of the fittest," the heart of the Darwin–Wallace theory remains sound: overproduction of offspring in nature, genetic variability and a sorting process, which result in both long-term stability and episodic divergence of populations. Increasingly, new research is focusing on gaining a deeper understanding of these mechanisms of genetic variation and differential sorting as they occur on various levels within the same organism.

7. *"Just History," Not Science.* Some assume that research and inquiry into living things must lead to the formulation of fixed "laws" like those of chemistry or physics. Dissecting the anatomical structures of extinct creatures, working out their distribution in space and time and reconstructing past climates and ecologies may be "just history" to a physicist or chemist but to most biologists it is certainly science.

The kind of scientific illiteracy that rejects evolution as a humanist religious belief can result in serious errors in understanding and even loss of human life. For example, in 1984, Dr. Leonard L. Bailey of the Loma Linda University (Seventh-Day Adventist) school of medicine tried to save the life of "Baby Fae," an infant born with a severely malformed heart. He surgically implanted a baboon's heart, but the organ was quickly rejected and the child died.

Soon after, he was asked why he hadn't used a chimpanzee's heart instead, which would have offered a far better chance of success because of the chimp's much closer evolutionary propinquity and genetic fit. Dr. Bailey replied that he "didn't believe in evolution," and, in any case, couldn't see what it had to do with the practice of medicine.

The word "evolution" in the 17th century meant an "unrolling" or "unfolding" of a plan that was already there, as in the theory of preformation. Similarly, it also meant embryological development of an individual.

When Darwin began turning his attention to "that mystery of mysteries, the origin of species," the idea that species change through time was called "transmutation" or the "development hypothesis." In the Darwin–Wallace papers read before the Linnean Society in 1858 announcing the principle of natural selection, neither author used the word "evolution," nor did it appear in the first edition of *Origin of Species*. (It first appears in the fifth edition of 1869.)

As with "survival of the fittest," it was British philosopher Herbert Spencer who originated the term "evolution" in this context. Wallace and Darwin later adopted it but, in the public mind, both phrases have become completely identified with the theories of Charles Darwin.

See also ADAPTATION; ALLOPATRIC SPECIATION; BRANCHING BUSH; COEVOLUTION, CONVERGENT EVOLUTION; DARWINISM; DARWIN'S FINCHES; EXAPTATION; FITNESS; GENETIC DRIFT; HOMOLOGY; NATURAL SELECTION; NEO-DARWINISM; ORTHOGENESIS; PIGGYBACK SORTING; SYNTHETIC THEORY; TANGLED BANK.

For further information:

Bowler, Peter J. *Evolution: The History of an Idea*. Berkeley: University of California Press, 1984.
Dawkins, Richard. *The Blind Watchmaker*. New York: Norton, 1986.
Futuyma, Douglas. *Science on Trial: The Case for Evolution*. New York: Pantheon, 1983.
Gould, Stephen Jay. *Ever Since Darwin*. New York: Norton, 1977.
———. *The Panda's Thumb*. New York: Norton, 1980.
Reader, John. *The Rise of Life*. New York: Knopf, 1988.

EVOLUTION, MYSTERIES OF
Major Unsolved Problems

Charles Darwin candidly admitted there were "great difficulties" and unsolved puzzles in our study of evolution. Science teachers often make the grave mistake of trying to gloss over these areas of ignorance, attempting to provide an answer for everything. Creationists rightly puncture such pretensions to knowledge, but then go on to scoff at evolutionary theory for what it *hasn't* solved, as if that disproves all of modern biological knowledge.

The areas discussed below pose the most profound challenges to evolutionary science. Despite years of research and discussion (and, in some cases, promising directions), there are still no answers:

1. *Origin of Life.* How did living matter originate out of nonliving matter? Was it a process that only happened once or many times? Can it still happen today under natural or artificial conditions? Did it evolve out of the kind of growth and replication processes we see in crystals or on an entirely different basis?

2. *Origin of Sex.* Why is sexuality so widespread in nature? How did maleness and femaleness arise? If it is necessary to maintain genetic variability, why can many microorganisms do without it? How can one account for such phenomena as parthenogenesis: frog eggs, for instance, can produce tadpoles if they are pricked by pins or stimulated by electric current, without having been fertilized by male sperm.

3. *Origin of Language.* How did human speech originate? We see no examples of primitive languages on Earth today; all mankind's languages are evolved and complex. Can the answer be sought in the structure of the brain, experiments on teaching apes, animal communication systems or is there no way to ever find out?

4. *Origin of Phyla.* What is the evolutionary relationship between existing phyla and those of the past? There is still no agreement on how many there are today, how many we know from the fossil past and which may have come out of which. Transitional forms between phyla are almost unknown.

5. *Cause of Mass Extinctions.* Asteroids are currently in vogue, but far from proven as a cause of worldwide extinctions. And though punctuated equilibrium theory helps account for the so-called sudden appearence of new groups and long persistence of others, it has raised many new questions about stability and extinction of species.

6. *Relationship between DNA and Phenotype.* Can small, steady changes (micromutations) account for evolution, or must there be periodic larger jumps (macromutations)? Is DNA a complete blueprint for the individual, or is it subject to various influences and constraints in its expression? Are there any circumstances under which environment or behavior can work "backwards," influencing changes in the DNA?

7. *How Much Can Natural Selection Explain?* Darwin

never claimed natural selection is the only mechanism of evolution. Although he considered it a major explanation, he continued to search for others, and the search continues.

To dismiss these wide-open questions with pseudo-answers just to fill in unanswerable gaps is intellectually dishonest and no service to science. When asked about the origin of life, for instance, some say it "probably came about when a spark of lightning hit a 'primeval soup' of organic chemicals." That research direction has been pursued for years but never proven; its mindless repetition only stifles students' creativity in coming up with new approaches to science's greatest mystery.

Does that mean we must abandon everything we have learned about evolution because of the great questions still unanswered? How much better to admit and identify areas of profound ignorance and challenge the next generation to explore them.

See also DINOSAURS, EXTINCTION OF; LIFE, ORIGIN OF; PHYLA; SEX, ORIGIN OF; THEORY, SCIENTIFIC.

EXAPTATION
New Uses for Old Structures

One of the great puzzles of adaptation is how did certain structures that work so well develop by stages? If birds' wings were not designed to fly from the first, for instance, why would their ancestors have started growing them? As Stephen Gould has phrased the problem, "What good is half a wing?"

Studies of anatomy, fossils and the behavior of living animals suggest an answer to this brain-teaser. Anatomical structures used in one kind of behavior often can be shifted or co-opted for another function.

A favorite example of evolutionist Elisabeth Vrba's is the African black heron, a wading bird with well-developed wings, that is perfectly capable of flying. But the bird has found another major function for its wings. It fans them around its head and holds them extended to eliminate glare from the surface of the water. Thanks to this built-in sunshade, it can get a much better view of the fish and tadpoles it seeks. (Also, the prey may seek the shade.) If most other birds should become extinct, someone might look at the African heron and conclude that flight was only a stage in the evolution of glare shields.

When birds first evolved from reptiles, it seems likely their half-wings served an entirely different purpose than flight. They may have been useful for heat regulation, for instance, or for steadying the animal as it ran along the ground.

Animals often find themselves in different environments than those to which they have become

FANNING OUT WINGS, these African black herons create glare shields so they can see fish and tadpoles in the water. Using their wings as shades more often than for flight, these birds illustrate how existing structures can take on new functions during evolution. (Photo by Alan C. Kemp.)

adapted. Sometimes they "make do" with the old structures under the new conditions. For instance, marine iguanas of the Galapagos swim underwater all day grazing on seaweed, but have never developed flippers or webbed feet. Recent observations of certain populations of deer in Oregon report them habitually feeding on fish that are abundant on the riverbanks, although their teeth and digestive systems are supposedly adapted to herbivorous diets.

Over time, new behaviors may create new selection pressures. Old structures may be "made over" as certain variations are preserved and modified by natural selection. Parts of the old reptilian jaw became the ear bones in mammals. In elephants, an organ of smell became a powerful, flexible "hand." And of course, Stephen Jay Gould's favorite example, the giant panda, retains meat-eaters' teeth, while the wristbone has evolved into a crude "thumb" for holding the bamboo stalks on which he now feeds.

Older writers used the term "preadaptation" for structures that were present earlier, then switched function. The term led to confusion, since it suggests foresight, design or "preplanning" of the ultimate use of that organ. Exaptation is consistent with the idea that each species is the result of a special, unique history, which at no point can be predicted.

See also ORTHOGENESIS; PANDA'S THUMB; TELEOLOGY.

For further information:
Gould, Stephen Jay, and Vrba, Elisabeth S. "Exaptation—a Missing Term in the Science of Form." *Paleobiology,* 8,1 (1982):4–15.

EXPRESSION OF THE EMOTIONS (1872)
The Evolution of Behavior

At a time when most evolutionists were dissecting and comparing muscles and bones, Charles Darwin was still ahead of the pack. In 1872, he published the first modern book on the evolution of behavior, *The Expression of the Emotions in Man and Animals,* the founding document of comparative psychology, ethology and sociobiology.

Darwin realized that studying the evolution of bodies without also studying behavior was meaningless. He also realized, as Konrad Lorenz wrote a century later, "that behavior patterns are just as conservatively and reliably characters of species as are the forms of bones, teeth, or any other bodily structures."

He described movements specific to species, which were later called "fixed action patterns": a species' particular way of courting, fighting, resting, or feeding. Darwin mentions, for instance, how dogs may turn circles and scratch the ground before going to sleep on a carpet, "as if they intended to trample down the grass and scoop out a hollow, as no doubt their wild parents did." (He credits his grandfather Erasmus as the first naturalist to describe such repeated, stereotyped behaviors in animals.)

In *Expression of the Emotions,* he also discusses general principles that seem to him "to account for most of the expressions and gestures involuntarily used by man and the lower animals, under the influence of various emotions and sensations." Some, he thought, originated in habits that became "fixed" somehow in the species, like the dog preparing the ground for sleeping.

Another was his principle of "antithesis": Opposite emotions produce opposite expressions. When a dog is aggressive his ears and tail stiffen and his body strains forward. When he is submissive, he crouches, with ears and tail folded down. Darwin's astute observation is still valid, though today's investigators focus on "reduction of ambiguity" in the messages. Opposite signals for opposite intentions are easy to read.

With characteristic thoroughness and curiosity, Darwin also explored the effects of electric stimulation on human facial muscles, the expressions of the

SIGNALS OF FEAR, AGGRESSION, submission or pleasure among animals, and the possible evolution of these behaviors, was the subject of Charles Darwin's book on *Expression of the Emotions.* Observing his own dogs and cats, as well as more exotic creatures, he founded the research tradition that led to ethology and sociobiology.

insane, emotional communication by actors and in works of art and the development of facial gestures in infants.

If human emotions and their expression formed an unbroken continuity with the natural world, it was only a step to the next conclusion: Animals have minds and thoughts as well as feelings. He was not willing to tackle that one directly, but served as mentor to the young George Romanes, who published his *Mental Evolution in Animals* (1883) shortly after Darwin's death. (That volume contains a chapter on "Instinct" written by Darwin.)

For *Expression of the Emotions,* Darwin not only collected voluminous observations of animals, but also wrote to many travelers and missionaries to find out if different races and cultures of humans had the same facial expressions for grief, surprise or anger. He discovered, somewhat to his surprise, "the same state of mind is expressed throughout the world with remarkable uniformity."

But the main thrust of this book was the attempt to carry the evolutionary argument a step farther than had been done before. Using the comparative method, he attempted to show that a human smile was derived from a monkey's grimace of submission, that animals and humans both tremble with fear; he was purposely knocking down the wall between human and animal behavior. It was a demonstration of the grand view he had written in a notebook years earlier: "We may all be netted together." And a challenge to future generations to ultimately understand the evolution of human behavior.

See also COMPARATIVE METHOD ETHOLOGY; LORENZ, KONRAD; ROMANES, GEORGE J.

For further information:

Darwin, Charles. *Expression of the Emotions in Man and Animals.* London: John Murray, 1872.

Ekman, P., ed. *Darwin and Facial Expression: A Century of Research in Review.* New York: Academic Press, 1973.

EXTERNALLY DETERMINED SEX

See CROCODILIANS.

EXTINCTION
Destruction of Species

The recent history of the past few hundred years records the destruction of hundreds of species of plants and animals, including the great auk, dodo and passenger pigeon. The American bison was pulled back from the brink of extinction when there were only a few hundred left alive, out of a population that had numbered 40 million. During a single five-year period (1870–1875), buffalo hunters were slaughtering 2.5 million of them annually.

Extinction has always been a fact of life. According to ecologists Paul and Anne Ehrlich in their book *Extinction* (1981), 98% of all species that have ever lived have become extinct. There are probably about 10 million species alive on the Earth today, of which only 1.5 million have been discovered and given scientific names. Many are now being exterminated before they are even discovered, particularly in the tropical rain forests, which are so rich in diverse life forms. (There may be one million species in the Amazon basin alone.)

Until the late 18th century, naturalists did not imagine extinction was possible. Each species was believed to be a distinct idea in the mind of God, a link in an unbroken cosmic chain that allowed no gaps. When fossils of strange animals, like mastodons, were discovered, it was assumed that there must be some still living in the vast wilderness areas that had not yet been explored.

As more and more fossils were found (the first dinosaur teeth were discovered only in 1825) and more wilderness settled, the evidence of extinct creatures began literally to pile up. Late in his career, the eminent French anatomist Georges Cuvier had to admit that fossil bones were the remains of extinct species.

In the intervening years, human activity has brought about many extinctions. Although we have yet to see a new species evolve in nature—a very slow process—we often see them end, which can happen very quickly. As British naturalist Sir Peter Scott has put it, "Living species today, let us remember, are the end products of twenty million centuries of evolution; absolutely nothing can be done when the species has finally gone, when the last pair has died out."

Paleontologists have documented several mass extinctions, which wiped out the majority of life on Earth, allowing new forms to radiate and develop. One such "mass dying" occurred after the Cambrian period, eliminating the once-numerous trilobites. Another is the famous and much-pondered Cretaceous extinction, which ended the 140-million-year reign of dinosaurs as the dominant form of life and ushered in the Age of Mammals.

Although extinctions have periodically swept away millions of species, there was also time and opportunity for new species to evolve. As the Ehrlichs point out, we don't know which species may be the key to our own survival. Perhaps we can get along quite well without the snail darter, whooping crane or California condor, but what if we inadvertently wipe out the organisms that recycle nitrogen? We simply don't know enough about the delicate global ecosystem with which we are tampering.

Man's acceleration of the extinction of contemporary species is accompanied by habitat destruction and industrial poisoning, which prevents new speciation. There is serious scientific concern that humans may destroy the ability of the planet to support life. If so, the human species itself may become extinct, as did other hominids whose skulls we study.

No doubt, they didn't imagine such an outcome was possible any more than we do.

See also DINOSAURS, EXTINCTION OF; GREAT DYING; LONESOME GEORGE; NEMESIS STAR; RAIN FOREST CRISIS.

For further information:

Ehrlich, Paul and Anne. *Extinction: The Causes and Consequences of the Disappearance of Species*. New York: Random House, 1981.

Stanley, Steven M. *Extinction*. New York: Scientific American Books, 1987.

FANTASIA (1941)
Disney's Epic of Dinosaurs

With his innovative feature film *Fantasia*, Walt Disney hoped to elevate animated cartoons to high art. Against a background of classical music, he filled the screen with such grand themes as the unleashing of hellish forces, man's eternal attempt to gain power over nature and the evolution and extinction of the dinosaurs.

An orchestral score conducted by Leopold Stokowski provided the framework for a series of imaginative (and sometimes grotesque) visual interpretations. In its most famous sequence, Mickey Mouse's meddling nearly destroys the natural order of the universe to the accompaniment of Paul Dukas's scherzo *The Sorcerer's Apprentice.*

Disney's remarkable vision of evolution and extinction, oddly enough, was set to Igor Stravinsky's stirring and dramatic *Rite of Spring.* (To Disney's dismay, the irascible maestro later called the musical interpretation "execrable," and "the visual compliment . . . an unresisting imbecility.")

Fantasia's dinosaurs are mighty but slow-witted behemoths caught in a change of climate. The Earth is heating up, and thirsty dinosaurs undertake a vast migration across bleak, parched landscapes in search of water. (Almost 50 years later, another animated dinosaur feature, *The Land Before Time*, not only copied the Disney style, but even took as its plot a long dinosaur migration across a parched planet in search of water!)

Geologic evidence points to world climates having gotten cooler, not hotter, at the time of the dinosaurs' demise. Disney's apocalyptic vision of dinosaurian dehydration had been inspired by the recent discovery of a spectacular "saurian graveyard" in Montana. Near the Bighorn Mountains, famed paleontologist Barnum Brown had excavated a site (Howe Ranch) where hundreds of sauropods apparently had become trapped in mud as their marshes and wetlands dried out. It was a local catastrophe, but in the Disney version it appeared as a global cataclysm.

Despite its scientific and artistic flaws, *Fantasia* was a milestone that stimulated the public imagination about prehistoric creatures, sympathetically presented reptilian monsters as tragic victims and established a lasting dinosaurian image in pop culture.

"FIGHT TO THE FINISH—Survival of the fittest is demonstrated by two dinosaurs in the 'Rite of Spring' sequence of Walt Disney's animated classic *Fantasia*." (The preceding sentence appears on the Disney publicity still, illustrating a popular misconception of Darwinian evolution.) (Copyright Walt Disney, Inc.)

Renowned conceptual artist-essayist Robert Smithson once traced the look of the Disney dinosaurs back to museum painter Charles R. Knight, who described Tyrannosaurus rex as "an enormous eating machine." "Violence and destruction are intimately associated with this type of carnivorous evolutionism." Smithson wrote that the memorable battle between a tyrannosaur and stegosaur in *Fantasia*:

> evoked a two-dimensional spectacular death-struggle, quite in keeping with the entire cast of preposterous reptilian "machines." No doubt Disney at some point copied his dinosaurs from Knight, just as Sinclair Oil must have copied directly from Disney for their "Dino-the-Dinosaur."

Fantasia's giant, bipedal "tyrant lizard" is shown with three clawed digits on its small "hands." Years later, Disneyland technicians fashioned a life-size animatronic robot tyrannosaur based on the same mistaken restoration.

When it was almost completed, Disney invited a distinguished paleontologist to the park to check its accuracy. "It's very good, Mr. Disney," said the scientist, "except for one thing. You'll have to change his forelegs; any schoolboy knows tyrannosaurs had two clawed digits, not three."

Disney stared at his dinosaur, thought for a long moment, then shook his head. "Nope," he replied, "I really think it looks better with three."

About thirty years later, in 1990, it turned out that Disney wasn't entirely wrong. Paleontologists discovered a new species of bipedal dinosaurs, *Nanotyrannus*, which were smaller relatives of *Tyrannosaurus*. They greatly resembled their larger cousins except for one thing—their forelimbs had three clawed digits.

See also BROWN, BARNUM; KNIGHT, CHARLES R.; SINCLAIR DINOSAUR.

FIRST FAMILY

See AFAR HOMINIDS.

FITNESS
What Price Survival?

Erasmus Darwin, grandfather of Charles, had speculated that in early tribes the biggest, most powerful males would attract more females, father the most children and thus perpetuate their attributes. Philosopher Thomas Hobbes had thought in a "state of nature" each man would be "at war with every other man," the stronger directly eliminating the weaker. Outside scientific circles, that is still the popular idea of "survival of the fittest."

Charles Darwin used "fitness" in several ways:

(1) in his grandfather's sense of robust individuals; (2) in terms of populations better adapted to particular environments, such as specialists at obtaining particular foods; and (3) those with more complex nervous systems or "improvements in organization." Today geneticists prefer phrases like "high respresentation of gene frequencies in descendant gene pools." While gaining a precision useful in certain kinds of research, this mathematical definition leaves behind any sense of the living creature.

Alfred Russel Wallace, in his classic *Darwinism* (1889), defined "fitness" from the viewpoint of a naturalist who had spent years observing animals in tropical forests. To Wallace the fittest are "those which are superior in the special qualities on which safety depends":

> At one period of life, or to escape one kind of danger, concealment may be necessary; at another time, to escape another danger, swiftness; at another, intelligence or cunning; at another, the power to endure rain or cold or hunger; and those which possess all these faculties in the fullest perfection will generally survive.

They are "the best organised, or the most healthy, or the most active, or the best protected, or the most intelligent, [which will] . . . gain an advantage over those which are inferior in these qualities." Wallace's "fitness" varies with the time of life, or even with the time of day!

Julian Huxley differentiates between "survival fitness" and "reproductive fitness." The first concerns the individual's success at growing to maturity, finding food and living long enough to reproduce. "Reproductive fitness" concerns efficiency in leaving offspring, such things as clutch or litter size, ratio of sexes in a population, competitive attractions for the opposite sex.

Fitness can also mean the range of variability in a population's gene pool, which allows flexibility in crisis situations, such as changes in climate, famines or diseases. Native populations are more "fit" to survive malaria in West Africa because of the high incidence of sickle-cell, a genetic disease that causes anemia in some individuals, but also confers some immunity to malaria. Thus many people with the trait live long enough to reproduce, while Europeans, with no resistance to malaria, used to call West Africa "the white man's graveyard."

Long-term fitness is also relative. A species may be very successful in one particular environment, but not in another. It may persist for a very long time, like the dinosaurs, who were certainly "fit" enough when they ruled the earth for 140 million years. But fitness is not eternal under all conditions. When mass extinctions claim more than 90% of the planet's life forms, fitness may be reduced to a matter of sheer luck. By far the great majority of all species that have ever lived have become extinct.

See also DIFFERENTIAL REPRODUCTION; "SURVIVAL OF THE FITTEST."

FITZROY, CAPTAIN ROBERT (1805–1865)
Explorer, Meteorologist, Administrator

It was Captain Robert FitzRoy's idea to take a young naturalist aboard H.M.S. *Beagle* during its five-year surveying expedition for the British Admiralty. After interviewing several applicants, he chose 21-year-old Charles Darwin, a well-recommended but inexperienced amateur who had never been to sea. This enthusiastic Cambridge-educated gentleman, FitzRoy thought, would make an intelligent companion and "messmate." Darwin was equally impressed with FitzRoy, describing him as "my beau ideal of a captain."

So greatly did the two young men affect the course of each other's lives that FitzRoy is often remembered only as "Darwin's captain." However, he would have made his mark in any event, as seaman, explorer, surveyor, mapmaker, meteorologist and governor of New Zealand.

A staunch believer in the literal interpretation of scripture, FitzRoy could not have imagined that his vessel would become famous as the birthplace of evolutionary biology. Years later, he vehemently repudiated his old friend and cabinmate Darwin, cursing him as "a viper in our midst." Although he fully acknowledged Darwin's great achievements, he could not forgive him for leaving the Creator out of his theories.

Born in Suffolk to the aristocratic Graftons, who descended from Charles II, Robert FitzRoy's immediate ancestors included courtiers, admirals, sea captains and a prime minister. (The name *fis roi* means "son of a king.") After graduating with top honors from the Royal Naval College at Portsmouth, he served on several vessels which took him from the Mediterranean to South America.

In 1828, then-Lieutenant FitzRoy was given command of H.M.S. *Beagle,* a 21-gun brig that had been sent to map the frigid and desolate southern coasts of South America, including Patagonia and Tierra del Fuego. Weather was rough, supplies short, and the crew was depressed and sick with scurvy. The previous captain had despaired and shot himself, but FitzRoy succeeded in taking over the survey and leading the expedition forward.

During this first command of the *Beagle,* FitzRoy became obsessed with the Indian tribes of Tierra del Fuego. Despite the harsh climate, these sturdy people went nearly naked, lived in rough huts, eked out a living by hunting and fishing, and appeared to the Englishmen almost as beings from another planet.

There were several skirmishes with these natives when they boldly invaded the English ship, and made off with its small whaleboats, which they much preferred to their own canoes. During one violent confrontation, a Fuegian was shot dead, and FitzRoy took some Indians as hostages for the stolen whaleboat. When the tribesmen refused to exchange the boat for their kinsmen, FitzRoy released them all except for a nine-year-old girl who wanted to stay aboard ship. FitzRoy named her Fuegia Basket and decided to teach her English.

Soon the captain "acquired" three other Indians, young men whom he considered "scarcely superior to the brute creation (animals)." As a human experiment, he brought them to England, "civilized" them at his own expense, then returned them to their homeland several years later—a major objective of the *Beagle's* second voyage. To FitzRoy's dismay, his philanthropic attempt to spread the light of Britannia ended in tragic failure. [See BUTTON, JEMMY.]

His other personal agenda was to find geological evidence confirming the biblical account of creation; he became increasingly upset when Darwin's researches seemed to be taking a quite different direction.

FitzRoy was imperious, arrogant and often pulled rank to win philosophical arguments. Once he told Darwin slavery was justifiable, as black people were child-like and better off under the plantation system. Darwin insisted that no man or woman would choose to be a slave.

During FitzRoy's visit to a Brazilian plantation, the owner gathered many of his slaves and asked whether they would rather be free. All agreed they were content with their lot, and FitzRoy crowed to Darwin that his case had been proved. "I then asked him," Darwin wrote in his *Autobiography* (1876), "perhaps with a sneer, whether he thought that the answers of slaves in the presence of their master was worth anything. This made him excessively angry, and he said that if I doubted his word, we could not live any longer together."

Despite these occasional flare-ups, relations between the two were good, and Darwin was chosen to accompany FitzRoy on many of his overland expeditions. During the five years (1831–1836), the *Beagle* survived storms, earthquakes, encounters with hostile Indians, sinister-looking gauchos and the infamous military dictator General Juan Manuel Rosas in the Argentine. Under FitzRoy's driving perfectionism, the ship made its way around Cape Horn,

ARISTOCRATIC NAVAL OFFICER, Admiral Robert FitzRoy's very name means "son of a king" *(fils roi)*. A brilliant navigator and commander, Charles Darwin called him "my beau ideal of a Captain"—though their friendship did not survive the evolution controversy.

circumnavigated the globe, pausing in the Galapagos, Tahiti, New Zealand and Australia. Upon returning to South America, the *Beagle* expedition made new explorations of its coasts, then returned to Falmouth, England with an immense body of new information.

FitzRoy's maps and charts are the basis for those still in use today. He wrote and edited *Narratives* (1839) of the *Beagle* voyage, though he was somewhat jealous when Darwin's account became far more popular and well-known than any other volume in the series.

His interest in meteorology led him to devise the popular, easy-to-use FitzRoy Barometer, which was widely distributed to British coastal fisherman. Also, he invented the system of raising storm warning flags, which was adopted all over the world. Odd as it may seem today, his most controversial contribution was his insistence that the weather could be foretold, or "forecast," as he called it.

He pioneered a system of gathering widely scattered reports into "synoptic" maps (the term still used for them today). As head of the Admiralty's meteorological office, he got *The Times* to begin publishing newspaper weather forecasts and maps, which had never before been done. For this he was widely ridiculed, as few believed weather could be predicted with any accuracy. His persistence carried the day;

eventually royal messengers called at the FitzRoy residence requesting a private weather forecast for Queen Victoria's vacation.

In later years, his imperious and moody personality—and conviction that he was right about matters he didn't understand—led him into many scrapes and difficulties. Nevertheless, he represented the British government in the South Seas, was promoted to admiral, and later in life served as governor of New Zealand.

The complex situation in New Zealand would have been difficult for even an excellent colonial administrator. Unfortunately, Fitzroy's temperament was particularly ill-suited for the intricacies of cross-cultural politics and diplomacy, and he left that country in worse turmoil than he found it.

Back in England, he nearly fought a duel of honor with a rival for political office, but trounced the fellow with an umbrella instead. This bizarre scene took place at the Admiralty, where his opponent—brandishing a whip—said: "Captain FitzRoy, I shall not strike you, but consider yourself horsewhipped." FitzRoy replied by knocking the man down with his umbrella, to the astonishment of various elderly officers.

Several times over the years he appealed to Darwin to recant his evolution theory and felt guilty his expedition had played a part in "undermining" scripture. At the famous Oxford debate between Thomas Huxley and Bishop Wilberforce, FitzRoy had held a Bible over his head and shouted, "The Book, The Book." [See OXFORD DEBATE.]

His inflexible traditionalism, perfectionism and need to control in the face of repeated failures and frustrations eventually proved too much for FitzRoy. He became a haunted, sickly man trying vainly to come to terms with a rapidly changing world.

With his glory days long behind him, he became increasingly despondent and withdrawn. Like his uncle, Viscount Castlereagh before him, FitzRoy became deeply depressed and took his own life at the age of 60.

His old friend Darwin had long feared this sad finale; FitzRoy, he thought, had a "hereditary disposition" to suffer from excruciating mood swings. But despite FitzRoy's stubborn fanaticism, Darwin also knew him as a kind and generous man with noble intentions. He remembered how the Captain had personally "fixed my hammock with your own hand," and the many considerations he had shown during the long, difficult, dangerous expedition.

Above all, Darwin always remained grateful that FitzRoy had chosen him for their historic voyage of discovery; he called it "the most fortunate encounter of my life."

See also (H.M.S.) BEAGLE; DARWIN, CHARLES; VOYAGE OF THE H.M.S. BEAGLE.

For further information:

Mellersh, H. E. L. *FitzRoy of the Beagle*. London: Mason & Lipscomb, 1968.

Moorehead, Alan. *Darwin and the Beagle*. London: Hamish Hamilton, 1969.

FIXED ACTION PATTERNS

See LORENZ, KONRAD; RITUALIZATION; TINBERGEN, NICKOLAAS.

FLAT–EARTHERS
Oldest "Bible Science" Sect

Hebrew scribes wove into the Old Testament their concept of a flat, stationary Earth—an ancient view they shared with older Babylonian and Egyptian cultures. Centuries later, when early scientists concluded the Earth was a spinning globe, orbiting the sun, Christians attacked astronomy as a Satanic attempt to destroy religion and morality. Long before scriptural literalists fought against the theory of evolution, they battled scientists over the shape of the Earth.

Right down to our own time, a handful of die-hard "flat-earthers" believe the Apollo moonwalk and astronaut photos of a round planet Earth are government lies fostered by a conspiracy of atheistic scientists.

Round-earth theories were often debated in the ancient world, even while the older notions prevailed. Aristotle had proposed a reasonable proof that the Earth is a globe, based on observations of lunar eclipses, differences in star patterns seen by travelers and the manner in which ships "descend" at the horizon.

But it was Ptolemy's *Almagest* (A.D. 140) that permanently established the global idea in Western culture, although he still considered Earth the center of the universe. It remained for Copernicus and Galileo to destroy that geocentric notion; and for the Catholic church to destroy Galileo.

In 1849, a modern flat-earth movement was founded in England by Samuel Birley Rowbotham. Under the pseudonym "Parallax," Rowbotham published a pamphlet in that year titled *Zetetic Astronomy: A Description of Several Experiments which Prove that the Surface of the Sea Is a Perfect Plane and that the Earth Is Not a Globe!* ("Zetetic" is from the Greek *zetetikos*, meaning "to seek or inquire.") "Parallax" tirelessly toured England for most of his life, lecturing on his own system of astronomy and attacking the idea of a spherical Earth.

In his view, the world is a vast flat circle, something like a pancake. The North Pole is at its center, and the perimeter is walled in by high sheets of ice that are mistakenly called the South Pole. Sun, moon and planets circle above at an altitude of about 600 miles. Phenomena like rising and setting and the apparent disappearance of ships over the horizon are explained by optical illusion and the "zetetic law of perspective."

About 1860, just after Darwin's *Origin of Species* (1859) appeared, the flat-earth movement and "Zetetic Astronomy" became household words in England and, a few years later, in America. As upholders of the literal word of the Bible, flat-earthers warned Christians that astronomy posed a more pervasive and established danger to their religion than the upstart geology. Flat-earther John Hampden wrote that Satan's mission "has been to throw discredit on the Truth of God . . . [through] that Satanic device of a round and revolving globe, which sets Scripture, reason, and facts at defiance."

In America, flat-earthism became a central doctrine of the Christian Catholic Apostolic Church, which claimed 10,000 members during the 1920s and 1930s. An entire town of about 6,000—Zion, Illinois—was founded by the sect and ruled by the iron hand of Wilbur Glenn Voliva. To its parochial schools, some families sent three generations to learn the flat-earth doctrine. Voliva even had his own radio station, on which he thundered against "the Devil's triplets, Evolution, Higher Criticism [historical biblical scholarship] and Modern Astronomy."

Voliva taught that the stars are much smaller than the Earth and rotate around it, the moon is self-luminous and the sun was a reasonably sized lamp, not a giant star:

> The idea of a sun millions of miles in diameter and 91,000,000 miles away is silly. The sun is only 32 miles across and not more than 3,000 miles from the earth. It stands to reason it must be so. God made the sun to light the earth, and therefore must have placed it close to the task it was designed to do. What would you think of a man who built a house in Zion and put the lamp to light it in Kenosha, Wisconsin?

And "Where is the man," Voliva asked, "who believes [if the earth really rotates] that he can jump into the air, remaining off the earth one second, and come down to the earth 193.7 miles from where he jumped up?"

After Voliva's death in 1942, the popularity of his flat-earth brand of fundamentalism started to wane, and his town of Zion has long since caught up with the 20th century. Still, some of its elderly members still believe in their hearts that, in their founder's words, "the so-called fundamentalists . . . strain at the gnat of evolution and swallow the camel of modern astronomy." Charles Darwin himself was aware of the connection when he wrote, "By far the greatest part of the opposition [to evolution] is just the same as that made when the sun was first said to stand still and the world to go round."

See also: FUNDAMENTALISM; HAMPDEN, JOHN; "SCIENTIFIC CREATIONISM."

FLOWERS

See ANGIOSPERMS, EVOLUTION OF; ORCHIDS, DARWIN'S STUDY OF; SPRENGEL, CHRISTIAN K.

FLUORINE ANALYSIS
Dating Bones of Contention

There are two basic dating techniques for establishing the age of ancient objects: relative dates and "absolute" or chronometric dates. An "absolute" date makes use of a natural "clock"—it can be radioactivity or tree ring growth—that allows a count backwards from the present. Potassium-argon and carbon 14 are known to change at a steady rate, and therefore give an age in years. Relative dating doesn't tell how old a fossil is, but simply confirms, or disproves, whether objects found together share the same age or come from different periods.

Fluorine analysis is a relative dating technique, which works only on bones, developed by Professor Kenneth Oakley of the British Museum. When bones are buried, quantities of fluorine from groundwater become part of their mineral structure. The longer they lie in the earth, the more fluorine they will absorb. Therefore, in any given site, bones of the same age contain about the same proportions of fluorine.

In the early 1950s, Oakley used this technique to make a sensational discovery: One of the long-established "fossil men," which had baffled anthropologists for more than 40 years, was a deliberate hoax! Oakley tested the famous Piltdown remains: pieces of skull and jaw dug up in 1912 at Sussex, England, along with crudely chipped stones. The fossil appeared to combine features of human and ape and had been hailed by such authorities as Professor Ray Lankester, director of the British Museum, as "the first Englishman."

But when Oakley made his analysis, he found the skull contained much higher levels of fluorine than the jaw, which meant it had lain in the ground a lot longer and was therefore much older. "It was unlikely," commented one anthropologist, "that the skull had met an untimely demise while the jaw lingered on for thousands of years."

This discrepancy led Oakley to closely reexamine the original material and to make other physical and chemical tests. While the skull was indeed an authentic *Homo sapiens* of ancient vintage, the jaw turned out to belong to a modern orang-utan and had been stained to match the skull. Its teeth had been filed to resemble human teeth, and the connections (condyles) between skull and jaw were cleverly broken. The identity of the hoaxer and the reasons for the scam remain a mystery, but Oakley's exposure of the Piltdown forgery established fluorine analysis as an important tool in authenticating the time frame of fossil bones.

See also CHRONOMETRY; PILTDOWN MAN (HOAX).

For further information:
Oakley, Kenneth P. *Frameworks for Dating Fossil Man.* Chicago: University of Chicago Press, 1964.

FLY ROOM
Thomas Hunt Morgan's Lab

A 16' × 23' room at New York City's Columbia University was known as "the fly room." In this makeshift laboratory Thomas Hunt Morgan and his associates used tiny flies to understand the workings of heredity.

Morgan had little faith in Mendelism when he came to Columbia in 1904, and so he was as surprised as any when his experimental work showed Mendel to be correct. At first, he tried to study transmission of characters in mice and rats, but had no success. He made progress only after he chose the humble fruit fly, *Drosophila melanogaster*, as his subject.

The fly could be bred by the thousands in milk bottles. It cost nothing but a few bananas to feed all the experimental animals; their entire life cycle lasts 10 days and they have only four chromosomes.

Between 1907 and 1917, Morgan and his team of seven worked in the fly room, attempting to encourage mutations with heat, X-rays and chemicals. Using the most laughably simple equipment, but a lot of scientific talent and perseverance, the "fly room" gradually provided evolutionary biology with a new foundation.

In 1933, Morgan won the Nobel Prize for Physiology and Medicine "for his discoveries concerning the function of the chromosome in the transmission of heredity."

For further information:
Allen, G. E. *Thomas Hunt Morgan: The Man and His Science.* Princeton: Princeton University Press, 1978.

FOOTPRINTS—OF EARLIEST HUMANS

See FOSSIL FOOTPRINTS, PALUXY RIVER; LAETOLI FOOTPRINTS.

FOOTPRINTS—OF DINOSAURS

See BIRD, ROLAND T.; FOSSIL FOOTPRINTS, PALUXY RIVER; "NOAH'S RAVENS."

FOSSEY, DIAN (1932–1985)
Gorillas' Greatest Defender

From an ordinary existence as an occupational therapist in Louisville, Kentucky, Dian Fossey's life was transformed into one of the great dramas of modern Africa. For 19 years, her beloved "family" was a band of wild mountain gorillas; human poachers became her sworn enemies. Ultimately her passion to preserve the great apes cost her life.

Fossey's lifelong interest in animals prompted her to visit Africa on a vacation in 1963, during which she met Dr. Louis Leakey at his fossil digs in Olduvai Gorge. The crafty paleo-anthropologist planted the seed in her mind of studying wild gorillas and later recruited her to carry out the project in the rain forests of Zaire.

Rare and elusive, the mountain gorillas live only on six extinct volcanoes, located in the border area of Rwanda-Zaire-Uganda. Previously, these gorillas had been tracked and killed by Paul du Chaillu in the 19th century and memorialized by sculptor-naturalist Carl Akeley, who mounted a family of them in the American Museum of Natural History. Field observers John Emlen and then George Schaller explored the area in the early 1960s. It was Schaller, working alone and without carrying weapons, who made the first long-term, close-up field study.

Fossey began her work in 1967, when there were about 600 animals left of the many thousands that had been discovered in the 1860s. Less than 400 remain today, of which 250 are found in Fossey's study area. It was very possible, she realized when she began her project, that these magnificent creatures might become extinct within 100 years of their discovery. When she began to establish her field station on the Zaire side of the mountains, the area was still remote, and the nearest village was several hours away down the mountain trail. Political turmoil soon caused her to relocate in Rwanda, which was more densely settled.

Even though the new study area was supposed to be in a wildlife preserve, the local people illegally farmed cattle, poached wildlife and cleared trees wherever they pleased in the park. In addition, the area was staked out with *sumu*, signs of African witchcraft scattered through the forest, which caused fear and uncertainty among Fossey's African staff.

In short, the Parc National des Volcans was a poor excuse for a national park or wildlife refuge. Hunting and trapping were rampant, and the few guards were easily bribed. To her dismay, Fossey found the dense foliage booby-trapped with wire nooses for the capture of antelope and forest hogs; these traps often snared gorillas, causing gangrene of hands and feet. Poachers killed entire troops so they could capture infants to sell to zoos and cut off the apes' massive hands and feet to sell as souvenirs. Alone with her love of gorillas in a war-zone between Africans and apes, she feared the gorillas would all be exterminated before she had a chance to study them.

For three years, she patiently stalked the apes, gradually habituating them to her presence. No previous researcher had spent so much time or been so patient in approaching gorillas. Where Schaller considered himself fortunate to find a vantage point yards away for a clear view with his binoculars, Fossey was not satisfied until the gorillas accepted her right in the middle of a family group.

At first she hiked many miles tracking groups, unsure of what to do when she got close. When she finally contacted them, she learned to drop to her knees, making gestures of submission and soft "contentment" noises that were nonthreatening. At the beginning, her courage was tested to its limits when huge, silver-backed males would run at her, roaring and chest-beating. At such times, she wrote in her book *Gorillas in the Mist* (1983), "I found it possible to face charging gorillas only by clinging to surrounding vegetation for dear life."

For her gorilla studies, Fossey earned a doctorate in zoology from Cambridge University. As she became accepted by the apes, she developed a personal empathy that was extraordinary in its depth and commitment. Although she consciously tried to avoid "humanizing" the gorillas in her writing, she was unsuccessful. Her detailed descriptions of family life are full of "proud fathers," and the gorillas conduct "valiant struggles" and wear "haunted expressions." This identification with her animals, though it may have blunted objectivity, made her a supremely committed conservationist, who may have single-handedly saved the species from extinction.

She trained her staff to go on "poacher patrols," dismantling traps and nooses and attempting to scare off hunters. Countering magic with magic, she left her own *sumu* in the forest—everything from Halloween masks to firecrackers. Sometimes it worked; many poachers were scared off, and animals were saved. But others regarded her as an intruder who had no right to tell them they couldn't hunt; their forefathers after all, had hunted in these lands since the dawn of time.

The most traumatic event in Fossey's life was the slaughter of one of her favorites, Digit, in 1977. She had spent hundreds of hours with Digit, had watched

TRIUMPH AND TRAGEDY characterize the life of Dian Fossey, who successfully entered into the world of Zaire's wild mountain gorillas, then defended them from extermination by poachers. On a rare visit to New York City, Fossey grudgingly admired Carl Akeley's mounted gorilla group at the American Museum of Natural History.

him grow from infancy to maturity, and regarded him as a personal friend. According to Fossey's account, her assistant had "found Digit's mutilated corpse lying in . . . blood-soaked vegetation. Digit's head and hands had been hacked off; his body bore multiple spear wounds." She agonizingly reconstructs his last stand:

> Digit, long vital to his group as a sentry, was killed in this service by poachers on December 31, 1977. That day Digit took five mortal spear wounds into his body, held off six poachers and their dogs in order to allow his family members, including his mate Simba and their unborn infant, to flee . . . Digit's last battle had been a lonely and courageous one. During his valiant struggle, he managed to kill one of the poacher's dogs before dying. I have tried not to allow myself to think of Digit's anguish, pain, and the total comprehension he must have suffered in knowing what humans were doing to him.

After that incident, Fossey suffered a nervous breakdown and "came to live within an insulated part of myself." Digit's death was publicized and became a rallying point for public interest in conserving the gorillas.

But Fossey became bitter, like a soldier whose buddy is killed by the enemy. She joined with police in military-type raids of local villages, rounding up poachers. She helped capture three men responsible for Digit's death, who were tried and given prison sentences, but several others eluded capture. Fossey continued to study the gorillas, and they allowed her to get as close as before, but "this was a privilege that I felt I no longer deserved."

Publicity brought more assistants, researchers and interested tourists to her Karisoke Research Center. She wanted little contact with them. Withdrawn and obsessed, she wanted only to be let alone with her gorillas and for her gorillas to be safe. Even one of the researchers who lived at her camp for two years hardly saw her; she communicated with him by writing notes.

Despite the growing success of the Digit Fund and the Mountain Gorilla Project in protecting the gorillas, Fossey kept aloof. She refused to take part in education programs that had been started in Rwandan schools, claiming that it was too little, too late. Her preference was to go out in the woods and destroy a few more traps or collar another poacher. On one occasion, she even kidnapped a poacher's ten-year-old son and held him hostage. She wanted all pursuit of the gorillas to stop immediately and forever and had little patience with educating the next generation of Africans.

"When you have any kind of rare species," George Schaller sympathized, "the first priority is to work for its protection. Science is necessarily secondary." Fossey even discouraged students from doing any further research, urging instead that they join in her battles with poachers.

On December 28, 1985, Dian Fossey was killed "by unknown assailants" at her forest camp.

The presumption that Fossey was murdered by vengeful poachers was widely believed as the finale to her legend; the truth is more difficult to accept. In her final years at Karisoke, her personality had deteriorated; she had isolated herself from researchers and students, spending weeks locked in her cabin. She had become resentful, suspicious of others and downright cruel to her staff. Fossey was tormented, dying of emphysema and finally friendless.

Those who were at Karisoke during her last years seem to agree that she was probably not killed by a village poacher, but by someone close to her, who had felt the full fury of her unjustifiable rages and merciless personal attacks. Though she remained on her mountain, she had descended into madness. She was buried in the gorilla cemetery in her camp, next to the remains of her beloved Digit.

See also AKELEY, CARL; "APE WOMEN," LEAKEY'S; DIGIT GORILLAS; KARISOKE RESEARCH CENTER.

For further information:

Fossey, Dian. *Gorillas in the Mist*. Boston: Houghton Mifflin, 1983.

Hayes, Harold. *The Dark Romance of Dian Fossey*. New York: Simon & Schuster, 1990.

Mowat, Farley. *Woman in the Mists*. New York: Warner Books, 1987.

FOSSIL FOOTPRINTS, PALUXY RIVER
Man and Dinosaur Tracks Together?

Creationist Robert Kofahl has claimed that human footprints and dinosaur tracks are found fossilized together in the same Cretaceous limestone. "They must have lived on earth at the same time, just as the Bible implies," he concluded. For years the famous fossil footprints along the Paluxy River, near Glen Rose, Texas, have been a favorite creationist exhibit to demonstrate that evolutionists are all wrong in their time table, which has the dinosaurs dying out 65 million years before man appeared.

In 1970, the Creation-Science Research Center and Films for Christ filmed the tracks for *Footprints in Stone*. They thought this movie clearly showed "children's prints, those of normal adults, and also very large prints up to eighteen inches long. The human characteristics were unmistakable. Some human prints were found overlapping three-toed dinosaur prints."

However, a new study, released in June 1986, has determined that the prints are not human at all. Glen J. Kuban, a computer programmer from Brunswick, Ohio, and an acknowledged expert on dinosaur tracks, began to reexamine the Paluxy tracks in 1980. Almost immediately, he found faint impressions of dinosaur toes that had gone largely unnoticed in the supposedly "human" prints. With Ron Hastings, a Texas science teacher, Kuban found that "almost every one of the alleged human tracks was accompanied by distinct colorations in the rock that, upon detailed analysis, revealed the pattern of dinosaur digits."

The different colors (blue-gray to rust, contrasting with the ivory-tan color of the limestone) suggest the tips of the impressions were later filled in by different sediments, thus altering their shape to resemble human tracks. On examining the evidence, professional paleontologists agreed with their conclusions.

In those tracks once proposed as human, the toes were always blurred or indistinct, and the impression of a flat-footed (plantigrade) walker was created by the rounded heels—which still raises the question of whether dinosaurs walked in a more flat-footed manner than has usually been assumed.

After several creationists visited the site, Dr. John D. Morris of the Institute for Creation Research at El Cajon, California admitted that the tracks can no longer be "regarded as unquestionably human," and added "It would now be improper for creationists to use the Paluxy data as evidence against evolution."

They also withdrew their film *Footprints in Stone* from circulation.

Interestingly, Mr. Kuban chose to publish his findings in the *Creation/Evolution* magazine of the secular humanists, but included the postscript: "I am a Christian and believe in the Creator, but have not yet formed definite conclusions about some aspects of the origins controversy . . . I chose to publish [here] not to attack creationism but to help set the record straight on the true nature of the Paluxy evidence."

See also BIRD, ROLAND T.; LAETOLI FOOTPRINTS; "NOAH'S RAVENS."

FOSSIL MAN

When Darwin published *Origin of Species* in 1859, scarcely a fossil of early men or near-men had yet come to light. His hypothesis that man evolved from primates (represented today by lemurs, monkeys and apes) was criticized for being "fossil-free." Opponents constantly taunted, "Where are the missing links?" Yet within a century of Darwin's death, there were enough fossils of early hominids to fill a large public gallery. (In 1984, for the unprecedented "Ancestors" exhibition at New York's Museum of Natural History, 40 famous early hominid skulls were flown in from all over the world.)

The first discoveries of early-man fragments were Neandertal crania, found in Germany and France beginning in 1856, described by Thomas Huxley in his *Essay on Man's Place in Nature* (1863). Although convinced they were the most ancient hominid fossils yet known, he thought them so similar to modern man they could not qualify as being demonstrably "closer to apes." During the 1860s, the prehistoric stone tools discovered by Boucher de Perthes near the Somme river in France were also attracting increased attention and intensified the search for the tool-maker.

After several more complete Neandertal fossils had come to light, the next major discovery was the Java Man (formerly known as *Pithecanthropus*) by Eugene Dubois in 1891, which some hailed as the long-sought "link." Many fossil-man hunters then shifted their focus from Europe to Asia, which they believed was the cradle of mankind. But despite a seven-year search (1921-1928) by the Gobi Desert Surveys, known popularly as the "Missing Link Expeditions," no "dawn man" remains turned up.

By the mid 1920s, there were two other finds: one real and revolutionary, the other an outrageous hoax. The fraud was Piltdown Man, a fossil forgery concocted of a human cranium and a fragmentary orang-utan jaw, "discovered" in Sussex, England in 1912.

(It was readily accepted because many scientists expected our ancestors to have a large brain combined with ape-like teeth and jaw.) The authentic discovery was ignored for years because it had the opposite configuration: human-like teeth and a small, ape-sized brain. That was the *Australopithecus africanus*, or South African man-ape, discovered by anatomist Rayond Dart in 1924, the first solid clue to an African genesis of mankind.

While European scientists continued to seek man in Asia, Dart, later joined by Robert Broom, continued to unearth more African man-ape fossils, including some larger, heavier-boned species known as *robustus*. (It was not until the 1960s that the significance of Dart and Broom's discoveries was fully appreciated.) During the 1930s, most paleoanthropologists were excited about the excavations in China that had yielded a dozen skulls of Peking man, then called *Sinanthropus*. Soon it became clear that Peking and Java man were the same species, now called *Homo erectus*, and eventually found also in North Africa, Europe, East Africa and Israel.

During the mid-to-late-20th century, the search intensified in East Africa, led by Louis and Mary Leakey and later by their son Richard. At Olduvai Gorge, Koobi Fora and other sites in Kenya and Tanzania, the record was enriched with some still-bewildering discoveries: the enigmatic, larger-brained *Homo habilis*, going back some two million years, as well as both smallish (gracile) and heavier (hyper-robust) australopithecines. An almost complete skeleton of a *Homo erectus* boy was found at Lake Turkana, estimated at 1.6 million years old.

Three or four different species appear to have existed at about the same time (older textbooks had arranged them in some ancestral-descendant series). It now seems clear there was an adaptive radiation or "branching bush" of ancient hominids in East Africa, of which only one "twig," *Homo sapiens sapiens*, managed to survive.

During 1974, Donald Johanson, Tom Gray and their colleagues from France and California made remarkable discoveries in the "Afar triangle" of Ethiopia. The most famous is "Lucy," the oldest partial skeleton of a hominid yet found, which Johanson assigned to a new species *Australopithecus afarensis*, estimated at three million years old. These upright-walking, small-brained hominids had apishly long arms not previously known from fragmentary skeletons.

A collection of 13 individuals, nicknamed "The First Family," were discovered at Hadar the following season; whether or not these bones represent more than one species is still a controversial question among experts.

No fossil australopithecines, *Homo erectus*, *Homo habilis*, or any other very early hominid, not even the comparatively recent Neandertals, are found in the Americas. Scattered remains of early AmerIndian populations may go as far back as 30,000 years; occasional reports of much older human or hominid fossils have not yet been substantiated in the Western hemisphere.

Anthropologists debate endlessly about which fossil hominid species were ancestral, which were "side branches" or "on the main line" leading to ourselves. The details of this complicated family tree are not likely to become clearer anytime soon. But the broad picture is encouraging, for as far as "missing links" go, there is not one but many. We have fossil remains of at least four early upright-walking man-like creatures with brains the size of modern apes dating back two or three million years. There are others that appear to have been closer cousins of ours (*Homo habilis*), some (*Homo erectus*) with a brain two-thirds the size of our own and others (Neandertal) with crania as large as ours or larger. We also have the remains of humans like ourselves dating from the last Ice Age (30,000 years ago), along with their stone and bone tools, weapons and some remarkable cave paintings of extinct large mammals.

While the evidence is still fragmentary and our understanding of it inadequate, theories of human evolution are no longer "fossil-free." The trouble is, given the contentiousness of some paleoanthropologists, the evidence is frequently overwhelmed by theories and expectations.

Thus, Dart's "Taung child" was misidentified as an ape by prestigious anatomists who for years insisted it had "too small a brain" to be hominid, while the fragmentary *Ramapithecus* jaws won instant and universal acceptance in the mid-1960s by top experts until it was discredited as a human ancestor two decades later.

Paleoanthropologist David Pilbeam, whose early career was based on helping successfully establish *Ramapithecus*, reached a humbling conclusion when the fossil was found to be far from the human lineage. "In paleoanthropology," he wrote in 1978, " 'theory'—heavily influenced by implicit ideas—almost always dominates 'data.' Ideas that are totally unrelated to actual fossils have dominated theory building, which in turn strongly influences the way fossils are interpreted."

Besides, Pilbeam told a scientific gathering, "I have become convinced that fossils by themselves can solve only parts of the puzzle . . . You are much better off using molecular evidence [for] timing of branching points . . . a difficult admission . . . for someone who was brought up to believe that everything we needed to know about evolution could be got from the fossils."

See also AFAR HOMINIDS; AUSTRALOPITHECINES; DESCENT OF MAN; GOBI DESERT EXPEDITIONS; HOMO ERECTUS; HOMO HABILIS; "LUCY"; NEANDERTAL MAN; PEKING MAN; PILTDOWN MAN (HOAX); RAMAPITHECUS.

For further information:

Day, Michael H. *Guide to Fossil Man: A Handbook of Human Paleontology.* 3rd ed. Chicago: University of Chicago Press, 1977.

Delson, Eric, ed. *Ancestors: The Hard Evidence.* New York: Alan R. Liss, 1985.

Johanson, Donald, and Edey, Maitland. *Lucy: The Beginnings of Humankind.* New York: Simon & Schuster, 1981.

Johanson, Donald, and Shreeve, James. *Lucy's Child.* New York: Wm. Morrow, 1989.

Leakey, Richard, and Lewin, R. *Origins.* New York: Dutton, 1977.

Lewin, Roger. *Bones of Contention.* New York: Simon & Schuster, 1987.

Reader, John. *Missing Links: The Hunt for Early Man.* London: Collins, 1981.

Willis, Delta. *The Hominid Gang.* New York: Viking, 1989.

FOSSILS
Reading the Rocks

Fossils (from the Latin *fossa*, a hole) originally meant any curious natural object that was dug up from the ground. That included not only petrified plants, shells or animal skeletons, but also crystals, gems and odd mineral ores. Early collectors debated whether the stones shaped like organisms were really once alive and how they might have been formed.

Conrad Gesner (1516–1565), an industrious early naturalist, wrote a book *On Fossil Objects, chiefly Stones and Gems, their Shapes and Appearances* (1565) that described and illustrated for the first time a large and varied collection of these mysterious objects.

Some of the most familiar shapes seemed to be mineralized seashells, though the puzzle was that they were found on mountaintops. Others appeared to be fish or animal skeletons, but did not resemble any creatures that now exist. Many had strange and bizarre geometric shapes unlike any living things.

For several hundred years after Gesner, attempts to understand fossils included theories that: (1) they are the remains of creatures that once lived; (2) they are a manifestation of "pure forms" that look like, but are not, once-living creatures; (3) they are some kind of mysterious, devilish trick to disturb the natural order; or (4) a divine test of faith from God. Intelligent investigators, at various times, have believed all these things.

Easily identifiable fossils, such as ancient oyster shells, appeared even to the ancient Greeks as having once been living creatures. Five hundred years ago

Leonardo da Vinci suggested a friend study fossil shell deposits to understand past changes in the Earth. But skeletons of unfamiliar organisms were puzzling. Some imagined that a mysterious force, or *vis plastica*, shaped rocks into forms that resembled strange animals or plants, but they had never been alive. Like geometric crystals or unusual gemstones, they were considered natural marvels and were prized by collectors.

It was only when an ever-increasing number of large animals and previously unknown fish were discovered that the older questions faded away, as paleontologists realized they had to explain a vast prehistoric zoo. By the early 19th century, the questions had become: How did so many creatures' fossil remains get into the cliffs and outcrops; How old were they; How did they fossilize; and What was the world like when these creatures were alive?

In one of the most bizarre Victorian theories, respected naturalist Philip Gosse argued that fossils were placed in the ground by God as part of the spontaneous creation of an ongoing world, just as he had included growth rings on trees that had never been saplings and navels on Adam and Eve, who were never born. Fossils were put in the earth to create it whole, complete with a past that never occurred. [See OMPHALOS.]

Gosse's views notwithstanding, most naturalists came to realize that fossils are the remains of once-living creatures, including many species that have become extinct. By the 1870s, fossil hunters of the "great bone rush" filled museums with dinosaurian giants and scoured the world for "missing links" between humans and apes.

Today, much less spectacular fossils are yielding the most important information about the past. Microfossils and drab remains of tiny sea creatures have provided keys to understanding such phenomena as cycles of global extinction, shifts in ancient climates, movements of continents, changes in the composition of sea water and the advance and retreat of glaciers.

See also BERINGER, JOHANNES; BURGESS SHALE; COPE–MARSH FEUD; FOSSIL MAN; ICE AGE.

FOUR THOUSAND AND FOUR B.C. (4004 B.C.)
Earth's Biblical Birthdate

Prior to the publication of Charles Darwin's *Origin of Species* in 1859, the first words in most English Bibles were "4004 B.C."—the date fixed for the creation of the world in the official view of the church. Until recently, many Bibles were printed with a special column of dates running alongside the text, which was meant to give the reader a precise chronology of events. These dates were not part of traditional scripture; they had been added in the 17th century. Most readers assumed, however, that these recently added dates were part of the ancient text.

According to the ecclesiastical timetable, the universe itself was created 2,000 years prior to that, or about 6,000 years ago. In our present understanding, that is about the time of the First Egyptian Dynasty, a complex civilization created by humans like ourselves. About 17,000 years ago cave paintings of mammoths and reindeer were made by artists, who were certainly human, when glacial ice covered Europe. Man-made stone tools and artifacts are well known from several hundred thousand and more years ago. To get to fossil hominids like *Homo erectus* we would have to go back to the half-million mark, while some African man-apes (australopithecines) are at least three million years old.

As geneticist Karl Pearson wrote in 1923, "If we have ceased to believe that the world and all its forms of life were created in 4004 B.C. it is because Darwin freed us from that cramping doctrine." For instance, Englishman John Woodward (1665–1728), who founded a geological museum at Cambridge to house the fossils he spent a lifetime collecting, might have groped his way toward great geological truths. But he was stymied by the scientific establishment of his day, which refused to publish his papers or lend credence to any human prehistory going back farther than 4004 B.C.

In the 18th century, some thought fossils were no older than the Creation itself, while others believed the Devil inserted them in the Earth to tempt man to disbelieve the church's teachings. "With a complete history of the world supposedly known from its creation in 4004 B.C. no real solution was possible" to the problems posed by fossils and flint tools. This "bondage of 4004 B.C.," as Pearson called it, continued in force even through the 1860s.

One of the reasons Darwin was so hesitant and cautious in presenting his views on the great age of the Earth was that he—like most Victorians—assumed 4004 B.C. was part of the original scripture itself. In 1861, Darwin was astonished to learn that the "official" biblical chronology had been calculated and added to the Bible by Ireland's Archbishop Ussher in 1650—just 200 years earlier.

See also CREATIONISM; USSHER-LIGHTFOOT CHRONOLOGY.

FRANKLIN, BENJAMIN (1709–1790)
A Darwinian Precursor

Every American schoolchild knows Benjamin Franklin as a printer, statesmen, scientist, diplomat and the author of *Poor Richard's Almanac*. One of America's great founding fathers, he was also a keen ob-

server and experimenter, inventor of the lightning rod, the Franklin stove and a pioneer in harnessing electricity. But it is not so well known that, like his friend Thomas Jefferson, his observations in natural history place him among the early intellectual forerunners of Charles Darwin. [See JEFFERSON, THOMAS.]

In 1750, Franklin wrote a letter about some pigeons he had observed nesting on the side of his house. It struck him that organisms increase to fill all the available space:

> I had for several years nailed against the wall of my house a pigeon-box that would hold six pair; and, though they bred as fast as my neighbours' pigeons, I never had more than six pair, the old and strong driving out the young and weak, and obliging them to seek new habitations. At length I put up an additional box with apartments for entertaining twelve pair more; and it was soon filled with inhabitants, by the overflowing of my first box and of others in the neighborhood . . .

For a man of Franklin's breadth of thought, it was but a short jump from theorizing about birds to people. In a short treatise written the following year, he concluded there is:

> no bound to the prolific nature of plants or animals but what is made by their crowding and interfering with each other's means of subsistence. Was the face of the earth vacant of other plants, it might be gradually sowed and overspread with one kind only . . . And were it empty of other inhabitants, it might in a few ages be replenished from one nation only; as for instance, with Englishmen . . . one million English souls in North America [doubling every 25 years] will in another century be more than the people of England . . .

Franklin's pamphlet was read by Thomas Malthus, an English clergyman and economist who took the ideas much farther. People increase at a geometric rate, he wrote, while their food supply increases only arithmetically. The result is that populations expand to the limits of their resources, followed by a crash, which could mean famine, war, or disease; man's tragic "struggle for existence," he called it.

Malthus's *Essay on the Principle of Population* (1789) in turn directly inspired Darwin and Alfred Russel Wallace, both of whom independently applied its principles to the natural world. Thus, Benjamin Franklin was not only one of the founding fathers of the United States, but of the theory of natural selection as well.

See also MALTHUS, THOMAS; NATURAL SELECTION.

FREEMAN, RICHARD BROKE (1915–1986)
Darwin Chronicler

Today there is a full-fledged "Darwin industry," made up of academics who pore over and write about

Charles Darwin's life and works for a living. But English zoologist Richard Freeman, one of the brilliant founders of the field, immersed himself in Darwiniana for the sheer joy of it.

Freeman's early career gave no indication that he would best be remembered as a Darwin scholar. An expert on insects with a doctorate from Oxford, he worked as a pest-control officer for the Ministry of Agriculture and Fisheries during World War II, doubling as a major in the Anti-Aircraft Home Guard, in command of 1,000 men.

In 1947, he joined the teaching staff of the University College London and taught zoology and taxonomy there for the next 35 years. Over the years he developed a passion for natural history books in general and books by and about Charles Darwin in particular. His book collecting and worldwide correspondence with scholars and bookdealers turned into a passion.

His first bibliographic book described more than 4,000 volumes on the plants and animals of Britain (*Natural History Books 1495–1900: A Handlist*, 1980). Once hooked, he went on to create unique guides to the Darwin literature, now considered indispensible to any serious Darwin scholar.

Freeman's *The Works of Charles Darwin: An Annotated Bibliographical Handlist* (1965; 2nd edition, 1977) lists every edition of every one of Charles Darwin's works. He set a new standard for scientific bibliography in the completeness and meticulousness of his survey of almost 2,000 items, including all editions, translations and reprints. With the increasing interest of book collectors in Darwin, Freeman's work is the first place to turn to determine the rarity and historical value of a Darwin work.

But Freeman's most remarkable compendium is *Charles Darwin: A Companion* (1978), an alphabetical listing of every person, place name and important topic mentioned in Darwin's books and letters. His *Darwin Pedigrees* (1984) brought information about the family up to date.

Although debilitated in his later years by partial stroke and confined to a wheelchair, Freeman continued his vigorous Darwin scholarship to the end. Those who venture into the immense Darwin literature find their path easier and more enjoyable as a result of Freeman's labor of love.

See also DARWIN CHARLES; DARWIN, CHARLES, OFFSPRING OF; DARWIN, CHARLES, WORKS OF.

For further information:

Freeman, Richard. *Charles Darwin: A Companion*. London: Folkestone: Wm. Dawson, 1978.

———. *The Works of Charles Darwin: An Annotated Bibliographical Handlist*. 2nd edition. Folkestone: Wm. Dawson, 1977.

FRERE, JOHN (1740–1807)
First Archeologist of Prehistory

The first published scientific evidence for prehistoric man was John Frere's "Account of Flint Weapons Discovered at Hoxne in Suffolk," read before the London Society of Antiquaries in 1797, a dozen years before Charles Darwin was born. Finely worked flint handaxes had been discovered in association with several large bones and a massive jawbone, probably of a mastodon, all sealed together in the same strata.

Frere was attracted to the site by workmen, who had been excavating clay for bricks. He took careful notes on the position of the bones and artifacts in the deposit, and made detailed drawings of the worked stones, which were published in *Archaeologia* in 1800.

These shaped flints were man-made tools, Frere concluded, "fabricated and used by a people who had not the use of metals . . . The situation [depth] at which these weapons were found may tempt us to refer them to a very remote period indeed, even beyond that of the present world."

Unfortunately, this pioneer of archeology was a lone voice, and the significance of his work went unnoticed until long after his death. Amazingly, the study of prehistory owes another important debt to its founder, John Frere—his great-great-great granddaughter is Mary Leakey! (She was born Mary Nicol; her mother's maiden name was Frere.) Along with her late husband Louis and son Richard, Mary Leakey has been a major force in the discovery of early hominids and their tools in East Africa.

See also LEAKEY, MARY.

FREUD, SIGMUND (1856–1939)
Evolution of Neurosis

On a table in his consulting room, Dr. Sigmund Freud kept a glass case full of the "antiquities" he enjoyed collecting: Egyptian scarabs, paleolithic stone hand axes, ancient figurines. What did this eclectic curio cabinet have to do with psychoanalysis? Freud was exploring how events of his patients' childhood had shaped their adult personalities. These prehistoric artifacts, he often told visitors, were clues to the "childhood of the human race."

Freud was trained as a biologist during the heyday of classical Darwinism, when German medical schools taught recapitulation in the womb as an accepted part of evolutionary theory. Promoted in Germany by Ernst Haeckel, the idea was that the development of each individual (ontogeny) is a speeded-up replay of the whole history of the species (phylogeny).

Based on general observations of the developing embryo, recapitulation theory became extremely influential outside of science. Pushed to explain many social phenomena, the misapplied analogy caused a great deal of mischief. Starting with the inaccurate notion that a human embryo at various times resembles an adult fish, reptile, monkey, etc. in the womb, it assumed all individuals go through the same evolutionary stages of development.

According to the extension of this idea, the minds of European children passed through a stage similar to adults of existing "lower" races (savages) or to our primitive prehistoric ancestors. Tribal peoples who had been enslaved or colonized by Europeans were regarded as "child-like primitives," requiring the firm supervision of a paternal missionary or colonial administrator.

To the triad of children, savages and early man, Freud added a fourth: the neurotic adult. (Other theorists of his time, such as Cesare Lombroso, had suggested that "criminals" and "morons" are stuck at an earlier stage of human development, or are "throwbacks" to a primitive type.)

Most writers have treated Freud as if his theories had arisen fully formed out of his own system of thought, with no scientific precedents. In fact, as Frank Sulloway has shown in *Freud, Biologist of the Mind* (1979) this "absolute originality" is a myth; 19th-century evolutionary ideas had an enormous influence in shaping Freud's thought.

Freud himself began his *Introductory Lectures on Psychoanalysis* (1916) with the statement of Haeckel's premise, which seemed to him self-evident: "Each individual somehow recapitulates in an abbreviated form the entire development of the human race." And, in 1938, he explained that "with neurotics it is as though we are in a prehistoric landscape—for instance in the Jurassic. The great saurians are still running around; the horsetails grow as high as palms."

Stephen Jay Gould notes that "recapitulation was both central and pervasive in Freud's intellectual development." His "oral" and "anal" stages represent not only the infant's early experiences, but also hark back to a four-legged animal ancestry.

Freud's famous *Totem and Taboo* (1913) is subtitled *Some Points of Agreement Between the Mental Life of Savages and Neurotics*. In this imaginative classic, Freud pondered the relationship of the worldwide taboo against incest, the "Oedipus complex" he thought he found in children and tribal totemism: clan identification with a sacred animal that must not be killed, except once a year when it is ritually eaten.

Here he concocted his own psychoanalytic myth of "the primal horde," the Freudian version of Original Sin. Freud imagined that the first prehistoric society was a patriarchal clan, ruled by a dominant father who monopolized food and sex. In order to get at the women, the sons murdered their father.

CHILDHOOD STAGES of development could underlie adult neuroses, taught Sigmund Freud, just as the "childhood" of the human species might explain certain religious practices and taboos. Like most biologists of his day, Freud was strongly influenced by the theory of evolutionary recapitulation.

But then they were too guilty to enjoy the women, which Freud thought was the origin of the incest taboo.

Later, the sons assuaged their guilt by merging the memory of their father with a symbolic totemic animal that it was taboo to kill. However, once a year, the sacred totem was symbolically slain and eaten. Freud's new "origin myth" was daring and imaginative, but there is no evidence these events ever took place.

Even more far-fetched speculations have recently surfaced in Freud's manuscript *A Phylogenetic Fantasy*, written in 1915, but forgotten and stored in an old trunk for 70 years and finally published in 1987.

This strange work traces hysteria, obsessions, anxiety neurosis and other modern disorders to the harsh life of our ancestors during the Ice Ages. Anxiety, for instance, arose because:

mankind, under the influence of the privations that the encroaching Ice Age imposed upon it, has become generally anxious. The hitherto predominantly friendly outside world which bestowed every satisfaction, transformed itself into a mass of threatening perils.

Under harsh conditions, people had to limit their numbers, which caused redirection of libidinal urges to other objects and resurface today as "conversion hysteria" or fetishism (sexual desires directed at objects like shoes or leather, rather than to the opposite sex). His idea was that behaviors that make no sense in today's world must have had a utility in the past, and have been handed down as a sort of inherited memory.

Freud's conclusions about the origins of dysfunctional behaviors are therefore based on two antiquated theories in biology: recapitulation combined with Lamarckian inheritance. Most present-day Freudians, unfamiliar with the history of evolutionary theory, cannot appreciate how deeply Freud's thinking rests on these two major 19th-century scientific fads, which have long since been abandoned by biologists.

See also BIOGENETIC LAW; COMPARATIVE METHOD; HAECKEL, ERNST.

For further information:
Sulloway, Frank. *Freud, Biologist of the Mind: Beyond the Psychoanalytic Legend.* New York: Basic Books, 1979.

FUEGIAN INDIANS

See BUTTON, JEMMY.

FUNDAMENTALISM
Conservative Religious Movement

Christian fundamentalism reached its zenith of popular support in the United States during the 1920s, when 37 antievolution resolutions were brought before the legislatures of 20 states. Seven states voted them into laws forbidding the teaching of evolution in schools. Groups such as the World's Christian Fundamentals Association and the Anti-Evolution League of America packed considerable power in their self-proclaimed war on Darwinism.

The term "fundamentalist" was coined by Curtis Lee Laws, editor of the Baptist *Watchman–Examiner* (July, 1920, p. 834) to denote "believers who cling to the great fundamentals and who mean to do battle royal" for them. Before the 1920s, there was a "fundamentalist" movement in America, but it was neither unified nor aggressive in attacking science.

In 1910, the General Assembly of the Presbyterian Church adopted a creed of "five fundamentals": (1) inspiration and infallibility of scripture, (2) deity of Christ, including virgin birth, (3) substitutionary

atonement of Christ's death, (4) literal resurrection of Christ from the dead and (5) the literal return of Christ in the Second Coming.

To these, the Reverend Jerry Falwell adds that modern fundamentalists also believe in a literal heaven and hell, the depravity of mankind, the importance of soul-winning (evangelism), the Holy Spirit, and the existence of Satan.

A series of 12 booklets called *The Fundamentals* was published by the Presbyterians in Chicago between 1910 and 1915 to "set forth the fundamentals of the Christian faith." They were sent free to ministers, missionaries and others engaged in "aggressive Christian work."

Although they contained some antievolutionary essays, the booklets also included writings of early fundamentalists who attempted to reconcile Christianity with Darwinian thought, which may seem surprising today. Such outstanding Christian thinkers as Augustus Hopkins Strong (1836–1921), the leading Baptist theologian, had no quarrel with evolution and said so in the pages of *The Fundamentals*.

Another eminent theologian, James Orr (1845–1913) published an article in *The Fundamentals* claiming that the Bible should not be read as a science textbook, that the world is much older than 6,000 years, and that "evolution is coming to be recognized as a new name for 'creation.'"

By the 1920s, however, the fundamentalist movement had turned into an aggressive, militant crusade, based on the fear that American children were being taught to deny God, subvert family authority, destroy the credibility of the Bible and base their morality on a crass Social Darwinism.

It was not until the inevitable confrontation with biology teacher John T. Scopes at the Tennessee "Monkey Trial" (1925) that fundamentalists began to lose their war, although they won that battle. Scopes was convicted, but Clarence Darrow's public airing of the issues won the day.

As historian James Moore puts it, Darrow "made [William Jennings] Bryan talk nonsense, confess ignorance, and, most important of all, admit that he did in fact 'interpret' the Bible." Under Darrow's relentless questioning, Bryan had given the lie to his own position that the "word of God" was unambiguous and not subject to human interpretation.

Over the years, the various state statutes banning the teaching of evolution have been struck down. Nevertheless, fundamentalists have made periodic shows of strength, most recently in the 1970s, when a conservative Christian constituency, led by the Reverend Jerry Falwell, succeeded in influencing a presidential campaign.

Previous presidents, such as former educator

Woodrow Wilson, had said "no intelligent person at this late date" denies evolution, and Theodore Roosevelt claimed to have studied natural history "at the feet of Darwin and Huxley." But President Ronald Reagan stated in 1976 that he thought many scientists were retreating from belief in evolution and that he saw nothing wrong with "other theories" of creation being taught in the schools.

In 1987, however, the Supreme Court struck down the last of the "equal time" laws for "Creation Science," ruling that it was an intrusion of religious instruction into the public educational system. Undaunted, fundamentalist groups have vowed to continue opposition to the teaching of evolutionary biology in schools partially supported by their taxes.

See also BUTLER ACT; CREATIONISM; "EQUAL TIME" DOCTRINE; MILITARY METAPHOR; REAGAN, RONALD; "SCIENTIFIC CREATIONISM"; SCOPES TRIAL; SCOPES II.

FUR CONTROVERSY

See ANIMAL RIGHTS.

GADARENE SWINE CONTROVERSY
A Debate About Demons

One of Thomas Henry Huxley's most colorful confrontations about biblical authority was not with a churchman, but with Prime Minister William Gladstone, who had publicly attacked the evolutionists. Scriptural teachings, he insisted, are revealed truth and should be believed without question. Somewhat whimsically, Huxley took his stand on the story of the Gadarene swine, one of the more bizarre episodes in the New Testament.

The story, in Mark 5:1–19 and Luke 8:26–39, begins when Jesus enters the country of the Gadarenes. [King James version, Gerasenes in later translations.] Two thousand pigs were feeding on a hillside above the Sea of Galilee when a madman approached Jesus on the road. He said he was possessed by so many devils that he called himself "Legion." Taking pity on the man, Jesus healed him on the spot and sent the flock of demons into the herd of swine. The possessed porkers then rushed wildly down a steep bank into the sea and drowned.

Huxley thought the story a particularly blatant example of primitive demonology grafted onto the authority of Jesus. If accepted uncritically, said Huxley, "the belief in devils who possess men, and can be transferred from men to pigs, becomes as much a part of Christian dogma as any article of the Creeds." If the Gospel tradition is "falsely coloured and distorted by the superstitious imaginings" of its tellers, "what guarantee have we that a similar unconscious falsification" of the sayings and doings of Jesus was not also handed down?

Besides, Huxley asked, what about the pigs' owner? Would Jesus wantonly destroy the property of an innocent farmer? Gladstone answered that the farmer was Jewish and shouldn't have been raising pigs.

This heated controversy dragged out for months in the public prints causing some critics to ridicule Huxley for devoting so much energy to an absurdity. Privately, he told friends, he was "sick and tired of these damned pigs." But these "too famous swine are not the only parties to the suit . . . The real question is whether the men of the nineteenth century shall believe the demonology of the men of the first century."

His last reference to the bedeviled herd was in a letter to his friend Sir John Lubbock in 1894, at a time when Huxley was very ill:

I am creeping about as well as a sharp attack of lumbago will let me, but for the most part horizontal. I wish there was a herd of swine for that [lumbago] devil to go into—only . . . they would not be able to rush violently anywhere or do anything but grunt. At least, that is *my* experience.

See also HUXLEY, THOMAS HENRY; OXFORD DEBATE.

GAIA HYPOTHESIS
Mother Earth's Stabilizers

Gaia, an ancient Greek name for the Earth goddess, has been applied to a new hypothesis about how our planet supports life. The controversial idea was developed in the 1970s by two English scientists: James Lovelock, an organic chemist, and Lynn Margulis, a molecular biologist.

They believe that our planet is not simply an insensible rock hurtling through space, but a sensitive system that operates according to its own laws to support, develop and preserve life.

The first scientific experiments on ecological life-support systems were done by the great English chemist Joseph Priestly in 1772. Exploring the question of whether plants breathe, Priestly placed a plant in an airtight bell jar; after a few days, it suffocated and died. Next, he enclosed a plant and mouse together in the same airtight chamber and found that both could breathe and live! Soon after, he realized that the animal inhales oxygen (O_2) and exhales carbon dioxide (CO_2), while the plant obligingly does just the opposite. It was the first scientific demonstration of a mutual life-support system.

In our own time, every school child learns that the Earth's atmosphere consists of a certain mixture of oxygen (21%), nitrogen (78%), carbon dioxide, and other gases. This ocean of air extends to a height of about 60 miles and envelopes the earth with a mantle of the same gases in the same proportions. But in 1972, James Lovelock wondered why these ratios remain constant over the whole planet and persist for vast periods of time with very little fluctuation.

He had found some exceptions, which help establish the more usual rule of stability. Evidence in the rocks reveals that eons ago, during the Cambrian era, there was an imbalance in the atmosphere. Oxygen levels dropped tremendously, causing mass extinctions among sea animals. However, by some unknown means, the balance was corrected, and the biosphere was rebuilt and stabilized once more for millions of years.

What caused the gases to make the necessary adjustment? We don't know. But Lovelock postulates the existence of a vast and delicate mechanism whose workings are still mysterious, but which seems in some way to react as if it were an organism. Earth's air, living matter, oceans, and land surfaces are all parts of the mechanism, the complexities of which are only now beginning to be unraveled.

According to Lovelock, in his book *Gaia* (1979), "the most essential part is probably that which dwells on the floors of the continental shelves and in the soil. The tough, reliable workers composing the microbial life of the soil and sea-beds are the ones who keep things going."

Lovelock is impressed, for example, by the millions of microbes that live in the guts of termites. Without this miniature universe of diverse creatures, termites would be unable to digest the wood (cellulose) on which they feed. But because an incredible living world has evolved inside—and along with—termites, the forests of the world are continually recycled, instead of being stacked miles high with billions of dead shrubs and trees.

Lovelock sees the entire range of living matter on Earth, from viruses to whales, from oaks to algae, as contributing to the Gaia system, which maintains and manipulates the Earth's atmosphere to suit its overall needs. It is also, Lovelock points out, a concept that is in tune with many of the world's ancient religious traditions, such as the Hindu, Buddhist and Native American.

Some worried scientists have welcomed evidence for the Gaia hypothesis. If a tough, self-regulating mechanism exists in Mother Earth, they reason, we can all rest easier, as it may compensate for such human tampering as destruction of rain forests or spread of radioactive waste.

It seems more likely, however, that we cannot expect such a far-reaching, delicate system to heal any and all wounds inflicted by man. Mary Midgley, a philosopher of science, has warned against this interpretation in her book *Evolution As A Religion* (1985). "If Gaia finds her system getting out of kilter because one element in it is insatiably greedy," Midgley writes, "she simply ditches that element as she has done so many others before. She is not in the least anthropocentric [man-centered] and has no special interest in intelligence . . . but resolute to remain in general green and alive, [resisting those who] turn the green, thriving world into a desert. And if this resistance fails, she herself can no doubt be killed with all her children. No universal fail-safe mechanism protects either her or us."

See also COEVOLUTION; ECOLOGY; ENDOSYMBIOSIS; EXTINCTION; RAIN FOREST CRISIS.

For further information:
Lovelock, J. E. *The Ages of Gaia: A Biography of Our Living Earth.* New York: Norton, 1988.
———. *Gaia: A New Look at Life on Earth.* Oxford: Oxford University Press, 1979.

GALAPAGOS ARCHIPELAGO*
Darwin's "Enchanted Islands"

"The natural history of these islands is eminently curious, and well deserves attention," Charles Darwin wrote of the Galapagos Archipelago (pronounced Ga-LOP-a-gus), a cluster of volcanic cones 600 miles west of Ecuador, in the deep Pacific Ocean. There are five main islands, and a scattering of smaller ones that pierce the water astride the equator. Early Spanish sailors called them *Las Islas Encantadas*, the Enchanted Isles, miniature self-contained worlds of desert, scrub, misty forest and black volcanic beach. "Here, both in space and time," Darwin wrote in his *Journal* (1839), "we seem to be brought somewhat near to that great fact—that mystery of mysteries—the first appearance of new beings on this earth."

Darwin and the *H.M.S. Beagle* spent only five weeks exploring the islands in September and October of 1835, but they made a lasting impression on him. The Galapagos Archipelago is a compact little model of how the whole planet may have been populated by radiating species, adapted to different ways of life and descended from common ancestors. In the first edition of his *Journal* he only hinted at such a conclusion, but in subsequent editions—as he sorted and pondered his collections—he began to realize that the Galapagos were key to the theory he was seeking to explain the origin of species.

Almost immediately, Darwin noticed that the Galapagan animals and most of its plants were unique—found nowhere else in the world. "There is even a difference between the inhabitants of the different islands," he wrote, "yet all show a marked relationship with those of [South] America, though separated from that continent by an open space of ocean . . . 600 miles in width.

The archipelago is a little world within itself, or rather a satellite attached to America, whence it has derived a few stray colonists . . . Considering the small size of these islands, we feel the more astonished at the number of their aboriginal beings, and at their confined range . . . the different islands to a considerable extent are inhabited by a different set of beings . . . One is astonished at the amount of creative force, if such an expression may be used, displayed on these small, barren and rocky islands . . .

The nature and diversity of animal life was a real delight to the ship's young naturalist. There were seals, dolphins, penguins, prehistoric-looking iguana lizards, bright red crabs, incredible giant tortoises (Darwin rode one) and a most interesting series of mocking birds and finches. Large reptiles dominated the landscape; aside from mice and rats, there were no mammals.

Despite their fearsome appearance, the marine iguanas were vegetarians that dove deep in the ocean to graze on seaweed. (Darwin cut one's stomach open to discover that fact.) As for the tortoises, they were so docile that sailors were in the habit of carting off hundreds of the hapless beasts, to be stored live

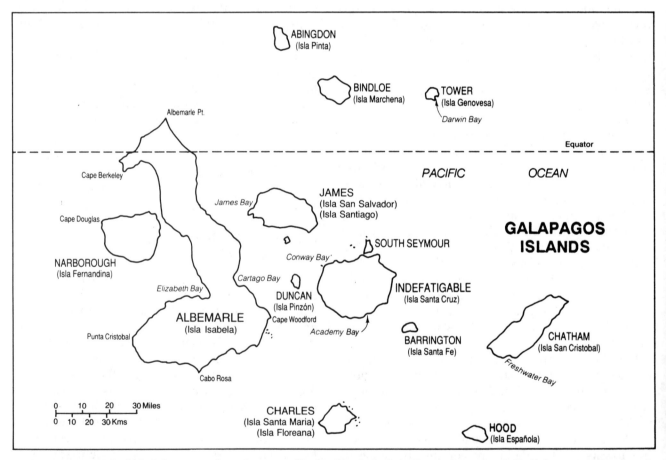

ENCHANTED ISLES of black lava seemed to take young Charles Darwin back in time as he explored the Galapagos Archipelago in the Pacific, west of Ecuador, in 1835. Marine iguanas and giant tortoises greeted the young naturalist, along with bold birds that had evolved in an environment free of predators.

in ship's holds awaiting their turn to feed meat-starved crews. One tortoise could yield 200 pounds of meat, and it took six or eight men to lift one and carry it away.

Because of their long isolation, the Galapagos animals had developed no fear of man. The finches, wrens, mockingbirds, doves and even carrion-buzzards—like the tortoises and seals—had an "extreme tameness." "A gun is here almost superfluous," Darwin wrote, "for with the muzzle I pushed a hawk off the branch of a tree." He was astonished at the contrast between this trusting disposition and the birds back in England, which had apparently learned to keep their distance from man thousands of years ago.

But it was the diversity of closely related species on the several islands that Darwin said "strikes me with wonder." He noted that each island had its own species of tortoise, mockingbird and finch. Indeed, 13 types of finches, each with a different type of beak, from delicate to heavy, from small to large, occupied different niches in the ecology. "Seeing this gradation and diversity of structure in one small, intimately related group of birds," he wrote, "one might really fancy that from an original paucity of birds in this archipelago, one species had been taken and modified for different ends."

Darwin wrote that statement some years later, after he had sorted out his collections of birds back in England, with the help of ornithologist John Gould. [See DARWIN'S FINCHES.] At the time, he almost missed that most important point: "that the different islands to a considerable extent are inhabited by a different set of beings."

> My attention was first called to this fact by the Vice-Governor, Mr. Lawson, declaring that the tortoises differed from the different islands, and that he could with certainty tell from which island any one was brought. I did not for some time pay sufficient attention to this statement, and I had already partially mingled together the collections from two of the islands. I never dreamed that islands, about fifty or sixty miles apart, and most of them in sight of each other, [of the same age and climate] would have been differently tenanted; but . . . this is the case . . . [I am] thankful that I obtained sufficient materials to establish this remarkable fact in the distribution of organic beings.

See also (CHARLES) DARWIN RESEARCH STATION; ISOLATING MECHANISMS; LONESOME GEORGE; VOYAGE OF THE H.M.S. BEAGLE.

For further information:
Steadman, David, and Zousmer, Steven. *Galapagos: Discovery on Darwin's Islands.* Washington D.C.: Smithsonian, 1988.

GALAPAGOS TORTOISE

See LONESOME GEORGE.

GALTON, SIR FRANCIS (1822–1911)
Eugenicist, Statistician

Charles Darwin's cousin, Sir Francis Galton, was the epitome of the compulsive Victorian genius. By the age of 22, he had earned degrees in both medicine and mathematics, then went on to become a famous travel writer (on Africa), statistician, experimental psychologist and meteorologist. But Galton saw a grander canvas for his talents: the future course of human evolution. Without his help, he came to believe, mankind would never evolve properly.

Galton is chiefly remembered today as the founder of the eugenics movement—a well-intentioned attempt to bring about social improvement through selective breeding. If horses and cattle could be improved through artificial selection, why not people? It was, of course, an unmitigated disaster, an ideology that caused untold suffering for more than 60 years. Eugenic principles were called upon as the "scientific" justification for restrictive immigration laws, compulsory sterilization, racist social policies in several countries and the nightmarish mass exterminations in Nazi Germany.

Galton had so many diverse interests, such an active mind and was so convinced of his own greatness that in later life he hired secretaries to follow him everywhere, lest some of his thoughts escape unrecorded. Few did.

Among his almost 200 scientific publications, are some intriguing oddball titles: "On a New Principle for the Protection of Riflemen" (based on the trajectory of spherical bullets) (1861), "Statistical Inquiries into the Efficacy of Prayer" (1872), "Visions of Sane Persons" (1882), "Measurement of Character" (1884), "Notes on Australian Marriage Systems" (1889), "Head Growth in Students at the University of Cambridge" (1889), "Arithmetic by Smell" (1894), *Fingerprint Directory* (1895), "Three Generations of Lunatic Cats" (1896), "Cutting a Round Cake on Scientific Principles" (1906).

Profoundly inspired by Darwin's evolutionary theories, Galton spent much of his life attempting to follow out their practical, mathematical and social implications. On first reading *Origin of Species* (1859), which is slow work for most, Galton recalled he "devoured its contents and assimilated them" quickly and easily, "which perhaps may be ascribed to an hereditary bent of mind that both its illustrious author and myself have inherited from our common grandfather, Dr. Erasmus Darwin." (Galton and

HEREDITARY GENIUS was more than the title of Sir Francis Galton's famous book; he believed it was the destiny of his own family. A cousin of Charles Darwin, Galton's ingenious applications of mathematics to biology laid the foundation for biometric research, but his "eugenic" program for breeding superior humans was a disaster.

Charles Darwin had different paternal grandmothers, however.)

These "new views" encouraged Galton "to pursue many inquiries which had long interested me, and which clustered round the central topics of Heredity and the possible improvement of the Human Race." At the time "even the word heredity was then considered fanciful and unusual."

Searching for patterns of hereditary talents among close relatives, he began to gather background information on a sample of accomplished Cambridge college students. Eventually, he pulled together data on 1,000 men from 300 families, for his study *Hereditary Genius* (1869). Some families, he found, produced an inordinate number of scientists, musicians, painters, judges, military commanders or even oarsmen or wrestlers.

These special talents, he believed, were innate. Galton thought the advantages of tradition, wealth, opportunity and education didn't count for much, compared with hereditary abilities present in the individual at birth. Thus, a person born with superior character and intelligence should inevitably rise to eminence, even if reared in the lowliest slum. Since few did, he concluded the lower classes remained poor because of genetic "inferiority."

Charles Darwin, on reading part of *Hereditary Genius,* wrote Galton, "You have made a convert of an opponent in one sense, for I have always maintained that, excepting fools, men did not differ much in intellect, only in zeal and hard work; and I still think this is an *eminently* important difference." Galton respectfully disagreed, insisting that "character, *including the aptitude for work,* is heritable like every other faculty."

Eugenics, the "science of breeding the best," became Galton's obsession. He wrote that "an enthusiasm to improve the race is so noble in its aim, that it might well give rise to a sense of religious obligation . . . its principles ought to become one of the dominant motives in a civilized nation."

> Man . . . has also the power of preventing many kinds of suffering. I conceive it to fall well within his province to replace Natural Selection by other processes that are more merciful and not less effective . . . This is precisely the aim of Eugenics. Its first object is to check the birth-rate of the Unfit . . . [and] the improvement of the race by furthering the productivity of the Fit by early marriages . . . of the best stock.

In addition to encouraging "the best" to produce more children, Galton became increasingly concerned about "reproduction of the unfit." Specifically, he feared the lower classes would take over society by having the most children; it would result in the "degeneration" of the "British race." His ideas were taken seriously, attracting such thinkers as playwright G. B. Shaw, novelist H. G. Wells and socialist Sidney Webb to join the Eugenics Education Society, which Galton founded in 1907.

The United States proved to be even more fertile ground than England for the adoption of eugenic policies. Sterilization laws for "imbeciles" were passed, and the usually compassionate Justice Oliver Wendell Holmes upheld its constitutionality in a decision of 1927: "We have seen more than once that the public welfare may call upon the best citizens for their lives. It would be strange if it could not call upon those who already sap the strength of the State for lesser sacrifices . . . It is better for all the world, if . . . society can prevent those who are manifestly unfit from continuing their kind." Later, it was found that many who were sterilized were not "imbeciles" at all.

Galton also established the tradition in England of biometrics, the application of statistical and populational techniques to the study of heredity and variation. His journals and laboratories laid the groundwork for population genetics, which, years later, fused with Darwinian theory and put the theory on a sounder basis. He also took up the unfinished work of Captain Robert FitzRoy, reestablishing his meteorological office and its practice of furnishing weather maps and predictions to the newspapers.

Galton was once asked, when in his seventies, why he had no deep wrinkles or furrows on his forehead. "Oh, that is easy," he replied, "it's because I am never puzzled."

See also BIOMETRICIAN VS. MENDALISM CONTROVERSY; COMPOSITE PHOTOS; DEGENERATION THEORY; EUGENICS; FITZROY, CAPTAIN ROBERT; GALTON'S POLYHEDRON; RACISM (IN EVOLUTIONARY SCIENCE).

GALTON'S POLYHEDRON
Equilibrium and Constraints?

Darwin's eccentric cousin, Sir Francis Galton, was the first to advocate the use of fingerprints in identifying criminals, founded the eugenics movement and wrote an influential book on *Hereditary Genius* (1869).

In that work, Galton presented an unusual analogy for the evolution of species. Suppose, he wrote, that you had a rough stone with many facets that was tumbling across a table. It would rest in a stable equilibrium on one facet until considerable energy was exerted to push it over onto the next facet, where it would again be quite stable. Its movement would thus consist of a series of abrupt progressions from stasis to stasis.

Galton did not use his idea in the modern sense of punctuated equilibrium in large populations. He imagined a kind of "buildup" in organisms, which could erupt in seemingly sudden changes of form. As Stephen Jay Gould puts it, "the polyhedron's response to selection is restricted by its own internal structure; it can only move to a definite number of limited planes . . . [an] "interaction of external push and internal restraint."

Naturalist and theologian St. George Mivart used Galton's polyhedron as the basis for his hypothesis of Specific Genesis, wherein an organism—though adapted to particular conditions—contains within itself certain well-defined possibilities for adjusting quickly to sudden environmental changes. Decades later such large rapid changes in individuals would be called "macromutations," but Mivart and Galton were writing years before the rise of genetics.

A pattern of rapid changes in species, followed by long periods of stability, is now seen by some paleontologists as a more accurate description of the fossil record than Darwin's insistence on slow, steady rates of change. In 1972, Gould and Niles Eldredge called attention to this model of evolution, and coined the term "punctuated equilibrium" to describe it.

Although an obscure historical curiosity, Galton's polyhedron can be useful in visualizing the limitations within organisms of the possibilities for rapid change. It is also a reminder that Darwin's view of gradual, uniform change always had its critics and that some earlier thinkers proposed models of abrupt change that were neither biblical deluges nor volcanic catastrophes. It is a problem that has not yet been solved; the question of rates of change in organisms is currently one of the most debated and investigated topics in modern biology.

See also EXTINCTION; GALTON, SIR FRANCIS; "NEW" CATASTROPHISM; PUNCTUATED EQUILIBRIUM.

GARDNER, BEATRICE and R. ALLEN

See WASHOE.

GARNER, RICHARD LYNCH (1848–1920)
Pioneer Primate Observer

In America, the only primates zoologist R. L. Garner had ever seen were locked in cages. Fascinated with human evolution, he decided to make pioneering studies of monkey and ape behavior in the wild during the 1890s. But since hunters insisted gorillas would invariably attack, Garner hit on a unique field-work method: He built a large cage in the deep forest and locked himself inside!

Garner's cage was no flimsy hunter's blind; it was strong enough to hold off any animal except an elephant, "giving the occupant enough time to kill an assailant." Armed with a rifle, revolver and bush-knife, he sat patiently for days and weeks on end, emerging each evening to return to base camp. Almost from the start, monkeys and other wildlife gathered in nearby trees to observe the caged human. Eventually, Garner did see passing gorillas and chimpanzees, but never tracked them or observed normal group behavior.

Nevertheless, he learned to imitate 10 chimpanzee sounds and identified 20 more. He also observed the whooping, frenzied, vegetation-tearing displays accompanied by drumming, which he described as a "chimpanzee carnival." Relying on local accounts, Garner correctly described the gorilla family group, with its several females clustered about a large male leader in *Gorillas and Chimpanzees* (1896).

Later, he became the Bronx Zoo's representative in Africa, buying animals and shipping them to New York. In 1906, Garner sent the first live gorilla to reach an American zoo (England's first had arrived in 1860). Another was sent in 1911, but neither survived for long. Zoo directors despaired of ever being able to keep gorillas alive.

Garner also studied captive monkeys, particularly some capuchins from South America. He found he could scare them by imitating their alarm call and

LOCKED IN A CAGE, zoologist R. L. Garner made an early attempt to study wild chimpanzees and gorillas in their home forest. Until a few decades ago, field observers thought free-living apes or baboons would rip them apart if they approached on foot.

that when physically dominated they made distinct "submissive" noises and gestures.

An overzealous Darwinian, he imagined he could discern distinct calls for "milk," "apple," "banana," "bread," and "carrot"—a whole "monkey language," which no one else could later verify. All of these sounds he transcribed as "melodies" in standard musical notation (*Apes and Monkeys*, 1900), concluding monkeys had "all the characteristics of speech."

Despite his limitations, Garner made the bold first attempt to observe wild apes, which inspired others (including Carl Akeley) to follow him. If the heavily armed, pith-helmeted Victorian scientist sitting in his jungle cage seems quaint, his successors did little better. For almost a 100 years they carried guns, tracked baboons from Land Rovers and watched gorillas through binoculars.

It was only in the 1960s that Jane Goodall began mingling with chimps and George Schaller tracked gorillas unarmed and on foot. By the 1970s, mountain gorillas had accepted Dian Fossey into their groups, Schaller had walked among tigers, Shirley Strum abandoned her van to mingle with baboons and Jacques Cousteau's divers swam alongside whales, completely dependent on the more powerful animal's tolerance.

Because human distrust of animals had been so strongly ingrained, it took incredible bravery for these naturalists to approach wild animals without walls or weapons. In doing so, they contributed more than scientific knowledge of the various species—they freed humanity from the mental cage that for so long had inhibited mutual trust between ourselves and other species.

See also APE LANGUAGE CONTROVERSY; BABOONS; CARPENTER, CLARENCE RAY; CHIMPANZEES; FOSSEY, DIAN.

GAUSSE, FRANZ J.

See PHRENOLOGY.

GAUTREY, PETER J. (b. 1925)
Darwinian Archivist

Peter Gautrey became a premier shaper, guardian and custodian of one of the most unique and valuable collections in the history of science: the thousands of books, manuscripts and letters that comprise Cambridge University's Darwin Library. But as a youth he had had no special interest in science; at eight years old his talent as a singer gained him a musical scholarship to the famous choir school at St. John's College in his native Cambridge, England. In 1940, he took a job at the University Library, unaware it would lead to a nearly half-century career (he retired in 1989).

In 1947, after four years' war service in Europe with the British Army, Gautrey returned to the library and worked in the manuscript reading room. There he met Sydney Smith, Bob Stauffer and other scholars who were assembling and organizing the long-scattered Darwin papers. Gautrey became fascinated, and soon the self-taught librarian became one of the world's outstanding experts on Charles Darwin's life and times, publishing articles on some of the rare documents he discovered in the archives. Among the "treasures" he brought to light were Darwin's notes on bees in his garden, questions about the facial expressions of tribal peoples he had sent to missionaries around the world, and parts of a previously unknown manuscript of the great biologist's autobiography.

Gautrey's work brought him into close contact with scholars, authors, historians, and scientists from all over the world, many of whom came to admire his mixture of wry wit, extroverted good humor, and British reserve. One of his dearest friends was Nora, Lady Barlow, granddaughter of Charles Darwin, who died in 1989 at the age of 103. Once Gautrey visited the Down House Museum with her; as they entered the old drawing room, Lady Barlow exclaimed: "I just felt as if Grandmama was coming out to meet us, just as she did when I was a girl." She was referring, of course, to Charles Darwin's wife Emma. Gautrey later recalled, "That was the closest I ever felt to personally touching life in the Darwin household of the last century."

See also BARLOW, LADY NORA; FREEMAN, RICHARD; SMITH, SYDNEY.

GENE MAPPING
DNA Inventory

In 1988, the United States government announced the genome mapping project—an all-out $3 billion effort to analyze and sequence all the subunits of human DNA. Recent advances in automated, computerized analyzers now make it possible to remove much of the drudgery from the millions of biochemical tests necessary to compile such a massive inventory.

Gene mapping means constructing a detailed written manual of the chemical instructions for making a given organism—in this case, a human being. Each of the 46 human chromosomes has an estimated 50,000 to 100,000 genes to be identified and listed. Then each of the three billion DNA subunits would be chemically identified and located in its proper order—a process known as sequencing. While gene maps have been made for simpler organisms, such as bacteria and fruit flies, the human genome is the most complex ever attempted.

The first complete genetic map of a living organism was completed at Columbia University in February, 1987, after 18 months of work by a research team. Its subject was a bacterium that lives in the human intestine, *Escherichia coli*. This microbe has a single chromosome, which is one-tenth the size of the smallest of a human's 46 chromosomes; it was broken into 23 pieces for the analysis.

Dr. James D. Watson, codiscoverer with Francis Crick and Maurice Wilkins of the structure of DNA, was appointed head of the human mapping project by the National Institutes of Health (NIH), which had come back to the project after initially turning it down. As director of the largest biological research project in history, the once-irrepressible maverick of

biochemistry was officially installed as its elder statesman.

When NIH had refused to take on the project, geneticists went to the U.S. Department of Energy, which had enormous resources and had already sponsored some projects on chromosome sorting. However, competition for control soon developed between the biologists and physicists, who had been carrying out experiments with nuclear energy and thought they could do any kind of lab science. Finally, a biochemist, Dr. Charles Cantor, was chosen to head the program and work was begun at the Los Alamos, New Mexico, labs and the Lawrence facility at Berkeley, California.

By 1986, Congress was persuaded to sponsor American leadership in deciphering the human genome. If the U.S. failed to take the initiative, Japan had the technology to tackle it. Repetitive testing and analysis will be handled by precision robots developed in Japanese companies during the 1980s. (One major piece of hardware, created at Hitachi, is an automatic scanner for reading DNA sequences, which allows a biologist to conduct experiments at a keyboard while robots perform the "benchwork.")

In Charles Cantor's phrase, "the genome is a historical text, with copying errors recorded and recopied in each generation." As with the bacterium, the research strategy is to break up the genetic instruction book into small "chapter-size" pieces for analysis. In addition to achieving its stated goal of mapping the genome, the project is expected to spin off new industries and make fundamental contributions to medicine.

"When we found the structure of DNA [in 1953] we never thought we were going to go this far or this fast," Dr. Watson mused upon accepting the appointment at NIH. But he was still impressed by the immensity and complexity of the task. "I'd like to get the answer," he told the *New York Times*, "while I am still alert enough to appreciate it."

GENETIC DRIFT
A Chance Factor in Evolution

Evolutionary change in breeding populations can happen in a number of ways. Mutations in genes can occur, natural selection can operate upon them and the frequency of genes in a particular population can shift in an adaptive, or directional, manner. There can be genetic recombinations, hybridizations and changes in the environment, which alter the survival value of existing gene frequencies for better or worse. But there's also a wild card in the deck: genetic drift.

Genetic drift is a random process, a chance factor in evolution, that is tied to the size of a breeding population. If a small population becomes reproductively isolated, its gene pool may no longer represent the full range of genetic diversity found in the parent gene pool. There may be a couple of albinos in the lot, or individuals with inheritable diseases or many blue-eyed individuals; it is all in the luck of the draw. Statisticians call it "sampling error." The smaller the breakaway population, the greater the effect of genetic drift.

Geneticist Sewall Wright first described the phenomenon, which can be measured among modern human populations, and called it the "founders effect" (others have called it the "Sewall Wright effect"). It operates when a very small group of individuals become founders of a new, larger population, producing a "genetic bottleneck." This founder's effect or genetic drift occurs in humans when a small tribal band splits up, and a few families migrate to colonize a new territory. Founder populations can also be the remnants of harsh winters, wars or famines.

Actual examples in human populations include some isolated Alpine villages, which have unusually high percentages of albinism, and a South Atlantic island (Tristan da Cunha) whose population—descended from a single Scots family and some shipwrecked sailors—has a high proportion of a rare hereditary eye disease. Anthropologist Joseph Birdsell, who has made extensive population studies among Australian aborigines, has concluded that very small breeding groups (under 500 and as few as 170) were the norm rather than the exception among hunters and gatherers. As with other creatures that became dispersed over a wide range, irregular and nondirectional changes in gene frequencies probably played an important role in human evolution.

GENETIC ENGINEERING
"Inventing" New Life-forms

Thousands of lab scientists go to work every day to tinker with life-forms, rearrange DNA and splice together genetic codes from distantly related creatures. While critics fear they may unwittingly unleash hordes of evolutionary monsters, "genetic engineers" believe they are doing what horsebreeders or corn farmers have always done—improve the usefulness of plants and animals for man.

Since the 1970s, when the technology for gene-splicing was first developed, genetic engineering has become big business. All over the world, biotechnology labs are making gene maps, snipping bits of chromosome from one organism and planting them into other species.

They have already caused bacteria to make human insulin and dissolve dangerous blood clots, and they will soon turn microbes into living factories for the production of serum and vaccines. Others are trying to "invent" bacteria that will eat up oil spills at sea or provoke human antigens to destroy organisms that decay teeth. By the mid-1980s, a piece of DNA from the gonorrhea organism was being used to make a quick test for the disease, replacing one that took several days.

Genetic engineers are revolutionizing world agriculture. In one experiment, a modified microbe was injected into corn to see if it could give the plants immunity from their archenemy, a moth larvae known as corn borer. The microbe had been genetically altered to attack alkaline chemicals, and it attacked the corn borers' stomachs, giving them ulcers—and protecting the plants.

Similarly, tobacco plants proved able to resist the budworm pest, after an implanted gene caused them to manufacture an insect poison—and they passed the gene on to their offspring. (It originally came from another plant, the cowpea, that produces a natural insecticide.)

In another startling development of the 1980s, tobacco plants—the experimental "white rats" of botany—received genetic material from fireflies. The result: a tobacco plant that glows in the dark. It is now much in demand for various studies of cell physiology, since the glowing "marker" cells can be easily traced.

Lured by immense profits, large corporations are in a race to develop the biotechnology for cancer cures and protein farms, pest-proof food crops, more efficient and productive farm animals and prevention of genetic diseases in humans. The U.S. Patent Office, in 1987, decreed that living things produced by recombinant DNA are patentable inventions. That same year a biologist patented a strain of mice that is guaranteed to develop cancerous tumors, for use in cancer-cure experiments.

Genetic engineering has its vehement critics, however. Environmental advocate Jeremy Rifkin, for example, has challenged scientists' rights to unleash "new" organisms into the environment for profit and promptly sued the U.S. Patent Office to challenge its new policy that living things can claim the legal status of "inventions."

One of the most intensive areas of current research is the application of genetic engineering to produce more useful farm animals: sheep with better wool and smaller appetites, hens that lay more eggs even when crowded together, cows that produce more milk and fast-growing fish.

At the University of Idaho, a state where fish-farming is a major industry, genetic engineers are attempting to create the ideal food fish of the future. So far, they have found fish much more easy to manipulate than mammals. Experimenters have been able to reverse the sexes of fish, multiply their chromosomes and even produce an "invented" variety made up of cells from two different species of trout.

Cattle embryos have been successfully cloned at Granada Genetics, Inc., a leading livestock genetic engineering firm in Marquez, Texas. Researchers there have already raised several small herds of genetically identical cattle and are preparing for the day when many cloned embryos of a single desirable animal can be routinely implanted in ordinary cows.

See also DNA; GENE MAPPING; LOEB, JACQUES.

For further information:
Yoxen, Edward. *The Gene Business.* London: Pan, 1983.
———. *Unnatural Selection? Coming to Terms with the New Genetics.* London: Heinemann, 1986.

GENETIC VARIABILITY
The Raw Material of Evolution

The incredible range of variation within species has always fascinated naturalists and frustrated classifiers. Anyone can see that dogs and cats are different species, but the population biologist, or practical breeder, becomes entangled in the great variations within each species, sometimes even within a single litter.

Although Darwin saw thousands of different species in the wild during his voyage on the H.M.S. *Beagle*, it was only after detailed scrutiny of a single group of organisms that he considered himself a finished naturalist whose theories could command respect. Upon his return home, he embarked on an eight-year study of barnacles, during which he studied intensely the variability in thousands of individuals.

Later, Darwin sought out local farmers who bred different types of pigs, gardeners who experimented with varieties of flowers and country gentlemen who bred fancy pigeons. He questioned them endlessly about strains, crosses and hybrids, conducted his own garden experiments and compiled the results in his two-volume work *Variation of Plants and Animals Under Domestication* (1868). In the 1870s, he corresponded with botanists who were breeding potatoes and encouraged them in their quest to find varieties immune to the blight that was ravaging Ireland, causing mass famine and emigration.

It was one of Darwin's great insights that if certain genetic characteristics could be combined or eliminated in domestic plants and animals by deliberate human selection, then selection could similarly work on genetic variation in the wild. In the *Origin of Species* (1859), Darwin's argument for natural selection was based on this amazing variability within species and on the analogy with artificial selection of domesticated organisms.

Today there is probably greater human genetic variability (polymorphism) than ever before. Since culture, diet and medical technology have lengthened our lifespans, more people than ever before are protected from being "weeded out" before they have a chance to reproduce. Therefore, today's human gene pool contains an astonishing range of biochemistries, skin colors, body structures, sizes, proportions and, unfortunately, genetically carried diseases. But at the same time, the genetic variability of the animals and plants with which we share the world, and especially the species we use for food, is rapidly shrinking.

Several hundred years ago, each area of the world had many kinds of rice, beans, corn, cattle, wheat, sheep or chickens, which local farmers nurtured and bred from the wild stock over thousands of years. Andean Indians, for instance, had scores of potato varieties adapted to different altitudes and levels of moisture.

During the past century, however, much of this variation has been deliberately bred out of domesticated plants and animals to produce "superior" strains. We have all benefited from the worldwide production of large, tasty and high-yield varieties of staples. But when farmers universally adopted the new "improved" types, often through the urging of government programs, to increase their yields, they stopped cultivating their ancient local varieties. Indian corn (maize) and Texas long-horned cattle became rare, bred, if at all, as curiosities.

By the mid-1970s, virtually all corn, wheat, rice and many other grains were each grown from a single strain. While there are great advantages to the new "super" varieties, there is a very real danger of putting all our eggs into one genetic basket. For instance, the better-tasting, faster-growing food crops are defenseless against wheat rust, corn blight and other fungal invasions. If they are genetically too uniform, they can all be wiped out in a single epidemic. And, in the 1970s, millions of acres of corn, from Iowa to Texas, succumbed. Near the end of the 19th century, a similar catastrophe destroyed the vast British coffee plantations on Ceylon (now Sri Lanka), which had to be replanted with tea, turning the English from a nation of coffee drinkers to tea drinkers almost overnight!

In nature, variability within a species is an insurance policy. During changing conditions, part of the population will be wiped out, but part will survive. This change in the composition of gene pools through time is a major mechanism of evolution. Contemporary destruction of rain forests and other habitats has been rapidly eliminating thousands of varieties within wild species, many of them tragically destroyed before they are even known to science.

Plant geneticists belatedly realized that some of the original, nearly lost varieties of food plants had to be preserved. In the late 1970s, various governments and the United Nations Food and Agriculture Organization (FAO) began creating seed banks to preserve genetic variability as a hedge against climatic changes, diseases and unforeseen situations. By the time researchers started scouring the world for seeds, many important varieties were close to extinction.

A major seed bank was established in the United States at Fort Collins, Colorado. It is the equivalent for genetic variability of food plants to Fort Knox. Instead of gold bullion, millions of seeds, representing thousands of plant varieties, are stored in its refrigerated vaults. Some are periodically germinated in laboratories to produce new seed to guard against extinction; many will remain viable for decades in cold storage.

However, plant geneticist Erna Bennett criticizes the government's preoccupation with quantitative storage. The plants need periodic grow-outs in fields, she argues, because long storage and laboratory germination kills a certain percentage. Seeds with valuable properties may be lost simply because of archival conditions, which create an unintended selection for survival in the artificial environment. "There is a gradual erosion of genetic diversity," she says, "Field areas must be preserved all over the world, and collecting must continue. The world's genetic inheritance is wasting away."

See also ISIS (INTERNATIONAL SPECIES INVENTORY).

For further information:
Schwartz, Anne. "Banking on Seeds to Avert Extinction." *Audubon Magazine*, 90,1 (1988):22–27.
Shell, Eleen Ruppel. "Seeds in the Bank Could Stave Off Disaster on the farm." *Smithsonian*, 20,10 (1990):95–105.

GENOME
Makeup of a Species

Genome is the total amount of genetic information in a species population. An individual carries genes on his or her chromosomes; the total of genetic instructions for that individual is the genotype.

A fundamental concept of modern biology is the distinction between this genotype (the individual's code) and the phenotype (the physical body or expression of the code.) But the genome applies to populations: It is the sum of all genotypes in a species. This populational blueprint for an entire species,

the genome, is the current focus of gene-mapping projects.

See also GENE MAPPING.

GENOTYPE
Blueprint for Individual

The genotype is the genetic makeup of an animal or plant—a coded plan for all the individual's characteristics. Its actual expression, the observable body that results from the genes, is known as the phenotype.

The distinction between genotype and phenotype is fundamental to Mendelian genetics. There may be many unexpressed or recessive characters in the genotype that do not show up in the phenotype. Changes that occur in the genotype, whether from natural mutation, action of several genes working together (pleiotropy), or artificial gene splicing or other tampering can show up in the phenotype. But changes made to the phenotype, such as mutilations or hypertrophy, will not affect the genotype.

See also MENDEL'S LAWS; "SELFISH GENE"; WEISMANN, AUGUST.

GEOFFROY SAINT–HILAIRE, ETIENNE (1772–1844)
French Naturalist

Etienne Geoffroy Saint–Hilaire, though he was no evolutionist, thought that animals' structures could undergo changes and transformations. He was therefore closer to Jean Baptiste Lamarck than to their mutual antagonist Baron Georges Cuvier, France's reigning master naturalist.

Cuvier had divided the animal kingdom into four branches, according to his idea of basic organization: *mollusca* (mollusks), *radiata* (starfish), *articulata* (insects) and *vertebrata* (backboned animals). Each species in nature, he had argued, was created perfectly adapted to its environment. If habitat or climate should change, the organisms are simply out of luck—they cannot evolve but simply become extinct.

The young upstart zoologist Geoffory Saint–Hilaire became the major French advocate of the German *Naturphilosophie* school of Johann Goethe and Lorenz Oken. Like his teachers (and Richard Owen in England), he believed in a mystical "unity of type," reflecting ideal forms throughout nature. Challenging Cuvier's classification, he mocked the idea that structure and environment were rigidly tied together. To Geoffroy Saint–Hilaire, animal structures could change with environmental disruptions that disturb embryonic development and produce the birth of "monsters" (novel forms, variants of species).

On the eve of the July revolution of 1830, Geoffroy

Saint–Hilaire and Cuvier held a public debate on their respective views of nature. Although the liberal regime of Louis Philippe won the seat of government, the conservative Cuvier successfully put down the attempted revolution in natural history. Their famous debate has often been wrongly characterized as Geoffroy's evolutionism versus Cuvier's fixity of type. That was not the issue; the real question pitted Geoffroy's idealist-structural interpretation of organisms against Cuvier's functionalist views.

See also CUVIER, BARON GEORGES; NATURPHILOSO-PHIE; OWEN, SIR RICHARD.

For further information:
Appell, Toby. *The Geoffroy–Cuvier Debate.* Oxford: Oxford University Press, 1987.

GEOGRAPHIC ISOLATION (IN SPECIES FORMATION)

See ALLOPATRIC SPECIATION.

GERTIE THE DINOSAUR
First Animal Cartoon

If animated cartoon creatures evolve, it seems appropriate that years before Mickey Mouse, the first animal "toon" star was a dinosaur. *Gertie the Trained Dinosaur* (1914) was the brainchild of innovative artist Winsor McCay, who had first introduced the friendly brontosaur in his Sunday comic strip *Little Nemo in Slumberland*. She proved so popular that he decided to feature her in an animated film.

But Gertie was more than a screen cartoon, she was McCay's partner in a clever vaudeville act that toured for years. McCay himself would appear on stage dressed in a formal suit, to put Gertie through her paces. Seemingly at his command, her projected image would eat, dance or laugh at his jokes. When she misbehaved, he scolded her and she cried. For the public, it was an important first view of a dinosaur "brought to life"—the first of many more to come in the movies.

A SINGULAR SENSATION, cartoonist Winsor McCay's vaudeville act featured "Gertie," his animated dinosaur who appeared to respond to his every command. She was the first big hit as an animal cartoon star, a niche later filled by ducks, mice and rabbits.

Near the end of the act, McCay would ask Gertie to lower her long neck so he could climb on. Then he would walk towards Gertie and, with split-second timing, slip backstage through a slit in the screen. At the same instant, a life-size cartoon of McCay in his tux would appear to be hoisted up by Gertie, smoothly crossing over into the cartoon universe. With her human friend perched on her back, Gertie would then turn and walk away.

See also DINOSAUR RESTORATIONS; FANTASIA.

GESNER, CONRAD

See FOSSILS.

GIGANTOPITHECUS
Extinct Giant Ape

Although King Kong was a phantom from Hollywood's dream factory, a real giant ape once actually existed during the human past. Not an ancestor of ours, he was perhaps a competitor and may well have been exterminated by stone-tool wielding *Homo erectus,* sometime during the mid-Pleistocene, about 300,000 years ago. This massive primate probably stood nine feet tall and weighed about 600 pounds, if the rest of the creature was in scale with its teeth and jaws. It was named *Gigantopithecus* (gigantic ape) because its jawbone and teeth are five times larger than that of modern man.

In 1935, remains of *Gigantopithecus* were accidentally discovered in a Hong Kong pharmacy by G. H. R. von Koenigswald, a Dutch paleontologist. Chinese apothecaries have always stocked unusual fossils, which they call "dragon's teeth," for use in ground-up medicines. Von Koenigswald regularly searched these drugstores for curiosities and was amazed to find an enormous tooth with an ape-like (Y-5) dental pattern. When more teeth began to show up, a field search began, which has since yielded hundreds of *Gigantopithecus* teeth and jawbones from various sites in China and Pakistan; other parts of the skeleton, however, have not yet been found.

At various times, *Gigantopithecus* was considered a candidate for an ancestral orang or even a hominid, but current opinion places the giant as a dead-end offshoot of the early pongids, before the branching off of the gorilla-chimp lines.

A site called Tham Khuyen in the Lang Son Province of northern Vietnam is the focus of current research. Located northeast of Hanoi, near the Chinese border, it was discovered by workmen mining bat guano from limestone caves. As fossils were unearthed, Vietnamese scientists became interested, as did Soviet prehistorians who were working in the area. The Vietnam War made work at the site impossible for several years. In 1986, the Hanoi government consented to let United States scientists John W. Olsen and Russel Ciochon join the excavations as part of a "cultural thaw" between the two nations.

Dr. Olsen commented that "some of our Chinese colleagues would say that some form of *Gigantopithecus* might have survived to this day, as the so-called yeti, or abominable snowman." Although he doesn't believe yetis exist, Olsen points out that if some giant primate did survive, the remote mountain ranges of western China or Central Asia would be the right place to look for it.

Olsen hopes his investigation will shed light on the relations between *Homo erectus* and *Gigantopithecus* for there are tantalizing reports that bones of the two species are mingled at the site. It would not surprise him if it turns out that *Gigantopithecus* was exterminated by early man, despite its enormous size and strength. "I believe that the mountain gorilla faces today the same fate at the hands of man as did *Gigantopithecus,*" Olsen says.

See also DRAGONS; KING KONG; Y-5 DENTAL PATTERN.

GILKY, LANGDON

See SCOPES II.

GISH, DUANE

See CREATIONISM.

GLACIATION

See AGASSIZ, LOUIS; ICE AGE.

GOBI DESERT EXPEDITIONS (1922–1930)
Search for Origins in Asia

Charles Darwin thought Africa was probably the "cradle of mankind," a clever hunch that would have been forgotten long ago, except that science now agrees with it. But Darwin was in the minority; from his day until more than 70 years after his death, the search for human origins focused on Asia. Between 1922 and 1930, the most audacious and expensive land expedition in history combed the far reaches of Outer Mongolia, then as little-known to science as the surface of the moon. Its mission: To find the birthplace of the human species.

Led by Roy Chapman Andrews of the American Museum of Natural History, the remarkable Central Asiatic Expeditions ranged over China and Mongolia, resulting in the first systematic study of the Gobi Desert (a redundancy, since *gobi* means desert). There were five seasons of exploration and collecting, which cost a total of $700,000, an outlandish sum during those years of the Great Depression. At the project's

"THE MISSING LINK" EXPEDITIONS, as they were called, set out from New York's American Museum of Natural History to explore Mongolia and China's Gobi Desert in 1924–1928. Scientists and equipment were transported by the most extensive caravan of motor cars, pack asses and camels ever assembled.

ADVENTUROUS ARCHEOLOGIST Roy Chapman Andrews, leader of the Gobi Expeditions, was a real-life "Indiana Jones," battling bandits and sandstorms in search of specimens and fossils. Although he found no "missing link," the East Asiatic Expeditions brought back many scientific treasures, including the first known fossilized dinosaur eggs.

zenith, in 1925, it had 14 scientists and technicians, 26 native assistants, five touring cars, two trucks and a support caravan of 125 camels.

Among the expedition's accomplishments were archeological excavations of Paleolithic and Neolithic sites, the first accurate maps of the Gobi, collection of thousands of new plants and animal specimens, detailed geologial surveys and the gathering of vast collections of fossils—including important early mammals and the first known dinosaur *(Protoceratops)* eggs. Amazingly, the mineralized eggs still lay in concentric rings, just as their mother had arranged them in her nest 150 million years before.

Although these exotic eggs, vast graveyards of "Shovel-tusked elephants" and other wonders brought fame and honor to the Gobi team, the expedition did not accomplish what it set out to do above all else: To find the evidence of museum president Henry Fairfield Osborn's "dawn man" in Asia. Just a few years before, some truly early hominid remains had been found by the South African anatomist Raymond Dart, but European and American scientists, convinced of Asiatic origins, had ignored them.

Scientist-explorers in the Gobi faced dangers and adventures worthy of the fictional archeologist Indiana Jones, who seems to have been patterned after Roy Chapman Andrews himself. On more than one occasion, they had to shoot their way out of ambushes or race their motor cars against brigands on horseback. When they ran out of motor oil, they kept the vehicles running on lamb fat, a local gooey cheese and Mrs. Andrews's skin creams.

Bands of armed bandits demanded cash for safe passage, corrupt politicians required payoffs and the scientists frequently wandered into local civil wars.

Reasoning that the foreigners were not crazy enough to spend a fortune collecting mere rocks, Chinese officials concluded they were looting fabulous national treasures and seized about 100 huge crates. To their amazed disappointment, they found nothing but fossils, which they eventually released for shipment to the museum in New York.

Andrews wanted to continue his explorations of the region's remarkably rich fossil beds, because he was convinced "we had barely scratched the surface of the Gobi's secrets." But despite his best efforts and even intervention by the U.S. State Department, the suspicious Chinese would not allow the expedition to return again after 1930. (After World War II, however, the work was followed up and expanded by Russian, Mongol and Polish scientists.)

Andrews had successfully met the enormous challenges of raising financial support, organizing and accomplishing impressive work in rough, dangerous country. But all these, he later wrote, were "infinitely less nerve-racking than trying to steer a safe course for an expedition's ship between the rock [sic!] of Oriental diplomacy. Disturbed internal conditions and fluctuating politics present an almost unsurmountable wall to the foreign explorer."

See also "CRADLE OF MANKIND"; DINOSAUR EGGS.

For further information:

Andrews, Roy Chapman. *On the Track of Ancient Man.* New York: Putnam, 1926.

Preston, Douglas. *Dinosaurs in the Attic.* New York: St. Martins, 1986.

GOBINEAU, COUNT ARTHUR DE

See ARYAN "RACE," MYTH OF.

GOETHE, JOHANN WOLFGANG VON

See NATURPHILOSOPHIE.

"GOLDILOCKS PROBLEM," THE
Conditions for Life

Life requires very special conditions for its existence. It can survive only between a rather narrow range of temperatures, and its environment must include proportions of oxygen, carbon, nitrogen and water. Within our solar system, some planets (Mercury, Venus) are too hot, some are too cold (Uranus, Neptune, Pluto), and most have unsuitable atmospheric gases or none at all, but Earth is "just right"!

Why some planets should be capable of supporting life while most cannot has been nicknamed "The Goldilocks Problem" by biochemists, after the fairytale heroine who tried out the three bears' porridge.

One was too hot, one was too cold, and only one was "just right."

In the 1970s, British chemist James Lovelock looked at the idea that only certain rare conditions can support life and turned it around. Instead, he thought, perhaps life itself creates and maintains proper conditions to insure its own survival. According to Lovelock, who has investigated such questions as why the proportion of gases in air is constant over long periods of time, all earthly life is part of a complex system that creates conditions "just right" for its own perpetuation.

See also BIOSPHERE; ECOLOGY; GAIA HYPOTHESIS.

GOLDSCHMIDT, RICHARD

See HOPEFUL MONSTERS.

GONDWANALAND
Paleozoic Supercontinent

Geologists believe a huge supercontinent, Gondwanaland, existed during Paleozoic times. During the Mesozoic it broke up to become South America, Africa, India, Australia and Antarctica. The outlines of these continents still fit roughly into one another, like pieces in a jigsaw puzzle.

In the late 19th century, Austrian geologist Eduard

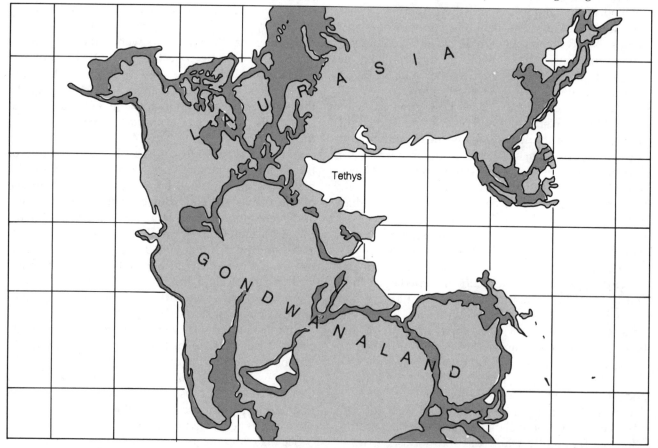

ANCIENTLY CONNECTED, the same geology and fossil creatures are found in what is now South America, Africa, India, Australia and South America. This map represents late Triassic times, about 200 million years ago, during the early dinosaurs' heyday.

Suess noticed amazing similarities in the geology of central India, Madagascar and southern Africa and concluded that they must have once been joined. He claimed a supercontinent had once existed and named it for the Gonds, an ancient Dravidian kingdom and people located in central India near the fossil beds. (Gonds still survive in India today as forest tribes, isolating themselves and their culture from the surrounding Indians.)

Many years later, a meteorologist named Alfred Wegener formulated his theory of continental drift, which remained controversial for many years. As more evidence accumulated, geologists began to form a picture of gigantic, movable continental plates that shaped the current geography of the planet. South America, Australia and Antarctica were also once part of Gondwanaland.

In the 1960s, Dr. Edwin H. Colbert of the American Museum of Natural History, found fossils of the tusked reptile *Lystrosaurus* first in the South African Karroo, then in central India (in the land of the present-day Gonds) and finally in Antarctica. Although the geology of continental drift had already been worked out, Colbert's series of discoveries was the icing on the cake. Gondwanaland's existence was confirmed.

See also BROOM, ROBERT; CONTINENTAL DRIFT; KARROO BASIN.

GOPI KRISHNA (1903–1984)
Evolutionary Yogi Philosopher

In the ancient Indian concept of Kundalini energy, Kashmiri mystic Gopi Krishna believed he had found the biological basis for religion, genius and evolution. Founders of great faiths, in his view, were more highly evolved individuals. By increasing the flow of Kundalini within their own nervous systems, Buddha and Jesus were able to inspire others with the ideal of evolving into superior beings:

> When we once accept that in the case of man, biological evolution toward a higher state of consciousness provides the springhead for all the phenomena connected with religion . . . it would then be clearly realized that it is not to propitiate a god or any other spiritual entity that one has to pay attention to the dictums of faith; but that the principle of evolutionary progress toward a spiritual goal has to be accepted in a way similar to . . . growth and development of a child towards a sane and healthy maturity.

Since Kundalini phenomena are not recognized by Western science, Gopi Krishna tirelessly campaigned for the systematic study of yogic and meditative states by medical physiologists. There is a unitary biological mechanism, he taught, that produces crea-

EVOLUTION OF THE HUMAN BRAIN to higher form of consciousness was quest of Kashmiri philosopher Gopi Krishna. In a dozen erudite books, this self-taught yogi urged Western scientists to investigate possible physiological mechanisms described in ancient Indian tantras.

tivity, genius, psychic talents and mystical or religious experience. This ancient Eastern concept of an evolutionary energy in humans, also known as "serpent power," has been symbolized for at least 5,000 years by two snakes coiled in a double helix.

Born in 1903 in a village near Srinagar, the capital of Kashmir, Gopi Krishna spent his life in the region, working as a minor civil servant to support his mother and sisters. At the age of 34, after practicing yogic meditation for three hours every morning for 17 years, an unexpected change came over him. In the midst of deep meditation in the lotus position, he "felt a strange sensation below the spine, at the place touching the seat . . . Suddenly, with a roar like that of a waterfall, I felt a stream of liquid light entering my brain through the spinal cord."

According to his autobiography *Kundalini: The Evolutionary Energy in Man* (1967), Gopi Krishna received that jolt without proper preparation or training, causing him ill health for a dozen years. After undergoing what he called a "highly accelerated evolutionary transformation," for the rest of his life everything he saw seemed to have a glowing, translucent quality. The final "transition" to a stable "cosmic consciousness" occurred in 1950 and lasted without interruption until his death at the age of 81.

While Gopi Krishna's ideas are nothing like scientific evolutionism, they are remarkably similar to those of the Jesuit paleontologist-mystic Pierre Teilhard de Chardin. In his last spiritual testament, *Le Cristique* (1955), Father Teilhard wrote that "everywhere on earth, at this moment, within the new spiritual atmosphere created by the appearance of the idea of evolution, there float . . . the two essential components of the Ultra-human: love of God and faith in the world.''

While Teilhard acknowledged that most people consider biological evolution and divine spirituality as very separate phenomena, he believed that "in me they fuse together spontaneously." Like Gopi Krishna, Teilhard's awakening to his "evolutionary consciousness" was accompanied by great pain:

> One would think that a single [spark] of this kind of light . . . would cause an explosion strong enough to transform [mankind's beliefs and behavior, unifying ethics, evolutionary biology, and religion.] How then can it be that I, looking around me, still intoxicated by what has been shown to me, should find myself alone, as it were, the only one of my species? The only one to have seen? . . . and so unable, when asked, to cite a single author, a single text, which might clearly describe the marvellous "Translucence" that has so transfigured everything I see?

Despite their vastly different backgrounds and cultural traditions, had Father Teilhard ever met Gopi Krishna, both would surely have been startled by the similarity of the visionary evolutionism they shared.

See also BEECHER, HENRY WARD; CADUCEUS; KUNDALINI; "OMEGA MAN"; TEILHARD DE CHARDIN, FATHER PIERRE.

For further information:
Gopi Krishna. *The Biological Basis of Religion and Genius.* New York: Harper & Row, 1971.
———. *Kundalini: The Evolutionary Energy in Man.* Berkeley: Shambhala, 1967.

GORILLAS
Largest Living Primates

Elusive creatures of the African forest, gorillas were practically unknown to Europeans a century ago. Often they were confused with chimpanzees in traveler's reports; both apes sometimes lived in the same forests. During the 1870s, French explorer and hunter Paul du Chaillu drew huge lecture audiences, hungry for his tales of encountering and killing wild gorillas in the Belgian Congo, now Zaire. The stuffed specimens with which he shared his lecture platform created a sensation at just about the time Darwin and Huxley put out their books on human evolution.

Du Chaillu described the great apes as creatures "from a hellish nightmare." The way to kill one was to walk up to it and stand your ground while it made the most hideous roars and mock charges. Then you waited until it grabbed hold of the muzzle of your rifle and lifted it to its mouth—that was the time to

fire. If you waited a moment too long, du Chaillu warned, the gorilla would bite the barrel closed or bend it into a curve, then kill the hunter. Thus was born the durable image of the gorilla as Gargantua and King Kong, which reached its zenith in the 1930s.

Gorillas were considered so ferocious that when zoologist R. L. Garner went to observe wild gorillas during the 1890s, he locked himself inside a specially built cage in the middle of the forest. No one in their right mind would try to get near a gorilla if he didn't intend to shoot it immediately.

Meanwhile, as a few real gorillas reached zoos and comparative psychology labs, another picture of the gorilla emerged. Wrenched from their forest homes and the social groups that nurtured them, individuals appeared sullen, intractable and were often described as "introspective"—meaning they would rather stare at the wall than participate in "intelligence tests" devised by their captors.

By the late 1950s, popular notions about gorillas were reshaped again by the field naturalist George Schaller, who tracked them from a distance. He carried no gun and was never attacked. Indeed, he found the apes shy, unaggressive and rather boring. They had few fights, made no tools, hunted no prey and were peaceful vegetarians. Despite their great strength and the male's massive canines, they mostly sat around eating wild celery.

While their days are spent feeding and resting on the forest floor, they often build night-nests in sturdy tree crotches where they may retire in safety.

They were intensely social; a typical group consisted of several females and young, a few subadult males and one large, experienced "silverback" male—the group's leader, protector and defender. Large silverbacks can weigh 400 pounds, and their charges and vegetation-tearing displays are frightening; however, they rarely follow through with an attack, preferring to drive intruders away by intimidating threats.

Schaller was followed by the remarkable and tragic Dian Fossey, who spent 17 years among the mountain gorillas, until her (still unsolved) murder at Karisoke, Rwanda, in 1985. Fossey not only studied these largest of living primates, but succeeded in drawing world attention to their precarious condition. Thanks to her local "war" on poachers in the 1970s and improved park security, the mountain gorilla populations were saved from extinction.

Fossey was impressed with the shyness and gentleness of many individuals, even fully grown males. She totally disproved myths about their violent nature and was able to approach close enough to become accepted in the middle of their groups, often playing, sunning and mock-feeding with them for days on end.

GENTLE GIANTS, wild gorillas are shy vegetarians who rarely attack other creatures. The great apes' ferocious appearance and bluff charges terrified 19th-century hunters, who described them as monstrous killer, as in the above image from explorer Paul du Chaillu. Seventy years later, the myth was perpetuated by Barnum & Bailey's 1938 circus poster (below).

When Uncle Bert and later Digit voluntarily approached her, Fossey felt she was "the most fortunate person in the world . . . my curiosity [about them] was rewarded with their curiosity about me." Gorillas eventually accepted and trusted her, forming "a bridge of chasms of millions of years between the two species." The distinction had blurred between "who was the observer and who was the observed."

Today, a score of gorilla groups have become so used to human visitors they have become important economic resources in Rwanda and Zaire, attracting substantial tourist dollars. Their celebrity status insures government measures for their continued protection, as they are now much more valuable to the local economy alive than as hunting trophies. Yet they survive under problematic conditions: Constant exposure to humans may irrevocably disturb their normal behavior and create a semi-domesticated ape. Science's understanding of gorilla behavior, ecology and evolutionary adaptations may still forever be lost, even if the apes themselves are given a reprieve from extinction.

See also APES; "APE-WOMEN," LEAKEY'S; DIGIT; FOSSEY, DIAN; KARISOKE RESEARCH CENTER; KING KONG.

For further information:

Du Chaillu, Paul. *Adventures in Equatorial Africa.* New York: Harper, 1871.

Fossey, Dian. *Gorillas in the Mist.* Boston: Houghton-Mifflin, 1983.

Schaller, George. *Year of the Gorilla.* Chicago: University of Chicago Press, 1964.

"GORILLA WOMAN"

See PASTRANA, JULIA.

GOSSE, PHILIP

See OMPHALOS.

GOULD, STEPHEN JAY (b. 1941)
Paleontologist, Essayist, Science Historian

When five-year-old Stephen Jay Gould first marveled at the towering *Tyrannosaurus* skeleton in the American Museum of Natural History, he decided to spend his life studying fossils. Although few children in Queens, New York, shared his early fascination for evolution, he never considered any other career but paleontology.

Now professor at Harvard University and a curator of its Museum of Comparative Zoology, Gould attended Antioch College, then returned to Manhattan, for graduate work in paleontology at Columbia University. For his doctoral thesis he investigated vari-

ation and evolution in an obscure Burmudian land snail, anchoring his later theorizing in intense scrutiny of a single group of organisms, as Darwin had done with barnacles.

At one point, he hoped to find correlations between variation and different ecologies within the creature's range, but the snails' sizes, colors and shell shapes seemed to vary quite independently of local environment. Impressed with the importance of non-selectionist factors in evolution, he also became interested in structural constraints: How slight changes in one feature must alter several others within definite limits—what Darwin had called "correlation of parts."

Gould also became interested in distinguishing incidental features from adaptive ones. He coauthored (with Richard Lewontin) an influential paper inspired by the spandrels of certain medieval cathedrals: Geometric architectural features decorated with impressive religious paintings. While art historians had analyzed their distinctive aesthetic, most had forgotten the spandrel's humble origin as an unavoidable engineering consequence of stress distribution—a structural byproduct of constructing that kind of dome. As a biological example, Gould points out that the human chin, often cited as "advanced" in comparisons with "lower" primates, holds no special corelation with higher intelligence. It is, like the spandrels, an incidental result of stress and growth factors in the human jawbone.

Although Gould has became closely identified with the influential idea of punctuated equilibrium, it actually originated with paleontologist Niles Eldredge and was developed by them jointly. Eldredge's detailed studies of trilobites brought home to him a pattern that had impessed Thomas Henry Huxley: The fossil record seems to show "bursts" of speciation, then long periods of stability. Darwin's reply was that the fossil record was then too sketchy and incompletely known to provide evidence of patterned rates. But vast accumulations of paleontological evidence over the last century do not support Darwin's case for a steady, gradual evolutionary rate. Eldredge's trilobite series suggested, instead, relatively short episodes of rapid evolution followed by long periods of stability, confirming Huxley's impression. Gould enthusiastically agreed; it was time to acknowledge that such episodic patterns in the rocks probably reflect the reality of life's history. By the 1980s, "Punctuationalism" had become widely adopted and was proving to be a fruitful hypothesis for generating new insights and research.

Although one of Gould's lifelong heroes is Charles Darwin, whose achievements he has celebrated in such books as *Ever Since Darwin* (1977) and *The Panda's Thumb* (1980), he is irreverent toward the orthodox Synthetic Theory of evolution that has prevailed in biology since the 1940s. Dissatisfied with the limits of its explanatory power, he is open to exploring other possible mechanisms and approaches to supplement traditional natural selection—to the dismay of more conservative colleagues.

One of his approaches has been to emphasize the hierarchy of levels on which evolution operates; biochemical, genetic, embryological, physiological, individual, societal, species, lineages. Sorting or selection on any of these levels, he believes, produces significant effects on the level above or below it—a promising and largely unexplored area for future research. Yale paleontologist Elisabeth Vrba and others agree with Gould about the importance of such hierarchical studies.

Among opponents of punctuationalism, a few (Richard Dawkins, Verne Grant) complain Gould has set up a "straw man" of Darwinian gradualism and that jumpy or "quantum evolution" was discussed years ago by Ernst Mayr and George Gaylord Simpson. Examples of Darwin's self-contradictory fudging are easy to find, but Gould maintains that change "by slow, insensible degrees" remained central to Darwinian thought. And while acknowledging his predecessors' insights, Gould argues that often it is a shift of emphasis and focus rather than a radically new idea that leads to deeper scientific understanding.

Gould's success as a popular author is another tempting target for critics; he does not shrink from public controversies. He appeared before Congressional committees on environmental issues, was a courtroom witness in the Arkansas Scopes II trial regarding teaching of evolution in the public schools and is prominent in speaking out against pseudoscientific racism and biological determinism. But even one of his most adamant detractors, Robert Wright, while recently attacking one of his books in the pages of the *The New Republic*, grudgingly began his diatribe: "The acclaim for Stephen Jay Gould is just shy of being universal . . . He is, after all, America's evolutionist laureate."

See also BARNACLES; BIOLOGICAL DETERMINISM; CONTINGENT HISTORY; PANDA'S THUMB; PUNCTUATED EQUILIBRIUM; SCOPES II.

For further information:
Gould, Stephen Jay. *The Mismeasure of Man.* New York: Norton, 1981.
————. *Ontogeny and Phylogeny.* Cambridge: Harvard University Press, 1977.
————. *The Panda's Thumb.* New York; Norton, 1980.
————. *An Urchin in the Storm.* New York: Norton, 1987.
————. *Wonderful Life: The Burgess Shale and the Nature of History.* New York: Norton, 1989.

GRADUALISM
Slow and Steady Change

One key feature of Darwin's original theory was that evolutionary change must have proceeded by "slow, insensible degrees"—a progression of tiny changes adding up to produce new species over immense periods of time. It was Darwin's attempt to apply Sir Charles Lyell's uniformitarian principles from geology to the world of life.

Lyell had been a voice of reason at a time when geologists invoked imagined violent and sudden catastrophes, convulsions, floods and debacles to explain the features of the Earth. Presently observable processes of wind, water, erosion and deposition, Lyell thought, should be applied to account for the facts of geology.

Darwin went so far as to adopt "Nature makes no leaps" as an axiom, or basic assumption. But from the first his friend and supporter, Thomas Henry Huxley, thought it "unnecessary to burden the theory" with an unproven gradualism, which he later described as an "embarassment."

When critics asked why the fossil record, though it showed change over time, did not demonstrate this presumed succession of small, gradual transitions, Darwin replied it was very "imperfectly" known, like a book with many pages, even whole chapters, missing. Another century of fossil discoveries, he believed, would fill in the picture. In fact, many transitional forms have since come to light, but they are still comparatively rare.

A similar challenge to gradualism came from studies of population genetics in the early 20th century, when geneticists studied small, measurable changes in populations (micromutations or microevolution). They assumed these would soon be easily shown to add up to the larger changes (macromutations or macroevolution) that produce new species. In fact, they did not.

By the 1970s, the concept of Darwinian gradualism came under increasing scrutiny and attack by biologists. Recent researchers have reopened the question of macromutation, or evolution by fairly rapid jumps. Such apparent discontinuities, or "jumpiness," in the fossil record led to theories of "punctuated equilibrium." Critiques of gradualism have not overturned Darwinian biology, or disproved evolution. They are part of a continuing attempt to build on its foundations and gain a clearer picture of how evolution works.

See also HOPEFUL MONSTERS; PUNCTUATED EQUILIB-
RIUM; UNIFORMITARIANISM.

For further information:

Eldredge, Niles. *Life Pulse: Episodes from the Story of the Fossil Record.* New York: Facts On File, 1987.

Stanley, Steven. *The New Evolutionary Timetable.* New York: Basic Books, 1981.

GRANT, MADISON

See ARYAN "RACE," MYTH OF.

GRASSES, EVOLUTION OF
New Evidence and Research Tools

Most museum paintings of life in the Miocene (23.5 million to 5.2 million years ago) show rhinos and early horses moving across grassy plains. But, until very recently, the animals were much better known than the vegetation, since fossils of ancient grass are often difficult to identify. Botanists speculate that grasses originated about 80 million years ago, but the oldest known fossil grass is from the late Oligocene, about 25 million years ago.

Only a few paleobotanists in the world specialize in the evolution of grasses. Significant research was done during the 1980s, for example, by Joseph Thomasson of Fort Hays State University in Kansas. In the northwestern part of that state is an extraordinary site with a lugubrious name: Minium's Dead Cow Quarry. It may turn out, according to Thomasson, to be a "Rosetta Stone" of grasses because of excellently preserved and unusually complete animal-plant associations.

Thomasson's discoveries were made possible by the scanning electron microscope, which enables him to photograph the actual cellular structure of the fossil plants. Previously grasses had been nearly impossible to identify and research. Ultramicroscopic analysis has already helped clarify the question of when grasses split off into two major groups: one that uses three-carbon photosynthesis (C3) and the other that uses a four-carbon (C4) system.

C4 grasses are more efficient, grow faster and are better adapted to heat and light. Their photosynthetic rate is twice or three times that of C3 grasses, which is why crabgrass (a C4) can so easily take over and destroy a lawn of Kentucky bluegrass (C3). Corn is C4; rice and wheat are C3.

A C4 grass leaf has a characteristic wreath-like pattern of cells surrounding its tube-like vessels, which enables them to separate energy-producing and energy-consuming reactions. Unlike C3 grasses, they can store carbon dioxide and are therefore able to close pores during high temperatures and retain water. Before the well-preserved fossils at the Dead Cow

Quarry site were known and studied, says Thomasson, "We had absolutely no idea when this split in those two physiologies occurred." Researchers had made crude "guesstimates" ranging from 10,000 to 100 million years ago.

Thomasson claims his study of Minium's Quarry "allows us to say that 7 million years ago, positively and without any doubt, that split had already occurred." And since the C4 grasses present are already well developed, it indicates the original split may have occurred much earlier.

Because C4 plants are adapted to warm, wet climates, they indicate a subtropical Miocene Kansas without extreme winters and an ecology like today's Serengeti Plains of East Africa. Prehistoric American grasslands were inhabited by huge herds of animals, including zebra, rhinos, pronghorn antelope, big cats, bears and horses, all of whose bones are found in clear association with the grass fossils at the site.

Although there are still huge blanks in the evolutionary history of grasses, including their origin and early development, the electron microscopy technique offers a new glimpse into the actual physiology of grass that fed rhinos and zebras seven million years ago in present-day Kansas. At last, there is a real basis for identifying the grass in those museum murals.

See also MINIUM'S DEAD COW QUARRY.

GRAY, ASA (1810–1888)
American Botanist, Evolutionist

Asa Gray was, in Charles Darwin's words, "a complex cross of lawyer, poet, naturalist, and theologian." He was also one of the greatest botanists of the 19th century and for years supplied Darwin with detailed information on the distribution of plant families and species throughout the world.

Although he was Darwin's greatest champion in American science, Gray had a theological bent and tried to defend evolution from charges of atheism. He wanted to make things easier for liberal clergymen, arguing that natural selection did not wholly do away with the "argument from design"; God accomplished his plan through evolution. When some insisted that new species could only be created by the direct intervention of God, Gray countered that they were putting limits on scientific investigation without enlarging the understanding of religion.

Despite his brilliance, Gray was very much in the shadow of Louis Agassiz, the most influential anti-Darwinian naturalist in America. (Both were professors at Harvard.) But Agassiz really did not understand Darwin's theories. Gray wrote that "he growls over it, like a well-cudgelled dog—is very much annoyed by it," but was just too old to change the entire view of nature he had learned from his teacher Georges Cuvier.

In 1857, Darwin had given Gray a sketch of his theory in a letter, making him one of the intimate circle sworn to secrecy about the coming bombshell. After *The Origin of Species* (1859) was published, Gray tried to please all sides and continued to equivocate for another dozen years. Finally, he published a clear statement of his position in a series of essays called *Darwiniana* (1876). Darwin liked it so well that the following year he dedicated his book on the forms of flowers to Gray.

Darwin was grateful for Gray's championing evolution in American science, but they parted company on the question of design (teleology). Gray thought variations in nature were provided by a good-natured God to be used by species for their benefit, like a friend who hands you pieces of a jigsaw puzzle in the proper order for you to fit them in.

In *Variation of Animals and Plants under Domestication* (1868), Darwin said natural selection worked more like a builder putting together a structure with odd bits of stone and rubble. The fragments were not shaped for particular ends, but were selected to fit in where they could be useful. He wrote to Gray that he had "no intention to write atheistically," but could not see "evidence of design and beneficence on all sides of us [because] there is too much misery in the world."

He could not believe, for instance, that a kind and loving God could purposely design the Ichneumid wasps, whose young feed on the insides of living caterpillars, "or that a cat would play with mice." He could not imagine variation of a particular species was "ordained and guided" any more than an astronomer would suggest that a particular meteroic stone fell according to "a preconceived and definite plan."

See also AGASSIZ, LOUIS; DIVINE BENEFICENCE, IDEA OF; NATURAL SELECTION; TELEOLOGY.

For further information:
Dupree, A. Hunter. *Asa Gray.* New York: Atheneum, 1968.
Gray, Asa. *Darwiniana: Essays and Reviews Pertaining to Darwinism.* New York; Appleton, 1876.

GREAT BONE WARS

See COPE–MARSH FEUD.

GREAT CHAIN OF BEING
Scala Natura

For hundreds of years before evolution became a part of the common thought of our culture, the concept of the "Great Chain of Being" dominated the Western view of nature. Everything in the universe was arranged in a hierarchy, from low to high, from noble to base. There were "base" metals like lead and "noble" ones like gold and silver. Man was the noblest of creatures; all other living things occupied different links in the chain below him. Above man were various spiritual beings, with God at the top of the cosmic order.

Arthur Lovejoy traced the history of the idea in his classic work *The Great Chain of Being* (1936), in which he called it "one of the half-dozen most potent and persistent presuppositions in Western thought." It lasted from medieval times right up until about a century ago.

The chain carried through to the social order, too. Peasants at the bottom, then servants to the gentry, then various grades of nobility and the monarch at top. Racism is embedded in the Chain of Being; the idea was used to rank the various races into "higher" and "lower." Of course, the white Europeans who devised it were at the top.

Darwin realized the chain concept simply didn't describe nature very well, and resolved not to use "higher" and "lower" in his descriptions of animals. Nevertheless, the mental habit was hard to break. The original title of his book on comparative psychology was to be *Expression of the Emotions in Man and the Lower Animals*, but he went through the manuscript striking out terms of rank and changed the title to *Expression of the Emotions in Man and Animals.* (1872).

See also ORTHOGENESIS; CONTINGENT HISTORY.

For further information:
Lovejoy, Arthur. *The Great Chain of Being: A Study of the History of an Idea.* Cambridge: Harvard University Press, 1936.
Tillyard, E. M. W. *The Elizabethan World-Picture: The Idea of Order in the Age of Shakespeare, Donne and Milton.* New York: Vintage/Random House, 1942.

GREAT DEVONIAN CONTROVERSY

See CAMBRIAN–SILURIAN CONTROVERSY.

GREAT DYING
Mass Extinctions

Extinction is a normal event in the history of life; every so often a species will be faced with the loss of its habitat, or the introduction of a new predator or be outcompeted for food and resources. But mass extinctions or "great dyings" are global catastrophes that wipe out tens of thousands of species at once when whole ecosystems collapse.

Mass extinction events have been used to define geological history. When geologist John Phillips first drew the horizontal lines that separate the Paleozoic, Mesozoic and Cenozoic eras in 1840, he was indicating that in the rocks above each line were fossils of different life forms than in those below it. The major "jumps" or "punctuations" record the wholesale extinction of a majority of living things, followed by a new radiation of different families and species.

Of the half-dozen or so mass extinctions, two of the most extensive and devasting were at the end of the Permian, just before the Age of Reptiles, and the Cretaceous "great dying" that ended it. In the Permian extinction, David M. Raup has estimated that 90% of all then-existing species (about five million kinds of living things) vanished. Thousands of kinds of dinosaurs were wiped out by the Cretaceous extinction, plus flying reptiles, plesiosaurs, ichthyosaurs—as well as many once-abundant creatures of the sea such as ammonites.

Causes of the global climatic and habitat changes that trigger these mass extinctions are not yet well understood, though current opinion veers toward scenarios of great meteors or asteroids crashing to Earth, at least in the Cretaceous devastation. Great dust clouds raised by such impacts, according to the theory, may have filtered the sun's rays, cooling the Earth's climates.

However, the world is experiencing a "mass dying" today caused by a human asteroid—the cataclysmic effects of man on the biosphere. During the past century, humans have reduced the white rhino from thousands to 20, currently slaughter 60,000 African elephants a year, and burn rain forest at the rate of 50 acres a minute. As the human population shoots up to 10 billion during the next century, the fate of thousands of species is sealed. Since man has appeared on Earth, we have increased the rate of extinction of other species by 1,000%.

Humanity's devastating impact on other living things, ecologists believe, could trigger a domino effect that could extend a new "great dying" to ourselves. We may be able to survive nicely without the tiny snail darter fish, the rare lemur known as the aye-aye or the rhinoceros hornbill bird, but we just don't know which organisms are crucial to maintain the biosphere for all living things.

In Paul Ehrlich's metaphor, it is like taking rivets out of the wings of an airplane. When there are thousands of rivets in place, we may still be perfectly safe if a dozen are removed. But if we keep on taking out the rivets, how can we know at what point there are too many missing until the wings suddenly fall off?

See also ALVAREZ THEORY; BIOSPHERE; DINOSAURS, EXTINCTION OF; EXTINCTION.

GREAT FLOOD MYTHS

See ATRAHASIS; CATASTROPHISM; NOAH'S FLOOD.

GREAT RIFT VALLEY

See OLDUVAI GORGE.

GREENHOUSE EFFECT
Global Warming of Atmosphere

Usually considered an environmental issue, the "greenhouse effect" may well play a major role in the current evolution of life on Earth. Past changes in climate have caused major disruptions of plant and animal populations, and possibly even the demise of the dinosaurs after 140 million years of success.

In a gardener's greenhouse, the sun's rays penetrate to warm the air; but the heat then becomes trapped by the glass, creating tropical temperatures even in winter. The same thing happens inside a car sitting in a parking lot all day with the windows rolled up.

In the Earth's atmosphere, a greenhouse effect is caused when thickened layers of carbon dioxide and other gases are built up by burning of fossil fuels (oil, gasoline, coal). Like the glass, they allow radiation (heat) to come through, but block its escape back into the upper atmosphere. In addition to industrial waste, the mass burning of tropical rain forests releases even more carbon and also destroys millions of trees that could convert carbon dioxide back into fresh oxygen.

We are now sending about 5.5 billion tons of carbon dioxide into the atmosphere each year; only half that much can be absorbed by oceans and forests. Some scientists predict that if the current level of fossil fuel use continues, by 2030 there could be a 3- to-9 degree rise in world temperatures. Such change could melt polar ice, raise ocean levels and seriously disrupt agriculture and ecosystems.

GREYWACKE FORMATION

See MURCHISON, SIR RODERICK IMPY; CAMBRIAN–SILURIAN CONTROVERSY; SEDGWICK, REVEREND ADAM.

GROUP SELECTION
Survival of the Social Unit

Cooperating groups of creatures may survive and reproduce better than individual members on their own. Social insects are the classic example. Hives of wasps or bees function almost as if they were themselves organisms, with individuals specialized to fulfill the functions of an immune or defense system, reproductive and food-gathering system and maintenance and repair system. Some bees even station themselves at air passages and cool the air by beating their wings, air-conditioning the hive.

Even those individuals who do not directly reproduce contribute their genes to the next generation via the interesting mechanism of kin selection, by which close relatives of breeding individuals are perpetuated. [See KIN SELECTION.] In other words, if the whole group is successful, its entire gene pool tends to reproduce itself.

Some observers of free-living monkeys, such as the Japanese primatologist Kinji Imanishi, have studied the possible survival advantage of individual social groups within a species. Monkeys may share certain habits of defense, food preferences or even food preparation traditions (such as the washing of vegetables to remove grit) that could have a selective advantage for the group as a whole.

Most evolutionists agree, however, that there is no support for the idea that individual animals are motivated to act "for the good of the group" or "the good of the species." To the contrary, it has been repeatedly demonstrated that the behavior may have evolved to *function* for the good of the group, yet, as far as anyone knows, individuals have no such noble aims in mind. It appears likely that they respond, even as asocial creatures do, purely as individuals acting and reacting on their own behalf.

See also FITNESS; NATURAL SELECTION; "SELFISH GENE."

For further information:

Gilpin, M. *Group Selection in Predator-Prey Communities.* Princeton: Princeton University Press, 1975.

Kummer, Hans. *Private Societies: Group Techniques of Ecological Adaptation.* Chicago: Aldine Atherton, 1971.

Wynne-Edwards, V. C. *Animal Dispersion in Relation to Social Behavior.* Edinburgh: Pliver and Boyd, 1962.

GUANACOSTE PROJECT
"Extinct" Forest Revived

Tropical biologist Dan Janzen's award-winning study of coevolution showed how a species of tree-dwelling ants acts as an acacia's immune system, protecting the tree from all pests and intruders. [See ANT-ACACIAS.] Not content to pursue academic research on how plants and animals evolve together, Janzen brought his expertise to a project no one thought possible: recreating an ecological community on the verge of extinction.

Tropical dry forest, of which less than 2% of the original habitat survives, is a special community of animals and plants that evolved together in Central America. Janzen is directing an unprecedented effort to restore dry forest over a 430-square-mile area of the Guanacoste National Park in Costa Rica. Once heavily forested, it was cleared for farms and overgrazed by cattle years ago, leaving only a tiny remnant of forest in an otherwise barren landscape.

Dry forests are so called because they evolved in a climate where there is no rainfall for six months of the year. When they were studied by the 19th-century naturalist Thomas Belt, says Janzen, they stretched in all directions like vast oceans; today, only tiny "ponds" of the unspoiled forest remain. Using seeds and plants from the few remaining stands, attracting (and sometimes importing) now rare animals, Janzen has attempted to reassemble the pieces of a shattered ecosystem.

Janzen read the old studies by Belt and others describing the forest ecology a century ago, studied fossils of its earlier inhabitants and learned what he could from the extant remnants. He wanted to know the role of each member of the forest, including the margay cat, howler monkey, coral snake, coati and agouti. (He estimates that the complete ecosystem contained 350 bird, 160 mammal, 3,500 plant and 30,000 insect species.)

One of the first things he had to find out was the relationship between some of the major plants and animals. He needed to know which animals help disperse the plants they feed on and which are destructive. The answers weren't always the expected ones.

Janzen found, for instance, that though monkeys eating wild figs appear to vandalize the trees, they are really its benefactors. Along with the fruit, they swallow the fig's small seeds whole, which pass unharmed through their digestive systems. Unwittingly, the monkeys spread the seeds throughout the forest, encased in packets of moist fertilizer. Small parrots, which may appear less destructive, crush and destroy all seeds they eat—a real menace to the trees.

Janzen's interest in plant dispersion led him to experiment with seeds found in animal droppings. (Charles Darwin would have approved; he soaked seeds in barrels of brine for months, then tried to germinate them to determine which species could float to oceanic islands and still sprout.)

One puzzle was the guanacoste tree itself, a major dry forest species that seemed to be disappearing. All the existing trees were very old; no new ones were being produced.

After investigating, Janzen concluded that the heavy guanacoste seeds must have been spread only by large mammals, perhaps originally by Pleistocene elephants or giant sloths that are now extinct. (Large seeds have been found in samples of their ancient dried dung, or coprolites.) In more recent times, the tapir may have filled the role, but there were now none in the area.

A few horses roam freely in the forest, and visitors often ask Janzen why he permits them to remain. His answer is they are needed as "substitutes," to fill the role of whatever large mammals once helped guanacoste trees reproduce. He found the large seeds can survive in a horse's digestive system for months, then are dropped miles away, where they often germinate. Possibly, the long soaking and rubbing also softens the seeds' tough outer coat, so they sprout more easily.

Over the years, Janzen has educated the local people about the forest ecology; their cooperation has been crucial to the project's success. Formerly destructive of plants and wildlife, they now have joined in the effort to rebuild the forest community.

Since he began the Guanacoste Project, Janzen has extended its area to six times its original nucleus. As the ecosystem reconnects relationships between living things, it gains in area and diversity. Endangered insects and birds are attracted to it as a safe haven; they settle in and multiply.

See also BELT, THOMAS; COEVOLUTION.

For further information:

Janzen, Daniel H. *Ecology of Plants in the Tropics*. The Institute of Biology's Studies in Biology No. 58. London: Edward Arnold Ltd., 1975.

———. *Guanacoste National Park: Tropical, Ecological and Cultural Restoration*. San Jose, Costa Rica: Editorial Universidad Estatal al Distancia, 1986.

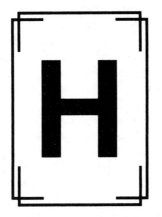

HAECKEL, ERNST (1834–1919)
German Evolutionist

Born in Potsdam in 1834, the son of a government lawyer, Ernst Haeckel was trained as a physician, graduating in 1859, shortly before the appearance of Darwin's *Origin of Species*, the book that dramatically changed his life. Here, he thought, was the answer to everything he had been seeking in science, philosophy, ethics, religion, politics—a unified, or monist, view of the world based on the creative properties of matter. His own fanaticized version of evolution—the "non-miraculous history of creation—became an obsession and guiding passion, with Darwin his greatest hero.

Abandoning his medical practice, he set out for the University of Jena, Germany, to begin a new career by studying anatomy under the great Carl Gegenbaur. From 1859–1860, he undertook a zoological expedition in the Mediterranean, during which he discovered 144 new species of radiolara. His monograph, *Die Radiolaren* (1862), and others on sponges, worms, medusae, and other sea creatures established his scientific credibility. Eventually, he described approximately 4 thousand new species of marine invertebrates. Haeckel became professor of zoology and comparative anatomy in 1862 and established himself for the next half-century as the famous "ape-professor of Jena."

Later, he worked his way up to man and the universe in a series of popular books on morphology and evolution. His *Generelle Morphologie* (1866) contains the first of the famous drawings of evolutionary "trees of life," as well as his exposition of the Monist philosophy.

That same year he made the pilgrimage to England, visited Darwin at Down House and was not disappointed. "It was as if some exalted sage of Hellenic antiquity . . . stood in the flesh before me," he later wrote. Darwin, too, was impressed. In his *Descent of Man* (1871), Darwin claims he would not have bothered to write the book if he had known that Haeckel had already begun one on the same topic.

Haeckel loved to invent scientific terms and created many still in common use. Ecology (the study of the relationship between organism and environment), Ontology (study of embryological development) and Phylogeny (study of evolutionary descent or lineage) are all of Haeckel's coinage.

He became convinced he had discovered the most basic law of evolution: "Ontogeny recapitulates phylogeny," or the development of an embryo (ontogeny) is a speeded-up replay of the evolution of the species (phylogeny). It was an enormously influential idea, utilized by both Darwin and Huxley, who were impressed with Haeckel's detailed illustrations comparing embryonic development in various animals and man. In their earlier stages, according to Haeckel's drawings, pigeons, dogs and humans looked identical.

This recapitulation theory enjoyed a tremendous vogue for a few decades, but eventually proved too vague to be of much use in research. Before it was discredited, however, it shaped scientific thought of the period, including the psychoanalytic theories of Sigmund Freud. [See FREUD, SIGMUND.]

When critics brought charges of extensive retouching and outrageous "fudging" in his famous embryo illustrations, Haeckel replied he was only trying to make them more accurate than the faulty specimens on which they were based. Although Darwin, his intellectual hero, believed that "caution is the soul of science," Haeckel was often reckless in his methods, as if he had some kind of direct line to the source of evolutionary truth.

In one of his bolder public pronouncements, Haeckel stated that the "missing link" between men and apes must have lived in Java or Borneo. Although there was no known fossil evidence for such a creature, he gave it a Linnaean name anyway—*Pithecanthropus alalus* (meaning "ape-man without speech")—and encouraged his students to go out and find it. It is one of the amazing stories of science that a young doctor who attended Haeckel's lectures, a Dutchman named Eugene Dubois, became so infused with his confident enthusiasm he went to Java and dug up what he called *Pithecanthropus erectus* (later renamed *Homo erectus*).

Elated, Haeckel promptly sent a congratulatory telegram "to the discoverer of *Pithecanthropus* from its inventor." Zoologists have since made a rule that scientific names cannot be given to species that have not yet been discovered.

Haeckel's book *Welträtsel* (The Riddle of the Universe, 1899) was one of the most incredible publishing successes of all times. It sold more than half a million copies in Germany alone and was translated into 25 languages, although its science was already outmoded.

In his *Natural History of Creation* (1868), Haeckel argued that the church with its morality of love and charity is an effete fraud, a perversion of the natural order:

"If we contemplate the mutual relations between plants and animals (man included), we shall find everywhere and at all times, the very opposite of that kindly and peaceful social life which the goodness of the Creator ought to have prepared for his creatures—we shall rather find everywhere a pitiless, most embittered struggle of all against all . . . Passion and selfishness . . . is everywhere the motive force of life. Man . . . forms no exception to the rest of the animal world . . . The whole history of nations

YOUNG ZOOLOGIST Ernst Haeckel (left, seated), with his net-wielding assistant, was the very image of the Romantic natural history tradition. In later life (right), as the famous "Ape Professor of Jena," Haeckel became one of Germany's most revered national heroes.

. . . must therefore be [a physio-chemical process] explicable by means of natural selection."

Haeckel's lectures drew huge, enthusiastic crowds at international conventions of freethinkers and monists. After one such gathering, held in Rome, the pope ordered a "divine fumigation" of the Holy City to clear the air.

Philosophers and scientists complained Haeckel's data were wrong, his methods impossible, his philosophy a mess. Such details fazed him not a bit. Haeckel remained enormously popular in Germany, even when his scientific reputation was all but gone. Because he steadfastly refused political office, attacked entrenched church authorities and promoted German nationalism, he was adored and honored by the government. In his old age, Haeckel had become a national hero, in one admirer's phrase, "a shining symbol that will glow for centuries."

He convinced masses of his countrymen to accept their evolutionary destiny and "outcompete" inferior peoples. It was right and natural that only the "fittest" nations should survive. He believed, literally, that "politics is applied biology"—an idea that later became a standard Nazi slogan. Despite the disastrous consequences of applied Social Darwinism in German history, Haeckel is best remembered as a charismatic biologist who surpassed even Thomas Huxley as a popularizer of evolution to mass audiences.

See also ARYAN "RACE," MYTH OF; BIOGENETIC LAW; DUBOIS, EUGENE; MONISM; MONIST LEAGUE.

For further information:

Boelsche, W. *Haeckel: His Life and Work.* London: T. Fisher Unwin, 1906.

Gasman, D. *The Scientific Origins of National Socialism in Germany: Social Darwinism in Ernst Haeckel and the German Monist League.* London: Macdonald, 1971.

Gould, Stephen Jay. *Ontogeny and Phylogeny.* Cambridge: Harvard University Press, 1977.

Haeckel, Ernst. *The History of Creation, or the Development of the Earth and Its Inhabitants by the Action of Natural Causes.* 2 vols. New York: Appleton, 1876.

HALDANE, J. B. S. (JOHN BURDON SANDERSON) (1892–1964)
Geneticist

J. B. S. Haldane was one of the great rascals of science—independent, nasty, brilliant, funny and totally one of a kind. Son of an Oxford professor of physiology, he began in science as his father's assistant. Eventually he taught genetics and biometry at University College, London, where he helped create the modern Synthetic Theory of evolution.

He learned Mendelian genetics while still a boy by breeding guinea pigs and often served as one himself

when he helped his father. In one childhood episode, the elder Haldane made him recite a long Shakespearean speech in the depths of a mine shaft to demonstrate the effects of rising gases. When the gasping boy finally fell to the floor, he found he could breathe the air there, a lesson that served him well in the trenches of World War I.

A physically courageous 200-pounder, Haldane continued the family tradition of using his own body for dangerous tests. In one experiment, he drank quantities of hydrochloric acid to observe its effects on muscle action; another time he exercised to exhaustion while measuring carbon dioxide pressures in his lungs.

Haldane was immensely cultivated; he had mastered Latin, Greek, French and German while still a student. Later, he wrote extensively on history and politics, and made important research contributions to chemistry, biology, mathematics and genetics. He is best remembered (along with E. B. Ford and R. A. Fisher) as an innovative pioneer in population genetics. Haldane's brilliance helped define the field, which reshaped modern evolutionary biology.

During World War I, Haldane volunteered for the Scottish Black Watch and was sent to the front. There he found, to his shock and dismay, that he liked killing the enemy. Twice wounded, he personally delivered bombs and engaged in sabotage behind enemy lines, prompting his commander to call him "the bravest and dirtiest officer in my Army."

In 1924, Haldane published a remarkable work of fiction, *Daedalus.* It was the first book about the scientific feasibility of "test-tube babies," brought to life without sexual intercourse or pregnancy. At the time, it was regarded as shocking science fiction. Haldane wrote it as a student might a century and a half into the future, looking back on how the production of "ectogenic" babies had transformed the world:

> The effect on human psychology and social life of the separation of sexual love and reproduction . . . is by no means wholly satisfactory. The old family life had certainly a good deal to commend it . . . On the other hand . . . the small proportion of men and women who are selected as ancestors for the next generation are so undoubtedly superior to the average that the advance in each generation . . . from the increased output of first-class music to the decreased convictions for theft, is very startling.

Daedalus was a popular and influential book, the original dose of "future-shock" for the 20th century. It inspired Aldous Huxley's novel *Brave New World* (1932), in which a society based on test-tube babies turns out to be not such a wonderful place after all.

Huxley also put Haldane in another of his novels, *Antic Hay* (1923), as Shearwater, "the biologist too absorbed in his experiments to notice his friends bedding his wife."

By the mid-1930s, leading geneticists such as Hermann Muller announced that *in vitro* ("in glass") fertilizations would soon be possible, and 40 years later they were a reality. Yet, though he predicted its feasibility, Haldane became an outspoken critic of eugenics. Genetic theory was being used for distorted political ends, he complained, by "ferocious enemies of human liberty."

Shortly before his death in 1964, the irrepressible Haldane wrote an outrageous comic poem while in the hospital, mocking his own incurable disease:

> Cancer's a Funny Thing:
> I wish I had the voice of Homer
> To sing of rectal carcinoma,
> Which kills a lot more chaps, in fact,
> Than were bumped off when Troy was sacked . . .

It was circulated among his friends, who savored the consistently witty irreverence with which Haldane had lived his courageous and productive life.

See also EUGENICS; HUXLEY, ALDOUS; KIN SELECTION; SYNTHETIC THEORY.

For further information:
Clark, Ronald W. *JBS: The Life and Work of J. B. S. Haldane.* New York: Coward-McCann, 1969.
Haldane, J. B. S. *The Causes of Evolution:* London: Longman, 1932.
———. *Heredity and Politics.* London: Allen and Unwin, 1938.

HAMILTON, W. D.

See KIN SELECTION.

HAMPDEN, JOHN (1819–1881)
Flat-Earther, Antievolutionist

John Hampden, the self-proclaimed champion of biblical "Flat Earthism," drew the great evolutionist Alfred Russel Wallace into one of the most bizarre episodes in the history of science. Wallace, Charles Darwin's partner in creating the theory of natural selection, expected to pick up some easy money by demonstrating that the world is round. Instead, he found himself enmeshed in what he later called "the most regretable incident in my life."

In 1870, Hampden publicly challenged any scientist to prove the Earth a spheroid and offered to bet £500 that no one could prove the "globular theory" to the satisfaction of independent referees. To collect, the challenger had to demonstrate the existence of a convex railway track or curving surface on a large body of water.

With his father a rector and his uncle a bishop, Hampden had sworn to destroy the "infidel pagan superstitions" of modern science. A zealous disciple of Samuel Birley ("Parallax") Rowbotham, author of *Earth Not a Globe* (1869), Hampden warned that if schools continued to teach the "Satanic globular theory" to the young, it would mean the destruction of all morality and religion. Evolution, he believed, was only the latest wrinkle in the more basic blasphemies of Copernicus and Sir Isaac Newton.

His preferred targets—wealthy, influential gentlemen like Charles Darwin or Sir Charles Lyell—would never rise to Hampden's bait. But Alfred Russel Wallace, with his openness, naivete and perpetual near-poverty, thought he might make a few points for science and win an easy five hundred. After all, he had started his career as a surveyor.

Hesitant, he asked the great geologist Lyell whether he thought it wise to accept such a challenge. "Certainly," Sir Charles replied, "it may stop these foolish people to have it plainly shown to them."

Wallace had chosen the worst possible advisor to protect his interests. Lyell had pushed Charles Darwin to publish quickly, warning that Wallace was also developing a theory of natural selection and might beat him into print. The same Lyell, Darwin wrote in his *Autobiography* (1876), "strongly advised me never to get entangled in a controversy, as it rarely did any good and caused a miserable loss of time and temper." While counseling Darwin to avoid public debates, the respected, gentlemanly Lyell was encouraging Wallace to lock horns with one of the most malicious, abusive crackpots in all of England.

Confident of his scientific prowess, Wallace entered the contest unconcernedly, like a lamb to the slaughter. Hampden asked him to pick an umpire, and Wallace chose a man named J. H. Walsh, who was a stranger to him. As the well-known editor of *The Field,* a newspaper for country gentlemen, Walsh had a reputation for fairness and objectivity. Graciously, Wallace offered to let Hampden appoint another impartial judge of his own choosing. He named a printer, who was a close friend and dedicated flat-earther, which Hampden did not think it necessary to mention.

On Saturday morning, March 5, 1870, the parties met near the north end of Old Bedford Canal, about 80 miles from London. The waterway ran straight and unobstructed for six miles between two bridges. Wallace affixed large cloth rectangles to the facing sides of each bridge, both bearing a bold black stripe running parallel to the ground and both placed exactly the same height above the water.

Halfway between the two bridges, Wallace stuck a tall pole in the bank, bearing two large discs as height markers. He measured the height of the lower disc, placing it exactly the same distance above the water level as the black stripes mounted on the bridges. After carefully lining up a surveyor's level-mounted telescope at the same height as the markers, Wallace asked the referees to sight through it from either bridge. From whichever vantage point, the pole's discs appeared much higher than the bars mounted on the bridges, which seemed to be on a downslope.

Walsh and another man looked through the telescope and both confirmed immediately that Wallace had proved his point. But Hampden's referee disagreed and actually jumped for joy, proclaiming the demonstration proved the flatness of the Earth. Hampden said it wasn't even necessary for him to look through the telescope, as he trusted his colleague. Wallace was stunned. He had not reckoned with the strange logic of "Zetetic" astronomy, as set forth in Rowbotham's book, which explains away such a result as a mere optical illusion, to be expected if the Earth is *flat*. Hampden immediately claimed victory.

Walsh took the drawings and interpretations of what was seen through the telescope, studied them for a week, then published them in his paper with an announcement that Wallace was clearly the winner. Furious, Hampden demanded his money back, but Walsh sent it to Wallace with congratulations. At this "perfidy," Hampden directed a barrage of letters and pamphlets at Walsh, Wallace and all their friends and colleagues, calling them liars, thieves, cheats and swindlers. He kept it up for the next 16 years, even haranguing the officers of all the scientific societies to which Wallace belonged.

When Walsh brought his own criminal action for libel, the court ordered Hampden to stop his flow of venom and "keep the peace" for a year. He managed it for a few months, then sent Mrs. Wallace a note:

MADAM—If your infernal thief of a husband is brought home some day on a hurdle, with every bone in his head smashed to a pulp, you will know the reason. Do you tell him from me he is a lying infernal thief, and as sure as his name is Wallace he never dies in his bed.

You must be a miserable wretch to be obliged to live with a convicted felon. Do not think or let him think I have done with him.

JOHN HAMPDEN

Wallace brought Hampden back to court, where he was fined and spent a week in jail, but over the next four years he kept repeating the offense and was convicted three times. He got several months in jail the next time, six more when he harassed Wallace again and was directed to pay damages and court costs, which he never did. Instead, he brazenly hid all his assets under a relative's name and declared bankruptcy.

Then Hampden began turning the tables in the law courts. He sued Walsh, the stakeholder, for his £500 when his barrister reminded him that English law did not recognize wagers. Losers of bets had no legal obligation to pay, and Hampden had asked for his money back before Walsh had paid it to Wallace. Hampden won, and Wallace bore Walsh's expenses in the suit. Although he had been declared winner of the challenge and wager, and several times winner in the courtroom, Wallace ended up deep in the red, disgusted with British justice.

For years, Hampden continued to attack and vilify Wallace in his various publications; he had thousands of supporters in England who belonged to the Flat Earth Society and regarded him as hero. In 1875, the incorrigible Hampden crowed that during the past 15 years, "no one but a degraded swindler has dared to make a fraudulent attempt to support the globular theory, [which] is ample and overwhelming proof of the worthless character of modern elementary geography."

Some 25 years after Hampden's death in 1881, Wallace, in his autobiography, expressed continuing amazement at his virulence, as if he were some strange specimen of natural history. "Seldom has so much boldness of assertion and force of invective been combined with such gross ignorance . . . And this man was educated at Oxford University!"

See also FLAT-EARTHERS; WALLACE, ALFRED RUSSEL.

For further information:

Godfrey, Laurie, ed. *Scientists Confront Creationism.* New York: Norton, 1983.

Wallace, Alfred Russel. *My Life.* 2 vols. New York: Dodd, Mead, 1905.

HAPPY FAMILY, THE
Survival of the Friendliest

"One of the most wonderful and unique sights to be seen in the world is that of the HAPPY FAMILY," trumpets the 1896 *Guidebook to the Consolidated P. T. Barnum Greatest Show on Earth.* Pictured is a large cage containing many different species of animals, some of them "natural enemies," living together in peace and harmony.

Evoking a zoological Garden of Eden rather than a Darwinian "struggle for existence," the famous exhibit demonstrated the possibility of:

that millenial time when the lion shall lie down with the lamb, and the wolf beside the sheep. In this cage will be found cats playing harmlessly with mice; dogs, foxes, monkeys and rabbits peacefully eating food out of the same dish without harming each other; owls, eagles, vultures watching the frolics of squirrels, bats and small birds, their usual prey, without making the least attempt to capture or kill them. In fact such a collection of animals in one cage was never seen since the days of Noah and the animals in the Ark.

Thousands of Barnum's paying customers found the exhibit strangely appealing as well as amazing. It seemed a reversal of natural law, a contradiction of the popular conception of "survival of the fittest." How did predators lose their cruelty and prey their fear? Barnum proclaimed that "a secret," but assured viewers that the animals were not unhealthy or drugged.

"We were the first to originate the idea of a Happy Family and our success caused numbers of our competitors to imitate us . . . but they lamentably failed," Barnum crowed. In fact, the great master of humbug shamelessly lifted the idea, which seems to have originated in England. In 1852, Francis T. Buckland wrote of two exhibitions of "Happy Families" in London. "One stands at Charing Cross and about the streets, the other remains permanently at Waterloo Bridge. They both claim to be the original 'happy family,' but I think the man at the bridge has the greatest claims to originality. He is the successor to the man who first started the idea, Austin by name," who opened his exhibit to the public around 1816.

There was no special trick to the effect. Combining animals in "unnatural" groupings takes great patience and requires choosing docile individuals, acclimating them gradually to each other, keeping them all well fed and obtaining the various species when young and rearing them together. Animals that cannot get along are excluded by the trainer until a workable balance is struck.

"LAW OF THE JUNGLE" seemed suspended in this famous live exhibit of peaceful predators and prey, as illustrated in P. T. Barnum's *Greatest Show on Earth Souvenir Book* for 1882. Barnum's claim that he invented the "Happy Family" was humbug.

Happy Family exhibits proved so popular, Mark Twain could not resist penning a parody—his own attempt, he claimed, to prove that man was of the "Lower Animals." First, he wrote, he experimented by placing some "Higher Animals," such as dogs, cats, and rabbits in a cage, and taught them to be friends. "In the course of two days I was able to add a fox, a goose, a squirrel and some doves . . . They lived together in peace; even affectionately."

"Next, in another cage," Twain continued, "I confined an Irish Catholic from Tipperary, and as soon as he seemed tame I added a Scotch Presbyterian from Aberdeen. Next a Turk from Constantinople; a Greek Christian from Crete; an Armenian; a Methodist from the wilds of Arkansas; a Buddhist from China; a Brahman from Benares. Finally, a Salvation Army Colonel from Wapping." When he checked on his experiment after a couple of days, he lamented, "the cage of Higher Animals was all right, but in the other there was but a chaos of gory odds and ends of turbans and fezzes and plaids and bones and flesh—not a specimen left alive. These Reasoning Animals had disagreed on a theological detail and carried the matter to a Higher Court."

Old-fashioned Happy Family exhibits are still seen occasionally at sideshows and resorts; one persists to the present day at Coney Island, New York. During 1985–1986, new, ecologically sophisticated versions appeared, containing various species in the same enclosure purporting to mimic a natural habitat: the immense "Deep Ocean" tank in Walt Disney World (Florida) and the "Rain Forest" exhibit ("Jungleworld") at New York's Bronx Zoo.

See also BARNUM, PHINEAS T; BIOPHILIA; TWAIN, MARK.

HARDY–WEINBERG EQUILIBRIUM
Genetically Balanced Populations

In 1867, Scottish engineer Fleeming Jenkin voiced an objection to Darwin's theory that touched off the famous "Paintpot Controversy." If "sports" (mutations) arise rarely in a population, Jenkin argued, how could they possibly have an opportunity to take hold? They would be "swamped by numbers and obliterated" in a few generations, like a few drops of black pigment thrown in a bucket of white paint.

Many others recognized this problem, none more so than William Bateson at the end of the 19th century, who clearly saw that swamping was a serious problem for his own theory of discontinuous evolution. In fact, before Bateson knew of Gregor Mendel's results, he had begun a program of research to investigate blending. Bateson's student E. R. Saunders conducted a series of experiments in cross-breeding and concluded that "in heredity the two characters do not completely blend, and the offspring do not regress to one mean but to two distinct forms. The variety, in short, is not swamped by intercrossing."

One of the main attractions of Mendel's work for Bateson was that Mendel provided not only evidence that blending did not occur but an explanation of *why* it did not. Yet, even after Mendel's paper was belatedly recognized in 1900, most scientists still thought dominant genetic characters would in time swamp the recessives.

The resolution of this problem began in 1908, when R. C. Punnett, Bateson's protege, addressed the Royal Society of Medicine on "Mendelian Heredity in Man." In the discussion, a statistician named Udny Yule suggested that a dominant allele (or variant form of gene) would increase in frequency until it reached a stability at 50%. This would produce the expected three-to-one Mendelian ratio among the offspring of hybrid crosses.

Punnett felt that was wrong, but could not prove it. He took the problem to his friend G. H. Hardy, professor of mathematics at Cambridge University, who is said to have then written the solution on his cuff at the dinner table. Because he considered it so elementary, Hardy refused to publish it in a journal his mathematical colleagues would normally read, so Punnett placed it in a biology journal. It was Hardy's only excursion into genetics.

What Professor Hardy came up with was that the simple binomial expression $(p^2 + 2pq + q^2 = 1)$ describes the proportion of each genotype in the population, where p represents the dominant allele (A), q the recessive (a) and $(p + q = 1)$. Punnett then drew a grid to demonstrate these proportions visually, which is enshrined in every basic biology textbook as "Punnett's Square."

Simultaneously, but unknown to Hardy, Wilhelm Weinberg (1862–1937) of Stuttgart, Germany, a physician whose hobby was genetics, had just published the same solution to the same problem. Although he shares the hyphen with Hardy on the equilibrium equation, Weinberg remained largely unsung and unknown. Like Gregor Mendel, the founder of genetics, Weinberg's many findings in theoretical population genetics were overlooked and only later rediscovered by science; but, by the time his work was recognized, population genetics had moved far beyond him.

See also "BEANBAG GENETICS"; MENDEL'S LAWS; PAINTPOT CONTROVERSY.

HAWAIIAN RADIATION
Diversity from Isolation

If Charles Darwin had explored the Hawaiian Islands (rather than the Galapagos, which so impressed him)

RAPID EVOLUTION from fairly recent common ancestor has produced scores of closely related but divergent honeycreepers in the Hawaiian Islands. Some species with sharp, heavy beaks can penetrate bark like woodpeckers, while delicate elongated curved beaks evolved in nectar-feeders. Species with parrot-like beaks crush seeds and pits.

he would have seen much more striking examples of diversity among closely related species. Evolutionists after him have found in these volcanic islands, long isolated from the major continents, an extraordinary natural laboratory of adaptive radiation.

Most famous are the 23 remaining species of finch-like birds known as Hawaiian honeycreepers and the more than 500 different fruit flies that have evolved, diverging from island to island and adapting to different habitats or foods on the same islands.

During the several million years since their ancestors reached the islands, some honeycreepers evolved into seedeaters with heavy beaks, others developed straight, thin beaks for spearing insects, while still others diverged into parrot-beaked species and delicate nectar-feeders with long curving bills and tubular tongues for probing flowers. More than half of the original 47 honeycreeper species have become extinct during the past 1,500 years, since the advent of humans and imported predators. (Some were wiped out by the original Polynesian settlers and others

only during the past few hundred years by Europeans.)

Under the former conditions of isolation, fruit flies (with their very short reproductive cycle) radiated far beyond the honeycreepers. Among the hundreds of species, some have become specialized for feeding on nectar or sugar; others eat decaying leaves; some are parasites on spider eggs; and some live only in a single valley on one island. They show a spectacular diversity in their body shapes, but even among those that appear pretty much the same (even to other fruit flies), they can be told apart by their sounds and behavior.

Hawaiian fruit fly populations have evolved scores of different courtship behaviors by which to recognize members of their own species, including elaborate airbone "dances." During the late 1980s, researchers also found the "songs" of Hawaiian fruit flies are as amazingly varied as their bodies. Some species make pulsing cricket-like sounds, while others sound more like cicadas than flies. Like body shapes or genes, these "songs" are providing more clues about how the various species diverged and spread throughout the islands.

See also ADAPTATION; DARWIN'S FINCHES; DIVERGENCE, PRINCIPLE OF; ISOLATING MECHANISMS.

For further information:
Carlquist, Sherwin. *Island Biology*. New York: Columbia University Press, 1974.
———. *Island Life: A Natural History of the Islands of the World*. Garden City, N.Y.: Natural History Press, 1965.
Gleason, H. A., and Cronquist, A. *The Natural Geography of Plants*. New York: Columbia University Press, 1964.
Good, Ronald. *The Geography of the Flowering Plants*. New York: Wiley and Sons, 1964.

HAWKINS, BENJAMIN WATERHOUSE (1807–1889)
First Evolutionary Artist

Best remembered for his giant models of dinosaurs sculpted for the Crystal Palace Exhibition (1853), Benjamin Waterhouse Hawkins pioneered the art of reconstructing extinct animals. Although his creatures are now considered fanciful and anatomically inaccurate, Hawkins was the first to stimulate tremendous public interest in life of the past.

Before he turned his hand to prehistoric monsters, Hawkins was an accomplished animal artist who contributed many illustrations to zoology monographs published by the British Museum (Natural History). He drew the plates (and transferred them to litho stone) for the volume on *Fish and Reptiles* of the *Zoology of the Voyage of H.M.S. Beagle* (1839–1845); his models were the specimens collected by the young Charles Darwin.

Hawkins made his dinosaur reconstructions under the supervision of Richard Owen, the irascible anatomist and director of the museum. An enemy of Darwin's, Owen coined the word "dinosaurs" to describe the newly discovered fossil creatures, but did not accept the idea of evolution. Hawkins's monsters were presented as "antediluvian." Later, Hawkins worked with leading evolutionists, including Thomas Huxley and Edward Drinker Cope.

The opening of his Crystal Palace exhibit was celebrated with a festive dinner held inside the huge *Iguanodon* model, attended by England's leading scientists. Hawkins's dinosaur park was such a huge success that he was summoned to America to build a grandiose Paleozoic Museum in New York's Central Park, a project that ended disastrously.

During his years in America, he reconstructed huge fossil skeletons for the great paleontologist Edwin Drinker Cope. He also painted haunting, nightmarish landscapes with grinning saurians reminiscent of gargoyles and hellish demons. Some of these works are still exhibited at the Princeton Museum, and his Crystal Palace statues have been restored to their former glory at Sydenham Park, near London.

Several generations of museum muralists, science fiction illustrators, movie animators, and modern makers of "dinorobots" owe a historical debt to Waterhouse Hawkins. He was the first of a distinguished line of artists to capture the popular imagination with reconstructions of extinct monsters.

OBSESSED BY EXTINCT MONSTERS, British artist Benjamin Waterhouse Hawkins is surrounded in his studio by full-sized models of ancient animals (left), the first major attempt to reconstruct the few dinosaur species then known. Two of his *Iguanodon* sculptures still adorn a small wooded park in the midst of a London suburb (right).

See also AMERICAN PALEOZOIC MUSEUM; DINOSAUR, RESTORATIONS; IGUANODON DINNER.

HEBERER, K. A.

See PEKING MAN.

HENNIG, WILLI

See CLADISTICS.

HENSLOW, REVEREND JOHN STEVENS (1796–1861)
Botanist, Geologist, Teacher

When the Reverend Professor John Henslow joined the teaching staff of St. John's College, Cambridge, in 1814, he had (at his father's insistence) given up his childhood dream of exploring African jungles. He was orthodox and devout in his religious beliefs, not very original as a thinker, and unambitious for advancement—an unlikely figure to help spark a scientific revolution. Yet, Charles Darwin said this unassuming cleric-naturalist was the greatest single influence on his career.

Henslow was a self-taught all-around naturalist, especially knowledgeable in geology, botany and entomology. When he came to Cambridge, no degrees were given in natural history subjects, a situation he determined to change. He revived the university's moldering herbarium and natural history museum, replanted its botanic garden and helped make scientific learning respectable in an institution long dominated by classics and theology.

Professor Henslow favored "hands-on" experience with living things. He took his students on high-spirited field trips into the countryside, where he showed them how to ask nature, rather than books, for answers. "Nothing could be more simple, cordial, and unpretending than the encouragement which he afforded to all young naturalists," his former student Charles Darwin later recalled. Never flaunting his vast knowledge, he put the young men perfectly at ease, treating their questions and blundering "discoveries" with the same respect he gave accomplished senior colleagues. Earnest mistakes were corrected "so clearly and kindly that one left him in no way disheartened, but only determined to be more accurate the next time."

By the time young Darwin returned home from college in 1831, Henslow's encouragement had transformed him from a boyish beetle collector into a budding naturalist. Although his father was urging him toward a life as a minister in the Church of England, he was busily making plans for a geologizing trip to Wales and the Canary Islands with Professor Henslow, which did not please Mrs. Henslow.

Everything changed, however, when Darwin received a fateful letter. Henslow had been asked to recommend a naturalist for a long round-the-world voyage of exploration and considered signing on himself, but this time his wife put her foot down. The professor's dream of exploring exotic lands and collecting strange plants and animals was thwarted again. Still, he hated to see the opportunity go to waste and urged young Darwin to seize it, even though he wasn't "a finished naturalist." Henslow would coach him from half a world away.

Dr. Robert Darwin did not relish the prospect of losing his bright, adventurous son to an ocean storm or tropical fever. Young Charles prevailed, however, and Henslow became his mentor and partner on this voyage of discovery. Although he never left England, Henslow assisted in the scientific explorations every bit of the way.

Just before the *Beagle*'s departure, Henslow advised Darwin to take along the just-published first volume of Charles Lyell's *Principles of Geology* (1830–1833). "By all means read it for its facts," he advised, "but don't believe any of the wild theories." Lyell's book became a crucial influence on the development of Darwin's thought as he geologized throughout his explorations.

During the entire five years of the voyage, Henslow corresponded regularly with Darwin, offering detailed advice and guidance to the inexperienced young naturalist. Darwin had taken on an enormous job and was at first unsure whether he was "collecting the right facts."

It truly became their voyage. From Brazil, Darwin wrote, "The delight of sitting on a decaying trunk amidst the quiet gloom of the forest is unspeakable and never to be forgotten. How often have I then wished for you." His teacher replied, "Your account of the Tropical forest is delightful, I can't help envying you."

Henslow took on the burden of receiving the endless crates and boxes of rocks, plants and preserved mammals, birds, insects and fish Darwin shipped home. Each one was carefully opened, damaged specimens removed and every item carefully arranged and stored.

"I firmly believe that, during these five years," wrote Darwin, "it never once crossed his mind that he was acting towards me with unusual and generous kindness . . . I owe more than I can express to this excellent man." Similarly, it seems never to have occurred to Darwin that he was giving his mentor great joy by allowing him to vicariously live out his boyhood dream of exploring the tropics.

Although the two men continued to correspond until Henslow's death, Darwin rarely discussed his evolutionary views, in an effort to spare Henslow's sensibilities. For his part, as the botanist Joseph Hooker wrote, Henslow was "a man who, with strong enough religious convictions of his own, had the biggest charity for every heresy so long as it was conscientiously entertained." He could not accept Darwin's conclusions, but never doubted his integrity.

When Richard Owen launched a vituperative, mean-spirited attack on Darwin's *Origin of Species* (1859), Henslow wrote his old pupil: "I don't think it is at all *becoming* in one Naturalist to be bitter against another any more than for one sect to burn the members of another." To the gentlemanly Reverend Henslow, torching heretics was simply not good manners.

See also DARWIN, CHARLES; LYELL, SIR CHARLES; VOYAGE OF THE (H.M.S.) BEAGLE.

For further information:

Barlow, Lady Nora, ed. *Darwin and Henslow: The Growth of an Idea; Letters 1831–1860.* Berkeley: University of California Press, 1967.

Russell-Gebbett, Jean P. *Henslow of Hitcham; Botanist, Educationalist, and Clergyman.* Lavenham, Suffolk: T. Dalton, 1977.

HETEROZYGOUS
Different-paired Genes

Alleles are different forms of the same gene. When each gene of an allelic pair produces different effects, they are said to be heterozygous (from the Greek *hetero* meaning different and *zygotous,* yoked); for instance, a person with one gene for type A blood and one for type B is heterozygous and will have AB blood.

If a person receives two alleles for type A blood (one from each parent), he would be homozygous for A and would have type A blood. It is possible to be heterozygous for a trait and yet have only one—the dominant—allele expressed in the body (phenotype). However, the other (recessive) allele may be expressed in succeeding generations.

See also DE VRIES, HUGO; HARDY–WEINBERG EQUILIBRIUM.

HINDU CONCEPT OF EVOLUTION

See AVATARS; GOPI KRISHNA.

HITCHCOCK, REVEREND EDWARD

See "NOAH'S RAVENS."

HOAXES

See BARNUM, PHINEAS T.; BERINGER, JOHANNES; BOUCHER DE PERTHES; CALAVARAS SKULL (HOAX); PILTDOWN MAN (HOAX); TASADAY TRIBE (HOAX).

HOLMES, SHERLOCK

See DEDUCTION; DIVINE BENEFICENCE, IDEA OF.

HOMINID
The Human Family

A hominid is an upright, bipedal primate of the family Hominidae (man-like creatures), which includes several species of *Australopithecus, Homo erectus, Homo sapiens* and perhaps others yet to be discovered.

The genus *Australopithecus,* known only from fossils, included at least four species. All were upright, bipedal and had brains about one-third the size of modern man's. The larger, heavier forms, *A. robustus* and *A. boisei,* and smaller or gracile species, *Australopithecus africanus* and *A. afarensis,* are all known from Africa. It is believed they lived between 1.5 and 4.5 million years ago.

A disputed type of early hominid, *Homo habilis,* is sometimes included with the australopithecines, but its status is not yet clear. Richard Leakey claims it belongs to the line leading to *H. sapiens,* and that all others are sidebranches.

Homo erectus belongs to the same genus as modern man, but was a different species. Java man (formerly *Pithecanthropus*) and Peking man (*Sinanthropus*) are now both considered *Homo erectus,* and the older names have been abandoned. This hominid was large (six feet tall), had heavy brow ridges, a brain half-to-two-thirds as large as modern man's, and is associated with Acheulean hand axes, fire and possibly cannibalism. Remains are found in Africa, Asia and Europe and are dated between half a million and 1.6 million years old.

Modern man, *Homo sapiens sapiens,* is the only hominid on Earth today; all living humans belong to this one species. An ancient variety was Neandertal man, sometimes classified as a subspecies *Homo sapiens neandertalensis,* associated with the Mousterian stone tool industry. However, the Neandertal population may have merged with *Homo sapiens.*

Hominids should not be confused with Hominoids. The Superfamily Hominoidea includes not only the Hominidae, but also the Pongidae, the family of great apes.

See also AFAR HOMINIDS; AUSTRALOPITHECINES; CANNIBALISM CONTROVERSY; HOMO ERECTUS; HOMO HABILIS; HOMO SAPIENS, CLASSIFICATION OF; MOUSTERIAN INDUSTRY.

For further information:
Lambert, David, and the Diagram Group. *Field Guide to Early Man.* New York: Facts On File, 1987.
Tobias, Philip V. *Hominid Evolution: Past, Present and Future.* New York: Alan R. Liss, 1985.

HOMINID GANG
Kenyan Fossil Finders

In the early days of fossil hunting, European or American scientists took credit for all discoveries even though actual finds were often made by anonymous native assistants.

Louis and Mary Leakey and their son Richard changed that by training a cadre of Africans who have become recognized as the most skilled fossil hunters on Earth. Informally known as "The Hominid Gang," these Kenyans are credited with many important discoveries.

Among the better-known members of the team are Bernard Ngeneo, who found the 1470 *Homo habilis* skull, and Kimoya Kimeu, the veteran leader of the team and discoverer of the remarkable "Turkana Boy." Kimeu began working with Louis and Mary Leakey at Olduvai Gorge and has continued for 25 years with Richard. For unearthing countless hominid fossils, Kimeu was presented with the prestigious National Geographic La Gorce Medal by Ronald Reagan at the White House in 1986.

The Hominid Gang has collected more than 14,000 fossils, including most of the East Turkana hominids. It consists of about a dozen individuals who search vast expanses of exposed sites full-time, six days a week, six months a year. They have been at it for 15 years and are credited with finding more than 200 hominid fossils, including a score of skulls.

In her book, *The Hominid Gang* (1989), journalist Delta Willis extends the nickname to include scientists and investigators engaged in trying to solve the enormous puzzle of hominid evolution.

See also HOMO HABILIS; LEAKEY, MARY; LEAKEY, RICHARD; TURKANA BOY.

HOMINOID GAP
Blank in Fossil Record

A 10-million-year "gap" in the fossil record tantalizes paleoanthropologists. It begins about 14 million years ago and covers a crucial period in the emergence of the human lineage.

By the Miocene period (5-to-24 million years ago), hominoids—the group encompassing man and apes—had already evolved in Africa. Ancestral apes, called dryopithecines, are very well represented by fossils, but it is not clear whether they may be considered on the line of possible early ancestors of humans.

(Anthropologists had accepted *Ramapithecus,* but in the 1970s it was ruled out as a human ancestor.)

From 14 million years ago until the australopithecine fossils appear at about four million years ago, the African fossil record is relatively blank. Dryopithecines were tailless, but had large canines and were not bipedal. Several hominid lineages apparently became bipedal during the Miocene, but we have no evidence of how or why. After this "gap" in the known sequence, hominid fossils begin with *Australopithecus afarensis* remains, which are dated at between four and five million years old. At about 3.5 million years, the first *Homo habilis* fossils appear; they persist and coexist with australopithecines for several million years. *Homo erectus* appears around a million years ago and overlaps the existence in time and geographical area with some of the older hominids, as well as with the most recently emergent species, *Homo sapiens.* (Today *Homo sapiens* is the only surviving genus and species of the complex radiation of hominids.)

Important changes occurred in African plants and animals between 7.5 and 4.5 million years ago, during the "Hominoid Gap." Geologic shifts may have caused the Mediterranean basin to dry up; drier climates changed forested area to open grasslands. Some paleontologists suspect these events were a crucial turning point in our ancestors' evolution.

A spectacular career awaits the discoverer of hominid fossils in the African Miocene, which would begin to bridge the yawning "Hominoid Gap."

See also DRYOPITHECINES; HOMINID; "MISSING LINK"; RAMAPITHECUS; TURNOVER PULSE HYPOTHESIS.

For further information:
Willis, Delta. *The Hominid Gang: Behind the Scenes in the Search for Human Origins.* New York: Viking, 1989.

HOMO ERECTUS*
Men of the Mid-Pleistocene

Homo erectus, a species of human somewhat different from ourselves, roamed Africa, Europe, China and Malaysia up until the mid-Pleistocene period, about 400,000 years ago, when it disappeared. We know these possible ancestors of ours mostly from their fossil skulls, some long bones, and the skeleton of an adolescent from East Africa, the "Turkana Boy." Estimated at 1.6 million years old, the boy's skeleton represents the earliest *Homo erectus* known. Most experts believe this hominid species originated in East Africa, from whence it dispersed northward, ranging widely throughout Europe and Asia during the next million years.

Adults are thick boned, with massive jaws and heavy eyebrow ridges. They were first called "ape-men" ("*Pithecanthropus*") or "missing links," but there is really nothing ape-like about them. They were a species of humans, with brains a bit smaller than our own: about 1,000 cubic centimers versus 1,400 for *Homo sapiens*. Pelvis and legbones show they were fully upright and bipedal, as we are.

Sometimes *H. erectus* is associated with crude stone hand-axes or picks; whether they had what we consider spoken language is not known. Remains from China were found with quantities of ashes from fires; some of the skulls were broken at the base, perhaps for extraction of the brains.

Homo erectus fragments have been found at widely scattered sites throughout the Old World: Java, East Africa, North Africa, China and Europe. Each discoverer made up a new name, convinced his find was very different from anything previously known to science. (A tendency to proclaim newly discovered hominid fossils to be "the oldest," "unique," or a "previously unknown species" still prevails in paleoanthropology.)

Dr. Eugene Dubois found the first *H. erectus* fossils in Java (1891) but he named the species *Pithecanthropus erectus*, or "ape-man who walks upright" (Java man). Forty-two years later (1933), excavations in an ancient Chinese cave yielded *Sinanthropus pekinensis*, or "Chinese man from Peking" (Peking man). A score of fragments from other sites were each given a separate species name.

In 1950, evolutionary biologist Ernst Mayr, an expert systematist (classifier of organisms), decided to tackle the mess. Unlike most anthropologists, to whom slight variations justified new species, he had the zoologist's wide-ranging familiarity with the animal kingdom. Mayr had unraveled complex relationships between families and species of birds at the American Museum of Natural History and was able to review the hominid fossils with detachment.

After comparative study, Mayr concluded there had been no scientific basis for splitting the so-called Pithecanthropines into a dozen invented genera, and instead lumped them all into the established genus *Homo*. The species name, *erectus*, was taken from the original find by Eugene Dubois.

Anthropologists gratefully accepted Mayr's reclassification, which resolved years of confusion. But there still is no agreement on whether *H. erectus* was a direct ancestor of modern man, a separate side-branch, which disappeared for unknown reasons, a rival we outcompeted—or none of these.

See also DUBOIS, EUGENE; MAUER JAW; PEKING MAN; TURKANA BOY.

For further information:

Delson, Eric, ed. *Ancestors: The Hard Evidence.* New York: Alan R. Liss, 1985.

Mayr, Ernst. "Taxonomic Categories in Fossil Hominids." Cold Spring Harbor Symposium on Quantitative Biology 15:(1950):109–118.

Reader, John. *Missing Links: The Hunt for Early Man.* London: Collins, 1981.

Shapiro, Harry. *Peking Man.* New York: Simon & Schuster, 1974.

HOMO HABILIS
Controversial Ancestor

While most anthropologists thought a convincing picture was emerging of the human lineage tracing back through *Homo erectus* to an ancestral australopithecine, the Leakey family dissented.

Back in 1961, Louis Leakey's son Jonathan had found a fragmentary skull at Olduvai Gorge, dated about 1.75 million years old, by the potassium-argon method, which seemed to be an "advanced" hominid. It was named *Homo habilis*, or "handy man," because Leakey believed it was associated with chipped stones found at the site. "This was the first evidence," wrote Richard Leakey, "that early members of the human lineage were contemporaries of the australopithecines, not descendants as was generally believed."

Debate swirled over whether this "habiline" was really a distinct species of *Homo*, an aberrant sapiens or a variety of australopithecine. In 1972, Bernard Ngeneo, of Richard Leakey's "Hominid Gang," found a similar but much more complete skull at East Turkana. It is generally known as the "1470" skull, from its accession number at the Kenya National Museum.

The 1470 skull was pieced together by Richard Leakey's wife Meave and several anatomists from dozens of fragments—a jigsaw puzzle that took six weeks to assemble. Dated at 1.89 million years old, with a cranial capacity of 750cc, Leakey believes it is the oldest fossil of a true human ancestor. In his view, the australopithecines and other hominid fossils were sidebranches.

Leakey fought hard to win a place for his 1470 (along with the previous habiline fragments found at Olduvai) because most anthropologists thought the skull was simply "too modern-looking" to be as ancient as he at first claimed. Initial potassium-argon tests had given an age of three million years, but later the deposits were restudied with improved techniques and found to be only half as old as Leakey and the geochronologists had thought. Even so, it is still fairly early: between 1.8 and 1.9 million years.

Leakey has backed off somewhat from his earlier assertion that his fossil ties in with the small but fully bipedal foot found at Olduvai or with the remarkable (earlier) footprints in hardened volcanic ash discovered at Laetoli by his mother, Mary, in 1976. But he still believes that an upright hominid—a truly man-like ancestor, which was not *australopithecus*—will be found to have lived in East Africa between three and four million years ago.

His published conclusion, similar to that of his father Louis, would push back the emergence of large-brained hominids several million years beyond the conventional view, separating the human line from other hominids early on. But there are other interpretations of the evidence, including the possibility that *Homo habilis* was part of a population intermediate between early australopithecines and *Homo sapiens*.

However, by 1989, Leakey sought to distance himself from his original theory, insisting any attempts at specific reconstructions of the human lineage were premature. He now stresses instead the diversity of coexisting early hominids, the great age of big-brained forms and the need for finding more fossils.

Until more evidence accumulates, the significance of the habiline material for understanding human evolution is still an open question. Even its association with stone tools (the basis of its name as "handy man") has not been firmly established. However, it is clear that *habilis* shows the largest braincase that old—almost two million years.

See also LEAKEY, LOUIS; LEAKEY, RICHARD; "OLDEST MAN."

For further information:
Leakey, Richard, and Lewin, Roger. *Origins: What New Discoveries Reveal about the Emergence of Our Species and Its Possible Future.* New York: E. P. Dutton, 1977.

HOMOLOGY
Similarities of Common Descent

"Homo" means "the same." The seven bones in the human neck correspond with the same seven, much larger, neckbones in the giraffe: They are homologues. The number of cervical vertebrae is a trait shared by creatures descended from a common ancestor. Related species share corresponding structures, though they may be modified in various ways.

All mammals—from mice to whales—share a common ancestor. And all have at least some body hair, are warm-blooded, suckle their young with milk from the female and have four-chambered hearts. A closely related subgroup or family, like the cats, share more specialized traits—such as retractable claws. Humans' close kinship with the great apes was first demonstrated by Thomas Henry Huxley in *Man's Place in Nature* (1863), in which he gathered overwhelming evidence for close homologies—muscle for muscle and bone for bone. Huxley's comparative anatomy demonstrated that man is more similar to the apes (gorillas, chimps, orangs) than the apes are to monkeys.

Some similarities between distant species may be caused by adaptation to similar environments, which is known as convergent evolution. Development of streamlined fins in fish (teleosts) and flippers in dolphins (mammals) are analogous: They function alike, but are very different in underlying structure. Wings of birds, insects, and bats are also analogous: similar in form and function, but with entirely independent evolutionary histories.

Linnaeus's original classification of animals does not distinguish between analogous and homologous structures. Creatures were often put in the same groups by resemblances to an imagined "divine plan" or "design." Since Darwin's impact on biology, species are classified to reflect the relative closeness or distance of their common ancestry.

For further information:
Owen, Richard. *On the Archetype and Homologies of the Vertebrate Skeleton.* London: van Voorst, 1848.

HOMO SAPIENS, CLASSIFICATION OF
Defining the Human Species

When the great Swedish naturalist Carl Linnaeus attempted to classify every living thing, he was not sure just how many species of humans existed. There were garbled reports from Africa and Asia about apes, "Jackos," "Pygmies," "wild men" and other near-men, but reliable information was scarce. In fact, few European scientists had even seen the carcass of a gorilla or chimpanzee, and the apparent variety of so-called savages was bewildering.

Of one thing Linnaeus was pretty certain: The earth contained creatures so anatomically similar to mankind that there was no reason to put them in a separate genus. (After Linnaeus, zoologists did separate them, placing man and each of the apes in distinct genera. Now, after 200 years, biochemical and genetic tests show they are much closer than anyone had thought. Humans and apes may soon share the same genus again.)

Despite the striking similarity of structure, Linnaeus was more concerned with how apes and men behaved and communicated. In fact, Linnaeus's definition of man was based not on his usual description of form (morphology) but on mind and behavior (psychology), particularly the use of language. Where he usually listed anatomical traits, he wrote an enigmatic line next to *Homo sapiens* (Man the wise): *Homo nosce Te ipsum* (Man, know yourself.) Or, to put it more flippantly, if you want to know what defines this creature, look in a mirror.

Under Linnaeus's system, one individual of each

species is described in detail, then preserved as the "type specimen," or holotype. Museum drawers are full of the skins, bones and pressed plants that represent each of the known living species. (The Linnean Society of London still has thousands from Linnaeus's own collection.)

In the case of man—a polytypic species where populations vary greatly—what individual shall represent us? Shall he or she be tall as a Watusi or short as a Mbuti pygmy? Dark-skinned or fair? With woolly hair or straight? In Africa, Asia and New Guinea, local populations vary more among themselves than averages of the groups differ from each other. More variation exists among black Africans, for example, than between Africans and Europeans.

Historian Wilfred Blunt argues that the type specimen of mankind must be the individual described in greatest detail by the namer of the species. After writing five versions of his autobiography, the man Linnaeus described most completely was himself. Blunt therefore concludes that Carl Linnaeus is the legitimate type specimen of *Homo sapiens*. Entombed in the floor of Uppsala Cathedral, his bones represent us all.

See also BINOMIAL NOMENCLATURE; LINNAEUS, CARL.

For further information:
Blunt, Wilfred, *The Compleat Naturalist: A Life of Linnaeus.* New York: Viking Press, 1971.

HONEYCREEPERS

See HAWAIIAN RADIATION.

HOOKER, SIR JOSEPH DALTON (1817–1911)
Botanist, Evolutionist

One of the most remarkable men of the 19th century, Joseph Dalton Hooker had wanted to be just like Charles Darwin (eight years his senior) when he grew up. As a young man, he kept advance proofs of Darwin's travel *Journal* of 1839 (obtained by his botanist father) under his pillow. Then, like his hero, he half-heartedly pursued a medical career, but left it to sign on as unofficial naturalist on the surveying ship *Erebus*. Among his contributions to the expedition was a remarkable collection of previously unknown plant species from the Antarctic.

Soon after the young botanist's return to England, he was elated when Charles Darwin sought him out. Within a few months, he became the great naturalist's closest friend, confidant, lieutenant and research partner—a relationship that was to last for almost 40 years.

Their coming together was no accident; throughout the *Erebus* voyage, long before they knew each other, Darwin had followed Hooker's adventures and achievements through the letters he sent home. Darwin got hold of them by the same route Hooker had received his *Journal* proofs: through the benign meddling of the older generation of naturalists.

After reaching England from the South Polar Seas, Hooker's letters followed a complicated path. Their intended recipient was his father, William Jackson Hooker (1785–1865), the famous botanist who was founder-director of the Royal Botanic Gardens and herbarium at Kew. Delighted with his son's accounts of strange places and wonderful new species, the elder Hooker proudly showed them to his wealthy amateur botanist friend, Charles Lyell of Kinnordy, who passed them on to *his* son, famed geologist Sir Charles Lyell—friend and mentor of Charles Darwin.

Hooker had taken Darwin's *Journal* along on his travels, consulting it constantly, just as Darwin had taken Lyell's *Principles of Geology* (1828–1832) on his own voyage. Upon Hooker's return, the usually reclusive Darwin sought him out and offered him the task of classifying the *Beagle*'s plant specimens, still languishing undescribed in the British Museum. Hooker jumped at the opportunity.

Darwin liked Hooker immediately and admired his vast botanical knowledge. (Hooker had, after all, been helping his father with thousands of plant specimens since the age of 12.) He knew much more than Darwin about identification and traditional botany, but Darwin's way of looking at plants ("philosophical botany") was a revelation to him. Unlike professional botanists of the day, Darwin focused on the living plant: its physiology, ecology, distribution, growth and reproduction.

It was not long before Darwin decided to take this excellent young man into his confidence about the theories he had secretly been developing for years. After all, Hooker could not help him marshal supporting evidence if he didn't know what to look for.

On January 11, 1844, Darwin revealed to his new friend that he had long been engaged in:

> a very presumptuous work . . . [most would] say a very foolish one . . . I was so struck with the distribution of the Galapagos organisms . . . I determined to [tackle the problem of] what are species . . . and I am almost convinced (quite contrary to the opinion I started with) that species are not (it is like confessing a murder) immutable . . . I think I have found out (here's presumption!) the simple way by which species become exquisitely adapted to various ends. You will now groan, and think to yourself, "On what a man have I been wasting my time and writing to." I should, five years ago, have thought so.

With this hesitant, self-satirizing letter, Darwin confided to Hooker the agenda of his life's work, which he had kept even from Lyell. Soon after, some fifteen years before publication of the *Origin of Species*, (1859) he entrusted him with a preliminary essay (the "1844 Sketch") of his theory of natural selection.

Darwin had picked the right man; Hooker was to become his indispensable collaborator and ally in establishing evolution, backing it up with pioneering studies of plant distribution throughout the world. In the introduction to his *Flowers of Tasmania* (1857), Hooker became the first respected naturalist to publicly take a stand with Darwin on natural selection.

Hooker became a frequent visitor at Down House, where Darwin would "pump him" each morning, handing him countless slips of paper asking for specific information on plant classification or distribution. Little by little, they began to piece together the outlines of succession and affinities, which came earlier and which later, and how plant families might have dispersed and spread over islands and continents. In the afternoons, they would stroll the Sandwalk, talking of "foreign lands and seas, old friends, old books, and things far off to both mind and eye."

Before Hooker left on grand new expeditions to India and Tibet, Darwin loaded him with hypotheses to test, samples to be taken, questions to be answered. Hooker was becoming the world's greatest expert on the distribution of plants, and all his knowledge, research, maps, and specimens were at Darwin's disposal. Sometimes Darwin feared he was selfishly taking all his friend's ideas, but Hooker always maintained he received more knowledge from Darwin than he gave.

Hooker went on three expeditions to India, all of them conducted in the grandest Victorian manner. On one exploration, he traveled by elephant 5000 feet up a sacred Himalayan mountain, attempting to reach Tibet. But the Rajah of Sikkim, paid by the Chinese to keep Englishmen out of the area, had sent 100 men to capture him, which didn't faze Hooker a bit.

He had taken the precaution of hiring his own private security force: 56 Gurkhas, who were "immense fellows, stout and brawny, in scarlet jackets, carrying a kookry [dagger] stuck in the cummerbund and heavy iron sword at their side." Within two hours, he had turned both his and the Rajah's men into a combined army of plant collectors.

On Hooker's expedition to the Sikkim Himalayas in 1848, he brought back hardy species of rhododendrons, which became an integral part of British gardens thereafter, an expression of the romantic Victorian notion of exotic beauty. He wrote feelingly of these danger-filled quests for exquisite, rare flowers amid rivers "roaring in sheets of foam, sombre woods . . . crested with groves of black firs, terminating in snow-sprinkled rocky peaks."

Hooker's massive party included armed guards, bird and animal shooters, cooks, porters, plant collectors and an herbarium crew. Weather was so wet in the Khasia Hills that it took special efforts to dry specimens and press them in papers. He rented "a large and good bungalow, in which three immense coalfires were kept up for drying plants and papers, and fifteen men were always employed from morning till night." His discovery of the rare blue vanda orchid in the Khasias, unfortunately, resulted in an invasion by ruthless commercial collectors who stripped the forests bare.

Hooker succeeded his father as director of the Royal Botanic Garden at Kew, serving for 20 years (1865–1885), at which time his son-in-law, William Thistleton-Dyer, took over. All three directors were knighted for making this institution a world center of plant knowledge. (His father's own collection of specimens had formed the herbarium's core, occupying 13 rooms of his private house before it was moved to Kew.)

In addition to his superb, voluminous contributions to botany, Hooker is best remembered for the assistance he gave his friend and role model, Charles Darwin, often at crucial times. For instance, during the legendary Oxford Debate over evolution, it was Hooker—not Thomas Huxley—who won the day with a sober recitation of evidence *after* Huxley and the bishop had traded barbs. And when Darwin's long hesitation to publish threatened his place in history, it was Hooker (with Lyell) who rescued his claim to fame. [See "DELICATE ARRANGEMENT."]

In the end, Joseph Dalton Hooker had gone his youthful daydreams one better—he had become one of Charles Darwin's heroes.

See also INDEX KEWENSIS; LYELL, SIR CHARLES; OXFORD DEBATE; SANDWALK.

For further information:

Allan, Mea. *Darwin and His Flowers: The Key to Natural Selection*. New York: Taplinger, 1977.
———. *The Hookers of Kew, 1785–1911*. London: Joseph, 1967.
Hooker, Joseph D. *Himalayan Journals*. 2 vols. London: John Murray, 1854.
Huxley, Leonard, ed. *Life and Letters of Sir Joseph Dalton Hooker*. 2 vols. London: J. Murray, 1918.
Scourse, Nicolette. *The Victorians and Their Flowers*. London: Croon, Helm, 1983.
Turrill, W. B. *Joseph Dalton Hooker*. London: Scientific Book Club, 1963.

HOOTON, EARNEST ALBERT (1887–1954)
American Physical Anthropologist

An early student of mankind's ways, the Roman philosopher Horace coined a maxim justifying his

FOUR STRANGE FIGURES inspired by Professor Earnest Hooton's claim that if fossil man walked the streets today in formal dress, no one would notice them. Chicago Museum of Natural History artist sketched Cro-Magnon (top, right), Neandertal (top, left), and (at bottom) Java man and Peking man, now known as *Homo erectus*.

The lemur is a lowly brute,
   its primate status, some dispute.
He has a damp and longish snout,
   with lower front teeth hanging out.
He parts his hair with this comb-jaw,
   and scratches with a single claw
That still adorns a hinder digit,
   whenever itching makes him fidget.
He is arboreal and omniverous
   from more about him, Lord deliver us.

Hooton was beloved by his students for his sometimes unintentional eccentricities as well. For instance, he so admired the great English physicist Sir Isaac Newton that he named a son after him, Newton Hooton. The boy took a lot of ribbing, so Hooton tried a more innocuous combination when he named his daughter Ima. Eventually, she married a Mr. Goody and became Ima Hooton Goody.

Hooton had begun his career as a classicist at the University of Wisconsin, where he won a Rhodes scholarship to Oxford in 1910. Although he had expected to continue his studies of ancient Greek and Roman literature, he became fascinated in London with the British archeologist R. R. Marett (1866–1943) and the anatomist Sir Arthur Keith (1866–1955), and returned to the United States a physical anthropologist. He taught the subject at Harvard from 1914 until his death in 1954, during which he trained the first generation of professional physical anthropologists in the United States.

For further information:

Hooton, Earnest Albert. *Man's Poor Relations*. New York: Doubleday, 1942.

———. *Up from the Ape*. 2nd ed. New York: Macmillan, 1946.

HOPEFUL MONSTERS
Macromutations

Richard Goldschmidt (1878–1958), a brilliant but unorthodox geneticist, did not believe that Charles Darwin's idea of slow, gradual changes could account for the origin of species. Forced out of his native Germany by the Nazis, he continued to develop his research at the University of California at Berkeley, where he wrote his magnum opus *The Material Basis of Evolution* published in 1940.

Although he recognized the constant accumulation of small changes in populations (microevolution), he believed they did not lead to speciation. Between true species he saw "bridgeless gaps" that could only be accounted for by large sudden jumps, resulting in "hopeful monsters."

Goldschmidt tried to explore possible genetic mechanisms of how rapid change might occur in lineages of organisms. He suggested that a relatively

wide-ranging curiosity: "I am human; nothing in human nature is alien to me." With characteristic humor, Professor Earnest Hooton of Harvard revised it to read: "I am a primate; nothing in primate nature is alien to me."

For 40 years, Hooton was the premier gatherer of material on monkeys and apes, popularizing knowledge about them and encouraging students to learn about man by studying the context of our evolutionary lineage. His very readable books, *Man's Poor Relations* (1942) and *Up From the Ape* (1946), were widely influential, helping inspire the wave of primate field studies that transformed the following generation's physical anthropology from an anatomical to a behavioral science.

His approach was quirky and entertaining, including such novel studies as comparing the measurements of a gorilla with those of a professional wrestler. (Hooton insisted he meant no offense to the wrestler, who was a "real gentleman.") Among zoological descriptions of the primates, he inserted such doggerel as:

small change might have a large effect on the phenotype, especially through "controlling" genes which mediate the expression of the organism's blueprint. Later, he thought macromutations or mutants (which used to be called "monsters") might arise in a single generation, and this biological novelty might enjoy a selective advantage under changing environmental conditions.

That was where the "hopeful" came in. One hope was that the mutation would prove so useful in the newly changed environment that it would become selected as a new norm. Another hope was that the variant would appear often enough in the population to allow several similar "monsters" to find one another and produce offspring. There is a grotesque humor about the unfortunate phrase "hopeful monsters" that lent itself to caricatures of Goldschmidt's ideas and obscured the theoretical issues.

See also DE VRIES, HUGO; GRADUALISM; SALTATION.

For further information:
Goldschmidt, Richard. *The Material Basis of Evolution.* New Haven: Yale University Press, 1940.

HORNER, JACK

See DINOSAUR HERESIES.

HORSE, EVOLUTION OF
Saddled With Errors

Professor Othniel C. Marsh of Yale, one of America's greatest paleontologists, set out to confirm Charles Darwin's evolutionary hypothesis by working out the evolution of the horse.

He collected a magnificent set of American fossil horses and published a paper in 1874 tracing its development from a small three-toed animal "the size of a fox" through larger animals with progressively larger hooves, developed from the middle toe. Darwin thought Marsh's sequence from little *Eohippus* ("Dawn horse") to the modern *Equus* was the best evolutionary demonstration anyone had produced in the 15 years since the *Origin of Species* (1859) was published.

When Darwin's friend and "general agent" Thomas Henry Huxley toured America, he visited Marsh at Yale and was mightily impressed with his progressive series of fossil horses. "Whatever I ask for, you can produce," he enthused.

Marsh's classic unilineal (straight-line) development of the horse became enshrined in every biology textbook and in a famous exhibit at the American Museum of Natural History. It showed a sequence of mounted skeletons, each one larger and with a more well-developed hoof than the last. (The exhibit is now hidden from public view as an outdated embarassment.)

Almost a century later, paleontologist George Gaylord Simpson reexamined horse evolution and concluded that generations of students had been misled. In his book *Horses* (1951), he showed that there was no simple, gradual unilineal development at all.

There were three complex radiations in the course of horse evolution, as they shifted from browsers on forest leaves to grazers on the grassy plains. Sixteen different genera developed, of which 15 became extinct. Several lineages had developed in each adaptive zone, producing complex branches rather than a single line. Some grazers had well-developed hooves; other retained their toes. Rates of development were not gradual, but "jerky." Teeth, toes and body size varied in different lineages, independently of each other.

It was an easy mistake to make, since only one genus of horse is left today, *Equus*. Marsh arranged his fossils to "lead up" to the one surviving species, blithely ignoring many inconsistencies and any contradictory evidence. Ironically, his famous reconstruction of horse evolution was copied by anthropologists. They, too, thought they saw a straight-line lineage "leading up" to the sole surviving species of a once-varied group: *Homo sapiens*.

See also BRANCHING BUSH; COPE–MARSH FEUD; ORTHOGENESIS.

For further information:
Simpson, George G. *Horses.* Oxford: Oxford University Press, 1951.

HOWELL, F. CLARK (b. 1925)
American Paleoanthropologist

All his life anthropologist F. Clark Howell of the University of Chicago has hoped to find spectacular fossils of early man. So far they have eluded him. But Howell developed methods for attempting to squeeze maximum information out of early-man sites whether they contain hominid fossils or not.

In the past, seekers of fossils might find hominid remains, and then wonder how to date them. So during their next season, they would call in a geologist or perhaps an expert on associated mammals. It was a piecemeal, haphazard approach, and much valuable information was lost or moved around at the site.

Howell pioneered the idea of bringing in a team of various experts to work in concert right from the start: a geologist to determine the history of the rocks, a surveyor to map the area, a photographer to document remains and artifacts as they are uncovered, a draftsman to record the positions of each item. Budget permitting, Howell would also include a palynologist (a specialist in fossil plant pollen) to figure out the vegetation of the period, a pedologist (or

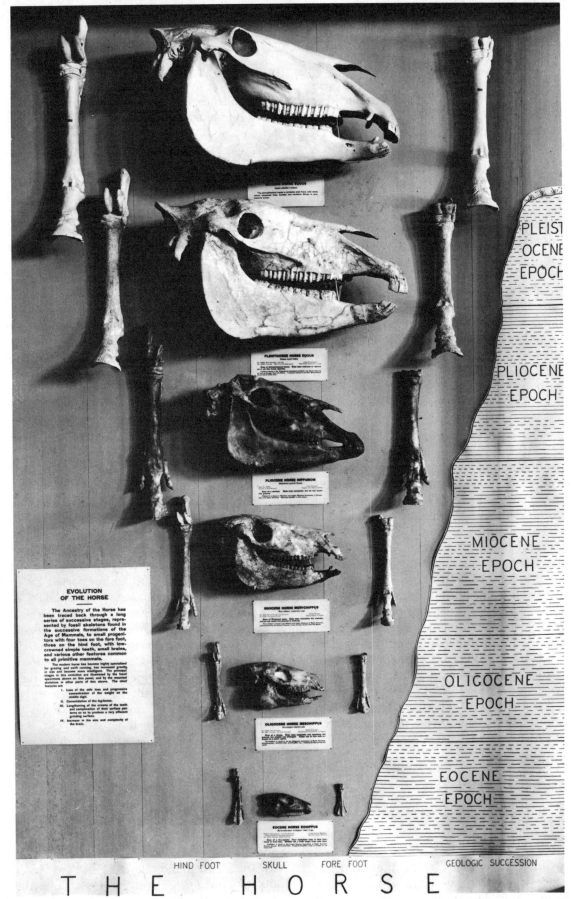

HIND FOOT   **SKULL**   **FORE FOOT**   **GEOLOGIC SUCCESSION**

THE HORSE

OUTMODED EXHIBIT depicted straight-line evolution of horse from small, five-toed animal, "leading up to" progressively larger hoofed *Equus* of today. Paleontologists later demonstrated that the real story was much more complex—a branching of many divergent populations over time and space.

223

expert on soils), a geochemist and a paleontologist, who could identify various animal bones found with the remains of ancient humans and their tools.

He used his team technique at Torralba and Ambrona, two sites in Spain that he excavated in the 1960s. The sites were first surveyed within a grid of small squares. Inch by inch, each found item was recorded and plotted, and a new map was made for every foot of excavation. Finally, Howell was able to put together a detailed three-dimensional picture of the entire site, which seemed to give a remarkably clear idea of what had occurred there almost half a million years ago.

Based on associated tools and the remains of ancient elephants, Howell concluded that these were places where *Homo erectus* killed and butchered the large mammals. Pollen and geology showed the area was once a swamp; an ash layer suggested the elephants had been driven into the mire by grass fires.

Stone tools were scattered around the elephants' bones, which were scratched where the tools had been used to butcher the carcasses. Sometimes half an elephant was missing, and was presumed by Howell to have been carried away by the ancient hunters.

Toralba and Ambrona became models of sophisticated excavation techniques, though Howell's interpretations were later criticized. There were, after all, no *Homo erectus* remains present which could link the site definitely with that hominid. The grass fires could not be proven to have been set by the hunters, who may have been scavenging animal carcasses that had become mired in the swamp.

Though Howell never found a hominid fossil his doctoral student, Donald Johanson, recovered some of the most spectacular hominids ever discovered: the *Australopithecus afarensis* fossils from Ethiopia. In his book *Lucy* (1981), Johanson expressed great respect for Howell's ability as an anthropologist and teacher, but could not resist adding matter-of-factly: "Clark's just not a finder."

See also HOMO ERECTUS; JOHANSON, DONALD; "LUCY."

HOYLE, FRED

See PANSPERMIA.

HUNTING HYPOTHESIS
How Apes Killed a Theory

Throughout the latter 1960s, with thousands of Berkeley students outside his window protesting a war, a smiling, amiable anthropologist taught that man is a killer by nature. Professor Sherwood Washburn of the University of California cited comparative field studies of primates and tribal peoples to support his "hunting hypothesis."

Washburn believed "the hunting way of life" separated humans from their relatives, the peaceful vegetarian apes. Aggressiveness, he thought, had been a positive basis of human evolution: the impetus for development of tools, technology, intelligence and social complexity. "If you don't believe we're naturally killers," he added, "put an air rifle in the hands of any young boy and watch how he loves to pop some birds or squirrels."

With the disappearance of forests, Washburn wrote, early hominids had to venture out onto the savannah grasslands. In that new ecology, there would be selection pressure for bipedalism—standing upright—to scan the flat ground for enemies or prey. There man made the transition to an aggressive hunter who killed for his living.

Cooperative hunting would shape social behavior, and the hunting psychology would select for aggressive behavior, which might eventually be turned against one's own kind. Tool use and finely tuned hand-eye-brain coordination would evolve with the development of weapons and butchering, an integral part of the hunting way of life.

Popularized by Robert Ardrey's bestseller *African Genesis* (1961), anatomist Raymond Dart's theory that "man was born with a weapon in his hand" had become part of the conventional wisdom. "Man the hunter" became newly reestablished in textbooks and popular culture, where the "caveman" who "brings home the bacon" never really went out of fashion. Now it was presumed to be supported by exciting new studies of chimpanzees (Jane Goodall) and gorillas (George Schaller) in the deep forest. There fieldworkers described the apes as peaceful vegetarians, who neither hunted nor made war on their fellows. To Washburn, they posed a major puzzle for anthropology: How could "aggressive" humans have been derived from gentle, fruit-eating ancestors? His solution: "adaptation to a hunting way of life."

But during the 1970s and 1980s, field studies of apes took a dramatic and unexpected turn. Continuing observations proved that chimpanzees *do* hunt; their favorite prey are monkeys, baby antelopes and the young baboons with which they frequently socialize.

Moreover, they hunt cooperatively; several chimps go out together to stalk and surround their prey, driving it into ambush. Killing and rending carcasses with their bare hands, chimps use no weapons or tools for butchering. Even more startling, the kill is not kept by the hunters; it is divided and shared among other members of the group. Chimps of all ages and sexes surround the hunters and beg for morsels with outstretched palms. Food-sharing is a leisurely and elaborate social ritual lasting several hours; dominant males do not grab meat, but wait patiently with the others for their doled-out portion.

Then came the biggest shocker of all: "war" within the species. Jane Goodall, in 1976, reported a "chimpanzee holocaust": the deliberate slaughter of all the males and infants in a neighboring group. Between three and five male chimps ganged up on each individual and brutally bashed him to death with fists and feet. Although they used no weapons, they did slam their victims against tree trunks. (They spared the females, who then joined the group and later mated with the victors.)

Goodall was heartbroken. She had thought chimps behaved "better than humans," but now realized her apes were no angels. "They are even more like us than I imagined," she sadly admitted. Later, she reported on chimp infanticide and cannibalism; two females kidnapped, killed and ate several infants within their own group.

Hunting, murder and mayhem can no longer be considered human specialties among the hominoids. Murderous aggression does not require upright posture, weapons-making or a life on the open plains; stooped-over, knuckle-walking chimps, mostly vegetarian, living in the tangled forest, without weapons or man-sized brains, can be as vicious and bloody as any carnivore.

Science remains ignorant of the basic adaptation that shaped the human species. Despite our bloody history, it is still possible that mankind is no more aggressive by nature than other primates.

See also AFRICAN GENESIS; CHIMPANZEES; KILLER APE THEORY.

For further information:
Ardrey, Robert. *The Hunting Hypothesis.* New York: Atheneum, 1976.

HURZELER, JOHANNES

See OREOPITHECUS.

HUTTON, JAMES

See ACTUALISM; STEADY-STATE EARTH.

HUXLEY, ALDOUS (1894–1963)
Novelist, Philosopher

Evolutionist Thomas Henry Huxley, friend and champion of Charles Darwin, founded an intellectual tradition in Western thought and in his own family. One of his grandsons, Sir Julian Huxley, became a distinguished evolutionary biologist and essayist in his own right and a founder of the modern Synthetic

Theory of evolution. Julian's brother, Aldous, followed a literary career, a poet's quest for new values and perceptions in the "brave new world" science had wrought.

Yet, echos of his evolutionist grandfather's major concerns reverberate throughout the fictional creations of Aldous Huxley: What kind of creatures are we? How can we know what is real? On what values should we base our actions? What can we honestly claim to know or believe? Into what kind of being will we allow ourselves to evolve?

Huxley's novels, casually saturated with the classical and historical traditions in which he was steeped, were addressed to an educated elite. But, like his grandfather, who enjoyed lecturing on evolution to working men, Aldous Huxley had no snobbery about writing for *Vogue, Life* or *Playboy* to popularize the results of his search for a view of life.

Huxley was a working writer, proud that he could support himself by the pen. During his early days the going was rough; his parents were "impecunious

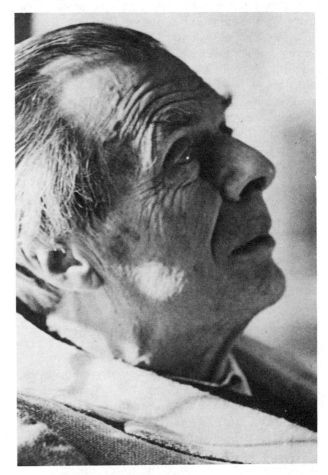

A BRAVE NEW WORLD of high-tech and low virtue is the best-remembered vision of Aldous Huxley. Novelist, essayist, futurist, mystic and sometime Hollywood screenwriter, he entertained, provoked and infuriated readers with consummate literary skill.

but dignified," as he recalled, "putting on dress clothes to eat . . . out of porcelain and burnished silver—a dinner of dishwater and codfish, mock duck and cabbage." From early youth he suffered from poor eyesight, which made a literary career especially difficult. He was blind by middle age.

He wrote everything from drama reviews to ad copy, plays, poetry, articles for the Sunday papers, even screenplays for Hollywood; but his fame rested on the essays and novels. During his last years he assumed the mantle of "philosopher, mystic and guardian of the public conscience." His pacifism during World War II and "inner explorations" with psychedelic drugs (*The Doors of Perception*, 1954) brought controversy and a measure of public scorn.

Among his best-known novels are *Chrome Yellow* (1922), *Antic Hay* (1923), *Those Barren Leaves* (1925), *Point Counter Point* (1928), and *Brave New World* (1932). The last is, perhaps, his most popular, despite its widely acknowledged flaws of one-dimensional characters and an unsatisfying "solution." It remains a vivid projection into the future of the eugenic ideas current in the early 20th century, when evolutionary biologists spoke seriously of the urgent need for mankind to take hold of its own evolution. Huxley saw and savagely satirized the kind of society that could result from biotechnology run wild; and the spectre of his "brave new world" continues to loom as an ominous warning, especially since the technology he predicted has actually arrived.

See also EUGENICS; HALDANE, J. B. S.; HUXLEY, SIR JULIAN S.; HUXLEY, LEONARD; HUXLEY, THOMAS HENRY.

For further information:

Clark, Ronald W. *The Huxleys*. New York: McGraw-Hill, 1968.

Huxley, Aldous. *Ape and Essence*. New York: Harper, 1948.

———. *Brave New World*. Garden City, N.Y.: Doubleday, Doran, 1932.

———. *Ends and Means: An Inquiry into the Nature of Ideals*. New York: Harper, 1937.

HUXLEY, SIR JULIAN SORELL (1887–1975)
Evolutionary Biologist, Administrator

"I like that little chap; he'll look you straight in the eye and deliberately disobey you," said Thomas Henry Huxley of his three-year-old grandson, Julian Sorell Huxley. The great Victorian evolutionist did not live to see it, but biologist Julian (and his brother, the novelist Aldous) were to extend the family's fame and achievements well into the 20th century.

While still a boy, Julian decided to follow his grandfather into the study of evolutionary biology. Later in life—also like Thomas Henry—he found a conventional scientific career too limiting and entered the public arena. Early on, Julian thought a commitment to achievement "implies constructiveness not only about one's own life, but about society, and the future possibilities of humanity." As Thomas's evolutionism had led to an interest in exploring new views of religion and ethics, Julian's own study of biology expanded into a rethinking of humanist values and the application of science to social problems.

As a young student on vacation, Julian went bird-watching and observed the remarkable courtship performances of the crested grebe. His description of their "plesiosaur-like dance," in which they entwine their long necks, jump out of the water, and utter a crescendo of excited cries, became a classic in the field study of animal behavior. His analysis of how various "normal" behaviors seem to become "ritualised" in the courtship sequence was later taken up as one of the influential concepts in the new study of animal behavior (ethology), developed by Konrad Lorenz and Niko Tinbergen. [See RITUALIZATION.]

But the first of Huxley's research to attract public interest was an experiment on the Axolotl salamander, which was believed to be a permanently juvenile species that retained external gills throughout life and never left the water. [See NEOTENY.] Huxley and a colleague fed two five-inch axolotls extracts of cattle thyroid. After two weeks of ingesting hormones, the animals changed color, absorbed their fins and gills and began walking on dry land. This "artificial metamorphosis" by an evolutionist created wild speculation in the press about its possible applications for humans, although Huxley patiently explained there were none.

In 1921, Huxley joined other naturalists on a field trip to Norway, where he studied more bird courtship rituals and marine invertebrates. As his reputation grew, he was invited to pursue research at the Wood's Hole Institute in the United States, where he investigated problems of growth and form.

Studying the fiddler crab, which develops an extraordinarily large claw it uses for signaling and courting, Huxley tried to understand the rates of growth and proportion of the creature. "I was here able," he wrote, "to find a definite mathematical expression relating the weights of the claws and the rest of the body. The two behaved like two sums of money put out at two different rates of compound interest; in other words, the ratio of their growth rates was a constant."

Later, Huxley weighed and compared deer antlers as he had fiddler crabs claws, and worked out their constant differential growth rates. His first major treatise, *Problems of Relative Growth* (1932), was followed by *The Elements of Experimental Embryology* (1934), coauthored with Sir Gavin de Beer.

AT HIS GRANDFATHER'S KNEE, Julian Huxley imbibed a tradition of scientific integrity, evolutionism, public service and the search for "religion without revelation" from Thomas Henry Huxley. Julian grew up to become a leader in forging the modern Synthetic Theory of evolution, and later served as the first director-general of the United Nations Educational, Scientific, and Cultural Organization (UNESCO).

Although his academic career was going well—he then occupied the chair of zoology at King's College, London—around age 40 Julian decided to abandon it. He wrote several influential books, including *Religion Without Revelation* (1927), *The Individual in the Animal Kingdom* (1912) and *Essays of a Biologist* (1923). With novelist H. G. Wells, who had been a student of Thomas Huxley, he collaborated on a monumental textbook *The Science of Life*, to make biology comprehensible to the average man, as Wells had tried to do with historical studies in his famous *Outline of History* (1919–1920). (Wells died during the collaboration, and Huxley finished the book with Wells's son.)

During the 1930s, Huxley was influential in helping bring together the modern, or Synthetic, theory of evolution: an update of Darwinism incorporating 20th century developments in paleontology, Mendelian genetics, population genetics, systematics and other branches of biology. Along with Theodosius Dob-

zhanksy, Ernst Mayr, George Gaylord Simpson, J. B. S. Haldane and others, Huxley was one of the architects of the new understanding, and it was he who named it with his book *Evolution: The Modern Synthesis* (1942).

For some years, Huxley served as director of the London Zoo and proved to be a dynamic and innovative administrator. After a succession of posts heading various governmental and cultural agencies, his stature had grown to the point that he was asked to serve as the first director-general of the United Nations Educational, Scientific and Cultural Organization (UNESCO).

Huxley relished the challenge; he thought the idealistic internationalism of the early United Nations could lead to an "evolutionary step up" (culturally, not biologically) from parochial warfare and the conflicts of antagonistic national governments. In his controversial pamphlet outlining his vision for UNESCO, he went on record as biologist turned social activist:

> Unesco must constantly be testing its policies against the touchstone of evolutionary progress . . . The key to man's advance, the distinctive method which has made evolutionary progress in the human sector so much more rapid . . . is the fact of cumulative tradition, the existence of a common pool of ideas which is self-perpetuating and itself capable of evolving . . . [Therefore] the type of social organization [is] the main factor in [advancing or limiting] social progress.

This attempt by an "evolutionist-humanist" to bring together the human species created immediate polarization. Huxley's notion that man was in charge of his own destiny shocked many religious leaders as impious. His reference to Marxism as "the first radical attempt at an evolutionary philosophy" made Americans bristle, while a remark about the value of contraception in world social planning angered Catholics and others. And Communist ideologues demanded to know "whether acceptance of Unesco required the abandonment of their philosophy in favour of Dr. Huxley's."

See also HUXLEY, ALDOUS; HUXLEY, LEONARD; HUXLEY, THOMAS HENRY; SYNTHETIC THEORY; WELLS, H. G.

For further information:

Clark, Ronald W. *The Huxleys*. New York: McGraw-Hill, 1968.

Harrison, G. A., and Keynes, M., eds. *Evolutionary Studies: A Centenary Celebration of the Life of Julian Huxley*. London: Macmillan, 1989.

Huxley, Julian. *Essays of a Biologist*. Cambridge: Cambridge University Press, 1912.

———. *Evolution: The Modern Synthesis*. London: George Allen and Unwin, 1942.

———. *Religion Without Revelation.* New York: Harper, 1927.
———. "The Tissue Culture King" (1929). In *Great Science Fiction by Scientists*, edited by Groff Conklin, 145–170. New York: Collier-Macmillan, 1962.

HUXLEY, LEONARD (1860–1933)
Biographer, Editor, Poet

Leonard Huxley, a quiet and unassuming literary man, not only chronicled the accomplishments of his father, Thomas, but inspired and shaped the continuation of a remarkable family tradition.

His own gifts were considerable, but he devoted them mainly to preserving the achievements of the previous generation. His classics *Life and Letters of Thomas Henry Huxley* (1900) and *Life and Letters of Sir Joseph Hooker* (1918) combine fine biography with judicious editorial work.

Leonard Huxley's arrangement and selection of personal papers always has point and purpose. Although his subjects' scientific careers are covered in meticulous detail, he never loses sight of them as individuals. His readers experience real intimacy with his subjects and gain a real sense of having known them personally.

Perhaps inhibited by living under the shadow of the great Victorian evolutionist, Leonard never exhibited the flamboyant Huxley inventiveness or intense commitment to the pursuit of new truths. Yet he not only chronicled his father's legacy, but successfully imbued his sons with the tradition. Leonard wrote:

> Unreason in every form is the enemy of scientific method, and the victory of science . . . means the gradual banishment of . . . many fancies . . . which survived to form a beautiful if misty background to everyday thought . . . Is it then true [that the world was robbed] of its illusions, its beauty, its aspirations, and given in their stead naked fact, mechanical order? . . . Beauty does not rest in untruth, nor is the loveliness of a landscape less appreciated by [understanding] perspective. The knowledge which destroys false beauties enthrones new ones . . .

Leonard's sons, to whom he passed on the family quest, were the biologist Julian and the novelist Aldous. Their grandfather had helped to locate man's place in nature, and now they were asking what man was going to do about it.

As scientists, writers and administrators, succeeding generations of Huxleys applied evolutionary knowledge to practical matters, continued to probe the question of mind and matter, plunged into controversial social and environmental issues and pondered the relation between science, ethics and religion—asking always what beliefs should govern men's actions.

See also HUXLEY, ALDOUS; HUXLEY, SIR JULIAN S.; HUXLEY, THOMAS HENRY.

For further information:
Clark, Ronald W. *The Huxleys.* New York: McGraw-Hill, 1968.
Huxley, Leonard, ed. *Life and Letters of Sir Joseph Dalton Hooker.* 2 vols. London: J. Murray, 1918.
———. *The Life and Letters of Thomas Henry Huxley.* London: Macmillan, 1900.

HUXLEY, THOMAS HENRY (1825–1895)
Evolutionary Biologist

Thomas Henry Huxley is remembered mainly as Darwin's combative champion and defending knight: a forceful speaker, witty writer and debater who gladly battled bishops and scientific foes to establish evolutionary biology. He is also known as the founder of an intellectual lineage, which includes his grandsons, novelist Aldous and biologist Julian. Not so well known is his lifetime of solid accomplishment in many fields.

Huxley was a first-rate comparative anatomist and paleontologist, an educator who established lab courses in colleges, a philosopher of science, invertebrate zoologist and popular essayist. He was also a medical doctor (though he never practiced), an anthropologist and a tireless campaigner for freedom of thought. Huxley often said he would rather not have students hold his ideas, if they believed them for the "wrong reasons," such as uncritical conformity. "Every great truth," he wrote, "begins as heresy, and ends as superstition."

Born at Ealing, a suburb of London, on May 4, 1825, he was the youngest of seven children. His father was a schoolmaster, but the family moved when Thomas was young, and he received little guidance in his early education. Although he wanted to be a mechanical engineer, his two brothers-in-law were doctors, and the family persuaded him to follow suit. One got him interested in human anatomy, and he soon became a highly skilled dissector and teacher. However, he believed he was infected at his first dissection (though he wasn't cut) and suffered afterwards from what was then called "a form of chronic dyspepsia."

After a stint at London University, he won a scholarship to Charing Cross Hospital medical school, where he gained a knowledge of comparative anatomy far beyond what was required of a physician. Later, he remarked that his boyhood interest in mechanical engineering had been transferred to organisms. He wanted to take them apart and analyze their structures as living machines. About that time (1845), he discovered a layer of cells in the root-sheath of hair, which is still known as Huxley's layer.

He applied for an appointment in the Royal Navy, and was made assistant surgeon on the H.M.S. *Rattlesnake*, which was about to survey the seas between Australia and the Great Barrier Reef. During the voyage (1846–1850), Huxley studied delicate ocean creatures (hydrozoa, tunicates, molluscs), which float near the surface, but do not last long for study. On the basis of his morphological work, he reclassified hydrozoa with sea anemones and corals, establishing them in a separate class.

Huxley's paper on the family of medusae (1849) also contained the important and original observation that two layers of cells in these primitive, hollow sea creatures were similar to an early stage of vertebrate embryos. This discovery took on added importance some years later, when evolutionists sought evidence for the "recapitulation" of evolution (phylogeny) in the development of individuals (ontogeny).

On his return to England, Huxley gained recognition and was made a fellow of the Royal Society. Although he continued to make important discoveries in comparative anatomy, he could not earn a living in science, and at times he despaired. Until the late 19th century, science was mainly the province of the wealthy or those subsidized by them, and few salaried positions existed.

Finally, Huxley was hired as a lecturer in natural history at the Royal School of Mines, and also was paid naturalist on a geological survey. In 1855, he felt settled enough to marry and sent for Henrietta Heathorn, whom he had met in Sydney, Australia while on the *Rattlesnake* voyage. She had been waiting for him for seven years during which they conducted a long-distance courtship by letter. Her health was so poor they both wondered if she'd still be alive by the time he was able to support a family. They were married in 1855, and Henrietta proved a strong and able wife.

Huxley first crossed swords with Richard Owen, the reigning anatomist and paleontologist at the British Museum, over the structure of the skull. Owen insisted the skull was developed from an enlarged vertebra, according to the theory of archetypal plans, which he had learned from Lorenz Oken and Johann Goethe. In a brilliant demonstration (1858), Huxley showed that the skull was different in structural development from the spinal column and that Owen had distorted the facts of anatomy in order to make his case. By decisively proving the senior anatomist wrong, he earned Owen's lasting enmity.

Huxley's life was transformed by a book: Darwin's *Origin of Species*, published in 1859. Prior to that time, he had not thought there was enough evidence to support a "development hypotheses" (evolution) or a plausible theory of how species could change.

Darwin's work hit him like a revelation. "How extremely stupid not to have thought of that myself," he remarked. Fortunately for the theory, it was Huxley who was asked to review Darwin's book for the influential *London Times*. His enthusiastic reception got the *Origin* off to a promising start, despite harsh critics—including the spiteful Owen.

That same year, Owen read a paper at Cambridge declaring that the human brain was structurally different from that of apes. Man's brain, he said, has a backward projection of the cerebral hemispheres, that covers the cerebellum and wraps around a "hippocampus minor," utterly different from other primates. Huxley rose at the meeting, stated that Owen was clearly wrong and that he would soon prove it.

In 1861, Huxley published two essays that showed Owen's statements to be contrary to well-known facts of anatomy; the human brain has no structures setting it apart from those of gorillas and chimps. Huxley expanded the work into *Zoological Evidence as to Man's Place in Nature* (1863), his most important book, which brought him into the public eye, where he remained for the rest of his life.

He became even more famous that year for his part in the famous "Oxford Debate" with Bishop Samuel Wilberforce, who had attempted to ridicule the new evolutionary ideas by asking Huxley if he was descended from an ape on the side of his grandfather or grandmother. Huxley's reply that he'd prefer an ape ancestor to a man "who used great gifts to obscure the truth" caused a sensation. (The inaccurate news went out that Huxley had said he would rather be an ape than a bishop.) Of the chemist Joseph Priestley, one of his heroes, Huxley once wrote: "There are men to whom the satisfaction of throwing down a triumphant fallacy is at least as great as that which attends the discovery of a new truth . . . and who are even more for freedom of thought than for mere advancement of knowledge." It was equally a description of himself.

An admirer of the philosopher Rene Descartes, Huxley believed doubt should have a high place in the life of the mind, not condemned as "grievous sin," and thought it appropriate that his given name was that of the doubting apostle. He wrote that it was "a man's most sacred act" to say what he believed to be true and that he would not lie where there was no clear evidence for belief. Accordingly, he coined the term "agnostic" to describe his lack of proof for or against the existence of God, and the term became part of the language. Some religionists have called agnosticism "a coward's way of being an atheist," but few have taken the trouble to appreciate Huxley's precision of thought. He well understood the limitations of the mechanistic philosophy and was not the advocate of "godless materialism" he has often been painted. Actually, Huxley had a strong tendency to mysticism about ultimate causes, but insisted materialism was simply the most workable assumption science could adopt for its purposes.

In all things, he advocated a healthy, active doubt, and—even in the depth of grief when his infant son died—refused to take comfort in "the hopes and consolations of the mass of mankind." In a famous letter to the Reverend Charles Kingsley, who had asked if he wasn't now sorry he didn't have religion, he replied, "My business is to make my aspirations conform to the facts, not to make the facts conform to my aspirations . . . I refuse to believe that the secrets of the universe will be laid open to me on any other terms."

Yet while he denounced literal belief in demons or miracles, Huxley clearly recognized that some eternal truths lay back of the prevailing religious mythology. "It is the secret of the superiority of the best theological teachers to the majority of their opponents," he wrote, "that they substantially recognize [certain] realities of things, however strange the forms in which they clothe their conceptions." He included among them:

> The doctrines of predestination; of original sin; of the innate depravity of man and the evil fate of the greater part of the race; of the primacy of Satan in this world . . . Faulty as they are, [they] appear to me to be vastly nearer the truth than the "liberal" popular illusions that babies are all born good and . . . that it is given to everybody to reach the ethical ideal if he will only try; [and] that everything will come right (according to our notions) at last.

Huxley's anatomical studies led him to unite reptiles with birds, a recognition of similarities between the two groups that has been enjoying a new vogue a century later. He did a great deal of paleontological work on dinosaurs and on classification of mammals, identified "crossopterigian" fishes and was active in ethnology and physical anthropology.

He also gave much attention to writing and thinking about the implications of evolutionary theory, not only for science but for philosophy and ethics. At a time when "Social Darwinists" insisted "natural law" sanctified elimination of the weak by the strong, Huxley came out with a different view. Because we are human, he argued, we have the intelligence and duty to fight against accepting "survival of the fittest" as our ethic. We have the option of being a nurturing, protective and compassionate species, deliberately turning our backs on the law of the jungle to create a humane society.

Debating the Bishop of Oxford shaped Huxley's

early reputation as a feisty, articulate scientist, but that was not his favorite type of contest. What he really liked was to master specialties far removed from his own fields of zoology and physiology and beat the experts at their own game. When theologian-naturalist St. George Mivart became one of Darwin's most adept and troublesome critics, Huxley took him on. But instead of arguing from a scientific base, he attacked with a virtuostic command of Roman Catholic theology, trouncing Mivart on his own turf. Similarly, with William Gladstone and others, he delighted not in biological demonstrations, but in refuting their fuzzy biblical, historical, and philosophical scholarship.

True to form, when Huxley responded to the insistent challenges of Spiritualists to disprove their claims of communication with the dead, he outfoxed them at their own tricks. He had learned that the Spiritualist movement started with two teenage sisters who produced phony "spirit raps" with their feet. Through long practice, Huxley became adept at loudly snapping his second toes inside his boots. Whenever his sense of humor moved him, he announced that he, too, had the power of summoning the spirits. Then, as if from nowhere, would follow a staccato of mysterious knocking sounds. Psychics were confounded, and true believers astounded.

In 1874, Darwin had asked Huxley to investigate fraudulent "psychics" who were swindling his brother-in-law. He expected Huxley to be a sitter or observer; it would never have occurred to him to suggest also learning how to become a *performer*. But such was the playful unorthodoxy of Huxley's rise to a challenge, he could not resist dissecting the seance from both viewpoints. "In these investigations," he concluded, "the qualities of the detective are far more useful than those of the philosopher."

In later years, Huxley took on many duties, including administrative positions in universities and government and, for a time, even served as director of the Royal Fisheries. But increasing bouts of ill health forced him into retirement; he died at Eastbourne on June 29, 1895. His witty and interesting letters were collected by his son Leonard in *The Life and Letters of Thomas Henry Huxley*, published in 1900; his great body of essays appears in a nine-volume work *The Collected Essays* of Thomas H. Huxley, which first appeared in 1893–1894.

Although Huxley is remembered as "Darwin's Bulldog," spreading the secular gospel of Darwinism, he actually had a few difficulties with Darwin's ideas. Not the basic idea of evolution, which he regarded as established fact, but Darwin's assertions about the universality of natural selection and gradualism.

He particularly told Darwin he thought his idea of

evolutionary change occurring everywhere at a slow, gradual rate was an "unnecessary burden" for the theory to bear. His reading of the fossil record impressed him with the fairly sudden appearances and rapid divergence of various classes and families follow by long periods of stability. Recent reexaminations of the evidence, in the light of "punctuational" theory, shows the acuity of Huxley's observations.

Also, a century of discussion, debate and experiment has still not settled some of Huxley's fundamental questions about the process of natural selection. Scientists are still trying to understand how it works and to what extent it satisfactorily accounts for a wide range of evolutionary phenomena.

See also AGNOSTICISM; GADARENE SWINE CONTROVERSY; MAN'S PLACE IN NATURE; MATERIALISM; OXFORD DEBATE; PUNCTUATED EQUILIBRIUM; SPIRITUALISM; WELLS, H. G.

For further information:

Ayres, Clarence. *Huxley.* New York: Norton, 1932.

Bibby, Cyril, ed. *The Essence of T. H. Huxley.* New York: Macmillan, 1967.

Clark, Ronald W. *The Huxleys.* New York: McGraw-Hill, 1968.

Clodd, Edward. *Thomas H. Huxley.* Edinburgh: Blackwood, 1902.

DiGregorio, Mario. *T. H. Huxley's Place in Natural Science.* New Haven: Yale University Press, 1984.

Huxley, Leonard, ed. *The Life and Letters of Thomas Henry Huxley.* London: Macmillan, 1900.

Huxley, Thomas H. *Collected Essays.* Macmillan, 1885–1890.

———. *Evidence for Man's Place in Nature.* London: Williams & Norgate, 1863.

Irvine, William. *Apes, Angels & Victorians; A Joint Biography of Darwin and Huxley.* London: Weidenfeld & Nicholson, 1956.

HYBRIDIZATION
Cross-Breeding of Species

A major characteristic of a species is that its members do not normally mate or produce fertile offspring with members of other species. Nevertheless, the fact that fertile hybrids are sometimes possible shows that species are populations capable in greater or lesser degrees of genetic compatability with related populations.

In the wild, species do not interbreed for many reasons besides genetic differences. For instance, they may be separated by rivers or mountain ranges and thus not have the opportunity. In fact, this enforced separation between populations is thought to be a major cause of species formation. When the geographic barriers are removed, the isolated populations may have grown so different that they are genetically incompatable. Where the populations have not diverged too widely, removal of barriers may promote hybridization, in some cases forming new species.

Charles Darwin made thousands of tests on hybrids of different varieties of plants and studied the results of crossing in domestic pigeons, hogs and horses. However, the underlying mechanisms of heredity, which were being discovered by the obscure monk Gregor Mendel during Darwin's lifetime, were never known to him. Nevertheless, Darwin groped toward a solution and published three books on variation and hybridization experiments.

One major puzzle he found insoluble was that he could not distinguish between what are today called "single-gene" traits, as in Mendel's peas, and other, more complexly caused (polygenic) traits, such as blue eyes or intelligence. Other mechanisms known today, such as recombination and pleiotropy, were not even suspected. Without a good model of genetic causes, Victorian evolutionists incorrectly combined the effects of several mechanisms in their so-called "blending inheritance" of traits from father and mother, an idea which was later disproved [See PAINTPOT PROBLEM.]

In the 1930s, experiments were done to breed organisms farther and farther away from a founding population until a "new species" was artificially created. A variety was produced that would "breed true" but would no longer cross with the closely related population from which it was derived. J. B. S. Haldane wrote in the 1940s that such man-made "new species" had been developed in tomatoes and fruit flies by British geneticists.

It is not known just how "distant" two related species can be and still produce hybrid offspring. Domestic and wild animals have produced interesting and sometimes useful (to man) hybrids. Successful crosses have been made between cattle and bison ("beefalo"), turkeys and chickens ("turkens") and horses and zebras. Usually, the male offspring of these unions are sterile, and females are either sterile, show reduced fertility or produce offspring that do not live long.

One of the most familiar of the hybrid domesticated mammals is the mule, a cross between a horse and a donkey. Mules have the sure-footedness of donkeys, the strength of horses, and a uniquely bright and stubborn personality all their own that has endeared them to generations of farmers. However, mules have to be cross-bred from horses and asses each time; as classic hybrids, they cannot reproduce themselves.

A compilation of *Mammalian Hybrids* (1987) by Annie T. Gray lists 573 crosses between different species and varieties of mammals that have produced offspring. Most crosses take place in zoos, since breeding between species rarely occurs in the wild. A most common zoo cross is the "ligon" or "tigler," offspring of a lion and tiger. The best known captive hybrid that produced fertile offspring was fathered by a male polar bear (*Ursus matrituimus*) who mated with a female kodiak bear (*Ursus arctos midenforffi*) at Washington's National Zoo. She gave birth to Pokodiak, a female cub, in 1936, and a male, Willie, in 1939.

Eleven years later, Pokodiak and Willie produced a male cub, Gene, which has lived for many years at the zoo.

See also "BEANBAG GENETICS"; MENDEL, ABBOT GREGOR; TRANSITIONAL FORMS.

HYPERROBUST AUSTRALOPITHECINE

See AUSTRALOPITHECUS BOISEI.

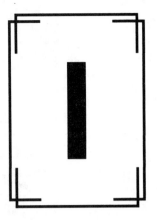

ICE AGE
Great Climatic Shifts

Harvard zoologist Louis Agassiz, the leading light of American biology, scorned Charles Darwin's theories of evolution as "impossible" when they were published in 1859. However, 20 years earlier, Agassiz himself had launched a major unorthodox theory to which Darwin was an early convert—the astounding idea that great continental glaciers had periodically advanced and retreated during prolonged Ice Ages.

Although at first skeptical, young Darwin was soon won over by Agassiz's radical ideas. He later wrote in his *Autobiography* (1876) that, as a student on holiday, he had gone "geologising" with Adam Sedgwick in Wales and "had a striking instance how easy it is to overlook phenomena, however conspicuous, before they have been observed by anyone." Following Sedgwick's lead, he had spent the day examining rocks for fossils:

> but neither of us saw a trace of the wonderful glacial phenomena all around us; we did not notice the plainly scored rocks, the perched boulders, the . . . terminal moraines [long heaps of pebbles and gravels]. Yet . . . a house burnt down by fire did not tell its story more plainly than did this valley. If it had still been filled by a glacier, the phenomena would have been less distinct than they are now.

As the Ice Age theory began to take hold in geology, it also posed a great mystery. What drives glacial cycles? In 1842, only a few years after Agassiz had announced his conclusions, Joseph A. Adhemar, a French mathematician, suggested cold periods might be caused by long-term changes in the position of the Earth relative to the sun. But it was not until the 1920s that Milutin Milankovitch, a Yugoslav astronomer, worked out the idea in detail.

Backed by a long and intricate series of calculations, Milankovitch proposed a curve of fluctuations in global climate governed by three factors: the slant of the spinning Earth relative to the sun (it regularly tilts a couple of degrees back and forth every 41,000 years), the shape of its orbit (which elongates and contracts every 100,000 years) and the precession, or wobble, of the planet (a circular movement completed every 23,000 years). Milankovich showed that these three factors could vary the amount of sunshine reaching Earth by about 20%—enough to account for the formation and melting of great ice sheets. It was an intriguing theory, but no one was able to test it against a complete record of past glaciations for many years. Over much of the Earth, the most recent advances and retreats of the ice had obliterated the evidence of previous episodes.

However, beginning in the 1950s, Cesare Emiliani, working at the University of Chicago, found a new way of obtaining such a record—not on land, but in seafloor sediments in the shells of tiny marine organisms called foraminifera. Emiliani showed that the proportion of radioactive oxygen isotopes in their remains reveals the composition of sea water at the time they were alive. Since sea water was chemically different than it is today at various periods in the past, the composition of these ancient foram shells can tell a good deal about glacial cycles.

Fluctuations in isotope levels within them, it turns out, correlate with the percentage of the world's water that is locked up in ice sheets. By the time the vapor from warmed surface water is carried by winds to colder regions, it has redeposited most of its heav-

EXCELLENT ARTISTS who lived during the last Ice Age, some 30,000 years ago, left paintings and etched drawings of bison, mammoths, European wooly rhinos, extinct horses and fat women. Neither their skeletons nor their works indicate any important differences from modern humans.

ier isotopes at sea as rain, leaving the lighter ones to fall as snow. Cores drilled from the sea-floor showed an isotopic pattern that correlated with Milankovitch's curves, and, as others demonstrated in the 1970s, the regular fluctuations in the Earth's orbital movements fit right into this emerging picture.

Many kinds of evidence corroborate existence of the cyclic pattern: global ice volume has peaked every 100,000 years during the past million or so. After each peak, the ice rapidly retreats for a few thousand years, causing a warming or interglacial period, such as the one we are in right now that began about 14,000 years ago.

However, the mechanisms that cause ice ages are not as simple as the interaction of regular planetary movements or differential sunlight or rainfall. Scientists are only beginning to discover the dauntingly complex relationships between oceans and atmosphere, living things and circulation of global currents, distribution of carbon and oxygen. The Earth is no dead ball of rock whirling through space, but a dynamic system whose intricacies we are just beginning to glimpse.

For instance, we now know that Ice Age air (some of which is still trapped in ice bubbles formed thousands of years ago) contains one-third less carbon dioxide than interglacial air. It has also been recently found that most (60%) of the world's carbon dioxide is in the ocean, captured by tiny ocean plants at the surface and "pumped" into the depths by the food chain. Since ocean currents profoundly affect the gas exchange process, it seems that, as geochemist Wallace Broecker and geologist George Denton put it, "only a major shift in the ocean's operation could account for such a dramatic change in atmospheric composition."

One recent geochemical study at MIT (again of foraminifera fossils, but this time of other chemicals in their composition) offers independent confirmation that ocean currents must have circulated differently during the last glaciation. Currents act like conveyer belts for heat, transferring vast quantities of solar energy gathered far from land to the continents, where oceanic winds strongly affect terrestrial climate and vegetation. These currents, in turn, are set in motion by complex relationships between the distribution of sea salt, differential formation of water vapor and the transport of fresh water from melting glaciers.

In Broecker and Denton's view, transitions between glacial and interglacial conditions represent jumps between two stable but very different modes of ocean-atmosphere operation. There is a "flip-over" effect that triggers the operation of a different system of forces; the complexities of these systems are still only dimly understood. But their basic proposition is that if there are really clear jumps between two states, "all climate indicators should register a transition simultaneously."

Evidence from different parts of the world all seems to tie in. Warmer North Atlantic surface water, melting of northern ice sheets as well as Andean mountain glaciers, reappearance of European forests, changes in plankton around Antarctica and the South China Sea—all point to around 14,000 years ago as the end of the last Ice Age.

Some scientists have expressed concern that the present interglacial phase may be artificially extended by the warming effects of human industrial activity, burning of fossil fuel and destruction of rain forests. It is still too early to calculate such effects, and,

indeed, real understanding of how Ice Ages are produced is still far from complete.

See also GREENHOUSE EFFECT; LASCAUX CAVES; MAMMOTH; NEANDERTAL MAN.

For further information:

Hadingham, Evan. *Secrets of the Ice Age: A Reappraisal of Prehistoric Man.* New York: Walker, 1979.

Sutcliffe, Antony J. *On the Tracks of Ice Age Mammals.* Cambridge: Harvard University Press, 1985.

White, Randall. *Dark Caves, Bright Visions: Life in Ice Age Europe.* New York: Norton, 1986.

ICHNOLOGY

See "NOAH'S RAVENS."

IGUANODON DINNER
Victorian Dinosaur Celebration

To celebrate New Year's Eve, 1853, 21 famous Victorian scientists were served a seven-course meal inside the belly of a giant *Iguanodon* statue. The occasion was the opening of the first dinosaur exhibit at the Crystal Palace exhibition at Sydenham. Dramatic dinosaur replicas did much to stir public interest in the approaching Darwinian revolution in sci-

THE IGUANODON DINNER, New Year's Eve, 1853. Anatomist Richard Owen (left, in head) coined the term "dinosaur," then gave a dinner at the Crystal Palace to celebrate the completion of his iguanodon reconstructions. The huge models, sculpted by Benjamin Waterhouse Hawkins, can still be seen at Sydenham Park in Southeast London.

ence, though at the time, many viewed the monsters as casualties of Noah's flood.

The life-sized iguanodon, constructed of metal and concrete, was the work of Benjamin Waterhouse Hawkins, the first great reconstructive dinosaur artist. Dinosaurs had been known only for a few decades; they had been discovered in Kent and given their name by the cantankerous chief anatomist of the British Museum, Richard Owen. At the dinner, with his usual charm and tact, Owen made a speech attacking Hawkins for "getting the iguanodon wrong," since recently discovered fossil tracks suggested the animal really walked upright on its two hind legs. Yet, Owen himself had supervised Hawkins while he built the dinosaur models, and so earned his place literally at the head of the table inside the iguanodon's head.

Invitations were inscribed on the wing-bone of a pterodactyl and a special song was composed for the occasion, with the chorus: "The jolly old beast/Is not deceased/There's life in him again." The dinner was a great success, although it was not the first of its kind. In 1801, the American artist and museum owner Charles Willson Peale had assembled the first skeleton of a mastodon at his Museum in Philadelphia. In December of that year, Peale invited 12 distinguished guests to join him for dinner inside the immense skeleton, where they made patriotic toasts and sang "Yankee Doodle."

See also DINOSAUR RESTORATION; HAWKINS, BENJAMIN W.; JEFFERSON, THOMAS; OWEN, SIR RICHARD; PEALE'S MUSEUM.

IMPACT THEORY

See ALVAREZ THEORY; DINOSAURS, EXTINCTION OF; NEMESIS STAR.

INDEX KEWENSIS (1892–1895)
Greatest Plant Catalogue

Charles Darwin's last written contribution to the study of natural history was not a book but a bank cheque—to underwrite the "botanist's bible," the *Index Kewensis*.

For years the Royal Botanic Gardens at Kew, near London, had kept up their lavishly illustrated catalogues of all known species of flowering plants. By 1880, however, they were hopelessly out of date. Victorian naturalists and botanists were adding newly discovered species so rapidly that a completely revised world inventory was needed.

Darwin's good friend Sir Joseph Hooker, director of the institution, could not raise government backing and appealed to the public for financial assistance. In January, 1882, Darwin sent him £ 250 to begin the

project and told his children to continue supporting the work with annual payments if he didn't live to see its completion. He died soon after, in April of that year, at the age of 73. His estate continued to underwrite costs of the work, and, within a few years, the manuscript of the Index weighed two tons.

The *Index Kewensis* is part of Darwin's lasting scientific legacy: Now computerized, it is still in use today and is continually kept current for more than 200,000 species. Some botanists have commented on the irony that the great evolutionist—who convinced the world that species are unfixed, changeable entities—should have funded an immense, definitive species list as his final gift to science.

See also HOOKER, SIR JOSEPH D.

"INDIAN RING" SCANDAL

See (CHIEF) RED CLOUD.

INDO-EUROPEAN (ROOT LANGUAGE)

See QUEST FOR FIRE.

INDUCTION
From Fact to Theory

Induction as a scientific method is often credited to Sir Francis Bacon. He believed that careful observation of many particular instances of a phenomenon would lead to a pattern, a general law or principle. While *deduction* follows a train of premises, *induction* emphasizes experience as the source of knowledge.

Charles Darwin claimed to have created his theory of evolution by means of natural selection "according to the true Baconian principles of induction." In fact, he did no such thing, but it was customary in his day that a true scientist worked by finding patterns or laws by gathering great quantities of "facts."

Even before Darwin's time, the Scots philosopher David Hume (1711–1765) had demonstrated that induction was not a surefire method of finding truth. One major problem is what is to be observed? If the scientist has no idea what he is looking for, he cannot know where to focus his attention or what facts to gather. Darwin himself once noted with amusement that "most people don't understand that all observations must be directed towards proving or disproving some particular view if they are to be of service."

See also DEDUCTION; SCIENCE; THEORY, SCIENTIFIC.

INFANTICIDE
Nature's Murder Mystery

Baby-killing is commonplace in nature. Fathers, mothers, siblings, close relatives, parent's new mates—all may destroy infants or eggs under certain circumstances. Infanticide has been observed among rotifers, ants, guppies, swallows, lions, rats, monkeys, acorn woodpeckers, apes, prairie dogs and humans, to name just a few. Since it removes genes from the next generation's population, the widespread phenomenon must affect evolution, but how?

Only a few years ago, scientists thought infanticide was an abnormal behavior produced by crowding or stress in captive animals. However, during the 1970s, field studies showed it to be common among wild populations. Can this behavior be beneficial to species? What is the effect on sex ratios, population structures, reproductive strategies? These are currently "hot" research topics in evolutionary biology.

One pattern has already emerged in species where social groups consist of "harems" of one or two males and many females, such as lions and chimpanzees. When males from an outsider group attack, they attempt to drive away or kill resident adult males. If successful, they slaughter the infants and mate with the females.

Lactating females usually become fertile again and quickly conceive. Also, they accept the situation in a manner foreign to human empathy, never rising up in outrage against the invading males who killed their offspring nor resisting their sexual advances.

Interest in the adaptive significance of infanticide was provoked by Sarah Blaffer Hrdy's work (1970s) on Indian Hanuman monkeys (langurs), but her theories turned out to be much more applicable to African chimps and lions. Invading males, after killing some infants, end up siring many more, thus perpetuating their genes. Often offspring of the invaders seem more vigorous than those produced by a closely inbred social group.

Infanticide in nature takes many forms. When a female poison-arrow frog comes upon a male brooding a nest for another female, she crushes the eggs, then mates with him. In black eagles, the first-hatched (larger) chick harasses its sibling, monopolizes the food, and the second starves to death. Among hyenas, a dominant female may harass a subordinate nursing mother until her cubs die, freeing her to help nurse the dominant female's litter.

In many species, the evolutionary mechanisms favoring infanticide are extremely puzzling. A seven-year study of black-tailed prairie dogs by John Hoagland (1985) found that more than half of all litters born were destroyed by lactating females killing their own sister's pups. Walter Koenig (1970s) observed that female acorn woodpeckers—another animal that lives in large colonies—smash eggs laid by close relatives. Whether there may be evolutionary or genetic benefits to the colony of such "kin-directed infanticide" is still entirely unknown.

For further information:
Hausfater, Glenn, and Hrdy, Sarah B. *Infanticide: Comparative and Evolutionary Perspectives.* New York: Aldine, 1984.

INHERIT THE WIND (1955)
Dramatization of Scopes Trial

A tense courtroom drama based on the famous "Tennessee Monkey Trial" of 1925, *Inherit the Wind* fascinated New York City theater audiences and went on to international fame. Based on a science teacher's clash with local religious beliefs, its overarching theme was freedom of thought.

At the historic trial, John T. Scopes was prosecuted for presenting Darwin's theory in a high school biology class in defiance of a state antievolution law. He was defended by Clarence Darrow, a flamboyant attorney with a passion for progressive ideas. An equally charismatic figure, William Jennings Bryan, represented militant fundamentalists for the prosecution. National press attention was focused on the little town of Dayton; the *Baltimore Evening Sun* sent a young journalist named H. L. Mencken.

Inherit the Wind was written in 1951 by Jerome Lawrence and Robert E. Lee. After a brief, successful tryout in Dallas, it opened in New York City on April 21, 1955 to rave reviews and settled in for an extended run. The Bryan character (called Brady) was played by Ed Begley, the journalist by Tony Randall and Darrow (Drummond) by Paul Muni.

Stanley Kramer made a classic motion picture version (1960) starring Spencer Tracy as the Darrow/Drummond character, Fredric March as Bryant/Brady, and Gene Kelly as the cynical newspaper man. In one of Drummond/Darrow's most memorable speeches, he argues that a loss of cherished beliefs may be the price of new knowledge:

> . . . progress has never been a bargain. You've got to pay for it. Sometimes I think there's a man behind a counter who says, "All right, you can have a telephone, but you'll have to give up privacy, the charm of distance . . . Mister, you may conquer the air but the birds will lose their wonder and the clouds will smell of gasoline! . . .

The play's title is from Proverbs 11:29, "He that troubleth his own house shall inherit the wind." Although the characters and conflict were clearly based on the Scopes case, only a few bits of dialogue were taken directly from the trial transcript. Playwrights Lawrence and Lee insisted in a preface that their play "does not pretend to be journalism." The stage directions set the time of the trial as "Not too long ago. It might have been yesterday. It could be tomorrow."

INHERIT THE WIND, the 1960 film classic based on the Scopes trial in Dayton, Tennessee, featured Spencer Tracy (in suspenders) as the Clarence Darrow character, Gene Kelly (left) as the cynical newsman based on H. L. Mencken, and Fredric March as the champion of biblical literalism.

See also BUTLER ACT; CREATIONISM; FUNDAMENTALISM; SCOPES TRIAL.

IN MEMORIAM

See TENNYSON'S IN MEMORIAM.

INSECTIVOROUS PLANTS
Darwin's "Disguised Animal"

> During the summer of 1860, I was surprised by finding how large a number of insects were caught by the leaves of the common sun-dew (*Drosera rotundifolia*) on a heath in Sussex. I had heard that insects were thus caught, but knew nothing further on the subject. I gathered by chance a dozen plants, bearing fifty-six fully expanded leaves, and on thirty-one of these dead insects or remnants of them adhered . . .

So begins Charles Darwin's 14th book, *Insectivorous Plants* (1875), the first scientific investigation into the fascinating behavior of carnivorous plants. Sometimes neglected among his many achievements in natural history, Darwin was the first to prove that Venus fly-traps, sundews and pitcher plants are ac-

tually meat-eaters. Like so much of his research, this "fine new field for investigation" began with "idling" observations while "resting" from other projects.

Darwin soon developed a special affinity—even defensive affection—for these remarkable organisms. When American botanist Asa Gray suggested he might better occupy his time, Darwin wrote (1863): "Depend on it you are unjust on the merits of my beloved *Drosera*; it is a wonderful plant, or rather a most sagacious animal. I will stick up for *Drosera* to the day of my death." Venus fly-traps, with their hinged, spiked leaves that snap closed on insect victims, Darwin thought "the most wonderful plant in the world." His wife, Emma, wrote a friend, "He is treating *Drosera* [the sundew plant] just like a living creature, and I suppose he hopes to end in proving it to be an animal."

Emma was not exaggerating. In addition to Darwin's stated aim of throwing light on possible transitional steps between the fly-trap and ordinary plants, he seems to have had bridging the plant-animal gap in the back of his mind. "By Jove," he wrote to his friend the botanist Joseph Hooker, director of Kew

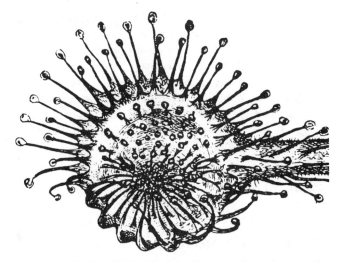

ROUND-LEAFED SUNDEW plant (Drosera) closes its tentacles over a bit of meat in an illustration from Charles Darwin's book *Insectivorous Plants*. He became so absorbed in the study he wrote a friend "at the present moment, I care more about Drosera than the origin of all the species in the world."

Gardens, "I sometimes think *Drosera* is a disguised animal."

Darwin's was the first extensive physiological-behavior study of insectivorous plants, covering 30 species belonging to eight genera. He wanted to get at three basic mechanisms: the power of movement of the plants when capturing prey, the secretions of glands for digesting insects and the absorption of the digested matter. His thousands of experiments resulted in the first definitive evidence—where before had been only speculation—that these plants really were getting nutrition directly from capturing animal life.

No novice to plant movement (he had published his book on *Movement in Climbing Plants* (1865) some years before), Darwin found both the general sensitivity of these plants and their finely coordinated movements to be remarkable. After all, they had neither muscles nor nervous system. "It appears to me that hardly any more remarkable fact than this has been observed in the vegetable kingdom," he marveled.

While admitting that the coordinated movement of plants was inferior to that of animals. Darwin envisioned them as a possible early stage in evolution of animal nervous systems. "But," he says, "the greatest inferiority of all is the absence of a central organ, able to receive impressions from all points, to transmit their effect in any definite direction, to store them up and reproduce them." In other words, they lacked a brain!

Experiments on the adaptive significance of carnivory were assigned to Darwin's son Frank, who grew two groups of plants. One he fed insects; the other had to take its sole sustenance from the soil. The insect-fed plants were more vigorous and produced more flowers and seeds. "The results," Frank concluded, "show clearly enough that insectivorous plants derive great advantage from animal food"—a finding confirmed many times since.

Carnivory in plants is now recognized in eight families, 15 genera, and roughly 500 species, or about 0.2% of all flowering plants. Thus it is a very rare occurrence; why this is so is not understood, and the best that has been done is to correlate habitat with carnivory. In all cases, insect-eating plants are found in moist, nutrient-poor, usually acidic soils, often in sunny locations.

Perhaps the most exotic habitats of carnivorous plants occur on the "Lost World" *tepuis* (pronounced tep-POO-eez) of Venezuela: high, flat-topped stone mountains or mesas, whose long-isolated plants and animals helped inspire Sir Arthur Conan Doyle's famous adventure tale. But they are also commonly found in north temperate peat and fen lands, the little known hillside seep bogs of western Louisiana and east Texas, and the *Cephalotus* bogs in southwestern Australia—all of them characterized by a poor supply of nitrogen and other plant nutrients.

See also DARWIN, CHARLES; GRAY, ASA; HOOKER, SIR JOSEPH D.; (THE) LOST WORLD.

For further information:
Allan, Mea. *Darwin and His Flowers*. New York: Taplinger, 1977.
Darwin, Charles. *Insectivorous Plants*. London: Murray, 1875.
Juniper, B., Robins, R. J., and Joel, D. M. *The Carnivorous Plants*. London: Academic Press, 1989.

ISAAC, GLYNN (1937–1985)
Paleoanthropologist

A native of Capetown, South Africa, Glynn Isaac spent his life absorbed in the drama of African prehistory, though he later made his home base in the United States. Intelligent, energetic, sincere, he was a teacher (University of California, Berkeley and, later, Harvard) but always had one foot in the field.

He spent his first professional years (early 1960s) as warden of prehistoric sites in Kenya and deputy director to Louis Leakey at the Center for Prehistory and Paleontology in Nairobi. Later, he and Richard Leakey were coleaders of the Kenyan excavations at Koobi Fora, where they unearthed 20 archeological sites, some as old as two million years.

Isaac was most interested in reconstructing patterns of living of early hominids and what could be learned of their dietary habits. Unfortunately, he died suddenly at the age of 47 while returning to Harvard from a conference in China.

He will perhaps best be remembered as a conciliator between powerful egos, a dedicated, even-tempered researcher with no ax to grind, who helped bring together disparate personalities in the common cause of understanding human origins. He made no enemies. In the field of paleoanthropology, that is a formidable achievement in itself.

ISIS (INTERNATIONAL SPECIES INVENTORY SYSTEM)
Computer Mating Endangered Species

With the extinction of so many wild populations of animals within the past few decades, the only hope for survival for certain species is in captivity. Siberian tigers, snow leopards, black rhinos may soon disappear completely in the wild. But matings in zoos (or between animals kept in different zoos) has presented problems.

A gorilla named Oscar at the Columbus Zoo was mated for four years, but by 1980 had fathered infants that died soon after birth, had heart defects or were miscarried. Geneticists eventually realized what was wrong. Zoo officials were unknowingly mating Oscar with his sisters, and the inbreeding was causing problems. Among close relatives, there is a higher probability that both parents will share some defective genes, and weaknesses will intensify in the offspring. (There also tends to be an overall reduction in fertility.)

To manage captive populations, evolutionary biologists have set up a program to coordinate activities among North American zoos, and, increasingly, with institutions throughout the world. Now zoo directors can be confident their breeding programs will not result in accidental inbreeding. Zoos in New York and California now routinely swap animals like the rare Przwalski's wild horse with zoos in the Soviet Union to keep up genetic variability and vigor in the stock.

ISIS, the International Species Inventory System, keeps computerized records of all individual animals in participating institutions. A few punched instructions, and the printout reveals the parents and grandparents of a given tiger or panda and its mating history. The ISIS catalog contains more than 53,000 living animals at 175 facilities. Birth, death and breeding statistics are kept on another 40,000 ancestors of these animals. By 1987, a complete printout of the total ISIS information weighed about 50 pounds.

Modern computer technology can locate suitable individual animals needed to keep captive populations healthily reproducing, even while their wild cousins become scarcer or disappear. This last-ditch effort is no substitute for conserving natural habitat, but it has greatly increased the effectiveness and efficiency of breeding attempts by zoos. Optimistic founders of the system chose the acronym ISIS because it is also the name of the ancient Egyptian goddess of fertility.

See also GENETIC VARIABILITY.

ISLAND LIFE

See WALLACE, ALFRED RUSSEL.

ISLAND OF DR. MOREAU

See WELLS, H. G.

ISOLATING MECHANISMS
Species Mate Identification

By their choice of mates (and rejection of others in closely related species), animals can tell us that their breeding population is a real species, not something arbitrarily defined by the zoologist.

Some of the most unusual, spectacular structures and behaviors in the animal world seem to function mainly for species recognition and sexual attraction. The riotous color patterns on the faces of African guenons (masked monkeys), the hundreds of different sounds and aerial "dances" of Hawaiian fruit fly species and the elaborate courtship rituals among birds of paradise help keep related breeding populations genetically distinct. Hence, they are known as isolating mechanisms, a term coined by geneticist Theodosius Dobzhansky.

Such signals are not fail-safe. Hybrid crosses between varieties or even species do occur, though they are comparatively rare. Some creatures even seem totally heedless of these genetic "traffic signals." Male dogs, for instance, are notoriously indiscriminate as to where they direct their procreative energies. And twice during the 1980s, male moose in New England attracted national media attention for their unrelenting attentions to dairy cows.

See also DOBZHANSKY, THEODOSIUS; GENETIC DRIFT; SPECIES, CONCEPT OF.

ISOLATION THEORY

See MAYR, ERNST; ROMANES, GEORGE J.; WAGNER, MORITZ.

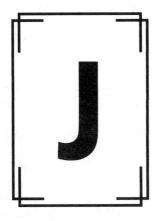

JANZEN, DANIEL

See ANT-ACACIAS; BELT, THOMAS; GUANACOSTE PROJECT.

JARAMILLO EVENT
Evidence for Plate Tectonics

In 1966, earth scientists first identified the Jaramillo Event, an ancient reversal of the Earth's magnetic field that took place 900,000 years ago. Recorded in rocks gathered at Jaramillo Creek, in New Mexico's Jemez Mountains, it was only the latest discovery in a long sequence of known magnetic polarity changes. But Jaramillo, as it turned out, was the crucial bit of evidence that triggered a revolution in geology as profound as that of Einstein in physics or Darwin in biology.

Competing American and Australian teams had been working for years to make sense of Earth's magnetic reversals and to connect them with some larger phenomena. To incorporate the evidence from Jaramillo, they had to construct an 11th version of the time scale based on these flip-flops, which had been pieced together from rocks all over the world.

At this point, American, Canadian, English, and French researchers (including Fred Vine, Lawrence Morley, Drummond Matthews, Victor Vacquier, Brent Dalrymple, Neil Opdyke, and others) realized there was a pattern to the sequence that exactly matched another they had seen: a recently compiled magnetic profile of the northern Pacific seafloor. It was like fitting in the one jigsaw puzzle piece that makes the whole picture understandable.

"I realized immediately," Fred Vine recalled, "that with the new time scale, the Juan de Fuca Ridge [in the Pacific, southwest of Vancouver Island] could be interpreted in terms of a constant spreading rate. And that was fantastic, because we realized that the record was more clearly written than we had anticipated. Now we had evidence of constant seafloor spreading."

Discovery of the Jaramillo Event, in conjunction with the data gathered in the East Pacific (1964) by the research vessel *Eltanin*, was the spark that set off a complete transformation of ideas in geology. If seafloors regularly spread, then "continental drift" was confirmed. As evolutionary theory had overthrown the fixity of species, so a century later plate tectonics overturned the ancient belief that continents and oceans were permanently locked into their present positions.

The Jaramillo/Eltanin pattern convinced many geologists that continent-sized slabs or plates of the Earth's crust are actually in constant motion. Since 1966, a rapidly growing body of research has documented that the Earth is made up of about 25 huge plates, which open and close ocean basins, thrust up mountains where they collide (as in the Himalayas) and even submerge under each other to be melted back into magma.

Plate tectonics is the grand unifying theory that made sense of a great body of previously unconnected geological data. It explains, for instance, why the shapes of continents—such as the outlines of eastern South America and western Africa—seem to fit together. (They *were* together, but split and drifted apart.) Similarly, it explains why some continents now widely separated in different climatic zones, contain identical kinds of rocks and fossil remains, such as the tropical *Lystrosaurus* found in India, South Africa, China and Antarctica.

See also CONTINENTAL DRIFT; LYSTROSAURUS FAUNA; PEER REVIEW; PLATE TECTONICS.

For further information:
Glen, William. *The Road to Jaramillo: Critical Years of the Revolution in Earth Science.* Stanford, Calif.: Stanford University Press, 1982.

JEFFERSON, THOMAS (1743–1826)
Presidential Paleontologist

Forty-nine Nobel Prize winners gathered in the White House on a spring evening in 1962 to be honored for their achievements in the arts and sciences. "I think this is the most extraordinary collection of human talent, of human knowledge, that has ever been gathered at the White House," said President John F. Kennedy, "with the possible exception of when Thomas Jefferson dined alone."

Jefferson was indeed a man of wide-ranging talents. He could design a graceful building, calculate an eclipse, tie an artery, argue a legal case, break a horse, survey an estate, dance a minuet or give one of the first scientific descriptions of a fossil mastodon skeleton found in America.

In the late 1700s, fossil bones of elephantine proportions were discovered in various sites in Europe and America, creating a sensation among intellectuals. What could they mean? How old were they? Benjamin Franklin and George Washington both expressed interest in the massive tusks and grinding teeth, but were puzzled. Could these bones, as Quaker naturalist Peter Collinson thought, be mixed remains of elephants and hippopotamuses, perhaps drowned in the Flood?

News of the finds greatly interested Jefferson, who had just published a description of ancient elephant teeth dug up near his Virginia property in his *Notes on the State of Virginia* (1784–1785). Then living in Paris as America's ambassador, he boldly contradicted the great French naturalists Georges Buffon and Jean Daubenton, who thought the bones were mixed up from different species. Jefferson argued that they represented one animal, in some ways similar to an elephant, but also in some ways different. However, he stopped short of concluding it was extinct. Irrevocable disappearance of one of God's creatures was still too irreverent a conclusion, even to a free-thinking deist. "Such is the economy of nature," Jefferson wrote in a 1799 paper describing the bones, "that no instance can be produced, of her having permitted any one race of her animals to become extinct; of her having formed any link in her great work, so weak as to be broken."

Even while engaged in his many activities as a planter, politician and statesman, Jefferson continued to seek additional fossils. In 1796, while serving as vice president under John Adams, he acquired some remains turned up by workmen excavating saltpeter in a western Virginia cave. These bones proved that still another previously unknown giant animal had existed in the area—a huge, clawed creature quite unlike the elephantine monster. Jefferson thought it a kind of lion and named it *Megalonix*, or great-claw, from its huge hooked nails.

A few years later, Dr. Caspar Wister and Jefferson jointly described it in the *Philosophical Transactions* of the American Philosophical Society. Wister correctly concluded that the animal was a giant sloth, similar to one that had been found in South America, and named the creature *Megalonix jeffersoni*. Jefferson again chose to believe that some of the creatures must still be alive somewhere, but extinction would soon become an unavoidable conclusion for him and all other fossil hunters.

Jefferson's fossils, among others, excited top anatomists, and inspired Georges Cuvier to tackle his comparative study of elephants, living and extinct. Over the next few years, the great French anatomist reviewed the evidence of fossil crocodiles, bears, rhinos and deer and concluded "they have belonged to creatures of a world anterior to ours, to creatures destroyed by some revolutions of our globe; beings whose place those which exist today have filled, perhaps to be themselves destroyed and replaced by others some day."

The craze was on for new digs, and farmlands across the country were eagerly dug up by many hands in search of "the great American incognitum," often with the interest and encouragement of then-President Thomas Jefferson. Finally, in 1801, Charles Willson Peale, curator of Peale's Museum in Philadelphia, successfully unearthed the first complete skeleton of a mastodon in New York state.

It was a patriotic triumph more than a scientific one. For years, American naturalists had smarted under the slurs of their European colleagues about the "degenerate" New World. Buffon and other Frenchmen had written that in America, dogs lost their bark, men their virility, and "All animals are smaller in North America than Europe." In reply, Jefferson had compiled a list of weights and measures of American bears, beavers and otters—all larger than their European counterparts. He had even sent Buffon a large stuffed American moose, but the Frenchman remained unimpressed. In 1803, his friend Peale crated up the complete, immense mastodon skeleton and took it to Europe for public exhibition. The awe-inspiring "mammoth" elephant came to symbolize

THOMAS JEFFERSON'S many interests included paleontology, which he helped establish in America after gigantic fossil bones were found on his estate. Since extinction of species was not yet accepted as possible, Jefferson believed these large unknown animals would eventually be discovered alive somewhere in the North American wilderness.

American pride. *Now* let the European naturalists dare to call North America a place of small, second-rate inhabitants! President Jefferson was well pleased.

See also EXTINCTION; MAMMOTH; SECULAR HUMANISM; PEALE'S MUSEUM.

JENKIN, FLEEMING

See HARDY–WEINBERG EQUILIBRIUM; PAINTPOT PROBLEM.

JENNINGS, HERBERT SPENCER

See PURE LINE RESEARCH.

JOHANNSEN, LUDWIG

SEE PURE LINE RESEARCH.

JOHANSON, DONALD (b.1943)
Discoverer of "Lucy"

A Chicagoan born to Swedish immigrants, Donald Johanson was determined to become a professional scientist, probably a chemist. Then one day, while still in high school, he read an article by Louis Leakey in *National Geographic* that was to change his life forever.

Leakey's account of his search for early-man fossils at East Africa's Olduvai Gorge stirred Johanson's sense of adventure. "The name Olduvai, with its hollow, exotic sound," he later recalled, "rang in my head like a struck gong." And the bold quest of the ambitious, competitive, audacious Leakey also struck a chord. It was the beginning of his intense love-hate relationship with the Leakey family, which was to have a dramatic effect on paleoanthropology for the next several decades.

Johanson decided to major in anthropology at Illinois University and went on to graduate work at the University of Chicago. While still a student, he wangled his way onto an international fossil-hunting expedition to Ethiopia and within two years had become world famous for discovering the spectacular "Lucy" fossils—hailed as the "oldest and most complete skeleton of an upright-walking hominid." By the time he completed his Ph.D. and became a curator at the Cleveland Museum of Natural History, he had discovered another treasure trove of hominid fossils ("The First Family"), and had taken some of the limelight from the Leakey family's long dominance of African paleoanthropology.

Ironically, the Ethiopian expedition, originally led by French geologist Maurice Taieb and American Jon Kalb had gained funding on Louis Leakey's recommendation, well before Johanson joined their team. And, in his early seasons in Ethiopia, Johanson had sought and won the cooperation of Mary and Richard Leakey. But Johanson's published views and statements instigated an ever-widening rift with the first family of paleoanthropology. This split was partly caused by his insistence that his Lucy (*Australopithecus afarensis*) was a more ancient human ancestor than anything the Leakeys had found, his knack for grabbing press attention at Leakey-organized conferences and his compulsion to play scholarly one-upmanship with Richard Leakey, a past master of that game.

When Johanson announced the proposed new australopithecine species *afarensis*, he did not use "Lucy" as the type specimen, but instead identified it with a jaw Mary Leakey had found in East Africa—a fossil she considered not an australopithecine at all, but more likely a *Homo habilis*. Offended at the "appropriation" of her discovery by an upstart with whom she did not agree, Mary made her displeasure with Johanson known.

With his talent for fossil-finding, fund-raising and political infighting, Johanson had managed to gain control of the Ethiopian project and emerged as a rival of Richard Leakey, at least in the popular press. In 1979, he announced his own family tree of the hominids, on which Lucy was ancestral both to other australopithecines and to the human line.

The Leakeys stuck to the beliefs of Louis (who had recently died) that *Homo habilis* was an "oldest" human ancestor and that australopithecines were merely sidebranches. When more fossils were discovered, Richard Leakey argued, they would show a distinct lineage leading through his *Homo habilis* to *Homo sapiens*. But beneath the scientific disputes was another question: Who would be King of the Hill?

Press accounts, including major articles in *Life* and the *New York Times*, eagerly seized on the growing rivalry between the two. Leakey tried to avoid responding to Johanson's challenges, instead promoting his own reading of the fossils, by this time imbedded in his own book *Origins* (1977) and his "Making of Mankind" television series.

Matters came to a head before millions of people in a televised confrontation in 1981, fomented by newscaster Walter Cronkite. Just before they were to begin filming, Leakey told both Johanson and Cronkite he did not want to debate their differences. Because of increasing creationist attacks on evolution, he said, it was more important to give the public some scientific education.

It was not to be. No sooner were the cameras rolling, than Cronkite asked each man for his view on what the fossils meant. Johanson at once pulled out a prepared chart of a "family tree" with Lucy as its matriarch, mankind evolving from the three-foot high australopithecine, with Leakey's *Homo habilis* as a sidebranch.

When asked for his own view, Leakey protested that he had brought no artwork, so he was handed a marker and asked how he would modify Johanson's chart. With a nervous chuckle, Leakey drew a big bold "X" through Johanson's entire scheme. "And what would you replace it with?" he was asked. Leakey then drew a large question mark. A few minutes later, he stormed angrily out of the building.

"When next I saw Richard," says Johanson in *Lucy's Child* (1989), "he looked right past me. That was four years ago. We haven't spoken since. In public, he has continued to insist that our 'rivalry' is largely the media's creation, and that beneath the hype there lie nothing more than minor professional disagreements. I wish it were that painless."

In 1987, Johanson and his colleague Tim White sought fossils at Olduvai Gorge in Tanzania, and found a fragmentary skeleton they identified as a female *Homo habilis* (OH65), estimated at 1.8 million years old. Seeing resemblances between this new hominid and his famed *Australopithecus afarensis* ("Lucy") from Ethiopia, Johanson nicknamed the find "Lucy's child" and made her the subject of his 1989 book. He argued that this Olduvai female was descended from the older *Australopithecus afarensis* population, and that both were direct ancestors of the genus *Homo*.

In 1981, Johanson founded the Institute for Human Origins in Berkeley, California, of which he remains director. The institute has served as a clearinghouse for information on fossil hominids and a launching base for field expeditions. It also houses a geochronology lab, headed by Garniss Curtis (coinventor of the potassium-argon method of dating), that carries on dating work with materials from East African sites.

See also AFAR HOMINIDS; "LUCY"; "OLDEST" MAN; OLDUVAI FEMALE.

For further information:

Johanson, Donald, and Edey, Maitland. *Lucy: The Beginnings of Humankind.* New York: Simon & Schuster, 1981.

Johanson, Donald, and Shreeve, James. *Lucy's Child.* New York: Wm. Morrow, 1989.

Lewin, Roger. *Bones of Contention: Controversies in the Search for Human Origins.* New York: Simon & Schuster, 1987.

Reader, John. *Missing Links: The Hunt for Early Man.* London: Collins, 1981.

REMARKABLE DISCOVERIES catapulted young Donald Johanson into the first rank of fossil hunters before he had completed graduate school. Tim White (right) helped Johanson fit the 3.5-million-year-old "Lucy" skeleton and "First Family" bones from Ethiopia into the puzzle of hominid evolution.

JOURNAL OF RESEARCHES INTO THE GEOLOGY AND NATURAL HISTORY OF THE VARIOUS COUNTRIES VISITED BY H.M.S. BEAGLE, ETC. BY CHARLES DARWIN

See VOYAGE OF THE H.M.S. BEAGLE.

JUKES FAMILY
Study of "Socially Unfit"

In the heyday of the influential eugenics movement, a scientific study was needed to demonstrate a genetic basis for criminality, poverty and feeblemindedness. If it could be shown that certain ethnic groups were innately inferior and "unfit," they could be sterilized or put away so that brighter, more moral, productive citizens could outbreed them, which was the core of the eugenics program.

If social pathology could be proven to result from heredity—not from social or economic conditions—mankind could confidently take charge of improving itself through selective breeding. The most well-known study of a family of "misfits" was Richard Dugdale's book on the Jukes family, published in 1877.

Dugdale traced the ancestry of the Jukes clan—a large family of prostitutes, criminals and drunkards—back through seven generations to a single founding couple in upstate New York. Although Dugdale actually recognized they were degraded more by their social situation than by heredity, his conclusion was ignored. The name Jukes became synonymous with biologically based criminality.

A follow-up study in 1915 by the Eugenics Records Office showed no improvement in the family's fortunes, and they were enshrined in biology textbooks as an example of Darwin's fears that modern society allowed unrestricted breeding of the "unfit." Today, the Jukes family is another kind of textbook case: of incompetent science, failure to distinguish between causes and effects, biased study design and the confusion of social and genetic phenomena.

See also BIOLOGICAL DETERMINISM; EUGENICS; SOCIAL DARWINISM.

"JUST–SO" STORIES
Fanciful Explanations of Origins

Rudyard Kipling, in addition to his journalism, adventure stories and chronicling of the British Raj in India, is remembered for a series of charming children's tales about the origins of animals. The *Just–So Stories* (1902) are fanciful explanations of how the elephant got his trunk (a crocodile bit it and stretched it) or how the camel got his hump (rolling around in lumpy sand dunes). Modeled on the folktales of tribal peoples, they express humor, morality or whimsy in "explaining" how various animals gained their special characteristics.

"Not long ago," writes science historian Michael Ghiselin, "biological literature was full of "just-so" stories and pseudoexplanations about structures that had developed 'for the good of the species.' " Armchair biologists would construct logical, plausible explanations of why a structure benefited a species or how it had been of value in earlier stages.

Charles Darwin knew that approach could yield no useful knowledge, because "it was the wrong question, and the wrong way of finding the answer," notes Ghiselin. "Evolutionary hypotheses have to be tested . . . Darwin worked out the methodology for testing such hypotheses and, in applying it to orchids, showed us how to use it."

Stephen Jay Gould has pointed out three basic fallacies in the attempt to reason out "logical" explanations for structures in organisms based on a vague survival value. First, it is "hyperselectionist"—seeking adaptation in every minute structure, whereas some traits could well be neutral. Second, plausible logic is no substitute for gathering sufficient evidence from fossils, behavior and embryology. And, finally, reconstructions cannot be made according to supposed "stages" of progression. Each organism is the product of a special, unique history, and we often have no way of knowing how early adaptations meshed with past environments.

See also ADAPTATION; ADAPTATIONIST PROGRAM; NEUTRAL TRAITS.

K/T BOUNDARY
End of Dinosaur Age

One of the important contemporary scientific debates is about the causes of the mass extinctions at the close of the Cretaceous epoch, about 65 million years ago. Volcanic activity, impacts of meteors or comets, drastic climatic change of unknown cause—whatever the reason, there is an abrupt change in numbers and types of fossils found at the close of the Mesozoic, or Age of Dinosaurs, which ended with the Cretaceous.

Scientists refer to this crucial, enigmatic transition in the history of life as the "K/T boundary." The Cretaceous epoch is abbreviated as K to distinguish it from the much earlier Carboniferous (coal-forming) epoch, abbreviated as C. Sedimentary rock layers above the Cretaceous, which include the fossil record of the Age of Mammals, are traditionally called Tertiary or T.

See also ALVAREZ THEORY; EXTINCTION; GREAT DYING.

For further information:

Hsu, Kenneth. *The Great Dying: Cosmic Catastrophe, Dinosaurs, and the Theory of Evolution.* New York: Ballantine Books, 1986.

KAMMERER, PAUL (1880–1920)
Controversial Evolutionary Biologist

In 1926, shortly after his research on Lamarckian inheritance was denounced as a fraud, Austrian biologist Paul Kammerer committed suicide. To this day, scientists speculate on his tragic life and argue about the integrity of his evolutionary experiments.

Kammerer is most often dismissed as a neo-Lamarckian who believed he had proved acquired characteristics could be genetically passed to offspring. In fact, for most of his career, he was a Mendelian who could offer no convincing explanations for his unusual experimental results. If anything, he thought they did not demonstrate novelty but atavism (reversion or throwback), an expression of existing genetic characters that had only been "suppressed" by the animal's mode of life.

Before World War I, he conducted experiments with salamanders, newts and an amphibian known as the midwife toad. Salamanders raised on different colored gravels changed their colors and breeding habits and passed the changes on to their offspring. Blind newts, bred in infrared light, "redeveloped" eyes. And male midwife toads regained the nuptial pads with which they grasp females when mating in water. Male midwife toads do not normally have the special nuptial pads (on their thumbs) that other, pond-spawning toads use for hooking onto swimming females. But when several generations are forced to live and breed in water, Kammerer claimed, the males (and their offspring) develop the pads.

After the war, Kammerer traveled to England and America to continue his work, but was met with cold hostility by Mendelian geneticists. They could not reproduce his results, and Oxford's Gregory Bateson accused him of incompetence or fraud.

Critics examined some of his pickled toads, and announced that their nuptial pads were more prominent because they had been injected with India ink. Kammerer protested that he had not tried to "fake" anything, but Bateson cried "hoax," the experiments were discredited and the depressed biologist killed himself a few months later.

Kammerer's story was the focus of *The Case of the*

Midwife Toad (1971) by Arthur Koestler, who suggests it was not Kammerer, but an overzealous assistant who attempted to restore faded markings by inking the 15-year-old specimens. Other theories blame an insanely jealous rival (there was one) or Nazi saboteurs displeased with Kammerer's politics.

His popular lectures about mankind evolving toward a species of "superman," a theme of great social importance in the late 1920s, caused a sensation. Kammerer preached against racism while many German, British and American scientists were using Mendelism to justify it. In their view, "inferior" individuals or races needed to be prevented from breeding if a "super race" were to evolve.

The optimistic egalitarianism of Kammerer's conclusions made front-page news. Lamarckian inheritance was now proven, he proclaimed, and therefore all peoples could make evolutionary progress through improvement of their social environment. Achievement of a species of "supermen" was possible without racism!

There was dark speculation that such political statements may have motivated Nazi agents to tamper with his specimens and so discredit him. On the other hand, Soviet scientists welcomed him with open arms. His views were compatible with socialist ideals of progress for the masses and would soon buttress the disastrous pseudo-science of Lysenkoism. Kammerer accepted a research position in Moscow, but died shortly before he was to begin.

According to Koestler's version, Kammerer was simply hounded to death: a victim of the narrow-minded scientific establishment. His work could not be replicated because few had his remarkable flair for breeding amphibians under artificial conditions. Other supporters insisted his earlier body of work could not all have been faked; it had been examined and praised by leading biologists. Yet, geneticists ceased quoting any of his previous experiments, and most were never reattempted.

Historians of science generally consider Koestler's account of the troubled biologist unreliable. Some even produced evidence that Kammerer's suicide was triggered by an unhappy love affair, rather than by acrimonious scientific controversies.

To this day, the debates and uncertainties continue. Historians and scientists agree on only one fact about Paul Kammerer: He was one of the most ambiguous, enigmatic and tragic figures in the history of biology.

See also LAMARCKISM; LYSENKOISM.

For further information:
Kammerer, Paul. *The Inheritance of Acquired Characteristics.* New York: Boni and Liveright, 1924.

Koestler, Arthur. *The Case of the Midwife Toad.* London: Hutchinson, 1971.

KANAM JAW

See "OLDEST MAN."

KANT, IMMANUEL

See RACE.

KARISOKE RESEARCH CENTER
Gorilla's Threatened Refuge

African mountain gorillas are making their last stand in the core of the Virunga Volcanos, a 25-mile-long chain of extinct volcanic cones that straddles the borders of Uganda, Zaire and Rwanda in East Central Africa.

Intense competition for land between wildlife and the exploding human population has forced the few surviving gorilla bands higher and higher into the mountains. Agriculture and cattle increasingly gobble up the lower habitats, leaving the gorillas no choice but to retreat to the more inaccessible slopes, like King Kong fleeing to the highest point in the landscape.

Political boundaries that slice across the volcanic chain are the legacy of a dispute between the colonial powers Belgium, Germany and Britain. By 1900, borders were agreed on, and the American sculptor-naturalist-taxidermist Carl Akeley convinced King Albert of Belgium to create the Albert National Park to protect the homeland of the gorillas. (The parts of the park divided among the modern states are called Parc National des Virungas in Zaire, Parc National des Volcans in Rwanda and The Kigezi Gorilla Sanctuary in Uganda.)

The acreage reserved for the apes has been steadily shrinking. In 1969, 22,000 acres were taken from the park for the cultivation of pyrethrum, a flower used for making commercial insecticides. Less than 30,000 acres remain in the park, but half of that is threatened by government cattle-grazing programs. The surrounding land contains 800 inhabitants per square mile, and the Africans go into the park to collect wood, poach game and graze cattle.

Kabara meadow (elevation 10,200 feet), near Mount Mikeno in Zaire, was an ideal place to find gorillas until just a few years ago. George B. Schaller, who made the pioneering study of the ape, had worked for several years in the Kabara area during the 1960s. Carl Akeley thought it "the most beautiful and tranquil spot in the world" and was buried there in 1926, but his bones were stolen by Zairoise vandals in 1979. (Some were returned and the grave resealed in

1990 by his American biographer, Penelope Bodry-Sanders.)

Dian Fossey also encountered her first gorillas at Kabara, but cattle and hunters had made them scarce. Because she needed a more undisturbed area, she reestablished her study site on the Rwandan side of the border. After weeks of searching, she found her "spectacular . . . impressive spot in all the Virungas . . . ideal for gorilla research":

> Exactly at 4:30 P.M. on September 24, 1967, I established the Karisoke Research Centre—"Kari" for the first four letters of Mt. Karisimbi that overlooked my camp from the south; "soke" for the last four letters of Mt. Visoke, whose slopes rose north some 12,172 feet immediately behind the 10,000-foot campsite . . . Little did I know then that by setting up two small tents in the wilderness of the Virungas I had launched the beginning of what was to become an internationally renowned research station.

Although sometimes lonely in the early years, Fossey felt she had "reaped a tremendous satisfaction that followers will never be able to know."

Fossey made the preserve her personal responsibility. She chased out cattle and poachers, even engaging in a bitter personal war during which gorillas were slaughtered and poachers jailed. Fossey died there on December 28, 1985, the victim of a still-unsolved murder. She is buried among the creatures she loved in the Karisoke gorilla cemetery.

Today the Center carries on a varied program of gorilla research, conservation of wildlife and habitat and education of local people to care for gorillas as a national treasure. In recent years the Karisoke gorillas have become the center of a growing tourist industry. Visitors who wish to watch gorillas must book in advance and pay a substantial fee, which is used to help protect and preserve the magnificent apes of the Virunga Volcanoes.

See also AKELEY, CARL; APES; "APE WOMEN," LEAKEY'S; FOSSEY, DIAN; GORILLAS.

For further information:
Fossey, Dian. *Gorillas in the Mist*. Boston: Houghton Mifflin, 1983.

KARROO BASIN
Boneyard of Mammal-like Reptiles

One of the most incredible and important fossil fields in the world is the Great Karroo, a long basin about 800 miles long and 400 miles across at the southern tip of Africa. It extends almost from coast to coast, just above Cape Town and Durban. The Karroo is famous for its vast accumulations of bones of mammal-like reptiles, strange creatures that were truly transitional between mammals and reptiles.

Karroo fossils were the first paleontological work of the talented and eccentric Dr. Robert Broom, a Scots physician who had immigrated to South Africa. Beginning in the 1890s, Broom sought clues to the origin of the mammals in the Karroo fauna. He discovered 70 new genera and more than 200 previously unknown species, worked out the detailed anatomy of most groups and straightened out their classification; in 1928, he received the Medal of the Royal Society for this work.

It was this same Robert Broom who, at the age of 69, decided to find more "Taung apes" (australopithecines) to fill out the story of human evolution. He was already the world's greatest paleontologist, he announced, and saw no reason why he shouldn't also become the world's greatest anthropologist as well. In fact, his discoveries of early hominids, along with those of Raymond Dart, changed the course of 20th-century fossil-man studies.

In 1962, vertebrate paleontologist Edwin Colbert came to the Great Karroo to search for mammal-like reptiles and found fossils of a creature called *Lystrosaurus*. "It was the beginning of my association with this fossil reptile that was to be of great consequence in my life," he wrote.

Over the next few years, Colbert was to also find remains of *Lystrosaurus* in India and then again in Antarctica! It was, he believed, the final proof of the debated theory of continental drift. South Africa, India and Antarctica had once been joined in the supercontinent of Gondwanaland, but they had broken apart and drifted to the far ends of the earth. The geology of the South African Karroo with its *Lystrosaurus* is unmistakable, even in Antarctica.

See also BROOM, ROBERT; GONDWANALAND; LYSTROSAURUS FAUNA.

KASPAR HAUSER EXPERIMENTS
Isolating Unlearned Behavior

One persistent problem in understanding evolution has been to what extent animals inherit behaviors as well as bodily structures. Can all birds fly without being taught? Will they sing their species' song without hearing other birds? If a child is never taught any language, will he or she be able to speak?

Scientists have attempted to find out what animals can do "instinctively" by raising them in isolation and seeing whether normal behavior develops. These experiments are called "Kaspar Hauser" experiments after a tragic little boy who lived in Nuremberg, Germany early in the 19th century (1812–1833). He was supposedly reared in complete isolation by a deranged man who kept him locked away from all other humans; his only contact with the world was receiving food and water passed to him through a hole.

When he was discovered as an adolescent in 1828, Kaspar became an object of curiosity among scientists. The cruel experiment they could not perform on purpose had been carried out by circumstance, and they were eager to study the results. They found Kaspar did not behave anything like a "normal" human being. According to some accounts, he spoke only a few words and had to be taught everything a small child picks up by imitating its elders. He was cared for and eventually "enculturated"—taught to speak, dress, eat and behave as a member of society, although he always bore the heavy scars of his years of social deprivation.

Four years of special tutoring, paid for by an aristocrat intrigued by his history, enabled Kaspar to function well enough to work as a court clerk in Ansbach. However, after a year of this employment, he died suddenly of a knife wound under mysterious circumstances. Whether murder or suicide, the case has never been solved. Some believe he was the illegitimate child of a noblewoman and was killed before he could learn the truth and claim a princely inheritance. German authors have remained fascinated with the political, social and psychological enigmas of Hauser's life for well over a century, inspiring many novels, poems, 20 plays and three major films.

Birds raised as "Kaspar Hausers" show some behavior that appears to be entirely innate and some that must be learned. Swallows kept in small cages where they could not flap their wings nevertheless flew normally when released. Other species seem to need practice to become good fliers. Chaffinches sing simple songs when raised as "Kaspar Hausers," but never develop the elaborate repertoires and local "dialects" of those individuals who hear other birds of the same species at an early age.

Scrutiny of the original Kaspar Hauser story shows the tragic child was not as totally isolated as he is depicted in most psychology texts. In fact, the man who kept him often spoke to him and provided a few toys, including a small model of a horse. Though uneducated, Kaspar was even taught to write his name while confined. That he was abused and stunted is undeniable; but that he was a valid example of total learning deprivation is apparently a myth that has become embedded in the literature of behavioral science.

For further information:

Singh, J. A. L., and Zingg, Robert M., *Wolf Children and Feral Man*. New York: Harper and Row, 1966. Contains reprint of Anseln von Feuerback, "Kasper Hauser: An Account of an Individual Kept in a Dungeon, Secluded from All Communication with the World from Early Childhood," 1833.

KEITH, SIR ARTHUR

See KEITH'S RUBICON; PILTDOWN MAN (HOAX); TAUNG CHILD.

KEITH'S RUBICON
Minimum "Human" Brain Size

British anatomist Sir Arthur Keith refused to accept the African australopithecine fossils as human ancestors because their brains were too small. Human qualities of mind, Keith proclaimed, can only appear when brain volume is at least 750 cubic centimeters, a point nicknamed "Keith's rubicon" (dividing line). And, at 450cc, *Australopithecus africanus* didn't qualify.

The Rubico river divided northern Italy from ancient Gaul. When Julius Caesar "crossed the Rubicon" with his army in 49 B.C., the Roman Senate immediately declared a state of war. Sir Arthur Keith's usage similarly means a boundary that, once passed, has inevitable and irrevocable consequences.

In Keith's day, the *Homo erectus* skulls at 950cc (actually 775–1,225cc) could comfortably be included as humans, since their range overlaps our own species (1,000–2,000cc). But the *Homo habilis* skulls discovered later measured about 640cc, just on the other side of the rubicon. Skulls of *Australopithecus* adults are about 500cc, which is larger than chimps but smaller than *Homo habilis.*

Keith's rubicon is no longer regarded as a valid yardstick for defining a "human" primate, if it ever was. How did he arrive at the "magic" number of 750cc? It was the smallest functioning modern human brain anatomists had seen at the time.

KELVIN, LORD (SIR WILLIAM THOMSON) (1824–1907)
Champion of Young Earth

Sir William Thomson, Lord Kelvin, codiscovered the second law of thermodynamics, and his Kelvin temperature scale is still the standard in physics labs. But the great 19th century physicist is chiefly remembered by evolutionists for an arrogant blunder, which Charles Darwin called "one of my sorest troubles."

Kelvin believed not enough time had elapsed since Earth's formation for evolution by natural selection to work. The immense time spans Charles Lyell and Charles Darwin imagined as the canvas of evolutionary creativity by slow, gradual degrees did not exist, said Kelvin. Earth was simply too young, and he could prove it with physics and mathematics.

In 1866, Lord Kelvin published a very short paper

titled "The 'Doctrine of Uniformity' in Geology Briefly Refuted." The Earth began as a molten body and has been steadily losing heat, he argued. Interior temperature measurements (from mines) could establish the rate of heat loss from Earth's crust. Then one could reason back to the time when the Earth was hot enough to be molten, which would give the age of the planet. In a series of elegant calculations, Kelvin then assigned an age of about 100 million years to the Earth, which he later whittled down to 20 million. (Today we believe the earth is 4.5 billion years old.)

Kelvin's assumptions were that heat dissipates at a constant rate, that the Earth began as a molten entity, that its composition is fairly uniform and that no renewable sources of heat exist in the Earth's crust. This last assumption ultimately proved to be his undoing, when decades later it was found that radioactivity in the crust constantly generates new heat. Rate of heat loss, therefore, became meaningless as a measure of age.

For 40 years, Kelvin campaigned for a young Earth and brought the prestige of the "hard" sciences to bear on geologists' thinking. To some, his "proof" presented an insuperable difficulty to evolutionary ideas. But most, including Thomas Huxley and Alfred Russel Wallace, accepted Kelvin's verdict on the age of the Earth and tried to cram all of evolutionary history into it, arguing that evolution must proceed at a quicker pace than Darwin thought.

In 1903, Pierre Curie discovered that radium salts constantly emit heat, and the following year Ernest Rutherford delivered a paper on radium that shattered Kelvin's "proof" forever. Rutherford later recalled that he hesitated when he spotted Kelvin in the audience.

> To my relief, Kelvin fell fast asleep, but as I came to the important point, I saw the old bird sit up, open an eye and cock a baleful glance at me! Then a sudden inspiration came, and I said Lord Kelvin had limited the age of the earth, provided no new source of heat was discovered. That prophetic utterance refers to what we are now considering tonight, radium!

The discovery of radioactivity not only destroyed Kelvin's argument. "It also," wrote Stephen Jay Gould, "provided the very clock that could measure the earth's age and proclaim it ancient after all!" The constant rate of radioactive decay proved to be an excellent measure of elapsed time, giving dates going back billions of years. Darwin's hunch about the great age of the Earth, based on his broad knowledge of geology, fossils and biology, was finally vindicated by the physics and mathematics that had so worried him.

See also CHRONOMETRY; FOUR THOUSAND AND FOUR B.C.

KETTLEWELL, H. B.

See PEPPERED MOTH.

KEW GARDENS

See INDEX KEWENSIS; HOOKER, SIR JOSEPH D.

KEYSTONE SPECIES
Shapers of Environment

When Robert T. Paine of the University of Washington coined the term "keystone species" in 1966, he meant species that are exceptionally important to the structure of communities, and help maintain their organization and diversity. When Paine removed certain predatory starfish (*Pisaster*) from intertidal rocks on America's northwest coast, one or two voracious mussel species soon devoured all other creatures and dominated the entire zone. Only when the mussel population was held in check by starfish could other types of bottom-dwelling creatures flourish there.

Some species are not merely adapted to environments—they create optimum living conditions for themselves or others. Coral colonies provide anchorage for sea anemones, sponges, and thousands of other invertebrates, which, in turn, support many species of fish. This varied community could exist without some of its members, but not without the coral. By creating a favorable environment, corals are the keystone of the ecosystem—just as an architectural keystone keeps a stone arch from falling apart.

Perhaps the most spectacular kind of keystone species are the *modifiers*: those, like corals or beavers, that can produce major changes in an environment. In East Africa, recent decimation of elephant herds by ivory poachers has had profound and unexpected effects on many organisms. Important modifiers of environments, elephants routinely clear thousands of acres of trees and underbrush, creating new grasslands. During droughts, they dig waterholes with their tusks and trunks, establishing oases that support dozens of other species.

A precise definition of keystone species is elusive, because effects on other organisms are relative. Where feasible, a species' importance to its ecosystem is best determined by removing it from habitat samples. For example, a field experiment conducted in the 1980s focused on the ability of a small saltwater snail to alter coastal habitats. Mark D. Bertness, a marine ecologist at Brown University, noticed that the advance of the periwinkle snail along the Atlantic coast of the northeastern United States seemed to be correlated with the conversion of coastal marshes to rocky beaches.

By removing the snails from small, caged-off plots, Bertness demonstrated that the algae-feeding periwinkles keep the beach rocks clean and smooth. Mud and algae layers cannot adhere or build up, grasses cannot take root, and the marsh habitat soon disappears—along with such marsh denizens as mud worms, mussels, and fiddler crabs.

Knowledge of keystone species is crucial for understanding how to preserve habitats and perhaps even how to restore some that have been destroyed. The idea of a natural balance is popular and reassuring, but many ecologists now view nature as unstable and in flux. Extinction of some species may cause little harm to the survivors, while others are absolutely crucial to the web of life.

See also CORAL REEFS; ECOLOGY; GAIA HYPOTHESIS; GUANACOSTE PROJECT.

For further information:
Bertness, Mark. 1984. Habitat and community modification by an introduced herbivorous snail. *Ecology* 65: 370–381.
Botkin, Daniel B. 1990. *Discordant Harmonies: A New Ecology for the Twenty-first Century.* New York: Oxford University Press.
Paine, Robert T. 1966. Food web complexity and species diversity. *American Naturalist* 100: 65–75.

KIDD, BENJAMIN (1858–1916)
Survival Value of Religion

For many social thinkers during the late 19th century, Darwinism was linked with optimism about social progress. Natural selection was evolving society toward a more rational, progressive way of life, perhaps a scientific utopia. So when Benjamin Kidd, in his book *Social Evolution* (1894), concluded that natural selection actually favors the preservation of "nonrational" institutions—such as religion—he provoked a critical storm.

Kidd's controversial idea was that the nonrational factor of religious belief had been immensely important in the evolutionary survival of societies. Reason and logic, he thought, were socially disintegrating forces, since each individual would act and plan for his own self-interest. Such mystical or "nonrational" beliefs as reward in an afterlife for selfless acts, the efficacy of prayer and ritual, and eternal punishment for immorality were the underpinnings of social solidarity and efficiency.

In providing supernatural sanctions against total selfishness, religion was an adaptive trait; it helped societies survive. As Kidd put it, "the nonrational factor" is an integrating principle:

> providing a sanction for social conduct which is always of necessity ultra-rational and the function of which is to secure in the stress of evolution, the continual subordination of the interests of the individual to the interests of the longer lived social organism to which they belong.

Darwinians had insisted morality did not need to be backed by religious belief and that the grand sweep of the history of life was toward ever-increasing rationality. Kidd's book, therefore, provoked wounded cries from secularist scientists, who accused him of reviving religion in a new evolutionary guise.

But Kidd defended his revisionist idea that society cannot survive on a pure diet of scientific rationality. He was impressed by the persistence in social life of nonrational beliefs and reinterpreted Social Darwinism to assign an adaptive advantage to religion. Or, as one of his nastier critics summed it up, Kidd had proposed a theory of "the necessity for a little stupidity in social life."

See also HUXLEY, ALDOUS; SECULAR HUMANISM; SOCIAL DARWINISM.

For further information:
Jones, Greta. *Social Darwinism and English Thought.* Atlantic Highlands, N.J: Humanities Press, 1980.
Kidd, Benjamin. *Social Evolution.* New York: Macmillan, 1894.

KILLER APE THEORY
Our Violent Heritage?

"Not in innocence, and not in Asia, was mankind born," begins the best-seller *African Genesis* (1961) by dramatist Robert Ardrey. We evolved in Africa "on a sky-swept savannah glowing with menace," he wrote, where "man was born with a weapon in his hand," descendant of a "killer ape" who hunted for a living and often turned murderous weapons upon his own kind.

On Ardrey's "personal odyssey into the human past," he had met the scientific mavericks of Africa, whom he saw as voices crying in the wilderness. While European anthropologists had long sought the oldest human fossils in France and England and looked to "civilized" Asia as man's probable birthplace, anatomist Raymond Dart and paleontologist Robert Broom had been finding more ancient hominids in Africa, but the European establishment wouldn't listen.

Unencumbered by such parochialism, Ardrey listened, not only to evidence that an African "southern ape" (*Australopithecus*) existed, but to Raymond Dart's theory that man was descended from a carnivorous ape. As one critic has put it, parodying Ardrey, the murderous ape had its roots "not in evidence and not in Africa, but in Ardrey's own emotions about Africa." He had first gone there to report on Kenya's Mau Mau uprising and experienced "in the terror-brightened streets of Nairobi the primal dreads of a primal continent . . . Africa scared me."

Dart, the South African anatomist who had discovered some of the first man-ape fossils, had painted a bloody picture based on his speculations about prehistory. In 1953, two years before Ardrey came on the scene, Dart published a paper called "The Predatory Transition from Ape to Man," in which he argued that a line of apes branched off from its fruit-eating cousins to become carnivores. With upright posture, the hand was freed to make weapons, and the hunting habit led to success. The new human creature, Dart thought, began turning his weapons on his own kind. Those with the better weapons, sharpest minds and strongest stomachs prospered by killing their rivals—and thus man was born with the mark of Cain upon him.

No doubt, the lingering popular fantasies of Hollywood's *King Kong* (1933) and the killer ape of Edgar Allan Poe's "Murders in the Rue Morgue" (1841) lurked somewhere in the background. From famed 19th-century gorilla hunter Paul du Chaillu to P. T. Barnum's Gargantua, the great apes were always depicted as fearsome killers, often with the implication they also carried off women for interspecies sex.

Dart offered battered and broken bones of fossil baboons as evidence and considered every object remotely associated with the early hominids a potential weapon. An antelope horn, a piece of long bone, a chipped rock: All were imagined by Dart as weapons, used for murder and warfare as well as hunting. Years later, C. K. Brain reexamined the hominid and animal bone accumulations, but concluded the holes and punctures were the work of leopards and hyenas.

The killer ape theory was picked up and echoed by animal behaviorist Konrad Lorenz, anthropologist S. L. Washburn and other authorities, although they were less dramatic than Ardrey in their writings. Nevertheless, the killer ape captured the academic as well as the public imagination, perhaps, offering an escape from guilt: Man was born aggressive, it is part of our biological nature, and we must accept killing as a fact of life. Soon, the pendulum of anthropological fashion swung the other way, and a rash of books were produced on such phenomena as the harmless, peaceful Tasaday forest tribe and cooperative social behavior in animals. A few years later, the women's movement would bring feminist interpretations of human evolution and studies of rank among females in baboon troops.

In the years since Ardrey's book was published, the number of new discoveries of fossil hominids has exploded and these fossils notably lack evidence of violent deaths. Meanwhile, George Schaller's decades of fieldwork among such true predators as lions, leopards and tigers show that man is, by comparison, a very peaceful species. In the United States, for example, there are about 10 killings for every 100,000 people. A field observer would have to watch 10,000 people for a full year in order to witness a single murder. If human beings were as violent as lions are among themselves, there would be a dozen or two slayings every year on every city block!

Richard Leakey, who has viewed and discovered more than the lion's share of early hominid fossils, has said, "The evidence for a predatory early hominid is perfectly valid, but a predatory ape is not a killer in the sense that the 'killer ape' was introduced and is popularly conceived of . . . Nobody would have heard of the killer-ape concept if it hadn't been for Robert Ardrey, but in the same breath I will say that a majority of the people now interested in supporting early-man research would never have heard of early man. . . . [His book generated tremendous interest] and if the price was the killer-ape hypothesis, then I think it's cheap at the price . . . [but] it should have been abandoned by everybody long ago."

See also AUSTRALOPITHECUS BOISEI; DART, RAYMOND ARTHUR; DARTIANS; LEAKEY, RICHARD; QUEST FOR FIRE.

For further information:
Ardrey, Robert. *African Genesis*. London: Collins, 1961.
————. *The Hunting Hypothesis*. New York: Atheneum, 1976.
Lorenz, Korad. *On Aggression*. London: Methuen, 1966.

KIMEU, KIMOYA

See HOMINID GANG.

KIMURA, M.

See NEUTRAL TRAITS.

KING KONG (1933)
Classic Ape Icon

"Well, Denham," says the policeman, "the airplanes got him." Impresario Carl Denham, who has just watched his "prehistoric Giant Gorilla" fall off the top of the Empire State Building, knows better: "Oh no, it wasn't the planes. It was *beauty* killed the beast." So ends *King Kong*, the 1933 movie masterpiece that has been called Hollywood's most perfect realization of a terrifying dream.

Horrific and fierce though he is, Kong's kinship

with ourselves is always painfully clear and evokes tremendous sympathy. He is not at fault for the nightmare that results from showman Denham's obsessive lust for glory. In his arrogance, Denham has attempted to subdue an elemental force of nature for the trivial amusement of paying crowds. His own downfall results, too, from blind refusal to acknowledge Kong's capacity for love—the ape's "humanity." Denham never imagined Kong would care more for Anne Darrow (Fay Wray) than for his own life or freedom, even destroying trains and buildings to be near her.

King Kong opened to packed houses on March 3, 1933, in the midst of the Depression, and quickly became the highest grossing film of its time. Writers have since analyzed its psychosexual overtones, Freudian imagery and status as parable, myth or nightmare. Merian C. Cooper, the producer-director, claimed it was "escapist entertainment pure and simple."

Many of the elements were drawn from the filmmaker's own life and interests and from the influence of current popularizations of evolution and natural history. Cooper had been a combat pilot in World War I, had filmed animals in remote jungles (with his partner Ernest B. Schoedsack) and had hoped to become a great explorer. Among other accomplishments, he helped pioneer technicolor film, organized major airlines and coproduced Westerns with famed director John Ford.

Like his screen alter ego, Carl Denham, Cooper was tough, adventurous, stubborn and a born showman with a sense of high drama. But his fascination with wild animals (especially baboons and gorillas) also made him something of an aspiring naturalist. While in Africa filming a "jungle adventure" (a very popular genre in the 1920s), he observed baboon troops near his locations, and kept systematic notes daily over a number of months.

After returning to New York City in 1929, he wrote an original 85,000-word monograph on baboon behavior, one of the first primate field studies. Unfortunately, there is no way to rate him as a naturalist. While cleaning his cluttered apartment, a maid mistook the manuscript for trash and threw it out. There were no copies, and Cooper never attempted to rewrite it.

Cooper lived a short distance from the American Museum of Natural History and became friendly with the wealthy explorer-naturalist W. Douglas Burden, one of the museum's trustees. Around 1925, Burden had financed and led a museum expedition to the remote and mysterious island of Komodo in the East Indies. Huge carnivorous lizards had been discovered there in 1912, soon after Sir Arthur Conan Doyle

published *The Lost World* (1912), a fantasy adventure set in an isolated land where ape-men and dinosaurs survived into the present.

After traveling 15,000 miles by ship, Burden, his crew, and lovely young wife reached Komodo Island. Later he wrote of its volcanic "gnarled mountains . . . that bared themselves like fangs to the sky . . . as fantastic as the mountains of the moon . . . We seemed to be entering a lost world." His first impression of a live Komodo dragon was of "a primeval monster in a primeval setting." (In imagination, Cooper had dreamed of capturing giant gorillas in Africa; at the time, they were almost as little known as the giant lizards.)

Burden managed to take the first motion pictures of the dragons, shot a few (which are still on exhibit at the American Museum of Natural History) and even captured two live nine-foot specimens for the Bronx Zoo. Years later, in a letter to Burden, Cooper recalled:

> When you told me that the two Komodo Dragons you brought back to the Bronx Zoo, where they drew great crowds, were eventually killed by civilization, I immediately thought of doing the same thing with my Giant Gorilla. I had already established him in my mind on a prehistoric island with prehistoric monsters, and I now thought of having him destroyed by the most sophisticated thing I could think of in civilization, and in the most fantastic way . . . to place him on the top of the Empire State Building and have him killed by airplanes.

Writing the script was now a matter of getting Kong atop what was then another marvel—the world's tallest building.

During 1929–1930, Cooper wrote the first drafts of his movie masterpiece in New York City. He frequently consulted with Burden and visited him at the museum, where he was also inspired by the murals of prehistoric life by Charles R. Knight, the "Father of Dinosaur Art." The idea of taking Fay Wray to Skull Island was inspired by Burden's wife, who had gone along on the dragon expedition. Burden had written a book about his great lizard, calling him "King of Komodo," which Cooper eventually transformed into "King Kong."

When David Selznick became head of RKO in 1931, he invited Cooper to assist him in Hollywood. Under a previous administration, the studio had begun a film called *Creation*, about prehistoric animals. It was an ambitious experiment in stop-motion animation by the talented artist Willis O'Brien, who had created similar scenes for *The Lost World* (1925). Cooper was to determine whether the project could be salvaged.

When he saw O'Brien's work, he was wildly enthusiastic, but not for *Creation*, which would be "just big beasts running around.'' What was lacking, he reported to Selznick, was a strong story featuring a ''ferocious menace.'' O'Brien's genius could bring to life the ''Giant Terror Gorilla'' Cooper had imagined. He suggested some scenes for O'Brien to sketch; the drawings were so compelling that Selznick gave approval to build the small gorilla model and shoot a few test sequences. The ''Eighth Wonder of the World'' was launched.

The first test sequence shot was Kong's death struggle with the pterodactyl. O'Brien's innovative crew, including the model-maker Jose Delgado and the young Ray Harryhausen became the core of a tradition. (In the scene where Kong wrecks the elevated trains, many of the miniature storefronts bore the names of the artists—''O'Brien's Florist,'' ''Delgado's Diner,'' etc.) Later, many of the same artists and their students worked on *Son of Kong* (1934), *Mighty Joe Young* (1947), *The Beast from 20,000 Fathoms* (1953) and the series of *Sinbad* films that began in 1958.

For years, Cooper kept Kong's real size a closely guarded secret. As late as 1964, at a Hollywood exhibit of classic special effects, visitors were still jarred by seeing the original Kong model on display—only about 15 inches high. Press photographers asked actress Fay Wray to hold up the miniature Kong, just as the giant ape had carried her in the film. Miss Wray refused the pose, explaining quietly, ''I could never do that to him.''

Somehow, what Merian Cooper has called an ''illogical'' fantasy compels belief, creating its own timeless reality. Although it appeared to be a novelty, it did not spring out of nowhere. King Kong's roots are keep, drawing on rich traditions.

Story elements came from the fairy tale ''Beauty and the Beast,'' from the ''fatal character flaw'' of Greek tragedies (both for Denham and Kong) and from the nightmarish tales of gorillas popularized by hunter Paul du Chaillu. It drew on the filmmaker's explorer friends and his own glory-seeking stubbornness, on Charles Darwin's *Descent of Man* (1871) and Alfred Russel Wallace's *Malay Archipelago* (1869), on museum expeditions of the 1920s to exotic islands, the romance of jungle movies, Tarzan, Conan Doyle's *Lost World* and animator Willis O'Brien's long-standing obsession with prehistoric monsters.

Remote from these sources, producer Dino de Laurentis attempted a remake of *King Kong* in 1975 using a man in an ape suit in place of stop-motion animation; the World Trade Center towers replaced the Empire State Building. The dismal result only reaffirmed the brilliance of the original.

See also APES; GORILLAS; (THE) LOST WORLD; O'BRIEN, WILLIS.

DREAMING OF PREHISTORIC MONSTERS, writer-producer Merian C. Cooper conjures up a cinematic vision of his giant ape King Kong in this 1933 publicity still. The Hollywood classic drew on Cooper's varied background as a combat pilot, wildlife photographer, filmmaker and student of natural history.

For further information:

Annan, David. *Ape-Monster of the Movies.* New York: Bounty Books, 1975.

Goldman, Orville, and Turner, George E. *The Making of King Kong.* New York: Ballantine Books, 1975.

Gottesman, Ronald, and Geduld, Harry. *The Girl in the Hairy Paw: King Kong as Myth, Movie, and Monster.* New York: Avon Books, 1976.

KINGSLEY, REVEREND CHARLES

See OMPHALOS; WATER BABIES.

KIN SELECTION
Altruism and the ''Selfish Gene''

Social insects, wrote Charles Darwin, present ''one special difficulty, which at first appeared to me insuperable, and actually fatal to my whole theory.'' How, he wondered, could the worker castes of bees or ants have evolved if they are sterile and leave no offspring?

Darwin solved the problem by devising an idea known today as "kin selection." Although the workers do not reproduce, their seemingly altruistic actions preserve and perpetuate their fertile relatives. As Darwin put it, "a well-flavoured vegetable is cooked, and the individual is destroyed; but the horticulturist sows seeds of the same stock, and confidently expects to net nearly the same variety."

Despite this important insight, which predated Mendelian genetics, Darwin's views were usually interpreted in terms of an individual's struggle to pass on its own genes. In the name of "survival of the fittest" (a phrase originated by Herbert Spencer), Social Darwinists argued that pity and altruism were contrary to nature and destructive to species.

Nevertheless, naturalists have recorded many instances of "altruism" in nature and not just in social insects. Some birds risk their lives by crying out to warn of predators approaching the flock, for instance. Recent students of Edward O. Wilson's sociobiological approach argue that such altruism, like that of the worker insects, is actually genetic selfishness. The bird who risks itself by warning its fellows protects close relatives that share many of the same genes, thus increasing the chance these genes will survive.

J. B. S. Haldane, the great British geneticist, was explaining natural selection to some friends in a pub and declared tongue-in-cheek that he would gladly sacrifice himself for his family if he knew that his genes would live on. "How many relatives would be enough?" asked his companion. Haldane seized the back of an envelope and did some hurried calculations. "I am willing to die," he announced, "for two brothers, four uncles, or eight cousins." That sort of theoretical problem—genetic survival through "kin selection," as analyzed by W. D. Hamilton—has since become a major focus of sociobiology.

See also "SELFISH GENE."

For further information:
Hamilton, W. D. "The Genetical Theory of Social Behavior." *Journal of Theoretical Biology* 7 (1964): 1–52.

KIPLING, RUDYARD

See "JUST-SO" STORIES.

KNIGHT, CHARLES R. (1874–1953)
Artist of Ancient Life

Charles R. Knight spent his life creating paintings and sculptures of a world no man had ever seen: a younger Earth of monstrous dinosaurs, birds with teeth and flying reptiles, all of which vanished millions of years ago. Combining skilled artistry with a mastery of anatomy and paleontology, Knight established a scientifically accurate vision of extinct creatures that is still unsurpassed. His enormous influence inspired generations of imitators in such diverse media as museum murals, scientific texts, comic books, toys and monster movies.

Born in Brooklyn, New York, on October 12, 1874, Charles was the only child of an American mother and English-born father who worked for the Morgan banking house, a connection that was to prove important. In later years, J. P. Morgan became a major patron and sponsor of Knight's murals for the American Museum of Natural History, of which Morgan was treasurer.

Knight's artistic talent and passion for animals showed itself early, but a childhood accident almost put an end to his career before it had begun. A playmate threw a pebble that struck his right eye, nearly destroying it. The eye remained weak throughout his life, contributing to Knight's lifelong fear of losing his sight.

During his teens, he began full-time training in art at both the Metropolitan Art School and the Art Students League in New York City. In 1890, he got his first salaried job, at Lamb's studios, a manufacturer of church windows. There he designed the birds and animals that appeared in their stained-glass panels. Even while working full-time, his love of drawing animals led him often to the Central Park Zoo. He still had no idea that his growing knowledge of living animals could lead to an important career painting creatures long extinct.

After his father's death in 1892, he decided to turn to book and magazine illustration, where perhaps there was a market for his animal drawings. He made the rounds of publishers and eventually began to get assignments for children's books, and from *McClure's* and *Harper's* magazines. To increase his knowledge of animal anatomy, he studied carcasses in the taxidermy department of the American Museum of Natural History and became friendly with staff scientists.

One day in 1894, the head taxidermist mentioned that the fossil department needed an artist to draw a lifelike restoration of an extinct mammal: the piglike *Elotherium*. Knight went upstairs, examined the fossil skeleton and made a skillful sketch of what the animal might have looked like in the flesh. Delighted with the results, the paleontologist started a flood of free-lance assignments, which established Knight at the museum.

His art soon came to the attention of the museum's aristocratic president, Professor Henry Fairfield Osborn (1857–1935), who was building its great dinosaur halls. Recognizing Knight's talent, he set him to work on a series of huge murals depicting prehistoric life, which were sponsored by J. P. Morgan.

Osborn had also hired Dr. William D. Matthew, who was mounting the fossil specimens in lifelike poses, a novel idea at the time. After questioning Matthew and Osborn for their interpretations of an animal and studying the mounted skeleton, Knight would then sculpt a small clay model and cast it in plaster. Painting from his model, he could pose it at any angle and see exactly where highlights and shadows would fall. In addition, his accumulated knowledge of muscles, attitudes and expressions of living animals was clearly reflected in his work. While the scenes and colors often reveal an imaginative romantic vision, the animals and plants were depicted with an unprecedented scientific accuracy.

Some of his most dramatic interpretations resulted from a brief, fruitful collaboration with Edward Drinker Cope, one of the founders of American paleontology. Cope had been the mentor of Henry Fairfield Osborn, and Knight had always wanted to meet him. They spent several months together at Cope's fossil-crammed home in Philadelphia, planning a series of paintings based on Cope's interpretations of his dinosaurs. Cope died soon after, but Knight continued the work, which culminated in a remarkable series of paintings based on their discussions.

Although he always worked on a free-lance basis, Knight is closely identified with the American Museum of Natural History, which has an extensive collection of his paintings and murals. Excellent examples of his work can also be seen at the Los Angeles County Museum and many other institutions; Knight did some of his finest work at the Field Museum in Chicago, where he spent four years decorating several entire halls with matching panels, murals and bas-reliefs. Among the private commissions he accepted was a special request from President Calvin Coolidge to paint his beloved collies.

Knight's reconstructions of dinosaurs, mammoths, saber-tooth cats and other prehistoric creatures established the images of these animals in popular culture. They were copied not only in scores of scientific and children's books, but have influenced

ARTIST WHO SAW THROUGH TIME, Charles R. Knight was the first to paint accurate, lifelike reconstructions of prehistoric animals, establishing modern images of the dinosaur. Knight is shown with clay model of stegosaurus he created as an aid to painting. Knight's renditions of tyrannosaur, triceratops, and brontosaurus are classics that were widely copied.

comic book artists, toymakers and filmmakers for almost a century.

Knight's dinosaurs were considered so definitive that even his mistakes were widely copied. In the 1890s, following a paleontologist's conjectures, he had painted *Agathaumus*, a creature that looked like a mixture of dragon and rhinoceros. Imagined from fragments (there was no skull or skeleton), it was later dropped by science. But that news failed to reach Hollywood; "Agathaumus" was animated by Willis O'Brien in the classic silent film *The Lost World* (1925), along with Knight's other dinosaurs. Firmly established in popular culture, it continued to appear in children's books on dinosaurs until the 1960s.

See also COPE, EDWARD DRINKER; DINOSAUR RESTORATIONS; KING KONG; (THE) LOST WORLD; OSBORN, HENRY FAIRFIELD.

For further information:

Czerkas, Sylvia, and Glut, Don. *Dinosaurs, Mammoths, and Cavemen: The Art of Charles R. Knight.* New York: Dutton, 1982.

Knight, Charles R. *Animal Drawing: Anatomy and Action for Artists.* New York: McGraw-Hill, 1947.

———. *Before the Dawn of History.* New York: McGraw-Hill, 1935.

———. *Life Through the Ages.* New York: Knopf, 1946.

———. *Prehistoric Man: The Great Adventurer.* New York: Appleton-Century-Crofts, 1949.

KOKO
First Signing Gorilla

During the 1970s a number of psychologists reported that they had successfully taught a modified form of American sign language to chimpanzees. Dr. Francine (Penny) Patterson, then a grad student at Stanford University, decided to see if gorillas could do as well as chimps. In 1972, she began working with Koko, a young female gorilla at the San Francisco Zoo. Four years later, Patterson purchased her.

Beginning when Koko was a year old, Patterson gave her an intensive education through countless hours of personal interaction and attempts to teach communication. Koko soon matched and then surpassed her chimpanzee rivals. By 1979, she was able to use more than 400 signs. According to Dr. Patterson, Koko was able to communicate about the past or future and was even able to argue, joke and lie.

She also began to learn speech sounds by punching the keyboard of an electronic speech synthesizer and by listening to human conversations. She knew the word "candy" so well that it had to be spelled out to keep her from looking for sweets.

When a young male gorilla named Michael was moved into her quarters, Koko began teaching him signs, especially *Koko* and *tickle*. When asked if she was an animal or a person, Koko signed "Fine animal gorilla."

According to Patterson, Koko can remember past emotions. Three days after Patterson was bitten by Koko, Patterson claims she had "a revealing conversation" with her.

Me: What did you do to Penny?
Koko: Bite. [Koko, at the time of the incident, called it a scratch.]
Me: You admit it?
Koko: Sorry bite scratch . . .
Me: Why bite?
Koko: Because mad.
Me: Why mad?
Koko: Don't know.

Patterson also believes Koko makes up "ingenious" lies. "Once, while I was busy writing," she related, "she snatched up a red crayon and began chewing on it. A moment later I noticed and said, 'You're not eating that crayon are you?' Koko signed 'Lip' and began moving the crayon first across her upper, then her lower lip, as if applying lipstick."

Work with Koko has continued for over two decades, during which Patterson has also been working with Michael, another gorilla she has taught to sign. Despite the famous "reconsideration" of ape language studies by Professor Herbert Terrace [see APE LANGUAGE CONTROVERSY; NIM CHIMPSKY] Patterson has stuck to her guns. Critics have fired away at her methods and controversial results, but she continues talking to her gorillas with unflagging conviction and devotion.

Alone among ape-language researchers, she claims that a gorilla can remember and communicate events from its infancy:

Using sign language, the 400-pound gorilla [Michael] can relate how he was captured—"big trouble"—when he was only two years old. He and his mother were "chased" by native gorilla hunters, who then killed his mother. He uses startlingly violent gestures to explain how the men "hit" him and his mother.

Some critics compare Patterson's historical accounts from apes to such researcher's delusions as "cold fusion" and the "canals" of Mars. She insists, however, that "Koko and Michael are forcing us to reexamine everything we've ever thought about animals."

See also CLEVER HANS PHENOMENON.

For further information:

Hahn, Emily. *Eve and the Apes*. New York: Weidenfeld & Nicolson, 1988.

Patterson, Francine, and Linden, E. *The Education of Koko*. New York: Holt, Rinehart & Winston, 1981.

"Conversations with a Gorilla." *National Geographic* 154, 4, pp. 438–465.

KROMDRAII

See BROOM, ROBERT.

KROPOTKIN, PRINCE PETER (1842–1921)
"Law of Mutual Aid"

During an era of industrial robber barons, militarism, *laissez-faire* economics and colonial exploitation, Social Darwinism became a handy excuse for ruthlessness. In the words of Prince Peter Kropotkin, ideologues twisted Charles Darwin's conception of nature into:

a world of perpetual struggle among half-starved individuals, thirsting for one another's blood. They made . . . *woe to the vanquished* . . . the last word of modern biology . . . [and] raised the "pitiless" struggle for personal advantages to the height of a biological principle which man must submit to as well . . .

Born a prince in pre-Revolutionary Russia, Kropotkin railed against the social system that offered him hereditary privilege. After holding a variety of diplomatic and military posts as a young man, he became a writer-philosopher and, in later years, turned his full energies to overthrowing his country's social system.

His early work took him to Siberia, and he later conducted a geological survey of Manchuria. A careful observer of both wildlife and local villagers, he became convinced that—even in such a cold, harsh environment, where one might expect competition to be keenest—survival depended more on cooperation than on competition. He observed horses forming defensive circles to resist wolf attacks, the cooperative hunting strategies of the wolves themselves, the social colonies of insects and birds.

Kropotkin published a remarkable book, *Mutual Aid* (1902), which was his corrective to the popular view of a "struggle for existence." Taking his cue from a lecture "On the Law of Mutual Aid" given in 1880 by the Russian zoologist Karl F. Kessler, Kropotkin spent years building his case for the survival value of compassion, nurturing and altruism. It was not until almost 70 years later, with the rise of sociobiology, that the role of altruism in evolution was seriously examined. Kropotkin was a lone voice calling attention to "mutual aid, mutual support, mutual defense" in the animal kingdom.

He clearly saw the implications for human politics and eugenics programs as well. Kropotkin criticized Darwin's remarks in the *Descent of Man* (1871) about the "alleged inconveniences" of maintaining what Darwin called the "weak in mind and body" in civilized society. Darwin seemed to think advanced societies were burdened with too many "unfit" individuals, Kropotkin scolded, "as if thousands of weak-bodied and infirm poets, scientists, inventors, and reformers, together with other thousands of so-called 'fools' . . . were not the most precious weapons used by humanity in its struggle for existence by intellectual and moral arms." It was Darwin himself, said Kropotkin, who had shown that "sociability" conferred an important evolutionary advantage. Therefore, Thomas Huxley's insistence that mankind must struggle against a harsh, competitive "law of nature" was unnecessary. To Kropotkin, it was social cooperation that gave a species its competitive edge.

As he grew older, Kropotkin became an anarchist-nihilist, doing everything he could to undermine a social system he saw as unjust, inhumane and "unnatural." If the corrupt political and economic institutions could be dismantled, he thought, mankind would return to its more "natural" state of harmony and cooperation.

See also KIN SELECTION; SOCIAL DARWINISM.

For further information:

Appleman, Philip, ed. *Darwin: A Norton Critical Edition.* New York: Norton, 1970.

Kropotkin, Peter. *Mutual Aid: A Factor in Evolution.* London: Heinemann, 1902.

———. *Memoirs of a Revolutionist.* New York: Atlantic Monthly, 1899.

KUBRICK, STANLEY

See 2001: A SPACE ODYSSEY.

KUNDALINI
"Evolutionary Energy"

Although Western science considers it a religious myth, to yogis and Hindu philosophers of India, *Kundalini* is a very real force: the life energy that drives evolution.

The Kundalini tradition, at least 5,000 years old, persists in India and Tibet, and is known from artworks of ancient Greece and the Middle East. Also called the Serpent Power, it is symbolized by the caduceus, two snakes intertwined in a double helix.

According to the ancient tradition, Kundalini lies coiled as a powerful force at the base of the human spine, near the sexual organs. Beginning as sexual energy, it is transformed and channeled from lower centers (*chakras*) to higher, traveling up the spinal nerves and passing through the seven centers of the body. When Kundalini reaches the highest (seventh)

TRAITOR TO HIS CLASS, Russian Prince Kropotkin preached anarchy to bring down corrupt aristocracy that ruled through brutal military force. After observing social behavior in animals, he concluded that evolution selects for cooperation rather than competition, the thesis of his book *Mutual Aid.*

center of the brain, bliss, or "sammadhi," results, and the consciousness is transformed.

In his book *Kundalini: The Evolutionary Energy in Man* (1967), the Kashmiri philosopher Gopi Krishna links the attainment of mystic states, artistic creation and genius with the rise of Kundalini. These elevated states of brain function, he believes, are real physiological phenomena, which can and should be studied by Western science. Yogis believe the evolutionary future of mankind depends on neural transformation promoted by specific breathing, sexual and meditative practices to raise Kundalini.

See also CADUCEUS; GOPI KRISHNA.

For further information:

Gopi Krishna. *The Awakening of Kundalini.* New York: E. P. Dutton, 1975.

Keiffer, Gene, ed. *Kundalini for the New Age: Selected Writings of Gopi Krishna.* New York: Bantam, 1988.

LA BREA TAR PITS
Asphalt Fossil Trove

In a small park near a busy shopping center in Los Angeles, California, seven miles west of the civic center, natural asphalt pits contain one of the richest collections of fossil Pleistocene animals in the world. This bituminous outcrop of petroleum shale—clay, sand and asphalt—contains hundreds of mastodons, bison, horses, camels, wolves, birds and saber-toothed cats. About 40,000 years ago, they were the only residents of Los Angeles.

Although fossils from the pit weren't described until 1875, the site had been mined for years, and thousands of tons of fossiliferous asphalt had already been shipped to tar roofs and roads in San Francisco. During the nineteenth century, it was part of Rancho La Brea, then on the western outskirts of the city.

University of California scientists began collecting fossils there in 1906, when the first saber-toothed cats (*Smilodon*) found anywhere in the world were discovered in the pits. Canine teeth of the saber-tooth resemble curved 10-inch knives and were used for stabbing and tearing prey. More than 700 skulls of the saber-toothed cat were found, and initial reports described densities of 20 saber-tooth and wolf skulls per cubic yard.

Most authorities believe that prey animals, such as the camels, horses and bison, may have become mired in the tar, and their struggles attracted crowds

TRAPPED IN ASPHALT POOLS, tens of thousands of animals left their bones in what is now Los Angeles, California. Saber-toothed cats, American lions, giant sloths and mastodons succumbed to the sticky tar. During the 1980s the Page Museum was built directly over one of the pits to exhibit fossil treasures right at the site of discovery.

of saber-tooth cats and wolves, which also were trapped in the sticky mass. A skeleton of an early American Indian was also found preserved under the bones of an extinct vulture.

The 23-acre park is open to the public free of charge. A large cluster of fossil bones has been left unexcavated and undisturbed, so the visitor can see the natural state in which the profusion of fossils occurs in the asphalt. It is not unusual to see a sparrow or squirrel wander from the park into the pit today and become entrapped in the gooey tar, the sad spectacle of a fossil in the making.

The extensive collections from Rancho La Brea (more than 565 species) are stored and exhibited in the George C. Page Museum of La Brea Discoveries in Hancock Park. Opened in 1977, the museum features life-sized models of mastodons trapped in the actual tar deposits.

For further information:
Harris, John and Jefferson, George. *Rancho La Brea: Treasures of the Tar Pits.* Los Angeles: Natural History Museum of Los Angeles County, 1985.

LAETOLI FOOTPRINTS
Earliest Fossil Man-Tracks

"They are the most remarkable find I have made in my entire career," said Mary Leakey in 1976, describing a series of fossil footprints she excavated near an ancient volcanic mountain in Tanzania. "When we first came across the hominid prints I must admit I was sceptical, but then it became clear that they could be nothing else. They are the earliest prints of man's ancestors, and they show us that hominids three-and-three-quarter million years ago walked upright with a free-striding gait, just as we do today."

These earliest man-tracks are at a site called Laetoli, in a wooded area about 25 miles south of Olduvai Gorge, where Mary Leakey, her husband Louis, and son Richard have made so many important fossil discoveries.

Preserved in the volcanic mud are tracks of various animals, including spring hares, guinea fowl, elephants, pigs, rhinos, buffaloes, hyenas, antelopes, baboons and a saber-toothed tiger. Among these are the tracks of three hominids—a large individual walking slowly north, a smaller one following behind and a youngster. Perhaps it was a family group, with the large one the male, the smaller the female and the juvenile their child. The young one seemed to skip by their side, at one point turning to look around to the left. The smaller individual for some reason walked directly in the tracks of the larger one.

Like nearby active volcanos in East Africa today, the ancient volcano "Sadiman"—very near the prints—

occasionally belched out clouds of grey ash over the surrounding countryside. This ash sets hard as cement when it is dampened slightly, then dried in the sun. A brief shower moistened the ash layer; tiny raindrop craters can be seen in its surface. Then, the sun came out and hardened it. Had there been no shower, the footprints would have blown away, and had there been a downpour, they would have been obliterated. But a brief, light shower prepared the ash to take impressions perfectly and then harden, leaving this extraordinary record of a few moments in the life of an upright-walking hominid group, moving slowly across the African landscape almost four million years ago.

See also LEAKEY, MARY.

LAMARCK, JEAN-BAPTISTE ANTOINE DE MONET, CHEVALIER DE (1744–1829)
Naturalist, Evolutionist

Pioneer evolutionist Jean-Baptiste Antoine de Monet (later known as the Chevalier de Lamarck) was a born fighter. His family's men had for countless generations been horse soldiers, imbued with honor, bravery, tenacity and a desire for glory. When Lamarck traded a military career for one in science, he had simply found a new field of combat, and, to this day, "Lamarckians" are embattled.

Lamarck's war-weary father, determined to shield his 11th child from becoming cannon fodder, sequestered him with the Jesuits as a priest-in-training. But at 19, young Jean-Baptiste fled his safe school to join a regiment defending a German town at the start of the Seven Year War. Within a few days, and without training or adequate preparation, Lamarck distinguished himself in the thick of battle. He seized a field command when his superior officers were killed and refused to let his men withdraw—though they were isolated and outnumbered—until official orders to retreat came through the lines. Immediately thereafter, he received a regular officer's commission.

After years of distinguished service as a soldier, the Chevalier de Lamarck traded his sword for a pen to seek glory in the battle for men's minds. Lamarck sought to do no less than revolutionize knowledge in all of science by creating his own system of chemistry, meteorology and biology (a term he coined).

Much of Lamarck's ambitious system was based on assumptions that were easily disproved during his lifetime. He took on the great experimental chemist Antoine Lavoisier (1743–1794), for instance, with fanciful "laws" of fluids and compounds based on logic and speculation. "In this respect," wrote his enemy Georges Cuvier, "he resembled so many others who spend their lives in solitude, who never entertain a doubt of the accuracy of their opinions, because they never happen to be contradicted."

After an unhappy stint as a bank clerk, Lamarck found he could sustain himself in cheerful poverty by writing encyclopedias of natural history. His *Dictionary of Botany* (1778) became quite popular, because (as Cuvier noted) he stole only from the best authors of the day.

For 11 years he published almanacs based on his meteorological system, predicting the year's weather in advance. The weather refused to cooperate, but people continued to buy a year's worth of inaccurate forecasts, just as they do today.

Lamarck also became a proficient botanist and put out the first "keyed" field guide to French flowers, making identification easy for casual hikers. He also created a new system of animal classification based on the fundamental distinction between animals with backbones (vertebrates) and all those without (coining the term invertebrates), which still stands today.

He struggled for years without a regular position, through four successive marriages that produced four children. Georges Jean, Comte de Buffon, one of the eminent naturalists of the day, recognized his abilities and hired him as tutor to his own son. For several years, Lamarck was able to pursue his natural history interests in that capacity, hoping one of the scarce positions in the then tiny scientific establishment might open up.

In 1794, when the dust from the French Revolution began to settle, the King's garden (*Jardin des Plantes*) was reorganized as the Museum of Natural History (*Musee d'Histoire Naturelle*), but most of the new posts were already promised to those who were professors and keepers under the old regime. Only the lowest, least desirable position was offered to him: keeper of insects, shells, and worms. Lamarck accepted.

Oddly enough, Lamarck's name has come to stand chiefly for an idea he did not originate: the development or atrophy of organs through "use or disuse" and their transmission to offspring who inherit these "acquired characteristics." In fact, just about everyone in Lamarck's day—ordinary people as well as scientists—believed that, and Lamarck claimed no credit for its invention.

Lamarck's classification work led him to several evolutionary speculations. First, he saw living things as tending to progress from simpler organisms to those with more complex nervous systems (man, of course, being the pinnacle of such perfection). Second, he tried to address the problem of extinction at a time when few could imagine "gaps" in the harmony of nature. Comparing fossil oysters in his collection with similar modern ones, he concluded that one had developed into the other. The ancient species had not really died out at all; they had simply changed into those of today.

Lamarck described no mechanisms for producing change ('Lamarckian" inheritance through "pangenesis" was an ill-fated invention of Charles Darwin). But he is remembered for several passages in which he asserts that bodies are shaped by habitual behavior caused by an animal's needs. Thus, generations of giraffes keep stretching to reach ever-higher branches, and wading birds keep raising themselves from the mud until their legs become stilts.

These much-quoted examples brought Cuvier's ridicule and later Darwin's exasperation ("God save us from the 'willing' of animals, volition etc.") Some of the criticism grew out of a semantic problem; the French word *besoin* can mean "lacks," "needs" or "wants," depending on context.

Darwin contemptuously knocked Lamarck for a vision of trees or worms "willing" themselves to adapt or progress—an absurdity of which Lamarck was not guilty. It was only the more advanced creatures, he thought, that had "sensibility" and could strive to meet their needs. Lamarck thought plants and invertebrates changed because of unconscious physiological responses to environments, rather than through volition.

Unlike most dissectors and collectors of his time, Lamarck always conceptualized organisms in relation to their behavior in nature and the challenges of changing environments. Denying then-accepted concepts of a steady-state Earth only some thousands of years old, he believed geology pointed to major shifts in climate, land and sea, which took place gradually over millions of years (deep time).

If species had been fixed at creation and remained static ever since, he realized, they could not survive environmental changes. Therefore, they must be constantly adapting, even if their appearance changed little. Further, he saw the development (evolution) of life as slow, smooth and gradual—a uniformitarian approach anticipating Darwinian gradualism.

But Lamarck's ideas did not take hold in France, though after the Darwinian revolution he was reclaimed by French biologists. "It is not enough to discover and prove a useful truth," Lamarck wrote, "it is also necessary to be able to get it recognized." During his lifetime, he became a pathetic figure, fighting for his ideas even after he went blind, attending scientific meetings on the arm of a devoted daughter.

Lamarck had argued that species of blind fish found in dark caves had lost their eyes through disuse. In one of the meanest remarks ever recorded in science, Cuvier publicly taunted the aged Lamarck after he once again defended his evolutionary views. "Per-

PIONEERING FRENCH EVOLUTIONIST Jean Baptiste de Monet, Chevalier de Lamarck tried unsuccessfully to convince his colleagues of the truth of evolution, but was not appreciated until after his death. Through a quirk of history Lamarck's name came to mean the inheritance of acquired characteristics, an idea he never invented.

haps," said the Baron, "your own refusal to use your eyes to look at nature properly has caused them to stop working." Even as his daughter led the frail genius from the room, she begged him not to despair: "Have no doubts, father, posterity will honor you."

See also LAMARCKIAN INHERITANCE; NEO-LAMARCKIAN; "NEW" LAMARCKISM.

For further information:

Lamarck, J. B. *Zoological Philosophy* (1809). Reprint. Chicago: University of Chicago Press, 1984.

Packard, Alpheus. *Lamarck, the Founder of Evolution.* New York: Longmans Green & Co., 1901.

LAMARCKIAN INHERITANCE
Passing On Acquired Characteristics

Practically every high school biology textbook for the past 40 years credits Jean-Baptiste, Chevalier de Lamarck (1744–1829) with developing an erroneous theory of inheritance that was corrected by Charles Darwin's theory of evolution. This piece of misinformation is wrong on both counts and represents a serious misreading of history.

Lamarck was a brilliant and original thinker, the first to develop a comprehensive and systematic theory of evolution. He stressed that all life forms change and believed they progressed to greater complexity by "striving upward." Also, unlike his more successful contemporary Georges Cuvier, he believed that Earth had gone through gradual natural changes, not a series of great catastrophes.

Lamarck's theory held that if an animal strived in a particular direction, for instance, a short-necked giraffe trying to reach leaves high on trees, its constant reaching would lengthen its neck, and its offspring would have slightly longer necks. (It is also known as the "use and disuse" hypothesis.) We know today that genetics doesn't work that way. If you worked out in a gym every day of your life, your larger muscles would not be passed on to your son or daughter. A child's body type is coded in the DNA of his and her parents' sex cells, unaffected by the bodily traits acquired during their lifetimes. Only mutations or recombinations in the genetic material will be passed on.

Lamarck did not consider the idea of "inheritance of acquired characteristics" an original or important part of his evolutionary theory. Everyone in his day accepted and believed it, and it continued to hold force until Mendel's ideas took hold around 1900. Darwin believed in the inheritance of acquired characteristics, although he realized that there was contradictory data, but he was never able to discover the mechanisms of inheritance.

Although it is a misnomer and historically inaccurate, the outmoded doctrine of "use and disuse" inheritance continues to be known as "Lamarckian." The reason is that the Russian Trofim Lysenko and other 20th-century "Lamarckian evolutionists" lifted this unimportant fragment of his thinking and extolled it as offering hope for the progressive improvement of societies.

See also LAMARCK, JEAN-BAPTISTE; LYSENKOISM: "NEW" LAMARCKISM.

LANA
Chimp Who Learned "Yerkish"

During the heyday of chimpanzee language projects in the late 1960s and early 1970s, some experimenters worked with hand signs and gestures. To reduce possible ambiguity in the apes' responses, psychologists at Yerkes Primate Research Center in Atlanta devised an artifical language: a simple "vocabulary" of abstract symbols and a few rules for stringing them together. In honor of Robert M. Yerkes, who pioneered behavioral research with apes, they named it "Yerkish."

Lana was the first of several chimpanzees at the Center to learn to "read and write" Yerkish, having begun her training in 1972. Almost immediately, she grasped that when she pressed a symbol key on her specially designed computer, its designated sign appeared on a screen. Psychologist Duane Rumbaugh of Georgia State University first taught her to associate symbols (names) with objects. If she pressed the symbol for candy, she got candy.

Soon after, Lana learned to put "please" before the noun and a period after, used verbs like "give" and "open" to make requests and even completed unfinished sentences like "Please machine open . . ." (door, window, etc.). By 1975, she had a vocabulary of 120 words and could arrange them to make statements or requests.

Mostly, she asked for food and drink. Rumbaugh recorded 23 different ways she had of asking for a cup of coffee. Other psychologists, such as Herbert Terrace, thought this extraordinary result was not a demonstration of the ape's subtlety of expression. Instead, he suggested, Lana was running through as many signs as she could, in hopes one of the combinations would eventually result in the cup of coffee.

Despite the carefully worked out visual symbol system, and the expensive electronic "language generator," Lana had managed to introduce an element of ambiguity, which undermined the project's validity. A few years later, Rumbaugh began again after completely redesigning his research methods to be (he hoped) tamper-proof by chimps.

See also APE LANGUAGE CONTROVERSY; NIM CHIMPSKY; WASHOE.

LANE-FOX, AUGUSTE

See PITT–RIVERS MUSUEM.

LANKESTER, E. RAY (1846–1929)
Evolutionist, "Ghost-Buster"

Charles Darwin knew Ray Lankester's father, the Kent county medical examiner, and inspired young Ray to become a naturalist. By the time he was in his teens, Lankester was a dedicated Darwinian, enthusiastically championing the scientific search for truth. A bit overzealous in his youthful crusade against superstition and dishonesty, he won international attention in 1876 by exposing the celebrated American "Spirit-Medium" Henry Slade—who had convinced many Londoners he was able to communicate with the dead.

Lankester graduated from University College, London with a degree in zoology, having favorably impressed his professor, Thomas Henry Huxley. Eager to please his idol Darwin, who detested professional "mediums" (today known as "channelers"), the young biologist hauled Slade into police court to expose him as a "common rogue." [See SLADE TRIAL.] Charles Darwin was so delighted he sent a generous contribution toward Lankester's court costs.

Curious and energetic, Lankester's interests ranged over a wide range of evolutionary biology. One now-forgotten area that interested him was the Degeneration Theory. Evolution, he thought, could lead in the direction opposite from progress, to decadence and degeneration.

"Any new set of conditions occurring to an animal which render its food and safety very easily attained", he wrote, "seem to lead as a rule to degeneration; . . . just as an active healthy man sometimes degenerates when he becomes suddenly possessed of a fortune." Fearing that England was becoming bloated by colonial wealth, he wrote:

> Rome degenerated when possessed of the riches of the ancient world. The habit of parasitism clearly acts upon animal organization in this way. Let the parasitic life once be secured, and away go legs, jaws, eyes, and ears; the active, highly-gifted crab, insect, or annelid may become a mere sac, absorbing nourishment and laying eggs.

He warned his fellow members of the "English race" against allowing themselves to be "overtaken" in the struggle for existence through sheer laziness.

Lankester wrote a long-running newspaper column titled (somewhat ironically), "Science From an Easy Chair"; these were later collected in several popular books, including *Diversions of a Naturalist* (1915), *Great and Small Things* (1923), and the *Kingdom of Man* (1907). For many years he was director of the British Museum (Natural History), and helped establish a dynamic Darwinian tradition in an institution formerly devoted mainly to classification.

His classic book *Extinct Animals* (1905) became the standard introduction to dinosaurs and ancient animals, illustrated with reconstructions he had helped create. It was used by Sir Arthur Conan Doyle as the source for prehistoric creatures in his classic adventure novel *The Lost World* (1912). (Sherlock Holmes's creator mentions Lankester by name in that work as the hero's "gifted friend.")

According to evolutionist-historian Ernst Mayr, Lankester was responsible for rescuing the idea of natural selection from limbo in British science. Despite the efforts of Alfred Russel Wallace, Sir Joseph Hooker, Walter Bates and a few others to promote it, natural selection became unpalatable to British scientists in the decades following Darwin's death.

Lankester had been impressed by the work of August Weismann, the German genetics pioneer. Weismann believed inheritance was carried by a "germ plasm" that could not be directly modified by behavior or environment. Lankester invited Weismann to visit England and enthusiastically welcomed him into the scientific community.

Lankester went on to found a school of thought at Oxford that attempted to combine Weismann's ideas on heredity with natural selection: a forerunner of the modern Synthetic Theory. Biologist-historian Ernst Mayr believes Lankester's influence was "of decisive importance," since until that time not a single experimental biologist was pursuing research based on natural selectionism. Darwin's favorite "ghost-buster" had reinvigorated the central proposition of his theory, marking a path later followed by R. A. Fisher, J. B. S. Haldane and Julian Huxley in shaping 20th-century biology.

Lankester was also influential in the development of prehistoric archeology; he coined much of the technical language still used to give accurate "word-pictures" of stone tools. He was also one of the major actors in the notorious "Piltdown man" affair.

Although at first skeptical of the purported "ape-man" fossil from Sussex gravel pits, within a few years he had joined his colleagues in hailing it as "the earliest Englishman." Abandoning earlier caution, he wrote the jaw was "the most startling and significant fossil bone that has ever been brought to light." And while he warned that association be-

tween the bones and chipped stones could not be established he had no doubt the flints were the workmanship of prehistoric man.

Although Lankester claimed to encourage the free flow of ideas at his lectures, he could not really tolerate any challenge to his expertise. Once a man in his audience politely said he doubted certain chipped flints had been shaped by an ancient man and gave his reasons. Lankester replied: "I have listened and I am amazed; one would expect to hear your remarks not from an educated Englishman, but perhaps from a member of an uncivilized race."

Stone and bone evidence seemed reassuringly "solid" to Victorian scientists, but sometimes proved to be as thin as the "spirits" Lankester had discredited in Slade's parlor. The great debunker of hoaxers had been taken in—some think by vindictive Spiritualists. It is still a mystery: Who created the Piltdown Man and why? Perhaps, as Professor John Winslow has suggested, it was a Spiritualist's revenge for Lankester's scientific arrogance and public prosecution (some said persecution) of the "medium" Henry Slade.

Lankester had played the hero of science for all it was worth; unlike the gullible followers of Slade, he insisted a true scientist is immune to shoddy evidence. But when the tables were turned, Lankester fell for a crudely faked fossil that appeared to support his own most cherished beliefs.

See also DEGENERATION THEORY; (THE) LOST WORLD; PILTDOWN MAN (HOAX); WEISMANN, AUGUST.

For further information:

Lankester, E. Ray. *Diversions of a Naturalist*. London: Methuen, 1915.

———. *Extinct Animals*. London: Constable, 1905.

———. *Science from an Easy Chair*. London: Methuen, 1919.

FEISTY, FEARLESS AND STUBBORN, Dr. Edwin Ray Lankester continued the Darwin-Huxley tradition in England as director of the British Museum (Natural History) and author of a long-running popular newspaper column on natural history.

LASCAUX CAVES*
Prehistoric Art Treasures

Among the great masterpieces of prehistoric art are the decorated limestone caverns of Lascaux, France. During the Pleistocene glaciations, our ancestors left a legacy of mysterious, magnificent paintings of mammoths, aurochs, wild horses and woolly rhinos. The artworks show careful observation and intimate familiarity with the great mammals that became extinct long ago. Now estimated at 17,000 years old, the pictures stood up to time and climatic changes, but finally almost succumbed to tourists.

In the half century since their discovery, millions of visitors have come to view the extraordinary cave art and unwittingly brought with them moisture, algae and bacteria, which began to attack the walls. By 1960, alarmed custodians of the prehistoric site

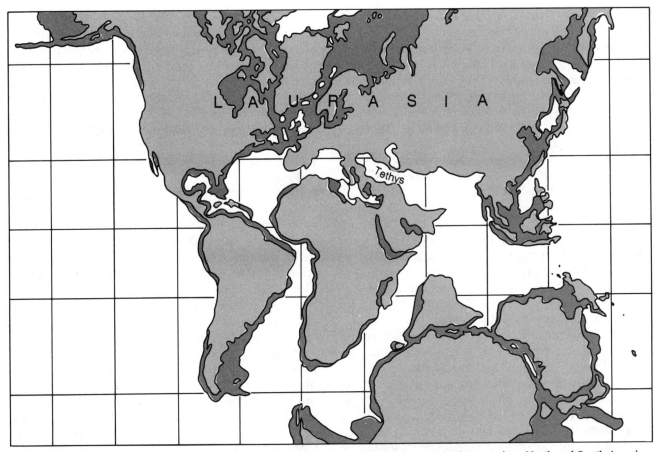

SUPERCONTINENT of Laurasia in mid-Cretaceous, about 100 milliion years ago. Later, it broke up to form North and South America, Greenland, and parts of Asia and Europe.

began to notice patches of green algae spreading and covering up the animal figures. Three years later, the cave was closed to the public and remains so, while conservators work with temperature and humidity controls and antibiotics.

About 200 yards from the cave is a man-made Lascaux II, a painstaking replica of the cave and its artwork, which was opened to the public in 1984. Within a huge blockhouse buried in a quarry, engineers recreated the contours of its walls. Artist Monique Peytral used the same pigments (red ochre, charcoal, sulphur) as Ice-Age artists and took five years to reproduce the paintings in exacting detail. More than 300,000 visitors now visit Lascaux II annually, while experts work to preserve the nearby originals.

For further information:
Vahn, Paul G., and Vertut, Jean. *Images of the Ice Age.* New York and Oxford: Facts On File, 1988.

LAURASIA
Northern Lands

Of the ancient former supercontinents, Laurasia was the northern land mass during the Mesozoic. When it broke apart, Laurasia formed North America, Greenland, Europe and Asia (except India).

See also CONTINENTAL DRIFT; GONDWANALAND; LYSTROSAURUS FAUNA.

LEAKEY, LOUIS (1903–1972)
Paleoanthropologist

Louis Seymour Bazett Leakey was a maverick, an independent fossil hunter, archeologist and anthropologist who, with his wife Mary, changed the entire picture of human prehistory. He is best known for his discovery of scores of fossils and stone tools relating to human evolution at Olduvai Gorge, Tanzania, where the Leakeys confirmed Charles Darwin's prediction that Africa was the probable "cradle of mankind."

Prior to the Leakeys' work, most anthropologists thought that man had evolved in Asia about a million years ago. Since his pioneering investigations—continued today by his son Richard—the focus has been on Africa, and the time depth has been pushed back to almost four million years.

Leakey was born in Kenya to Church of England missionaries, but the Africans they were "converting" in turn made a Kikuyu of young Louis. He grew up among them, was initiated into the tribe at 13 and often thought and dreamed in Kikuyu, which was his first language. He became known as the

"White African," and used that nickname as the title of his first book of reminiscences.

Early education came from his father's tutorials, as well as initiation into the culture and hunting methods of his African teachers. Plans to send him abroad for education were delayed by World War I, but at 16 he entered a school at Weymouth and eventually made his way to St. John's, Cambridge. There he persuaded dubious university officials to accept Kikuyu as his foreign language requirement, submitting as proof of his competence a dictated letter signed by a Kikuyu chief's thumbprint.

At 20, he took a year off to lead a fossil-hunting expedition to Tanganyika for the British Museum, after which he returned to Cambridge and took a degree in anthropology. His first independent "archeological expedition" to East Africa, consisting of himself and a friend, yielded many late Stone Age bones and artifacts from the Nakuru–Naivasha area of Kenya, to which he would return.

In 1931, he began his work at Olduvai Gorge, and established the famous FLK site, which stands for "Frida Leakey Korongo," a gully named after his first wife. (Nearly 30 years later, his second wife, Mary, found the remains of *Zinjanthropus* at FLK.) At the Kavirondo Gulf area of Lake Victoria, Leakey found early *Homo sapiens*, the earliest then known, in 1932. And, at nearby Kanam, he found a lower Pleistocene jaw, which he always considered an ancestral hominid distinct from *Australopithecus*.

For more than 40 years, Leakey (later joined by Mary) worked in the African sun under harsh conditions to search for remains of early man. He had to haul water hundreds of miles in his Land Rover to the site and often had to dodge charging rhinos and prowling lions to do his work. Not only did he find many early hominid remains, including *Homo habilis*, the still-debated fossil he considered the oldest known human ancestor, but also what he believed were early "living sites" strewn with Acheulean stone tools and a crude, early assemblage of worked pebbles he called the Olduwan culture.

During the 1950s, Kenya was in turmoil because of the Mau Mau uprising, a bloody revolution to win freedom and independence for the Africans. Two thousand Kikuyu who refused to take the Mau Mau oath were murdered, along with 32 white civilians. One of the victims was Leakey's cousin Gray Leakey, who was buried alive after having been forced to watch his wife strangled. Leakey campaigned for an end to the violence and terror, although he fully supported the Africans' goal of self-government.

When Jomo Kenyatta, a Kikuyu and a Western-trained anthropologist, was tried for conspiracy by the white government, Leakey was asked to serve as interpreter at the trial. Defense lawyers claimed his interpretation of certain Kikuyu words was inaccurate and prejudicial, and the episode created years of enmity between him and Kenyatta, who later emerged victorious as the first president of an independent Kenya. A price was put on Leakey's head by the Mau Mau; he tried to explain his position in two books *Mau Mau and the Kikuyu* (1952) and *Defeating Mau Mau* (1954), in which he sought to "heal the mental wounds that have been inflicted on all races in Kenya."

In 1959, Mary discovered *Zinjanthropus* at Olduvai while Louis was in camp with a fever. It was the first early hominid to which the new potassium-argon method of dating could be applied, and the resulting determination of 1.75 million years radically changed scientific views on the time scale of human evolution. More than 2,000 stone tools and flakes were also patiently gathered at the site and studied.

Leakey brought an unconventional, experimental approach to his theorizing about early man. When investigating the role of weapons in human survival, he attempted to subsist on small game caught with his bare hands alone and succeeded. To determine the use and efficiency of prehistoric tools, he skinned and butchered large game animals with them. He even became quite proficient in chipping his own stone tools in order to understand how they were made by early man.

Once he raised eyebrows at a graduate seminar in America by suggesting that students try eating whole mice, frogs and lizards to see how their skeletons looked after they had passed throught the human digestive tract. (He had found the remains of small creatures at prehistoric sites and wondered if they had got there by accident or as a food item of early man.)

In addition to his fossil hunting, he was also a tireless fund-raiser, a founder of the Kenyan National Game Park and the first director of the Coryndon Museum, now the Kenya Museum where his son Richard succeeded him as director.

His interest in the behavior of early man also led him to select and encourage several women to conduct field studies of the great apes. They all made important contributions to understanding the way of life of chimps, gorillas and orangs, which had never before been observed so closely and for such long periods in the wild. [See "APE WOMEN," LEAKEY'S.]

During the last years of his life, slowed down by heart disease and arthritis, he followed with intense interest the career of his son Richard, who was making important discoveries of his own. After a few years of father-son feuding, the two became reconciled, and Richard became a successor rather than a rival. Leakey died of heart failure in 1972, but Richard, Mary and a host of others carry his work forward.

See also FOSSEY, DIAN; HOMO HABILIS; LEAKEY, MARY; LEAKEY, RICHARD; "OLDEST MAN"; OLDUVAI GORGE.

For further information:
Leakey, Louis. *Adam's Ancestors.* New York: Harper & Row, 1960.
———. *By the Evidence: Memoirs, 1932–1951.* New York: Harcourt Brace Jovanovich, 1974.
———. *White African: An Early Autobiography.* Cambridge, England: Shenkman, 1966.

LEAKEY, MARY (b. 1913)
Archeologist, Paleoanthropologist

Mary Nicol Leakey is the great-great-great-granddaughter of John Frere, the first British prehistorian, who in 1800 published descriptions of flint tools and mammoth bones found sealed together in a cave deposit in the Suffolk countryside. Frere attributed them to "a remote period, even beyond that of the present world," an interpretation far too radical for his contemporaries to even consider. For 60 years his discovery was ignored.

Everyone paid attention to Mary in 1978, however, when she announced that she had uncovered a layer of fossilized hominid footprints estimated at 3.5 million years old at the site of Laetoli in Kenya. By that time she had established herself as one of the 20th century's most credible explorers of the distant past, the matriarch of a remarkable family of fossil hunters who together focused world attention on East Africa as the "cradle of mankind."

Born in London, she became interested in early artifacts when her painter father took her to visit archeological sites in France. In 1933, she met her future husband, Louis Leakey, and became his dedicated partner in the search for early man in Africa. One of the spectacular finds usually attributed to her husband's efforts, the *Zinjanthropus* or "Nutcracker man" was actually discovered by Mary and her dalmatians in 1959 at Olduvai Gorge. The "Zinj" skull has since been assigned to *Australopithecus boisei*, a hyper-robust hominid with heavy bones, powerful jaws, a jutting ridge of bone (the sagittal crest), which anchored heavy jaw muscles, and enormous molar teeth.

A talented and meticulous artist, Mary drew most of the illustrations of bone and tool specimens for Louis's works and published her own drawings of ancient African rock paintings. Although she is often depicted spending many years working with her husband at Olduvai Gorge, in fact Louis was not there nearly so often as she. While he was raising funds abroad, it was Mary Leakey who did much of

FINDING SPOTS to dig for early hominid fossils has been a specialty of Mary Leakey and her dalmatians. Discoverer of the "Zinjanthropus" skull (a hyperrobust australopithecine), Mrs. Leakey also unearthed a series of footprints three million years old—the remarkable trackway of early man at Laetoli.

the systematic development of the site and who pioneered the archeological methods that have since become widely adopted in the field.

Louis was excitable, impulsive and intuitive, while Mary is careful, systematic and logical. She recalled that he would get strong hunches about where fossils might be and "would go off for no good reason. He was very often right, but very often wrong as well." The combination of his rushing to conclusions and her painstaking attention to detail worked very well. "If we'd both been the same kind of person," she has said, "we wouldn't have accomplished as much."

In 1948, Mary discovered a partial skull of an early ape-like creature, which was named *Proconsul*, dated at 16 million years old. Discovered at Rusinga Island in Lake Victoria, Louis named it for Consul, a famous chimp in the London zoo, since he believed it to be ancestral to the chimpanzee. Although Louis is popularly given the credit for the Olduvai fossils, he never actually found a skull by himself. Mary, the African "hominid gang" and sometimes the Leakey children were the actual discoverers.

Among Mary's important finds was a jaw that Donald Johanson named (against her wishes) the type specimen of *Australopithecus afarensis*, linking it with his "Lucy" discovery from the Afar region of Ethiopia, 1,000 miles to the north. However, Mary never accepted the "honor" and maintains her find is a species of *Homo* and not the same as Johanson's Hadar hominids.

At a site called Laetoli in Kenya, 30 miles south of Olduvai Gorge, in 1976–1978, she made what she considers the most exciting discovery of her career: preserved footprints of three hominid individuals who had left their tracks in soft volcanic ash more than three million years ago. It is a remarkable record of "fossilized" behavior, establishing that very ancient man-like creatures walked exactly as we do.

Mary is also, of course, mother and archeological mentor to Richard Frere Leakey, whom she had the foresight to tag with her ancestor's family name. Despite his youthful disavowal of the Frere–Leakey legacy, Richard perpetuated family tradition by becoming a pioneering, cantankerous, world-class prehistorian. The discoveries of his "hominid gang" include the oldest, most complete skull of a large-brained hominid (the 1470 *Homo habilis*) and the most complete skeleton of an ancient hominid (WT 1500, or the Turkana boy, *Homo erectus*). Ten years after his mother found the first hyperrobust *australopithecus* at Olduvai, Richard discovered another complete skull (406) of the same *boisei* hominid.

See also FRERE, JOHN; LEAKEY, LOUIS; LEAKEY, RICHARD; "LUCY."

For further information:

Leakey, Mary. *Disclosing the Past: An Autobiography.* Garden City, N.Y.: Doubleday, 1984.

———. *Olduvai Gorge: My Search for Early Man.* London: Collins, 1979.

Lewin Roger. *Bones of Contention.* New York: Simon & Schuster, 1987.

Reader, John. *Missing Links: The Hunt for Early Man.* London: Collins, 1981.

Willis, Delta. *The Hominid Gang.* New York: Viking, 1989.

LEAKEY, RICHARD ERSKINE FRERE (b. 1944)
Paleoanthropolgist, Museum Administrator, Wildlife Conservationist

When Richard Leakey was a boy in Kenya, he grew to hate the sight of old stones and bones. Since his father Louis and mother Mary had dedicated their lives to the search for human origins, each important new scrap of fossil skull competed with him for their attention. If there was one thing he was absolutely *not* going to be when he grew up, he decided, it was a paleoanthropologist.

Given the option of attending college or going to work when he was seventeen, he recalls in his book *One Life* (1981) "there was no doubt in my mind that I should avoid at all costs an academic life, and . . . [I was] determined to distance myself from my parents and their work."

Instead, he learned the ways of the bush, of organizing expeditions and tracking wildlife. He decided to become a photo safari guide, leading tourists to remote wilderness areas. Several times he organized equipment and vehicles for film crews covering his parents' work and was impressed by the worldwide interest in their search for fossils of human ancestors.

More like his father than he would be comfortable to admit (even to himself), the brilliant renegade from a family of brilliant renegades began to dream of following the track of early humans. But it would have to be on his own terms, not as his father's assistant or apprentice. "I wanted a show of my own," he later recalled.

Following his own hunches, in 1964 he (and his friend, archeologist Glynn Isaac) prospected a few promising sites near Lake Natron. Early on, his assistant Kimoya Kimeu found the only known lower jaw of *A. boisei*, a species previously known only by Mary Leakey's fragmentary find at Olduvai in 1959 (when it was called *Zinjanthropus* or "Nutcracker Man"). Known as the "Peninj jaw," it was the first of many exciting discoveries by Leakey and his associates. In 1968, he accompanied his father on a fund-raising visit to the National Geographic Society in Washington, D.C. and made his first bold step out of his father's shadow.

After Louis had received continued support for his and Mary's work at Olduvai Gorge, young Leakey seized the floor to describe his own plans. Although untrained as an anthropologist or archeologist, he asked the startled board members to bankroll him at Lake Turkana (then Rudolf). Startled, they agreed to sponsor one season. "But if you find nothing," they warned, "you can never come back begging here again."

His "Leakey luck" held, for he turned out to be a fossil-finder of the first rank. But as Leakey's reputation grew, so did the tensions between him and his father, rival anthropologists and government bureaucrats involved in administering the Kenya National museums and archeological sites. His book *One Life* details the convoluted daily battles and infighting at which he proved to be proficient. Richard Leakey was determined to do things his own way, and, for the most part, he succeeded.

In 1972, Leakey's team discovered the famous 1470 skull—a hominid with a large brain (750 cc) he identified as *Homo habilis,* a species of man rather than *Australopithecus.* His father Louis was delighted; it was the kind of evidence he had been seeking for an archaic type of human older than *Australopithecus.*

Unfortunately, shortly after that reconciliation, Louis died of a heart attack in London. With his mother's support Richard became his successor in running the search for early man in East Africa. Spectacular fossil finds continued to emerge from Kenyan sites, including the remarkable skeleton of an adolescent *Homo erectus,* known as the Turkana Boy, and the "Black Skull" of a robust australopithecine.

Over the years, Leakey has helped overthrow simple linear schemes of hominid evolution and demonstrated that several distinct species of near-men simultaneously inhabited ancient East Africa. Yet, he has simultaneously fought to establish *Homo habilis* as our earliest known ancestor, with all australopithecines as sidebranches. On this issue, Louis Leakey and his family were in remarkable agreement.

During decades of leadership in the search for human origins, Leakey earned the respect of scientists as a top paleoanthropologist, despite his open contempt for formal training. Now a prominent scientist, administrator, conservationist, author of several popular books and burdened by numerous responsibilities, he still flaunts a Huck Finn-like delight in having skipped school and gotten away with it. He is fond of telling interviewers he has "never been to a university—except to lecture." His public talks invariably draw huge, paying crowds.

In 1979, Leakey almost died of kidney failure and appeared to face the end of his active career. His brother, Philip, with whom he had never gotten along, offered to donate one of his own kidneys. Richard agonized over old resentments, the dangers to his brother, and the enormity of this gift of life, and finally accepted the organ transplant. Philip, a career politician, was widely quoted as telling a reporter, "Well, I can't really say I hate my brother's guts any more."

Richard Leakey is fond of insisting he is not an anthropologist—a profession he swore he would never enter. As director of the museum, he has supervised projects in botany, zoology, geology and many other fields. Proud of his Kenyan nationality, he has worked hard to bring Africans into the scientific enterprise, which for so long was monopolized by Europeans.

Controversy still follows him everywhere, and he seems to thrive on it. He tells his rival Donald Johanson that his famous "first family" of *Australopithecus afarensis* is really not one, but two species, neither of which are ancestral to humans. His wars with bureaucrats, departments and agencies over jurisdictions, funds and permissions are endless. In 1988, he accused high officials in the Kenyan government of being "soft" on elephant poachers, charging they were profiting from wildlife slaughter. He was promptly chastised, pressure was exerted and he resigned his posts as head of the museum and Department of Archeological Sites. Within a week, the president of Kenya personally saw to it that Leakey was reinstated. In 1989, he left the museum to accept a government post as National Parks Director of Wildlife Management. He began with a vigorous campaign to save the elephant herds, which are being pushed to extinction by ivory poachers.

See also HOMINID GANG; HOMO HABILIS; JOHANSON, DONALD; LEAKEY, LOUIS; LEAKEY, MARY; "OLDEST MAN"; TURKANA BOY.

For further information:

Leakey, Richard. *The Making of Mankind.* New York: Dutton, 1981.

———. *One Life: An Autobiography.* London: Michael Joseph, 1981.

Willis, Delta. *The Hominid Gang.* New York: Viking, 1989.

LEBENSBORN MOVEMENT
Attempt to Create "Master Race"

Evolutionary studies inspired the eugenics movement, founded by Charles Darwin's cousin Sir Francis Galton in the 19th century. Galton proposed programs of selective breeding to "improve" the human species and discourage genetically "inferior" individuals from reproducing. However, no governments actually attempted to artificially evolve a new, "superior" breed of humans until the disastrous experiments of Nazi Germany.

In 1936, Heinrich Himmler and his Stormtroopers (S.S.) founded an institution called *Lebensborn* or "Fountain of Life." Its purpose was to create millions of blond, blue-eyed "Aryan" Germans as the genetic foundation of the new "Master Race." Lebensborn children would be raised to be obedient, aggressive, patriotic and convinced their destiny was to dominate or destroy all "inferior" races or nations. Galton's well-intentioned dream of human improvement had become a nightmare in reality.

The stated objective of Lebensborn was "to enlarge Germany's existing blood basis of 90 million to 120 million." This was to be achieved by forcibly taking "Aryan" or "Nordic" (blond, blue-eyed) children from their parents, wherever found, and raising them as good Nazis, either with childless Nazi couples or in foundling institutions. Another method was to mate

Aryan-looking single women with "superior" fathers, some of whom would impregnate many women. Not surprisingly the men chosen to perform this patriotic task were almost all S.S. officers.

According to the Lebensborn report for 1938, "eight hundred thirty-two valuable German women decided despite their single state the sacrifice entailed, to present the nation with a child." Part of the report emphasized the economic windfall of the program. If each new child during his lifetime contributed 100,000 Reichsmarks to the economy, Lebensborn could claim to have enriched Germany by 83 million Reichsmarks.

On his first inspection tour of occupied Poland, Heinrich Himmler had been amazed by the Nordic appearance of many "Slavic" children and had decided it was his duty to kidnap them from their parents and send them to Lebensborn foster homes and orphanages in Germany. Two special S.S. agencies—the Volksdeutsche Liaison Office and the Race and Settlement Office—searched Europe for "human specimens considered suitable for Germanization." They also "cleared" vast areas of land of "inferior" Poles and Jews by killing or enslaving the inhabitants, and practiced euthanasia (mercy killing) on thousands of Germans who were feebleminded, infirm, mentally ill or crippled. (The gas chambers were originally invented to weed out their own "unfit" Germans, and were later applied to the genocide of entire "undesirable races," particularly Jews.)

Nazi eugenics had two aspects: the extermination of millions of "undesirables" and the selection and breeding of preferred "Aryan" types. It was an article of faith that the blond, blue-eyed "Nordic-looking" children would also prove intellectually and morally superior and that they would "breed true" when mated. Neither assumption was correct. Instead of changing evolutionary history, the Lebensborn experiment produced a group of perfectly ordinary, confused orphans, who had to cope with the chaotic aftermath of a society gone mad.

See also ARYAN "RACE," MYTH OF; EUGENICS; GALTON, SIR FRANCIS

LEIDY, JOSEPH (1823–1891)
Founder of American Paleontology

Dr. Joseph Leidy, who founded American paleontology, was one of the last great naturalists to tackle the entire range of biology. While teaching human anatomy at the University of Pennsylvania's medical school, he became a self-taught expert in the structures of all kinds of living animals. But, above all, Joseph Leidy is remembered as the first scientist to discover, assemble and describe an American dinosaur.

During the summer of 1858, workmen digging a house foundation in the small town of Haddonfield, New Jersey, uncovered some puzzling fossils. Smaller chunks were carried off by curious onlookers, who used them as unusual paperweights and doorstops. Some of the larger bones were sent to Professor Leidy, then director of the Academy of Natural Sciences in Philadelphia.

Leidy immediately grasped their importance. He knew it had been hardly a decade since European scientists had first recognized the existence of extinct giant reptiles, named "dinosaurs" (terrible lizards) by the British Museum's Sir Richard Owen. Rushing to New Jersey, Leidy collected the rest of the bones as they were dug up, then went from house to house in the neighborhood, offering to buy up all the scattered teeth, foot bones and fragments that were decorating the villagers' desks, gardens and window ledges.

Back in Philadelphia, he was able to reconstruct the skeleton as a reasonably complete duck-billed dinosaur, which he named *Hadrosaurus foulkii*, the first known dinosaur from America. He thought it was a relative of England's *Iguanodon*, the very first dinosaur recognized anywhere. This "great extinct herbivorous lizard," he thought, "may have been in the habit of browsing, sustaining itself, kangaroo-like, in an erect position" on hind legs and tail—a judgment still considered accurate.

Leidy's dinosaur turned out to be a more complete specimen than any at the British Museum, enabling him to make the most accurate dinosaur reconstruction then known. His Hadrosaur, arguably the first mounted dinosaur skeleton in any American museum, was put on exhibit at Philadelphia's Academy of Science, where it still can be seen.

After this dramatic success, Leidy became increasingly interested in American prehistoric animals, which were entirely unknown and beginning to be discovered in increasing numbers. Using his great knowledge of living creatures, he was to reconstruct, describe, classify and name a whole new world. He was entirely up to the great task. Unlike his successors—the militant paleontologists Edward Cope and Othniel Marsh—Leidy was a patient and peaceable man. It was he who first found and named many of America's early mammals, but he bowed out of paleontology, unwilling to participate in the frenzied "bone wars" of his students.

Sadly, in the later bitter rivalry between Cope and Marsh, Leidy was all but forgotten. Paleontologist Henry Fairfield Osborn, director of the American Museum of Natural History, recalled that many of the Eocene and Oligocene animals had been given three names in the scientific literature:

> the original Leidy name and the Cope and Marsh names. It has been the painful duty of Professor [William Berryman] Scott and myself to devote thirty of the best years of our lives trying to straighten out this nomenclatural chaos.

Even as he grew older, Leidy maintained his youthful enthusiasm for unraveling nature's secrets. When in his mid-sixties, after a lifetime of prodigious labor, someone asked if he were ever tired of his work. "Tired!" he said, "Not so long as there is an undescribed intestinal worm, or the riddle of a fossil bone, or a rhizopod new to me."

See also COPE–MARSH FEUD; IGUANODON DINNER; "NOAH'S RAVENS"; PEALE'S MUSEUM.

For further information:
Howard, Robert West. *The Dawnseekers: The First History of American Paleontology*. New York: Harcourt, Brace, Jovanovitch, 1975.

LEMURS

See PROSIMIANS.

LESQUEREUX, LEO (1806–1889)
Pioneering Paleobotanist

One of the most tragic and remarkable of earth scientists, Leo Lesquereux helped found the study of fossil plants in North America. His beautifully illustrated volumes of American *Coal Flora* (1879–1884) rivaled the best that were being published in Europe, yet he was unable to earn a living from teaching or research. A Swiss immigrant to America, he learned English after he became deaf and so never heard the language in which he wrote his books.

Lesquereux's family pursued a typically Swiss occupation—they manufactured parts for watches. As a boy, he loved to climb difficult mountains to gather rare flowers from their peaks. At the age of 10, a misstep took him over the edge of a cliff and he bounced down a mountainside, unconscious most of the way. No bones were broken, but he lay in a coma for two weeks. His hearing was permanently damaged; it would grow progressively worse over the years.

He tutored French, married a noblewoman who was one of his students and became principal of the local high school. His spare time was spent studying mosses and collecting rocks in the mountains, but this idyllic life was not to continue. His hearing grew worse, and an operation in Paris intended to restore it, instead extinguished it completely. He had to give up his career as an educator and followed his friend August Agassiz (brother of Louis, the Harvard paleontologist) to America.

Lesquereux was 41 when he arrived in Boston in 1847. He had a wife and five children to support, was now totally deaf and spoke no English. Agassiz hired him to classify a plant collection. Soon his reputation grew, prompting an invitation to Ohio to assist in studies of mosses. Other invitations and commissions followed, which allowed him to make important and original contributions to his science, but none ever paid well enough to support his family.

No university would hire a deaf professor or researcher, so Lesquereux worked as an unpaid assistant on various geological expeditions in order to pursue his passion for paleobotany. In between field trips, he had to give a great deal of time to nonscientific pursuits in order to support his family. He set up a small watchmaking and jewelry business; later, he trained his three sons in that business as well.

Nevertheless, Lesquereux plugged away tirelessly at his work, overcoming every obstacle. On expeditions, he sat day after day on waste slate heaps, turning over every rock in his search for the plant fossils that fascinated him. Cold winds, noonday heat or pouring rain made no difference. Over the years, he made a number of important discoveries and produced a stream of excellent publications. And despite the hardships he faced, he survived and continued to be productive well into his eighties.

Colleagues marveled at his ability to operate under conditions that would have driven a lesser man to despair. Writing in 1895, one recalled, "I have been present when Lesquereux talked with three persons alternately in French, German, and English by watching their lips. The interview would begin by each one saying what language he intended to use." Then Lesquereux would answer each in their own tongue. However, since he had never heard the sounds of English, "his pronunciation of it was curiously artificial and original."

Lesquereux's faith in revealed scripture was strengthened rather than weakened by the *Origin of Species* (1859). Charles Darwin, in a letter to his friend Joseph Hooker in 1865, expressed amazement at "an enormous letter from Leo Lesquereux on Coal Flora." Lesquereux had originally been a trenchant critic of the *Origin of Species*, but said Darwin:

> he says now after repeated reading of the book he is a convert! But how funny men's minds are! He says he is chiefly converted because my books make the Birth of Christ, Redemption by Grace, etc., plain to him!

For further information:
Andrews, Henry. *The Fossil Hunters: In Search of Fossil Plants*. Ithaca: Cornell University Press, 1980.

LEWONTIN, RICHARD

See RACISM (IN EVOLUTIONARY SCIENCE); RED QUEEN HYPOTHESIS.

LIBBY, WILLARD F.

See RADIOCARBON DATING.

LIFE, ORIGIN OF

Scientists cannot agree on a single formal definition of life, though most people think they can recognize a living thing when they see it. At the level of chemical reactions (and far back in time), the distinction between life and nonlife gets fuzzy. Living things are built of nonliving carbon, hydrogen, nitrogen, oxygen, phosphorus and sulphur, but just how they got organized into something alive remains one of biology's deepest mysteries.

In 1953, at the University of Chicago, Stanley L. Miller and Harold C. Urey mixed ammonia, water vapor, hydrogen and methane to simulate Earth's early atmosphere, then crackled lightning-like electrical sparks through it. Amino acids, the building blocks of proteins, appeared, along with other organic compounds usually produced only by living things.

Over the years, continued experiments with gases and chemicals bombarded with heat, ultraviolet and electrical energy produced almost every chemical component (nucleotides) of DNA and RNA, but the parts never assembled into the long molecules that make life possible. More recently, longer DNA-like molecules did form spontaneously from simpler carbon compounds and zinc salts at California's Salk Institute.

Given the tendency of these chemicals to form nucleotides and of nucleotides to build long-chain molecules, according to biologist Lynn Margolis, the process of life emerging from nonlife is "hardly surprising" and "shockingly straightforward." Since an RNA molecule's structure resembles "one-half of an open zipper . . . when the right ingredients are present, the components of the missing half simply line up and fit, thus forming a perfect copy of the original."

Unfortunately, as Margolis admits, "no cell has yet crawled out of a test tube," and thousands of similar experiments have produced goopy organic tars, but no recognizable life. Decades of persistent failure to "create life" by the "spark in the soup" method (or to find such productions in nature) have caused some researchers to seek other approaches to the great enigma.

A few, including geneticist Francis Crick and the astronomer Fred Hoyle, think life may have arrived on Earth from elsewhere in the universe, perhaps hitchhiking on a meteor or comet. But this notion of wandering celestial rocks "seeding" the Earth (Panspermia), even if true, does not solve the riddle of how life originated: it simply pushes the question father back to another time and place.

Another recent theory looks to the structure of clays and minerals as possible templates, trapping and holding molecules in patterns, which might make them susceptable to the process of natural selection.

But even the most promising, technically sophisticated attempts to demonstrate the origin of life from nonliving chemicals are still guesses and gropes in the dark. For almost a century, many scientists have taught that some version of the "spark in the soup" theory "must" be true. Repetition of this idea as fact, without sufficient evidence, has done a disservice to new generations by capping their curiosity about a profound and open question.

See also BIOSPHERE; CLAY THEORY (OF ORIGIN OF LIFE); EUKARYOTES; PANSPERMIA.

For further information:
Cairns-Smith, A. G. *Seven Clues to the Origin of Life.* Cambridge: Cambridge University Press, 1985.
Halvorson, H. O., and Van Holde, K. E., eds. *The Origins of Life and Evolution.* New York: Alan R. Liss, 1980.
Oparin, A. I. *Life: Its Nature, Origin and Development.* New York: Academic Press, 1964.
Shapiro, Robert. *Origins: A Skeptic's Guide to the Creation of Life on Earth.* New York: Bantam, 1986.

LIGHTFOOT, REVEREND JOHN

See USSHER/LIGHTFOOT CHRONOLOGY.

LINNAEAN "SEXUAL SYSTEM"
Erotic Botanica

Carl Linnaeus has come down in history as a dry-as-dust classifier who gave thousands of Latin names to plants labeled in museums. What is often forgotten is that some of those labels created a sensation in the mid-18th century and thrust their author into fame and infamy.

As a student, Linnaeus had taken a keen interest in the sexuality of plants and peppered his natural history writings with erotic imagery. In his thesis, "Praeludia Sponaliarum Plantarum," presented in 1729 at the university in Uppsala, he wrote of the pistils and stamens of flowers:

The actual petals of a flower contribute nothing to generation, serving only as the bridal bed which the great Creator has so gloriously prepared, adorned with such precious bed-curtains and perfumed with so many sweet scents in order that the bridegroom and bride may therein celebrate their nuptials . . . When the bed has thus been made ready, then is the time for the bridegroom to embrace his beloved bride and surrender himself to her . . .

Taking his cue from the earlier work of Camerarius (1694) and le Vaillant (1718), Linnaeus had no hesitation in describing pollination as a sexual act. Pollen was sperm, seeds were ova, and there is no fertilization if the anthers are removed (castration).

Although his teacher, Professor Celsius, was impressed with the thesis, and it was instantly popular with his fellow students, Linnaeus was treading on dangerous ground. Churchmen looked to nature for morality: the industrious ant, the persistent tortoise, the faithful dove. To equate plant fertilization with the sex life of animals was to call attention to "polgamy, polyandry, and incest" in nature.

By the time he brought out his *Systema Naturae* in 1735, Linnaeus had devised an entire "sexual system" for classifying plants. Flowering plants were grouped according to the number of stamens, while flowerless groups were defined by the form of the female organs.

Monandria is like "one husband in a marriage," *Diandra* like "two husbands in the same marriage," and the *Polyandria*—a group that includes the linden and the poppy—"twenty males or more in the same bed with the female." The counting of stamens and pistils was a practical way of determining the group to which it belonged and the erotic imagery a strong aid to remembering the family name.

The system proved to be very practical in identifying species and also greatly helped to popularize the study of plants and flowers. It even inspired Charles Darwin's grandfather Erasmus to write a controversial book of erotic poetry. His *The Botanic Garden; or Loves of the Plants* (1790), illustrated with drawings of sensual human lovers, celebrated the sexuality of vegetation with decorous lustiness.

Many of Linnaeus's fellow naturalists admired the system's usefulness and were amused by its explicitness, but others were outraged at this "botanical pornography." Long after the great taxonomist's death, the bishop of Carlisle wrote that "nothing could equal the gross prurience of Linnaeus' mind." Another churchman protested that such botanical names as *Clitoria* (a pea plant) "is enough to shock female modesty." Even a century later, the great German poet-naturalist Goethe advocated censorship of botanical textbooks used by university students.

See also BINOMIAL NOMENCLATURE; LINNAEUS, CARL; LINNEAN SOCIETY; SPRENGEL, CHRISTIAN K.

LINNAEUS, CARL (1707–1778)
Founder of Taxonomy, Botanist

In Sweden, Karolus (or Carl) Linnaeus is still a great name, as important as Shakespeare is to England or Dante to Italy. During his lifetime, he was a national celebrity who dominated European biology, although churchmen were shocked by his "sexual system" for classifying plants.

Linnaeus's classification of living things still forms the basic language of world science. When he created his method of naming species 200 years ago, he believed all species were unchanging creations of God; yet his system was easily transformed into a vast chart of evolutionary relationships.

He devoted his life to naming and classifying every living thing according to similarities in structure. Before Linnaeus developed his system, natural history classification was in chaos. Without regard for systematic anatomical comparison, some put flying fish and birds together (because both fly) or turtles and armadillos (because both have shells.)

As European colonization of far-flung lands reached its zenith, immense numbers of creatures and plants new to science were sent back to England, Holland and France. The need to bring order and system to this rapidly expanding body of knowledge was great.

Since natural science grew out of the religion of his time, Linnaeus regarded each species as a distinct idea in the mind of God. He saw natural groups as divine "plans" of organization and thought his mission was to discern that plan. As he modestly put it, "God created, but Linnaeus classified." (Late in life, however, experience with hybrids and problematic varieties made him doubt whether species were quite so "fixed" after all.)

Churchmen believed that man was God's final work, the crowning glory of Creation. Linnaeus said that he was following Adam's example by spending his life giving names to all the other living things, though he was also interested in their behavior, reproduction and distribution. His system of classification became so successful that most later naturalists became preoccupied with finding and naming new species to the exclusion of studying the lives of plants and animals in nature.

The surname Linnaeus was coined by his father Nils, after a lime tree near his birthplace. Carl was therefore born Linnaeus and did not Latinize "von Linne" later in life, as many books erroneously repeat. (Actually, he used both names without preference.) Born in the southern Swedish province of Smalland near a beautiful lake, he grew up surrounded by gardens and wildflowers. His parents had hoped he would follow his father into the church, but, to their disappointment, he showed no talent for metaphysics and other priestly subjects.

Carl loved to work with plants instead, and a kindly professor steered him toward medical study. (A botanist was almost a physician, for all physicians had to be botanists. Most drugs came from the rare plants they grew themselves in special gardens of *materia medica*.).

But Carl was much more interested in the plants than in their medical uses and set out to be a botanist. It was a difficult road, paved with hardships and poverty; he bounced from one teacher to another, never losing sight of his goal: to be professor of botany at the university at Uppsala.

Little by little, his wide knowledge, unceasing labor and personal charisma brought him to the top of his field. He developed unsurpassed collections of

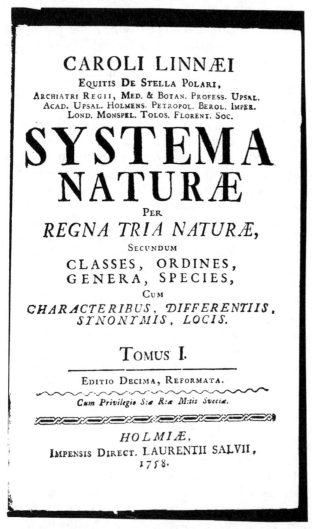

CAROLI LINNÆI

EQUITIS DE STELLA POLARI,
ARCHIATRI REGII, MED. & BOTAN. PROFESS. UPSAL.
ACAD. UPSAL. HOLMENS. PETROPOL. BEROL. IMPER.
LOND. MONSPEL. TOLOS. FLORENT. SOC.

SYSTEMA NATURÆ

PER

REGNA TRIA NATURÆ,

SECUNDUM

CLASSES, ORDINES,
GENERA, SPECIES,

CUM

*CHARACTERIBUS, DIFFERENTIIS,
SYNONYMIS, LOCIS.*

TOMUS I.

EDITIO DECIMA, REFORMATA.

Cum Privilegio S:æ R:æ M:tis Sveciæ.

HOLMIÆ,
IMPENSIS DIRECT. LAURENTII SALVII,
1758.

GROUNDBREAKING work on classification, or taxonomy, Linnaeus' *Systema Natura* revolutionized botany and zoology. His system was superior to any that had been used before and proved adaptable to the Darwinian transformation of biology into an evolutionary science.

animals and plant specimens, containing thousands of species he had personally discovered, classified and named.

Eventually, he devised scientific names for the roughly 4,200 species of animals and 7,700 plants then known. Botanical names published before 1763 have no standing unless they were adopted by Linnaeus in his *Species Plantarum* (1752) or *Genera Plantarum* (5th edition, 1754). His *System of Animate Nature* (1735) has gradually been expanded, until today it includes 350,000 plants and more than a million animals.

See also BINOMIAL NOMENCLATURE; LINNAEAN "SEXUAL SYSTEM"; LINNEAN SOCIETY.

For further information:
Blunt, Wilfred. *The Compleat Naturalist: A Life of Linnaeus.* New York: Viking Press, 1971.
Jackson, Benjamin D. *Linnaeus.* London: Witherby, 1923.
Linnaeus, Carl. *Systema naturae. Regnum animale.* 1758. Facsimile reprint. London: British Museum (Natural History), 1956.

LINNEAN SOCIETY
British Science Organization

Carl Linnaeus (1707–1778), the inventor of the scientific classification system of living things, spent a lifetime amassing a priceless museum of specimens. Many still exist: 2,000 are in the Natural History Museum in Stockholm. But his personal collection—including all his books, manuscripts, correspondence, dried fishes, shells, insects and herbarium—became a legacy for his widow. She would have preferred cash.

In 1784, Mrs. Linnaeus happily sold the collections to a wealthy English naturalist named James E. Smith. They were his pride and joy until he died 30 years later, at which time Mrs. Smith put them up for sale. She could not find an individual who wanted the entire collection and decided to sell it piecemeal.

In 1829, a public-spirited group of gentlemen decided to form an association to keep the collection and library intact and out of the hands of widows. After raising the necessary contributions, they bought the great taxonomist's book and specimens, and founded the Linnean Society of London, which still owns and maintains them.

The Linnean Society soon became a major force in British science, particularly during the 19th century, as it was a place where a naturalist could present new discoveries to the country's scientific elite.

In April, 1858, the distinguished geologist Sir Charles Lyell and the botanist Sir Joseph Hooker "communicated" to the Society a joint theoretical statement by Charles Darwin and Alfred Russel Wallace. Although their versions were separate (Darwin's had been written 14 years earlier), their theory proposed that species gradually arose through "natural selection"—a directional force operating upon a source of random variation.

That occasion marked the beginning of modern biology, but neither Darwin nor Wallace appeared at this famous non-event. It attracted little attention and no discussion at the Linnean Society, whose namesake had championed the permanence of species. In the Society's annual report for 1858, its president noted that it had been a dull year in biology, with no important new theoretical developments of the sort that occasionally revolutionize a scientific field.

See also "DELICATE ARRANGEMENT"; LINNAEUS, CARL.

LIVING FOSSILS

See CRYPTOZOOLOGY.

"LIZZIE BORDEN" EVOLUTION
Replacing "Parental" Species

While puzzling over patterns in the fossil record in the 1980s, paleontologist Robert Bakker thought of Lizzie Borden, the accused ax murderess of a century earlier. Her name seemed the perfect nickname for his evolutionary interpretation. As Bakker reads the evidence, pockets of modified populations keep extending their ranges until they wipe out their "parent" species.

Miss Borden, according to the old rhyme, "took an ax and gave her mother forty whacks," after which she also chopped up her father. The truth, despite popular tradition, is that the evidence was deemed insufficient, a jury acquitted her, and the crimes were never officially solved.

Bakker had been struggling with Darwin's view that species are modified by accumulating minute, gradual variations over time. Many paleontologists agree that, in most cases, fossils just don't support that model.

Instead, species seemed to be remarkably stable over long periods, some changing hardly at all over millions of years. When there were changes, they appeared in "jumps," which were widespread through populations over a large geographic area. Sometimes "parent" and modified "descendant" populations appeared very close together in time.

Bakker believes the fossil species he studied did remain stable over long periods and throughout a wide range. But then a population in one part of the range produced a new adaptation, perhaps to local climatic conditions that later became more widespread. This would mean that a "parent" population and a modified descendant population existed at the same time, rather than the whole species evolving gradually.

As the small, newly modified population extended its range, moving into areas occupied by the older species, it would gradually outcompete and replace the older species throughout the habitat. Eventually, the "parental" population would be wiped out.

However, as in the case of its namesake, there is still plenty of room for argument about the evidence for "Lizzie Borden evolution." Perhaps her name is even more appropriate than Bakker realized; the jury is still out.

See also ANAGENESIS; GRADUALISM; PUNCTUATED EQUILIBRIUM.

LOCH NESS MONSTER
Legendary Surviving Dinosaur

According to ancient Scots tales and testimony from scores of modern "Nessie-sighters," a species of large unknown animal lives in the deep, narrow body of water known as Loch Ness, near Inverness. Many who believe in the "monster's" existence point out that the Loch has special conditions, which could have allowed the survival of a relict population of aquatic dinosaurs or the closely related plesiosaurs. Discoveries of other supposedly extinct creatures or "living fossils" include the coelocanth, the tuatara and the okapi within the past century.

Clearly, there could not be one "creature" that has persisted over thousands of years; it would have to be a small breeding population containing males, females and young. Distinguished British naturalist Peter Scott became so convinced Loch Ness contained living plesiosaurs, he even attempted to give them a "scientific" Latin name. However, zoologists made a firm rule some years ago not to allow names for species that have not yet been found.

Cryptozoologist (student of "hidden animals"), Robert Rhine believes he snapped an underwater photo of the creature and its diamond-shaped flipper in 1977, which made the front pages of newspapers around the world. (Murky water made the shot extremely blurry.) Rhine also claimed he had the bulk of the animal's body in view for 20 minutes in the presence of six witnesses.

Local residents had been sighting "Nessie" for years, but world interest in the story intensified in 1933, when a new road around the Loch opened up the area to automobiles. Since then, there have been more than 4,000 reported sightings, though very few are taken seriously by experienced "Nessie" hands.

Many sightings can be discounted as wakes of boats, floating masses of vegetation, oddly shaped logs or giant eels. Nevertheless, after most have been

A SURVIVING PLESIOSAUR HERD has been a favorite candidate for the legendary "beasties" of Scotland's Loch Ness. This version of the creature was sculpted by special effects artist Bill de Pauolo for a film to be titled *Legend of the Loch*.

debunked, there remains a sprinkling of good observations from credible witnesses and even unexplained sonar blips of moving shapes at great depths. Thousands have searched for Nessie, although actually capturing the creature is forbidden by local law (the first ordinance making it a crime to kill or trap an animal no one is sure exists).

If there were a group of plesiosaurs in the Loch, the ecology might well be favorable to their survival. The Loch is 22 miles long, 2 miles wide and at least 1,000 feet deep. Its submerged cliff sides are crenelated with unexplored caverns, some perhaps containing airshafts. The deep waters of the Loch hold a steady temperature (42–44°F) year round because of its great volume and it never freezes over. Abundant eels and fish would provide a continuous food supply.

In the fall of 1988, the most ambitious, coordinated attempt to locate the Loch Ness creature was mounted: Operation Deep Scan. Twenty-four boats equipped with deep sea fish-finders lined up to form a sonar curtain, then swept the Loch from end to end. A remote control mini-submarine with mounted television camera was also used to probe the peat-darkened waters. Of course, the noisy activity of these motorized "monster-hunts" might be enough to drive any shy aquatic creature into hiding.

Unexplained sonar blips could have been caused by huge eels or fish. However, as Sir Richard Owen and the "sea-serpent" watchers of a century ago realized, one can't prove a negative. Science cannot say there is no such creature just because none have yet been caught.

As one veteran monster-hunter put it, "It's much more fun to believe there is a plesiosaur here than that there isn't." Each year, one million visitors to Loch Ness prove the point, arriving with hopes of a sighting and leaving with Nessie mugs, key chains, and T-shirts. Spending about $250 million, they have turned this possibly imaginary "beastie" into the third top tourist attraction in the United Kingdom. See also CRYPTOZOOLOGY; DRAGONS.

For further information:
Costello, Peter. *In Search of Lake Monsters.* New York: Coward, McCann, 1974.

Mackal, Roy P. *The Monsters of Loch Ness.* Chicago: Swallow Press, 1976.

Witchell, Nicholas. *The Loch Ness Story.* Hammondsworth, England: Penguin Books, 1974.

LOEB, JACQUES (1859–1924)
Early Bioengineer

Jacques Loeb, a German biologist who immigrated to America, stands at the head of the bioengineering tradition. His main interest was in manipulating or controlling the processes of life, a focus that put him at the heart of some raging controversies.

Loeb is best remembered for his experiments on sea urchins demonstrating parthenogenesis: the production of offspring from a female without fertilization by a male. By using mild electrical shocks or pinpricks, he found he could get unfertilized urchin eggs to hatch. Later, Loeb produced tadpoles from unfertilized frog eggs using similar techniques. The resulting worldwide publicity around 1900 stirred interest in a scientific basis for "virgin birth" in more complex creatures.

From his teachers at Strassburg in the 1890s, Loeb became convinced that the prevailing "materialist" or "reductionist" view of organisms was incomplete; nature was far more marvelous and unpredictable than most scientists had thought. But Loeb didn't really care how organisms worked; as far as he was concerned, they might as well be "black boxes" with unknown workings. What interested him most were what effects could be produced and what controls he could discover as "an engineer of living substance."

After coming to America, he worked on his experiments first at Bryn Mawr, then at the University of Chicago, and later at Berkeley. He thought the inventive, practical Yankee spirit would find his work congenial and was shocked when he found himself surrounded by American critics.

Progressive evolutionists thought he was tampering with a Divine plan by "meddling" with life. "Mechanistic" physiologists disowned him because he couldn't and wouldn't offer explanations for his experimental results. Finally, he became intellectually isolated, pigeonholed by American science as an archreductionist, espousing the philosophy he had fought so hard to oppose.

For further information:
Pauley Philip. J. *Controlling Life: Jacques Loeb and the Engineering Ideal in Biology.* Oxford: Oxford University Press, 1988.

LONESOME GEORGE
Symbol of Endangered Species

There is something deeply affecting about a creature who will spend the next 100 years searching for a mate and has no way of knowing he is the very last of his kind.

Newspapers around the world carried his picture: a 200-pound giant land tortoise of the Galapagos Islands, nicknamed "Lonesome George." He became a symbol of endangered species everywhere, evoking sympathy as "the loneliest creature on Earth." The San Diego zoo erected a bronze statue in his honor, and he inspired thousands to contribute to the cause of threatened wildlife.

George is the last of a saddle-backed species (*Geochelone elephantophus abingdoni*) originally native to Pinta Island, one of the driest in the Galapagos. An upturned shell shape enables him to raise his neck and reach high for cactus pads, unlike the dome-shaped varieties found on other islands.

However, since George became a celebrity, he doesn't have to reach for food anymore. He has been moved to Santa Cruz island, 100 miles from his homeland and is kept in a spacious pen, where his food is spread on the ground. When he was discovered in 1971, giant tortoises had not been seen for years on Pinta Island and were believed extinct there. A male without a mate, "the last of a doomed species," he was removed for safekeeping, and a $10,000 reward was offered to anyone who could find a female on Pinta. Although the island was combed by bounty hunters, no other tortoise was ever found.

Darwin visited the Galapagos in 1835 and was fascinated by its life forms, including the tortoises. He was intrigued to learn that each of the 14 islands had its own species and that many islands had their own related varieties of mockingbirds and other creatures, which led him to conclude that the populations had diverged from common ancestors and were not the result of multiple creations.

Tortoises did not fare well in Darwin's day. Sailors, pirates and whalers carried off more than 200,000 of them to help feed crews on long voyages. Hundreds were dumped, live, into ship's holds for months, awaiting their turn to become soup. So badly were they abused that mariners, perhaps to assuage their own guilt, liked to say that wicked naval officers were doomed to return as giant tortoises.

Those remaining on the islands were decimated by the goats, cats, pigs, dogs and rats that had been introduced by the sailors. These imported intruders ate thousands of the tortoises eggs as well as their vegetable food. Giant tortoises take their time to move, to breed and to live—some for more than 200 years. In fact, giant tortoises are the longest-lived land animals of all; some still alive today were youngsters of 50 when Darwin visited the islands.

Major efforts have been made to preserve some of the other related tortoises, such as the once-threatened Wolf variety, which now numbers in the thousands again. But what will become of old Lonesome George? Some have suggested crossbreeding him with a female of another variety to preserve his genes. Others would like to freeze his semen or clone his cells or simply turn him loose on his home island. But for now, he's going to remain a pampered prisoner, living in comfortable captivity where he can be accessible in the unlikely event that a female of his kind is discovered some day.

Why is Lonesome George the recipient of so much attention when he may already be a "lost cause"? William Beebe, an American naturalist, once wrote: "When the last individual of a race of living things breathes no more, another heaven and another earth must pass away before such a one can be again."

See also EXTINCTION; GALAPAGOS ARCHIPELAGO.

LORENZ, KONRAD (1903–1989)
Ethologist, Behavioral Evolutionist

Konrad Lorenz's entertaining account of his animal acquaintances, *King Solomon's Ring* (1952), is a natural history classic. According to myth, Solomon's magical ring grants man the gift of speaking with birds and beasts; in reality, Lorenz pioneered new scientific understanding of animal communication. A naturalist in the Darwinian tradition, he also attempted to understand how animal "displays" or social signals have evolved.

Born the son of a prosperous physician in Altenberg, Austria, Lorenz took a medical degree at Vienna, but spent most of his life pursuing a childhood fascination with animals. Working with creatures common about his country home, such as jackdaws and geese, he was able to unravel behavior patterns that had eluded generations of comparative psychologists, with their mazes and "puzzle boxes."

With Nickolaas Tinbergen and Karl von Frisch, he founded ethology, a science of behavior that considered the whole animal. Ethologists were not interested in some arbitrary measure of "intelligence" or "learning ability," but in the animal's natural behavior: its courtship displays, social signals, species isolating mechanisms, nesting and territorial behavior. The influence of such a viewpoint revolutionized the study of animal behavior and placed it back into the context of an animal's life cycle, ecology and evolution. (By the 1980s, the term "behavioral ecology" had largely replaced "ethology.")

Lorenz's approach was to watch what animals did naturally, rather than stick them into preconceived experiments. He noticed, for instance, that a brooding greylag goose retrieved rolling eggs with a particular combination of two motions. She would bob her bill vertically, while making steadying horizontal movements. Lorenz tried a "natural" experiment: remove the egg after the goose had rolled it halfway back to the nest and see what she would do.

Amazingly, the bird continued to draw her neck in while bobbing her bill up and down, but ceased the sideways steadying movements. The bobbing vertical action, Lorenz concluded, was "fixed" and innate, while the other movements needed the stimulus of the rolling egg. Through such experimental "dissection," Lorenz believed, one could identify real "units of behavior." Such genetically determined behavior patterns, he thought, were adaptive traits evolved by natural selection.

In the popular press, Lorenz became famous for his research on "imprinting." Previously, scientists had found that for a day or so after hatching, young ducks and goslings would follow their mother to food and away from danger. They were, in Lorenz's terms, "imprinted" upon the mother duck.

Lorenz showed he could imprint the young waterfowl on something very unlike their natural mother: for example, on himself. He raised many broods of geese and ducks, even leading them daily to the water and showing them where to forage. Some scientists found his methods bizarre. "It's an amazing psychological phenomenon," commented one critic, "he believes he's their mother." Yet Lorenz was astute enough to realize his own parental instincts were triggered by the duckling's "cuteness"—the round, big-eyed face common to the young of many vertebrate species. [See CUTENESS, EVOLUTION OF.]

Sometimes "King Solomon's ring" gave only the illusion of working. While insisting that ethology must objectively describe observed behavior patterns, Lorenz found it difficult not to consider his study subjects little people, with human-like emotions and motives.

Pair formation among greylag geese, he wrote, contains all the emotions of human marriage. It "follows exactly the same course as with ourselves." In his book *The Year of the Greylag Goose* (1979) Lorenz spoke of one bird's "scorned mistresses," another's "unfaithful mate" and a goose's "dumbstruck grief" at the loss of his beloved. Reviewing the book in *Natural History* magazine, a zoologist wondered, "How did this soap opera get into a book about geese?"

In attempting to reconstruct the evolution of behavior, Lorenz tried to compare his "units" of behavior the same way 19th-century anatomists had compared bones and muscles. During courtship, for instance, waterfowl show a range of different, stereotyped behaviors peculiar to various species: One gives an exaggerated preening display, while another combines a few of the same movements with "ritualized" feeding or aggressive behavior.

Lorenz tried to arrange such "displays" into a family tree of behaviors, deducing how the more elaborate had evolved from the simpler. Although zoologists were intrigued with his conclusions, ultimately they were abandoned as conjectural and unprovable—much like the 19th-century arrangements of human technology in evolutionary series. [See COMPARATIVE METHOD; PITT–RIVERS MUSEUM.]

With his fellow ethologists, Tinbergen and von Frisch, Lorenz was awarded the 1973 Nobel Prize for Physiology or Medicine. But his brilliant, influential work was flawed by a biological determinism that lent itself to too-facile "jumps" from animals to man. It was an attractive bit of circular reasoning. After projecting human motivations on animals, he jumped back from animals to man to prove his case. And in *On Aggression* (1966), he argued that wars and political turmoil stemmed from an innate need for humans to attack and fight each other over mates and territory unmediated by "genetic mechanisms for social cooperation." Human fierceness was built into us by evolution, he concluded, unlike what he considered the gentler carnivores—wolves and lions!

Even more upsetting to his supporters, it came out that Lorenz—widely admired in Europe and America—had bolstered Nazi racial theories with his writings on biological "purity of type." In an infamous paper published in 1940 ("Disorders Caused by the Domestication of Species-Specific Behavior"), Lorenz argued that domesticated animals lose their inborn preference to mate with the "pure wild type." Humans, he thought, had become "domesticated" by race-mixing in Europe, resulting in the loss of pure "types" and "degeneration" of the "higher" races—a situation that could only be corrected by state-controlled breeding. (Only five years before, German Jews and gentiles were forbidden to intermarry based on such pseudoscience, a flimsy mask of reason for the politics of destruction.) During the Nazi era, Lorenz wrote:

> The high valuation of our species-specific and innate social behavior patterns is of the greatest biological importance. In it as in nothing else lies directly the backbone of all racial health and power. Nothing is so important for the health of a whole *Volk* as the elimination of "invirent types": those which, in the most dangerous, virulent increase, like the cells of a malignant tumor, threaten to penetrate the body of a *Volk* . . .
>
> [By studying animals] which are easier and simpler to understand, [we] discover facts which strengthen the basis for the care of our holiest racial, *volkish*, and human heredity . . . [Where civilization causes] imbalance of the race, then race-care must consider an even more stringent elimination of the ethically less valuable than is done today . . . to replace all selection factors that operate in the natural environment.

Some of Lorenz's defenders have claimed he was politically naive or misunderstood. Certainly, he was

appalled when, in 1943, he personally witnessed a group of gypsies carted off to the death camps. But as biologist Garland E. Allen has written:

> Ideas can become deadly weapons when they provide a supposedly objective and rational description for human social behaviors. Scientists need to understand the role their work can play in a social system where the control of the flow of ideas lies in the hands of those elite who have their own social aims in mind. Lorenz's case exemplifies this point dramatically . . . whether Lorenz saw the implications of his work in this direction or not [he] contributed to, and was used by . . . the Nazis.

During the 1970s, there was an exaggerated reaction among biologists and anthropologists to E. O. Wilson's book *Sociobiology* (1975), a new attempt to explain the biology of behavior from insect and fish to man. To Wilson's astonishment, scientific meetings became stormy and hostile; epithets of "biological determinist" quickly turned to "fascist" and "Nazi." It was, unfortunately, part of the legacy of Konrad Lorenz, who left an equally strong image as the charming papa of a line of baby ducks.

See also BIOLOGICAL DETERMINISM: ETHOLOGY; TINBERGEN, NIKOLAAS.

For further information:

Lorenz, Konrad. *Evolution and Modification of Behavior.* Chicago: University of Chicago Press, 1965.

———. *King Solomon's Ring: New Light on Animal's Ways.* New York: Thomas Crowell, 1952.

———. *On Aggression.* London: Methuen, 1966.

Singer, Peter. *The Expanding Circle: Ethics and Sociobiology.* New York: New American Library, 1981.

(THE) LOST WORLD (1912)
Doyle's Prehistoric Adventure

Sir Arthur Conan Doyle, the creator of Sherlock Holmes, had a love-hate relationship with evolutionary scientists. On the one hand, he was intrigued by their visions of ape-men and prehistoric monsters, but he was repelled by their narrow materialism and rejection of his Spiritualist religion.

His "boy's adventure novel," *The Lost World*, expressed both attitudes. Originally begun as a serial novel in the *Strand* magazine in April 1912, it introduced his raucous maverick scientist, Professor Challenger, who was to star in several subsequent adventures.

Challenger, a robust, dark-bearded explorer, leads an expedition to the deepest unknown jungles of South America, where his party stumbles into a "lost world," which had remained unchanged for millions of years. They encounter stegosaurs, flying reptiles and tribes of red-haired ape-men in adventures Doyle admitted were inspired by the novels of Jules Verne.

When Challenger returns to civilization, he has a hard time convincing his fellow scientists he is not a liar. What evidence will satisfy them? A bone? A photograph? "A bone can be as easily faked as a photograph," says one of the characters—a sly reference to the controversy that surrounded Doyle's belief in "spirit photography."

In an appreciation of the book, John Dickson Carr noted that its satiric descriptions of scientific debates are more entertaining than the dinosaurs. Conan Doyle paints a comic picture of:

> grave-bearded men of science, where some abstruse theory is concerned, behav[ing] exactly like temperamental prima donnas and . . . fully as jealous of each other . . . the zoologists are as interesting as the zoo.

Doyle identified closely with his hero and even donned a fake beard to pose as Challenger for the frontispiece when the story was published in book form.

In 1983, Dr. John H. Winslow proposed that *Lost World* contained hidden clues to the Piltdown hoax and suggested that Conan Doyle was the hoaxer. Among other things, he noted that a map of the "Weald" in the novel was similar to the terrain of the Piltdown site, that there were references to "faked bones" and "prehistoric practical jokers" in the book and that the ape-men were red-haired. (The Piltdown "man" was a human skull with the jaw of an orangutan, the only red-haired ape.)

Conan Doyle's motive, Winslow claimed, was to embarass the "materialist" scientists who denied any role to "spirit" in human origins. Some Darwinian evolutionists had embarrassed Doyle's Spiritualist friends, claiming they were incapable of separating genuine evidence from fraud or forgery. In particular, Winslow singled out Edwin Ray Lankester, director of the British Museum (Natural History).

Lankester had become famous while a young man for exposing the medium Henry Slade and prosecuting him as a "common rogue" in police court. [See SLADE TRIAL.] It was the first time a scientist had charged a psychic with fraud in a court of law and did much to publicly discredit the Spiritualist movement.

Doyle never forgave Lankester and later spent much of his fortune promoting Spiritualism, which he believed was an "all important revelation" for mankind. Still, Doyle was impressed with Lankester's book *Extinct Animals* (1905) and drew heavily on its descriptions of prehistoric dinosaurs for *The Lost World*.

Although the *New York Times* ran a headline AR-THUR CONAN DOYLE IS PILTDOWN SUSPECT (1983) in reporting Winslow's views, Winslow was never able

APE-MEN AND DINOSAURS survive into the present day in an isolated "lost world" discovered by Professor Challenger, a hero of science created by Sir Arthur Conan Doyle. First published in 1912, the classic science fantasy-adventure inspired hundreds of imitative books and movies about "dinosaur islands."

to muster more than circumstantial evidence that Doyle was in fact the culprit. Other, equally plausible suspects have been advanced by others and have been summarized in *The Piltdown Inquest* (1987) by Professor Charles Blinderman.

One of the first and most influential dinosaur adventure movies was made of *The Lost World* in 1925, starring Wallace Beery as a memorable Professor Challenger. Although the special effects, including one of the first uses of miniature models in stop-frame photography, are crude by current standards, they caused a sensation in 1925.

The process was so new that Conan Doyle created a stir by previewing the dinosaur sequences without explanation. At the American Club of Magicians in New York on June 2, 1925, he announced that he would show the audience a glimpse of something not exactly "psychic," but certainly "not nature as we can now observe it."

After building an atmosphere of expectation and mystery, he showed prehistoric iguanodons, tyrannosaurs and brontosaurs fighting and rearing their young on the big screen.

Next day the *New York Times* headlined: DINOSAURS CAVORT IN FILM FOR DOYLE. SPIRITIST MYSTIFIES WORLD-FAMED MAGICIANS WITH PICTURES OF PREHISTORIC BEASTS—KEEPS ORIGIN A SECRET.

Conan Doyle's monsters had been animated by a young genius in the new medium named Willis O'Brien. Some years later, O'Brien created and animated the models for the original *King Kong*.

See also KNIGHT, CHARLES R.; LANKESTER, E. RAY; O'BRIEN, WILLIS; PILTDOWN MAN (HOAX).

LOVELOCK, JAMES

SEE GAIA HYPOTHESIS.

LUBBOCK, SIR JOHN (LORD AVEBURY)
(1834–1913)
Pioneering Prehistorian

When John Lubbock was a boy, his father, a wealthy banker, astronomer and mathematician, told him that he had some wonderful news. Young Lubbock guessed he was going to get a pony to ride. "No, much better than that," Lubbock Senior replied, "Mr. Darwin is coming to live in Down village." "I confess I was much disappointed," Lubbock recalled, "though I came afterwards to see how right he was."

The Lubbocks lived on a grand estate called High Elms in the Kentish countryside (it burned down in the 1970s), about a mile down the road from Darwin's Down House, which still stands and is open to the public. "Insofar as one could be born and bred to Darwinism before 1858," writes historian George Stocking in *Victorian Anthropology* (1987), "John Lubbock was." He found in Darwin a teacher, mentor and father figure who greatly influenced his life and career. From an early age, he became part of Darwin's select inner circle, which included such brilliant men as Thomas Huxley, Joseph Hooker and Charles Lyell. At the celebrated Oxford Debate at which Huxley confronted Bishop Wilberforce, Lubbock gave a long, effective defense of Darwinism using evidence from embryology.

The eldest of 10 children, Lubbock attended Eton, joined the family banking business at 14, entered Liberal politics and simultaneously pursued a scientific and literary career. He was the epitome of the overachieving Victorian gentleman, remarkably talented and active in many fields. As a young man, he told Darwin his three goals: to be Lord Mayor of London, Chancellor of the Exchequer and president of the Royal Society. Darwin said he could be any one if he gave up the other two. Lubbock ignored this advice and did not reach any of those positions.

Nevertheless, his accomplishments in finance, politics, and science are astonishing. While a teenager, he discovered the first fossil musk-oxes in England, thus helping to establish the existence of a cold glacial period, which delighted Darwin. He went on to publish many original papers in natural history and botany and discovered a new species of crustacean in Darwin's collection, which he named after his mentor. But his most lasting contributions were in the fields of comparative psychology, prehistory and the behavior of social insects.

One of the leaders of what was then known as the "prehistoric movement," he focused on exploring a time period thousands of years before what was generally considered knowable "history." To the pre-

historians, ancient Greece was a late development. They were intrigued with the stone "hand-axes" and other evidences of early man coming to light, largely through the excavations of the Frenchman Boucher de Perthes, whose work was met with skepticism and disbelief for 25 years. Lubbock toured the Somme River gravels in 1860, escorting the geologist Sir Joseph Prestwich and others to the remarkable stone tools sites. All were impressed by the abundant evidence of extinct mammoths, woolly rhinos and other cold-weather animals living at the same time as man. Lubbock published accounts of these sites in the *Natural History Review,* of which he was an editor, and in 1865 published his classic work *Prehistoric Times.*

Lubbock decorated the walls of High Elms with hundreds of primitive tools and weapons from ancient digs and current tribal peoples. He coined the terms for the divisions between the Old Stone Age and New Stone Age still used today: Palaeolithic and Neolithic. As historian George Stocking points out, the full title of Lubbock's book *Prehistoric Times, as Illustrated by Ancient Remains, and the Manners and Customs of Modern Savages* "embodies the basic principle of 'the comparative method' of sociocultural evolutionism," an approach that was to dominate anthropology for a century. In 1867, at the urging of Thomas Huxley, Lubbock became the first president of the Royal Anthropological Institute.

While he admired the cleverness of early peoples, as reflected in their artifacts, his view of contemporary tribes was distorted by upper-class Victorian condescension. In general, he thought non-Western tribes had beastly manners and no "real" religion. Far from painting the savage as noble and free, Lubbock thought tribal man was "a slave to his own wants [and] passions . . . [suffering] from the cold by night and the heat of sun by day; ignorant of agriculture, living by the chase, and, improvident in success, hunger always stares him in the face and often drives him to the dreadful alternative of cannibalism or death."

But in his later book *Origin of Civilisation* (1870), published within a year of Darwin's *Descent of Man* (1871), Lubbock revised his view of savages to allow for evolutionary progress. If tribes were considered degenerate descendants of superior ancestors, they could not be "missing links" in evolution. He had written in *Prehistoric Times* that tribal peoples were inferior, stupid and disgusting, but essentially rational. In *Origin of Civilisation*, he tried to show how "higher and better ideas of Marriage, Law, and Religion" had developed over time as the various races raised themselves from "utter barbarism." Although his archeological writings broke new ground, Lub-

NATURALIST, PREHISTORIAN, POLITICIAN, AND BANKER, Sir John Lubbock became Charles Darwin's "scientific son" when the Darwins moved into a nearby property. Lubbock's mansion became home to captive colonies of ants and wasps; its walls were decorated with fossils and prehistoric tools.

bock's naive comparisons of primitive cultures quickly became outmoded. [See COMPARATIVE METHOD.]

As an observer of social insects, Lubbock was on the cutting edge of research. One special room in his home contained more than 30 ants' nests of many species. His classic *Ants, Bees, and Wasps* (1882) detailed many experiments on the behavior, social organization and "mental activity" of ants. Indeed, his observations on social insects were sometimes kinder and more sympathetic than his descriptions of tribal peoples. But then he never observed "savages" at first-hand.

Among Lubbock's other important discoveries, he was the first to document color vision in bees, which confirmed Darwin's view that flower forms and colors were adaptations to attract pollinating insects. [See DIVINE BENEFICENCE, IDEA OF.] Ants are mindless "automata," he thought but occasionally show altruistic feelings for their fellows. Lubbock describes them "tenderly" carrying off wounded comrades.

He was also intrigued by so-called slave-owning ants, which were so dependent on their "slaves" that their mouth parts degenerated, and they had to be fed by them. This proved, said the Liberal Lubbock, that slavery caused "degeneration" among slave-holding ants "as it did among humans who become dependent on slave labor." (Recent studies of these

species by Howard Topoff and others have shown that the biology of ant behavior is so dissimilar to human institutional slavery as to make such comparisons meaningless.)

After several unsuccessful tries, Lubbock was elected to Parliament in 1869; his public list of supporters included John Stuart Mill and Charles Darwin. He is still remembered in England as "Saint Lubbock" for creating the first secular bank holiday in England (August 7, "St. Lubbock's Day"). It was the first time the masses were given a long weekend from their labors without a religious excuse. He also fought for shorter working hours, the introduction of scientific education into the schools and laws protecting native birds and forests.

In later life, he wrote popular books on travel, free trade, botany, *The Senses of Animals* (1891), economics and *The Pleasures of Life* (1887–1889), which included his celebrated list of the "One Hundred Best Books." He outlived many of his great scientific friends who had belonged to the "X Club"—an intimate circle of Darwinian scientists—and served as a pallbearer for Darwin and Huxley. One of the last of the original evolutionists, he was fooled during his final days by the Piltdown skull, which he admired as "the most simian yet found."

During the 1890s, Lubbock became interested in a group of ancient mounds containing the remains of a great circle of ancient stones. The site, which he called "the finest megalithic ruin in Europe . . . older and much grander than Stonehenge" was near Avebury, in Wiltshire. When he learned they were going to be demolished for commercial development, he promptly bought up the land and preserved the prehistoric mounds. When he was rewarded for his life of public service with a peerage in 1899, he chose the title Lord Avebury, after the site of his beloved prehistoric ruins.

See also "X" CLUB.

For further information:
Avebury, Lord (Sir John Lubbock). *Ants, Bees, and Wasps.* London: Kegan Paul, Trench, 1882.
———. *Origin of Civilization and the Primitive Condition of Man.* London: Longmans, Green, 1870.
———. *Prehistoric Times. As Illustrated by Ancient Remains and the Manners and Customs of Modern Savages.* London: Williams & Norgate, 1865.
Hutchinson, Horace. *Life of Sir John Lubbock, Lord Avebury.* 2 vols. London: Macmillan, 1914.

"LUCY"
Early North African Hominid

According to anthropologist Donald Johanson's account *Lucy: The Beginnings of Humankind* (1981), this ancient near-man (or near-woman) is "the oldest, most complete skeleton of any erect walking hominid found anywhere in the world." Discovered by Johanson, Tom Gray and several colleagues at Hadar, in the Afar region of Ethiopia on November 30, 1974, "Lucy" captured the attention of the world press.

While celebrating the find, Johanson's team sat around their evening campfire playing the Beatles tune "Lucy in the Sky with Diamonds" on a tape machine. Since the shape of her pelvis appeared female (although there are no known "male" pelvises for comparison), they nicknamed her "Lucy." Her scientific name, *Australopithecus afarensis*, means "southern ape from the Afar," and Johanson has championed the species as our earliest known direct ancestors. (The name is a strange contradiction because the genus was first named from South African fossils, but Afar is in northern Ethiopia.)

Although Lucy was proclaimed to be a 40% complete skeleton, in fact less than 25% of Lucy's complete skeleton was actually recovered (47 bones out of 206). Only by "doubling" many bones (for instance, assuming a missing left rib would be a mirror-image of its known mate on the right side) did the specimen appear about 40% complete. Nevertheless, with long bones of arms and legs, ribs, sacrum and half a pelvis, Lucy remains the most complete skeleton ever found of such an early hominid, three million years old.

Despite excited press accounts, Lucy was not the first specimen to establish that australopithecines were bipedal and erect. Thirty years earlier, South African anatomist Robert Broom had found australopithecine leg bones and pelvises at Sterkfontein showing that early hominids were upright walkers, but Lucy certainly provides dramatic, clinching confirmation. As one paleontologist put it, "the South African fossils are excellent, but they have not as good a salesman as a Leakey or Johanson."

Although the Lucy fossils were initially dated at three million years, Johanson had announced them as 3.5 million because he said the species was "the same" as a skull found by Mary Leakey at Laetoli, Tanzania. By proposing Mary Leakey's find as the "type specimen" for *Australopithecus afarensis*, he was identifying Lucy with another fossil 1,000 miles from the Afar and half a million years older! Mary thought the two not at all the same and refused to have any part of linking her specimen with *afarensis*. Convinced her own specimen belongs to the genus *Homo*, she announced that she strongly resented Johanson's "appropriating" her find, her reputation and the older date to lend authority to Lucy. Thus began the bitter, persistent feud between Johanson and the Leakeys.

Richard Leakey argued that the designation *Aus-*

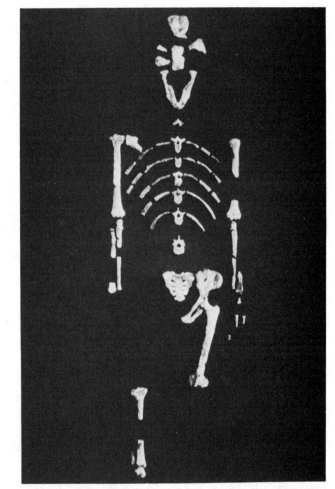

MOST COMPLETE SKELETON from 3.5 million years ago, "Lucy" was discovered by the Johanson-Tieb field team in the Afar region of northern Ethiopia in 1976. Classified as a new hominid, *Australopithecus afarensis*, "Lucy" stood less than four feet tall, but has the pelvis, leg bones and feet of a fully bipedal walker.

tralopithecus afarensis was not justified, that the Afar fossils represented two different species, not simply large males and smaller females, as Johanson believes. Also, to some, Lucy looked human enough to be considered part of the genus *Homo*. In fact, Johanson himself originally described the fossils as *Homo*, a species of man, but soon after changed his mind based on the assessment of his colleague, Tim White. They now describe the bones as too ape-like in the jaws, teeth and skull to be considered *Homo*, yet also sufficiently distinct from other, later australopithecines to warrant their own species.

Lucy certainly is one of the most remarkable relics ever discovered. She has irrevocably established in the public mind a view of our ancestors as diminutive hominids, with a small-brained head, fully erect body and pelvis and legs adapted for bipedal walking, not stooped over or walking with the aid of the hands.

Lucy has also helped overthrow the old theory that erect posture, which would free the hands for tool-making, evolved with the enlargement of the brain. Upright walking had evolved in Lucy and other early hominids long before they had very large brain size. We did not "stand tall" as result of some special human cleverness.

See also AFAR HOMINIDS; AUSTRALOPITHECINES; JOHANSON, DONALD; TURKANA BOY.

For further information:
Johanson, Donald, and Edey, Maitland. *Lucy: the Beginnings of Humankind.* New York: Simon & Schuster, 1981.
Johanson, Donald, and Shreeve, James. *Lucy's Child.* New York: Wm. Morrow, 1989.
Lewin, Roger. *Bones of Contention: Controversies in the Search for Human Origins.* New York: Simon & Schuster, 1987.
Reader, John. *Missing Links: the Hunt for Early Man.* London: Collins, 1981.

LUNAR SOCIETY

See DARWIN, ERASMUS.

LYELL, SIR CHARLES (1797–1875)
Geologist

"I really think my books come half out of Lyell's brain," said Charles Darwin, "I see through his eyes." But Sir Charles Lyell, the great naturalist's friend and mentor, had a hard time returning the compliment. Although he privately encouraged Darwin's evolutionary work for years, Lyell could not bring himself to endorse his friend's theories in his own popular geology books. Much to Darwin's disgust, Lyell was a past master of the art of coming down squarely on both sides of an issue.

Lyell's father, Charles Lyell of Kinnordy, was a Scots laird who was torn between scientific and literary interests. He was a keen amateur botanist, but also managed to produce a well-known English translation of Dante's *Inferno*. The younger Charles was also divided between two professions; he started out as a lawyer, but his strong interest in geology finally won out. He had admired the works of James Hutton and was instructed by the geologist William Buckland, with whom he examined volcanic layers on field trips to Italy.

Just before Darwin was to leave on his five-year voyage aboard H.M.S. *Beagle,* his Cambridge professor Reverend John Henslow recommended he take along the first volume of Lyell's *Principles of Geology* (1830), which had recently been published. "By all means read it for the facts," Henslow instructed, "but on no account believe the wild theories."

In this founding document of modern geology, Lyell had argued: (1) that the geologic past can best

ARGUING HIS CASE for the "Uniformitarian" view of geology came easily to Sir Charles Lyell; he had been trained as a lawyer. Lyell's *Principles of Geology* revolutionized earth science, and greatly influenced young Charles Darwin's theories.

be understood in terms of natural processes we can actually observe today, such as rivers depositing layers of silt, wind and water eroding landscapes, glaciers advancing or retreating (actualism); (2) that change is slow and steady (gradualism), rather than quick and sudden; (3) that natural laws are constant and eternal, operating at about the same intensity in the past as they do today. (Sometimes slower or faster, but averaging about the same overall rate of change.)

Darwin devoured the book, which was brilliantly written, thoroughly grounded in fieldwork and seemed to place the study of geology on a new and sensible footing. Although Lyell believed living species were fixed and not related by common descent, he gave Darwin the means of seeing what wonders could be wrought in geology by slow, small forces operating over immense spans of time. "I am tempted to extend Lyell's methods even farther than he does" Darwin wrote.

Although he inspired Darwin and became his lifelong friend and mentor, Lyell had great difficulty accepting "the descent of man from the brutes," because, he confessed, "it takes away much of the charm from my speculations on the past relating to such matters." Nevertheless, later in life he had to grudgingly acknowledge the growing evidence. Darwin was frustrated and angry with Lyell's refusal to support him wholeheartedly in print, though he did so in private conversations. He simply could not, as he put it, "go the whole Orang."

What was really peculiar to Lyell are two ideas rarely associated with his *Principles of Geology*: the older ideas that earth and water trade substances and shape each other, maintaining some kind of long-range balance (the steady-state Earth), and that time and life proceed in cycles. It was conceivable to Lyell that man and our familiar animals could all become extinct, only to be replaced by dinosaurs again in a subsequent creation, followed, in some distant age, by a "new creation" of man. Aside from historians of science, Lyell's belief in cyclic time has been all but forgotten.

Lyell wrote that only a few extinctions occurred at a time, which "gradually" added up, rather than wholesale extinctions. New species were "called into being" to replace them. Somewhat cynically, he refused to specify how or what he meant, leaving the interpretation open, so as not to ruffle the feathers of the theologians.

Caution about antagonizing anyone and an extreme desire for social acceptance curbed Lyell's adventurousness in the realm of ideas. Darwin thought it ridiculous how Charles and Lady Lyell would spend hours poring over dinner invitations, making it a matter of great importance which to accept and which to decline. It was no accident that Lyell's determination to offend no one of importance resulted in his receiving a knighthood, and later, being named baronet. Darwin received no national honors during his lifetime.

Lyell saw his *Principles of Geology*, which first appeared during 1830–1833, through 13 revised editions. It had started out as a long, connected argument, but later became a jumble of bits and pieces added to include newer research. Finally, the book itself became a career and produced substantial revenues for its author.

Charles Darwin had Lyell in mind when he wryly remarked that scientific men should be put to death at the age of 60, so their inflexible habits of mind could not interfere with the progress of the newer generations. After discussing Darwin's theories at length in his *"Geological Evidences As to the Antiquity of Man"* (1863), Lyell asked Darwin if "now he might be allowed to live." However, he always hedged his presentations of his friend's evolutionary ideas and never explicitly embraced them in his public writings.

See also CATASTROPHISM; GRADUALISM; STEADY-STATE EARTH; UNIFORMITARIANISM.

For further information:
Bailey, Edward. *Charles Lyell*. London: Nelson & Sons, 1962.

Eiseley, Loren. *Darwin's Century: Evolution and the Men Who Discovered It*. New York: Doubleday, 1958.

Gould, Stephen Jay. *Time's Arrow, Time's Cycle*. Cambridge: Harvard University Press, 1987.

Greene, John C. *The Death of Adam*. Ames: Iowa State University Press, 1959.

Lyell, Charles. *On the Geological Evidence of the Antiquity of Man*. London: John Murray, 1863.

———. *The Principles of Geology*. 3 vols. London: John Murray, 1830–1833.

LYSENKOISM
Ideological Genetics

During the 1930s, Trofim D. Lysenko rose to power in the U.S.S.R. by convincing government ideologues that he could create a genetic science consonant with their political philosophies. They wanted science to support the view that Soviet society could be literally transformed in a few generations, and that the Russian people were progressively evolving. It might take too long to create the socialist utopia if each generation had to be separately educated without cumulative inheritable improvements.

Lysenko proclaimed that Mendelian genetics was a "tool of bourgeois society" in teaching that mutation is random and that genes are usually passed on unchanged. His reading of Marxist principles was that man and nature are improvable and perfectable. Stalin gave him free rein, and, within a few years, his rivals (and any honest scientists who openly supported Mendelism) were out of work, imprisoned or even executed.

Among his claims, Lysenko said he could alter species of wheat by changing their environment and he intended to transform Soviet agriculture. He said he could change winter wheat into spring wheat merely by altering the temperatures at which they were grown. He even claimed that environmental manipulation could change wheat into rye in one generation!

According to Lysenko, Charles Darwin's concept of the "struggle for existence" was a bourgeois notion used to justify competition in a capitalist society. Nature was altruistic. Seeds should be planted in clusters, so that all except the most perfect and vigorous one would sacrifice themselves for the good of the species.

Lysenko became so powerful he was never forced to produce evidence for his assertions. His policies retarded Soviet biology and genetics for 30 years and devastated the agricultural production of the country. Repeated crop failures and shortages finally caused his ouster in 1965.

After the death of Russian dictator Josef Stalin, Lysenko revealed that the head of state had helped prepare his famous speech of 1948. "Comrade Stalin found time even for detailed examination of the most important problems of biology," Lysenko declared in his eulogy for Stalin in *Pravda* (1953). "He directly edited the plan of my paper, 'On the Situation in Biological Science,' in detail, explained to me his corrections, and provided me with directions as to how to write certain passages in the paper."

The term "Lysenkoism" is used in a restricted sense to describe his notions of environmental manipulation of genes, coupled with inheritance of acquired characters. In its broader meaning, Lysenkoism has come to symbolize the disastrous consequences of making science subservient to political ideology.

See also LAMARCKIAN INHERITANCE.

For further information:

Joravsky, David. *The Lysenko Affair.* Cambridge: Harvard University Press, 1970.
Zirkle, Conway, ed. *Death of a Science in Russia: The Fate of Genetics as Described in Pravda and Elsewhere.* Philadelphia: University of Pennsylvania Press, 1949.

LUCY'S CHILD

See OLDUVAI FEMALE.

LYSTROSAURUS FAUNA
Evidence for Ancient Continent

Lystrosaurus is a large mammal-like reptile (synapsid) of the Triassic period, originally known from South Africa, where it is associated with other characteristic fossils. Often it is found in the same rocks as another mammal-reptile called *Thrinaxodon*, some small ancestral lizards (eosuchians), the little reptile *Procolophon* and a particular type of amphibian—collectively known as the *Lystrosaurus* fauna.

Squat, tusk-toothed *Lystrosaurus*—and the unique array of animals always found with it—helped prove Antarctica, South Africa and India were once joined together. Also, oceanographic map-makers have found that details of their submerged coastal outlines today, at a depth of 1,000 fathoms, would form a remarkably close fit.

According to paleontologist Edwin Colbert, one of the discoverers of the *Lystrosaurus* fauna on all three continents, these animals once occupied a continuous range on ancient Gondwanaland. When the supercontinent broke up, one segment—once a land of lush forests and strange tropical animals—drifted southward to become frozen Antarctica.

See also CONTINENTAL DRIFT; GONDWANALAND; PLATE TECTONICS.

MacCREADY, PAUL

See PTEROSAUR, MACCREADY'S.

MALARIA

See RESISTANT STRAINS.

MALAY ARCHIPELAGO

See WALLACE, ALFRED RUSSEL.

MALTHUS, THOMAS (1766–1834)
Inspiring Economist

Because of theorists like Thomas Malthus, economics came to be known as "the dismal science." Yet the dreary conclusions of Malthus's famous *Essay on the Principle of Population* (1798) gave hope to two great naturalists—Charles Darwin and Alfred Russel Wallace—and directly sparked their development of the idea of natural selection.

Malthus, a British clergyman and political economist, was deeply distressed by the teeming slums of preindustrial England. Human misery, he perceived, arose from human fertility combined with irresponsibility. Taking his cue from natural populations of plants and animals, he showed that reproductive potential in all creatures, including man, far exceeds the ability of resources to support unchecked production of offspring.

Biologists have since discovered that overproduction in nature goes far beyond what Malthus had supposed. For instance, it is now known that every female sturgeon produces six million eggs per year, each mayworm 40 million, and each tapeworm 60 million. Without enormous mortality, the Earth would shortly be overrun by the descendants of a single breeding pair of almost any creature. Put mathematically, Malthus argued that organisms reproduce in a geometric ratio, while food resources tend to remain constant. Even granting technological improvement, man's food resources would only increase at an arithmetical rate, which could not keep up with the geometrical increase in population.

Therefore, Malthus concluded, humans will become increasingly miserable as starvation becomes inevitable among the poorer social classes. His own solution was to advocate sexual abstinence for the poor and punitive laws against parents producing more children than they are able to support. His ideas are being put into practice today in China, where rigid government quotas on family size are backed by strong economic and social sanctions. Of course, in Malthus's day large families were the norm. Throughout most of the world where population growth is rapidly exceeding food supply, the Malthusian time-bomb is still ticking.

Malthus's principle that excessive fertility in nature produces competition for survival (and large numbers of "expendable" individuals) gave Darwin and Wallace, acting independently of each other, the foundations they needed for the theory of natural selection. In his *Autobiography* (1876), Darwin wrote:

> In October 1838, that is, fifteen months after I had begun my systematic inquiry, I happened to read for amusement Malthus on *Population*, and being well prepared to appreciate the struggle for existence which everywhere goes on from long-continued observation of the habits of animals and plants, it at once struck me that under these circumstances favourable variations would tend to be preserved, and unfavourable

ones to be destroyed. The results of this would be the formation of new species. Here, then I had at last got a theory by which to work.

Wallace had also read Malthus, and apparently made a similar connection as in 1858 he mulled over his observations on the competition for resources among tribal peoples of the Malay Archipelago. During a malarial fever, Wallace conceived the theory of natural selection, and realized it would apply both to human and animal populations.

Later, Social Darwinists would attempt to return the biological theory of natural selection to the human economic sphere, applying it to the conditions in urban slums that had been its point of departure. The whole episode is an impressive example of how social and biological theory draw from each other, as well as a striking instance of independent invention by similarly prepared minds.

See also DARWIN, CHARLES; NATURAL SELECTION; SOCIAL DARWINISM; WALLACE, ALFRED RUSSEL.

For further information:
Chase, A. *The Legacy of Malthus.* New York: A. Knopf, 1977.
Malthus, Thomas. *An Essay on the Principle of Population.* London: Johnson, 1798.

MAMMAL-LIKE REPTILES

See KARROO BASIN.

MAMMOTHS
Siberian Fossil Elephants

When great fossil elephants began to come to light in the later 1700s, there was a great deal of confusion in sorting them out. Those found in Siberia had first become known to science, since travelers had been amazed at the quantities of fossil ivory found there

since the early 1600s. A Dutchman named Cornelius Witzen was the first to print the word "mammoth" in 1694, noting that "the Inlanders [Russian settlers in Siberia] call these teeth *mammout-tekoos*, while the animal itself is called *mammout*."

Contradictory stories about the fabulous "mammout" trickled out of the north country for years, including debates over whether the animal was still alive. In the 1770s, immense quantities of fossil bones and tusks were found; demand for them soon prompted a flourishing trade. During the 19th century, the tusks of perhaps 50,000 Siberian mammoths were unearthed and sold for ivory in Europe.

When elephant-like tusks and fossils were discovered in North America, President Thomas Jefferson took a keen interest in them and called them "mammoths." But he failed to notice, as the French expert Georges Cuvier pointed out, that the teeth of the American species were entirely different. While the Siberian mammoths had flat grooved grinders like those of elephants, the American fossils had many curve-shaped bulges all over their grinding surfaces.

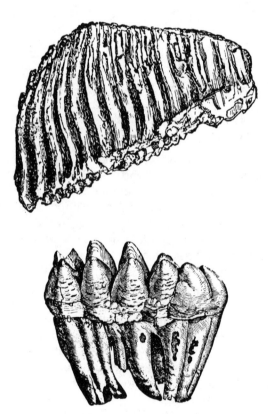

MAMMOTH'S GRINDING TOOTH (top) shows a series of long grooves and ridges. The American fossil elephant has a very different pattern (below); its many long, rounded shapes inspired the name "breast-shaped-tooth" or Mastodon.

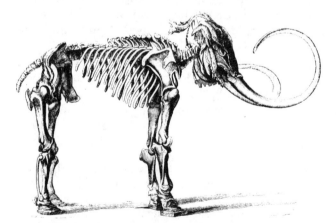

PEALE'S MASTODON became world famous as America's pride in 1799, symbolizing the new country's great size and vigor. Benjamin Franklin and Thomas Jefferson had encouraged Charles Willson Peale's efforts to unearth the fossil skeleton.

Cuvier suggested calling the Ohio Valley animal a "mastodon," meaning "breast-tooth," in view of these breast-shaped bulges on their grinding surfaces.

Nevertheless, Jefferson (and almost everyone else) kept referring to the American fossil tuskers as "mammoths," and "mammoth fever" swept the country. Giant loaves of bread were advertised as "mammoth bread" in Philadelphia bake shops. Some of Jefferson's admirers in Massachusetts made a "mammoth cheese," which weighed 1,235 pounds, and shipped it to the White House. He put it on exhibit in the East Room, which he nicknamed "the Mammoth Room."

See also JEFFERSON, THOMAS; PEALE'S MUSEUM.

For further information:
Silverberg, Robert. *Mammoths, Mastodons, and Man.* New York: McGraw-Hill, 1970.

MAN'S PLACE IN NATURE (1863)
Huxley's Most Influential Work

Evidence as to Man's Place in Nature, a scientific classic, was written by Thomas Henry Huxley (1825–1895) when he was 38 years old. His first and most important book, published in January 1863, it established the close kinship of man and apes in science as well as in the popular mind.

It was the logical follow-up to Charles Darwin's *Origin of Species* (1859), which had conspicuously left out any discussion of the origin of man. All Darwin would allow himself was the simple sentence "Light will be thrown on the origin of man and his history." Darwin detested public combat, Huxley loved it. He was ready to tackle the problem of man's kinship to the apes head-on and told Darwin, "I am sharpening my beak and claws in readiness."

It was audacious to even talk about man's "place in nature," implying that he was grouped with other animals in the study of natural history. Up to that time, prevailing thought held that man was somehow "outside" of nature, imbued with spiritual qualities setting him totally apart from apes and monkeys. A famous remark by a bishop's wife during that period says it all: "Descended from the apes! My dear, let us hope that it is not true, but if it is, let us pray that it will not become generally known." It was only after Huxley took the plunge and made it "generally known" that Darwin was willing to put out his own *Descent of Man* in 1871.

The major thesis of *Man's Place in Nature* is that comparative anatomy shows man and apes are much more similar than are apes and monkeys. First, Huxley describes the anatomy of the man-like apes and recounts the fragmentary evidence of their behavior as it was known. Then, after an exhaustive analysis,

muscle for muscle and bone for bone, he shows "It is quite certain that the Ape which most nearly approaches man, in the totality of its organization, is either the Chimpanzee or the Gorilla."

Although Huxley was often misrepresented as having said that man is descended from a chimp or gorilla, he nowhere makes that claim. His point was that we share a common ancestry. Replying to those who think it degrading or demeaning to acknowledge man's cousins as apes, Huxley finds hope rather than despair in the prospect. With a flourish of Victorian optimism, he sees progress in man's "ascent":

> Thoughtful men, once escaped from the blinding influences of traditional prejudice, will find in the lowly stock whence Man has sprung the best evidence of the splendor of his capacities, and will discover in his long progress through the Past reasonable ground of faith in his attainment of a nobler Future.

See also DESCENT OF MAN; HUXLEY, THOMAS HENRY; OXFORD DEBATE.

For further information:
Huxley, Thomas H. *Evidence as to Man's Place in Nature.* London: Williams and Norgate, 1863.

MANTELL, GIDEON ALGERNON (1804–1892)
Discovered First Dinosaur

With the thousands of types of dinosaurs known today, it seems incredible that less than 200 years ago giant reptiles were only the dragons of myth. Yet not until 1822 was the first discovery of dinosaur bones made—by Dr. Gideon Mantell and his wife Mary Ann in England.

It was actually Mrs. Mantell who spotted the first huge fossil tooth as she accompanied her husband on one of his "geologising" walks through Tilgate Forest, in Kent. Dr. Mantell was an avid amateur geologist, constantly searching for unusual relics of "antediluvian" creatures. He found a few more fragments and attempted to get the creature identified.

When local authorities seemed baffled, he sent his precious discovery to the world's greatest fossil expert, Dr. Georges Cuvier in Paris. The teeth belonged to a rhinoceros, Cuvier replied, and the bones were those of a hippopotamus.

But an English naturalist who saw the teeth disagreed. They were lizard-like, he thought, and convinced Mantell with the help of an iguana skull from the American tropics. Mantell concluded that he had found a giant lizard, named it *Iguanodon* ("iguana-tooth"), and read a paper describing it before the Royal Society of London in 1825. Cuvier, on second thought, admitted he was wrong and agreed that it was a previously unknown type of monstrous reptile.

Not long after another dinosaur was found near Oxford by Reverend William Buckland (1784–1856), although the term "dinosaur" had yet to be invented by the British Museum's anatomist Richard Owen. Mantell redoubled his efforts at Tilgate Forest and was rewarded with a new saurian—this one bearing the cutting teeth of a meat-eater.

In 1832, Mantell dug up an armored *Hylaeosaurus* covered with bony plates and spikes. By now, Mantell, completely obsessed with his fossil hunting, was ignoring his medical practice. His house became so crammed with rocks and bones that his wife left him, probably regretting that she had sparked his obsession by finding that first tooth. Finally, he had to sell off his dinosaurs to the British Museum in order to survive.

Maidstone, the county seat of Kent, has an official coat of arms containing an iguanodon that can be seen on all public buildings. Although it looks authentically Old English, the design was actually adopted by the town in 1949. It is the only municipality in the world with a dinosaur on its official seal, commemorating the discovery of an iguanodon in a local quarry by the first dinosaur hunter, Dr. Gideon Mantell.

See also DINOSAUR; DINOSAUR RESTORATIONS; IGUANODON DINNER.

For further information:
Colbert, Edwin H. *Men and Dinosaurs: The Search in Field and Laboratory.* New York: E. P. Dutton, 1968.
Mantell, Gideon. *Medals of Creation: or, First Lessons in Geology, and in the Study of Organic Remains.* London: Henry G. Bohn, 1844.
———. *The Wonders of Geology.* New Haven, Connecticut: A. H. Maltby, 1839.

MARGULIS, LYNNE

See BIOSPHERE; LIFE, ORIGIN OF.

MARSH, O. C.

See COPE, EDWARD DRINKER; COPE–MARSH FEUD.

MARXIAN "ORIGIN OF MAN"
Dialectical Darwinism

"Labor is the prime basic condition for all human existence, and this to such an extent that, in a sense, we have to say that labor created man himself . . . First labor, after it and then with it, speech—these were the two most essential stimuli under the influence of which the brain of the ape gradually changed into that of man."

So wrote Friedrich Engels in his 1876 essay, "The Part Played by Labor in the Transition from Ape to Man." Engels collaborated with Karl Marx in developing what they called "scientific socialism," which exalted labor as the definitive characteristic of *Homo sapiens.*

Marx had defined labor in *Capital* (1867) as a process in which man "mediates, regulates and controls his material interchange with nature by means of his own activity . . . Acting upon nature outside of him, and changing it, he changes his own nature also." Marx distinguished human labor from the work of spiders or bees, which he thought instinctual or automatic. To man alone he attributed a conscious purpose, a mental picture of the results of his labor. Since labor is the crucial factor in human evolution, an industrial system that "alienated" workers from their tools and labor is literally inhuman.

Engels discussed how chimpanzees build tree nests and shelters and even wield sticks and stones. But though ape anatomy is similar to man's, "no simian hand has ever fashioned even the crudest of stone knives." [See ABANG.]

During the vast period of prehistory, Engels speculated, the hand, freed from locomotion, "could henceforth attain ever greater dexterity and skill." Thus, "the hand is not only the organ of labor, it is also the product of labor. Only by labor, by adaptation to ever new operations . . . [muscles, ligaments, and bones adapt and improve until the hand can] conjure into being the paintings of a Raphael . . ."

But when social organization diversified to where some individuals merely planned the labor that would be carried out "by hands other than their own," at that point bourgeois philosophy originated. "All merit for the swift advance of civilization was ascribed to the mind, to the . . . activity of the brain."

Essences, ideals and religious thought dominated human cultures and "still rules them to such a degree that even the most materialistic natural scientists of the Darwinian school are still unable to form any clear idea of the origin of man, because under the ideological [bourgeois religious] influence they do not recognize the part that has been played by labor."

This Marxian idea of an early feedback loop (or dialectic) between the evolution of the hands and labor doubtless seems bizarre to non-Marxians, unaccustomed to seeing the loaded word "labor" used in discussions of human evolution. However, if one re-reads Engel's passages and substitutes the phrase "tool-use" for "labor" every time it occurs, the theory becomes identical to the anthropological orthodoxy of the past few decades.

For further information:
Engels, Friedrich. *The Origin of the Family, Private Property and the State.* New York: Pathfinder, 1972.

MATERIALISM
Everything's the Matter

In modern everyday speech, "materialism" means a special fondness for wealth, cars, houses, possessions—material "stuff" you can see and touch. In 19th-century science and philosophy, Materialism was the belief that matter is the only "stuff" there is. Evolutionary theory—like classic chemistry and physics—is based on that materialist assumption. Science itself is often defined as "the study of the properties of matter."

The underlying postulate of Materialism (and science) is the unprovable idea that everything—the Earth, the stars, animals, even our minds, dreams and personalities—is a product of physical matter, which is everywhere the same. Although the idea goes back to the ancient Greeks, its acceptance in science really began in the 18th century with Newtonian physics.

Only a century ago, Materialism was still a very hot topic, provoking passionate debate. Because its view of nature excludes demons, gods, ghosts and fairies (they are immaterial), Materialism was considered by many to be an enemy of religion, morality and ethical values. Scientists sought personality in brain rather than soul and focused on the creative properties of matter, leaving a creator God out of their theories. [See BELFAST ADDRESS.]

But most scientists who adopted the Materialist postulate did not abandon Judeo-Christian values and many professed a belief in God. Newton himself spent more time studying the Book of Revelation than he did physics. Yet, Materialist scientists shocked Victorian lecture audiences by asserting that "thought is as much a secretion of brain as urine is of kidneys." Thomas Henry Huxley liked to compare the mind to the whistle on a steam engine—a noisy adjunct to the body, driven by the same force.

Opponents of Materialism asked "Is that all there is?" and complained science was reducing man to a thing devoid of spirit. Vitalists insisted there must be an undiscovered "life force" pushing evolution forward; Spiritualists argued for "an invisible guiding hand" from another dimension, while fundamentalists looked no further than "The Creator" for explanations of natural phenomena.

Although perhaps more comforting than "cold" Materialism, these nonphysical approaches proved very poor tools for advancing biological understanding. Rejecting all these unseen, unmeasurable "forces," paleontologist George Gaylord Simpson wrote in his *Tempo and Mode in Evolution* (1944):

the progress of knowledge rigidly requires that no non-physical postulate ever be admitted in connec-

tion with the study of physical phenomena . . . the researcher who is seeking explanations must seek physical explanations only . . .

Simpson's proscription excludes not only consciousness, spirit and God, but also Platonic ideals—patterns, types, archetypes—that such naturalists as Richard Owen and Lorenz Oken had believed to be as real as bones. [See ESSENTIALISM; NATURPHILOSOPHIE.]

Physical explanation as the only scientific one is a legacy of Newtonian physics, which by the 19th century was the only accepted scientific model of the universe. However, modern physics has changed the picture drastically.

Critic of anthropology William R. Fix writes: "quantum physicists have been led more and more to consider models of consciousness and theories of perception as part of the 'stuff' that the new physics is about." Subatomic particle behavior, uncertainty principles and other recent developments have led quantum physicists to "describe 'reality' in terms that are often restatements of Buddhist metaphysics."

Fix's critique is that modern evolutionary biologists accept as their "reality" a "naive realism" based on outdated Victorian science: an old-fashioned Materialism that was daring and appropriate in Darwin's day but is no longer the model used in physics. (In one of his private notebooks, Darwin had written to himself: "Why is thought being a secretion of brain, more wonderful than gravity a property of matter? . . . Oh, you materialist!")

Thomas Henry Huxley, who fought to establish Darwinian theory, lamented that "there are numbers of highly cultivated and indeed superior persons to whom the material world is altogether contemptible; who can see nothing in a handful of garden soil, or a rusty nail, but types of the passive and the corruptible." Before they could appreciate Darwin's work, Huxley would have to raise their estimation of "matter," and his sophisticated understanding of Materialism was anything but naive.

"To modern science," he wrote, "the handful of soil is a factory thronged with swarms of busy workers [microorganisms]; the rusty nail . . . an aggregation of millions of particles, moving with inconceivable velocity in a dance of infinite complexity yet perfect measure; harmonic with like performances throughout the solar system . . . these particles [and the energy that stirs them] have [always] existed and will exist . . . Form incessantly changes, while the substance and the energy are imperishable."

In Huxley's view, matter was no less remarkable and mysterious than what some would call "consciousness." In 1878, he wrote that what we call

["Mind" and "Matter"] in our little speck of the universe are only two out of infinite varieties of existence which we are not competent to conceive—in the midst of which . . . we might be set down, with no more notion of what was about us, than the worm in a flower-pot on a London balcony has of the life of the great city . . .

Although Huxley has often been criticized for "reducing" everything to matter, in truth he never forgot that Materialism was only a useful working assumption. He was always careful not to consider it an ultimate explanation or a provable fact, and it never diminished his sense of wonder.

He was resentful that some scientists "talk as if . . . [accepting] matter as the substance of all things cleared up all the mysteries of existence. In point of fact, it leaves them exactly where they were."

One Victorian wit, despairing of ever finding a solution to the Mind–Matter question, summed up his perplexity in this epigram: "What is mind? No matter. What is matter? Never mind."

See also MECHANISM; SPIRITUALISM; THEORY, SCIENTIFIC.

MATTHEW, PATRICK (1790–1874)
A Darwinian Predecessor

Eccentric Scots naturalist Patrick Matthew believed that good ideas must eventually surface, no matter where they appear. A tree expert and fruit grower by trade, he published a concise summary of natural selection 26 years before Charles Darwin and Alfred Wallace, but his achievement was unnoticed and ignored. That was because, in 1831, Matthew chose to reveal his theory of evolution in an appendix to an obscure work on *Naval Timber and Arboriculture*, where it made no impact whatsoever.

After Darwin shook the scientific world with his epochal *Origin of Species* (1859), Matthew wrote a letter to the *Gardener's Chronicle*, claiming priority not only for "the organic evolution law," but also for the idea of the "steam ram . . . and a navy of steam gun-boats as requisite in future maritime war"—other prophetic ideas for which he received no credit.

Upon reading Matthew's protest, Darwin replied that he had not heard of the earlier publication, but would now "freely acknowledge that Mr. Matthew has anticipated by many years the explanation which I have offered of the origin of species [by] natural selection." He also noted that, considering the strange place in which it was published, "I think that no one will feel surprised that neither I, nor apparently any other naturalist, has heard of Mr. Matthew's views." In subsequent editions of the *Origin*, however, Darwin duly credited Matthew and other forerunners in a historical preface.

Not satisfied with Darwin's acknowledgement, Matthew tried to belittle his achievement by noting that he had thought natural selection was perfectly obvious, and could not see what all the fuss was about:

To me the conception of this law of Nature came intuitively as self-evident fact, almost without an effort of concentrated thought. Mr. Darwin here seems to have more merit in the discovery than I have had—to me it did not appear a discovery . . . it was by a general glance at the scheme of Nature that I estimated this select production of species as . . . fact—an axiom, requiring only to be pointed out . . .

The difference, as Stephen Jay Gould has written, was that Matthew had indeed stated the theory first, but had failed to appreciate and work out its vast implications. In contrast, Darwin amassed huge amounts of data and tackled such diverse problems as plant and insect coadaptations, human evolution, variation in domesticated animals and plants and hundreds of other topics. "He established a workable research program for an entire profession," notes Gould.

Even though Darwin was not the first to "discover" natural selection, it is what he did with it that history remembers. He gets full credit for devoting a lifetime to carefully refining the idea, patiently exploring its application to thousands of observations and experiments, grasping its full significance and conducting a tireless campaign to convince scientists to adopt it as a productive framework for biology.

Darwin himself was somewhat amused when an even "earlier" discoverer of natural selection came to light. A scientist named William Charles Wells—famous for his "Essay on Dew"—had previously published a statement of natural selection applied to the races of man. On learning of Wells in 1865, Darwin wrote a friend, "So poor old Patrick Matthew is not the first, and he cannot, or ought not, any longer put on his title-pages 'Discoverer of the Principle of Natural Selection.' "

See also NATURAL SELECTION; WELLS, WILLIAM C.

MATTHEW, WILLIAM D.

See KNIGHT, CHARLES R.

MAUER JAW (1907)
"Heidelberg Man"

Where are the fossils? Where are the "missing links"? Around the turn of the century, skeptics, anti-Darwinians and even professional paleontologists were confronting the lack of known human and near-human fossils. During all the time that Charles Darwin was writing about human origins, no fossil evidence had come to light except a scrap of Neandertal skullcap. He inferred that man had originated in Africa because our closest living relatives, the gorilla and chimpanzee, were found there. (Another 100 years would pass before his shrewd guess was backed up by dozens of African hominid fossils.)

In the 1890s, after Charles Darwin's death, the Dutchman Eugene Dubois discovered the *Pithecanthropus* fossils in Java, which were later so hotly contested as human ancestors. [See DUBOIS, EUGENE.] But there was still precious little from Europe, and so each new jaw or scrap of skull that shed light on early man created a sensation.

The Mauer Jaw caused such an uproar. Discovered on October 21, 1907 by two workmen digging in a huge commercial sand pit near Mauer, Germany, it was a human-looking jawbone that seemed much too large to belong to a modern man. Geologists from Heidelberg University had been watching excavations at the pit, since other fossils had been unearthed there, and they asked the owner to be on the lookout for possible human remains. The jawbone was 80 feet below ground level and was split in half by the workman's shovel.

Professor Otto Schoetensack of the University excitedly claimed the jaw, which he cleaned, restored and described. It appeared to him heavy and wide enough to belong to an ape, but the teeth were clearly human. Schoetensack assigned it to a new species of man, *Homo heidelbergensis*, or Heidelberg man. Although he couldn't tell much about what the complete skeleton looked like, he could give it a rough date—the jaw was found in association with the bones of extinct animals of the Early Pleistocene, the first hint than man had lived in Ice-Age Europe.

Today the Heidelberg man is considered simply a European representative of *Homo erectus:* a once widespread species that half a million years ago lived in Africa, then migrated to Asia and Europe. Dubois's Java man or *Pithecanthropus* was the same *Homo erectus* (Chinese examples were not found until 1927).

See also DUBOIS, EUGENE; HOMO ERECTUS.

MAYR, ERNST (b. 1904)
Evolutionary Biologist, Ornithologist, Historian

On March 23, 1923, a young German medical student named Ernst Mayr chanced to spot a pair of very rare ducks—the first fateful "accident" of a brilliant, unplanned and unexpected career. Recently graduated from the Dresden secondary school, he had thought his life was set: to follow the four-generation family tradition of successful physicians. He certainly did not expect to become an explorer, naturalist, ornithologist, philosopher-historian of science, Har-

vard professor and one of the 20th century's greatest evolutionary biologists. The ducks changed everything.

A birdwatcher since childhood, Mayr had bicycled into the countryside on spring break to hike around the wooded parklands and lakes of Moritzburg, a former hunting preserve of Saxon kings. When the two ducks with brilliant red bills and crests swam into view, he realized they were extraordinary. Immediately he cycled back to Dresden, but was unable to find anyone to come and confirm his sighting. Upon checking bird books, he discovered they were red-crested pochards, a species not seen by anyone in nearly 80 years.

When he told his birdwatching friends of his discovery, no one believed him, and the young man felt totally crushed. At a party soon after, he poured out his heart to a stranger, a pediatrician, and told him the duck story. Improbably, the man knew the greatest ornithologist in Berlin, one Professor Erwin Stresemann, and promptly wrote a letter of introduction. But when Mayr journeyed to Berlin, he received a rough reception. Stresemann quizzed him mercilessly, probed his knowledge of natural history and asked to see his prior notes and journals of field observations. At last convinced that Mayr was a first-rate observer, he published the sighting as genuine.

Impressed with the young man, Stresemann invited him to work in the Berlin Museum that summer as a volunteer, classifying bird specimens received from the tropics. Fascinated by the rain-forest wildlife, Mayr thought he was "given the keys to heaven" and continued to work at the museum during breaks from medical school. Just before he was to receive his degree, his mentor Stresemann offered to send him to the tropics, if he delayed his medical career and earned a doctorate in ornithology.

By age 21, Mayr had earned that doctorate and accompanied Stresemann to the International Zoological Congress at Budapest in 1927, where he was introduced to Lord Walter Rothschild, titular head of the wealthy European banking family. At his own private museum at Tring, in Hartfordshire, England, Rothschild was assembling the world's largest and most comprehensive bird collection. Again, Mayr benefited from a well-timed accident of circumstance. Rothschild's staff naturalist in New Guinea had suddenly died after many years of service, and he was desperately seeking a new bird collector. Mayr was hired on the spot.

Within the year, Ernst Mayr had traveled through six unexplored New Guinea mountain ranges, eventually collecting 3,400 bird skins and discovering 38 new species of orchids. In 1930, while suffering from malaria and dysentery in his mountain camp, he received an urgent invitation to join an expedition to the West South Seas sponsored by the American philanthropist Harry Payne Whitney. Again, Mayr was in the right place at the right time—a week before departure, the expedition had suddenly found itself without a leader. He accepted.

The Whitney South Seas Expedition was an epic scientific adventure, which made important contributions to biology, discovered scores of new species and provided the American Museum of Natural History with the materials for a new hall. In 1931, after collecting in the Solomon Islands, Mayr was hired to come to New York and work with the bird specimens at the museum.

He asked his chairman, Frank Chapman, what he should do first. Chapman replied, "You've been cracked up to me as an expert on South Sea island birds. *You* should know what you should be doing." "I had been raised in the Central European tradition where the boss tells you what to do next," Mayr recalled years later, "and I was shattered by such freedom."

In his first year at the museum, Mayr published a dozen papers, describing scores of new species and subspecies. The following year Rothschild's curator retired; Mayr was invited to take charge of the collection at Tring. There he worked up a series of related species from different islands, a convincing and dramatic demonstration of geographic speciation.

In 1936, he invited the evolutionary geneticist Theodosius Dobzhansky to study the series. He was impressed and Mayr's study influenced Dobzhansky's important book *Genetics and the Origin of Species* (1937), the founding work of the Synthetic Theory of evolution.

During the 1930s and 1940s, Mayr collaborated with Dobzhansky, Julian Huxley, and George Gaylord Simpson to help formulate the modern evolutionary synthesis, incorporating new discoveries by naturalists and population geneticists into the framework of Darwinian theory.

Mayr might have stayed on as Rothschild's curator at Tring, but for another "great accident." Rothschild had been involved with a married, titled woman who was now blackmailing him with threats of a family scandal. Her merciless and increasing demands for large sums of money ruined Rothschild, forcing him first to cut back on staff and finally to sell off his beloved and precious collection. New York's American Museum of Natural History purchased 280,000 bird skins from Rothschild, courtesy of the Whitney family, in the hardest year of the Depression. Mayr helped pack and ship 185 cases, each containing 7,000 skins of rare birds.

Mayr then returned to the New York museum to continue his work on speciation, and published an influential study on organic diversity, based on a series of bird species collected in the Solomon Islands and Fiji. With the island's robins, he was able to show that different colors among sexes are not determined by sex hormones, but by geographic isolation. On one island, both sexes of the robin species were drably colored; on the next island they were both brightly colored; on still another the male was bright and the female drab. His lectures on this research were well received at scientific meetings.

But Mayr has claimed he was then invited to give the prestigious Jessup Lectures at Columbia University only because at one meeting at which he spoke, the preceding speaker (the brilliant geneticist Sewall Wright) had seemed particularly dull. According to Mayr, Wright had turned his back to the audience and mumbled inaudibly as he filled a blackboard with mathematical equations, thus making Mayr's presentation appear even more interesting.

Mayr was asked to give two lectures on "Speciation in Animals" at Columbia, while a botanist, E. Anderson, was to give two on plants; the lectures were to be published together as a book. But Anderson, who suffered from manic-depression, was unable to submit his lectures, and Mayr was asked to fill in the rest of the volume. It became *Systematics and the Origin of Species* (1942), an unplanned but influential classic that redefined species in terms of breeding populations. If two subpopulations of geese that look alike are in contact but do not interbreed, they are considered separate species. On the other hand, the snow goose and the blue goose look very different and were considered different species. But when naturalists found that they flock together and interbreed, they were reclassified as color phases belonging to a single species.

Building on his decades of familiarity with island populations of birds, Mayr advanced a general theory (in 1954) of how species evolve. Either through the appearance of geographic barriers or by a few "founders" settling in a new area beyond the species' customary range, a very small population can become established—the first step toward reproductive isolation. [See GENETIC DRIFT.] Over time the little colony inbreeds, local conditions exert their selective pressures, and descendants become increasingly different from their ancestral population. If they are ever reunited, the two populations may no longer be capable of interbreeding or producing viable offspring. This kind of rapid evolution at the edge of a species' range (technically called "peripatric evolution") was emphasized by Mayr in the 1950s, and became one of the foundations for the punctuated equilibrium theory of Niles Eldredge and Stephen Jay Gould in the 1970s. [See PUNCTUATED EQUILIBRIUM.] The apparent quickness of such evolutionary change, however, is only from the geologist's long perspective. In fact, it is gradual and occurs over many thousands of generations. [See TURKANA MOLLUSKS.]

Amused at the chancy, undirected path of his own career, Mayr enjoys describing it in terms of contingent history—a process that parallels evolution itself. Had Rothschild's former lover not blackmailed him, Mayr might have spent his life at Tring in England and never had contact with Columbia and Harvard, which eventually led him from his work as taxonomist (classifier) to historian of science, to philosopher of evolution. Had Wright not been such an inaudible speaker at the meetings, had Rothschild's field collector not died, had Mayr not seen those rare ducks or met the pediatrician who knew the ornithologist, he claims he would not have had his remarkable career at all.

See also ALLOPATRIC SPECIATION; CONTINGENT HISTORY; DOBZHANSKY, THEODOSIUS; ROTHSCHILD, LORD WALTER; SYNTHETIC THEORY; WAGNER, MORITZ.

For further information:
Dobzhansky, Theodosius. *Genetics and the Origin of Species.* New York: Columbia University Press, 1937.
Mayr, Ernst. *Animal Species and Evolution.* Cambridge: Harvard University Press, 1963.
———. *The Growth of Biological Thought.* Cambridge: Harvard University Press, 1982.
———. *Systematics and the Origin of Species.* New York: Columbia University Press, 1942.

MECHANISM
Materialist Philosophy

Like all sciences, evolutionary biology is based on the idea that nature's workings, like those of a machine, can best be understood without reference to consciousness, gods or spirits. Mechanism (and its companion concept "Materialism") has always had profound limitations as a mode of explanation, yet its application to scientific problems has yielded extraordinarily fruitful results. Materialism assumes that everything, living and nonliving, is formed of the same kind of measurable, physical matter.

Previously, it was believed man was somehow set apart from nature; even his chemical bodily processes were thought to be special. Physiological studies over the last century, however, have established that our bodies operate according to the same biochemical processes found elsewhere in nature. Even the human heart, for so long considered the repository of love, poetry and courage can be replaced for months by a plastic pump, with no discernible loss of "spiritual" qualities in the recipient.

Among the most important thinkers in establishing mechanistic explanations in science are the British physicist Sir Isaac Newton (1642–1727) and the French philosopher Rene Descartes (1596–1650). Newton's

laws of motion and gravitation proved indispensable for building machinery in the industrial era and also encouraged the concept that the universe itself was like a machine. Its components could be analyzed, and the way they worked together was assumed to be constant, predictable and without supernatural interference.

Descartes believed there was a sharp division between mind (spirit, possessed only by man) and body, shared by man and animals. Nonhuman creatures, in Descartes' view, were simply "automata," mindless machines like robots or what were then called "clockworks mannikins." (The title of Anthony Burgess's novel *A Clockwork Orange* [1962] is a reference to the human brain as a programmable machine encased in a soft, fruitlike rind.)

Religious philosophers were outraged at science's blind, "clockworks" universe. Where was the place in this design for a caring God who could intervene in human affairs? A mechanistic cosmos devoid of plan, purpose or consciousness seemed devastatingly cold and unappealing. Many scientists, including Newton himself, did not replace their personal religious beliefs with a mechanistic worldview, but found it a necessity when doing science. It wasn't a matter of comfort, hope or morality; mechanistic models simply worked best at mimicking and predicting the behavior of natural phenomena.

Without Newtonian physics and the later advances of a "mechanistic, materialistic" chemistry and physiology, Darwin's evolutionary model would have been neither possible nor plausible. It is no accident that he was buried in Westminster Abbey just a few feet away from the tomb of Sir Isaac Newton.

See also BELFAST ADDRESS; CARTESIAN DUALITY; MATERIALISM.

MEGALADAPIS
Largest Lemur

One of the strangest collections of living and fossil primates comes from the island of Madagascar, now known as the Malagasy Republic, off the east coast of Africa. Its rain forests today still hold the last remnants of the prosimians, the group ancestral to monkeys, apes and man. Its lemurs, indrids and the amazing nocturnal aye-aye, which pokes a long, clawed middle finger into logs to spear insects, are found nowhere else in the world.

Strange as these rare, endangered species are, even more remarkable creatures lived on Madagascar until only a few hundred years ago. Among them were several species of giant lemurs, some with skulls almost two feet long. They were named *Archeoindris*, *Palaeopropithecus* and *Megaladapis* by the English scientist Charles Forsyth-Major, who first described their remains in 1894. Apparently, lemurs had reached Madagascar in the Pliocene and, being isolated, had radiated into many species, some tiny and some huge.

Radiocarbon dating of the giant lemur skulls has revealed that they are between 1,000 and 2,800 years old—comparatively recent as fossils go and well within the range of known human history. In fact, the migration of people to Madagascar about 1,500 years ago is probably responsible for the extinction of the creatures. *Megaladapis* remains are often found with bones of the giant, recently extinct flightless bird *Aepyornis*, which may be the fabled "roc" of Arabian legend. Also known as the "elephant bird," it was about three times the size of an ostrich—its eggs weighed 20 pounds.

Evidence that *Megaladapis* survived until very recent times comes from the journals of a French explorer named Francois de Flacourt, who visited the island in 1758. He described a large living creature called "tre-tre-tre-tre" by the natives, presumably after its call. It was the size of a two-year-old calf, he wrote, with a round head and human-like face, monkey-like hands and feet and man-like ears. According to de Flacourt, "the natives of the region flee him as he does them." If this was *Megaladapis*, one paleontologist recently lamented, "we can only regret having so closely missed a chance to observe what must have been one of the most curious primates ever to have existed."

MENCKEN, H. L. (1880–1956)
Journalist at Scopes Trial

By 1925, Henry Louis Mencken was a tough, witty, cynical cigar-chomping newspaperman who had spent 15 years covering big-city crime and politics. That year the *Baltimore Evening Sun* sent him to the little rural town of Dayton, Tennessee to report on the trial of a science teacher named Scopes, whose crime was teaching Darwin's theory of evolution. The prosecutor was the nationally known politician William Jennings Bryan, a biblical fundamentalist. Defending Scopes was famed liberal lawyer Clarence Darrow.

Mencken's dispatches were alternately sensible, satirical, condescending and cruel; they were widely reprinted all over the country. Mencken painted Bryan as a rabble-rousing hypocrite. "If the fellow was sincere," he wrote, "then so was P. T. Barnum . . . He was, in fact, a charlatan, a mountebank, a zany without sense or dignity." Recent historians of the trial have been much kinder to Bryan. [See BRYAN, WILLIAM JENNINGS.]

Although technically Bryan won the case, by the end of the trial he was a broken man. Darrow had

assaulted him with a barrage of ridicule that left him utterly worn out and defeated. A few days later he died suddenly.

But even when Darrow had finished, Mencken did not let up. If Bryan had survived his first stroke, Mencken's "memorial" article would have given him another. Bryan's whole career, he wrote:

> . . . was devoted to raising half-wits against their betters, that he himself might shine . . . One day [Darrow] lured poor Bryan [into repeating] his astounding argument . . . that man is [not] a mammal . . . I am glad I heard it, for otherwise I'd never believe it. There stood the man who had been thrice a candidate for the Presidency of the Republic—in the glare of the world, uttering stuff that a boy of eight would laugh at . . .

Mencken's adventures as a correspondent at the Scopes trial are told in his book *Heathen Days* (1943). He later established himself as a popular "critic of ideas," and published a distinguished study of *The American Language* (1919).

In *Inherit the Wind* (1951), the drama about the Scopes trial, the character of the abrasive, cynical reporter is undisguised Mencken. He was played by Tony Randall in the Broadway stage version and by Gene Kelly in the Spencer Tracy film.

See also INHERIT THE WIND; SCOPES TRIAL.

MENDEL, ABBOT GREGOR (1822–1884)
Father of Genetics

The son of peasant farmers, Gregor Mendel was born in 1822 in the Silesian village of Heinzendorf in what is now Czechoslovakia. In an autobiographical sketch written in 1850, Mendel described his youth as "sorrowful" largely, as he says, because of his repeated failure to secure for himself the means to continue his education: "The distress occasioned by the disappointed hopes, and the anxious, dreary outlook which the future offered me, affected me so powerfully at that time that I fell sick, and was compelled to spend a year with my parents to recover." This was apparently the first of a series of nervous breakdowns that Mendel suffered. To assure his future, he "felt compelled to step into a station of life" that would free him "from the bitter struggle for existence." This station was the Catholic Church, and, in 1843, Mendel entered the Augustinian monastery in Brunn as a novitiate.

Freed from immediate necessity, Mendel began a career as a teacher, but was required to pass a state examination in order to receive a permanent appointment. He took the exam in 1850, but failed. The following year, at the monastery's expense, Mendel enrolled as a student at Vienna University to study natural sciences and continued there until the summer of 1853. In 1856, Mendel once more attempted the state examination, but this time could not even finish it. He suffered another nervous breakdown and in depression returned to Brunn, where he remained ill for some months. He never attempted the examination again, but worked for years as a temporary uncertified teacher.

Mendel's scientific fame rests on a single paper published in 1965 in the journal of the Brunn natural science society. The paper summarizes the results of his lengthy hybridization experiments on peas. It did not attract attention during his lifetime, and Mendel's few scientific correspondents saw nothing remarkable in his work though it was, in fact, the long-sought key to understanding heredity.

"Mendel's Laws," though they were not explicitly stated in that 1865 paper, implied that parental characters do not blend in their progeny but are transmitted as discrete factors. Each trait is controlled by a pair of factors in which only one of the pair enters each mature sex cell or gamete. When joined with a gamete from the other parent a new pair is formed: the genetic inheritance of the offspring. A mating between a pair that differ in a single character only, $AA \times aa$, produces only heterozygous Aa offspring. If two of this generation mate, $Aa \times Aa$, the series AA, 2Aa, aa results. Half of this generation is therefore identical to the grandparents (AA and aa) and half identical to the parents (Aa). Each pair of differing traits in a hybrid association, for example, flower color and seed color, are inherited independently of all other differences in the parental plants.

Until recently, historians traditionally viewed Mendel as being scientifically isolated, discovering these "laws of heredity" by a set of brilliant experiments conducted in a small garden plot at the monastery then spelling them out clearly. According to this tradition, his achievement was never grasped by his contemporaries and was ignored until 1900, when three researchers—Hugo de Vries, Carl Correns and Erick von Tschermak—independently rediscovered the laws of heredity and Mendel's paper as well. They then brought the paper to the attention of the scientific community, where it was immediately recognized as the foundation document in the new science of genetics.

During the 1980s, this view of the history of genetics has been challenged: The rules referred to as "Mendel's Laws of Heredity" were never clearly formulated by Mendel but were read into his work by his rediscoverers. As Alan Bennett puts it, "There is no statement [in Mendel's paper] of the simple hypothesis that a given character is controlled by a pair of factors, and that a gamete carries only one member of the pair." In Mendel's notation, the result of a cross between Aa and Aa would be A, 2Aa, a, not AA, 2Aa, aa. This latter interpretation was appropriate for Mendel's rediscoverers, whose knowledge of cytology was far greater than Mendel's. In fact, Mendel's paper gives evidence that he envisioned the possibility of the simultaneous inheritance of three or more traits of a single character.

The realization that Mendel did not quite discover the modern interpretation of genetics raises the question of what precisely did Mendel discover and what were his aims? Finding the answers to these questions has become the focus of recent historical research leading to different interpretations among experts. Because of a paucity of historical documents, the answer may never be fully uncovered. But whatever historians may ultimately decide about Mendel's discoveries as he himself envisioned them in 1865, the point for the history of science is that in 1900 his results were immediately reinterpreted and extended and that this reinterpretation has guided genetics ever since.

Mendel's experiments in plant hybridization lasted only about 15 years. After 1868, when he was elected abbot at the age of 46, he ceased teaching; his position as head of the monastery gave him public duties and administrative responsibilities that left little time for scientific work. Mendel died in 1884 of cardiac degeneration and kidney disease made worse by chronic nicotine poisoning—he was a heavy cigar smoker in the latter part of his life.

After his death, the new abbot had his experimental garden destroyed and all his notebooks, papers and scientific records burned. Mendel's only claim on the world's attention or gratitude was the single published paper of 1865 in an obscure journal of natural history, which had remained unknown throughout his lifetime.

It was not until 16 years after his death that Mendel's ideas were resurrected, and his key to understanding the age-old question of heredity at last appreciated. Ironically, among the few relics of Mendel's life to survive are two poems from his youth, both honoring the power of the printed word. One concludes:

> May the might of destiny grant me
> The supreme ecstasy of earthly joy . . .
> To see, when I arise from the tomb,
> My art thriving peacefully
> Among those who come after me.

See also MENDEL, REDISCOVERY OF; MENDELIAN RATIOS CONTROVERSY; MENDELISM, ACCEPTANCE OF; MENDEL'S LAWS.

MENDEL, REDISCOVERY OF (1900)

Tradition holds that the belated recognition of Gregor Mendel's classic work in genetics occurred during the winter of 1899–1900 when Hugo de Vries, Carl Correns and Erik von Tschermak, after discovering the basic 3:1 hybrid ratios for themselves, independently tracked down the original paper through references in the scientific literature. According to statements made by these researchers many years after the event, each concluded that Mendel, 35 years earlier, had not only discovered the same hybrid ratios, but had come up with an explanation identical to the one they had independently formulated.

Recent historical interest in the multiple "rediscovery" does not confirm this traditional account. Instead, it appears that each of the rediscoverers, while working on hybrid crosses, read Mendel's paper and that it was only through this reading that they understood the meaning of their own results.

Of the three, Tschermak has the least claim to the title "rediscovery." Although he found Mendel's paper independently of the others, historians claim his writings show little understanding of what Mendel said. Nonetheless, he did reference Mendel's work in his 1900 publication and later went on to champion the importance and priority of Mendel's contribution.

De Vries, the most enigmatic "rediscoverer," maintained that he found Mendel's paper referenced in L. H. Bailey's pamphlet *Cross-breeding and Hybridizing* (1892), published after he had performed his experiments on hybridization. But after de Vries's death, his close colleague and successor Theo J. Stomps told a quite different tale. According to Stomps, de Vries actually received a copy of Mendel's paper from Martinus W. Beijerinck early in 1900, while still in the midst of his research.

Stomps's account fits with the fact that until that time de Vries showed no signs of understanding the hybrid ratios he was obtaining from his crosses. It now seems likely that de Vries received Mendel's paper, read it, saw that his many hybrids could be explained in Mendel's terms and wrote the first of his famous Mendelian papers *Sur la loi de disjonction des hybrides* (1900), a paper that, strangely enough, did not mention Mendel's name.

Carl Correns had seemed, until recently, the least problematic of the "rediscoverers." He had also seemed the one who understood Mendel best. In his account of how he came to find Mendel's paper, he says as he lay awake in bed toward morning in October of 1899, mentally running over his hybrid data, that the interpretation came to him "like a flash"—but that he did not become aware of Mendel's paper until a few weeks later. Correns certainly knew of Mendel's paper by December of 1899, since he referenced it in a publication at that time on another subject. But his casually tangential reference to Mendel in that paper leaves grave doubts that he understood the importance of Mendel's work at that time.

This evidence has led Onno G. Meijer of Free University, The Netherlands, an historian who specializes in this period, to question the "flash" episode of Correns's recollection 20 years later. In fact, Correns published on heredity only after he received a copy of de Vries's first paper in March of 1900. If he had independently discovered Mendelian ratios and principles in October 1899, as he later claimed, why did he ignore Mendel in his December publication and then wait until March (directly after reading de Vries's paper) to publish on heredity?

In short, the traditional rediscovery account is probably false. These three researchers did not rediscover what Mendel had discovered 35 years earlier, but were able by means of Mendel's paper to interpret their results in Mendel's fashion. In other words, Mendel told them what they were looking at! Without Mendel's paper, it is unlikely that Tschermak or de Vries would ever have discovered the famous 3:1 ratio. Correns might have eventually discovered it, but seems to have also to have been enlightened by Mendel, perhaps via de Vries.

Scientifically (but perhaps not historically or intellectually), it matters little whether these three researchers independently rediscovered Mendel's conclusions or whether Mendel was instrumental in directing their interpretations of their data, since they created a large enough stir that the science of Mendelian genetics took root at last.

Tschermak saw that Mendel's paper was reprinted in Europe in 1901, and William Bateson in England also had it translated and published in 1901. Highly visible publications by Tschermak, Correns, Bateson, and de Vries in German, French and English around the turn of the century insured that Mendel's paper would not be lost again.

See also DE VRIES, HUGO; MENDEL, ABBOT GREGOR; MENDELIAN RATIOS CONTROVERSY; MENDEL'S LAWS.

For further information:

Corcos, A. F., and Monaghan, F. V. "Role of de Vries in the Rediscovery of Mendel's Paper." *Journal of Heredity* 78 (1987): 275–276.

Meijer, O. G. "Hugo de Vries no Mendelian?" *Annals of Science* 42 (1985): 189–232.

Olby, Robert C. *Origins of Mendelism*. London: Constable, 1966.

Roberts, H. F. *Plant Hybridization before Mendel*. New York: Hafner, 1965.

Stern, C., and Sherwood, E. R. *The Origin of Genetics.* New York: W. H. Freeman, 1965.

MENDELIAN RATIOS CONTROVERSY
Did Mendel Fudge His Results?

About 70 years after Father Gregor Mendel described his pioneering genetic work with pea plants in a monastery garden, a respected English statistician went over his data and came to an astonishing conclusion. Professor R. A. Fisher, a leader in modern evolutionary and population genetics, announced in 1936 that Mendel's experiments had been falsified.

Fisher did not contend that Mendel's theories were wrong, only that he fudged the numbers he used to prove it. By applying a mathematical test known as "chi-square" analysis, he concluded that Mendel's results with thousands of pea plants were literally too good to be true. Because there is usually a range of variation in large numbers of trials, Fisher argued, it is incredible that Mendel's numbers came out almost "right on the button."

Although he did not accuse Mendel of outright fraud, Fisher expressed the opinion that assistants "bent" the data to fit Mendel's expectations. Because Fisher had a top reputation in biomathematics, his accusations that Mendel's results were "too perfect" were widely accepted and became incorporated into textbooks. Recently, Mendel was even listed in a "rogue's gallery" of scientific frauds in the prominent journal *New Scientist.*

During the 1980s, however, several geneticists, including Ira Pilgrim, Alain Corcos and Floyd Monoghan took a fresh look at Mendel's data. Could Mendel's results really be "too good to be true?" Pilgrim wrote:

> The implication of this concept is that if a scientist is testing a theory that predicts a 50:50 ratio; and, on testing, gets results of 500:500, that he should repeat the experiment in order to get *worse* results, lest he be accused of cheating. Mendel's results agreed with his theory. Why shouldn't they, since his theory was correct?

If Mendel's theory describes nature, then 500:500 is more probable than any other result, though it is unlikely he will hit it exactly. Yet Mendel had enough experience with statistics to make a large series of trials, which insured that the results would be pretty close to expected ratios. (The numbers would be much more likely to vary from the mark in smaller samples.)

On a more technical level, Pilgrim and others have given mathematical demonstrations of the correctness of Mendel's methods and the spuriousness of Fisher's "demonstration." For one thing, they have shown that truth or falsity of data cannot be deduced by statistical analysis of results. Only repetition of the experiments can prove whether the results are reproducible. By this criterion, Mendel's work has been validated over and over, while the genetic experiments of Trofim Lysenko, for instance, have not.

Also, Fisher's use of the chi-square test was entirely inappropriate. The test can determine the probability that experimental data fit the hypothesis. It cannot tell whether the data represent a truly random sampling, which is what Fisher was trying to find out.

From 1983–1986, papers on Mendel's "too good to be true paradox" have been flying fast and furious in genetics journals. But the last word on the subject belongs to Ira Pilgrim, who found "no evidence that Mendel did anything but report his data with impeccable fidelity. It is to the discredit of science that it did not recognize him during his lifetime. It is a disgrace to slander him now."

See also LYSENKOISM; MENDEL, ABBOT GREGOR; MENDEL, REDISCOVERY OF.

MENDELISM, ACCEPTANCE OF
The Lost—Found Theory of Inheritance

"My time will come," said the brilliant scientist-monk Gregor Mendel. Working with thousands of pea plants in the monastery garden at Brunn, Austria, he made the crucial breakthrough (1865) in understanding how heredity works. But no biologist recognized his achievement until thirty-five years later, though it was the key to evolution they had all been seeking. Why?

To begin with, Mendel published very little, only two short papers in the journal of a local natural history society. Although he sent reprints to some of the important botanists of his day, many ignored it simply because he wasn't at a major university, had no prestigious friends in science and did his work out in the boondocks.

Charles Darwin, in contrast, had taken infinite pains to cultivate a friendly circle of influential biologists. Quick acceptance of his evolution theory followed years of ingratiating letters and a carefully managed campaign to convince the scientific community. In addition to being a great scientist, Darwin was also a consummate diplomat, tactician and public relations man.

Mendel put his faith in the merit of his work, naively believing qualified scientists would surely recognize its worth. He sent a reprint of his paper to the day's leading botanist, Carl Nageli, who was also working with hybrids. Although Nageli then conducted a long correspondence with Mendel, encouraging him to pursue his work, in reality he was no friend to him.

First, he encouraged Mendel to continue testing his theory on hawkweed, instead of pea plants. Hawkweed has a complicated type of inheritance, which was certain to lead to confusing results. And when Nageli published his big book on inheritance and evolution in 1884, there was not a single reference to Mendel.

Ernst Mayr has pointed out that Nageli had a complex theory of his own and was one of the few biologists of the time who believed in pure blending inheritance. "To accept Mendel's theory," Mayr concludes, "would have meant, for Nageli, a complete refutation of his own."

The cloistered cleric could have used some of the worldliness of the young Thomas Henry Huxley, who made certain to keep one of his papers out of the hands of the great authority in his field, Richard Owen. Huxley wrote a friend:

> You have no idea of the intrigues that go on in this blessed world of science. Science is, I fear, no purer than any other region of human activity; though it should be. Merit alone is very little good; it must be backed by tact and knowledge of the world to do very much.
>
> For instance, I know that the paper I have just sent in is very original and of some importance, and I am equally sure that if it is referred to the judgement of my "particular friend" Professor Owen that it will not be published. He won't be able to say a word against it, but he will pooh-pooh it to a dead certainty.
>
> You will ask with some wonderment, Why? Because for the last twenty years [he] has been regarded as the great authority on these matters, and has had no one to tread on his heels, until at last, I think, he has come to look upon the Natural World as his special preserve, and "no poachers allowed." So I must manoeuvre a little to get my poor memoir kept out of his hands.

Huxley also used his savvy to "manoeuvre" his friend Darwin's work from rebel theory to scientific respectability. If Mendel had had a Huxley, his "time" would certainly have come a lot sooner.

See also MENDEL, ABBOT GREGOR; MENDEL, REDISCOVERY OF; PEER REVIEW; THEORY, SCIENTIFIC.

MENDEL'S LAWS

In genetics, the principles of inheritance are known as Mendel's Laws, after the Austrian abbot who conducted thousands of breeding experiments with pea plants in the garden of the Brunn monastery. The two basic formulations are:

The Law of Segregation: Sometimes called "particulate inheritance," this states that a hybrid or heterozygote (Aa) transmits to each mature sex cell (gamete) only one factor (A or a)—not both—of the pair received from its parents.

The Law of Independent Assortment: Different characters (e.g., shape, Aa, and color, Bb of peas) are recombined at random in the gametes (e.g., AB, aB, Ab, ab).

Although tradition attributes these laws to Gregor Mendel, he never actually formulated them. Instead they were read into his writings by his "rediscoverers" around 1900 and have been attributed to Mendel ever since. In Mendel's famous 1865 paper, independent assortment is only briefly alluded to, while segregation is never even mentioned—it is present as an assumption only.

See also BIOMETRICIAN–MENDELISM CONTROVERSY; MENDEL, ABBOT GREGOR; MENDEL, REDISCOVERY OF.

For further information:
Monaghan, F. V., and Corcos, A. "The Origins of the Mendelian Laws." *Journal of Heredity* 75 (1984): 67–69.

MESOZOIC (Geological Epoch)
Age of the Dinosaurs

Some paleontologists spend a lifetime fascinated by ancient fish or fossil snails but the museum-going public favors the dominant reptiles of the Mesozoic period, better known as dinosaurs.

The Mesozoic, or "time of the middle animals," began about 248 million years ago and lasted about 183 million years. For most of that time dinosaurs of all shapes and sizes populated the planet—from creatures the size of a chicken to the largest land animals that ever walked the Earth. During the mid-Mesozoic, the first birds appeared, and mammals developed somewhat inconspicuously throughout the era. Mammals did not "out compete" the great reptiles, though they did replace them as the dominant land species.

There are three major divisions of the Mesozoic: the Triassic (248–213 million years ago), Jurassic (213–140 million years ago) and Cretaceous (140–65 million years ago). During the Triassic, small dinosaurs began their spectacular radiation into many families and species, adapting to many varied niches.

Sauropod dinosaurs and flying reptiles lived during the Jurassic, named for deposits from the Jura Mountains. The remarkable fossils of *Archaeopteryx* are found in the late division of the Mesozoic. During the late Cretaceous, small mammals evolved, but were inconsequential in the dinosaurian landscape.

A mass dying, or extinction, at the end of the Cretaceous wiped out not only the dinosaurs, but also sea-going plesiosaurs, mollusks and ammonites in the oceans; indeed the majority of all animal life. Despite a great deal of research, and even more speculation, what happened at the close of the Mesozoic is still a tantalizing mystery.

No traces of men or other hominids are found in Mesozoic deposits; none appear until 62 million years later. Despite pop culture images like *One Million B.C.* or "The Flintstones," humans and dinosaurs never walked the Earth at the same time.

See also ALVAREZ THEORY; ANGIOSPERMS, EVOLUTION OF; DINOSAURS, EXTINCTION OF; GREAT DYING; NEMESIS STAR.

METAPHORS IN EVOLUTIONARY WRITING

Poet Robert Frost thought every scientist could learn a practical lesson from poets—skill in the use of metaphors. A good writer, said Frost, knows just how far to extend a metaphor before it becomes strained. Imagery, unexpected comparisons, seeing similarities in the dissimilar are the poet's stock-in-trade; they are also the hallmark of exciting scientific writing. Charles Darwin would certainly have agreed; the metaphors he included in his exposition of evolutionary theory retain their vigor long after his careful collections of "facts" became outdated.

Explaining natural selection, Darwin pictured nature as a stock breeder, selecting variations that improve or adapt species and discarding others. But whereas a human breeder could only select characteristics he could see, like size or color, nature could act on the composition of the blood, the size and shape of internal organs, and on every part of the organism. "What might she not achieve?"

Another of Darwin's metaphors is the "thousand wedges" he pictured pressing in on every part of a plant or animal to fit it to its environment and way of life. Here the image suggests a great tree split open—a job requiring tremendous force—achieved by many small wedges driven along its length. It is a picture that evokes not only the concept of many consistent and directional pressures, but great effects resulting from cumulative small causes.

The final paragraph of the *Origin of Species* (1859) contains Darwin's metaphor of the tangled bank, a jumble of plants and animals, worms and birds and vegetation. It is a commonplace image, observable in any wood or garden, yet confusing to the casual observer. Beneath the apparent disarray, Darwin sees orderly forces that evolve species in relation to others, each with its role to play in the ecology.

These life forms "most beautiful and most wonderful" are being evolved "whilst this planet has gone cycling on according to the fixed law of gravity"—Newton's metaphor of the universe as a machine. In his final sentence of the *Origin of Species* Darwin ties the "tangled bank"—with all the beautiful messiness of life—to the logical, predictable order of a whirring engine.

Perhaps Darwin's most widely known metaphor is his depiction of evolutionary history as a tree of life branching from a broad common trunk to finer limbs, slender branches and delicate twigs. Stephen Jay Gould has slightly modified the metaphor into a branching bush to emphasize the fullness and bushiness of the delicate lineages and to minimize the inflexibility and straight directionality of the trunk.

Darwin thought we had better learn the laws of nature and fit in with them, since we are part of the community of life. He lamented, for instance, that human society might be breeding too many of the "unfit" and too few "superior" individuals. But he never created a metaphor for man's relationship to the cosmos, beyond this casual word-cartoon: "A dog might as well try to understand the mind of Newton."

His friend "Doubting" Thomas Huxley, premier agnostic, seemed better at creating metaphors for ignorance than for knowledge. His own dark image of man's relationship to nature was that of the hidden chess player. In seeking to learn nature's laws and live in harmony with them, he wrote, we have "for untold ages" been playing a game of chess on which our happiness, fortunes and very lives depend:

> The chessboard is the world, the pieces are the phenomena of the universe, the rules of the game are . . . the laws of Nature. The other side is hidden from us. We know that his play is always fair, just and patient. But also we know, to our cost, that he never overlooks a mistake, or makes the smallest allowance for ignorance. To the man who plays well, the highest stakes are paid . . . and one who plays ill is checkmated—without haste, but without remorse.

As for any real apprehension of the universe around us, Huxley wrote, we understand about as much "as a worm in a flower pot on a balcony in London knows of the life of the great city around him."

See also BRANCHING BUSH; MATERIALISM; RED QUEEN HYPOTHESIS; TANGLED BANK.

For further information:
Beer, Gillian. *Darwin's Plots: Evolutionary Narrative in Darwin, George Eliot and Nineteenth-Century Fiction.* London: Routledge & Kegan Paul, 1983.

Hyman, Stanley E. *The Tangled Bank: Darwin, Marx, Frazer and Freud as Imaginative Writers.* New York: Atheneum, 1962.

Leatherdale, W. H. *The Role of Analogy, Model and Metaphor in Science.* New York: New Holland, 1974.

METAPHYSICAL SOCIETY

One of the most remarkable collections of English scientists, theologians, priests, philosophers and iconoclasts, the Metaphysical Society was founded on April 21, 1869, 10 years after publication of Charles Darwin's *On the Origin of Species.* Its purpose was to explore and discuss all points of view on faith, belief, science, God, morality, miracles, truth and the basis of "knowing" anything.

Founded by the poet Alfred Tennyson, the Reverend Charles Pritchard (an astronomer) and R. H. Hutton, the group eventually grew to about 60 members. At various times, discussion participants included the Prime Minister William Gladstone, Archbishop Manning, the Duke of Argyll, Sir John Lubbock, Dr. William B. Carpenter, Professor St. George Mivart, Henry Sidgwick, the Bishop of Gloucester, the Archbishop of York, Professor Thomas Henry Huxley, Walter Bagehot, John Ruskin, and Professor John Tyndall—a "who's who" of British intellectual and religious life.

Members sometimes read formal papers for discussion, on a wide variety of topics: Is God unknowable? What is Death? Has a Frog a Soul? On the words Nature, Natural, and Supernatural. What is Matter? How do we come by our Knowledge? The Verification of Beliefs. What is a Lie?

At the first meeting, a member remarked that "if we hung together for twelve months, it would be one of the most remarkable facts in history." In fact they continued to meet once a month for 12 years.

The Metaphysical Society finally disbanded not because of violent disagreements, but because the dialogue was no longer producing new truths. It had become a gathering of gentleman who agreed to permanently disagree. "After twelve years of debating," one of its founders recalled, "there seemed little to be said which had not already been repeated more than once."

MIDWIFE TOAD

See KAMMERER, PAUL.

MILITARY METAPHOR
"Warfare" of Science *vs.* Religion

Many people assume Darwinian evolution must always conflict with Judeo–Christian beliefs. Some describe science versus religion as an all-out war between two opposing camps: an "army of scientists" against "soldiers of the Lord." Some scientists and fundamentalists see it that way, too.

But historian of ideas James Moore has amassed an impressive body of evidence that this military metaphor is a misleading simplification of the complex relationship between religion and science.

In his book *The Post-Darwinian Controversies: A Study of the Protestant Struggle to Come to Terms with Darwin* (1979), Moore points out that many of the top Victorian scientists who embraced evolution (including Asa Gray, Charles Lyell, Joseph Hooker) were believers in traditional Christianity and most were conservative. Moreover, he shows that the vast majority of early naturalists and geologists were actually churchmen themselves. Even Charles Darwin originally studied and prepared for a career as a clergyman-naturalist.

Popular and influential theologians and preachers, including Charles Kingsley and Henry Ward Beecher, did not see any conflict in accepting both science and the Bible. The great basic discoveries of fossil formations and ancient life were made by "scriptural geologists," whose major aim was to reconcile the geological record with the Bible.

What is perhaps most significant is that among Victorian thinkers "scientific" and "religious" ideas coexisted in the same people. Some were troubled with internal conflicts (Philip Gosse, Hugh Miller), but others were quite comfortable as religious scientists (Henry Drummond, Alfred Russel Wallace, Asa Gray).

Mythologist Joseph Campbell has pointed out that any arguments about time, shape of the Earth, geological history and origins of species were never really between religion and science. The basic conflict is between modern science and the science of the ancient Hebrews.

While the intellectual combat epitomized by Thomas Huxley's famous clash with Bishop Wilberforce [see OXFORD DEBATE] is memorable for its drama, Moore demonstrates that "behind the scenes" there was always more of a complex give-and-take between tradition and innovation. Evolutionary biology grew out of the pioneering research of churchmen and was quickly accepted by many religionists. Scientific genetics was developed in a monastery garden by Father Gregor Mendel, an Austrian monk.

Darwin himself was a devoted family man, pillar of his community, court magistrate, philanthropist and contributor to his local church. Nicknamed by some "the Saint of Science," his exemplary life was often cited by Victorian writers as proof that an evolutionist was no danger to society and religion. When he died, the church had no objection to Darwin's burial in Westminster Abbey, England's most sacred shrine.

See also FUNDAMENTALISM; GRAY, ASA; MILLER, HUGH; OXFORD DEBATE; PRESTWICH, SIR JOSEPH; "TWO BOOKS," DOCTRINE OF.

For further information:

Gillispie, Charles C. *Genesis and Geology.* Cambridge: Harvard University Press, 1951.

Moore, James R. *The Post-Darwinian Controversies: A Study of the Protestant Struggle to Come to Terms with Darwinian Great Britain and America, 1870–1900.* New York: Cambridge University Press, 1979.

White, Andrew D. *A History of the Warfare of Science with Theology.* New York: Appleton, 1896.

MILLER, HUGH (1802–1856)
Champion of Scriptural Geology

What uplifting scientific pursuit, conducted out in the fresh air of the countryside, can lead a sane man to madness and death? In 1856, most literate people would quickly have answered, "Geology." For that year a remarkable Scots author, who had singlehandedly made studies of the rocks not only respectable but widely popular, came to a tragic end. His name was Hugh Miller, a writer of such remarkable power he turned a book about fossil fish into an international bestseller.

Despite the 19th century craze for natural history, geology was considered dangerous. Fossils of ancient life discovered in rock strata were telling a fascinating but disturbing story. As evidence of prehistoric events, they were steadily pushing back the Earth's age; skeletons of monstrous reptiles and giant ground sloths not mentioned in scripture were piling up.

Although geologists reassured everyone there could be no disharmony between God's word and his record in the rocks, they fought each other tooth and nail over the meaning of each, while most people were bewildered and put off by the whole subject. Many clergymen recommended they avoid it.

Hugh Miller was a writer before he was a geologist, but as the son of working-class people—blacksmiths and harness makers—he had to get a trade. Short, stocky and muscular, he decided to try his hand at being a stonecutter at a quarry and hoped to also find time to write. As he went to work the first day, he dreaded becoming a drudge, too exhausted to pursue dreams of literature, "working to eat, and eating to work."

To young Miller's amazement, he enjoyed physical work in the beautiful wooded setting, soon became fascinated with the rocks and eventually was able to create a spectacular career writing geology books. His works, *The Old Red Sandstone* (1841), *Footprints of the Creator* (1849) and *Testimony of the Rocks* (1868), were admired by scientists and literary critics alike and gained him a tremendous following on both sides of the Atlantic.

Miller exemplified an ideal of the Victorian age: the self-taught working man who could uplift himself and join the "aristocracy of merit" in science. Gentlemen scientists of more privileged classes, like Sir Roderick Impy Murchison, Dean William Buckland and his idol, Louis Agassiz, welcomed Miller as a valued colleague.

Miller's discoveries were plentiful, and his descriptions of the strata and their fossils were meticulous, stylish and even entertaining. But readers were really won over by Miller's cheerful, unpretentious accounts of life as a laborer-scientist-philosopher. Above all, he conveyed a deeply felt piety, which his love for science seemed never to disturb.

By example he showed the ordinary man need not be threatened by geology; it was as joyful a contemplation of God's work as any other branch of natural history. And it was no act; his personal friends and colleagues found him as delightful, sincere and enthusiastic as he appeared to his thousands of readers.

He had no tolerance at all for the theories of evolution or "development" of one species into another. Almost gleefully, he attacked the writings of "Telliamed" (Benoit De Maillet), Jean-Baptiste Lamarck, and especially Robert Chambers's anonymously published *Vestiges of Creation* (1844). (Miller knew and liked Chambers, never suspecting he was secretly "Mr. Vestiges.")

Evolutionary ideas, Miller wrote, have "become popular among mechanics and tradesmen and are causing much damage, leading them to atheism and immorality." Belief in evolution rather than special creation, he wrote, could only come from "sciolists and smatterers," by which he meant ignorant dabblers in pseudo-science.

Yet, there was a dark, brooding side to this "meditative stonemason," whom historian Lynne Barber imagines "holding a Bible in one hand and a fossil fish in the other." Despite his valiant attempts to bolster biblical accounts of creation with geological evidence, it became an ever more difficult battle.

In his early essays, he had insisted there was no contradiction in the Earth having been created in six 24-hour days, but in his last posthumously published work, *Testimony of the Rocks* (1868), he had to admit the biblical "days" might each represent millions of years. Also, he had become increasingly morbid, convinced the record of the rocks pointed to inevitable extinction for mankind—part of a pattern no different than what had gone before.

While writing *Testimony*, he suffered from horrible dreams and visions, awakening convinced he had wandered the streets all night. (At such times, he insisted on checking his clothing for mud stains, but none were found.) He often wrote all night and day, with a knife and gun at his side to repel imagined burglars or intruders. There were searing headaches:

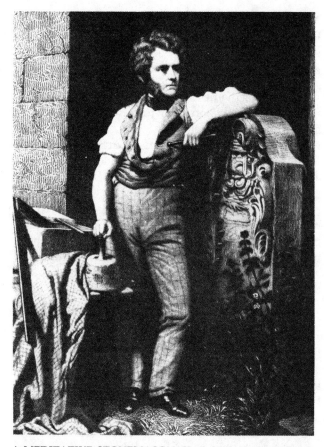

A MEDITATIVE STONEMASON, Hugh Miller exemplified the unpretentious workingman as natural philosopher. His internationally popular books popularized geology, while defending biblical creationism against the new theory of evolution.

He thought his brain was burning up, his mind had "given out" and he "couldn't put two thoughts together." (Yet the writing is perfectly lucid, showing no deterioration in quality.)

Finally, on December 23, 1856, Miller used his pistol to take his own life, leaving this bizarre note worthy of a Victorian horror story:

> DEAREST LYDIA—My brain burns. I *must* have *walked;* and a fearful dream rises upon me. I cannot bear the horrible thought. God and Father of the Lord Jesus Christ have mercy upon me. Dearest Lydia, dear children, farewell. My brain burns as the recollection grows. My dear, dear wife, farewell.
>
> HUGH MILLER

His doctor ascribed his descent into madness and suicide as the result of "overworking the brain." In our century, one could plausibly diagnose depression, nervous breakdown, chemical mood swings, migraines, drug abuse or brain tumors. (Miller did suffer from the quarryman's disease, silicosis, caused by breathing in rock dust.)

But public opinion had its own answer: Hugh Miller died of trying to reconcile fossils with scripture and blew out his circuits. Geology became "dangerous" once more, and ministers again warned their flocks against studying it. A few years later, in 1859, Charles Darwin published the *Origin of Species,* which also became a scientific bestseller.

Miller's arguments against evolution, which were raised against the *Vestiges,* looked hopelessly feeble when weighed against Darwin's massed evidence. Although he had once been acclaimed among the 19th-century's most interesting and persuasive writers, no one—except historians of science—would ever read Hugh Miller's books again.

See also AGASSIZ, LOUIS; SCIOLISM; "TWO BOOKS," DOCTRINE OF; VESTIGES OF CREATION.

For further information:
Barber, Lynne. *The Heyday of Natural History.* Garden City, N.Y.: Doubleday, 1980.
Miller, Hugh. *Footprints of the Creator.* Boston: Gould & Lincoln, 1849.
———. *The Old Red Sandstone.* Edinburgh: Nimmo, 1841.
———. *The Testimony of the Rocks.* Boston: Gould & Lincoln, 1868.

MIMICRY, BATESIAN*
Imitating Poison Prey

Henry Walter Bates (1825–1892) is chiefly remembered for proving that nature can come up with some very deceptive packaging. One of the most gifted tropical geologists of the 19th century, Bates was a good friend and traveling companion of the remarkable Alfred Russel Wallace, Darwin's "junior partner" in the discovery of natural selection. He and Wallace also shared a fascination with unusual patterns and colors in the animal kingdom.

Wallace and Bates made many contributions to the natural history of the tropics, separately in Asia and Africa and together on the Amazon in South America. Like Wallace, Bates discovered scores of new species and was a particularly avid collector of butterflies. He was intrigued by Wallace's discovery that harmless insects were usually drably camouflaged, while poisonous, bad-tasting species were dressed in gaudy colors as a warning to predators.

Bates noticed something else—the coloration of some innocuous species copied or mimicked the poisonous ones. Thus, predators might mistake them for noxious prey and leave them alone. He took care to point out that this was not conscious imitation, but a gradual, growing resemblance between species brought about by natural selection. His writings on the phenomenon convinced other naturalists, who began to call it "Batesian mimicry." Some critics scoffed at the idea that some species evolve elaborate markings and colors to imitate others, but Charles Darwin did not. "I am rejoiced that I passed over the whole subject in the *Origin,*" Darwin wrote Bates, "for I should have made a precious mess of it. Your paper is too good to be largely appreciated by the mob of naturalists without souls."

Recently, experimenters have put Bates's idea to the test. One researcher put scrub jays in with poisonous black-and-orange monarch butterflies and its harmless mimic, viceroys. Sure enough, inexperienced young birds gobbled up the viceroys. But birds that first ate what Alfred Russel Wallace called "the disgusting morsels" of monarchs quickly learned their bitter lesson and would not touch the harmless viceroy mimics.

When various orange butterflies were introduced, even very approximate look-alikes went untouched by the birds—demonstrating that even slight resemblances confer survival advantage. Over generations of selection, the resemblance could become increasingly fine-tuned, as Bates suggested 100 years ago. "Mr. Bates further observed," wrote Darwin, "that the [mimics] are comparatively rare, while the imitated swarm in large numbers; the two sets being mingled together."

See also BATES, HENRY WALTER; CONSPICUOUS COLORATION; MIMICRY, MULLERIAN; WALLACE, ALFRED RUSSEL.

MIMICRY, MULLERIAN*
Resemblances of Poisonous Species

Henry Walter Bates developed a theory of mimicry to explain why harmless butterflies he saw in Amazonia evolved to resemble poisonous ones. If predators can be fooled into avoiding the harmless mimics, he reasoned, then those that look similar to noxious ones will leave more offspring.

But Bates noticed another kind of mimicry that puzzled him. Several poisonous species of butterflies in the Amazon also have evolved similar colors and patterns. What adaptive advantage could be gained by poisonous prey resembling one another?

An acceptable explanation was proposed by Fritz Muller (1821–1897), a German zoologist. He assumed that young, inexperienced predators learn to avoid certain prey through trial and error—by killing and eating some. If the foul-tasting species varied widely in appearance, predators would have to kill many of each before they learn which to avoid.

But if the noxious prey came to share coloration, the predator learns to avoid one basic pattern and those it samples are spread out over many species. The survival value of resemblance is that it cuts the loss to each population. Mullerian mimicry is common among poisonous tropical butterflies, both in Africa and the Amazon.

See also CONSPICUOUS COLORATION; MIMICRY, BATESIAN.

For further information:
Owen, R. *Camouflage and Mimicry.* Chicago: University of Chicago Press, 1982.
Wickler, Wolfgang. *Mimicry in Plants and Animals.* New York: McGraw-Hill, 1968.

MINIUM'S DEAD COW QUARRY
Remarkable Miocene Site

Located in northwestern Kansas, a Miocene deposit with the outlandish name "Minium's Dead Cow Quarry" is yielding many new insights into the ecology of mid-America seven million years ago.

While most of the fossil species are known from other sites, Minium's quarry is remarkable for the rich animal-plant associations and the extremely good state of preservation of the fossils. Grass fossils, for example, are capable of being analyzed for their cellular structure and physiology. [See GRASSES, EVOLUTION OF.] Paleobotanist Joseph Thomasson has studied the intact vessels in leaf cross-sections and marvels, "It's as if you found a fossilized rabbit and it still had its eyelashes."

A picture of rich, subtropical grasslands teeming with large mammals emerges from studies of the site. But it is the close association of animals and plants that makes it so exciting to researchers. Paleontologist Richard Zakrzeski says, "In other places, if you find the seeds you don't find the bones; if you find the bones you don't find the seeds. As far as I know, it's unique to find a diversity of both plants and animals in the same sediments."

Paleontologists usually have to infer diet from tooth structure or other clues. But here plants and animals can be directly correlated, which will throw light on other less complete sites. It gives paleoecologists "a much better handle on who was eating what," Zakrzeski reports. His colleague, Joseph Thomasson, gives an example: "I've got a vial of seeds that I took out of that rhinoceros jaw. You can't get any closer association than that."

MIOCENE GAP

See "HOMINOID GAP."

"MISSING LINK"
Mythical Ape-Man

The most widespread misconception about human evolution seems to be the myth of the "missing link." For 100 years, many have accepted the cliche that Charles Darwin's theory is not "proved" because "no one has yet found the missing link between ape and man."

"Links" in a Great Chain of Being is an idea derived from medieval theology. Church philosophers ranked all creatures as "higher" or "lower," with man at the top, the crowning glory of creation. Above man were the angels, archangels and other spiritual beings, leading up to Almighty God. The Chain of Being was also reflected (or perhaps based) on the earthly order, with those of "low degree" at bottom, an aristocracy above and royalty at the apex of the social pyramid. That one species might develop or evolve into another was as unthinkable as a servant aspiring to move into the ruling class—a violation of the natural order.

A second source of the "missing link" idea lies in the comparative method developed by 19th-century naturalists. When Thomas Huxley wrote his famous essay on "Man's Place in Nature" (1863) he made an exhaustive comparison of the anatomy of monkeys, apes and man. Humans and apes showed overwhelming, detailed similarities in structure from which the kinship between them was inferred, but at the time there were no known fossil men, fossil apes or near-men, with the exception of a few fragmentary Neandertal crania.

Darwin was careful to say in his *Descent of Man* (1871) that humans are not descended from anything like a modern ape or monkey, but that these groups were related through common ancestry. He also guessed (correctly, it turned out) that ancient near-men fossils would be found in Africa, since it appears to be the ancestral home of the chimp and gorilla. Despite his caution, however, many of his contemporaries imagined a half-man, half-ape as the ancestor of both men and apes. Ernst Haeckel, the influential German evolutionist, went so far as to include this hypothetical "missing link" in his books and even gave it a Latin species name. He called it *Pithecanthropus alalus*, "ape-man without speech."

Since Haeckel's day, taxonomists have made a strict rule against giving scientific names to species before they've been discovered. Otherwise, unicorns and mermaids might return to catalogues of zoology. But Haeckel's insistence that *Pithecanthropus* existed and would be found had a remarkable effect. It inspired a Dutch army surgeon named Eugene Dubois to set out in search of *Pithecanthropus*—and he found it! In 1893, after several dedicated, difficult years, Dubois dug up the fossil hominid remains he called *Pithecanthropus erectus*, popularly known as the Java ape-man. It later turned out that they were not very "apish" after all, and were really an ancient species of human; whether or not they were "without speech" no one knows. (They have since been renamed *Homo erectus*, to reflect the conclusion they were not apes at all.)

In 1912, a "missing link" hoax fooled British anthropologists and was not exposed for 40 years. An unknown prankster combined an ape jaw with a human skull and planted the forgery in Sussex, England, at a site called Piltdown. One reason it was eagerly accepted as authentic was that it certainly combined ape and human characteristics—half and half. One famous anthropologist, Sir John Lubbock, said he was impressed that it was "the most simian" fossil man yet found. The Piltdown hoax was not only a cruel joke on the scientists who analyzed it, but a devastating comment on the simplistic concept of a "missing link."

There have been other famous "missing link" jokes that were not so malicious. When Charles Darwin went to Cambridge in 1877 to receive public acclaim for his life's work and an honorary doctorate, the students had decorated the hall with an effigy of a monkey hanging from several huge, interlocking hoops representing the "link." And 50 years later, at the conclusion of the Scopes "Monkey Trial" in Tennessee, lawyer Clarence Darrow—who lost his client's right to teach evolution in the schools there—received a telegram from a friend in California who was trying to cheer him up. It read, HAVE FOUND MISSING LINK—PLEASE WIRE INSTRUCTIONS.

Over the past few decades, paleoanthropologists have unearthed a bewildering variety of human and near-human fossils going back more than three million years. There are erect, bipedal, near-men (australopithecines) with small brains and somewhat apelike teeth. There are larger, more robust hominids with huge molars and powerful jaw muscles. Known variants on the human stock include *Homo habilis*, *Homo erectus* and populations of so-called Neandertals. Their evolutionary relationships to each other and to ourselves are still a mystery and may never be completely known. None of them is "half-way" between apes and men, but they are all links to our complex ancestry. Fossil hominids are no longer "missing."

See also AUSTRALOPITHECINES; DART, RAYMOND ARTHUR; DUBOIS, EUGENE; HOMO ERECTUS; PILTDOWN MAN (HOAX); TAUNG CHILD.

For further information:

Dart, Raymond. *Adventures with the Missing Link*. New York: Viking, 1959.

Delson, Eric, ed. *Ancestors: The Hard Evidence*. New York: Alan R. Liss, 1985.

Reader, John. *Missing Links: The Hunt for Early Man*. London: Collins, 1981.

"MISTER VESTIGES"
The Unknown Evolutionist

In 1844, some 15 years before Charles Darwin's *Origin of Species* (1859) appeared, a controversial book put forward a "Hypothesis of the Development of the Vegetable and Animal Kingdoms." Entitled *Vestiges of Creation*, it argued that species were not created suddenly, but developed gradually according to natural laws. Its author kept his name a closely guarded secret.

The unknown evolutionist had correctly anticipated the outcry and vilification that followed. Darwin later admitted the man he called "Mr. Vestiges" prepared the way for his own work and drew off some of the early hostility of critics. Although it sold 10 editions in its first decade, the anonymous author refused to reveal himself, and speculation on the identity of "Mr. Vestiges" became a popular topic at Victorian dinner parties. Rumors circulated that the author was the geologist Sir Charles Lyell, Countess Ada Lovelace (Lord Byron's daughter) or even Prince Albert.

Within a few years, many English scientists had guessed that the mystery man was Robert Chambers (1802–1883), a well-known Scots writer and publisher. An eclectic intellect, Chambers was the respected author of well-known encyclopedias, biographical dictionaries and works on history and folklore. However, his real name never appeared on the *Vestiges* until the posthumous 12th edition.

Chambers took elaborate precautions to hide his authorship. He sent recopied manuscripts and proofs to his printer through several intermediaries, even disguising their country of origin. Chambers entrusted only two close friends, his wife, Anne, and brother William, with the secret; it was never revealed during his lifetime.

In a *Preface* to the 1884 edition of *Vestiges*, published a year after Chambers's death, his old friend Robert Cox at last revealed the truth. Chambers's wife and brother had died, both of them content to let future historians continue the guessing game. But Cox felt that would be an injustice to his friend's achievement. As the last man alive who knew the secret, he wrote, "I am unwilling that it should die with me."

See also CHAMBERS, ROBERT; VESTIGES OF CREATION.

MITOCHONDRIAL DNA
Genetic "Family Archives"

Genetic blueprints for an individual, coded in DNA, reside in the cell nuclei of most plants and animals: half from the individual's mother and half from the father. But some DNA also exists outside the nucleus, in tiny organelles called mitochondria, which are crucial to a cell's metabolism. Mitochondrial DNA (MtDNA) is passed on only through the female line; the father's genes do not affect it at all.

Quite apart from its function in cell physiology, recent discoveries about mitochondrial DNA have opened up new possibilities for tracing evolutionary lineages of living populations. During the 1980s, such diverse creatures as desert tortoises, red-winged blackbirds, eels and humans have been studied on the assumption that complex similarities in mitochondrial DNA are attributable to shared matrilineal ancestors.

In some cases, the studies clarified relationships between many local populations (subspecies) that had puzzled zoologists because of variable external characters. Mutational differences in MtDNA can be more clearly read, free of the untraceable complications of male contributions. One much-publicized study of human females from African, Asian and American populations pointed to the existence of a recent common African "mother" for all mankind—a "Mitochondrial Eve" who lived only 200,000 years ago.

Despite some problems in calibrating the "mitochondrial clock" to give reliable time frames, geneticist John C. Avise thinks of mitochondrial DNA as "nature's family archives." If properly read, he believes this "evolutionary bookkeeping" could help unravel the tangled history of species dispersal over the Earth.

In 1988, Avise obtained bittersweet results in a study of Florida's seaside dusky sparrow. Because of habitat destruction, the species was reduced to only a few surviving birds. They were captured and bred with individuals of what was thought to be a closely related species, so some of their genes, at least, would live on. But after the last dusky sparrow died, Avise and colleagues studied its MtDNA and arrived at two startling conclusions.

First, the dusky local population had fooled zoologists by external color; it was not a genetically distinct species at all. Second, and even more distressing, the "closely related" species with which the last survivors were bred were not their nearest relatives. Another living sparrow population, differently colored and hundreds of miles to the north, was actually a much closer match.

Avise warns that the geographic distributions and population structures of species are changing at unprecedented rates because of human interference. "This may be the last century," he writes, "in which we can hope to rescue the natural historical records of most species before that information is hopelessly garbled and irretrievably lost."

See also MITOCHONDRIAL "EVE."

MITOCHONDRIAL "EVE"*
Biochemical Ancestress

In 1986, newspapers featured a startling science story: MODERN'S MAN'S ORIGIN LINKED TO A FEMALE ANCES-

TOR and BERKELEY SCIENTISTS FIND EVE IN AFRICA. University of California geneticists Allan Wilson, Mark Stoneking and Rebecca Cann had announced the results of a remarkable study utilizing mitochondrial DNA as a marker to trace the ancestry of modern human groups. And they believed the family tree led back to a woman who lived in Africa about 200,000 years ago.

Mitochondrial DNA is the "extra" genetic material found outside the cell nucleus and is only traceable through the female line. It appears to be a fast-mutating DNA that changes 10 times faster than the DNA in cells' nuclei.

The Berkeley researchers collected and analyzed this material from 150 women from five geographical regions of the world: Africa, Asia, Europe, Australia and New Guinea. Despite an overall similarity, two main groupings appeared in the data: one from Africa and the other from everywhere else.

Wilson and colleagues believe that their mitochondrial DNA technique supports the conclusion that "the common ancestor of modern humans lived in Africa, about 200,000 years ago," relatively recently by the evolutionary timescale. When individuals from this population moved out of Africa into Europe and Asia, it appears, "they did so with little or no mixing with existing local populations of more primitive humans."

See also CLADISTICS; MITOCHONDRIAL DNA.

MIVART, ST. GEORGE J. (1827–1900)
"Heretical" Biologist

English biologist St. George J. Mivart, a devout convert to Catholicism, embraced the Darwinian theory of evolution with one qualification: After the human body evolved from ape-like ancestors, God intervened to infuse it with a soul. In many ways a forerunner of Father Pierre Teilhard de Chardin (1881–1955), Mivart tried to bring scientific truth to the church and a religious perspective into biology. Like Teilhard, he succeeded mainly in drawing heavy fire from both quarters.

Originally graduated as a barrister, Mivart returned to his boyhood interest in natural history and studied comparative zoology on his own. Eventually, he produced first-rate contributions, such as his monumental 557-page anatomy *The Cat* (1881), which guided generations of students. For many years he was an instructor in biology at St. Mary's, a Catholic college, where he taught that evolution was entirely compatible with church dogma.

Ironically, it was his friend and scientific mentor Thomas Huxley who challenged Mivart's published arguments that evolution had been anticipated in church doctrine. Meeting him on his own grounds, Huxley quoted Catholic history and theology with the ease of a bishop to demonstrate that Mivart's claim was founded entirely on wishful thinking. Huxley would have liked to believe the church was congenial to evolutionary ideas, he wrote, but could no more condone "unfaithfulness to truth" by fudging ecclesiastical rulings than by stretching science beyond its proper boundaries.

For almost a decade (1861–1869), Mivart was Huxley's devoted student ("my constant reader"), attending almost every single anatomy lab and evolutionary lecture. But his increasing "theological fervor," fanned by a priestly colleague, eventually led him to an anti-Darwinian stance. In 1869, he told Huxley, whose friendship he treasured, that he intended to write a strong critique on the "insufficiency of Darwinism." It was a painful, emotional confrontation, though Huxley neither became angry nor argued. Mivart later recalled:

> As soon as I had made my meaning clear, his countenance became transformed as I had never seen it. Yet he looked more sad and surprised than anything else. He was kind and gentle as he said regretfully, but most firmly, that nothing so united or severed men as questions such as those I had spoken of . . .

Now Mivart determinedly churned out articles for Catholic journals on difficulties of the theory of natural selection and wrote a popular book (*On the Genesis of Species*, 1871) as a rebuttal to Darwin's *Descent of Man* (1871). His stated object was "to show that the Darwinian theory is untenable, and that natural selection is not *the* origin of species . . . upon scientific grounds only. My second object was to demonstrate that nothing . . . in evolution generally, was necessarily antagonistic to Christianity." He followed up with *Apes and Men, an exposition of structural differences bearing upon questions of affinity and origin* in 1873.

One of his major arguments against natural selection—one that has never gone away—is the question of how complex adaptive structures could originate. What good is one-quarter of a wing or half an eye? Why would such incipient structures be selected before they were fully useful? Mivart thought this stumper proved the "logical insufficiency" of the theory of natural selection. [For one answer to Mivart's objection, see EXAPTATION.]

Darwin was "mortified" by Mivart's private expressions of friendship even while he was attacking him in the public press. "You never read such strong letters Mivart wrote to me," Darwin complained to a friend in 1871, "about respect [and] begging I would call on him . . . yet [in his published articles] he makes me the most arrogant, odious beast that ever lived. I cannot understand him; I suppose that accursed religious bigotry is at the root of it . . . It has mortified me a good deal."

Although he was praised by cardinal-to-be John Henry Newman, who admired his competence in science, Mivart's increasingly liberal theological views led to his excommunication in 1900, shortly after which he died. Years later, his friends argued to church authorities that Mivart was not willfully heretical, but that the diabetes that caused his death had also unbalanced his mind. Accepting this "mechanistic" interpretation of Mivart's heresy, the church allowed him a Christian burial.

See also TEILHARD DE CHARDIN, FATHER PIERRE.

For further information:

Gruber, J. W. *A Conscience in Conflict: The Life of St. George Jackson Mivart.* New York: Columbia University Press, 1960.

Hull, David L., ed. *Darwin and His Critics.* Cambridge: Harvard University Press, 1973.

Mivart, St. George Jackson. *On the Genesis of Species.* London: Macmillan, 1871.

MOLECULAR CLOCK

See CLADISTICS; RAMAPITHECUS.

MONAD
Particles of Life

Gottfried Leibniz (1646–1716) the German philosopher, had proposed the monad as an elementary particle of life. Impressed by the new world of previously invisible creatures that was being revealed by the microscope, he defined them as the smallest living units, just as atoms were theoretically the smallest particles of matter.

Monads were thought to be "units of force" that could develop into plants or animals. From the first, the concept of monads as entities was confused with the simple microscopic organisms that were as yet unidentified.

Sir Charles Lyell thought monads might exist in great numbers at the present time, perhaps in warm ponds, where they could develop into new species to replace those few that became extinct from time to time. (He thought there was a slow, steady rate of species replacement and regarded the idea of rapid mass extinctions as a "catastrophist's" daydream.) Thus, monads were thought to be not only the original source of species, but an enduring resource that was continually renewing the Earth with species as needed in the "economy of nature."

By the end of the 19th century, the "monad" joined the "pangene" and the "homunculus" on the scrap heap of nonexistent theoretical entities.

MONBODDO, LORD (JAMES BURNET) (1714–1799)
Man Creates Himself

A hundred years before Charles Darwin, Scots jurist James Burnet, Lord Monboddo, looked to the apes and not to Adam for the origins of humanity. Man's body in the state of nature, he thought, was the work of God, "but as we now see him, he may be said . . . to be the work of man." The difference was language and human invention—what today is called culture. In retrospect, his insistence that apes can acquire humanity seems to anticipate an evolutionary perspective.

Monboddo believed "that no species of thing is formed at once, but by steps and progression from one state to another." He devoured traveler's reports of far-flung lands inhabited by orangs and other man-like creatures. The orangutan, according to 18th century writers, had a human form, walked on two feet, lived in society, defended itself with sticks and stones, constructed shelters and buried its dead. Despite its lack of language, Monboddo considered the orangutan a half-man, which could develop its intellect and learn to reason if given sufficient time.

A pioneer thinker on prehistory in the skeptical tradition of his day, Monboddo was not really an evolutionist but a philosopher of human nature. He championed the wisdom of the ancients and was famous for giving "classical suppers" attended by the most learned men in Edinburgh. Dressed in togas and sipping wine from garlanded vessels, they discussed and debated the enigmas of language, science and philosophy.

Many of his contemporaries considered him an eccentric or worse. "It is a pity to see Lord Monboddo publish such notions," wrote Dr. Samuel Johnson, "a man of sense and so much elegant learning . . . in a fool doing it, we should only laugh; but when a wise man does it, we are sorry."

Johnson was referring to the anthropological speculations Monboddo published in two six-volume treatises: *Of the Origin and Progress of Language* (1773–1792) and *Antient Metaphysics* (1779–1799). Language, Monboddo throught, was "necessarily connected with an inquiry into the original nature of man." Aristotle's definition of man as "A rational animal capable of intellect and science," was correct. Orang-utans were near-men not because of any "outward" or purely anatomical resemblances, but because they showed the rudiments of "rational" behaviors, such as using tools and burying their dead. (In fact, though they do throw sticks and stones, modern studies have not found orang-utans to be notable tool-users in the wild and certainly not grave-diggers.)

Monboddo and his contemporary Jean-Jacques Rousseau believed mankind's chief characteristic is the plasticity of human behavior. God did not place man at the pinnacle of creation, but instead gave him and her the means to get there by their own efforts. In rising above brute creation, therefore, man creates himself. By emphasizing humanity's dependence on learned and shared behavior, Monboddo foreshadowed the premise of cultural anthropology.

Between the ape and the civilized man he imagined the "noble savage," so dear to the 18th-century philosopher's heart, an unspoiled, unaffected being who could hold his own with the animals while still possessing the full range and potential of a human being. Thus, Lord Monboddo's writings paved the way not only for the theories of Darwin, but also for the *Tarzan* books of Edgar Rice Burroughs.

See also NOBLE SAVAGE; TARZAN OF THE APES.

MONISM
Social Darwinist Philosophy

In late 19th- and early 20th-century Germany, Monism (from Greek meaning "one") emerged as a popular, influential cultural-political expression of Social Darwinism. Its founder, Ernst Haeckel (1834–1919), the eminent German evolutionary biologist, characterized it as the union of matter and spirit, based entirely on the creative properties of matter. Haeckel described Darwin's theory of descent as "the non-miraculous history of creation," and rebelled against the religious or "dualistic" viewpoint within which he was raised. Haeckel developed "Monism" into a nationalistic, romantic, and anticlerical movement. As historian Daniel Gasman has shown in *The Scientific Origins of National Socialism* (1971), Monist ideas later served as one of the main supports for ideologists of Hitler's Third Reich.

When Haeckel first read the *Origin of Species* (1859) in 1860, he felt a flash of revelation and inspiration. Ironically, Charles Darwin's sober exposition of a materialist, scientific view of nature affected the young German biologist so strongly that his zeal equalled that of a religious fanatic. Haeckel said he "found in Darwin's great unified conception of nature . . . the solution of all the doubts which had bothered me." More than merely a scientific theory, writes Gasman, Haeckel saw in Darwinian evolution:

> a complete and final rendering of the nature of the cosmos. Through evolution he studied the world and everything in it including man and society as part of an organized and consistent whole. He

therefore called his new evolutionary philosophy "Monism" and contrasted it with all of traditional thought, which he rather disdainfully labelled "Dualism," condemning the latter for making distinctions between matter and spirit, and for invidiously separating man from nature.

Haeckel taught that since man's social existence is governed by the laws of evolution, there can be no "split" between morality and religion, man and animal, altruism and competition. His combination of Social Darwinism with a deeply felt mystical nationalism led him to preach "regeneration of the German race," extinction of the church and traditional religion, triumph over non-Germans and fulfillment of a "higher evolutionary destiny" for the nation.

See also ARYAN "RACE," MYTH OF; HAECKEL, ERNST; MONIST LEAGUE; SOCIAL DARWINISM; WOLTMANN, LUDWIG.

MONIST LEAGUE
Darwinian Politics

In 1904, Professor Ernst Haeckel (1834–1919), the philosophical father of Monism, proposed the formation of a Monist League at the International Free-Thought Congress, which was meeting in Rome. Germany's foremost evolutionist, Haeckel had advocated Charles Darwin's ideas in Germany and had written several popular and influential books on evolution. Now he proposed to translate his version of Social Darwinism into political action.

In 1906, the League became a reality in his university town of Jena. Despite the fact that he was past 70, Haeckel was its guiding spirit, though its first president was Dr. Albert Kalthoff, a radical theologian. Within five years, the Monist League had 6,000 members meeting in 42 cities and towns throughout Austria and Germany, published a weekly journal (*The Monist Century*) and had developed enormous influence, both in the international Free-Thought Movement and among German intellectuals.

Haeckel depicted Germany as on the brink of social disaster unless it could "bring itself into harmony with the laws of biology" as Haeckel believed them to be. He sought to remove the "contradictions" within both church teachings and liberal philosophy, with their supernatural or moral sanctions for altruism and compassion. Darwinism for Haeckel was a harsh "struggle for existence," and the laws of nature had to become the laws of society.

Any mixture of races, he thought, would lead to the deterioration of the German people. They needed to be brave, cunning and willing to be armed in every way against "biological decay." In Haeckel's hands,

Darwinian thought became a volatile mixture of anticlericalism, rationalism, materialism, racism, patriotism, eugenics and Aryanism. Darwin himself was not a Social Darwinist and thought it a bitter joke that his scientific theories were being used, as he put it, to "prove Napoleon was right and every cheating tradesman is also right."

See also ARYAN "RACE," MYTH OF; HAECKEL, ERNST; MONISM; SOCIAL DARWINISM.

MONKEYS

See BABOONS; CATARRHINES; CAYO SANTIAGO; PLATYRRHINES.

MORGAN, THOMAS HUNT

See FLY ROOM.

MORREN, GEORGE

See CANNIBALISM CONTROVERSY.

MORRIS, DESMOND

See CONGO; QUEST FOR FIRE.

MORTON, SAMUEL

See RACISM (IN EVOLUTIONARY SCIENCE).

MOSAIC EVOLUTION
A Patchwork Process

Organisms do not change "all over" by gradual degrees. Sometimes one part of the system may remain stable over long periods, while another evolves rapidly. Many birds, for example, seem to be a "mosaic" of adaptations acquired at different times. Most show little variation in the ancient body plan while wings, feet or beaks have become specialized for different methods of obtaining food.

One of the most striking examples of mosaic evolution is our own species, *Homo sapiens*. Different parts of the human body did not evolve toward our present form at the same time or at the same rate, but form a mosaic of complexes, evolved at a different times. First came the fully rotating shoulders and upper torso, which developed early (perhaps 15 million years ago) and which we have in common with the chimps, gorillas, orang-utans and gibbons. Anthropologists believe our common ancestors developed this arm-shoulder-chest complex in a forest habitat, as an adaptation to climbing and moving through the trees, before the hominid line diverged from the apes.

Next, about four-to-five million years ago, the human pelvis, legs and feet became adapted to upright walking and running. This shift to a bipedal posture was accomplished by the time of *Australopithecus*, who was still small-brained but very like humans from the waist down.

The most recent phase of hominid evolution—the expansion of the brain and further reduction of the jaws—first appears during the *Homo erectus* stage, whose fossils are dated at between half a million and 1.5 million years ago.

Life in the trees shaped our hands, arms, shoulders and chest. Ground-living, possibly on the African plains, produced our double-arched feet, long legs and basin-shaped pelvis. And perhaps an increasing reliance on language, culture and tools is responsible for the most recent changes: the accommodation of the skull to a larger brain and smaller dental apparatus. We are composite creatures; our bodies and behaviors have been shaped by a particular sequence of events in our evolutionary history.

MOTH, INDIAN GRAIN

See PEPPERED MOTH; RESISTANT STRAINS.

MOUSTERIAN INDUSTRY
Neandertal Tool Kit

A characteristic assemblage of chipped stone tools is associated with Neandertal man in many sites and is assumed to have been produced by them. Most of the tools are flakes, characterized by scrapers and denticulates. There are no fine points or blades, which are thought to be associated only with *Homo sapiens*.

The Mousterian stone tool industry takes its name from Le Moustier, in the French Dordogne, where it was first discovered in the mid-19th century. Since then, Mousterian tools have proved to be remarkably widespread in both time and geographical distribution, showing little change from Europe to the Middle East, from Russia to Africa, over a period of perhaps 100,000 years.

Limitations of skill and tool-types and long persistence of unchanging Mousterian traditions over such wide areas is one of the most baffling enigmas in paleoanthropology.

See also NEANDERTAL MAN.

MURCHISON, SIR RODERICK IMPY (1792–1871)
Geologist

Perhaps no other geologist is more typical of his era than Roderick Murchison. Ambitious, tyrannical, possessed of seemingly limitless energy (some have hinted that this might have been drug-induced) and with a towering ego, Murchison could also be urbane and charming. He was, in fact, the perfect example of the "gentlemanly specialist," that curious breed of Victorian scientist who, having independent means, an inquisitive nature and little formal training, did much to advance our knowledge of the world.

Murchison was born in 1792 into a wealthy Scots family and was educated at Great Marlow, an important military school, which perhaps accounts for his later pugnacity. His military ambitions were stifled, however, by a posting in Ireland, far from the field of valor during the Peninsular Campaign. He spent his early years traveling in Europe, then settled in a large country house and indulged his passion of foxhunting six days a week. In 1823, however, under the influence of Sir Humphrey Davy, president of the Royal Society, and his own wife, who was fed up with his lavish expenditures, Murchison sold his hounds and horses to his hunting cronies and decided to take up geology. His friends were much amused when one of his earliest discoveries, during a "geologising" trip to Germany, turned out to be a fossil fox.

After studying for a time under Reverend William Buckland, the great geologist from Oxford, and then William Lonsdale, Murchison came under the spell of one of the preeminent scientists of the 19th century—the Reverend Adam Sedgwick. Sedgwick led him into the study of the Graywacke, a very ancient, distorted series of rocks about which little was known. Murchison took it up with his usual enthusiasm and made the Upper Graywacke his personal domain.

He named these strata the Silurian after an ancient kingdom of Britons. In 1838, he published his mammoth two volume work *The Silurian System*, which he dedicated to Sedgwick. In it, and his succeeding book *Geology of Russia in Europe and the Ural Mountains* (1845), Murchison delineated the succession of Paleozoic rocks, sweeping aside the claims of other geologists. Thus was born the long-contested Cambrian–Silurian Question.

Although now all but forgotten, this knotty problem divided the scientific world for nearly two decades as Sedgwick, Murchison, Thomas de la Beche and other key figures sought to determine the true succession of Europe's ancient rocks. Although the problem was eventually resolved, with Sedgwick's Cambrian, de la Beche's Devonian and Murchison's Silurian all retaining their sovereignty, it was Murchison, as head of both the Geological Society and the British Geological Survey, who captured the public's imagination.

It was he who was seen as the leading light of British geology, and Murchison was loath to give credit to his peers. This episode taken from the

FAMOUS AS THE KING OF SILURIA, Sir Roderick Impy Murchison gave up full-time fox hunting to devote his energies and resources to geology. He advanced understanding of the earlier rock strata, but arrogantly suppressed the work of other talented geologists.

Midland Naturalist of 1849 typifies his incredible popularity among the general public:

> An excursion was made to Dudley, when Sir Roderick Murchison, in the great cavern of the Castle Hill, briefly explained the system of strata to which he had given the name of Silurian . . . In proposing a vote of thanks to him, the Bishop of Oxford [Dr. Samuel Wilberforce] said that although Caractacus was an old king of [Siluria] . . . yet Sir Roderick had extended the Silurian domain almost illimitably, and it was only just and proper that there, upon a Silurian rock, he should be acknowledged the modern King of Siluria.
>
> The Bishop, then taking a gigantic speaking-trumpet, called upon all present to repeat after him . . . one word at a time—Hail—King—of—Siluria! The vast assembly thrice [repeated the words] with stentorian voices and most hearty hurrahs, . . . and ever afterward, Sir Roderick was proud to be acknowledged "King of Siluria."

Although he shied away from scientific controversies later in his life, Murchison retained his vigor until the end and dictated his last presidential address from his deathbed at the age of 79.

In 1871, by his will, Murchison bequeathed £1,000 to the Geological Society for the founding of the

Murchison Geological Fund, which, in addition to funding worthy scientists, was to present a bronze Murchison medal every two years. The artwork, commissioned by its founder, features a bust of Murchison on the obverse; on the reverse are two crossed geological hammers, surrounded by fossil trilobites and brachiopod shells. Across the top is the single word: SILURIA.

See also CAMBRIAN–SILURIAN CONTROVERSY; SEDGWICK, REVEREND ADAM.

For further information:

Geikie, Archibald. *The Life and Letters of Sir Roderick Impy Murchison*. 2 vols. London: John Murray, 1875.

Murchison, Roderick I. *Siluria: The History of the Oldest Known Rocks Containing Organic Remains*. London: John Murray, 1854.

Secord, J. A. *Controversy in Victorian Geology: The Cambrian–Silurian Dispute*. Princeton: Princeton University Press, 1986.

NAPI
Native American Creation Story

According to one version of a Plains Indian creation story first written down in the 19th century, in olden times the Sun was a great fiery chief who lived in his lodge in the sky. His principal servant was Napi, an immense being who did the Sun's bidding, so he wouldn't be distracted from keeping the Earth warm.

Napi was usually occupied with the Sun's many tasks, but one day he found himself with some free time and sat down to smoke his pipe near a spring. As he sat, he noticed some damp clay next to him, so he picked up some lumps and started to form them into shapes. He made a great many little sculptures out of the clay, then let them all dry in the sun. As they hardened, Napi smoked and studied them.

Finally, he picked one up, blew his breath on it, and said, "Go you now, my son. Be a Bighorn Sheep and live out on the plains." And the sheep galloped off.

Then Napi blew on the others, giving life to the Bear, the Antelope, the Beaver, the Badger and many more. To each animal he gave a name and then sent them to where they were each supposed to live.

One strange little clay shape was left, one with two legs instead of four, and Napi smoked and looked at it for a long time. After awhile, he blew the breath of life into it, and said, "Go you now, my son. Be a Man. Live with the wolves, and hunt meat on the plains."

Napi thought he had done well, and that all the creatures would be happy. But a few days later, when he went to the spring again to smoke his pipe, all the animals came to complain. First, the Buffalo said "Grandfather, I cannot live in the mountains where you sent me. The hills are too steep, the rocks break my hooves, and there is no grass. I cannot live there."

The Bighorn Sheep complained that he could not live on the plains. "Grandfather," he said, "my hooves grow too fast there and curl up. There is no moss, no hills to climb, and my legs get weak." And the Antelope had similar complaints about living in the mountains with the Buffalo.

"All right," said Napi, "I have thought how to fix this, and here is what we will do. My sons, I will give you each a home suited to you. You, Bighorn Sheep, go up to the mountains, and take the Goat to live there, too. Bear, my son, you go and live among the forested hills; Cougar, you go there also. Buffalo, my son, go and take Antelope and live in the plains and eat the grass there. Badger and Prairie Dog, go also to the plains and dig burrows in the earth, where you will find food. And Wolf, you will share the meat of the plains with Man."

So all the animals listened to Napi and went where he told them to live, and they have lived there and been content ever since. All except Man, "who is never satisfied anywhere and always wants everything."

See also ATRAHASIS; ORIGIN MYTHS (IN SCIENCE).

NATURALISTIC FALLACY
Nature is "Right"

When humans seek answers to questions of ethics, behavior or custom, they sometimes turn to observations of plants and animals to see what is "natural." That is the Naturalistic Fallacy, the idea that what appears to be the case in nature is right or correct for mankind.

ORIGIN OF MAN in Plains Indian legend tells of Napi, the Sun's helper, forming humans and animals out of clay. Many origin myths from all over the world begin similarly; in the biblical version, God creates man "from the dust of the earth."

Aesop's fables, dating from ancient Greece, taught that the "industrious" ant was more to be admired than the "lazy" grasshopper who fiddled all day and would have no food for the winter. Or that one could learn from a tortoise that "slow and steady wins a race." (One could equally argue a tortoise proves you can't get anywhere unless you stick your neck out.)

Medieval "bestiaries" detailed the supposed behavior of animals (largely inaccurately) and offered good or bad examples for mankind. The moralistic tradition was carried over into 19th-century science and still survives, despite great strides in accurate observation and the realization that nature can provide examples for almost any kind of behavior.

Some early anthropologists argued that primitive promiscuity was natural to man after observing chimpanzees, while others pointed to the monogamy of gibbons. In some baboon species, one male keeps a "harem" of females, while others have temporary "consorts."

Among animals there is every conceivable kind of system—from monogamy to socially cooperative mating and rearing of young. There is also frequent cannibalism and infanticide in nature, as well as "altruism" and parasitism. Ants have a social system of "slavery" (though in no way resembling our meaning of the term), seagulls have "lesbian" female pairs, which court and nest together (though again, the human label is misapplied).

We don't even know what is "natural" for our own species. Every few years a new theory emerges on what is our "natural" diet, our "natural" life span, our "natural" sexual practices, our "natural" social system or our "natural" relationship with nature.

Nature is endlessly fascinating, but offers no "natural" way of life for humans to copy. Even in evolution, there is no "natural" tendency toward "progress," "perfection" or "ascent." Most of the time, we don't even know what is going on in nature.

NATURAL LAW
Formulation of Regularities

Natural law was one of the basic ideas that led to the ascendancy of science in Western thought. English physicist-astronomer Sir Isaac Newton was one of its founders, demonstrating that complex physical phenomena could be reduced to discoverable "rules." Since a godless universe had no place in his religion, however, Newton had little patience for those who insisted miracles were impossible.

Most agreed that what was "natural" was what was fixed, regular, expected. But from the first, critics complained that scientists had made of natural law a god, producing all things from the properties of matter: the materialist blasphemy. Theologians saw no logical difficulty in supernatural events. If God could start the world spinning, he could temporarily stop it. If he could make one kind of beetle, why not 100,000? He was, by definition, infinite and beyond human comprehension.

If God had set natural law in motion, they argued, why should it be "unnatural" for him to intervene directly every now and then? Miracles and interventions left an open field for the speculative imagination. Scientists realized that research was only possible with a commitment to finding regular "secondary causes." Such was the "revolution" of Charles Lyell in geology and Charles Darwin in biology.

See also NATURPHILOSOPHIE; "PHILOSOPHICAL" NATURALISTS; THEORY, SCIENTIFIC.

NATURAL PRODUCTS CHEMISTRY
Evolutionary Pharmaceuticals

Living things have evolved thousands of chemical substances during 3.8 billion years in the world's oceans. Now scientists are "prospecting" the biochemistry of exotic sea creatures, hoping to find cures for such deadly diseases as AIDS and cancer.

Homely, boneless sea creatures like corals, sponges and marine worms have had eons to develop chemical defenses against rivals and predators. "Some of these remarkable substances will also destroy viruses and cancers in humans," says Professor C. Robert Petit, of the University of Arizona's Cancer Research Center. He and his colleagues are trying to find out which ones.

Dr. Petit is part of an effort to develop "natural products chemistry," a field that began around 1812 in Europe. Biologists since then have searched molds, microorganisms and plants for new medicines, leading to such discoveries as quinine, tetracycline and penicillin. Many were derived from plants of the tropical rain forest, a repository of many still-unknown medicinal chemicals that may be destroyed before anyone has a chance to find them. [See RAIN FOREST CRISIS.]

Since the 1970s, a small but dedicated group of scientists have turned to the ocean. There, in the two-thirds of the planet covered by water, 400,000 species of animals and plants still exist—more than 80% of the Earth's organisms. If only 10% of them produce potentially useful chemicals, 40,000 unique substances are waiting to be discovered.

One major research effort in the U.S. is conducted by a nonprofit institute called SeaPharm, Inc., of Princeton, New Jersey. Using a "funnel approach," its scientists collect thousands of different organisms from the sea, then analyze each species' composition in the laboratory and test its effects on viruses and tumors. Each organism is analyzed by liquid chromatography; computerized instruments reveal the presence of different chemicals, and even draw diagrams of their molecular structures. Only a handful of useful chemicals emerges from the small end of this research "funnel," but these few substances may be capable of saving many millions of human lives.

Despite the great evolutionary distance between humans and eyeless, limbless sea animals, their bodies produce substances that have remarkable effects in our own systems—biochemical testimony to Charles Darwin's conclusion that all life is one great, related family. Recently, SeaPharm scientists discovered an effective cell growth inhibitor for mammals which a marine worm uses for defense. Its biochemical structure is utterly unlike anything the researchers had expected.

The long history of these supposedly "simple" creatures includes hundreds of millions of years for evolution to try more chemical experiments than man can conceive. When a substance looks promising, pharmacologists try to isolate its active component, modify it to get rid of undesirable side effects and synthesize it—so they don't have to keep taking living creatures out of the ocean.

One promising cancer-fighting drug is being developed from a New Zealand sponge. Substances from other ocean-dwelling organisms are being tested as treatments for herpes and other viruses. In their imaginative quest for new "natural products" medicines, marine pharmacologists rely on both evolutionary diversity and the underlying kinship of all life forms.

For further information:

Krieg, Margaret B. *Green Medicine: The Search for Plants That Heal.* New York: Rand McNally, 1964.

Plotkin, Mark J. "The Outlook for New Agricultural and Industrial Products from the Tropics." In *Biodiversity,* edited by E. O. Wilson, National Academy Press, 1988.

NATURAL SELECTION
A Mechanism of Evolution

Charles Darwin complained his critics said what was good in his theory was old and what was new in it was wrong. The "old" part was simply the fact of evolution: that species had developed over time and that all life is linked through common ancestry. The new part was how it worked: the mechanism of natural selection.

That living things evolve is as certain as a scientific fact can be. The evidence is overwhelming and continues to accumulate. Just *how* evolution occurs—entirely through natural selection, or in other ways as well—is still an open question. However, despite a century and a half of criticism and attack, natural selection continues to be one of the most fruitful organizing principles of biological research.

Natural selection starts with two observations:

1. There is vast overproduction of new individuals in nature. Every organism produces many more offspring (or eggs or seeds) than will survive to reproduce themselves, as anyone who walks through the woods can see.

2. There is a great amount of variation between individuals, which a casual observer may *not* see. All zebra foals or bullfrog tadpoles may look alike at first glance, but a naturalist who spends years studying them is struck by the wide range of variability within the same species.

In each generation, many individuals will not reproduce. Selection pressures may include predators, climate, other members of their own social group, competition for space, food or mates, parasites and disease. The popular notion that "survival of the fittest" means simply that strong "winners" kill off weak "losers" is ideology, not biology.

Darwin used the term "natural" as opposed to what he called "artificial" selection, the deliberate breeding of varieties of domestic animals or plants by man. If pigeon breeders could select for fancy plumage and horse breeders for speed or disposition, he thought, surely organisms would be similarly shaped by relentless selection in nature.

Creation of new life forms, according to current theory, requires only a source of genetic variation and a "sorting sieve," which lets some alleles through to the next generation and blocks others. (An allele is a variant of a gene.)

Paleontologist George Gaylord Simpson's famous image is a hat containing several sets of the 26 letters of the alphabet on slips of paper. Supposing you were to draw out letters at random, chances are poor that they would spell C, A, and T, in that order. But if each time you pick a slip, you throw it away if it is not C, A, or T, and return it to the hat if it is one of those three letters, soon you will have mostly Cs, As, and Ts in the hat. Your chances of drawing out C–A–T keeps improving; eventually you must draw them in the proper order.

Darwin first wrote down his theory of natural selection in 1842; he sent a brief statement of it in an 1844 letter to his botanist friend Joseph Hooker. He was not the first to invent it, but he was the first to convince the world of its validity.

Bit by bit he assembled an enormous amount of data from geology, zoology, botany, animal husbandry, paleontology, biogeography and scores of other disciplines. Patiently and brilliantly he marshaled his evidence, giving his readers the feeling they had discovered the theory themselves, almost against Darwin's objections.

Another great naturalist, Alfred Russel Wallace, who was 14 years Darwin's junior, had come to the same conclusions. Working in the jungles of Malaysia, Wallace lay on a hammock with malarial fever thinking about Malthus's *Essays On Population* (1798)— the same book Darwin said led him to the theory— when he independently conceived the idea of natural selection. It is one of the most striking cases of parallel discovery in the history of science.

In 1858, Wallace mailed his own paper on natural selection to Darwin, sending the older naturalist into a panic to finish his "big book" on the subject. Wallace and Darwin received joint credit for the theory, which was communicated to the Linnaen Society in 1858. Although Wallace's name is little known today, Darwin wrote him he considered natural selection "as much yours as it is mine."

Darwin was handicapped because there was little real understanding of inheritance in his day. Sometimes he spoke as if random variations were selected, eliminated or preserved in populations; at other times he adopted the so-called "Lamarckian" theory that an individual's habitual use or disuse of organs would be passed on to offspring. (It wasn't just Lamarck; practically everyone in his time believed in use-in-heritance.)

So uncertain was Darwin on the question of inheritance that, under fire from critics, he began to retreat from natural selection as his main evolutionary mechanism. In later editions of the *Origin of Species*, he suggested possible alternatives and special cases. By the last (6th) edition, Darwin had shrunk natural selection in importance to one of several possible mechanisms of evolution.

Meanwhile, his junior partner, Alfred Russel Wallace, continued to maintain the validity of their original vision, insisting that natural selection was the major force driving evolution. In 1889, some years after Darwin's death, he wrote his own detailed exposition of evolutionary theory, crammed with the latest evidence for natural selection. Ironically, he titled it *Darwinism*. But, although Wallace remained a convinced selectionist about plants and animals, he thought the evolution of the human species needed another explanation.

Influenced by his Spiritualist beliefs, Wallace could not see why early humans should have evolved a brain "so much better" than seemed strictly necessary for survival unless some "unknown agency" had intervened. [See WALLACE'S PROBLEM.] Our physical bodies evolved by natural selection, he concluded, but at some point "the Unseen World of Spirit" had injected a higher consciousness and intellect into the human species.

He had advanced that view some years earlier (1868), sending Darwin into melancholy. How could

Wallace apply natural selection to every living creature except man? He was abandoning science to bring special creation in the back door. "I only hope," Darwin wrote Wallace, "that you have not murdered too completely your own and my child."

Ever since Darwin's day, biologists have debated the mechanisms, but not the fact, of evolution. Does selection work only on individuals, on natural societies and social groups, possibly at the species level? Should we seek selection in every aspect of organisms or are some characteristics "neutral" or random? Does selection work slowly and gradually or by discontinuous jumps?

Natural selection has been misused as a political analogy, overextended into "just-so" origin stories and has undergone radical shifts in meaning as evolutionary theory broadened to take in population genetics and DNA. Yet despite more than 100 years of attempts to dislodge it, natural selection remains a central idea in biology, still generating new theories, observations and fruitful research designs.

See also ORIGIN OF SPECIES; "SURVIVAL OF THE FITTEST"; "TERNATE PAPER"; WALLACE, ALFRED RUSSEL.

For further information:

Haldane, J. B. S. "Natural Selection." In *Darwin's Biological Work: Some Aspects Reconsidered,* edited by P. R. Bell, 101–149. Cambridge: Cambridge University Press, 1959.

Kottler, Malcolm. "Charles Darwin and Alfred Russel Wallace: Two Decades of Debate over Natural Selection." In *The Darwinian Heritage,* edited by David Kohn, 367–432. Princeton: Princeton University Press, 1985.

Sober, Elliot. *The Nature of Selection.* Cambridge: MIT Press, 1984.

NATURAL THEOLOGY
Religious Science

From the time of the Renaissance until the late 18th century, natural theology was the prevalent system of thought in Euro–American studies of the natural world.

As developed by such outstanding practitioners as the Reverend John Ray and the Reverend William Paley, it was a synthesis of science and religion that admitted no conflict between the two. Any regularities, perceived patterns or designs in nature were taken as evidence of a supreme being or designer.

Modern "creationism" and "creation science" spring directly from the old natural theology tradition: Their aim is to discern the hand of the Creator in nature. An underlying premise is that there can be no conflict between observed nature and revealed religion, since both are traceable to the same source of truth.

Science, according to this view, is in the service of providing evidence for religious beliefs. But the history of knowledge shows that whenever science restricts its inquiries to bolster political, economic or religious ideologies, it cannot long survive. As Thomas Henry Huxley put it, "where science adopts a creed, it commits suicide."

See also LYSENKOISM; PALEY'S WATCHMAKER.

NATURA NON FACIT SALTUM (NATURE MAKES NO LEAPS)

See SALTATION.

NATURPHILOSOPHIE
German Idealist Movement

For almost a century, *Naturphilosophie*—a combination of Platonic idealism with a search for aesthetic purity—was the prevailing philosophy of natural history in Germany. Ushered in by the poet-philosopher Johann Wolfgang von Goethe (1749–1832), it was the search for archetypes; ideals of pure form and design.

Promulgated by a group of professors at Munich, especially Freidrich W. J. von Shelling, Lorenz Oken and Ignatius Dollinger, it became extremely influential as students carried it all over the world.

Naturphilosophie aimed to encompass all nature in an absolute, unified system of ideas about plan, pattern and type. As historian A. Hunter Dupree put it:

> Where an empiricist [experimentalist] might infer the Creator from the hand and eye, the Germans had it the other way round. The Creator and the design were the ultimate reality, of which the hand and the eye were but manifestations. They could see evolution everywhere, but it took place in the mind of God.

Despite the excesses of their philosophical system and the barrier it posed to acceptance of Darwinian ideas, some naturalist adherents nevertheless managed to conduct brilliant and precise research. The German embryologist Karl von Baer made many important discoveries, as did the English anatomist-paleontologist Richard Owen and the Swiss-American Louis Agassiz, who elucidated the Ice Ages and fossil fish. All of them were staunch and influential opponents of Charles Darwin's view of form resulting from natural processes and history, rather than imperfect manifestations of an ideal type or plan.

See also AGASSIZ, LOUIS; ESSENTIALISM; OWEN, SIR RICHARD.

For further information:

Oken, Lorenz. *Elements of Physiophilosophy.* (Translation of *Lehrbuch der Naturphilosophie*). London: Ray Society, 1847.

NAZIS, EVOLUTIONARY PROGRAM OF

See ARYAN "RACE," MYTH OF; LEBENSBORN MOVEMENT; MONISM.

NEANDERTAL MAN
First Known Fossil Man

"Fossil man does not exist!" declared the great French paleontologist Georges Cuvier, expressing the scientific world's long-standing conviction. But in 1856—three years before Charles Darwin published his *Origin of Species*—German quarrymen working a limestone deposit accidentally discovered Europe's first known human fossil. Science was not prepared to accept it.

A heavy skullcap and 15 parts of a skeleton, showing obvious differences from modern *Homo sapiens*, were excavated from a limestone cave on a rock face high above Germany's Dussel River, near the village of Neander. Fortunately, the quarry owner called the local teacher, who sent the bones to Professor D. Schaaffhausen of Bonn.

After careful study, he published his opinion (1857) that the Neander valley skeleton was not a German, but perhaps belonged to a "wild northern tribe" they had conquered long ago. Certainly, he thought, this man with the very extraordinary skull shape was older than the Celts.

Schaaffhausen's conclusion that the Neandertaler was a normal adult from an early tribe was essentially correct, but more prestigious "experts" insisted it was some kind of pathological freak.

Rudolph Virchow (Fear-COUGH), Germany's most influential anatomist, insisted the skeleton was not an ancient man at all, but merely a recent cripple with a deformed skull who suffered from arthritis. Another prominent anatomist pronounced it to be an abnormal idiot who probably had been a recluse living in the cave. No, it was a perfectly normal individual, wrote another learned professor: a normal Mongolian Cossack who had served in Napoleon's army 30 years before and had died during the retreat from Moscow.

Over the next century, more and more remains of these strange "deformed" men and women with the heavy jaws and eyebrow ridges kept turning up. One came from Gibraltar, several from France, and later specimens from the U.S.S.R., France, Israel, China, Italy and north Africa. Eventually, science recognized Neandertalers as a distinctive and wide-ranging population, which had inhabited much of

"STORMBOUND" was how artist Charles R. Knight envisioned Neandertal folk during the last Ice Age (left). Their heavy-browed skulls (right) and characteristic Mousterian stone tools remain to pose unresolved questions about who they were and what became of them.

the Old World, from Europe to the Near East, Africa, and Asia during the final phase of the last Ice Age.

Neandertal skeletons were discovered ritually buried, some with animal horns and red ochre, others with evidence that the body had been covered with flowers. Many of the sites contained a distinctive stone tool industry called Mousterian by archaeologists, after a rich site at Le Moustier in France.

Neandertals are now well known, but they present one of the mysteries of human evolution. Anthropologists cannot agree on how they were related to *Homo sapiens* populations, which apparently coexisted with them during the same period. At one time, Neandertals were considered our ancestors; but their contemporaniety with modern-looking people raises new questions.

Their Mousterian tool tradition persisted for more than 40,000 years and appears remarkably unchanged over time and geographical area. In contrast, the *Homo sapiens* of the Ice Age left a remarkably varied and changing stone technology, plus a rich legacy of cave paintings and carved figurines, some of them remarkable works of art. No carvings or paintings are associated with Mousterian sites.

A final unsolved mystery is what became of them. Did they die out for some reason, or merge with the gene pool of *Homo sapiens*? Or did they disappear because we outcompeted or simply exterminated them?

See also FOSSIL MAN; MOUSTERIAN INDUSTRY.

NEBRASKA MAN
A Swinish Missing Link

One of the most singular and embarrassing incidents in the history of evolutionary science began in 1922, when a solitary molar tooth was found in Nebraska. First-rank paleontologists, anthropologists and anatomists examined the cusp pattern, and all agreed with its discoverer that the tooth belonged to an ancient ape-man: a "missing link" of tremendous importance, to which they gave the name *Hesperopithecus* or "Western ape."

The tooth was certainly ancient; it was embedded in million-year-old Pliocene deposits. But what else could be said about it? For starters, English anatomist Sir Grafton Elliot Smith and a museum artist collaborated to produce a painting of both male and female *Hesperopithecus* for the *Illustrated London News*. Their "reconstruction" featured full figures of a well-muscled, slope-browed pair in a prehistoric landscape complete with early horses and camels.

Professor H. F. Osborn, head of the American Museum of Natural History, welcomed the news.

Antievolutionist politician William Jennings Bryan was a Nebraskan, and Osborn rubbed it in. "The Earth spoke to Bryan from his own State," he crowed, "this little tooth speaks volumes . . . evidence of man's descent from the ape."

In 1925, when John Scopes was tried for breaking Tennessee's state law against teaching Charles Darwin's theory of evolution in the public schools, the *Hesperopithecus* tooth was introduced as evolutionary evidence along with other fossils of early man then accepted by science (including Piltdown, which was later revealed as a fossil forgery).

Two years after the "Monkey Trial," a team of paleontologists returned to the Nebraska site where *Hesperopithecus* had been discovered five years earlier, determined to find more of this mysterious creature. To their joy, weathering had exposed parts of a jaw and skeleton on the precise spot. Eagerly, they brushed away dust and sand until the ancient fossil emerged to tell its truth—the infamous molar had once belonged to an extinct pig!

See also GIGANTOPITHECUS; "MISSING LINK"; PILTDOWN MAN (HOAX).

NEMESIS STAR
Source of Cyclic Extinction?

In the "new catastrophism" of the 1980s, evidence from astronomy, geology and paleontology points to a tantalizing possibility that evolution is more than the history of competition and adaptation among earthly life forms: It may be a response to cyclic cosmic events.

According to this hypothesis, the development of life has been punctuated by violent cataclysms that caused mass extinctions. Whole groups of creatures were wiped out all over the Earth, leaving unoccupied niches that were filled by later radiations.

First Luis and Walter Alvarez at the University of California at Berkeley announced in 1979 that they had found deposits of the metallic element iridium at Cretaceous boundaries in rock strata, which would coincide with the extinction of the dinosaurs. These deposits are not normally found on the Earth's surface, except at the impact craters of ancient meteors. Next, in 1984, two University of Chicago scientists, David Raup and John Sepkoski, published a compilation of evidence that ancient mass extinction occurred roughly every 26 million years. The question then became: What could possibly cause such massive, regular, traumatic oscillations in the history of life which seem to coincide with depositions of a space metal onto the Earth?

Two groups of scientists independently proposed

THE EARLIEST MAN TRACKED BY A TOOTH: AN "ASTOUNDING   DISCOVERY" OF HUMAN REMAINS IN PLIOCENE STRATA.

A PREHISTORIC COLUMBUS WHO REACHED AMERICA BY LAND?—AN ARTIST'S VISION OF   HESPEROPITHECUS (THE APE-MAN OF THE WESTERN WORLD) AND CONTEMPORARY ANIMALS.

AN EMBARRASSMENT to evolutionists, "Nebraska Man" or *Hesperopithecus* became part of the scientific literature on the basis of a single tooth, which actually belonged to a fossil pig. This painting, commissioned by distinguished anatomist Sir Grafton Elliot Smith, appeared in a 1922 issue of the *Illustrated London News.*

a similar answer in the April 1984 issue of the respected journal *Nature.* Their theory was that our sun has a smaller unrecognized companion star, which is now more than two light-years away and has eluded observation because of its proximity to the overpowering brilliance of the sun. Its entire swing takes about 26 million years, and each time it nears Earth, it disrupts the belt of comets (the Oort cloud, beyond Pluto) in our outer solar system. Freed comets and meteors are attracted to Earth and impact here, causing severe disruptions to life. Not to fear, though—it's not due to happen again for another 13 million years. (The proposers of this idea are Marc Davis, Piet Hut, Richard A. Muller and—in an independent group—Daniel P. Whitmore and Albert A. Jackson IV.)

Their expectation of finding a companion star to our sun results from various discoveries and observations going back more than 100 years. They have named the sun's hypothetical sister "Nemesis," after the Greek goddess who personified divine retribution. While it is unusual to name something before it has been discovered, the reality of a predicted "death star" can be explored and tested. Astronomers think that if Nemesis does exist, they have a 50% chance of finding it within the next three years.

Paleontologist Stephen Jay Gould points out that the naming before the finding worked out in the case of Eugene Dubois's discovery of *Pithecanthropus.* However, Gould is disturbed by the name Nemesis, which connotes a force of destruction poised to exterminate life. Ever the optimist, Gould has proposed the name "Siva" (pronounced SHE-va) for the hypothetical star, after the Hindu god. Although Siva represents the universal force of destruction, Gould reminds us that he "forms an indissoluble triad with Brahma, the creator, and Vishnu, the preserver. All are enmeshed in one—a trinity of a different order—because all activity reflects their interaction." Statues of Siva show him holding the flame of destruction in one hand and in another carrying a drum to regulate the rhythm of the cosmic dance, which represents creation. As Gould sums up the symbolism of ancient Indian mythology:

> He moves within a ring of fire—the cosmic cycle—maintained by an interaction of destruction and creation, beating out a rhythm as regular as any clockwork of cometary collisions. "In this perpetual process of creation and destruction," [Indian author] Parthasarathy writes, "the universe is maintained." Unlike Nemesis, Siva does not attack specific targets for cause or punishment. Instead, his placid face records the absolute tranquillity and serenity of a neutral process, directed toward no one but responsible for maintaining the order of our world.

See also ALVAREZ THEORY; DUBOIS, EUGENE; EXTINCTION; "NEW" CATASTROPHISM; PUNCTUATED EQUILIBRIUM.

For further information:

Muller, Richard. *Nemesis, the Death Star.* New York: Weidenfield & Nicolson, 1988.

Raup, David. *The Nemesis Affair: A Story of the Death of Dinosaurs and the Ways of Science.* New York: Norton, 1986.

NEO-DARWINISM
"Hard" Inheritance

George Romanes, Charles Darwin's disciple in animal psychology, coined the term "Neo-Darwinism" in 1905 to describe the theory of natural selection without a belief in the inheritance of acquired characters ("Neo-Lamarckianism" or "soft" inheritance). Romanes was convinced that Gregor Mendel and August Weismann were correct about the discontinuity of "germ plasm."

Unlike most scientists of the time, who saw the new genetics as the death-knell of Darwinism, Romanes expected the two theories to merge into an improved understanding of evolution. Alfred Russel Wallace, for instance, fought Mendelism tooth and nail, believing it to be utterly incompatible with the theory of natural selection he had coauthored with Darwin.

Some writers (incorrectly) use Neo-Darwinism as a more general term, to cover all of 20th-century biology, including the combination with population genetics in the 1930s that produced the Synthetic Theory.

See also DARWINISM; LAMARCKIAN INHERITANCE; NATURAL SELECTION; ROMANES, GEORGE J.

For further information:

Bowler, Peter J. *Evolution: The History of an Idea.* Berkeley: University of California Press, 1984.

Mayr, Ernst. *The Growth of Biological Thought.* Cambridge: Harvard University Press, 1982.

NEO-LAMARCKIAN

From the late 19th century until well into the 20th, some adopted the idea of evolution, but not the mechanism of "random variation" and "natural selection." As Samuel Butler said, it "banished mind from the Universe." Intellectual or moral progress made by individuals would die with them, rather than be passed on to their offspring for the improvement of the species, except by teaching. Each generation had to learn anew.

Neo-Lamarckians saw themselves as bringers of hope and inspiration, the antidote to what physicist Sir John Herschel called Darwin's "law of higgelty piggelty." Among them were the novelist Samuel Butler, playwright George Bernard Shaw and philosopher Henri Bergson. Fifty years later, two renowned scientists took up Lamarck's banner: the Austrian experimentalist Paul Kammerer and the Soviet ideologue Trofim Lysenko, who set Russian genetics and agriculture back 30 years.

All believed Lamarckian inheritance to be the true basis of evolution and stood fast against the new Mendelian genetics. But Lamarck had proposed no theory of *how* acquired characteristics could be inherited; he simply shared the common belief of his time that changes in parents can be passed to offspring. Even Charles Darwin shared that belief and suggested a mechanism for how it might work: his futile theory of pangenesis.

Developments in 20th-century genetics have put the crude form of "Neo–Lamarckianism" to rest, but the problem of inheritance of acquired characters remains important. The "Central Dogma of Genetics" says genetic information goes only one way—from DNA to expression in the organism, and never the reverse. But scientists continue to discover exceptions and complexities to this general rule.

Possible mechanisms for the inheritance of acquired characters include epigenetic regulators, which control embryonic development, and Barbara McClintock's "jumping genes," which may take up different positions on a chromosome to help adjust an organism to its environment. Retroviruses can be acquired during an individual's lifetime, enter the DNA and be inherited by offspring.

These research directions have produced a "New Lamarckism" (as opposed to the dead-end Neo-Lamarckianism): the current scientific search for mechanisms by which acquired characters or responses to environment may be inherited.

See also BUTLER, SAMUEL; LAMARCKIAN INHERITANCE; "NEW" LAMARCKISM; SHAW, GEORGE BERNARD.

For further information:
Hitching, Francis. *The Neck of the Giraffe: Where Darwin Went Wrong.* New Haven, Conn.: Ticknor & Fields, 1987.

NEOTENY
Are Adults Big Babies?

Neoteny is the retention of juvenile characteristics in the adult form of an animal. A classic example is the axolotl, a Central American salamander that spends its entire life underwater, never shedding the gill structures similar species outgrow when they develop lungs. The axoltotl (its name is from a Nuhuatl Indian word meaning "servant of water") remains in a permanent larval stage.

This curious phenomenon seems rare in nature and would be of little interest except that it may have been an important factor in human evolution: *Homo sapiens* has many characteristics of a fetal or infant ape. Adult chimps and gorillas, for instance, have elongated faces, heavy brow ridges, powerful jaws, small braincase in relation to overall skull and other characteristic proportions. Baby apes have flat faces, rounded brain case, light brow ridges, proportionately smaller jaws, and many other bodily features strikingly like adult humans. (See CUTENESS, EVOLUTION OF.)

This idea—that man is in some way a fetalized ape—has tantalized students of evolution for a century. Although it remains a puzzle, some have suggested it might help account for the apparently small genetic difference between humans and apes. One of the crucial differences may simply be genetic instructions to juvenalize an anatomical blueprint that is held largely in common.

INFANT APE AND ADULT HUMAN have strikingly similar facial proportions, such as high forehead, smallish nose, and flattened face (left), while mature ape developed low forehead and protruding snout (right). Neoteny theory suggests some species evolve genetic instructions to retain juvenile features.

For further information:
Gould, Stephen J. *Ontogeny and Phylogeny.* Cambridge: Belknap/Harvard University Press, 1977.

NEURAL DARWINISM
Theory of "Mind"

A new debate on the nature of "mind" began in 1987 with the publication of *Neural Darwinism: The Theory of Neuronal Group Selection* by Nobel laureate Gerald Edelman. After years of dissatisfaction with computer models of the brain, "black boxes" and "wiring diagrams," Edelman attempted to explain the workings of mind in terms of Darwinian selection.

In a long, difficult and complex argument, which very few neurologists seem to grasp and have yet to confirm, Edelman sets forth his theory that perception, action, and learning are based on a selection process. Groups of cells that respond successfully to stimuli from the environment are preserved and their connections strengthened. Those that do not are eliminated: thus a self-correcting process of adaptation continues throughout life.

Memories, rather than being stored in neat, local compartments, are continually reworked and recategorized. Every experience in our lives, from before birth up until death, alters and shapes our brains.

Edelman's "neural Darwinism" is an attempt to apply selection theory to one of the most intractable of all phenomena: the workings of the mind. Time and testing will tell whether it is scientifically productive or will itself be "selected out" in the struggle for existence of useful explanations.

As Ambrose Bierce wrote in *The Devil's Dictionary* (1906), the mind is "a mysterious form of matter secreted by the brain. Its chief activity consists in the endeavor to ascertain its own nature, the futility of the attempt being due to the fact that it has nothing but itself to know itself with."

For further information:
Edelman, Gerald M. "Neural Darwinism: Population Thinking and Higher Brain Function." In *How We Know*, edited by Michael Shafto. New York: Harper & Row, 1986.
———. "Group Selection as the Basis for Higher Brain Function." In *The Organization of the Cerebral Cortex*, edited by F. O. Schmitt. Cambridge: MIT Press, 1986.

NEUTRAL TRAITS
Nonadaptive Characters

The Darwin–Wallace theory of natural selection assumed "useful" variations would become established in a population, while all others would be eliminated. But some naturalists insisted that many traits in plants and animals had no demonstrable positive *or* negative advantage—they were nonadaptive or "neutral."

Although he easily demolished arguments for plan or "design" in nature, Charles Darwin seemed stumped by the critics who suggested evolution might be even *more* random than he had supposed. After the third edition (1861) of the *Origin of Species*, he began to allow more room for a pluralistic view of evolution, until natural selection became only one mechanism among many.

Distressed by this turn of events, Alfred Russel Wallace published his own version of evolutionary theory in 1889, a few years after Darwin's death. Wallace urged a return to the original vision of their joint theory—a thoroughgoing selectionism. Ironically, he titled the book *Darwinism* and rejoiced when reviewers called him "more Darwinian than Darwin." (They really meant "more selectionist," more insistent on the importance of adaptation in shaping organisms. Wallace always assumed that persistent traits had some "selective advantage," even if their value was not immediately obvious.)

But Wallace did not have the last word. After his death, the 20th-century synthesis of Darwinism with Mendelian genetics renewed interest in the possibility of neutral traits, especially among population geneticists.

By 1932, geneticist J. B. S. Haldane had concluded "that innumerable characters [of animals and plants] show no sign of possessing selective value, and moreover, these are exactly the characters that enable a taxonomist to distinguish one species from another." A few decades later, however, Oxford biologists proved the opposite—that many characters then considered "neutral" do, in fact, have selective value, which one revealed by experiment, not by speculation.

In the 1960s, with new discoveries about the enormous genetic variability within natural populations, a "neutralist school" arose once again. Researchers J. L. King, M. Kimura and others argued that if evolution was not like a preplanned journey with an intended destination, it might be more like a "random walk"—taking a turn in this direction or that, for no particular reason except the contingencies of its history. Both "chance" or selection shape organisms, but in what proportions? The debate continues to inspire new research designs.

See also ADAPTATIONIST PROGRAM; GENETIC DRIFT.

For further information:

Kimura, M. *The Neutral Theory of Molecular Evolution.* Cambridge: Cambridge University Press, 1982.

"NEW" CATASTROPHISM
Modern "Cataclysmic" Theories

Sudden global extinctions of almost all earthly life, a rain of giant asteroids from outer space, rapid prolif-

eration of new species—all appear to belong to the "catastrophic" era of science, a vogue that flourished about 200 years ago. Yet, jumps ("discontinuities") and cataclysms, although based on very different understandings, are back in favor today. Some have called their renewed scientific respectability "the new catastrophism."

Earlier catastrophists reached their conclusions differently. Some claimed that processes or "natural laws" observable today could not be used to interpret the past. Many tried to reconcile Earth's history with miraculous accounts given in the Bible, such as Noah's flood. At the time a poor understanding of processes allowed imaginations to run wild, picturing waves of poison gases or rampant volcanoes, unsupported by evidence.

By the 1860s, indiscriminate applications of "multiple creations," "catastrophes" and divine interventions had lost credibility. A growing weight of evidence supported the idea that that slow, steady processes currently observable shaped the Earth in times past. Geologists stopped inventing unknown causes and events. [See OCCAM'S RAZOR.] In biology, the Darwin–Wallace theory stressed evolution of species by gradual change rather than sudden leaps or special creations.

By 1870, "uniformitarian" geology and "gradualist" evolution had become the guiding principles of science. In an important sense, they still are. But a shift in emphasis occurred a century later, during the 1970s and 1980s, as the realization grew that the fossil record and living populations show patterns of fits and starts, not smooth directional change.

The Synthetic Theory of evolution of the 1940s, which melded the findings of genetics, populational mathematics, and paleontology still stands, but its emphasis on a long-term picture of steady, gradual change has been challenged. A "punctuationalist" model of the history of life has gained credence postulating long periods of stability interrupted by brief bursts of intense change. ("Brief" being a relative term, involving hundreds of thousands of years.)

The puzzle of global mass extinctions—such as the disappearance of the dinosaurs—points to great climactic changes, possibly caused by gigantic asteroids crashing into the Earth. [See ALVAREZ THEORY.] During the 1970s, some astronomers suggested global changes might be linked to cyclic events in outer space, occurring at regular intervals of 26 million years. (Data continue to be collected that will either confirm or discredit the idea.)

In studies of populations and cellular genetics, too, slow, gradual accumulation of small mutations does not now appear to be a major process in the formation of new species. A new search for "discontinuous" mechanisms is well underway.

This "new catastrophism" is not a return to the old, either in method, evidence or assumptions. Nor is it any kind of sweeping explanation, but an interesting confluence of particular developments in various fields. Yet, late 20th-century evolutionism has unmistakably swung away from a complacent, untested belief in "gradualism." Current research has adopted a new focus: the search for mechanisms of rapid, dramatic changes in the history of life.

See also ACTUALISM; CATASTROPHISM; SIR CHARLES LYELL; NEMESIS STAR; "NEW" LAMARCKISM; PUNCTUATED EQUILIBRIUM; UNIFORMITARIANISM.

For further information:
Eldredge, Niles. *Life Pulse: Episodes from the Story of the Fossil Record.* New York: Facts On File, 1987.
———. *Time Frames.* New York: Simon & Schuster, 1985.
Stanley, S. *The New Evolutionary Timetable.* New York: Basic Books, 1981.

"NEW" LAMARCKISM
The Responsive Gene

The old idea that behaviors or experiences acquired in an individual's lifetime can be passed on to offspring became associated with the name of the French biologist Jean-Baptiste Lamarck, though he did not originate it. The rise of Mendelian genetics disproved this "old" Lamarckism because changes (mutations) in the genetic material appeared unaffected by environment or individual experience.

According to the "central dogma" of genetics, protein synthesis can occur in one direction only: from DNA (the chemical code) to RNA (the messenger that interprets it as blueprint) to protein (body cells). The individual's genetic make-up (genome) remains fairly static, affected only by chance variations caused by replication mistakes or exposure to radioactivity or certain chemicals.

However, as researchers probed more deeply into heredity during the 1970s, some became dissatisfied with the prevailing model of a static genome, totally unresponsive to environment except through the selection of chance mutations. Geneticist Barbara McClintock, working independently with maize for 30 years, discovered the unexpected phenomenon of "jumping genes"—genes that may abruptly change their positions in relation to others on the chromosome. Such changes in position, she found, can profoundly affect how the blueprint is expressed in the plants. In experiments made during the late 1980s, researchers were able to move genes to different positions in a fruit fly's chromosomes. The artificial re-positioning resulted in a different eye-color for the fly's offspring; this new color was subsequently inherited by its descendants although the genes remained the same.

Genomes, it now appears, are not at all static. Within the ocean of genetic material, there is constant movement and reorganization. Whether some of this activity can be a direct response to environmental factors has recently become an open (and very controversial) question.

One research team during the 1980s reported an apparent responsiveness to environment in the genomes of hungry bacteria. Some seemed to evolve more rapidly when placed on lactose, for instance, and produced lactose-digesting mutants improbably often. Also, bacteria appear to pass among themselves bits of genetic material (plasmids) that can be incorporated into the recipient's own genome. Furthermore, some genes can be "switched" on and off as needed for certain situations.

The central dogma notwithstanding, in the early 1970s, David Baltimore and Howard Temin found that certain viruses use RNA rather than DNA as genes and are able to reverse-transcribe the DNA from RNA. By 1987, reverse transcription had been found to be widespread, not only in these "retroviruses" (which infect RNA and travel to DNA to cause heritable diseases) but in yeasts, plants and even mammals.

Some researchers (J. V. McConnell followed by Holger Hyden, Georges Ungar and others) are attempting to explore whether RNA may actually be changed by learning. If there were a mechanism for the changed RNA to transcribe back into DNA, a molecule with such easy access to the genome would play a major role in evolution. An idea so antithetical to the currently orthodox Synthetic Theory of evolution would not find easy acceptance. Still, RNA as a mediator of evolution, responding to environment and experience, remains an untested hypothesis that fascinates proponents of the "new" Lamarckism.

See also "CENTRAL DOGMA" OF GENETICS; LAMARCKIANISM, NEO-LAMARCKIAN.

For further information:
Sheldrake, Rupert. *A New Science of Life.* Los Angeles: Tarcher, 1981.
Steele, E. J. *Somatic Selection and Adaptive Evolution.* Chicago: University of Chicago Press, 1981.
Wills, Christopher. *The Wisdom of the Genes: New Pathways in Evolution.* New York: Basic Books, 1989.

NEWTON, SIR ISAAC

See NATURAL LAW.

NEW WORLD MONKEYS

See PLATYRRHINES.

NIETZSCHE'S "SUPERMAN" (UBERMENSCH)

See "OMEGA MAN."

NIM CHIMPSKY
Ape Communication Experiment

A few years after linguist Noam Chomsky of MIT had written that language is biologically unique to humans, an ape named Nim Chimpsky talked back.

Playfully named by psychologist Herbert S. Terrace of Columbia University, Nim was the subject of an experiment begun in December 1973. Its purpose was to teach American Sign Language to the two-week-old chimp, in hopes that he could learn a language, if he didn't have to *speak* the words.

It was an idea that went back to Robert M. Yerkes, who had tried to teach chimps to talk in the 1930s, and concluded that they were better at mimicking visual signals than vocal ones. By 1975, Nim was putting two signs together, and a year later, combinations of three, such as "you-tickle-me" or "me-more-eat."

Nim was given lots of affectionate attention by human trainers, who treated him like a small child. Almost five years later he was returned to the Institute for Primate Studies in Norman, Oklahoma. At that point, it was thought that Nim understood 300 signs, could produce 125 of them and had put thousands of "sentences" together.

Professor Terrace and Nim became well known for their achievements, but soon a controversy began to swirl around them. Was Nim really "speaking" or was he signing without comprehension to gain the approval of his handlers? Were his responses triggered by subtle cues from Terrace and his staff? Critics brought up the lesson of a famous German "talking" horse. [See CLEVER HANS PHENOMENON.]

In 1979, Terrace wrote a book, *Nim*, in which he disavowed his previous results. After reviewing the photos and videotapes, he concluded that he had vastly overrated Nim's language abilities. Word combinations had not increased in length, and much behavior could indeed be attributed to unconsciously given cues by humans. The extent to which "signing" apes can really master symbolic language is still a tantalizing question for research.

See also APE LANGUAGE CONTROVERSY; KOKO.

For further information:

Chomsky, Noam. *Language and Mind.* New York: Harcourt, Brace, Jovanovich, 1968.

Terrace, Herbert. *Nim: A Chimpanzee Who Learned Sign Language.* New York: Knopf, 1979.

NINTH BRIDGEWATER TREATISE
God as the Master Programmer

Charles Babbage (1792–1871) was a genius of Victorian mathematics who is generally considered the father of computers for his ingenious and elaborate "calculating engines": intricate mechanical marvels that could solve numerical problems.

A rationalist who looked to science for formulations of natural laws, he was dismayed when the eight volumes of the *Bridgewater Treatises* appeared in the 1830s. Commissioned by the Earl of Bridgewater's bequest, these purportedly scientific works often resorted to divine "miracles" to explain the facts of natural history. Babbage was especially disturbed by Dean William Buckland's view of catastrophes, successive creations and the occasional miraculous intervention of God.

In 1838, Babbage published a well-considered reply, his own, unauthorized *Ninth Bridgewater Treatise*. While the "official" treatises had the avowed purpose of cataloguing the inexplicable workings of the creator, Babbage argued that what seemed to be miracles might really be explained by some higher natural law, rather than by supernatural acts.

His calculating machine, a forerunner of the modern computer, could be programmed to change its mode of operation according to a predetermined plan. As historian Peter Bowler puts it, "Surely God could build such a preordained pattern into the universe which could change the normal laws of nature from time to time in such a way that would *appear* miraculous to the casual observer."

Babbage's ideas were adopted by Robert Chambers in his influential and pseudonymous work *Vestiges of Creation*, published in 1844. Although it was an evolutionary theory that in some ways anticipated Darwin, *Vestiges* put forward a very different notion of *how* things change. Rather than a consistent, uniform operation of natural processes, Chambers thought the rules of nature kept changing in a series of gear-shifting progressions.

Like Babbage, Chambers viewed God as a great programmer, who inputted this "law of progression" into the universe. Through a series of transformations, it unfolds different processes until it reaches the laws of nature we know. But, in Bowler's words, "the scientist is helpless to investigate the cause of such changes; he can see the overall pattern [but] cannot understand the programming mechanism itself."

See also BRIDGEWATER TREATISES; VESTIGES OF CREATION.

"NOAH'S ARK PROBLEM"
Survival of Those That Fit

One of the problems that bothered those naturalists who attempted to reconcile scripture with nature was how many animals were there on Noah's Ark? In the 18th century, only hundreds of species were known,

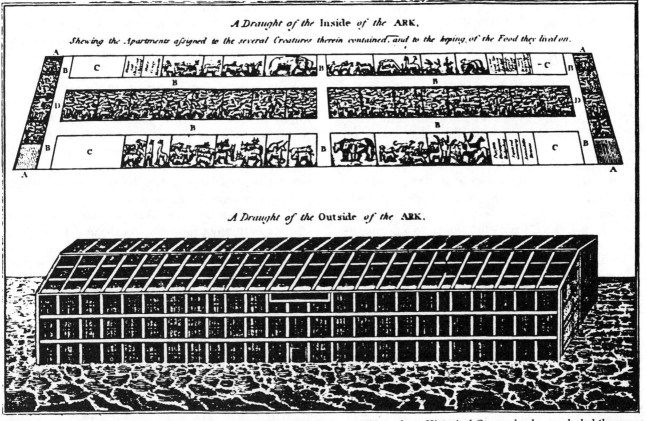

A Draught of the Inside *of the* ARK.

Shewing the Apartments assigned to the several Creatures therein contained, and to the keeping of the Food they lived on.

A Draught of the Outside *of the* ARK.

CAPACITY OF NOAH'S ARK was calculated by Dr. Edward Wells in an 1801 work on *Historical Geography;* he concluded there was room enough for seven individuals of all clean (kosher) beasts, two of each unclean beast, with enough hay, food and water for all.

and there was no difficulty in imagining they could all fit on the Ark. But with rapid expansion of knowledge during the voyages of exploration, early Victorian science recognized 1,000 species of mammals, 6,000 birds, 1,500 reptiles and amphibia, and the numbers were increasing year by year. How could two of each of these species, with enough food for more than a month, be crammed into a three-story Ark that measured "300 cubits long by 50 cubits broad"?

Serious scholars addressed themselves to this "Noah's Ark Problem." One worked out the area of the ark (if the cubit was about one and a half feet) to be 450 feet by 75 feet. Multiplied by three (the number of storys), that gave an area one-seventh the floor area of the Great Exhibition at the Crystal Palace. Another writer, a Dr. Hamilton of Mobile, Alabama, wrote that since most of the species were small, many of them could be crowded together in small compartments—almost shoulder to shoulder. Historian Lynn Barber, in *The Heyday of Natural History* (1980), writes:

> [By the 1850s most realistic thinkers] . . . agreed that it was not possible to fit all the known animals into the Ark, and the best solution was to make Noah's

flood purely local in extent, and to confine it to an area somewhere around the Caspian. The "Noah's Ark problem" now seems the most absurd of all the nineteenth-century religio-scientific debates, but to the Victorians it was the most real . . . A visitor to any zoological garden could see the difficulty of fitting so many animals into an area one-seventh the size of the Great Exhibition, and for many laymen this was the first hint that the Book of Genesis could not always be taken at face value.

In the 1890s, a major expedition set out to find the "actual" Noah's Ark, which was believed to have been sighted atop Mount Ararat in the Zagros Mountains of Turkey. If they could just prove with physical evidence that such a craft existed, creationist supporters of scripture believed they could establish once and for all the literal truth of biblical accounts. Although earlier searchers failed, a few diehards (including a former astronaut) continue to seek the Ark late into the 20th century.

See also ATRAHASIS; NOAH'S FLOOD.

For further information:

Barber, Lynn. *The Heyday of Natural History 1820–1870.* Garden City, N.Y.: Doubleday, 1980.

Montgomery, John W. *The Quest for Noah's Ark.* Minneapolis: Bethany House, 1972.

Teeple, Howard. *The Noah's Ark Nonsense*. Evanston, Ill.: Religion and Ethics Institute, 1978.

NOAH'S FLOOD
The Great Deluge

For centuries, Noah's flood was offered as a serious explanation for the existence of all fossil-bearing rocks. Scriptural literalists believe fossils, rather than a complex record of the history of life, were deposited when millions of animals drowned in the 40-day downpour. Some carcasses sank to the bottom (lower strata) while others floated up and landed on higher ground.

Young Charles Darwin—no slouch as a geologist— was delighted to find a giant ground sloth skeleton in South American fossil beds and noted thousands of seashells high and dry on mountainsides. Such observations stimulated his interest in theories of geologic succession, which he tried to discuss with the *Beagle* expedition's commander.

But Captain Robert FitzRoy had a ready explanation. The shells had remained on the mountain when the waters of the worldwide Great Deluge had receded; and gigantic extinct animals had perished because they were too large to fit in the doorway of Noah's Ark and so drowned in the flood.

For decades, scriptural geologists tried to reconcile the record of the rocks with the Great Flood, but the growing mass of evidence just didn't fit. Some parts of the world showed no signs of having been inundated, while others had clearly been submerged not once, but many times. Even the great reconciler of the Bible with geology, Hugh Miller, was forced to conclude in his last book, *The Testimony of the Rocks* (1868), that perhaps Noah's flood was locally confined to the Middle East.

Other problems with the serious consideration of one universal deluge covering the Earth were: (1) the source of such a great volume of water, (2) the kind of fossil layers that actually occur and their order (3) the sheer quantity of fossils, and (4) petrified footprints, dinosaur nests, and rain spatters, which are necessarily formed in surface mud, yet were found beneath layers of rock.

As for the water, a Quaker "Bible-scientist" named Isaac Newton Vail (1840–1912) proposed in 1874 that Earth once had "rings" like Saturn, made of water vapor, which broke up and fell. Geophysicists over the past century have found no evidence to support the idea, which Vail published in his book, *The Waters Above the Firmament* (1886).

If all animals had existed at the same time and their remains had been submerged in a single flood, a hodgepodge of life forms would be mixed together in the rocks, with no orderly pattern of succession. Smaller trilobites would have floated up while heavier ones remained on bottom, along with the large land animals, which would sink; but that's nothing like the actual record of the rocks.

Fossil layers (strata) everywhere show a pattern of succession that is completely incompatible with sorting by water action. Throughout the world, sea creatures and trilobites are in the lower rock strata, fish in the next higher, then amphibians, reptiles, birds and mammals. Nowhere does one find reptiles on bottom and trilobites on top, birds before reptiles, or primates mixed with dinosaurs.

It is rare for an individual animal's remains to become naturally preserved, yet the quantity of fossil skeletons is enormous. Such accumulations must have occurred over immense periods of time. The Karroo formation in South Africa alone contains fossil remains of about 800 billion animals. If all the trillions of fossilized creatures had been alive at the same time, they would have blanketed the entire Earth.

Many older geology books (1800–1860) continued to accept the Great Deluge, even after the consistent patterns in fossil evidence had become well known. When the British Museum's Richard Owen coined the term "dinosaur" and commissioned the great life-size Iguanodon models for the Crystal Palace in 1853, neither he nor his sculptor Benjamin W. Hawkins had evolution in mind. Owen considered them "antediluvian monsters"—perhaps from a "pre-Adamic Creation"—as did the Victorian public, which flocked to see them.

See also ATRAHASIS; CREATIONISM; IGUANODON DINNER; "NOAH'S ARK PROBLEM"; "NOAH'S RAVENS"; PROGRESSIONISM.

For further information:

Custace, Arthur. *The Flood: Local or Global?* Grand Rapids, Mich.: Academic Books, 1979.

Godfrey, Laurie. *Scientists Confront Creationism*. New York: Norton, 1983.

"NOAH'S RAVENS"
Bird-like Tracks in Stone

Reverend Edward Hitchcock, president of Amherst College during the mid-19th century, has a unique place in history. There is no doubt that, beginning in 1835, he conducted the first sustained research on dinosaur evidence in North America. But another man, Joseph Leidy, gets the credit for being first, because Hitchcock had no idea what he was studying; he thought they were giant birds.

When he first became aware of the abundant petrified trackways preserved in Connecticut Valley sandstone, dinosaurs were still virtually unknown to

AMISH FARMERS in Pennsylvania were used to seeing the giant "bird tracks" imprinted in stone, which they assumed were evidence for Noah's Flood. Reverend Charles Hitchcock, the president of Amherst College, devoted his life to studying and collecting them, never suspecting they were the footprints of dinosaurs.

science. The footprints had always been local curiosities, which farmers referred to as "tracks of Noah's ravens."

Fascinated, Hitchcock spent most of his time and energy collecting, studying and describing them for the next 30 years, and, in 1858, published his splendid book on the *Ichnology of New England*. To the day of his death, he believed they were the footprints of large birds. Years later, paleontologists were able to link them with Upper Triassic bipedal dinosaurs.

Joseph Leidy, who found and described a *Hadrosaurus* in 1858, has gone down in history as the first to discover a dinosaur in North America. Paleontologist Edwin H. Colbert acknowledges Hitchcock's priority for his "unknowing research." But the fact remains, Colbert writes, "that Leidy was the first to describe a dinosaur on this continent, *knowing* it to be a dinosaur."

See also BIRD, ROLAND T.; FOSSIL FOOTPRINTS, PALUXY RIVER; LEIDY, JOSEPH.

NOBLE SAVAGE
Ideal of "Natural" Man

People have often been dissatisfied with themselves and their society, wishing to believe in another time and place where folks were pure, unspoiled and "naturally" good. Scientists as well as philosophers have searched eagerly for man in the natural state: Adam and Eve before the Fall at a simpler "stage" of social evolution.

This ideal primitive was an obsession of philosopher Jean-Jacques Rousseau (1712–1778), the most influential and famous admirer of the "noble savage." An acute observer of artfully mannered, parasitic and devious French aristocrats, he concluded it is civilization that corrupts. When people lived in the wilderness, in simple tribes next to nature, he thought, they were straightforward, self-sufficient, democratic and honest.

Rousseau went so far as to insist that everything that distinguishes civilized man from the "natural" is evil, including arts, science, printing and complex institutions of law and government.

But where were these noble savages? With the exploration of the New World, some thought they had found them in the American Indian. Rousseau idealized the Indians as "children of the wilderness . . . nature's noblemen of the Plains," a Romantic notion that inspired European painters and sculptors to depict graceful, muscular superheroes, completely in tune with nature. It was a powerful image, this "natural man," which remains vivid in the collective imagination of Western culture.

Twentieth-century adventure hero "Tarzan of the Apes" is very much Rousseau's noble savage—explicitly stripped of his European title, "Lord Greystoke," as well as his clothes. Reared by apes in an isolated jungle, uncontaminated by "corrupting institutions," he reverts to man's basic goodness and heroic virtue. Tarzan has remained popular for almost a century, incarnated in scores of popular books and movies.

This Romantic ideal of the uncorrupted natural man (and woman) was revived in full force by "hippies" during the 1960s. Like Rousseau, they openly opposed and disobeyed laws they considered unjust, thought the current war a perversion of human nature and gave lip service to a purer, simpler way of life. And, like them, Rousseau had thrown away his watch 200 years earlier, advocated natural foods, long hair for men and public breast-feeding of babies.

But Rousseau also spoke of his state of nature as "a state which exists no longer, perhaps never existed, probably never will exist, of which none the less it is necessary to have just ideas, in order to judge well our present state." Utopian novelists and, later, science fiction writers have continued to explore his premise.

When Rousseau sent his *Discourse on Inequality* (1754) to Voltaire, the elder philosopher sarcastically replied, "One longs, in reading your book, to walk on all fours. But . . . I have lost that habit for more than sixty years . . . Nor can I embark in search of the savages of Canada . . . because the example of our actions has made [them] nearly as bad as ourselves." Nevertheless, Rousseau's ideas became so

popular in the French court, sophisticated ladies [in a vogue to appear more "natural"] began to breast-feed their infants at formal social functions.

Yearning for the noble savage spilled over into science, both in anthropologists' descriptions of tribal peoples and in their imaginative reconstructions of our far-distant ancestors.

There has been much debate, for instance, as to whether the early descriptions of Samoan culture by Margaret Mead in the 1920s were not idealized as a natural paradise free of sexual guilts, conflicts and jealousies. Some decades later, seemingly idyllic descriptions of life among African bushmen or the BaMbuti forest people were called into question by later investigators.

Whatever the resolution of these examples, certainly many anthropologists have favorably compared "their" people with Western civilization. Along with the tribal people's alleged superiority in their "simplicity," it is often presumed that they live with more respect for nature, purposely preserving the land and its creatures from permanent destruction, even as they use it for sustenance.

On close examination, this assumption seems mere wishful thinking. Certainly, some tribal peoples use natural resources with understanding and wisdom, but most simply lacked the technology or population size to create destructive impact comparable with our own. Nevertheless, recent studies show that Polynesian peoples managed to exterminate scores of bird species, before the arrival of Europeans, eating their way through the island's fauna. Many agricultural tribes practiced wanton burning of forests (and still do), while some American Indian groups—contrary to popular belief—were needlessly destructive of habitat.

Oddly enough, when it seemed the idea of the noble savage had just about been abandoned by sadder and wiser anthropologists, animal behaviorists picked it up. Dian Fossey thought her gorillas were "superior" to humans in their gentleness, loyalty and group solidarity. Jane Goodall once said chimpanzee mothers could teach humanity a thing or two about nurturing youngsters with devotion and tolerance. (She was shocked and disillusioned when she later observed chimpanzee "warfare," infanticide, and cannibalism, concluding the apes were "more like humans than I at first supposed.") Recent observers of mongooses and meerkats cite their "devoted sentry duty" and "voluntary babysitting" as "more responsible" than that of most humans.

Early man is often described now in similar terms: sharing food with members of their kin group, caring for cripples, cooperating in hunts, dividing labor between the sexes, living in harmony with their

ROMANTIC TRADITION lingers in this 1870 reconstruction of the paleolithic hunter as a "noble savage," inspired by stone tool discoveries in the French Dordogne region. The clothes, hair and stance are the artist's conception, as is the hafting of the stone tool in a wooden handle.

environment. When a cynical hoaxer offered a contemporary vision of such an idyllic people—the alleged "Tasaday tribe" of the Philippines—the noble savage was hailed once more as a lesson and example for civilized man. Few ever heeded Rousseau's disclaimer that his vividly portrayed natural man "exists no longer, perhaps never existed, probably never will."

See also CHIMPANZEES; SOCIAL BEHAVIOR, EVOLUTION OF; TASADAY TRIBE (HOAX).

NOOSPHERE

See TEIILHARD DE CHARDIN, FATHER PIERRE.

NOPCSA, BARON FRANZ (FRANZ BARON NOPCSA VON FELSO–SZILVAS) (c. 1875–1933)

One of the most bizarre characters in the history of evolutionary science was the Transylvanian nobleman Franz Baron Nopcsa von Felso-Szilvas. While still a young student, Nopcsa [NOPE-sha] contributed a brilliant paper on *Limnosaurus*, a new species of dinosaur he had discovered on his sister's estate. Dinosaurs would become his obsession, second only to his special passion for the land and culture of Albania.

Fascinated by the abundant remains of Cretaceous reptiles in his native Transylvania, Nopcsa discovered and described dozens of new species. Later, he turned to theoretical questions about evolutionary relationships of dinosaur families. His major book on fossil reptiles, published in 1922, established the classification of the major groups upon which all later work in the field was based.

Nopcsa's system was actually a refinement of an earlier analysis by the German paleontologist Friedrich von Huene, who had first proposed two major divisions of dinosaurs: the *Saurischia* (lizard-hipped) and *Ornithischia* (bird-hipped). Greatly influenced by von Huene, the flamboyant, polyglot Nopcsa popularized their collective knowledge of dinosaurs in five languages.

Before World War I, when pockets of feudalism still persisted, Nopcsa was able to live like a baronial lord. As he drove through Hungary, between his various estates, country folk bowed low and men snatched off their caps. (Some years later, in a changed Europe, the imperious Nopcsa was beaten and bloodied with pitchforks by some of these same peasants.)

One historian described Nopcsa's young manhood as "a semiroyal person in a sort of dreamland out of an operetta." Restless, with only his noble rank and scientific studies to occupy him, Nopcsa began having grander ambitions—like becoming king of his beloved Albania.

He had long identified with the Albanian people, and their backward, story-book culture isolated in the Balkan mountains. On frequent, adventurous travels throughout the country, he made excellent ethnographic studies of its language and customs. When Albania was freed from Turkish rule in 1913, the Austro–Hungarian leaders decided the country should have a king who would be friendly to Vienna. Nopcsa confidently applied for the job.

Since he knew the dialects and culture of the country, he argued, he could take it over in short order. All he required from the central powers was a small army of about 500 soldiers with artillery. (He himself guaranteed to provide a couple of steamships and a white horse, on which he intended to ride triumphantly through the streets, arrayed in a splendid uniform.) He also proposed to bolster the country's economy by marrying an American millionairess—after he was king, of course. He seemed genuinely astonished when the Austro–Hungarian leaders turned him down, and instead installed an Austrian prince, who had to flee the country six months later.

Nopcsa became an officer during World War I,

most of the time serving as a spy along the dangerous Hungarian border. When peace came, most of his lands and estates were confiscated, since they were now inside the new borders of Rumania. For awhile, Nopcsa headed the Hungarian Geological Survey, but his arrogant ways antagonized all his colleagues. Disgusted, he gathered his last monies and set off with Bajazid, his devoted male lover and secretary, on a motorcycle trip from Eastern Europe to Italy.

Unable to find a way to recapture the wealth or glories of his youth, Nopcsa sank into depression in his later years, crippled by "a complete breakdown of my nervous system." In April of 1933, he put a pistol to his head and killed himself, but only after first murdering Bajazid while he slept.

Nopcsa believed it was a mercy killing, though he apparently (as usual) made the decision for both of them. "My old friend and secretary . . . [had not] the slightest idea of my deed," wrote Nopcsa in his suicide note. "I did not want to leave him ill, miserable and poor, for further suffering in this world."

See also DINOSAUR.

For further information:
Colbert, Edwin. *Men and Dinosaurs: The Search in Field and Laboratory.* New York: E. P. Dutton, 1968.

"NUTCRACKER" MAN (ZINJANTHROPUS)

See AUSTRALOPITHECUS BOISEI.

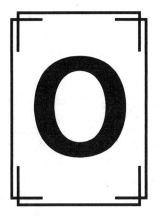

OAKLEY, KENNETH

See FLUORINE ANALYSIS; PILTDOWN MAN (HOAX).

O'BRIEN, WILLIS (1886–1962)
Special Effects Film Pioneer

Restorations of prehistoric animals had begun with huge sculptures of dinosaurs at the Crystal Palace exhibit of 1853 and were carried forward 50 years later by the magnificent paintings of Charles R. Knight. It was inevitable that movies, the dominant art form of the 20th century, would bring the great creatures of the past to life on the screen. But how could extinct animals be made to move and appear real?

Willis O'Brien developed the answer: a special-effects technique called stop-motion animation. Today it is used in all kinds of fantasy pictures, including the George Lucas *Star Wars* epics, to create spaceships and robots. But the development of stop-motion owes its existence to O'Brien's obsession with re-creating scenes of the prehistoric Earth.

All his early efforts were devoted to giving the moviegoer a glimpse of a vanished world populated by tyrannosaurs, flying pterosaurs and other creatures known to man only as fossil bones.

O'Brien's technique seems simple today, but when he started working on it in 1910, the process was unknown and revolutionary. Miniature models about a foot high are made of the animals in clay, and these are then cast in flexible rubber. At the core of the finished model is a metal armature construction with ball sockets for joints.

Stop-motion animation is a slow, painstaking process. Each joint is moved a fraction of an inch by hand and the scene photographed. When the succession of individual frames is projected, the illusion of movement is achieved.

O'Brien first developed his art for a picture called *Creation*, which was to depict scenes from the grand pageant of the evolution of life on Earth. He com-

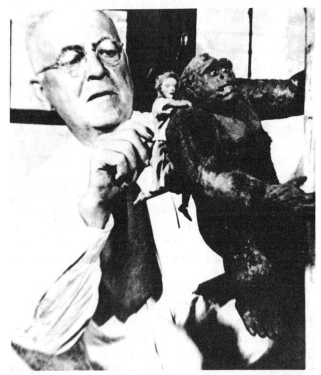

STOP-MOTION ARTIST Willis O'Brien never appeared before the movie cameras, but he was actually the "star" of *King Kong* and *Mighty Joe Young,* imparting movements and personality to the miniature "giant apes" and dinosaurs. O'Brien's career grew out of a lifelong interest in creating animated images of extinct animals and their prehistoric world.

335

pleted several scenes but work on the film was discontinued. However, his work had impressed the studios, and they hired him to animate Sir Arthur Conan Doyle's novel *The Lost World* (1912), the story of a scientific expedition's adventures in an isolated jungle still populated by dinosaurs and ape-men.

After years of perfecting his technique, O'Brien was hired by Merian C. Cooper to create the scenes for what was to become the masterpiece of stop-motion: the original *King Kong* (1933). Among the first scenes filmed for *Kong* were O'Brien's animations of Kong's fight with a pteranodon and the attack of a giant dinosaur on the expedition's raft.

Among its innovations was the combination of live actors with the miniature models to create exciting multiple images. Today, such effects are achieved with great precision through the use of computers. Willis O'Brien did it all by hand.

See also KING KONG; LOST WORLD.

OCCAM'S RAZOR
Principle of Parsimony

William of Occam (c. 1295–1350) was a renowned scholastic philosopher, second in influence only to St. Thomas Aquinas. Since there is no reliable account of his life, it is even uncertain whether he came from Ockham in Surrey, or Ockham in Yorkshire. He was often embroiled in political controversies and once had to be rescued from the pope's wrath by the German emperor.

He is best remembered in science for the principle that became known as "Occam's Razor": "Entities are not to be multiplied without necessity." In other words, a well-constructed theory about nature is the simplest possible explanation consistent with the facts. The "razor" shaves off any unnecessary flourishes or complications.

Its usual formulation, as given above, is found nowhere in Occam's writings. Philosopher-historian Bertrand Russell supplies the actual version: "It is vain to do with more what can be done with fewer." Or as modern minimalist designers would say, "Less is more."

Applied to geology and paleontology over the years, "Occam's razor" has raised (and resolved) key questions: Why postulate that God continually intervened in changes of climate, variations in species, floods, etc., when it is simpler to assume establishment of uniform laws (proximate causes) that operate to produce all subsequent events? Why assume God made fossils of creatures that never existed in order to test man's faith, when there is the much simpler explanation that they are the remains of animals that once lived?

Why imbue life with a "vital force" or "innate drive towards progress," when such "multiplied entities" are neither demonstrable nor necessary to account for evolution?

And why assume (as Charles Darwin did) that patterns of apparently long stability and abrupt change are functions of "imperfections" in the fossil record, which may someday be corrected by new discoveries? The simpler explanation is that these widespread geological patterns reflect actual events (punctuated equilibrium).

In 1966, Maurice G. Kendall proposed an update of Occam's principle specifically as applied to theories about how the mind works. "We should not invoke any entities or forces to explain mental phenomena," he wrote, "if we can achieve an explanation in terms of a possible electronic computer." Kendall called this special application "a kind of Occam's Electric Razor."

OKEN, LORENZ

See NATURPHILOSOPHIE.

"OLDEST MAN"
Quest for Earliest Humans

Although Louis Leakey enjoyed turning up fossil apes, his greatest desire was to discover the "Oldest Man." To the end of his life, he believed the human lineage was far more ancient than anyone thought, and that australopithecines were all sidebranches.

Leakey's first "Oldest Man" was a skeleton discovered at Olduvai Gorge in 1913 by the German geologist, Hans Reck. The bones were dug from very old deposits, which also contained fossils of extinct animals. After Leakey and Reck published papers proclaiming "Olduvai man" the "oldest" known human, later work proved the skeleton belonged to a recent African, whose grave had been dug into ancient deposits—an "intrusive burial."

Undaunted, Leakey continued to search for the earliest recognizable humans and titled one of his books *Adam's Ancestors* (1934). His belief in the great antiquity of a distinctive human line established a new fashion in paleoanthropology: a quest for the "Oldest Man" similar to the search for the "Missing Link" 70 years earlier. It has carried into the next generation in the debates between Richard Leakey and Donald Johanson over who found the earliest ancestors of the genus *Homo*.

In 1931, the elder Leakey found a fossil mandible at a site called Kanam; he announced this bone as "The world's earliest *Homo sapiens*." This was Louis Leakey's second "Oldest Man." Despite initial scientific interest, the material proved problematical. The site had been accidentally disturbed, and the fossils could not be dated.

In 1961, the Leakey family found skull fragments of an apparently unknown species, which was named *Homo habilis* ("handy man") because Louis believed it had made the very early stone tools found nearby. Since the fossil was contemporary with australopithecines, he announced that *Homo habilis* was the earliest known human ancestor, a separate line not descended from the australopithecines and Leakey's third "Oldest Man."

A decade later, when Richard Leakey's team found the 1470 skull (assigned by him to *Homo habilis*), young Leakey announced that it represented the "earliest suggestion of the genus *Homo*." Louis Leakey—who was ailing and had been estranged from his son for years—was excited by the find. He saw Richard's discovery as a confirmation of the antiquity of true man, vindication of his discredited Kanam jaw and evidence that *Australopithecus* was a hominid sidebranch, not a human ancestor at all. Father and son reconciled, agreeing that the "habilines"—not australopithecines—were the ancestors of humans.

But the Leakey family's battle to establish *habilis* as the "Oldest Man" was just beginning. First there was a prolonged controversy over the age of the 1470 deposits; geologists, paleontologists and geochemists disputed the original estimate determined by geochronologists working with Richard Leakey. (It was a million years younger, they concluded.) Although Leakey stubbornly maintained that his original, earlier dates were correct, he eventually had to back down and accept the evidence.

Next, there were the spectacular finds of Donald Johanson's team in the Afar region of Ethiopia of upright-walking australopithecines. Johanson claimed they were the earliest known human ancestors and even nicknamed a group of skeletons "The First Family." Controversy continues, but seems as incapable of resolution as anthropology's former preoccupation with the "Missing Link." It turned out to have no answer, because it was the wrong kind of question.

Commenting on Leakey's initial recalcitrance in the 1470 dating controversy, historian of science John Reader notes that "modern paleo-anthropologists are no less likely to cling to erroneous data that support their preconceptions than were earlier investigators. Dubois and the 'Missing Link,' Leakey and the 'Oldest Man'—both dismissed objective assessment in favour of the notions they wanted to believe."

See also AFAR HOMINIDS; HOMO HABILIS; LEAKEY, LOUIS; LEAKEY, RICHARD; "LUCY"; "MISSING LINK."

OLDUVAI FEMALE
Female "habiline" OH62

A national science magazine named her "Woman of the Year" in 1987, but the honoree didn't care—she had been dead for about 1.8 million years. Found by Tim White and Don Johanson in May, 1987 at Olduvai Gorge, Tanzania, the fragmentary skeleton was labeled "OH62," for the 62nd Olduvai hominid to be discovered.

Most of the 302 fossil fragments are from skull, teeth, thigh, arm and shin bones. They are consistent with a female about 30 years old who stood about three feet tall. White and Johanson think she belongs to the species *Homo habilis*, the rather large-brained upright hominid some believe is our earliest ancestor and the first maker of stone tools.

Although human-like in posture and enlarged brain, OH62 appears more "ape-like" in having very long arms (estimated from fragments), reaching down almost to her knees. The upper arm is 95% as long as

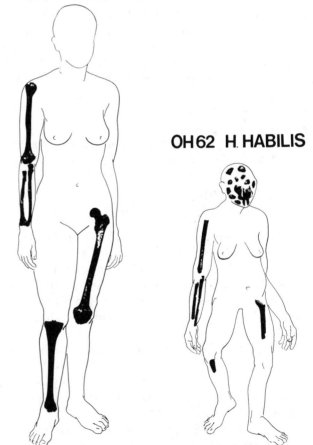

COMPARISON OF OLDUVAI FEMALE leg bone (femur) with that of modern woman shows early hominids were quite small, about three and a half feet tall. According to this reconstruction by Donald Johanson, their arms were proportionately longer than *Homo sapiens*, with hands dangling almost to the knee.

the thighbone, while in modern humans, it is only 70–75% as long.

Johanson was struck with the resemblance of OH62 to his famous fossil "Lucy," the *Australopithecus afarensis* he found in Ethiopia. (Lucy is estimated at three million years, older than OH62 by 1.2 million years.) He and White also agree that she looks very little like *Homo erectus*, the presumed human ancestor into which she may have evolved a scant 200,000 years later. If *Homo erectus* and OH62 are both on the human line, there was "an abrupt transition," says White.

To explain her similarity to much older forms and lack of resemblance to the more recent *erectus*, they invoke the model of punctuated equilibrium—long evolutionary stability, followed by a period of swift change. Until more samples are found, however, it is equally plausible to suppose that these species arose as diverging branches and represent distant cousins, rather than a single line leading to modern humans.

This Olduvai fossil is the subject of Johanson's book—*Lucy's Child* (1989), a sequel to his bestseller *Lucy: The Beginnings of Humankind* (1981). Johanson's critics think he is straining to connect the two discoveries. OH62 was compared not to similar discoveries of *Homo habilis* from the same time frame, but to another species and another genus, a full million years earlier.

See also HOMO HABILIS; "LUCY"; "OLDEST MAN."

For further information:
Johanson, Donald, and Shreeve, James. *Lucy's Child.* New York: Wm. Morrow, 1989.

OLDUVAI GORGE
Early Man Site Extraordinaire

Olduvai Gorge lies along the Great Rift Valley, a geological discontinuity that runs for hundreds of miles through East Africa and north to Ethiopia. It is stark, arid country that until recently was remote and inaccessible. When Louis Leakey began visiting sites there in 1931, he had to haul in his own water by Land Rover.

The gorge exposes an ancient lake basin in Tanzania. Its 100 meter cliffs expose deposits covering a timespan from over two million to 10,000 years ago. Louis and Mary Leakey worked various sites at Olduvai over a period of 40 years.

Their painstaking, pioneering work has changed the face of prehistory and made numerous contributions to the study of human evolution. Olduvai Gorge has yielded fossils of several types of ancient hominid, including *Australopithecus*, *Homo habilis*, and *Homo erectus*. In addition, Louis believed they had

discovered "living sites" containing assemblages of stone tools from different time depths, but the evidence is ambiguous. At Laetoli, 30 miles south of Olduvai Gorge, Mary Leakey found a remarkable series of footprints (1974) fossilized in volcanic ash and dated at 3.6 million years old. Teams of experts in geology, dating, ancient climates, fossil mammals and other fields have worked with the Leakeys to round out the picture of early environments at Olduvai. In 1987, Don Johanson and Tom White found their "female hominid" (OH62) at Olduvai. As fossils continually weather out of the cliffs, new discoveries are being made at this African treasure trove of human prehistory.

See also LEAKEY, LOUIS; LEAKEY, MARY; LEAKEY, RICHARD; OLDUVAI FEMALE.

For further information:
Leakey, Mary. *Olduvai Gorge: My Search for Early Man.* London: Collins, 1979.

OLD WORLD MONKEYS

See CATARRHINES.

"OMEGA MAN"
Future Evolved Being

Father Pierre Teilhard de Chardin, the Jesuit biologist, was a churchman who looked back to the dawn of prehistory and forward to the next step in the evolution of mankind. Although he made many contributions to the study of Pleistocene mammals, early man and paleolithic archaeology, Teilhard's lifelong quest was to reconcile his beliefs in Christianity and evolution and to project their final synthesis. Christianity was the hope of making men better and evolution was the means.

Teilhard believed there is a divine plan for Earth: Human destiny was to break out of the biosphere and into the *noosphere*, the realm of a higher thought, consciousness and self-knowledge, "When for the first time in a living creature instinct perceived itself in its own mirror," he wrote, "the whole world took a pace forward." The next breakthrough, he thought, would be the conquering of this new dimension by "Omega Man," a future being who would surpass *Homo sapiens* both intellectually and spiritually.

Omega Man is quite unlike Nietzsche's self-contained and aloof "Superman," who evolves through ruthless competition and triumph of will. Instead, Teilhard projected Christianity's ideal of brotherly love, a mystical union of individuals in species-wide cooperation, for wise and loving stewardship of the Earth. Omega Man is an expression of cosmic optimism by a hopeful philosopher.

See also TELEOLOGY; TEILHARD DE CHARDIN, FATHER PIERRE.

For further information:
Teilhard de Chardin, Pierre. *The Phenomenon of Man.* New York: Harper & Row, 1959.

OMPHALOS (1857)
Creationist book by Philip Gosse

In the opinion of even his most sympathetic friends, *Omphalos* (1857), Philip Gosse's unique defense of creationism, was a bizarre failure. A Victorian author of popular books on birds, insects and ocean life, Gosse (1810–1888) was a good naturalist who had exchanged informative letters with Charles Darwin on natural history subjects. But he was increasingly distressed by what he considered the conflict between evolutionary ideas in natural science and the biblical account of creation.

Gosse sought a theory that would reconcile the evidence of Earth's past with a literal interpretation of Genesis. The result was *Omphalos* (Greek for "navel"), which his son Edmund called "this curious, this obstinate, this fanatical volume." His theory was that fossils—like Adam's navel—were evidence of a natural birth that never occurred.

This paradox was explained by Gosse's strange idea of "prochronism." Since nature is an ongoing, cyclical process, he argued, the act of creation had to start *somewhere* in the cycle. A chicken implies an egg in its past, just as an egg implies it was laid by a chicken. If God had created a chicken in an instant, we could imagine the egg from which it hatched, but that egg would exist outside time (a "prochronic" egg).

Fossils and other geologic evidence of Earth's past, Gosse argued, are prochronic. For example, just as Adam was created with a navel, there were growth rings on the trees in the Garden of Eden. They were brought into being "complete" by the creator, so they could be part of the ongoing natural process—evidence left by God to suggest a past that never existed. Thus, Gosse could accept that strata and fossils exist, yet still insist that the Earth was actually created in six days, only 4,004 years before Christ.

In short, Gosse was actually maintaining that God had put fossils in the ground to fool geologists! "Who will dare to say," he asked, "that such a suggestion is a self-evident absurdity?"

The answer to that was "everyone," friends as well as foes. The Reverend Charles Kingsley, himself a naturalist who wanted to believe in Genesis, was horrified. He simply could not give up what he knew of geology, he wrote Gosse, to "believe that God has written on the rocks one enormous and superfluous lie." Later, Kingsley publicly and sadly called *Omphalos* a "desperate" attempt:

> . . . more likely to make infidels than to cure them. For what rational man, who knows even a little of geology, will not be tempted to say, "If Scripture can only be vindicated by such an outrage to common sense and fact, then I will give up Scripture, and stand by common sense."

After the dismal reception given his magnum opus, Gosse retreated from the world, a bitterly disappointed man. After years of being the public's darling for his natural history books, his son related, "He could not recover from amazement at having offended everybody by an enterprise which had been undertaken in the cause of universal reconciliation."

See also CREATIONISM; "SCIENTIFIC CREATIONISM"; "TWO BOOKS," DOCTRINE OF.

ONTOGENY RECAPITULATES PHYLOGENY

See BIOGENETIC LAW; FREUD, SIGMUND.

OPTIMAL FORAGING THEORY (OFT)
Controversial Efficiency Model

A current evolutionary controversy revolves about whether animals' "fitness" can be measured by how efficiently they capture energy. If a seed-eating species has survived for eons, the argument goes, they must be maximally efficient—outcompeting others by finding the most highly nutritious seeds while expending the least energy to get them.

John R. Krebs of Oxford and David W. Stephens of Amherst apply mathematical optimality theory to living things in their book *Foraging Theory* (1987), arguing that natural selection has made animals "optimally" efficient. Optimal Foraging Theory (OFT) measures "fitness" in terms of "the average rate of energy gain per unit of foraging time," assuming that better foragers will leave more offspring.

Field observers of animal behavior have noticed that they often act as if they were trying to optimize available energy. Songbirds abandon a waning food source when "diminishing returns" set in, and they are spending more energy than they are taking in. Other species commonly seek higher energy foods when given a choice between various items.

Krebs proposed the mathematical model in the belief it would provide numerical values for the effects of natural selection. "You only have to look around at things in nature, even bits of your own body," he says, "to see how incredibly beautifully designed they are to do the job they do."

To critics, notably Stephen Jay Gould and Richard C. Lewontin of Harvard, the premise smacks strongly of 19th-century teleology: The notion that nature contains "perfection," as if creatures were "designed" to be the best possible foraging machines they could be. In fact, they argue, animal behavior is shaped by many factors other than efficiency in finding food: for example, defense against predators, courtship displays, or random accidents of history. Moreover, an "optimum" situation implies some kind of final condition, whereas animals may still be evolving or adapting to recently changed environments.

The debate continues. Krebs has backtracked somewhat from what he originally supposed was "a law of nature." Lewontin finds concrete demonstrations of a predictive optimality model elusive, and several ecologists have publicly condemned it as "A complete waste of time."

But the fate of the theory will ultimately depend on whether simple models give good results for a range of species or whether each concrete instance becomes a "special case" for which the theory must be adjusted. The OFT controversy demonstrates once again the difficulty of applying neat mathematical models to the behavior of living things, which are the product of unique histories, and are still evolving and adapting to changing environments.

See also ECOLOGY; FITNESS; RED QUEEN HYPOTHESIS; TELEOLOGY.

ORANG–UTAN
The Red Ape

Two-hundred years ago, it was the red-haired Asian ape, the orang-utan, and not the African chimpanzee, that was considered man's closest kin. During the 18th century, both Scots philosopher Lord Monboddo and the French naturalist Georges Buffon had marveled at the orang's approximation to human form, "lacking only in the art of speech." Samuel Taylor Coleridge, the poet, had accused evolutionist Erasmus Darwin (Charles's grandfather) of espousing "an Orang-outan theology." A generation later, geologist Sir Charles Lyell told Charles Darwin himself he simply couldn't "go the whole Orang" in accepting his theory of evolution.

One of the reasons Alfred Russel Wallace, Darwin's partner in founding evolutionary biology, went to Malaysia was in hopes of studying the orang and also finding fossil clues to early man. Many then believed Asia was the cradle of humanity, and Wallace was the first European to study a species of ape—the orang—in the wild.

Orangs figured prominently in the imagination of early-19th-century writers, as in this extraordinary passage from Henry Hallam's *Literature of Europe* (1818): "Every link in the long chain of creation does not pass by easy transition into the next. There are necessary chasms, and, as it were, leaps from one creature to another . . . If Man was made in the image of God, he was also made in the image of an ape. The framework of the body of Him who has weighed the stars and made the lightning his slave, approaches to that of a speechless brute, who wanders in the forests of Sumatra. Thus standing on the frontier land between animal and angelic natures, what wonder that he should partake of both!"

Orang-utans in Malaysia today are dangerously close to extinction, for most of their original forest habitat has been cut down. They are unique among apes and monkeys for their solitary habits; orangs in the wild do not congregate in social groups, but travel alone through the trees, occasionally coming down and walking on the forest floor.

They have never been observed to use tools spontaneously in the wild, but those reared in field stations take great interest in such human gadgets as latches or water pumps, which they easily master. A captive orang-utan in the Basel zoo is the only ape that has learned to make stone knives by chipping with another stone, when the need arises for a sharp blade. [See ABANG.]

When orphans are gathered in "ape refugee camps," orangs seek the company of their fellows and seem to hate to return to the forest. Watching films of their misery at being turned out to forage after being "spoiled" by social life makes one wonder whether the "solitary" behavior is basic to the species, or was a secondary necessity since the Malaysian forests cannot support great concentrations of these large apes in one area.

Anatomical comparisons, along with DNA and serum albumin tests, indicate the orang-utan is much more distantly related to man than are the African apes and probably split off and went its own way millions of years before chimps and gorillas diverged.

"Orang-utan" is a Dyak word meaning "old man of the woods"; tribal peoples of Malaysia believe orangs can speak, but never do while people are around, for fear they will be put to work. When Alfred Russel Wallace shot, skinned and prepared the skeletons of several orangutans for shipment to the British Museum (Natural History), the local tribesmen were terrified Wallace would go after their heads and skins next.

See also: APES; BUFFON, COMTE GEORGES; MONBODDO, LORD; WALLACE, ALFRED RUSSEL.

For further information:

De Boer, L. E., ed. *The Orang-utan: Its Biology and Conservation.* The Hague: W. Junk, 1982.

Peterson, Dale. *The Deluge and the Ark: A Journey into Primate Worlds.* Boston: Houghton Mifflin, 1989.

ORCHIDS, DARWIN'S STUDY OF
Analyzing Adaptations

During the mid-1800s, an amazing craze for orchids swept England. Collectors braved tropical fever, snakebite and hostile Indians to supply the greenhouses of gentlemen and ladies back home. (No one yet knew how to grow orchids from seeds.) In the midst of this popular mania, Charles Darwin chose orchids as the subject of his first book after the *Origin of Species* (1859). Published in 1862, *The Various Contrivances by which Orchids are Fertilised by Insects* was revolutionary, though it is almost unread today.

Orchids is devoted to dissecting flowers of many species, exploring their structure and understanding the evolution of their special adaptations. Although most of the botanical details are technical, its major arguments are still crucial to understanding Darwin's thought and continuing influence.

Darwin's approach completely changed existing ideas about plant sexuality, while breaking new ground for understanding coevolution and pollination ecology. Previously (with the obscure exception of Christian Sprengel), naturalists thought of flowers as beautiful, pleasing, bizarre or useful to man—examples of divine artistry. Darwin set out to show that the strange and intricate structures "beyond what the most fertile imagination of man would invent" were the result of sex and history. Structures were analyzed strictly as adaptations to the pollinating insects that evolved along with them.

In Darwin's words, "Nature tells us, in the most emphatic manner, that she abhors perpetual self-fertilisation." From field reports, experiments in his greenhouse and data from botanist friends, he found "an almost universal law of nature that the higher organic beings require an occasional cross with another individual . . . no hermaphrodite fertilises itself for a perpetuity of generations." Since pollen from flowers on the same orchid plant was ineffectual, he demonstrated that "varied and beneficial . . . contrivances have [as their] main object the fertilisation of the flowers with pollen brought by insects from a distant plant."

This idea that flowers evolved to attract insects to aid them in sexual reproduction was still new and somewhat shocking. [See DIVINE BENEFICENCE, IDEA OF.] Darwin documented case after case where flowers evolved in tandem with particular kinds of insects, in some cases to lure them inside a pollen chamber, in others to provide landing platforms. Following his lead a century later, even more amazing adaptations have been discovered. Some orchids give off the sexual smells of female wasps to attract the males, others even mimic the body of the female

wasp. Fooled into copulating with the flower, the wasp rolls in its pollen, which he will unwittingly carry to another flower.

So convinced was Darwin that the structures of flowers had coevolved with insects that he even predicted the discovery of a bizarre moth that no one believed could exist. The beautiful white Christmas Star orchids of Madagascar sport foot-long structures containing nectar only at the bottom. "What can be the use," Darwin wondered, "of a nectary of such disproportionate length?" Then, with Sherlock Holmesian logic, he concluded "in Madagascar there must be moths with proboscides capable of extension to a length of between ten and eleven inches! This belief of mine has been ridiculed by some entomologists . . ."

Forty years later, a night-flying moth with a 12-inch coiled tongue was discovered on the island. It was named *Xanthopan morgani praedicta*—Morgan's yellowish moth that was predicted! No one has observed it fertilizing the orchid, which is not strange since the moth comes out only after dark. But, in the lyrical words of Luis Marden, "the orchid is there, the moth is there, and doubtless they find each other in the soft tropical night of the Great Red Island."

As for his approach to sorting out adaptations, Darwin explained his views in this passage from *Orchids*:

> Although an organ may not have been originally formed for some special purpose, if it now serves for this end, we are justified in saying that it is specially adapted for it. On the same principle, if a man were to make a machine for some special purpose, but were to use old wheels, springs, and pulleys, only slightly altered, the whole machine, with all its parts, might be said to be specially contrived for its present purpose. Thus throughout nature almost every part of each living being has probably served in a slightly modified condition for diverse puposes, and has acted in the living machinery of many ancient and distinct specific forms.

With this approach, Darwin veered away from the natural theologian's concept of "perfect adaptation" by a "Designer." [See PALEY'S WATCHMAKER.] Adaptations are not perfect; they are often demonstrably makeshift. It was much more fruitful to focus on the "contrivances" and contraptions, evidence of made-over parts showing the pathways of an organism's specific, unique history. [See PANDA'S THUMB.]

Darwin would have been amazed to see the great strides made in orchid raising during the past 100 years. About 18,000 wild species and 35,000 recorded hybrid crosses are now known. (No hybrids were known until 1856, when the first "orchaceous mule"—to use the wonderful Victorian term—was produced.)

Growing them from seeds is easy now, since it was discovered some years ago that they will sprout in an agar-sugar mixture. In the wild, sugar is provided by a symbiotic fungus, a fact that was unknown to the Victorians.

Orchid commerce has been recently revolutionized by meristem culture, a type of cloning developed by frenchman Georges Morel in 1956. By growing bits of apical cells (from the tip of new shoots) in rich nutrients, an indefinite number of "copies" of an orchid can be produced. These plants have not "reproduced" in the normal sense—numbers of identical blooms are really all part of one genetic individual.

Although most of us imagine them as sensuous tropical exotics, there are plenty of tough, drab little orchids in the cooler climes. One of Darwin's hopes was that his book would stimulate interest in the homely little orchids of England; he derived great pleasure from an "Orchis Bank" of local species that grew near his home in rural Kent. During the late 1960s, Dr. Kenneth Marsh and Joan Marsh—who currently occupies Darwin's old magistrate's bench in the Bromley court—headed a feisty effort to save Darwin's orchid bank from developers. Alas, without success.

See also COEVOLUTION; ORIGIN OF SPECIES.

For further information:

Allan, Mea. *Darwin and His Flowers.* New York: Taplinger, 1977.

Darwin, Charles. *On the Various Contrivances by which Orchids are Fertilized by Insects.* London: Murray, 1862.

Scourse, Nicolette. *The Victorians and Their Flowers.* London: Croon, Helm, 1983.

OREOPITHECUS
Mysterious "Swamp Ape"

One of the most puzzling primate fossils, found in 12-million-year-olds beds of brown coal in northern Italy, is a four-foot-high creature called *Oreopithecus*. Its face is short and flat, its teeth are like those of early monkeys and its pelvis and feet are man-like. It has been variously thought to be an ancestral gibbon, an ape-like monkey or a relative of man. At present, it is placed in a separate family within the hominoids.

Part of the Miocene primate radiation, the first specimens were found in 1870, and many parts of skeletons have since come to light—all from the Italian coal beds. A Swiss fossil hunter named Johannes Hurzeler became fascinated with the creature and raised funds to reopen a flooded Tuscany coal mine in 1964 to search for better fossils. Just before he ran out of funds, he discovered an almost complete skel-

eton of the creature in the ceiling of the pumped-out mine.

Despite the monkey-like teeth, it seems to have been much closer to developing a bipedal gait than any known apes or monkeys. Some anthropologists believe it developed upright posture parallel and independent of the human line and may have been outcompeted by our own ancestors.

Geologists conclude that the fossil-bearing coal was formed from swamp vegetation, and it is a mystery why so many *Oreopithecus* individuals should be associated with the swamp habitat. Did they fall from trees while trying to leap over swampy areas? Or did they find an important food item in swamps? There is no good explanation for what attracted so many of these ancient primates to the often-fatal swamps.

ORIGIN MYTHS (IN SCIENCE)
Evolutionists as Story-Tellers

Origin myths are stories people tell about how they came to be who they are. The Zuni Indians, a farming people, claim to have emerged from a mystical hole in Mother Earth, establishing their special kinship with the land. A tale from the nomadic Plains Indians explains that Father Sun's helper created men from clay along with animals. But since humans were not as well designed to get food as other creatures, they became perpetually dissatisfied wanderers. [See NAPI.] Unfortunately, when scientists put forth their best guesses about the origin of our species, they, too, become myth-makers, however unintentional.

While a student at Yale University in the early 1980s, anthropologist Misia Landau studied some of the older descriptions of human evolution written by such 19th-century evolutionists as Henry Fairfield Osborn, Sir Arthur Keith and Elliot Smith. Analyzing the passages as literature rather than for scientific content, she found they almost always took the form of a classic European "hero" myth, with its typically recurring elements.

The epic usually begins with a hero (a tree-dwelling ape, man-ape, early hominid, etc.) who lives contentedly in a stable environment (the trees, the forest), but is expelled from his happy home (climatic change, retreat of forest) and is forced to set off on a dangerous journey. (What women were doing during this saga of "man" now seems conspicuously absent.) Then he must overcome a series of difficult challenges (new conditions, competing species) and emerge triumphant (develop language, tool use, intelligence). Often, he undergoes still more hardships (Ice Age), but finally emerges erect and victorious. However, as in classic myths, the scientist-narrator typically warns that if this hard-won success results in arrogance, the inevitable outcome will be self-destruction.

Landau's professor, paleontologist David Pilbeam, was quick to agree that "our theories about human origins have often said far more about the theorists than they have about what actually happened . . . relatively unconstrained by fossil data." Pilbeam should know. It was he who, with Elwyn Simons, built a reputation by elevating some very shaky fossil evidence into a celebrated human "ancestor" in the *Ramapithecus* debacle, then spun elaborate stories about the alleged creature's social organization and tool-using ability. [See RAMAPITHECUS.]

Roger Lewin, in his book *Bones of Contention* (1987), cites another example: "The emphatic shift in theoretical stance between the 1950s and 1960s, when the specter of Man the Hunter, Man the Killer Ape dominated paleoanthropology, and the 1970s and 1980s, when peace and cooperation were stressed instead, with the emergence of Man the Social Animal." In the intervening years, Lewin points out, there was no new fossil evidence to support such a dramatic shift in reconstructions of human origins. The change, says Lewin, reflected "a shift in current social attitudes away from a time when war was an acceptable instrument of international policy."

Assumptions about the abilities and attributes of men and women and the traditionally subordinate status of women, has also influenced science's "origin myths." For one thing, woman was all but invisible in the male-dominated accounts of the origin of "man." Nineteenth-century male scientists depicted aggressive males who were mighty hunters; females were attracted to the strongest dominant males, who in turn selected the most pleasing mates. In primate studies, the troop was believed to be centered around domineering males, who competed for status with other males, while females vied for the males' protection. This view, already present in Charles Darwin's *Descent of Man* (1871) persisted for a century and was subsequently adopted by such writers as Sherwood Washburn, Robert Ardrey, Raymond Dart and Konrad Lorenz.

In the 1970s, with the influx of women into anthropology (and, perhaps not coincidentally, with the rise of the feminist movement), a new theme sounded in reconstructions of early humankind. Adrienne Zihlman of the University of California, Santa Cruz, wrote of "Woman the Gatherer" as the economic mainstay of primitive society, providing the everyday roots, grubs and vegetable foods, which were only occasionally supplemented by the men's hunting activities. Prehistorian Glynn Isaac, that ami-

able master of compromise between antagonistic viewpoints, declared he read the evidence of early hominid sites as indicating "food sharing" at a home base, with both sexes contributing equally to the larder.

David Pilbeam has adopted an epigram, which sums up what he has learned about scientists' pronouncements on the behavior and intelligence of fossil man (and woman):

> We do not see things the way they are;
> We see them the way *we* are.

He used to tell his students at Yale this trenchant quotation comes from the *Talmud*, but with his renewed devotion to truth, Pilbeam now admits he found it in a Chinese fortune cookie.

See also BABOONS; KILLER APE THEORY; PILTDOWN MAN (HOAX).

ORIGIN OF SPECIES (1859)
Epochal Essay on Evolution

One of the great classics of science, Charles Darwin's *On the Origin of Species by Means of Natural Selection, or the Preservation of Favoured Races in the Struggle for Life* was published on November 24, 1859; it is a dense, difficult book, which even Darwin's best friends found rough going. Uncertain of its appeal to nonspecialists, he offered to cover his publisher's losses. To his amazement, however, the entire first printing of 1,250 copies was snapped up by booksellers the first day. The *Origin* continued selling through six revised editions over the next 17 years and has remained in print ever since.

Origin of Species is one long demonstration that the best way to account for thousands of facts about the world of life is by "descent with modification"—or what later came to be known as "evolution." (The word appears nowhere in the first edition; Darwin used it first in his *Descent of Man* (1871), and in the 6th edition of the *Origin* (1872). However, "evolved" is the last word in all editions of the *Origin*.)

Similarities and differences between animal species, Darwin showed, pointed to common origins rather than separate, independent creations. He did not originate that idea (it had been stated before by, among others, the French biologist Jean-Baptiste Lamarck, and his own grandfather Erasmus Darwin, but Charles had patiently marshaled the evidence that convinced the scientific world that evolution is a fact. Having done that, he went on to propose natural selection as its mechanism. The *Origin* sparked a revolution in thought and kindled intense controversies; in the words of Thomas Huxley, it was considered a "decidedly dangerous book by old ladies of both sexes."

Previous evolutionary theories depended, in part, on destinies, goals toward which organisms must irresistibly develop, or untestable vital forces. Darwin approached living things as natural phenomena, shaped by natural causes that could be probed by experimental research and systematic field observations. For instance, Darwin doesn't merely speculate that certain plants might have reached far-flung oceanic islands as floating seeds; he soaks many kinds of seeds in barrels of salt water for months, then plants them to see which species will sprout. He does not merely speculate about whether crossing different varieties of domestic pigeons will improve the stock—he performs the breeding experiments. He invents a whole new science, then shows how to apply it.

Building his case in a disarming, indirect manner, Darwin brings the reader along almost in spite of himself. He constantly raises possible criticisms and objections, readily admits there are many difficulties, appears to yearn for an alternate explanation. Yet he seems to be inexorably led by the evidence to his stated views, despite his best efforts to resist.

So convincing were Darwin's diffident arguments for evolution and natural selection that one contemporary reader, who was skeptical at first, wrote a friend: "I don't know what's so brilliant about your Mr. Darwin. If I had access to all the facts he does, I'm sure I would have come to exactly the same conclusions."

Having established evolution as the most plausible explanation for the facts of geology, paleontology, comparative anatomy, embryology and other specialties, Darwin then proposes a mechanism for the way it works. Natural selection is a two-step process: (1) overproduction and variation within a species, and (2) greater survival and reproduction of those individuals with any slight advantage over their fellows; "fitter" traits are preserved and accumulated in successive generations. "Multiply, vary, let the strongest live [and reproduce] and the weakest die [leaving few progeny]."

In addition, the *Origin* is also a founding work on the interpretation of fossils, distribution of plants and animals (biogeography), taxonomy or classification, comparative morphology and many other fields of modern biology. However, it skirts the question of human evolution, which Darwin tackled a dozen years later, in the *Descent of Man*. Here, he allowed

AN INTELLECTUAL BOMBSHELL, *The Origin of Species* by Charles Darwin appeared in 1859, immediately sold out its first edition and has remained continuously in print ever since. Darwin's study at Down House, where the epochal work was written, is still visited by thousands every year. Portraits of botanist Joseph Hooker, geologist Sir Charles Lyell and other scientists helped inspire him.

himself only the single remark that "Light will be thrown on the origin of man and his history."

Darwin had been incubating the major ideas in the *Origin of Species* for about 21 years. He had opened his first "transmutation" notebook in 1837, a year after returning from the *Beagle* voyage, and continued recording ideas on the subject for the next several years while working on other projects. He wrote a 13-page "pencil sketch" of natural selection in 1842 and a much expanded summary (231 pages) in 1844; both were later submitted to the Linnean Society in 1858, to be published along with Alfred Russel Wallace's independent invention of the theory.

Until Wallace forced Darwin's hand by mailing him the "Ternate" essay ("On the Tendency of Species to Depart Indefinitely from the Original Type," 1858) from the Malaysian jungle, Darwin had intended to set forth the theory in a massive "Species Book," bulging with overwhelming quantities of relevant facts. Although he had made some headway and was a strongly convinced evolutionist and natural selectionist by the 1840s, Darwin kept on collecting data for another 15 years. Then, springing into action to avoid being scooped by Wallace, he dashed off the 155,000-word *Origin* in only 13 months. Since it was intended only as a summary of the proposed longer work, he originally titled it *An Abstract of an Essay on the Origin of Species etc. etc.*, but his wise publisher cut the first three words. His "big book," of course, was never completed.

Scientific content alone cannot explain the special excitement generated by the *Origin of Species*. According to critic-historian Stanley Hyman, it is also "a work of literature, with the structure of tragic drama and the texture of poetry." It conveys the urgency of a personal testimony, and an evangelical sense of mission, exhorting the reader to discover "this view of life."

Through Darwin's eyes we no longer see just a sparrow or a cactus, but a roiling drama of conflict and competition, a dynamic landscape of organic beings caught in a relentless struggle for existence. Nature is a frenetic, omniscient goddess, "daily and hourly scrutinising, throughout the world, the slightest variations; rejecting [some] . . . preserving [others] . . . silently and insensibly working" to choose and reject, to bestow favor or toss into the pit of extinction.

"With a book as with a fine day," Darwin wrote to Huxley in 1863, "one likes it to end with a glorious sunset," as he had done with the *Origin*'s concluding paragraph:

> Thus, from the war of nature, from famine and death, the most exalted object which we are capable of

conceiving, namely, the production of the higher animals, directly follows. There is grandeur in this view of life, with its several powers, having been originally breathed into a few forms or into one; and that, whilst this planet has gone cycling on according to the fixed law of gravity, from so simple a beginning, endless forms most beautiful and most wonderful have been, and are being evolved.

See also DARWIN, CHARLES; NATURAL SELECTION; VOYAGE OF THE H.M.S. BEAGLE; WALLACE, ALFRED RUSSEL.

For further information:

Darwin, Charles. *On the Origin of Species by Means of Natural Selection*. London: John Murray, 1859.

Darwin, Francis, ed. *The Foundations of the Origin of Species by Charles Darwin*. (Includes Charles Darwin's earlier sketches of the theory.) Cambridge: Cambridge University Press, 1909.

Simpson, George G. *The Book of Darwin*. New York: Washington Square Press/Simon & Schuster, 1982.

Stauffer, R. C., ed., *Charles Darwin's Natural Selection, Being the Second Part of His Big Species Book Written from 1856 to 1858*. Cambridge: Cambridge University Press, 1975.

ORTHOGENESIS
"Goal-Directed" Evolution

Orthogenesis (meaning an origin that is spelled out from the beginning) is the idea that evolution follows a preordained path, especially in the case of man, guiding us on to a higher destiny. It is wonderfully compatible with many religious beliefs about the soul's inevitable progress, but causes much confusion to those who seek a biologist's understanding of evolution.

Some philosophers have imagined a guiding force or "vital principle," which acts like a spiritual magnet, drawing mankind toward "higher" grades or levels. These orthogenetic ideas are derived from the medieval Chain of Being, an ascent from lowly rungs upwards toward the infinite.

Modern evolutionary theory holds that evolution is "opportunistic," in the word of paleontologist George Gaylord Simpson. At any point, it goes in the direction that is advantageous, often reshaping old structures for new uses. It does not know its destination, nor is it impelled to follow one particular direction. Sometimes the move is "sideways" into a different adaptive zone, as horses shifted from forests to grasslands. Evolution can also be "degenerative," as in the development of parasites whose organization goes from complex to more simple.

Humans cannot claim to be perched on an inevitable pinnacle of development, the ultimate "goal" of apes and near-men. We are simply creatures shaped by a particular and unique history. As Simpson put it, "evolution doesn't move in straight lines, but the minds of some scientists do."

See also CONTINGENT HISTORY; GREAT CHAIN OF BEING; HORSES, EVOLUTION OF; "OMEGA MAN"; TELEOLOGY.

OSBORN, HENRY FAIRFIELD (1857–1935)
Paleontologist and Museum Director

Henry Fairfield Osborn, the legendary president of the American Museum of Natural History, was a great man and not shy about admitting it. When young scientists came to see him, he sometimes suggested they try out his chair. "Young man," he would say, "when I was your age I studied with Huxley, and once met Darwin. They inspired me. I want you to know what it feels like to sit at the desk from which I preside over this institution. Perhaps you, too, will be inspired to become great some day, as I am."

Osborn even published an entire book that lists his articles and speeches and is illustrated with pictures of his many awards and medals. Not out of pride, he explained, but "to help inspire young men." Egotistical and autocratic though he may have been, there is no denying Osborn's solid accomplishments as paleontologist, evolutionist and dynamic director of the museum during its formative period.

When he took over guidance of the museum in 1908, it had not one single dinosaur. Osborn bought up various collections and hired the young Barnum Brown to go out into the field for years of dinosaur hunting. The result was the greatest collection of dinosaur fossils in the world. He also hired the pioneer painter of prehistory, Charles R. Knight, to create the famous murals for the dinosaur halls—and develop a whole genre of art in the process.

Osborn not only could recognize talent and select superb scientists and artists, he was also a master fund-raiser who twisted the arms of New York's social elite for the millions he needed to build the growing institution.

Also a serious paleontologist, he spent decades working up a massive history of elephants, one of his lasting contributions to paleontology. His work with Charles Knight and William Matthew created a new concept for museums by mounting the skeletons of extinct animals in realistic poses. Up until then, most museums merely exhibited their slabs of fossils, just as they came from the quarry. A bronze bust of Osborn at the museum bears the legend: "For him the dry bones came to life, and giant forms of ages past rejoined the pageant of the living.

It was Osborn who agreed to send Roy Chapman Andrews out on the East Asian Expedition to the

IMPERIOUS INNOVATOR Henry Fairfield Osborn was an immense influence in spurring evolutionary studies. As director of the American Museum of Natural History, he initiated field expeditions, paleontological research, public education and the dramatic presentation of prehistoric animals.

Gobi Desert to find the earliest ancestors of mankind, which were then assumed to be in Asia. Although they guessed wrong on the fossil men, there were treasures enough, including the first known nests of fossil dinosaur eggs. It was one of the most ambitious natural history expeditions ever attempted. Like Osborn himself, it was on a grand scale.

See also ARISTOGENESIS; BROWN, BARNUM; GOBI DESERT EXPEDITIONS; KNIGHT, CHARLES R.

For further information:
Hellman, Geoffrey. *Bankers, Bones, and Beetles: The First Century of the American Museum of Natural History.* Garden City, N.Y.: Natural History Press, 1968.

OSTEODONTOKERATIC CULTURE
Early Tools or Carnivore Rubble?

After South African anatomist Raymond Dart discovered the first australopithecine fossil in 1924, he spent years sifting through the jumble of animal bones found in the same limestone deposits, known as breccia. His statistical study of more than 4,500 fossil animal bone fragments convinced him that early man-apes had selected certain parts—bones, teeth and horns—to use as the first tools and had based their culture on it.

Dart was impressed, for instance, by the very high proportion of the jawbones of small antelope in the breccia, as well as by long bone-ends and horn cores. He made imaginative reconstructions of australopithecines using the horn cores as daggers, the long bones as bashers, the small jaws as cutters and scrapers. In a series of articles published in the 1950s, Dart coined the term *osteo-donto-keratic*, meaning "bone-teeth-horn" culture.

Dart's conclusions have never been accepted by scientists, even after he was given worldwide attention because of Robert Ardrey's bestseller, *African Genesis* (1961). The bones are not found in stratified deposits, but in irregular conglomerations of miscellaneous material. Some think the bone accumulations were made by hyenas or other carnivores; most believe there is no convincing way to tie in the bone parts with the culture of early hominids.

See also AFRICAN GENESIS; AUSTRALOPITHECINES; DART, RAYMOND ARTHUR; DARTIANS; KILLER APE THEORY.

OVERTON, JUDGE WILLIAM

See "SCIENTIFIC CREATIONISM."

OWEN, SIR RICHARD (1804–1892)
Zoologist, Paleontologist

Victorian England's most prestigious zoologist and paleontologist, Sir Richard Owen, held every scientific honor of his time. A master of classification, comparative anatomy and paleontology, he disliked his nickname "the British Cuvier" because he considered himself far superior to the great French naturalist. His fellow scientists were unanimous in their opinion of his character: Owen was the coldest, most arrogant, spiteful back-stabber any of them had ever encountered. And his most venomous, underhanded attacks were directed at the young evolutionists Thomas Huxley and Charles Darwin, who had once admired him.

Owen was the first director of the British Museum (Natural History) and the premier expert consulted by Her Majesty's government on everything imaginable: from evidence in murder cases to the authenticity of sea serpent sightings.

Young Darwin had sought Owen's friendship and advice and had even asked the senior naturalist to write the official description and classification of the fossil mammals he had collected on his voyage aboard H.M.S. *Beagle*. Owen undertook the task, and his monographs, published between 1838 and 1840, occupied four volumes of *Zoology of the Voyage of H.M.S. Beagle*. But, as Darwin later recalled, "after the publication of the *Origin of Species*, he became my bitter enemy, not owing to any quarrel between us, but as far as I could judge out of jealousy at its success . . . Certainly his power of hatred was unsurpassed."

Owen had a chilly politeness that masked his un-derhanded attacks. As historian Lynne Barber put it, "if Owen could steal the credit for someone else's achievements, he would always do so; if he could not, he would strive to discredit the achievement. Hugh Falconer, the elephant expert, warned Darwin that Owen was 'not only ambitious, very envious and arrogant, but untruthful and dishonest.' "

In addition to his 400 technical papers on anatomy, Owen is remembered for several noteworthy contributions to reconstructing extinct creatures, which, ironically, later helped establish the Darwinian theory he detested.

In 1839, a sailor brought him a puzzling six-inch fragment of bone from New Zealand, which Owen identified as an unknown flightless bird. His colleagues were skeptical—could any bone so massive possibly have belonged to a bird? But Owen stuck by his original determination. A decade later, several crates of fossils reached the museum, filling out the skeleton of the same creature he had named *Dinornis*. The existence of New Zealand's giant extinct moa bird was confirmed.

Similarly, Owen was the first to correctly reconstruct the giant ground sloth *Mylodon* from South American fossils and was even able to deduce something about its way of life. But his most famous reconstructions were the series of life-sized prehistoric animals, including *Iguanodons*, which were sculpted under his direction by the artist Benjamin Waterhouse Hawkins. Originally created for the Crystal Palace exhibition in 1853, they remain in the gardens of Sydenham, near London, to this day. [See IGUANODON DINNER.]

Dinosaurs will always be linked with Richard Owen. He was the first to understand and publish the significance of the remarkable fossil reptile bones that had recently been discovered by Dr. Gideon Mantell and others. It was Owen who coined the term "dinosauria," meaning "terrible lizards," for this previously unknown group of extinct giants. He also described and named *Archaeopteryx*, the early reptile-like bird, when he purchased the world-famous specimen for the British Museum.

In his reviews of Darwin's books, Owen was consistently inaccurate and unfair in representing what had been said. Privately, he opined that Charles "was just as great a goose as his grandfather." When he was shown advance proofs of the *Origin of Species* (1859), he complained it contained too many statements of Darwin's opinions, beliefs or conjectures. When Darwin immediately offered to cut them out, Owen sternly advised him not to change a thing, or he would "spoil the charm" of his book. Then, in his review, Owen promptly singled out Darwin's use of such phrases as "I am convinced" or "I believe"

HOLDING BONE OF MOA, a giant extinct bird, the British Museum's great anatomist Richard Owen savors a moment of glory. Owen identified the fragment as belonging to a monstrous bird, despite ridicule from other experts. More complete fossils of the creature proved him correct.

and attacked them as unscientific. Darwin felt not only abused, but tricked and cheated as well.

Owen adhered, for the most part, to the idea of "archetypes" he had learned from the German zoologist Lorenz Oken, a follower of the *naturphilosophie* school. He clearly saw and studied in detail the widespread pattern of homologies in structure—the principle that a bat's wing, human foot and whale's flipper are made of the same structural elements, bone for bone. He wrote a treatise *On the Archetype and Homologies of the Vertebrate Skeleton* in 1848, in which he illustrated homologies (corresponding similarities) in the anatomy of diverse animal species.

But to Owen, these were variations of a divine plan, not transformations based on common descent with modification. Nevertheless, Owen was the first to make the crucial distinction between analogies (differently structured parts with similar functions, such as insects' wings and birds' wings) and homologues (the same underlying structures performing a

wide variety of functions, such as horses' hooves corresponding to apes' third finger.) Although the horse walks on an enlarged nail of a single digit, the other four are present as "splints" in the leg.

It was also Owen who prepared Bishop Samuel Wilberforce with the ammunition to attack Darwin at the Oxford debates in 1863. [See OXFORD DEBATE.] It was not the first of his "undercover" attacks on the evolutionists. Owen had anonymously published a scathing, spiteful review of the *Origin of Species* in the *Edinburgh Review* (1860), then lied to Darwin's face that he had not written it.

At other times, Owen seemed to want to take credit for developing an evolutionary perspective in biology. Darwin noted, with annoyance, that Owen later taught that perhaps "all birds are descended from one, and advances as his own idea that the oceanic wingless birds have lost their wings by gradual disuse. [But] he never alludes to me, or only with bitter sneers."

Owen, in fact, sometimes hinted that he had his own theory of evolution, which was better than Darwin's and which he had originated earlier—but he never revealed just what that idea might be. In effect, he was claiming evolution was wrong, but that, if it was right, he had invented it.

Despite all his efforts, however, Owen was unable to stem the Darwinian wave that swept the life sciences. He went on to establish and direct the new British Museum of Natural History at South Kensington from which he finally retired at the age of 80. His arrogant prediction that Darwin's work "would be forgotten in ten years" followed him into old age, for it was his own theories that were discredited and forgotten. Although he outlived Darwin by several years, his archetypal ("essentialist") ideas about species had become as extinct as the dinosaurs.

See also ARCHEOPTERYX; ESSENTIALISM; HOMOLOGY; NATURPHILOSOPHIE.

OXFORD DEBATE (1860)
"Darwin's Bulldog" *vs.* The Bishop

Soon after Charles Darwin's *Origin of Species* appeared in 1859, the theory of evolution was under attack by churchmen. The first confrontation of opposing forces took place at Oxford University on June 30, 1860, during the week-long meetings of the British Association. "Darwin's Bulldog," Professor Thomas Henry Huxley, debated Samuel Wilberforce, the Bishop of Oxford, on the bishop's home ground.

Wilberforce was a wily and confident orator. Although he didn't know much science, he had been coached for several days by Richard Owen, the British Museum's paleontologist who was a rival of Darwin's and an enemy of Huxley's. Only a few days before, Owen and Huxley had clashed about the comparison of human and gorilla brains, with Huxley accusing the renowned Owen of not knowing what he was talking about. Now Owen was seeking revenge on the Darwinians by prepping the bishop for the Saturday session.

Huxley, after a week of meetings, was going to skip the last day. He knew Samuel Wilberforce was an accomplished speaker, capable of sliding around any argument or confrontation—which had earned him the nickname "Soapy Sam." (According to the bishop, he got the name because he was always in hot water and always came out of it with clean hands.) Besides, Huxley was not a scheduled speaker.

On Friday, Huxley ran into Robert Chambers, the author of *Vestiges of Creation* (1844), and told him he planned to miss the next day's session since he "did not see the good of giving up peace and quietness to be episcopally pounded." Chambers accused him of deserting, and Huxley replied, "Oh! If you are going to take it that way, I'll come and have my share of what is going on."

The new University Museum was crowded with 700 men and women who sensed an impending battle of giants. On the platform were the botanists J. S. Henslow and Joseph Hooker, Bishop Wilberforce, Dr. Draper of New York, Sir John Lubbock, Sir Benjamin Brodie (the Queen's physician and president of the Royal Society) and Huxley.

Dr. Draper droned on for an hour with a mediocre paper discussing Darwin's views, and several clerics and other members of the audience rose to respond. Among them was Admiral Robert FitzRoy, captain of the *Beagle* 30 years before, who held a Bible over his head and denounced Darwin for having been a viper in his midst. Finally, there were calls for the bishop, who started off fluent and florid, with ruminations about how disquieted he would be if one could prove to him that he had a "venerable ape" in his family tree. According to Hooker's account, Wilberforce "spouted for half an hour with inimitable spirit, ugliness and emptiness and unfairness" without any grasp of the scientific issues despite Owen's briefing.

Overconfident that the audience was his, Wilberforce pulled out all the stops and, with a turn to Huxley, demanded to know whether he would prefer to be descended from an ape on the side of his grandfather or his *grandmother*? At this, Huxley turned to Brodie, slapped his knee and whispered "The Lord hath delivered him into mine hands!" The Bishop had violated Victorian propriety by getting personal, attacking his opponent's family, and—worst of all—insulting womanhood!

"SOAPY SAM" was the nickname students gave Bishop Samuel Wilberforce of Oxford (left), a skilled speaker who could slide around any question. Wilberforce attacked Darwin's theories in the famous "Oxford Debate" of 1860, but his audience was won over by Thomas Huxley (right), Joseph Hooker, and other evolutionists.

Huxley waited to be called to speak by the audience, then rose slowly and deliberately. When we talk of descent, he calmly explained, we mean descent through thousands of generations. He had listened carefully to the bishop's speech, he said, but could not discover any new facts or new arguments—except questions about his personal predilections in the matter of ancestry. "It would not have occurred to me to bring forward such a topic as that for discussion myself," Huxley continued, "but if the question is put to me, would I rather have a miserable ape for a grandfather, or a *man* highly endowed by nature and possessed of great means and influence, and yet who employs these faculties and that influence for the mere purpose of introducing ridicule into a grave scientific discussion, I unhesitatingly affirm my preference for the ape."

Women fluttered their handkerchiefs and Lady Brewster fainted, while the room rebounded in laughter. Huxley continued with serious, sober arguments, and Lubbock and Hooker spoke after him. All were persuasive, well-informed Darwinians. Hooker pointed out that, from the bishop's remarks, it was clear he could never have read Darwin's book and was completely ignorant of botanical science, which had benefitted greatly from Darwin's work.

In writing to a friend of the incident, Huxley noted "I happened to be in very good condition and said my say with perfect good temper and politeness—I assure you of this because all sorts of reports were spread about . . . that I had said that I would rather be an ape than a bishop."

The importance of the Oxford debate was that it was the first time open resistance was made to the church's authority over the question of human origins, and science's right to pursue investigations touching on the nature of man. Instead of being crushed under ridicule, Huxley and his colleagues won a wider interest and fair hearing for the new theories.

Although he thereafter treated the bishop with perfect courtesy, Huxley could not resist a cruel parting shot at the time of the bishop's death. Many years later Wilberforce was riding through a field, when he was thrown from a horse and hit his head on a stone. "His end has been all too tragic for his life," Huxley commented. "For once, reality and his brains came into contact and the result was fatal."

See also CHAMBERS, ROBERT; FITZROY, CAPTAIN ROBERT; HOOKER, SIR JOSEPH D.; HUXLEY, THOMAS HENRY; OWEN, SIR RICHARD.

PAEDOMORPHOSIS

See NEOTENY.

PAINTPOT PROBLEM
"Swamping" of Mutations

In the mid-19th century, most biologists believed in "blending inheritance." When a dark-skinned person took a light-skinned mate, for example, they observed that the children would often be of an intermediate color. Similarly, a tall father and a short mother sometimes produce children of medium height. Because of such apparent "blending," scientists believed offspring would tend to average the parental traits.

Mendelian genetics was still unknown. Father Gregor Mendel, breeding pea plants in his monastery garden, was working out his principles of particulate inheritance, which would eventually show these visible "blends" were special, misleading examples, caused by complex factors. Common single-gene traits do not blend—they come out in different proportions in succeeding generations, depending on whether they are "dominant" or "recessive." But the behavior of genes was not known to Charles Darwin and his contemporaries and was appreciated by science only after 1900.

Henry Fleeming Jenkin (1833–1885), a brilliant Scots engineer who collaborated with the physicist Lord Kelvin, was an outspoken critic of Darwin's theory of natural selection. Darwin had proposed that large mutations (he called them "single variations" or "sports") that conferred a selective advantage would spread through the population. Jenkin pointed out that such variations could not spread, but instead would be diluted and disappear. If each succeeding generation was an "average" of its parents, the new variation would be "swamped" after a few crosses. He used the analogy of a drop of black paint added to a bucket of white. The white paint would completely overpower the black as if it had never existed. Hence, Jenkin's objection became known as "the Paintpot Problem."

Darwin said he was convinced by Jenkin's argument as it referred to large changes or "sports," but small scale recurrent changes were another matter. Natural selection could work on a range of small, gradual variations that kept recurring in a population, and these would not be "swamped."

When Mendelian genetics was rediscovered, its demonstration of particulate inheritance made "blending inheritance" obsolete. The "Paintpot Problem"—which once seemed so challenging—was put on the shelf of retired ideas.

See also MENDEL, ABBOT GREGOR; MENDEL'S LAWS.

For further information:
Hull, David, ed. *Darwin and his Critics*. Chicago: University of Chicago Press, 1973.

PALEOLITHIC
Old Stone Age

Sir John Lubbock, who was Charles Darwin's neighbor and protege, was a pioneer of what was known as the "Prehistory Movement." Discoveries of fossil man and his tools had not yet been established; Lubbock brought top geologists to France to see for themselves the sites excavated by Boucher de Perthes.

Thousands of finely chipped flint tools were soon accepted as the work of early man. Lubbock named the earliest layers the Paleolithic (100,000 to 1 million+), meaning "Old Stone Age." Later, he discerned a Mesolithic ("Middle Stone Age"); eventually a "Neolithic" (9,000 years ago) was added—the period of ground and polished axes associated with the beginnings of agriculture. These labels are still used by archeologists. When metals were found, a "Bronze Age" and an "Iron Age" were added but these were later dropped as meaningless.

One of the fallacies associated with the Victorian labeling of "ages" was that they thought the tool types represented definite stages in human history. When tribal peoples were discovered with stone axes, it was said that they were "still in the stone age," which of course they were not. No people's history has to pass through some predetermined number of stages, and humans who use stone implements today are not necessarily anything like those of a million years ago in their language, thought or social organization.

Also, what we find today is not necessarily the only technology used by the ancient people. Where only stone tools survive, the people may also have used ropes, nets, wooden implements and other perishable materials. The basic joke of the "Fred Flintstone" cartoons is that *everything* is made of stone—including books, combs, furniture—because they're supposed to be living in the Stone Age.

See also LUBBOCK, SIR JOHN; MOUSTERIAN INDUSTRY; PITT-RIVERS MUSEUM.

PALEY, REVEREND WILLIAM

See NATURAL THEOLOGY; PALEY'S WATCHMAKER.

PALEY'S WATCHMAKER
Design Implies Deity

Everyone with a passing interest in natural history in the 19th century—which included the majority of educated Victorians—had heard of William Paley's famous "Watchmaker." If they did not read his widely studied book *Natural Theology*, first published in 1802 and reprinted almost annually thereafter, they "would have had its ideas dinned into them from the cradle up, and heard them expounded from every pulpit in the land," as Lynn Barber has phrased it.

Paley's book begins with "The Watch on the Heath," an analogy that illustrates the "argument from design," which he thought proved the existence of a creator:

> In crossing a heath, suppose I pitched my foot against a *stone*, and were asked how the stone came to be there . . . [I could answer] it had lain there for ever . . . But suppose I had found a *watch* upon the ground . . . I should hardly think of [the same answer because] when we come to inspect the watch, we perceive . . . that its several parts are framed to point out the hour of the day . . . The inference, we think, is inevitable, that the watch must have had a maker; that there must have existed, at some time, and at some place or other, an artificer . . . who formed it for the purpose . . . and designed its use.

According to natural theology, everything in nature reflects the beauties and perfection of the God who designed it. Paley marshalled the same kinds of facts Darwin later did: the construction of the human eye, the physiology of plants, the relationships between organs. But to Darwin, the intricate mechanism of nature came about through the gradual fitting and honing of natural selection. To Paley, all had been designed in one stroke by a "Master Designer"—and his watchmaker argument is still often advanced by today's creationists.

See also DIVINE BENEFICENCE, IDEA OF; NATURAL SELECTION; NATURAL THEOLOGY.

For further information:

Dawkins, Richard. *The Blind Watchmaker: Why the Evidence of Evolution Reveals a Universe without Design.* New York: Norton, 1986.

Paley, William. *Natural Theology: or, Evidences of the Existence of the Deity, Collected from the Appearance of Nature.* London: Baynes, 1802.

PALUXY RIVER

See BIRD, ROLAND T.; FOSSIL FOOTPRINTS, PALUXY RIVER.

PANDA'S THUMB, THE
An Evolutionary "Contraption"

That zoo favorite, the giant panda, does not have a true opposable thumb, since its five toes long ago evolved into a bear-like paw. But when it adapted to feeding on tasty bamboo leaves, one of its wrist bones became an extra digit for holding and stripping the stalks. Thus, pandas have six digits—but the extra thumb is a made-over part.

Harvard paleontologist Stephen Jay Gould used the panda's "thumb" as the central image in a popular book of evolutionary essays (*The Panda's Thumb: Reflections in Natural History*, 1980). During a period when creationists were repopularizing the old idea that adaptation implies perfect design, Gould emphasized the sometimes quirky history of life.

Evolution uses what is available; it doesn't always produce neat, flawless creatures. When we examine jerry-built structures—organs that have shifted from one function to another—he argues, we see the twists and turns of history, not the planning of a divine designer.

Since the evolutionary pathway leading to oppos-

GIANT PANDA EVOLVED an "extra" thumb from one of its wrist bones, a secondary specialization for feeding on bamboo stalks. It also retains its actual first digit, but its "true thumb" is immobilized in the structure of its paw.

able thumbs was not taken by the panda's ancestors, its thumb was "committed" to the paw structure. So the radial sesamoid bone developed into an efficient, but inelegant thumb. Gould reminds us that Charles Darwin consistently turned to those organic parts and geographic distributions that seemed to make the least sense, for in the "leftovers" and "contraptions" could be read a particular history.

The panda's thumb is Gould's favorite example that the strongest proof of evolution is not optimal design, but "odd arrangements and funny solutions . . . paths that a sensible God would never tread but that a natural process, constrained by history, follows perforce."

See also ADAPTATION; NATURAL SELECTION; NATURAL THEOLOGY; PALEY'S WATCHMAKER.

For further information:
Gould, Stephen Jay. *The Panda's Thumb: More Reflections in Natural History.* New York: Norton, 1980.

PANGEA
Original Supercontinent

Pangea was a single, huge equatorial continent, which coalesced about 225 million years ago. Enormous plates moved together in conjunction with climatic shifts and other drastic geophysical changes. The result, apart from the formation of the single land mass, was the Permian extinction: the most extensive

"great dying" in the Earth's history, during which whole families of species disappeared forever.

Eventually, the great moving plates began to tear apart, pulling Pangea into a northern and a southern half. The northern supercontinent, Laurasia, was formed by 65 million years ago. It, in turn, broke up to form North America, Europe and Asia. Gondwanaland, the southern portion, later drifted apart to form the continents of South America, Australia, India and Antarctica.

See also GONDWANALAND; LAURASIA; PLATE TECTONICS.

PANGENESIS
Darwin's Wrong Guess

Because Charles Darwin realized evolutionary theory was incomplete without an understanding of heredity, he devised a theory of how parents pass on their characteristics to offspring. He called it pangenesis; it turned out to be wrong.

According to pangenesis, a trait acquired by a parent during his or her lifetime could be passed on to children (Lamarckian or "soft" inheritance). If a man worked to develop large muscles, for instance, the repeated habit of weight-lifting would somehow leave a lasting record in the cells of his body. Particles carrying this information were called "gemmules." They would migrate from all parts of the body to the sex cells, whereby they could be inherited by the offspring.

Within 20 years of Darwin's death, Mendelian genetics was belatedly recognized by science, and August Weismann demonstrated that what he called the "germ plasm" is distinct from the rest of the body.

Sex cells, we now know, produce genetic material independently, without input from arms, intestines or any other area of the body. Acquired characteristics, such as lopped off fingers, trained muscles or a mastery of French, are not passed on through "gemmules." However, Darwin was on the right track; he correctly surmised the existence of tiny biochemical structures that contain coded blueprints for the offsprings' characteristics. We call them genes, a term coined by Wilhelm Ludwig Johannsen in 1909.

See also MENDEL, ABBOT GREGOR; WEISMANN, AUGUST.

PANGLOSSISM

See TELEOLOGY.

PANSPERMIA
Life from Outer Space

British astronomer Fred Hoyle and his colleagues first suggested in 1978 that the seeds of life may be constantly raining down on earth from outer space—a theory he calls Panspermia.

According to Hoyle, the conditions for the origin of life may have been better among the vast amount of organic matter he believes floats through interstellar space. Unimaginably immense quantities of chemical molecules colliding in space might make the rare and improbable combinations more likely, almost inevitable. Simple life forms or amino acids may have ridden to Earth on comets or meteors. Of course, Hoyle recognizes this is no explanation for the origin of life; it simply moves the problem to another time and place.

Although biologists ridiculed panspermia for years, in 1989 it acquired some supporting evidence: amino acids were found within meteor craters. And, contrary to expectations, the molecules were not damaged by their rough landings. On learning of these discoveries, astronomer Carl Sagan commented: "The impact giveth and the impact taketh away!"

See also CLAY THEORY (OF ORIGIN OF LIFE); LIFE, ORIGIN OF.

For further information:
Crick, Francis. *Life Itself.* New York: Simon & Schuster, 1981.
Hoyle, F., and Wickramasinghe, N. C. *Evolution from Space.* New York: Simon & Schuster, 1981.

"PARALLAX" (SAMUEL BIRLEY ROWBOTHAM)

See FLAT-EARTHERS.

PARROTS, LANGUAGE ABILITY OF

See APE LANGUAGE CONTROVERSY.

PARTHENOGENESIS

See LOEB, JACQUES.

PARTICULATE INHERITANCE

See "BEANBAG GENETICS"; MENDEL'S LAWS.

PASTRANA, JULIA (mid-19th century, dates unknown)
Victorian "Gorilla Woman"

Early evolutionists thought a hirsute Spanish dancer named Julia Pastrana was a "throwback" to an ape-like stage of humanity. Charles Darwin, Ernest Haeckel and Alfred Wallace all wrote of her unfortunate birth defects as scientific curiosities, and the poor woman was endlessly studied and measured.

With his customary tact, Darwin referred to Pastrana as "a remarkably fine woman, but she had a thick masculine beard and a hairy forehead." In both

her upper and lower jaw, she had a double set of teeth, one row being placed within the other.

"From the redundancy of her teeth," wrote Darwin, "her mouth projected, and her face had a gorilla-like appearance." Darwin wondered if large teeth were correlated with hairiness: many hairless animals had small teeth or none. (He was thinking of armadillos and porpoises, but seemed to have forgotten that bare-skinned elephants and walruses grow the biggest teeth of all.)

Julia Pastrana became famous as the "gorilla woman"; she appeared in freak shows and sold photographs of herself. When she died, a heartless showman had her skin stuffed and mounted; he did a brisk business charging the public to see it.

PATTERSON, PENNY

See KOKO.

PEALE'S MUSEUM (1784–1845)
Early Natural History Museum

In July, 1786, America's most famous portrait painter, Charles Willson Peale (1741–1827), put an ad in the *Pennsylvania Packet* asking for "Natural Curiosities" and "Wonderful Works of Nature." Artist, amateur scientist, entrepreneur and showman, Peale was putting together the world's biggest and best private museum of natural history. (No such public institutions yet existed.)

To his delight, the public's interest matched his own, and he could barely keep up with the avalanche of specimens: minerals, shells, fossils, feathers, creatures dead and alive. Benjamin Franklin sent a French angora cat and President George Washington a brace of pheasants.

It all started when a local naturalist asked the artist to illustrate some puzzling, gigantic bones he had found. After drawing these "curiosities," Peale decided to exhibit them among his portraiture. Thus, the first real museum of natural history grew out of an art gallery, rather than a science laboratory.

Peale's shifting interests were reflected in the naming of his sons. While the older boys were all named for artists—Raphael, Rembrandt, Titian, Reubens and Vandyke—those born after he became interested in science were called Charles Linnaeus and Benjamin Franklin Peale. (Now best remembered as a philosopher-politician, Franklin was also famous as a scientist during his lifetime.)

By trial and error, Peale developed methods of preserving specimens (often dousing them in arsenic to keep away insects) and found an eager public willing to pay the admission fee to his establishment. It soon became one of the most celebrated in the world, the envy of European naturalists and a Philadelphia landmark.

Intrigued with the gigantic bones that had begun his natural history collection, Peale decided to find a complete skeleton of this unknown monster ("the

great American incognitum," as it was nicknamed) for his museum. After following several leads, he found what he was looking for at the farm of John Marsten, near Newburgh in New York state. Marsten had excavated some large fossil bones from a water-filled marl pit and he sold them to Peale for $200.

Convinced the rest of the skeleton was buried there, Peale launched a full-scale excavation, rigging elaborate block and tackles and a specially built pump, operated by a man-powered tread-wheel. Costing Peale $1,000, it was the first full-scale scientific expedition in American history and a complete success. Although a few bones (including the jawbone) were missing, Peale obtained an almost-complete skeleton of a huge mammal never before seen, except for fragments, and had it shipped back to Philadelphia.

A few months later, his crew dug up a similar skeleton at another farm, and Peale pieced the two individuals together to form the first mounted mastodon (or "mammoth," as he called it). When Americans got their first look at the great extinct elephant that once roamed the present Hudson River valley, it became an immediate sensation. It was also a new symbol of national pride, since European naturalists had insisted that "everything is smaller in America." For half a century, the word "mammoth" meant the

PEALE'S MASTODON was discovered in this water-filled marl pit near Newburgh, New York, in 1799. Charles Willson Peale (holding drawings) designed and built a bucket-wheel powered by a treadmill. Spectators took turns running inside the wheel to bail out the site while workmen quarried the massive fossil skeleton, which was later exhibited at Peale's Natural History Museum (above).

biggest and best. (It was later supplanted in popular parlance by the name of P. T. Barnum's huge circus elephant, "Jumbo.")

Peale's discovery of the mastodon was an important and original contribution to science, though the French anatomist Georges Cuvier was first to accurately describe and classify it. After the massive fossil bones had been mounted in the museum's Mammoth Room in December of 1801, Peale and a dozen jubilant patriots celebrated by holding a formal dinner inside the skeleton, where they sang "Yankee Doodle." (Some years later, not to be outdone, British scientists and museum artists held an elegant dinner inside the world's first life-sized dinosaur model—an *Iguanodon* at the Crystal Palace Exposition.)

Peale always claimed his main interest was scientific and took pride in being the first museum-keeper to arrange his large and diverse collection according to the Linnaean system of classification. But detractors were fond of pointing out that his exhibits also included such crowd-pleasers as waxworks, stuffed monkeys wearing suits and gowns, a cow with five legs and two tails, a machine for drawing silhouettes and clever mechanical dioramas that he called "moving pictures."

Raised among such versatile interests, one of Peale's sons, Rubens, became a shameless showman who built the museum into an even more popular (and unscientific) entertainment; another (Titian) became a serious naturalist; and still another (Rembrandt) a fine painter who incorporated superbly observed flowers and animals into his compositions. Charles Willson Peale had provided sufficient inspiration and precedent for them all.

The "descendants" of his museum were comparably diverse, ranging from the Grand Colossal Museum and Greatest Show on Earth of P. T. Barnum to the American and British Museums of Natural History. Part of Peale's legacy to the great natural history museums is their perennial problem of determining who is in charge of creating exhibits: scientists, showmen or artists. The answer each museum gives determines its own distinctive "flavor." In Charles Willson Peale—as in the best museum people since—these three distinct talents were highly developed in a single individual.

See also BARNUM, PHINEAS T.; IGUANODON DINNER; JEFFERSON, THOMAS; MAMMOTH.

For further information:
Barber, Lynn. *The Heyday of Natural History.* London: Jonathan Cape, 1980.

PEARL, RAYMOND

See PURE LINE RESEARCH.

PEER REVIEW
Science's Referee System

Scientific innovations, when first presented, are often unrecognized by experts in the field. When Alfred Russel Wallace published his brilliant "Sarawak Law" on species in 1855, it was almost completely ignored by zoologists. In 1858, the epochal Darwin–Wallace papers on evolution by natural selection were not greeted with acclaim by the majority of members when they were read at the Linnean Society of London. Some months later, the president of the Society, in his annual report, lamented that no significant new theories had been presented that year in biology.

One objective of science is to generate novel breakthroughs: something never seen, done or understood before. Yet scientists have a strong tendency to reject new facts or theories on those very grounds—that they have never been seen, understood or done before. Gregor Mendel's experiments on hybrid pea plants is another notorious example from evolutionary biology. Although published in a scientific journal and disseminated to the leading botanists of Europe, his work was not recognized as the foundation of a new science of genetics until 30 years later.

Charles Darwin was well aware that scientists are not so much convinced of new ideas as that younger men eventually replace their senior colleagues. All scientific men should have the grace to die at the age of 60, he wrote, so they could not "oppose all new doctrines" with their "inflexible" brains and impede the progress of the next generation.

Initially plagued by doubts as he began writing the *Origin of Species* (1859), Darwin first "fixed in my mind three judges, on whose decision I determined mentally to abide"—the botanist Sir Joseph Hooker, the comparative anatomist Thomas Huxley and the geologist Charles Lyell. He would put aside his "awful misgivings" if they could agree with his approach and conclusions. Lyell alone of the three was afraid to "go the whole Orang"; his years of fence-sitting greatly upset Darwin.

After the initial acceptance by his self-chosen referees, Darwin mounted a personal campaign to convince about a dozen other top men in natural history of the truth of evolution. He even picked and targeted them and kept running lists of who was still "unconverted." If these colleagues could be won, he thought, "my theory will be safe."

Thomas Huxley early on had no illusions about the "objectivity" of peer review. He had written an original anatomical monograph but knew the leading journal was then controlled by Sir Richard Owen, the antievolutionary director of the British Museum.

Owen was the top expert in the field, but "he thinks natural history is his private preserve, and no poaching allowed," Huxley wrote a confidante.

Alfred Russel Wallace, the coauthor of the theory of evolution by natural selection, was exasperated by the difficulties geologists had convincing their colleagues that many man-made stone tools had been found in sealed layers with the bones of extinct elephants and other Ice Age mammals. In his book *The Wonderful Century* (1904), Wallace wrote:

> In 1840 a good geologist confirmed these discoveries, and sent an account of them to the Geological Society of London, but the paper was rejected as being too improbable for publication! All these discoveries were laughed at or explained away . . . These, combined with numerous other cases of the denial of facts on *a priori* grounds, have led me to the conclusion that, whenever the scientific men of any age disbelieve other men's careful observations without inquiry, the scientific men are *always wrong*.

Ambitious scientists consider themselves to be intelligent, keen observers, meticulous and diligent. To acknowledge a novel discovery may be to admit that a colleague or rival is smarter, more methodical, a better observer—or even just luckier than oneself. If the new knowledge should come from an amateur, the resistence can be especially intense. [See CHAMBERS, ROBERT; SCIOLISM.]

During the early 19th century, science was still mostly in the hands of gentlemen amateurs who supported their journals with subscriptions and made decisions in small cliques. A century later, scientists constitute a huge, professional class and produce an enormous number of papers. Moreover, pressure to publish regularly is intense, even though few papers are actually read.

At least Wallace and Mendel got their ideas published, even if they were ignored. One of the founders of the modern revolution in geology wasn't so lucky. In 1963, Canadian geophysicist Lawrence W. Morely, working independently, made a synthesis of current data and worked out one of the foundations of modern plate tectonics: The sea floor is continually spreading, causing continents to move, split apart or crunch together.

Morely had been studying magnetic bands in rocks retrieved from the ocean floor, a record of alternating polarity reversals in the Earth's history. When he correlated them with matching evidence from inland rocks, he came up with a theory that seafloor spreading is the engine that moves the continental plates.

Morely was no weekend amateur, but chief of the Geophysics Division of the Geological Survey of Canada. In February, 1963, he wrote a brief letter to the journal *Nature* outlining his theory, but two months later received word they "did not have room" to print it. Next, he tried the *Journal of Geophysical Research*, which sent Morley's paper to an anonymous referee, who advised that "such speculation makes interesting talk at cocktail parties, but it is not the sort of thing that ought be published under serious scientific aegis." By the time they returned it to him, five months later, two other researchers had independently published the theory before him. (Fortunately, there was a happy ending; it did eventually become known as the Vine–Matthews–Morely hypothesis, one of the foundations of modern geophysics.)

In 1977, psychologist Michael J. Mahoney, of Pennsylvania State University, decided to test the objectivity of peer reviewers in judging scientific merit. He sent two "cooked-up" papers on a disputed topic to 75 experts in the field whose opinions about this controversy were known. The vast majority praised the dummy papers whose conclusions agreed with their own positions, while rejecting those that appeared to offer evidence for the opposing view. When he published these results, Mahoney was threatened with loss of his job and called "unprincipled and unethical" for daring to experiment on the scientific system itself without the participants' knowledge. (Getting their permission, of course, would have made the study impossible.)

A few years later, two other psychologists, Douglas P. Peters of the University of North Dakota and Stephen J. Ceci of Cornell, decided to follow up on Mahoney's innovative research. Their strategy was a bit different. They took many papers that had already been published by top scientists from prestigious institutions, had them retyped and substituted the names of unknown authors from low-prestige institutions. Then they sent them to the journals in which they had originally been published for peer review.

Their results: All of the papers were rejected, though none of the expert reviewers caught on that they had previously appeared in the same journal. When they published their results, these investigators, too, were vilified and harassed for attempting to find out how well the scientific system is working.

In 1983, two biologists summarized all the scientific studies of peer review and submitted their review article, "On the Probity of Peer Review," to the *American Scientist*. The journal editor replied that their paper had been sent to referees for peer review. The final verdict: "I could not concur more with the importance of your subject [but] we have very little space available for articles that would be interesting to scientists generally . . . Our contents are limited primarily to reviews of recent scientific research."

See also MENDEL, REDISCOVERY OF; MENDELISM, AC-
CEPTANCE OF; SARAWAK LAW; SCIENCE; TAUNG CHILD.

PEI, W. C.

See PEKING MAN.

PEKING MAN
Homo erectus from China

Beginning in the 1920s, from a limestone quarry at
Chou Kou Tien, or "Hills of the Dragons," near
Peking (now spelled Beijing), an international team
of scientists excavated the spectacular fossil evidence
of Peking man.

Although the finding of this important hominid,
now called *Homo erectus*, is often attributed to Canadian Davidson Black, the German-American Franz
Weidenreich and the Chinese archeologist W. C. Pei,
these investigators were only the last and best funded,
by Rockefeller grants, in a complicated history of
discovery.

That story begins in Chinese apothecary shops,
where fossils said to be "dragon bones" have been
sold since ancient times as ingredients for folk medicines. Around 1900, a German naturalist, K. A.
Heberer, began to search the drugstore bins for interesting specimens, identified 90 fossil mammals
and came upon what looked like a fossil ape or
human tooth (which was later assigned to the extinct
ape *Gigantopithecus*).

Germans, Swedes, Austrians, Frenchmen and
Americans were all poking around China in the early
years of the 20th century, when the idea of an Asian
"cradle of mankind" was at its most popular. The
first major expedition came from Sweden's Uppsala
University, at the instigation of Johan Gunnar Andersson, a mining expert advising the Chinese government on ore deposits.

Professor C. Wiman of Uppsala, who had built a
Paleontological Institute around the fossils Anderss-
son had already shipped to Sweden, put together a
field team headed by Otto Zdansky, a young Austrian paleontologist. Zdansky quickly found the first
two fossil teeth of Peking man (later called *Homo
erectus*), but he cautiously kept them under wraps for
several years.

Scientists and eccentrics, adventurers and bum-
blers continued to poke around the Chinese "dragon
bone" quarries, hoping for a lucky strike. Important
early plants and strange dinosaurs began coming to
light, and Zdansky's hominid teeth were finally announced in 1926. Other paleontologists were drawn
to see for themselves what was going on.

The American Museum of Natural History's East
Asiatic Expedition (known as "The Missing Link Ex-

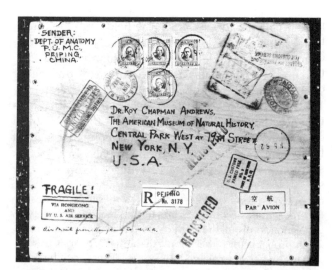

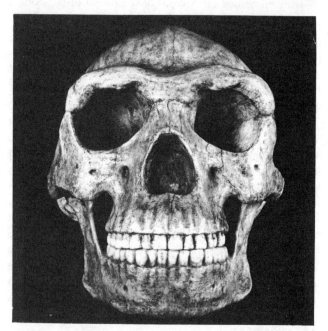

CRATES OF PEKING MAN'S BONES were lost during World
War II, as they were en route from China to the American Museum of Natural History in New York. This particular crate did
reach its destination, but it contained plaster casts—not the original *Homo erectus* fossils.

pedition") had agreed to limit its explorations to the
Gobi desert and northern territories, but its paleontologist Walter Granger visited Chou Kou Tien for a
look. So did French paleontologist Teilhard de Chardin, supposedly in China as a missionary. (He later
served as interim director at the site and analyzed
associated faunal deposits.)

Zdansky had retired from the field to a comfortable
university post, when a new round of Swedish activities began in the 1920s. In 1927, they were headed
by Birger Bohlin, a young expert on fossil giraffes,
who was assigned an army of Chinese laborers and
"ordered to find man." He did turn up a few more

jaws and teeth, which begin Bohlin's obsession with the image of Peking man (and woman). He drew charming reconstructions of their faces and has filled the walls of his home with endless variations of it over 50 years' time.

After Bohlin's tenure, still another, rival Swedish team continued the work, and the Americans entered on a large scale, funded by the Rockefeller Foundation. Canadian anatomist Davidson Black had persuaded the Rockefeller trustees to finance several seasons of intensive excavation and establish a Cenozoic Research Laboratory. After two years of moving thousands of tons of rock from the hillside—and shipping about 1,500 cases of mammal, plant and dinosaur fossils to Beijing—the frustrating result was only a few more hominid teeth.

On the basis of Wiman's few teeth and other fragments, Davidson Black had described and published in the journal *Nature* what he called the *Sinanthropus* ("Man from China") 1926. It was not until the very end of the December 1929 season, however, that Black's Chinese colleague, W. C. Pei found the first cranium deep in a limestone cave. The heavy-boned hominid skull featured prominent brow ridges and a somewhat smaller braincase (about 1,000cc) than modern humans (1,500cc). After his death in 1934, the site was directed for a year by Father Teilhard de Chardin, who was followed by Franz Weidenreich. By 1936 when all work had to stop because of the Japanese invasion of China, 14 skulls in varying conditions had been discovered, along with 11 jawbones, 147 teeth and a couple of small armbone and femur fragments. There were also stone tools and ambiguous evidence of a layer of carbon ash, which led to the name "The cave of hearths."

But just what had been excavated? A living site? A burial ground? A place of ritual cannibalism? Theories ran rampant. Peking man was represented mainly by skulls—hardly any postcranial material. Not a pelvis or a rib. Just skulls. And the openings at their bases, the foramen magnums, had been widened and smashed, as if someone had wanted to scoop out the brains.

Some concluded Peking man had lived in the Chinese caves, used fire and was a cannibal. Others thought they had been captured and their heads taken by *Homo sapiens*, who left no trace of themselves but the remains of their ghastly feast. Did they have language? Were these our ancestors? Or a sidebranch of human evolution our ancestors helped exterminate? One interpretation or another becomes fashionable every few years, but these enigmas remain unresolved.

At the time of discovery, it was clear that Peking man (then put in its own genus and species, *Sinan-*

thropus pekinensis) was virtually identical to the Java man (then called *Pithecanthropus erectus*), which had been found by Eugene Dubois in the 1890s. Over the next 50 years, similar remains turned up in the Middle East, Europe, North Africa and East Africa, including a fairly complete skeleton of an adolescent [see TURKANA BOY]. During the 1950s, Ernst Mayr reexamined the various classifications, and "lumped" these widespread hominids in the single species *Homo erectus.*

The Peking man fossils also pose a much more recent mystery: No one knows what happened to them. When war broke out in 1936, Weidenreich had hesitated to ship them out of the country since they belonged to the Chinese government. (Fortunately, he had sent excellent casts and notes back to New York. They are all that remain to science of the precious, hard-won fossils.)

Amidst the uncertainties of war-torn Beijing, it proved impossible to store them safely with Chinese authorities, so Weidenreich finally packed them for military shipment to the United States. They were believed to be aboard the marine ship *S.S. President Harrison*, which was sunk in the Pacific in mid-November 1941. So Peking man's bones may now be resting on the ocean's bottom.

However, there have been sporadic reports that the crate never made it onto that ill-fated ship, but was left behind in a railway station, where it was confiscated by the Japanese, stolen by looters or simply lost in the confusion. Tantalizing tips and rumors through the years have kept a spark of hope alive that the priceless relics were not destroyed after all and might yet resurface to reveal some of their long-kept secrets.

See also BEIJING MAN; DUBOIS, EUGENE; GOBI DESERT EXPEDITIONS; HOMO ERECTUS; TEILHARD DE CHARDIN.

For further information:

Reader, John. *Missing Links: The Hunt for Early Man.* London: Collins, 1981.

Shapiro, Harry. *Peking Man.* New York: Simon & Schuster, 1974.

PEPPERED MOTH
Adapting to Pollution

A homely little insect living in a polluted woods has been evolving almost fast enough to watch. During the 1930s, a group of British biologists seeking to demonstrate evolution in action found their ideal subject in the peppered moth.

This small, drab moth has three color phases: a whitish, a brownish mottled and a dark, almost black type. All three colors exist in the population, but in Victorian insect collections the lighter and mottled types predominate. Then, in the late 19th century, the formerly rare dark phase began to appear more frequently, until by the 1920s the species was almost all black.

H. B. Kettlewell and his colleagues correlated the moth's color change with the industrialization of the Manchester area. Soot and sulphur dioxide from mills and factories coated tree trunks, turning them black, and killing mottled tree lichens. Lighter and mottled moths now stood out clearly against the trunks, making them easy prey for predatory birds, while the dark moths became hard to see.

As trees darkened over the years, formerly rare dark moths survived in greater numbers, reproduced, and their frequency in the population dramatically increased. Within a few decades, the majority of the moths were dark. Countless biology textbooks have used the rapid evolution of the peppered moth from light to dark as a classic demonstration of natural selection in the modern world, observable within one human lifetime.

But the peppered moth's adaptation refuses to remain frozen in insect collections or textbook illustrations; the resilient insect continues to evolve before our eyes. Because of antipollution laws, English skies have been clearing over the past few decades. Trees have been getting lighter again, and new studies show that the lighter-colored moth is making a big comeback.

Laurence Cook, a biologist at the University of Manchester, published a study in 1986 of 1,825 moths collected throughout Great Britain. Graphs of the two forms show that the area dominated by the black moths is steadily shrinking toward England's northeast corner. Meanwhile, a scientist who has been trapping moths for years near his home in Merseyside reports that dark peppered moths comprised 90% of the total before 1975, but had fallen to 60% in 1984.

Has the white moth's coloration become protective again, now that its environment is cleaner and lighter colored? It would be a neat explanation of the recent reversal, adding to the tidiness of the peppered moth's story. Nature, however, is full of mysteries and ambiguities. Sir Cyril Clark, who collected the data, "cannot find significant changes in the prevalence of the lichens that supposedly hide light moths." Also, he has not been able to discover where the moths spend most of their daylight hours, but he seldom sees them on trees or walls (perhaps they're now too well camouflaged or Sir Cyril's eyesight is failing).

An alternate theory is that the white moths have another, unknown genetic advantage in the cleaner air. It would be satisfying for scientists to believe white moths have recurred because they are again difficult for birds to find, but so far conclusive proof

is lacking. One thing is certain: Scientists will be watching the colors of the little rascals for many years to come.

See also BUMPUS'S SPARROWS; GENETIC VARIABILITY; NATURAL SELECTION.

For further information:
Kettlewell, H. B. *The Evolution of Melanism.* Oxford: Oxford University Press, 1973.

PERMIAN EXTINCTION

See GREAT DYING.

"PHILOSOPHICAL" NATURALISTS
Searching for "Nature's Laws"

Beginning in the late 18th century, when most zoologists were concerned with collecting or discovering new species, the more curious sought regularities or "laws." They referred to themselves as "philosophical naturalists."

Natural philosophy—a system of general principles describing the nonliving world—had already been established by physicists and astonomers. But "natural history," the study of rocks, plants and animals, was still largely an exercise in description and classification. "Philosophical naturalists" may have enjoyed collecting, but they also shared a faith that the "laws" of life itself would soon be discovered.

They tried many approaches to find such laws. German poet Johann Wolfgang von Goethe thought he glimpsed a "unity of plan" behind living structures. In 1790, he analyzed flowering plants as variations on an "ideal" leaf and animal skulls as transformed vertebrae. Zoologist William Sharpe MacLeay invented a "quinary system" of classification (1819), based on his idea that all nature was built around the number five. In their day, these were worthy attempts by brilliant men, and there were many others.

All shared the conviction that discoverable, underlying principles would "make sense" of the facts of anatomy, geology, botany, animal distribution and fossils. Without philosophical naturalism, there could have been no evolutionary biology.

See also BIOLOGY.

For further information:
Rehbock, Philip. *The Philosophical Naturalists: Themes in Early Nineteenth-century British Biology.* Madison: University of Wisconsin Press, 1982.

PHRENOLOGY
Reading Character from Crania

During the 19th century, thousands came to believe in phrenology, which purported to "read" human character with "scientific" accuracy from bumps on the skull. Some went so far as to base hiring decisions—or even marriage plans—on the results of analyzing shapes of heads or facial features.

Phrenology was the "sister" to physiognomy (fiz-ee-OG-no-me)—the reading of facial features. Among its enthusiasts was young Captain Robert FitzRoy, who almost rejected Charles Darwin as ship's naturalist because his nose lacked angular definition. "The Captain doubted," Darwin reported, "whether anyone with my nose could possess sufficient energy and determination for the voyage." (Later in life, Darwin was amused when a phrenologist told him he had a "bump of reverence developed enough for ten priests.")

Franz Joseph Gall (1758–1828), an influential German physician and neuroanatomist, first developed the theory that scores of personality traits, talents and attributes are detectable by relative sizes of their corresponding "brain centers." Phrenologists believed they had mapped the mental "organs" that controlled "Ideality," "Sublimity," "Hope," "Veneration," "Firmness," "Secretiveness," "Destructiveness," "Amativeness" and "Love of Home"—to name a few.

Gall and his collaborator Johann Spurzheim had become famous by the 1830s, when their ideas were considered an important scientific breakthrough. Books and magazines based on phrenology continued to be popular for almost 100 years; occasionally there are still modern revivals.

During the 1870s, a family firm in New York, Fowler and Welles, made a fortune offering individual head-readings, training programs for phrenologists, traveling lecturers and publications. (The woman in that partnership, Lydia Folger Fowler, was America's second woman doctor.) Their *Phrenological Journal* was popular and progressive, often carrying articles on anthropology, animal behavior, evolution, psychology and social reform.

As a method of self-analysis, skillful use of phrenology was certainly as revealing as Tarot cards, personal horoscopes or other symbolic representations of character traits. Unfortunately, as a rationale for science it led to a great deal of social mischief. "Experts" gravely decided on "racial capabilities" on the basis of head shape; "craniologists" collected and measured skulls endlessly with no useful result. Criminal "types" were identified and classified by facial features, which we now know to be irrelevant. Worst of all, people were judging one another by cranial "bumps" or ear shapes rather than by their actual deeds and real capacities.

Darwin's codiscoverer of the principle of natural selection, Alfred Russel Wallace, staunchly defended

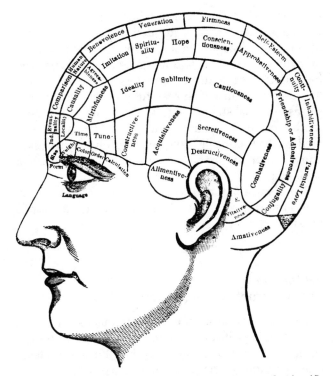

READING BUMPS on the skull was a popular pseudoscientific fad of the 19th century, originated by Gall and Spurtzheim. Despite excesses that now seem ridiculous, their system of studying "head bumps" was the first to encourage research on localization of brain functions.

phrenology. But by the beginning of the 20th century, biologists had rejected it so emphatically they denied science owed anything to it, even as they made advances in localizing brain functions through electrode stimulation.

"But this very fact of the connection of certain definite brain-areas with muscular motion is no new discovery, as modern writers seem to suppose," Wallace wrote in 1898, "but was known to Dr. Gall himself, although he [lacked] modern appliances for the full experimental demonstration of it." Gall knew full well, said Wallace, that stimulation at a brain site could cause a muscular action "as though you are drawn by a wire."

Almost a century later, historians admit he was right; phrenology was a necessary prelude to more sophisticated studies of localized brain function. Unfortunately, Wallace's enthusiasm led him to boldly predict that it would make a dramatic comeback in the 20th century:

Phrenology will assuredly attain general acceptance [as] the true science of mind. Its practical uses in education, in self-discipline, in the reformatory treatment of criminals . . . and the insane, will give it one of the highest places in the hierarchy of the sciences.

He believed it was rejected because his scientific colleagues had "narrow-minded prejudices" against his favorite ideas: phrenology, Spiritualism and mesmerism. For years he had railed against doctors who had fired a brilliant surgeon (also a phrenologist) because the man had successfully employed hypnosis as an anaesthetic. But Wallace himself fought tooth and nail against widespread adoption of a medical technique he believed "absolutely useless" and a "dangerous delusion"—vaccination against smallpox.

See also BIOLOGICAL DETERMINISM; FITZROY, CAPTAIN ROBERT; JUKES FAMILY; SCIOLISM; WALLACE, ALFRED RUSSEL.

PHYLA
The Systematist's Nightmare

When the great systematist Carl Linnaeus set up his system of classification in the 18th century, he encompassed all living things in the "Empire Biotes," a designation that is rarely mentioned today.

Within that "Empire" he set up the "Kingdoms" of plants and animals, and within them the Phyla—the great groups of organisms with differing basic structures or organization.

Most biology students are familiar with the Phylum Chordata—the animals with backbones. But when we turn to the invertebrate phyla, the ground gets a little shaky. Most biologists group the fungi in another phylum and blue-green algae is still another.

But to attempt to go farther is to lose even the experts in uncertainty and confusion. There are unicelled parasites like trypoanosomes (the carriers of malaria), which some put in a phylum all their own. Another phylum is represented only by one-celled creatures that digest wood and live only in the guts of termites, where they have evolved in a closed universe for eons. And there are many unclassifiable microscopic creatures that have been considered plants by botanists and animals by zoologists. Many of them even have two names—a zoological and a botanical. No one can really say if they are animals or plants.

While this is a very active field for contemporary research, at present scientists are not at all in agreement on how many phyla exist, how they are related to each other and which evolved from which. Adding to the confusion are a number of extinct phyla known only from fossils—such as some of the strange Burgess Shale creatures—whose relationship to other basic groups of organisms is entirely unknown.

See also BURGESS SHALE; CLADISTICS; EUKARYOTES.

PIGGY—BACK SORTING
Gene "Hitchhiking"

Sometimes natural selection favors a particular trait, and another, unrelated one gets carried along with it. The trait that comes along for the ride may seem puzzling from the standpoint of evolutionary "fitness"—it may appear to be nonadaptive or even harmful.

This hitchhiking process can be imagined as a child's toy that contains marbles of three sizes: small, medium and large. Inside the container are three sieves arranged in layers, each punched with different sized holes, corresponding to each of the three diameters of marbles. Turn the toy over, and the marbles will sort themselves out by size at each layer—for this is a selection machine.

But now suppose that the largest marbles are colored red, the medium-sized blue, and the smallest yellow. They will be sorted and separated out by color—red, blue and yellow. But the sorting mechanism, the sieve, is color-blind and cannot know it has sorted the marbles by color. It has selected for size only, but in the process the marbles have also been grouped by color.

An actual example of gene hitchhiking has been recently studied among a human population on the

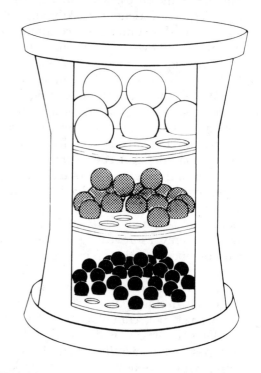

CHILD'S TOY "selection machine" groups balls by size, demonstrating how some attributes can ride "piggy-back" on those being sorted. Smallest balls happen to be black, next smallest gray, while others are white. Although the sieve selects for size only, the colors are sorted out as well.

island of Sardinia. The islanders have a very high proportion of a particular type of color blindness; researchers wondered why it is so common in this somewhat isolated population. Certainly, defective vision confers no advantages according to the theory of natural selection.

Genetic studies showed this kind of color blindness is "linked" with genes that confer a hereditary immunity from malaria, which is the crucial selective factor in leaving more offspring. Darwin made a list of such seemingly unrelated correlations among fruit trees, where "wrinkled skin," for instance, went along with immunity to certain insects, while certain colored fruits appeared along with resistance to various plant diseases.

Often, genes are passed on as a "package deal," linked together on chromosomes. Natural selection acts on the whole "package," even if it should contain a deleterious gene mutation. The success of such linked genes has been compared to a football game in which the whole team wins the trophy, including the jerk who kept fumbling the ball.

PILTDOWN MAN (HOAX)
Famous Fossil Forgery

On December 18, 1912, newspapers throughout the world blared sensational headlines: MISSING LINK FOUND—DARWIN'S THEORY PROVED. The source of all the excitement was a gravel pit at Piltdown, Sussex, in the southern English countryside, where a local amateur archeologist had found "The Earliest Englishman."

William Dawson, the discoverer, had been searching for fossils and stone tools for many years. Assisted by workmen and the young French Jesuit priest Teilhard de Chardin, he found pieces of a skull (cranium) and jaw, along with remains of ancient mammals, and some stone and bone tools. Proudly he proclaimed a new species, *Eoanthropus dawsonii*, "Dawson's Dawn Man," which was named and authenticated by experts at the British Museum.

Remarkably, the skull appeared entirely human, but the jawbone fragment was ape-like. A good deal of interpretation was necessary: The face was missing and so were the parts of the jaw hinges that join the skulls. While a few scientists questioned whether they really belonged together, most rushed to embrace Piltdown as a genuine intermediate between humans and apes.

Artists made imaginative reconstructions of his face, and statues of his presumed physique graced museums. In the U.S. there was even a popular comic strip in the Sunday papers called *Peter Pilt-down*, precursor of *Alley Oop* and "The Flintstones."

Most prestigious British anthropologists put their names and reputations on the line in authenticating Piltdown. When Sir Arthur Keith was challenged on his reconstruction, he gave a dramatic demonstration of his prowess. A known skull was smashed into pieces, and he correctly reconstructed its shape and cranial capacity from a few fragments. Others who championed the authenticity of the "great discovery" included anatomist Arthur Smith Woodward and Sir Ray Lankester, director of the British Museum.

Forty years later, the famous bones again made world headlines: PILTDOWN APE-MAN A FAKE—FOSSIL HOAX MAKES MONKEYS OUT OF SCIENTISTS. Back in 1911, someone had taken a human cranium and planted it at the gravel excavation together with a doctored orang-utan jaw. The orang teeth had been filed to make them look more human, and the jaw was deliberately broken at the hinge, to obscure correct identification. All the fragments had been stained brown with potassium bichromate, which made them appear equally old.

Piltdown's authenticity was repeatedly questioned over the years, but conclusive proof of fraud came in the 1950s, when the British Museum's Kenneth Oakley devised a new method for determining whether ancient bones were of the same age. His radioactive flourine test proved the skull fragments were many thousands of years older than the jaw. They could not be from the same individual unless, as one scientist put it, "the man died but his jaw lingered on for a few thousand years."

Critics pounced on Piltdown as an example of the weakness of evolutionary anthropology. Looking back, it appears that British scientists, fed up with news of sensational fossil men found in Germany and France, strongly craved an ancestor of comparable age. Besides, Sir Arthur Keith had theorized that the "big brain" came first in human evolution and was the hallmark of humanity. Piltdown filled the bill. Because of that bias, he had pooh-poohed Raymond Dart's australopithecine discovery in Africa—a creature with an ape-sized brain and "human" teeth and jaw, exactly the opposite of Piltdown.

Successful hoaxes generally share two characteristics: they prop up questionable but cherished beliefs while supporting local pride or patriotism.

In their enthusiasm, few thought it strange that the "earliest man" should have been found about 30 miles from the home of Charles Darwin or that a strange, paddle-like bone implement found in the pit resembled nothing so much as a prehistoric cricket-bat—a bit *too* appropriate an artifact for the "first Englishman."

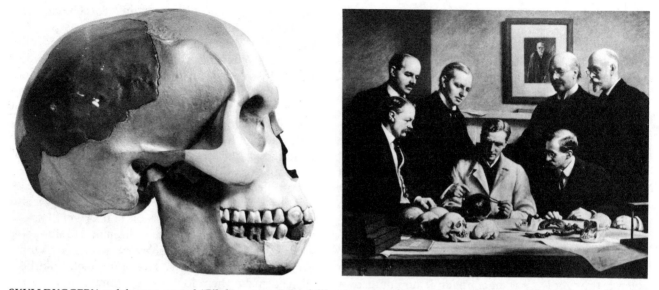

SKULLDUGGERY and forgery created "Piltdown man," the deliberate combination of fragments of a human skull and orang-utan jaw that fooled anthropologists for decades (left). Painting of principal scientists who authenticated Piltdown (right) shows Sir Arthur Keith in center, reconstructing a test skull from fragments—as Charles Darwin's portrait gazes down from the wall.

When the hoax was finally discovered, there were cheers and jeers. Creationists proclaimed all evolutionary science as phony. Anthropologists said the exposure proved their discipline is self-correcting and eventually roots out frauds. Advocates of the African australopithecine fossils felt vindicated in their view that hominids had small canines and walked upright very early on and only developed expanded brains much later.

A member of the British Parliament proposed a vote of "no confidence" in the scientific leadership of the British Museum. The motion failed to carry when another M. P. reminded his colleagues that politicians had "enough skeletons in their own closets."

The Piltdown hoax remains one of the most intriguing mysteries in the history of science. Who was the culprit, and what was his motive? For years, the finger of accusation pointed at Dawson. His motive would have been the fame and attention he gained as discoverer of England's most ancient inhabitant. During the past decade, historians of the hoax have implicated others.

Many accept the conjecture of Harvard's Stephen Jay Gould, who believes Dawson initiated the hoax as a prank, then enlisted the cooperation of his some-time-assistant, Pierre Teilhard de Chardin, who was then a young seminarian.

A minority view, advanced by Professor John Winslow, pins the real-life mystery on Sherlock Holmes's creator, Sir Arthur Conan Doyle, who lived a short distance from the Piltdown site. A fanatic Spiritualist, Doyle's motive might have been revenge against the scientists who ridiculed both his beliefs and his credulity in accepting the evidence of professional spirit-mediums.

E. Ray Lankester, for instance, who later became director of the British Museum (Natural History), incurred Conan Doyle's lasting resentment as the first scientist to prosecute a "spirit-medium" for fraud in police court. Conan Doyle had believed in the integrity of the accused, Henry Slade, and never forgave Lankester for "persecuting" him. [See SLADE TRIAL.] Winslow has also suggested that perhaps Conan Doyle wanted to demonstrate how easily scientific experts would uncritically accept evidence for their own beliefs, even as they scoffed at the authenticity of his photographs of "spirit-beings."

So far, there is still no proven solution to the mystery. Gould sticks with Teilhard and has amassed considerable circumstantial corroboration. After publishing one article, Winslow never followed up on his case against Conan Doyle—presumably for lack of hard, clinching evidence. Other possible candidates have been offered, as have stories of mysterious tapes-recorded accusations by aged survivors, and apocryphal anecdotes that muddy the waters.

Professor Charles Blinderman, in *The Piltdown Inquest* (1987), furnishes a detailed review of the full roster of about 10 of the "usual suspects." After reviewing all the tantalizing theories, he suggests that the fictional "missing link" was planted by a fictional culprit: the ubiquitous Victorian villain, Professor Moriarty. His motive? To discredit the creator of his nemesis, Sherlock Holmes!

See also CALAVARAS SKULL (HOAX); FLUORINE

ANALYSIS; LANKESTER, E. RAY; SPIRITUALISM; TEILHARD DE CHARDIN, FATHER PIERRE.

For further information:
Blinderman, Charles. *The Piltdown Inquest.* Buffalo: Prometheus Books, 1986.

Weiner, J. S. *The Piltdown Forgery.* Oxford: Oxford University Press, 1955.

PITHECANTHROPUS ERECTUS

See HOMO ERECTUS.

PITHECOPHOBIA

See ARISTOGENESIS.

PITT–RIVERS MUSEUM
Evolution of Technology

Colonel Auguste Lane Fox, a gentleman of good family but modest means, was a leader of the "prehistoric movement" within the Darwinian revolution. Not only did he believe that the races of man had evolved in "stages," but so had cultures and even tools and artifacts. An avid collector of bows, clubs and stone tools, he dreamed of creating a museum where they could be exhibited in evolutionary series. It would show, step-by-step, the development of human technology.

Lane Fox got his opportunity in 1880, when he inherited the huge 29,000-acre estate of his great uncle, George Pitt, the second Baron Rivers. (Twelve intervening heirs had died through illness or accident.) Convinced he was carrying out the will of God to reveal the laws of evolution, he took the name Pitt-Rivers and began excavating numerous prehistoric sites on the estate.

Within a short time, the immense collection of prehistoric and tribal artifacts of now-Major General Lane Fox Pitt-Rivers had outgrown two small museums. He offered to donate the entire collection to the British Museum—with one catch. The museum must guarantee to keep his artifacts permanently exhibited in the "evolutionary series" he had arranged, according to "laws of technological evolution" he had discovered. Trustees refused to meet that condition, so the Major General took his legacy to the world elsewhere.

In 1882, Oxford University agreed to accept the collection on Pitt-Rivers's terms, and he underwrote construction of a separate building to house it, as an annex to the university museum. He also provided funds for the preparation and preservation of the collection and stipulated that the university establish regular lectures in anthropology.

Pioneer anthropologists E. B. Tylor and Baldwin Spencer helped arrange exhibits in the new Pitt-Rivers Museum, which was later directed by Henry Balfour. A typical publication of Balfour's was *The Natural History of the Musical Bow* (1899), in which he chose specimens from the collection to demonstrate the evolution of a weapon into a musical instrument. He believed he could classify all such artifacts, like plants or animals, into a system of "genera," "species," and "varieties."

One tragic result of Pitt-Rivers's obsession with evolutionary classification of tools was what historian George Stocking calls "the paleolithic equation." Stone tools from the ancient past of Europe were exhibited side by side with those of contemporary Tasmanians or Australians. The implication was that because flint axes, knives and spearheads appeared similar, the modern tribes were "stone-age people," biologically equivalent to long-extinct ancestral Europeans.

As Stocking has written in his book *Victorian Anthropology* (1987):

> Left behind long since by the ancestors of the Europeans, [Tasmanians] had outlived their time by many thousands of years . . . extinction was simply a matter of straightening out the scale, and placing [modern tribal peoples] back into the dead prehistoric world where they belonged. Not only did the paleolithic equation help to distance the horror of the Tasmanian's extinction; it seemed even to set the seal of anthropological science upon their fate.

In recent years, uncertainty and debate about the future have surrounded the Pitt-Rivers Museum. Some believe Oxford should preserve the donor's Victorian "evolutionary" sequences of spears, bows, fire drills, "primitive" musical instruments and stone axes intact, which was, after all, the pledge it made to Pitt-Rivers. Besides, the entire museum itself is now a Victorian cultural artifact of historical significance.

Others feel the exhibits are misleading, outdated, racist and should be rearranged according to geographical or cultural areas or possibly along functionalist principles. One tongue-in-cheek notice in the *Royal Anthropological Institute's Newsletter* claimed that a group of "radical Museo-Marxists" want the tribal artifacts to be classified "according to their means of production."

See also RACISM (IN EVOLUTIONARY SCIENCE); "VERCORS" (JEAN BRULLER); WELLS, H. G.

For further information:
Stocking, George W. *Victorian Anthropology.* New York: Free Press, 1987.

PLANET OF THE APES
Endless Ape Epic

"May I kiss you?" a grateful Charleton Heston asks the female chimpanzee who saved his life. She offers her cheek, trying to suppress her obvious disgust at such intimate physical contact with a human. "I'm sorry," she says, "it's just that you're *so* ugly."

A wildly popular series of "Ape Planet" action movies began with a witty satirical novel by French author Pierre Boule (who had already scored a major movie hit with *Bridge On the River Kwai*). His novella *Planet of the Apes* (1963) was a humorous parable, tied loosely to an action-adventure plot. The screen version shifted its major focus to the violent action, which continued through the numerous lucrative sequels.

Orangs, chimps and gorillas in Boule's original parable act not so much like different species as takeoffs on our own social and vocational subcultures. Chimps are the true intellectuals. They see matters clearly, try to act decently and are flexible and innovative, which often lands them in trouble with the ape establishment. Orthodoxy in thought and policy is maintained by overly dignified orangutans, who serve on all scientific boards and committees. Gorillas are brutal, stupid and follow orders well; they form a state police or military caste. Among the film's most striking images are the legions of armed, leather-clad gorillas mounted on horseback.

Humans on their world have "degenerated" to the status of defenseless animals. Although anatomically like ourselves, they have lost their powers of language and sociocultural domination. Looks are not what make the apes and humans different; their bodies are close enough to function interchangeably. The crucial difference (as Lord Monboddo proposed in the 18th century) is in their thought, communication and behavior. That subtext of the *Planet of the Apes* series is a reflection in popular culture of the changing status of apes in science. There, but for the grace of language and symbol-using, go we.

Boule's original plot makes us reexamine our own treatment of apes. In *Planet*, the apes speak and wear clothes, while naked humans are rounded up and caged for medical experimentation. When the astronaut Taylor (Charleton Heston) tries to protest, he finds that no ape will believe humans can speak, since most of the individuals they've seen have degenerated into mute animals.

During a jury trial to determine whether Taylor is capable of intelligence, the three ape judges form a "no-see, no-hear, no-speak" tableau, which sci-fi writer Douglas Murray suggested may be "homage to the Scopes Monkey Trial and its continuing aftermath."

In 1968, the first film version of *Planet of the Apes* was released, with a screenplay by Rod Serling (creator of the classic television series "Twilight Zone") and Michael Wilson, a screenwriter who had been blacklisted during the McCarthy era. Produced by

HUMAN LAB ANIMALS are captured for study by chimpanzee scientists in the popular fantasy-adventure series *Planet of the Apes,* based on Pierre Boule's anthropological satire. Here actor Charlton Heston and fellow specimen are herded to research center by gorilla security forces.

Arthur Jacobs, the film starred Heston, Roddy McDowall, Kim Hunter, Maurice Evans, James Whitmore and Linda Harrison.

Special prosthetic makeup for the apes' faces was created by John Chambers, who had originally developed the process to help disfigured war veterans. His masks were so successful (and so expensive) that the molds were used many times over in the sequels, thus amortizing their cost. The designs also became the basis of hundreds of popular toys, models and other merchandise.

Beneath the Planet of the Apes, the first of four sequels, was released the following year (1969). Although based on a flimsy premise (a spaceship sent to rescue the human astronauts falls into the same time warp), it was one of the most successful of any genre sequels.

The rescue expedition discovers apes using humans as practice dummies and experimental creatures while preparing for a great war. A cult of surviving intelligent humans worships The Bomb— the god that destroyed their species. At the end, both humans and apes try to set the thing off and succeed. (In the original screenplay, a mutant ape was to emerge from a partially wrecked world and immediately kill a nesting dove. The producers changed it to something they considered less heavy-handed: a general and complete holocaust.)

From here on, the sequels became hopelessly confused and painfully convoluted (*Escape from the Planet of the Apes,* 1971; *Conquest of Planet of the Apes,* 1972; *Battle for the Planet of the Apes,* 1973). Characters are resuscitated, time warps jump from past to future and back again, and Roddy McDowall ends up playing his own son.

Despite puerile plots that are quickly overwhelmed by battle scenes and "action adventure," the popularity of these fantasies marks a change in mass perception of apes. No longer is the movie ape a monster (King Kong), a clown (Cheetah, Bonzo), or a "primitive man" (*Greystoke, 2001: A Space Odyssey*). The apes hold social roles, but are also individuals—mostly conforming and brutal, but sometimes rebellious and heroic; they demand to be taken seriously.

See also APES; HOMO SAPIENS, CLASSIFICATION OF; KING KONG; MONBODDO, LORD.

PLATE TECTONICS
Traveling Continents

As late as 1987, some scientists still doubted whether the "new geology" was here to stay. That Africa or North America moved around, migrating from one part of the world to another, seemed incredible. Yet, evidence that the idea of continental drift was valid kept piling up. And then a space-age technique convinced even the skeptics—by actually measuring the speed of plate movement with a "ruler" from beyond the stars.

Earth's crust is a thin outer shell, made up of about 20 tectonic plates. Each plate picks up molten volcanic rock from the Earth's deeper layers, which spews up through rifts on the ocean floor. As a layer of magma cools, its metallic particles align themselves with the Earth's magnetic poles, which change direction a couple of times every million years. Crustal layers, therefore, are made of alternating bands of rock with different magnetic directions, "freezing" a history of their formation. Thickness of the bands may indicate length of intervals or varying speed of Earth's movements.

Another line of evidence comes from similarity of rocks and fossils that appear in continents that are widely separated today. For example, Edwin Colbert's discoveries of *Lystrosaurus,* in South Africa, India and Antarctica, supports the idea they had once formed part of a single land mass now known as Gondwanaland.

Dozens of other studies added support to the theory. And the U.S. National Aeronautics and Space Administration sponsored a Crustal Dynamics Project that, in 1987, reported it had actually measured the rates at which continents move.

Since continental movement is very slow, it was necessary to compare their positions to a point that seemed almost perfectly fixed. James Ryan of Goddard Space Flight Center and his colleagues chose immensely distant quasars that emit radio noise from a far side of the universe.

Huge radio antennas were built in Alaska, Hawaii and Japan; the millisecond differences in their reception of the same signals gave a very precise way of measuring the antennas' distances from one another. After three years of tracking changing reception times, a clear pattern of continental movement emerged.

Hawaii and Alaska are moving about 52.3 millimeters closer each year, while Hawaii is moving toward Japan at a rate of about 83 millimeters per year. The Pacific plate is moving much faster than its Atlantic counterpart; North America and Europe are moving farther apart by 17 millimeters each year.

Collisions between plates are thought to thrust up mountains. Relatively recent ramming of the Indo-Australian plate into Asia, for example, is believed to have created the Himalayan range. When the edge of one plate dives beneath the edge of another, some crust returns to the mantle; the phenomenon is called a subduction zone.

Earthquakes occur along "fault lines," which are now seen as meetings of plate boundaries. The famous San Andreas fault in California, responsible for the infamous San Francisco quake of 1906, is really a boundary between the North American and Pacific plates. A medium-range quake rocked northern California in October 1989, destroying freeways and a section of the Bay Bridge.

Geologists have debated for years how severe the next California quake will be and how predictable. Until very recently, there was really no way to know. But with measurements from satellites and outer-space signals, it is now possible to calculate the precise rate of plate movement—an important step toward predicting the timing and severity of a quake.

Studies of plate tectonics have begun to explain many diverse phenomena, but some basic questions are still wide open: What drives the plates to move? Why does the Earth's magnetic field reverse itself? Do "pulses" of Earth activity cause changes in climate? As always in science, new answers bring with them newer questions.

See also CONTINENTAL DRIFT; GONDWANALAND; LAURASIA; WALLACE'S LINE.

For further information:
Glen, William. *Continental Drift and Plate Tectonics.* Columbus, Ohio: Charles E. Merrill, 1975.

Seyfert, Carl K. *The Encyclopedia of Structural Geology and Plate Tectonics.* New York: Van Nostrand Reinhold, 1987.

Wyllie, P. J. *The Way the Earth Works: An Introduction to the New Global Geology and Its Revolutionary Development.* New York: John Wiley, 1976.

PLATYRRHINES
New World Monkeys

New World monkeys are a large group that appear to have separated from their Old World ancestors in Oligocene times (24-to-37 million years ago). Once established in Central and South America, they radiated into many divergent species. The group is named from its "flat" (platy) nose (rhine), as opposed to the higher nostrils of the Catarrhines.

Much more distant from man than the Catarrhine (Old World) group, they are mostly arboreal creatures of the rain forest, feeding on plant and insect food (some on leaves) in the canopy. Among the best known are the howler monkey, whose resonant vocalizations can be heard for miles; the spider monkey, which can swing from arms and tail alike; the little squirrel monkey, often bought as pets, although they do not make good ones; and the Capuchin, which used to be the favorite companion of organ grinders.

See also CATARRHINES; PRIMATES.

"PLEISTOCENE OVERKILL" HYPOTHESIS
Extinctions Caused by Man?

At the very end of the last Ice Age, which extended from 70,000 to 10,000 years ago, 35 types of large mammals became extinct in North America. After surviving the rigors of this glacial period, mastodons, sabor-toothed cats, giant ground sloths and armadillos, native horses and camels, and giant beavers, bears and dire wolves all suddenly disappeared. For 150 years scientists have wondered why these hardy animals all died out about the same time—11,000 years ago.

Ecologist Paul S. Martin of the University of Arizona first advanced his "Pleistocene overkill" hypothesis in 1967, based on evidence from climatology, prehistoric archeology and paleontology. He believes the best explanation is that early human hunters were a major cause of the mass extinctions. Such a possibility flies in the face of the "conventional wisdom" that hunting man lived in harmony with a state of nature and was part of its "natural balance," or that populations of early man were too small to cause such widespread destruction of large mammalian species.

Early man in North America certainly hunted the great beasts. Fluted stone spear points of the Clovis type—named from a site near Clovis, New Mexico—have been found with the remains of mastodons, early horses, tapirs and camels. When hunting tribes migrated across the Bering land bridge from Asia to Alaska, Martin suggests, they entered a "hunter's garden of Eden"—a land of 100 million large game mammals with no experience of or adaptation to human predation.

An archeological colleague of Martin's, C. Vance Haynes, agrees that man may have played a major role in the Pleistocene extinctions, but thinks climatic changes might also have contributed. As glaciers melted, sea levels rose, which lowered water tables over much of North America causing large mammals to congregate in great numbers around the few remaining springs. If so, it would have made them even more vulnerable to cooperative hunts by human groups. Other scientists give even more weight to climatic changes, which may have caused wholesale destruction of habitats. But Martin maintains that changes in climate would not explain the heavy extinctions in the Americas and Australia, while the toll was much lighter in Asia and Africa.

Other ecologists and archeologists have joined in the debate; the "Overkill" hypothesis is far from proven. The timing of Ice Age extinctions is still not well understood, and radiocarbon chronologies have yet to be worked out. A neat "kill site" containing datable remains of human hunters with extinct North American mammals has still not been found. The Cutler Site recently excavated in southern Florida reveals 65 animal species, charred animal bones, human skeletons and projectile points in a cave, but the association among these various remains is not clear.

The questions still remain: Did most of the large North American mammals become extinct at the same time, about 11,000 years ago? Were climatic or other factors responsible, or were early human hunters the main reason for their disappearance during the last Ice Age?

PLENITUDE
Biotic Saturation

One of the old, traditional ideas that Darwin inherited from natural theology was the notion of "plenitude," which went along with the *Scala Naturae*, the Great Chain of Being. In a world created by a beneficent God, every possible niche in nature was filled by a species that "belonged" there. Only 100 years before, belief in plenitude even ruled out the concept of extinction, for it was considered sacrilege to believe that any gaps or imperfections in the natural order existed. When fossil remains of giant sloths or mastodons turned up in 18th century America, most naturalists assumed the unknown creatures must still exist somewhere in the wilderness, since God's "ideas" (species) could not be lost or destroyed.

It was Georges Cuvier, the great French paleontologist, who first dared to establish that some species had actually disappeared from the face of the earth forever. However, even after extinction was accepted as a reality, it was believed that the roles occupied by species that had disappeared were replaced by new species, so that the "economy of nature" retained its balance.

Steven M. Stanley, in *The New Evolutionary Timetable* (1981) points out that the continuing influence of "plenitude" on Charles Darwin caused him some thorny problems with his theory of natural selection. First, how could species vary significantly in a world into which they were packed niche against niche, occupying every possible place in nature? If selection winnows slight variations among individuals, there would be no room for it to operate if all niches were filled with species adapted to them. And how could these species move into new niches? [See RED QUEEN HYPOTHESIS.]

If Darwin's belief in plenitude gave natural selection little room to operate, he thought perhaps small waves might travel through the whole system, causing slow "insensible" changes over long periods of time in very large populations. Stanley argues it was Darwin's commitment to plentitude that forced him to adopt that model of slow, gradual change, in which nature has no gaps. Since the weight of modern evidence indicates plenitude is unfounded, Darwin's "gradualism" seems recently to have given way to the punctuational model.

In fact, recent studies seem to support the view that "the world is packed loosely enough with species that much variation is tolerated." Rather than very small, gradual changes in large worldwide populations, sometimes small populations diverge rapidly into distinctive new species. Evolution may be more jerky and "experimental" than Darwin thought, trying out new local species, which come into existence rather quickly. Then they may either move into empty niches, create new ones, or outcompete already existing species in similar habitats.

See also DIVINE BENEFICENCE, IDEA OF; GRADUALISMM; GREAT CHAIN OF BEING.

POIKILOTHERM
Cold-Blooded Animals

For most of evolutionary history, most animals were "poikilothermic"—their body temperatures dependent on the air or water that surrounds them. Insects, fish and reptiles are living poikilotherms.

"Cold-blooded" is a somewhat inaccurate description, since many are able to keep their body temperatures fairly high. Lizards, for instance, raise their metabolism by basking on sun-heated rocks. Creatures that produce and regulate their own internal temperatures, such as birds and mammals, are known as "homeothermic."

POPULATION CYCLES

See COEVOLUTION.

POPULATION THINKING
Shift in Biological Concepts

According to Harvard evolutionist and historian of biology Ernst Mayr, Darwinian evolution is based on a major shift of thought from "essentialism" to "population thinking." It is the difference between seeing a species as the "ideal type specimen" once favored by collectors and the "reproductive population" observed by field naturalists.

The older view of "types" was compatible with divine creation, "archetypes" and a zoology based on study of museum-drawer specimens. But Charles Darwin and Alfred Wallace had deliberately gone beyond being "mere collectors" (though they never forgot its allure) and shifted to trying to understand species in the wild, natural breeding populations. Variability among individuals was the natural condition of a species—the raw material of natural selection—not a "departure" from some ideal individual.

Darwin and Wallace were familiar with the studies of Belgian anthropologist Lambert Quetelet, who in 1842 had showed how normal frequency distribution curves applied to human population measurements; they were also inspired by Thomas Malthus's discussions of population growth and resources. Although the coauthors of natural selection theory were both untalented at mathematics, their basic grasp of population thinking opened biological problems to mathematical analysis. Darwin's cousin, Sir Francis Galton, founded the biometrics movement, which eventually resulted in sophisticated studies of population dynamics and population genetics—the foundations of current research into biological evolution.

See also BIOMETRICIAN-MENDELISM CONTROVERSY; ESSENTIALISM; SPECIES, CONCEPT OF.

For further information:
Crow, J., and Kimura, M. *An Introduction to Population Genetic Theory*. Minneapolis: Burgess, 1970.
Mayr, Ernst. "Typological versus Population Thinking." In *Conceptual Issues in Evolutionary Biology: An Anthology*, edited by E. Soper, Cambridge: Bradford/MIT Press, 1984.
Hutchinson, G. Evelyn. *An Introduction to Population Ecology*. New Haven: Yale University Press, 1965.

POSITIVISM
The "Scientific" Religion

Positivist philosophy was a peculiar beast—on the one hand helping define science and, on the other, creating a screwball religion of its own. Its founder, the Frenchman Auguste Comte (1798–1857), disavowed his countryman Jean-Baptiste Lamarck's idea of progressive evolution in biology, but insisted on the evolution of human understanding.

Comte (and his mentor Claude Saint-Simon, 1760–1825) had drunk deeply from the cup of the French Enlightenment philosophers. They believed in inevitability of human progress, the perfectability of man and the irresistible advance of scientific knowledge—all tending to an enlightened, benevolent, egalitarian society in the near future. Saint-Simon had written, "The golden age, which a blind tradition has hitherto placed in the past, is before us."

Human knowledge, Comte taught, must evolve through three stages. First, it is theological: Workings of nature are considered to be mysterious, the unpredictable will of God, the result of intervention by a First Cause. In the next stage of human understanding, philosophers look for nature's regularities or "second causes," although they may still cling to the idea of an underlying supernatural creator. Finally, Comte proclaimed people will refuse to speculate about any ultimate causes, which are incapable of proof. They should be concerned only with "positive facts" and "proven laws," which will be discerned in all phenomena and can form the only "positive" basis for human action and belief.

English logician and philosopher of science John Stuart Mill became interested, and wrote an influential book on *Auguste Comte and Positivism* (1865), which raised Comte's stock considerably among English intellectuals. Mill, Darwin and other scientists were sympathetic to the Positivist's attempts to find "laws" and regularities in nature; to establish morality based on reason, not revelation; and to question nature directly, without intereference from churchly assumptions or authority.

As Positivism gained ground in Britain, something amazing happened among adherents of this "rational," "scientific" philosophy—they tried to turn traditional religion inside out. Led and inspired by a kind of scientific priesthood, Positivists were going to remake society from top to bottom. Humanity itself, as the source of truth, beauty and knowledge, was to become the object of worship and veneration. Members would wear little statuettes of a "Goddess of Humanity" around their necks in place of the Virgin, familiar trappings "to ease the transition," as they put it.

Comte himself decided to begin this glorious "Age of Positivism" with a new calendar. Year One of the Positive Era (P.E.) was to be 1789, the beginning of the French Revolution. Henceforth, a year would be divided into 13 months of 28 days each. Months were to be renamed after great men of Western civilization, such as Moses, Homer, Shakespeare, Descartes, etc. Each day of the month would commemorate a lesser luminary in that field. (For instance, in the month of Moses, days were named for prophets; in Homer, for poets; in Descartes, for philosophers.)

Positivism championed a "scientific" view of the world in all departments of life. In time, Comte believed, it would be possible to discern the "laws" of human social behavior and sociology would acquire the precision of physics or chemistry. Social sciences would then become the crowning glory of all sciences and the best guide for human destiny. (Despite a modest shortfall on these aims, Comte *did* help to found and establish sociology.)

While Mill had supported the Positivist quest for "natural laws," and even their attempts to discredit traditional interpretations of "God's Will," he was appalled when Comte tried to transform himself into the high priest of the New Religion of Humanity. Comte's quest for unity, Mill complained, had become the imposition of uniformity. What had been proclaimed an open-minded search for new truth was rapidly hardening into narrow dogma.

Positivist leaders, such as Richard Congreve and Frederic Harrison (both former Church of England theologians) had become obsessed with detailing the new rational rituals and secular sacraments. Master schedules were being drawn up for the scientific improvement of society. In France, under Comte himself, the plan was to replace Napolean with a Positivist triumvirate, which would eventually empower a kind of Positivist pope—all in the name of devotion to Humanity.

Thomas Henry Huxley, who had at first applauded Positivist scientific values, quickly distanced himself from their attempt at a secular religion. With characteristic wit, he noted they were simply "re-inventing Catholicism minus Christianity."

Despite its initial splash, Positivism lasted only about 25 years and never became a widely popular movement. Nevertheless, it was an instructive, and perhaps inevitable, product of the century of Charles Darwin and Lewis Carroll. Some of Comte's ideas on science and sociology continue to be taught in universities today—usually with no reference whatsoever to his role as high priest of the New Religion of Humanity.

For further information:
Smith, Warren S. *The London Heretics 1870–1914*. New York: Dodd, Mead, 1968.

PRE–ADAMITE CREATIONS
Before the "Present World"

In the 18th and early 19th centuries, geologists and paleontologists commonly used the term "pre-Adamite" for animals and plants they believed existed before the time of Adam.

When they could not reconcile new fossil discoveries to Scripture, geologists assigned them to "previous creations" outside the scope of the Bible. All animals of the "present world," they believed, were created as fixed species at the same time as Adam and Eve. (We now consider *Homo sapiens* a very recent arrival, evolving later than most of the animals with which we share the Earth.)

Books published right up to the time of Darwin's *Origin of Species* (1859) refer to "Pre-Adamite" creatures inhabiting a "Pre-Adamite World." There were many successive creations, it was thought, separated by catastrophes (floods or fires), after which new kinds of animals and plants were separately created. Scriptural geologists saw no conflict in accepting many previous creations before the one described in Genesis. Professor D. Lardner of University College, London, writing in 1856, claimed there is "no discordance between Scriptural history and geological discovery, [only] the most remarkable and satisfactory accordance with natural phenomena."

By Lardner's time, an immense body of detailed information had accumulated on stratigraphy, glaciation and fossil organisms. Such dramatic large vertebrates as plesiosaurs, pterodactyls, giant sloths and sea-going ichthyosaurs were well known. Disturbing as the growing mass of evidence may have been to some, many geologists continued to accept innumerable "pre-Adamite" creations side by side with biblical accounts.

At least one artist of the day slyly wondered if the present creation would be succeeded by still another, in which *post*-Adamite giant reptiles would reign again. A popular cartoon shows a Professor Icthyosaurus lecturing his reptilian students on a human skull. "With such small teeth and weak jaws," says the professor, "it is altogether wonderful that this creature was able to procure food."

PREFORMATION THEORY
Created Once and Forever

A popular traditional Russian toy is a doll that contains a smaller doll inside it; that one contains a still smaller doll; and so on until, after 10 or 15 progressively smaller versions, one reaches a tiny miniature doll. The 18th century theory of how life produced copies of itself, "Preformation," asserted that the tiny seeds or "germs" in a woman's body contain miniature versions of her offspring already fully formed, but too small to see, although some who looked through crude microscopes did believe they saw tiny, preformed humans or "homunculi." All they do during pregnancy is grow larger; they were present in the mother's body when she was born. And her own "germ" was contained within her mother, "and so on back through the generations to the first woman, Eve."

Historian of ideas Peter J. Bowler says the theory "held that the whole human race literally was created by God in the beginning, enclosed one within the other, like a series of Russian dolls, waiting to be unpacked generation after generation." Similarly, "evolution" was used in the sense of "unrolling" or "unfolding" something already present, which of course implies a predetermined direction or goal.

See also BIOGENETIC LAW; EVOLUTION; MENDEL'S LAWS; ORTHOGENESIS.

PREHISTORIC MOVEMENT

See BOUCHER DE PERTHES; LUBBOCK, SIR JOHN.

PRESTWICH, SIR JOSEPH (1812–1896)
Confirmed Prehistoric Man

Clergyman-geologist Sir Joseph Prestwich did not discover any stone tools or fossil men and never publicly suggested mankind was older than church teachings acknowledged. Yet he is credited with establishing the fact that man coexisted with long extinct mammals and that humans lived on Earth much earlier than anyone had previously suspected.

When a French gentleman from Abbeville, Boucher de Perthes, announced in 1846 that he had excavated ancient man-made flint implements sealed with undisturbed bones of rhinos and mammoths in local gravel beds, no one paid him the slightest attention. Three years later he found many more tools with more remains of extinct mammals, and again he was ignored and ridiculed. Only one French scientist came to see for himself, became convinced and came to be scorned along with Boucher de Perthes.

In 1858, Darwin's friend Dr. Hugh Falconer saw Boucher de Perthes's flint collections and wrote to Sir Joseph Prestwich, asking him to visit Abbeville. A respected churchman, Prestwich was renowned as a highly competent geologist who had done careful studies of English gravels. He also was known as a scrupulously honest observer who avoided conjecture, speculation and theological controversy.

Prodded by Falconer, Sir John Evans and Sir John Lubbock, Prestwich went to Abbeville and Amiens in 1859 (the year Darwin's *Origin of Species* was published) to see the gravel beds. After examining the evidence, he said he was convinced: (1) that the flints showed unmistakable human workmanship; (2) that they occurred in undisturbed sediments; (3) that they were associated with remains of extinct mammals; and (4) that the time period was late in Earth history, but before the formation of the present land surface.

Prestwich reported his observations, but made no attempt to account for them or explain their significance. As a devout Anglican, he abstained from drawing the obvious conclusion about extending the church's currently held chronology. He gave a truthful account of what he saw and left the inferences to others.

Nevertheless, only after Prestwich verified the claims of Boucher de Perthes did anyone in England, France or the rest of the world take them seriously. He, more than any other geologist, established prehistoric man as fact.

The impact of Prestwich's immense credibility as a clergyman-geologist illustrates the complex relationship that existed between science and religion in the 19th century. Science owes much to the churchmen who developed geology, natural history and genetics—the two ways of thinking were not always "at war." Frequently, they coexisted in the same individuals.

See also BOUCHER DE PERTHES; LUBBOCK, SIR JOHN; MILITARY METAPHOR.

PRIMATES
Lemurs, Monkeys, Apes and Man

Within the Class of hairy, warm-blooded, milk-nurturing creatures known as mammals is the Order *Primates* (in Latin, pronounced Pri-MATE-eez). It was so named by Linnaeus, meaning "the highest" or "first rank" because it contains the creatures most closely resembling man, presumed to be "the pinnacle of creation." (This great classifier ranked all other mammals as "secundates," and the rest of the animal kingdom "tertiates.")

Today there are about 180 living species of primates, many of them in imminent danger of extinction as their forest habitats are being destroyed by their human relations. South American monkeys, for instance, have evolved in their rain forests for more than 40 million years, but vast ranges have been ravaged and destroyed during the last 40 years—one-millionth of their history.

Primates have hands with flat nails rather than claws, and many show a cluster of features evolved in the trees—a dangerous, three-dimensional world where one slip can mean a fatal fall. Many have flattened faces with the eyes on a single plane, allowing three-dimensional ("binocular") vision, crucial for navigating the complex lattice of vines and branches.

In animals with longish skulls, the eyes are positioned on either side of a snout; they cannot perceive depth, but rely more on odors for information. Tree-living puts similar pressures on nonprimates as well. Squirrels among rodents, racoons and cats among carnivores have all evolved flatter faces, binocular vision, and more adept "handedness" than their ground-dwelling cousins. Behavior of tree-climbing creatures also tends to be more flexible and quick.

One major division of the Order, the Prosimians (meaning "before monkeys"), is in some ways similar to very early primates. It includes all lemurs, which are now entirely confined to the island of Madagascar, off the eastern coast of Africa, where isolation preserved them from competition with monkeys. They are vaguely monkeylike, but with protruding, foxlike snouts, long, bushy tails and retain some claws along with flat nails on their digits.

American monkeys evolved independently of Old World primates from very distant common ancestors. Unlike their African counterparts, New World forest monkeys have prehensile tails, with which they can hang from branches or grasp objects. These Platyrrhines ("flat-noses") include the wooly, spider, howler, squirrel and capuchin monkeys, which not long ago ranged widely throughout the forests of Central and South America. (North America has no native monkeys.)

Old World monkeys, apes and humans form a super-family within the primates. With the exception of man, most are confined to Asia and Africa. (The last "European" monkey is a remnant colony of macaques on the Rock of Gibraltar.) Among the monkeys are tree-livers such as the African colobus and "masked" guenons, mixed-habitat Indian rhesus and langurs and ground or rock-dwellers such as baboons, drills and mandrills.

Old World monkeys are mainly quadrupeds. Even the tree-dwellers run along broad branches on all fours and cannot hang or swing. With shoulder joints "locked" as those of dogs or cats, they are unable to rotate their arms through a circle like apes or baseball pitchers.

There are four living apes: the gorillas and chimpanzees in Africa and the gibbons and orang-utans in Asia. All are tailless, have fully flexible shoulders and (gibbons excepted) are among the largest and strongest primates. The lighter gibbon is an acrobat, an arboreal specialist farthest removed from the others. Gorillas and chimps live mostly on the ground, usually walking with the aid of their hands, by shifting the weight forward onto folded-over fingers. (This has been misleadingly called "knuckle-walking"; apes do not walk on their knuckles, which would soon ruin the joints.) Orang-utans live in the high trees, climbing carefully and expertly, but can manage quite well on the forest floor, though they place the edges of their hands on the ground rather than the folded fingers.

In size, primates range from the two-ounce pygmy marmosets to 500-pound gorillas. There are solitary species and obsessively social species, part-time meat-eaters and full-time leaf-eaters. Some species are sexually monogamus, some promiscuous; others form male-dominated "harems," while still others are controlled by alliance-forming matriarchs. Several primate species never leave the trees. Some spend their whole lives on open plains, rocky cliffs, snowy mountains or among ruins of ancient stone temples. The order *Primates* contains some of the shyest, most inoffensive creatures on Earth; it has also produced the one species in evolutionary history that is alone capable of destroying all the rest.

See also APES; BABOONS; CARPENTER, CLARENCE RAY; CHIMPANZEES; DRYOPITHECINES; GORILLAS; ORANG-UTAN.

For further information:

Bourne, Geoffrey H. *Primate Odyssey.* New York: Putnam's 1974.

Kavanagh, Michael. *A Complete Guide to Monkeys, Apes and Other Primates.* New York: Viking Press, 1983.

Peterson, Dale. *The Deluge and the Ark: A Journey into Primate Worlds.* Boston: Houghton Mifflin, 1989.

Smuts, B. et al. eds. *Primate Societies.* Chicago: University of Chicago Press, 1987.

PROCONSUL
Possible Chimp Ancestor

Proconsul africanus, a possible Miocene ancestor of chimpanzees, is one of the Dryopithecines, a family of early apes that lived 12–15 million years ago. Although fragmentary remains of this hominoid are known from Europe, Asia and the Middle East, the most complete skull was found on Rusinga Island in Lake Victoria, Kenya, by Mary Leakey. Its lower molar teeth have the distinctive Y-5 cusp pattern, which is shared today only by apes and humans.

The genus name *Proconsul* whimsically memorializes a "celebrity" chimp named Consul who was a longtime favorite at the London Zoo. "Proconsul" means "before Consul"—in other words, Consul's great-great-grandaddy.

See also DRYOPITHECINES; Y-5 DENTAL PATTERN.

PROGRESSIONISM
Pre-Evolutionary Geological Idea

As geologists of the early 19th century dug more and more fossil animals out of the rocks, they began to notice a pattern in the strata or layers. At bottom were mostly invertebrates, above them a gap, then mostly fish, a gap, then amphibians, reptiles, and finally mammals—always in that order. In their sys-

tem, the idea of evolution did not yet exist, so they interpreted the series of animal groups (faunal succession) as a creation, followed by destruction, followed by a "higher" creation, again destruction, and so on.

Although the great French paleontologist Georges Cuvier had sidestepped the issue, some of his followers insisted that a brand-new creation had occurred after each "catastrophe." As world conditions changed, there was a series of separate creations, each one an improvement on the last. That idea was known as "progressionism." It was different from the later ideas of evolutionary "progress" because it did not maintain that all life through the ages is related and connected.

However, the idea of progressionism—the appearance on Earth of progressively higher forms—was one of the geological concepts that were forerunners to Darwin's evolutionary ideas. Most textbooks credit the geologist Charles Lyell, Darwin's mentor, with inventing a simple concept of "uniformitarianism," which Darwin then applied to living things. Actually, the history of that idea is much more complex. Progressionism was rejected by Lyell; he believed in recurring *cycles* of creation and destruction. In his view after a series of creations, the whole process was liable to repeat itself over again. For most of his life, he strongly resisted the idea of a single, unique history for all living things.

Louis Agassiz, the Harvard paleontologist who worked out the sequence of glaciers, staunchly opposed Darwinian evolution to the last. Agassiz taught a brand of progressionism that assumed each new creation reflected God's current conception, a step-by-step revision of all life.

"It did not occur to him what a blasphemy this interpretation really was," comments Harvard evolutionist and historian Ernst Mayr. "It insinuated that God, time after time, had created an imperfect world, and that he completely destroyed it [each time] in order to do a better job"—hardly a flattering picture of an omniscient Creator.

See also: AGASSIZ, LOUIS; CATASTROPHISM; LYELL, SIR CHARLES; UNIFORMITARIANISM.

For further information:

Bowler, P. J. *Fossils and Progress: Paleontology and the Idea of Progressive Evolution in the Nineteenth Century.* New York: Science History Publications, 1976.

PROSIMIANS
"Primitive" Primates

Prosimians (which means "before monkeys") are primitive only in the sense that they are more like early primates than any other group of living animals. But many of them have developed special adaptations, which make them quite different from their ancestors.

Among the prosimians are the lorises, lemurs and such oddities as the tarsier and the aye-aye. Most are confined today to the island of Madagascar, off the east coast of Africa. Most are shy forest animals, coming out only at night and are more often heard than seen, earning them the Latin name *Lemures*, or "ghost." Destruction of the forest by an increasing human population is making prosimians scarce, and their chances for survival precarious.

Older books include the little tree-shrew of Southeast Asia among the prosimians, but several studies in the 1970s established that they are not primates after all, but generalized mammals similar to those that first emerged during the late Cretaceous period. They are not now classified in any established group of mammals, but appear to be a "living fossil" with no close living relatives.

See also PRIMATES.

PROTOPLASM
Material Basis of Life

Many high schools still teach that "protoplasm," an undifferentiated organic compound, is the physical basis of life that fills the cells of every living creature. The term originated around 1840 and became dogma in biology. It was thought to be the material basis of all organisms, composed of chemicals such as carbon, amino acids, proteins, trace metals—a uniform substance that filled the cells of all creatures.

A hundred years ago the idea that life had a "material basis" made of common chemicals was controversial. Many thought there must also be some "vital" or "animate" component that chemists and biologists could not find under their microscopes. Thomas Henry Huxley jolted audiences as much with his lectures on protoplasm as with his expositions of the close relationship between man and apes.

In his classic lecture on "The Material Basis of Life (1868)," he explained that the protoplasm of one animal, when eaten by another, is transformed from lifeless matter back into living tissue.

After the lecture, he said, he intended to dine on mutton, and that "a singular inward laboratory, which I possess . . . will convert the dead protoplasm into living protoplasm, and transubstantiate sheep into man." Moreover, Huxley noted:

> If digestion were a thing to be trifled with, I might sup upon lobster, and . . . the crustacean would undergo the same wonderful metamorphosis into humanity. And were I to return to my own place by sea, and undergo shipwreck, the crustacean might . . . return the compliment; and demonstrate our

common nature by turning my protoplasm into living lobster.

Science still operates on the principle that there is a material basis to life, but the word "protoplasm" has been abandoned. During the past half century, as microscopes improved, cytologists have dissected numerous "substructures" with different functions within the cell. These minute bodies play roles in metabolism and other life processes, much like organs in multicellular animals. Earlier notions of a uniform, undifferentiated protoplasm were false; the instruments and techniques had not yet been developed to observe the wonderful complexities of single cells.

After the advent of high-power electron microscopes in the 1940s, whole new worlds (and new questions) came into view. Researchers began to find more differentiation, more particles with different functions and varying compositions of cells in different kinds of organisms. Some cells are now thought to be tiny communities composed of entities that may have had independent origins. Nineteenth-century concepts of a simple, undifferentiated "protoplasmic material" became antiquated.

See also EUKARYOTES; MATERIALISM.

PTEROSAUR, MacCREADY'S
Radio-Controlled Replica

Paul MacCready, an aerodynamics expert who built his reputation designing "impossible" flying machines, decided to recreate a sight that hadn't been seen by a living being on earth for 65 million years— a giant flying reptile winging through the skies. He envisioned whole flocks of them, as part of an outdoor prehistory museum filled with full-sized mechanical dinosaurs.

A former soaring champion, MacCready had already met extraordinary challenges in designing unorthodox flying contraptions. He built a human-powered airplane that flew a prescribed course in 1975 and won the $100,000 Kremer Prize for it. In 1979, he constructed another small plane powered only by the energy of its pilot, the *Gossamer Albatross*, and won the prize again when it crossed the English Channel. Since then, he has produced several more that run on solar cells and even flashlight batteries.

This time he was inspired by the 1972 discovery of a giant pterosaur fossil in a west Texas quarry. It was an astounding specimen of *Quetzalcoatlus northropi*, which had a wingspan measuring 36 feet! Its shape almost defied flight: a large head, long slender beak and no tail.

"Most airplanes are very static, 'dead' shapes," said MacCready. "Pterosaurs had many moving parts in complex relationships." His replica took his company, AeroVironment Inc. of Monrovia, California, two years to build, at a cost of $700,000. It is constructed of fiberglas, plastic and metal, has an on-board computer to coordinate its operating components and is radio-controlled from the ground.

Destined for the Smithsonian Institution's Air and Space Museum, the wing-flapping marvel performed successfully in a private flight over the Mohave Desert in early 1986 and is featured in the museum's film, *On the Wing*, about natural and mechanical flight.

However, its first public performance was a disaster. At the annual armed forces open house at Andrews Air Force Base outside Washington, D.C. on May 17, 1986, the pterosaur was lofted at 9:14 A.M. by a towline. Seconds later, its 18-foot wingspan twisted into a nosedive. After a brief realignment by the on-board computer, the replica's head detached from its body and the "creature" plunged to Earth. Although an emergency parachute opened behind it, descent was too rapid for the chute to buffer the hard crash landing.

"I'm chagrined, of course, but I'm not surprised," Mr. MacCready was quoted by the press. "As I've said before, just trying to make this thing fly was like trying to shoot an arrow with the feathers in front. The plate holding the head on had a crack in it, and apparently that was the weak spot . . . Now we know why pterosaurs are extinct."

Live creatures with the mechanical pterosaur's aerodynamic did in fact live and thrive for several million years during the Cretaceous era. They varied in size from giants like the Texas fossil to small species no larger than bats and ranged widely over the world. Whatever the reason for their extinction, an inability to stay aloft was not one of them.

MacCready's pterosaur was well publicized, and its disappointing debut was amplified on national television news broadcasts. The normally staid *New York Times* carried pictures of the crash on its Sunday edition front page (May 18, 1986) with a wry headline: PTERODACTYL REMAINS EXTINCT.

See also DINOSAUR RESTORATIONS.

PTOLEMY

See FLAT-EARTHERS.

PUNCTUATED EQUILIBRIUM
Episodic Evolution

When paleontologists Stephen Jay Gould and Niles Eldredge proposed to amend the established "Synthetic Theory" of evolution in 1972, they ignited a continuing controversy. Many biologists welcomed their new model of "punctuated equilibrium" as a valuable contribution to understanding evolution, while others bitterly resented its instant popularity.

Punctuated equilibrium (or "punk eek," as it has been nicknamed) views species populations as systems that display recurrent patterns of evolution. Rather than the smooth, gradual change imagined by Darwin—now known as "gradualism"—Gould and Eldredge suggested species tend to remain stable, changing little over long periods of time. (The system is then in "stasis" or "equilibrium.") Eventually, that stability is "punctuated" by an episode of rapid change. [But see TURKANA MOLLUSKS.]

Ever since Charles Darwin's day, the fossil record has posed a difficulty for evolutionists—and an arguing point for creationists—because it did not appear to confirm his notion of a slow, uniform development of species. Instead, some fossil organisms seem to persist through millions of years relatively unchanged, then disappear, after which a spate of new ones "suddenly" springs up. Sometimes (as with the dinosaurs) an entire group dies out, and then the rocks reveal the subsequent rapid appearance of many new forms (as with radiation of the mammals).

Darwin thought the rarity of complete evolutionary sequences could be explained because the fossil record was sketchy and incomplete. So few animals become fossilized and conditions for preservation are so rare, he argued, that the rocks present only a fragmentary sampling of gradual transitions. Most of the "in-between" fossils are lost, giving the appearance of a jumpy picture. Creationists interpreted the discontinuities to mean there has been no evolution at all—merely a series of successive creations.

A century later, a great deal more is known about the fossil record, and its "incompleteness" no longer seems so convincing. Masses of additional data only reinforce the same story and the worldwide record is consistent. It no longer seems to be an "artifact" of chancy preservation that we see little change in species for long periods, followed by extinctions, followed by rapid radiations. More likely, the rocks reflect real patterns in the history of life.

But what known processes could produce such rapid divergence of species? Ironically, it was Ernst Mayr of Harvard, one of the founders of the Synthetic Theory, who had proposed a plausible mechanism back in 1942. Mayr had suggested that speciation could occur fairly rapidly in small, isolated populations. Cut off from the larger gene pool by geographic barriers, a small amount of variation would be amplified by selection: the "founder's effect" of the biometricians.

Mayr's contribution was part of the Synthetic Theory, but for years paleontologists ignored it. It was a special case of evolutionary process, they thought, rather than a major explanatory principle. When punctuationists realized that conditions in small breeding isolates could account for rapid radiations, they gave Mayr's speciation model more weight. This shift in emphasis has already proved productive in interpreting distribution patterns of species.

Critics charge Gould and Eldredge with oversimplifying, claiming the Synthetic Theory has always encompassed both gradual and more rapid change. "Gradualism has become something of a straw man and whipping boy for the punctuationists," writes Verne Grant. He also argues that their emphasis on macromutation (large genetic jumps) and "founder's effect" is nothing new. However, he grudgingly admits that applications of the punctuationist perspective—for instance Steven Stanley's important study *The New Evolutionary Timetable* (1981)—have produced many new insights.

But scientific theories are never proved or disproved by argument or logic. Over the long run, the real test is whether they generate fruitful research, explain previously puzzling data and lead to important new discoveries. By these criteria, in the brief interval since it was proposed, the theory of punctuated equilibrium seems to be holding its own.

See also EXTINCTION; GOULD, STEPHEN JAY; MAYR, ERNST; "NEW" CATASTROPHISM; THEORY, SCIENTIFIC.

For further information:

Eldredge, Niles. *Life Pulse: Episodes from the Story of the Fossil Record.* New York: Facts On File, 1987.

———. *Time Frames: The Rethinking of Darwinian Evolution and the Theory of Punctuated Equilibria.* (Includes reprint of original Eldredge and Gould paper.) New York: Simon & Schuster, 1985.

Stanley, Steven. *The New Evolutionary Timetable.* New York: Basic Books, 1981.

PUNNETT, R. C.

See HARDY–WEINBERG EQUILIBRIUM.

PURE LINE RESEARCH
Limits of Variability

Alfred Russel Wallace and Charles Darwin had insisted that through gradual, continuous change, species could (in Wallace's phrase) ". . . depart indefinitely from the original type." Around 1900 came the first direct test of that proposition: the "pure line research" of Wilhelm Ludwig Johannsen (1857–1927). What would happen, Johannsen wondered, if the largest members of a population were always bred with the largest, and the smallest with the smallest? How big or how small would they continue to get after a few generations? Would they "depart indefinitely" from the original type, or were there built-in limits and constraints?

Experimenting on self-fertilizing beans, Johannsen selected and bred the extremes in sizes over several generations. But instead of a steady, continuous growth or shrinkage as Darwin's theory seemed to predict, he produced two stabilized populations (or "pure lines") of large and small beans. After a few generations, they had reached a specific size and remained there, unable to vary further in either direction. Continued selection had no effect.

Johannsen's work stimulated many others to conduct similar experiments. One of the earliest was Herbert Spencer Jennings (1868–1947) of the Museum of Comparative Zoology at Harvard, the world authority on the behavior of microscopic organisms. He selected for body size in *Paramecium* and found that after a few generations selection had no effect. One simply cannot breed a paramecium the size of a baseball. Even after hundreds of generations, his pure lines remained constrained within fixed limits, "as unyielding as iron."

Another pioneer in pure line research was Raymond Pearl (1879–1940), who experimented with chickens at the Maine Agricultural Experiment Station. Pearl took up a problem with important implications for a nation that loves its breakfast eggs. Could you evolve a hen that lays eggs all day long?

He found you could breed some super-layers, but an absolute limit was soon reached. Hens could lay only so many eggs, above which no amount of selective breeding could increase the quantity. In fact, Pearl produced some evidence indicating that production might actually be increased by *relaxing* selection—by breeding from "lower than maximum" producers.

Johannsen and other pure line researchers, who once seemed the nemesis of evolutionary theory, in fact ultimately rescued it from vague Darwinian speculations on fossils or museum-drawer specimens. Selection and variation had been brought into the laboratory and opened to experimental examination. Johannsen's beans, Spencer's paramecia and Pearl's hens shaped a new research tradition, which brought science a bit closer to perceiving the evolutionary process.

See also GALTON'S POLYHEDRON; GRADUALISM.

QUEST FOR FIRE
Ultimate Hollywood Hominids

For 60 years, most Hollywood "caveman" films belonged to a sleazy genre—curvey starlets wearing scanty skins and monosyllabic musclemen beaning each other with papier-maché boulders on backlot sets. Sometimes rubber dinosaurs terrorized the tribe, despite science's conclusion that giant reptiles had disappeared millions of years before humans appeared on Earth. Nevertheless, there has always been an audience hungry for images of how our ancestors might have lived.

In 1982, director Jean-Jacques Annaud completed *Quest for Fire,* a conscientious high-budget film about early man, against which all future efforts will be measured. An ambitious re-creation of life in the Pleistocene era, it was shot in harsh wilderness locations, featured real animals and attempted to depict Ice Age language and culture as authentically as possible.

Four types of hominids were shown coexisting in the same time and place. Three protagonists (played by Gary Schwartz, Naseer El-Kadi and Frank Olivier Bonnet) were men of the Ulam tribe—a wandering band of *Homo sapiens* who wear animal skins, take shelter in caves and eat almost anything but human flesh. They are depicted nonchalantly munching on insects and leaves, but really get excited (and literally drool) when they see meat on the hoof.

While the Ulam know how to use fire for warmth and cooking, they do not know how to create it. If their carefully guarded fire goes out, they must find a smouldering tree that was struck by lightning or steal fire from another tribe. (There was no possibility in the film that they might buy or barter it.)

A second tribe, the Wagabou, use clubs made from bones. Covered with thick body hair, they are a different species—primitive, aggressive hunters who prey on weaker tribes. Near the start of the film, they stage a brutal dawn raid on the Ulam's cave shelter—bashing, raping, killing.

The Wagabou seem clearly to be based on the early "killer ape" reconstructions of Raymond Dart. They represent a more "primitive" type of hominid, something like what *Australopithecus* was imaged to be by Raymond Dart, Robert Broom, and Robert Ardrey in *African Genesis* (1961)—a murderous man-ape who was born with a weapon in his hand and the mark of Cain on his brow. They are described as "semi-cannibalistic," resorting to human fare only when game is scarce.

A third group, the Kzamm, are fierce, cannibalistic Neandertals. Human flesh is their preferred food, and it is from their camp that our intrepid heroes must steal their fire. At first, the hungry Ulams nibble at the remains of a Kzamm meal, but when they realize they are gnawing on human bones they recoil in disgust and horror.

While stealthily raiding the cannibals' camp, they rescue a young woman (Rae Dawn Chong) who was slated to be barbequed. She joins the Ulams and ultimately becomes their savior when they reach the land of her people, the Ivaka. Her tribe lives in settled villages of mud and straw huts and has mastered the art of making fire with a friction device, the fire-drill. Ivakas wear little clothing, but paint their bodies with blue clay and wear reed masks. Throwing-sticks give their arms greater leverage and their spears longer flights; they also use animal skin containers. In general technology and appeerance, the Ivakas are modeled on the modern Australian aborigines, the "mud

SHAGGY ANCESTORS roam a vast wilderness landscape in Jean-Paul Arnaud's *Quest for Fire*, a $20 million attempt to visualize the life of early man. Noted ethologists and linguistic consultants made the film a summary of 1980s conjectures about the prehistoric world.

people" of New Guinea and the Noubas of Africa. The Ulam leader takes the Ivaka girl as his mate, and they return to the Ulam, where she teaches the pathetic, shivering band how to make fire—causing them to erupt in primal screams of joy.

Scorning artificial studio sets, producer Michael Gruskoff shot the film in wilderness locations: the Badlands of Alberta, Bruce Peninsula of Oregon, Tsavo wildlife reserve in Kenya and the Cairngorms of the Scots highlands, where tundra and Scotch pines date back to the Pleistocene. The hardships and deliberate choice of vast, remote locations proved essential to enable Annaud and his actors to feel how early men may have dealt with their world when they themselves were so small and weak. The scale of the humans against the immense open landscapes lends a pictorial dimension no studio set-up could match.

Masks and body costumes for the different characters were elaborate and expensive. Outfits for the Wagabous (complete customfitted masks and body suits, individually sculpted and finished with hand-hooked hair) were estimated at $10,000 each. One hundred and fifty latex masks had to be constructed for 80 actors, to allow for wear and tear during filming.

Even more difficult was the task of "making up" the live animals to look like their ancient counterparts. Lions were fitted with plastic tusks to resemble saber-tooth cats and a herd of Indian elephants was transported to Scotland to play mammoths. Each elephant was fitted with shoulder humps, curved tusks, and hundreds of pounds of wooly hair—truly a mammoth make-up job! In a crucial sequence, the Ulam take refuge among the herd of mastodons to elude their Kzamm persuers.

To create a credible communications system for the early men, the filmmakers hired as consultants zoologist-behavioral theorist Desmond Morris (author of *The Naked Ape*, 1967) and novelist-linguist Anthony Burgess (author of *A Clockwork Orange*, 1962.) Morris and Burgess, in a unique collaboration, worked out a "primitive" language, which combined words, gestures and primate communication signals.

Burgess developed his language for early man from the "Indo-European" theory used by linguists. By comparing words in various related language groups, linguists have come to believe English developed from "Primitive Germanic," a common family to which German, Dutch, and the Scandinavian languages also belong. These Germanic languages, in turn, are thought to have evolved from an even older base, Indo-European, the common ancestral language from which groups are also derived.

For "fire," one of the most crucial words in the film, Burgess uses "atra," which is tied to several related concepts in English. There is "art," from the old meaning of scientific skill, as in making fire. Also, "hearth," the core of the home or camp, related to "heart," the central, warming principle.

Indo-European "tir" is used for "animal," which became "deer" in English. (Up until recent times, "deer" meant all animals, as in Shakespeare's "Mice and rats and such small deere/Have been John's food for many a year.") Burgess created such compounds as "tir preng" (hunt), "tir garsna" (lion), and "tir meg" (mastodon).

Morris fit gestures to Burgess's words, borrowing from tribal people's mimicry of animals and current field observations of living primates. For example, his early men use monkey-ape signals for dominance (staring), submission (looking away), lip-smacking and teeth-chattering and practice social grooming by picking bugs from each other's hair.

If *Quest for Fire* ultimately does not present a true vision of our ancestors, it certainly offers a good picture to future historians of the state of our own theories about them. Like most serious attempts to visualize the prehistoric past, it tells us something about the state of our present knowledge and the limitations of our imagination and forces us to pin down our best guesses of the moment.

See also AFRICAN GENESIS; DART, RAYMOND; ORIGIN MYTHS (IN SCIENCE).

RACE
Geographic Variability

In zoology, a race commonly means a variety or subspecies: a partially isolated breeding population with some differences in gene frequencies from related populations, yet still entirely capable of interbreeding. When dealing with local "races" of differently pigmented frogs, no zoologist is inclined to praise the intellectual superiority of a "two-spotted" over the "three-spotted" population.

In describing their fellow humans, however, scientists have historically propounded the most bizarre theories, based mainly on perceived differences in culture, social traditions and their own ideals of beauty, which they, more often than not, have confused with biology. Several considered various tribal peoples on a par with separate species; others confidently categorized particular "races" with such pseudo-evolutionary terms as "primitive," "advanced," "degenerate" or "superior."

Nineteenth-century evolutionary ideas about "race" continue to be tossed about, though scientists never found a "pure race" anywhere among human subpopulations. And anthropologists long ago gave up working on definitive classification of "racial types," because it cannot be done. (Various learned professors "identified" between four to 400 races.) Human variability is fluid, complex and its evolutionary significance remains largely unknown.

Traditional measurements and "classifications" of hundreds of supposed "races" and "subraces"—along with estimates of their intelligence or "character" in relation to Europeans—is outmoded baggage from the days of colonialism and "Aryan" anthropology.

Yet many of these archaic concepts linger on, perpetuated by those who would justify social prejudice or political oppression.

It was only around 1900 that anthropologists established the crucial distinction between a population's biological inheritance and its cultural tradition. For instance, "black music" is a rich Africa-rooted cultural tradition that must be learned by each generation, not a genetic ability that comes along with skin pigment. English is not a "white" language; anyone of any color can learn to speak it. A European child raised among East African Kikuyus will speak, think and dream in Kikuyu.

A few centuries back, German philosopher Immanuel Kant (1724–1804) made a very general observation about human variability and geography. Light-skinned, wavy-haired people with prominent noses inhabited Europe, a cold, wet climate. Straight-haired people with flatter faces and eyelid folds inhabited the cold, dry reaches of Asia. And dark-skinned, flat-nosed peoples with wooly hair lived in the hot, wet climates in Africa. However, Kant's rough but suggestive attempt to correlate "racial" variability with ecology was forgotten and had little influence on biology or anthropology.

From an evolutionary point of view, the adaptive value of such visible features as skin color or facial structure is still unproven. Whether differences evolved in response to climate or through genetic drift in small, isolated populations is simply not known.

Charles Darwin thought human variability showed no discernable adaptive advantages, and so explained diverse skin colors and hair textures by his theory of "sexual selection." Local ideals of beauty, expressed in selection of mates, he believed evolved different appearances among diverse populations.

Over the past half-century, the idea of traditional "races of man" has completely broken down in science. Recent genetic studies have shown that there is more variation *within* geographic populations than there is between them. Also, there are genetic characters (blood proteins, for instance) that vary independently of so-called racial groups. When plotted on maps, their distribution patterns (clines) cut right across populations with different skin colors or hair textures. Yet the frequencies of these hidden, biochemical traits are just as valid markers of gene flow as skin pigment or hair form.

Recent studies of mitochondrial DNA, traced through generations of a worldwide sample of females, indicate that the divergence and spread of human populations from Africa was comparatively recent—just a few hundred thousand years ago, an eyeblink in evolutionary time. It seems likely human subpopulations have simply not been genetically isolated long enough to evolve very significant differences. (As recently as 50 years ago, it was debated whether various human groups were interfertile; all of them are.) We may perceive such variations as skin color or hair texture as large because of our own social conditioning; biologically and genetically, they are insignificant.

See also MITOCHONDRIAL "EVE"; RACISM (IN EVOLUTIONARY SCIENCE); SEXUAL SELECTION.

For further information:

Count, Earl W., ed. *This Is Race*. New York: Schuman, 1950.

Dobzhansky, Theodosius. *Mankind Evolving*. New Haven: Yale University Press, 1962.

Lewontin, R. C. *Human Diversity*. New York: Scientific American Library, 1982.

Montague, Ashley. *Man's Most Dangerous Myth: The Fallacy of Race*. 5th ed. Oxford and New York: Oxford University Press, 1974.

RACISM (IN EVOLUTIONARY SCIENCE)
Biased Research History

Science is a human enterprise, like art, literature or music; all are are embedded in their time, place, and culture. Although biologists and social scientists claim to seek "objective" truths, theirs is a history of repeated failure to see their own biases. A particular blind spot has been the investigator's attitudes and assumptions regarding social groups, cultures or races other than his own—and their evolutionary "position" relative to white Europeans.

Almost any 19th or even mid-20th century book on human evolution carries illustrations showing the progression: monkey, ape, Hottentot (or African Negro, Australian Aborigine, Tasmanian, etc.) and white European. Few of the early evolutionists were free of such arrogance, not even the politically liberal Charles Darwin and Thomas Huxley. Both campaigned passionately against slavery, not because they believed all human groups had the same potential, but because they opposed cruel mistreatment, even of one's social or racial "inferiors."

They were raised, after all, to believe light-skinned European males were the lords of the Earth, which easily translated into the "highest products of evolution." Women were considered somewhat inferior beings, yet capable in their own limited spheres. But the colonialized countries of Africa and India were "the white man's burden"; their inhabitants, subjugated by military force, were stigmatized as underlings by nature, less evolved, closer to children and the primitive world.

Darwin, however, was impressed and puzzled by the "Christianized" Fuegian Indians aboard his ship and how different they had become from their rough tribal relatives. Among evolutionists, the freest from racism was Alfred Russel Wallace, who remarked that he neither carried a gun nor locked his doors during years surrounded by Amazon Indians, Malays and Asian tribes. Wallace had developed "a high opinion of their character and morals."

But most European scientists were only too eager to construct a scale or rank order, always with dark-skinned peoples on the bottom and whites on top. Ideologists Comte Joseph de Gobineau in France and Houston Stewart Chamberlain in England, evolutionists Ludwig Woltmann and Ernst Haeckel in Germany, and paleontologist Henry Fairfield Osborn in America—all were entirely agreed on their own evolutionary superiority.

Tribal peoples, they thought, were like the ancestors of Europeans of thousands of years ago who used stone rather than metal for their tools and were described as "still living in the Stone Age": the "Paleolithic Equation," which held living tribal peoples to be identical with primitive humans of prehistory.

Their brains were small, according to measurements of cranial capacity by Samuel George Morton, who devoted his life to studying the dimensions of skulls. But recent reexaminations by Stephen Jay Gould showed an unconscious bias on Morton's part. His measurements were way off—though Gould thinks not deliberately. His expectation that "Negro" brains would be smaller than those of whites influenced his methods and conclusions. Gould was unable to duplicate Morton's results, using the same skulls and measurement techniques.

In 1989, anthropologist Napolean Chignon's stud-

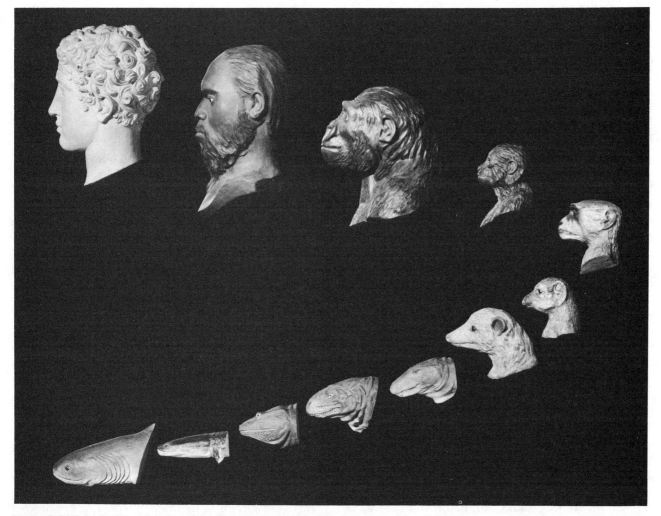

FALLACIOUS LADDER, "Our Face from Fish to Man" at the American Museum of Natural History, arranged living species in ascending series, from sharks through opposums, to monkey, gorilla, and man. Racist assumptions of the 1930s mandated placement of an African or aboriginal Australian "one rung below" a European white man—represented by a classical Greek god! (By the 1960s the exhibit had become an outmoded embarrassment and was removed.)

ies of warfare among Amazonian Indians were being used by the Brazilian government to argue that these tribesmen are murderous, primitive people, incapable of being absorbed into the life of a modern nation. Not so coincidentally, gold has been discovered in their province, and "sociobiological" arguments about their hereditary fierceness gives a convenient "scientific" excuse for treating them as less than human.

Racism (theories about the "natural" superiority and inferiority of human populations) carried to its limit usually leads to some group of people being systematically robbed of their land, labor, property—or exterminated altogether. Earlier in history, it was enough that they were of a different culture, religion or skin color for them to be branded animals, heathens, savages or infidels. But since science has won increasing acceptance as a source of truth, its "authority" has been cited to justify exterminations of Tasmanians and Australian aborigines by the British,

Native Americans by "Anglos," Jews and gypsies by Germans, and Africans by British, French, and Belgians.

One early, persistent controversy was whether all races were minor varieties within a single species (Monogeny, or "single origin" theory) or distinct kinds of humans amounting to separate species (Polygeny, or "multiple origin" theory). The nonracist view, holding for the unity of mankind, came from religious fundamentalists, who traced all humans back to the single creation of Adam and Eve. Early evolutionists, on the other hand, saw nonwhites as "primitive" peoples who were late to evolve or geographically specialized.

As late as the 1960s, the respected American physical anthropologist Carleton Coon published a massive, long-awaited study, *The Origin of the Races* (1962), which was taken seriously by a wide public, despite its racist assumptions and nonsensical biology.

Coon had proposed that each of the "five races" (by his classification) had evolved separately in Asia, Europe, America, Australia, and Central and Southern Africa. He thought a "pre-sapient" population of hominids (near-men) had first spread over the world; then, each group "crossed the threshold of humanity" separately and at different times.

He vastly overemphasized group differences, which he confused with cultural behaviors, and imagined some kind of inevitability for "pre-human" populations to evolve into *Homo sapiens* wherever they might be. Can one similarly imagine "pre-moose" hoofed animals dispersing throughout the world, then each local population evolving to "cross the threshold of mooseness" at different times?

Although acute at examining skulls and skeletons, Coon was particularly inept as an observer and reporter of human behavior. In one particularly obnoxious passage, he contrasted what he perceived as the wide range of subtle nuances in an "Italian Caucasian's" facial expressions with the "limited repertoire" of an "African Negro." The black man, he said, could only widen his eyes or show his white teeth for emphasis in conversation. Amazingly, such was the state of anthropological science in the United States in 1963, that the book was seriously debated instead of being dismissed as pseudoscience.

Recent evolutionary studies, aided by sophisticated new biochemical techniques, have demonstrated the genetic closeness and unity of our species. Richard Lewontin has pointed out that what we may perceive as great differences (skin color, hair texture) because of social attitudes, are correlated with negligible genetic differences between widely separated human populations.

Lewontin has also demonstrated that there is more variability between individuals in a given local population than between the averages of widely separated populations. For instance, there is much more variability between black Africans than between some kind of "average" African and European. Further, if most of the world's humans were wiped out, and the only surviving group was a thousand tribesmen in a remote mountain forest of New Guinea, that small, isolated population would contain 99% of the human genes and variations that exist.

See also COMPARATIVE METHOD; DESCENT OF MAN; EUGENICS; PITT–RIVERS MUSEUM; RACE.

For further information:

Coon, Carelton. *Living Races of Man.* New York: Knopf, 1969.

Gould, Stephen Jay. *The Mismeasure of Man.* New York: Norton, 1981.

Grant, Madison. *The Passing of the Great Race.* New York: Scribners, 1918.

Kamin, L. *The Science and Politics of IQ.* Potomac, Md.: Lawrence Erlbaum, 1974.

Morton, S. G. *Crania americana.* Philadelphia: John Pennington, 1839.

Stanton, William. *The Leopard's Spots: Scientific Attitudes toward Race in America.* Chicago: University of Chicago Press, 1960.

RADIOCARBON DATING
The C14 Clock

Nobel-prize-winning physicist Willard F. Libby conceived the technique of radiocarbon dating in 1946 and announced his first dramatic result in 1950. Deep within the world-famous painted cave at Lascaux, in France, some ancient charcoal had been found. Libby put the charred wood through his process and arrived at an age of about 15,000 years and so dated the Ice Age hunters who had lit a fire in the cave, perhaps to make their paintings of wooly rhinos, mammoths and other animals now extinct.

How did he do it? Radiocarbon dating is based on measuring the proportion of a rare, radioactive form of carbon (C^{14}) in a sample of bone, wood or other substance. Because of continual bombardment from the sun's radioactive rays, a fairly constant level of carbon 14 is usually maintained in the Earth's atmosphere. Living plants and animals therefore maintain known, constant levels; when they die, the carbon 14 begins to decay into nitrogen at a steady rate. Since it can be reliably predicted how long a given quantity takes to change, Libby reasoned, the proportion of carbon 14 that remains in bone or charcoal can tell approximately how long ago the organism died.

Various factors can throw off the correct calculation of age. For instance, a sample my be contaminated by contact with ancient organic material such as coal. Past fluctuations in the atmosphere, such as would occur with a volcanic eruption, may also throw off the estimate and make the tested material appear older.

Accuracy is constrained, also, by statistical limitations. Since the method counts radioactive emissions over a standard period of time, individual emissions are assumed to conform to the norm. But since each emission is a random event, there may be some deviation, expressed as a plus/minus range, which allows for error. For instance, if an ancient sample of basketry is assigned the age of 19,600 ± 2,400 years, which means there are two chances in three that the sample is between 22,000 and 17,200 years old.

Since no single date can be considered really definitive, archeologists take a series of 50 or so samples from a site to establish consistency and reliability.

Radiocarbon dating has a limited range, because

after 50,000 years not enough C^{14} remains in the sample to make an age determination. Therefore, the method is useful in prehistoric and archeological sites, but not for dating older fossils. Other techniques, such as potassium-argon and fission track dating, are used for dating more ancient materials, such as early hominid fossils and artifacts.

See also CHRONOMETRY.

RAIN FOREST CRISIS*
Contemporary Mass Extinctions

Tropical rain forests, the evolutionary wonderlands that so fascinated Charles Darwin and Alfred Russel Wallace, are being destroyed by man at the rate of 60 acres a minute. By the end of the 20th century, unless drastic measures are taken to reduce the damage, they will be almost gone—and with them many irreplaceable pages of the world's evolutionary history will disappear forever.

"Among the scenes which are deeply impressed on my mind," Darwin wrote as a young naturalist, "none exceed in sublimity the primeval forests undefaced by the hand of man . . . no one can stand in these solitudes unmoved." In his *Journal of the Voyage of H.M.S. Beagle* (1845), he recorded his delight in exploring a rain forest in 1832:

> Delight itself, however, is a weak term to express the feelings of a naturalist who, for the first time, has been wandering by himself in a Brazilian forest . . . The elegance of the grasses, the novelty of the parasitical plants, the beauty of the flowers, the glossy green of the foliage . . . brings with it a deeper pleasure . . . If the eye was turned from the world of foliage above, to the ground beneath, it was attracted to the [thick beds of ferns and mimosae] . . . wonder, astonishment, and devotion . . . fill and elevate the mind . . . in these fertile climates, teeming with life, the attractions are so numerous, that [I was] scarcely able to walk at all.

Teeming with life, indeed. Within 25 acres of Bornean rain forest, botanists recently identified 700 different tree species, as many as exist in all of North America. One single tree studied in Peru was home to 43 different species of ants—more than are distributed throughout the British Isles. It took 60 million years for the present diversity of species to accumulate in these forests, and the various insects, frogs, snakes and plants have evolved complex webs of interdependent relationships.

For example, fig trees have evolved to depend on a species of wasp that matures in its fruits and also pollinates it. Certain species of ants live only in acacias; in return for food and shelter, they clear other insects and debris out of the tree.

There are also complex relationships between water and gases; great amounts of oxygen are generated in rain forests, and the vegetation holds and recycles water and nutrients. Studies are currently underway on the important role played by these forests in creating and maintaining weather conditions over vast portions of the atmosphere.

Oddly enough, the soil in which rain forests grow is neither thick nor rich. Most of the valuable nutrients are trapped in the cycle of life and decay that goes on between a few inches from the surface and the treetops or canopy area. When forests are leveled and cleared for farming, the soil itself is very poor, capable of supporting crops for only a few years. Rain then washes away the shallow topsoil, and erosion and floods follow.

Within the past 40 years, rain forests in Brazil, Indonesia, Central America and Hawaii have shrunk from 5 million to 3.5 million square miles. More than 25,000 square miles are lost each year. If these forests disappear, as they will if destruction goes unchecked, thousands of species of plants and animals will disappear forever before they have even been discovered, named or studied.

The loss of these species and their ecosystems may be of great importance to humanity because:

1. Loss of species causes unpredictable disruptions in recycling of nutrients and production of clean, fresh air and water and triggers an increase in crop pests and disease-carrying insects.
2. Loss of such unglamorous species as soil bacteria and beneficial insects that are ecologically vital in maintaining the gas and nutrient cycles may affect climate and quality of life worldwide.
3. Rain forests are storage banks of genetic diversity or variability. One hundred acres of tropical forest contain more species than 1,000 square miles of Maine woods.
4. Each species, as the product of a long evolution, carries a special genetic code unique to its population. When a species becomes extinct, its genes are lost for human use. And, in this age of genetic engineering, such genes may have crucial importance to mankind for agriculture and medicine in the near future.
5. When forests are burned, the released carbon builds up atmospheric gases that tend to heat up the atmosphere. This "greenhouse effect" could cause a warming of temperate regions and melting of polar ice, which would raise sea levels and submerge coastal areas.
6. Many tropical plants and animals could be an immediate source of foods or medicine if they were studied rather than destroyed. Coffee, qui-

nine, pecac and reserpine have come from tropical rain forest plants. The Madagascar periwinkle, a rain forest flower, produces two drugs (vincristine and vinblastine) that have recently proved highly successful in treating Hodgkin's disease and leukemia.

Edward O. Wilson, the Harvard sociobiologist and expert on social insects, says the "current reduction of diversity seems destined to approach that of the great natural catastrophes at the end of the Paleozoic and Mesozoic eras [the extinction of the dinosaurs]. In other words, the most extreme for 65 million years."

Perhaps the greatest loss of all is knowledge. The total number of living species on Earth may be upward of 30 million, of which only 1.6 million have been identified. Tropical forests, which occupy only 7% of the planet's surface, may contain as many as half of all life forms. When the habitat is plundered, a delicate system of plants and animals that have evolved over millions of years can be wiped out in one day. Such a loss has been compared to the burning of a vast, rare library or a complex mechanism smashed before we even know what its parts are or how it works.

See also COEVOLUTION; ECOLOGY; EXTINCTION; GAIA HYPOTHESIS.

For further information:

Alameda, Frank, and Pringle, Catherine M., eds. *Tropical Rainforests: Diversity and Conservation.* San Francisco: California Academy of Sciences, 1988.

Canfield, Catherine. *In the Rainforest.* Chicago: University of Chicago Press, 1984.

Wilson, E. O., ed. *Biodiversity.* Washington, D.C.: National Academy Press, 1988.

RAMAPITHECUS
A Temporary "Ancestor"

For more than 20 years of the mid-20th century, a primate called *Ramapithecus* was considered the ancestor of all man-like creatures. Monkey-sized, with hominid-like teeth and jaws, it lived in forests from Egypt to India some 14 million years ago. Anthropologists thought it may have used tools and lived in social groups.

Its discoverer, Elwyn Simons of Yale's Peabody Museum, had been largely unknown outside his special field of early, obscure fossil primates. Up to that time, the evolutionary history of hominids—back beyond a few million years—was a blank. Simons "found" (1960) *Ramapithecus* in a drawer filled with primate fossils at the Yale Museum. It consisted of a couple of long-neglected upper jaw fragments from

India's Siwalik Hills, collected by paleontologist G. Edward Lewis in 1932.

When Simons held the two fragments of the upper jaw (maxilla) together; they seemed to form a smooth, rounded "U-shape," like a hominid palate rather than a monkey's. And the small "eyeteeth" sockets were consistent with a hominid's; monkeys and apes have large, sharp canines. Since there were no other known candidates from that period 14 million years ago, Simons decided *Ramapithecus* must have been an ancestral hominid, the earliest known forerunner of man.

Despite the lack of a complete skull, face or any skeletal bones, Simons set out to convince the world that *Ramapithecus* was the earliest human ancestor yet found—older by far than *Homo erectus, Australopithecus,* or even *Homo habilis.* As is the custom when scientists want to establish the importance of discoveries, Simons made the rounds of major universities. His charming, polished lectures soon enhanced his reputation.

An ambitious graduate student, David Pilbeam, joined his mentor Simons in promoting the "good news" of *Ramapithecus,* and the pair rode a heady wave of academic acclaim. They had filled in a troubling blank on man's family tree; within a few years, most textbooks routinely listed "Rama's ape" as the earliest known human ancestor. Discussions at scientific meetings covered its food habits, ecology, lifestyle and possible social organization.

Few anthropologists troubled themselves over the extremely fragmentary nature of the evidence, and fewer still examined the original fossil. But Vincent Sarich, of the University of California at Berkeley, was approaching human ancestry from another direction—molecular chemistry.

Sarich believed his comparisons of serum albumens from various living primates showed whether they were related in the recent or the distant past. More closely related species had less time to evolve differences; species that diverged from one another in the very distant past showed greater differences.

According to Sarich's "molecular clock," apes split off from hominids only five million years ago, not the 15–30 million that had been supposed by Simons and Pilbeam. *Ramapithecus,* he concluded, was simply too early to be hominid. And he made the famous statement: "Anything earlier than 8 million years cannot be hominid, no matter what it *looks* like."

His comment provoked howls from paleontologists, who thought he was saying fossils are irrelevant to reconstructing evolutionary history. Even his mentor, Professor Sherwood Washburn, is reported to have upbraided him with: "That's the dumbest thing you ever said."

But Sarich stuck to his guns; he trusted what his test tubes of blood serum proteins were telling him more than he trusted an expert's interpretation of a very few fossil fragments. He urged a reexamination of the fossil evidence.

Before Sarich had made his tests, anatomists Alan Walker and Peter Andrews had used the existing fragmentary material to make skilled reconstructions that cast further doubts on *Ramapithecus* as a human ancestor. And, in 1976, a new fossil of the creature, a complete and well-preserved jaw, was unearthened by Pilbeam's field team, working in Pakistan near the original site. This fossil jaw had a clear sharp ("V-shaped") angle in its contour, not a gently curved parabola ("U-shape"), as in hominids.

Now Pilbeam went back to the original fossils and noticed for the first time that the fragments could easily be pieced together differently, making the dental arch look angular rather than curved. And the teeth, on closer inspection, were not so hominid after all. When fragments of a face came to light soon after, it was seen to resemble an ancestral orang-utan rather than a hominid.

After being our "ancestor" for more than two decades (1961–1982), *Ramapithecus* was unceremoniously dumped.

Pilbeam publicly avowed he would "never again cling so firmly to one particular evolutionary scheme." Simons declared that all the pronouncements about tool use and social life came from Pilbeam, for he (Simons) certainly would not "allow myself the luxury of speculating about an animal that doesn't any longer exist."

Besides, by graciously accepting the new evidence and discarding his world-famous "discovery," he demonstrated a true scientific attitude; which again enhanced his reputation.

See also CLADISTICS.

For further information:

Lewin Roger. *Bones of Contention*. New York: Simon & Schuster, 1987.

Simons, Elwyn. *Primate Evolution: An Introduction to Man's Place in Nature*. New York: Macmillan, 1972.

"RANDOM WALK" EVOLUTION

See NEUTRAL TRAITS.

RAY, JOHN

See BRIDGEWATER TREATISES; PALEY'S WATCHMAKER.

REAGAN, RONALD (b. 1911)
Anti-Evolutionary President?

Ronald Reagan may be the only president of the United States who has ever used opposition to Darwinian evolution as part of a campaign strategy. His distant predecessor Teddy Roosevelt wrote books on wildlife and was an enthusiastic evolutionist. Another president, Woodrow Wilson, when asked about Darwin in 1913, replied: "Of course, like every other man of intelligence and education, I do believe in organic evolution. It surprises me that at this late date such questions should be raised."

At a much later date—August 22, 1982—President Reagan remarked during a speech in Texas that he "had a great many questions" about whether evolution was a proven fact. In Dallas to address a conference of conservative fundamentalist Christians, he stated evolution was "only a theory [which] is not believed in the scientific community to be as infallible as it was once believed" and that "recent discoveries down through the years have pointed up great flaws in it."

In the political view of his fundamentalist audience, teaching evolution in high school biology classes had become a sensitive issue, threatening their deepest values. By the mid-1970s creationists had persuaded school boards in Texas, California, and some southern states and some textbook publishers to omit any reference to human evolution in science programs. According to press reports, Reagan told the fundamentalists he agreed with their view that if schools teach evolution, they should also teach the biblical version of creation.

But several years after the election, Reagan quietly dropped any public opposition to the teaching of evolutionary theory, after courts and educators in several states prevailed over the religious pressure groups and had evolution reinstated in biology textbooks. However, Reagan never issued a public statement retracting his antievolutionary remarks.

In 1984, paleoanthropologist Richard Leakey of the Nairobi Museum and Dr. Alan Walker, anatomist at Johns Hopkins University, attended a White House luncheon with the president. Their purpose was to explain the recent evidence for human evolution to Mr. Reagan and to lobby for more governmental support of the quest for human origins. They assured him that while the mechanisms of evolution were certainly still in question, the evolution of our species from other primates was considered a scientifically established fact.

They also asked if Kenyan fossil hunter Kimoya Kimeu could receive the prestigious National Geographic Society's La Gorce medal at the White House. The president said he would be honored, but aides reportedly stepped in and revised the invitation so that only very limited press coverage would be permitted. Kimeu did receive his medal at a White House ceremony on October 22, 1985.

ANTI-EVOLUTION statements were featured in Ronald Reagan's successful campaign for the U.S. presidency in 1980, though he later honored paleoanthropologists at the White House. At dinner with the scientists, the president praised the lively intelligence of Bonzo the chimp, with whom the former actor had appeared in several movies.

Leakey and Walker described the latest skulls and skeletons from East Africa for Mr. Reagan, as well as the fossilized hominid footprints found by Mary Leakey at Laetoli. They also discussed recent studies of the great apes, citing new biochemical evidence of man's close genetic affinities to chimpanzees.

"You don't have to tell *me* how smart chimpanzees are," the President is reported to have replied. In his Hollywood acting days, he recalled, he appeared with a chimp in the cult classic *Bedtime for Bonzo* (1951), and was greatly impressed with his costar's intelligence. In the film, Reagan played a psychology professor who was rearing the ape in his home, unprepared for the chaos a chimp could cause on a college campus. At one point in the movie, Reagan speaks the line: "Even if Bonzo gave a lecture on

Darwinism, it wouldn't help me now!" Warming to his subject, the president regaled his guest evolutionists with stories of Bonzo's clever and malicious antics on the movie set.

According to accounts of the meeting, the two scientists found Mr. Reagan "personally very sympathetic" to evolutionary studies, but he had apparently followed his speechwriters' advice in appealing to the anti-Darwinian bias of his fundamentalist audience during the campaign.

See also: FUNDAMENTALISM; HOMINID GANG; SCOPES II.

RECAPITULATION THEORY

See BIOGENETIC LAW; FREUD, SIGMUND; HAECKEL, ERNST.

(CHIEF) RED CLOUD (c. 1830–1880)
Ally of Dinosaur Hunters

Nineteenth-century fossil hunter Othniel C. Marsh (and his arch-rival Edward Cope) viewed the "Wild West" as the world's greatest graveyard of evolution. To those dedicated dinosaur diggers, the bloody war between whites and Native Americans was merely an annoyance to fieldwork.

If necessary (and it was), the evolutionists would simply negotiate their own separate peace with the Sioux Nation. The outcome was a museum full of dinosaurs, and a major public scandal known as the "Red Cloud Affair," which rocked the presidency of U.S. Grant.

Red Cloud, an Oglala Sioux chief, led his warriors in scores of successful battles with the U.S. Army, until the government finally offered to trade land concessions for peace. According to a treaty of 1868, all of western South Dakota and the sacred Black Hills were reserved to the Sioux. But the chief couldn't know how badly white men would want what was *under* the Indian lands—gold and dinosaur bones.

Red Cloud kept his word, but the U.S. government did not. Three years later, gold was discovered in the Black Hills, and greedy white prospectors swarmed over Indian land in complete violation of the treaty. Army forces protected the gold-seekers instead of the Sioux who had laid down their weapons. In 1874, General George Armstrong Custer led 1,000 men into the area on an "exploring expedition" for gold.

Native Americans in the Badlands, near the Black Hills, were in a justifiably ugly mood. The Interior Department's "Red Cloud Agency," which administered the territory, was the first target of their discontent; they began by hacking the Agency's flagpole to splinters.

Ignoring official warnings, Yale Professor Othniel C. March entered the area with his field crew, bent

CHIEF RED CLOUD'S warriors sometimes trounced the U.S. Army, but protected Yale Professor O. C. Marsh's dinosaur diggers. Returning the favor, Marsh exposed widespread corruption in the Bureau of Indian Affairs. In 1882 the two battlers posed in a New Haven, Connecticut photo studio with wampum belt and pipe of peace, symbolizing their friendship.

on extracting dinosaur skeletons from the Black Hills. In the midst of 12,000 Native Americans ready for battle, the paleontologist set up camp and asked to meet with Red Cloud and his council.

Marsh's request to dig up bones of extinct monsters on Sioux land was a new one to Red Cloud, who was trying desperately to keep the peace. Still, the scientist's earnest quest for knowledge of ancient animals intrigued him. Other tribal leaders thought it was yet another ruse to dig for gold and that a government that cheated them with rotted rations and tattered blankets could not produce a truthful white man.

Marsh's earnest obsession won the day. "Help me get fossil bones," he promised the Sioux council, "and I'll take all your grievances to the highest levels in Washington." Red Cloud took the gamble. With an escort party of Sioux Warriors, Marsh and his workers set out for the Badlands, despite threats by hostile dissidents and dangers from other bands not bound to Red Cloud's word.

After weeks of frantic labor, two tons of crated fossils were loaded onto wagons, and the field crew departed a day before a massive Indian war party swooped across their site. Marsh now investigated Red Cloud's complaints; he was shocked at the rotten

pork and beans, filthy flour and scrawny beef the Agency was providing. Many of the people were sick and starving because of callous profiteering within the corrupt Grant administration.

When Marsh tried to bring the situation to the attention of Washington officials, he was met with evasiveness and deceit, which only strengthened his resolve. Patiently, he began to unravel the brazen "Indian Ring" headed by Secretary of the Interior Columbus Delano. Despite Delano's best efforts to discredit his and Red Cloud's complaints, Marsh personally confronted President Grant, the cabinet, and finally took his case to the newspapers.

After a long and bitter battle, Marsh and Red Cloud swayed public opinion, and Marsh gained national attention. Editorial writers hailed him as a truthful scientist-reformer who had taken on the lying "big boys" and won, support and publicity that helped him gain an edge in his rivalry with Edward Cope. Marsh was now the most famous paleontologist in America.

Chief Red Cloud and Marsh remained friends for life. "I thought he would do like all white men, and forget me when he went away," said the old warrior. "But he did not. He told the Great Father everything just as he promised he would, and I think he is the best white man I ever saw."

See also COPE, EDWARD DRINKER; COPE–MARSH FEUD.

For further information:

Hyde, George E. *Red Cloud's Folk.* Norman: University of Oklahoma Press, 1937.

Plate, Robert. *The Dinosaur Hunters: Marsh and Cope.* New York: McKay, 1964.

RED QUEEN HYPOTHESIS
Environmental "Tracking"

In Lewis Carroll's *Through the Looking Glass* (1872), the Red Queen instructs Alice that she has to keep running just to stay in the same place because the giant chessboard keeps moving beneath their feet. Leigh Van Valen of the University of Chicago sees a similarity to the Red Queen's problem in the paradox of organisms that must keep evolving just to hold their own in a constantly changing environment.

Van Valen believes that there can never be "perfect" adaptation of organisms since environments do not remain static. Natural selection therefore enables creatures to maintain, rather than improve, their adaptations.

His evidence for the "Red Queen hypothesis" is based on a study of extinction rates in various lineages. Families and species that have been around the longest would seem to be best adapted and therefore least likely to become extinct. But that is not at all the case. Extinction seems to operate independent of how ancient or recent a species or family may be. If a species can't "track" or keep up with a constantly changing environment, it becomes extinct.

Successful environmental tracking depends on having a viable range of genetic variability in the species gene pool. For instance, if a region becomes drier, a plant population will need to draw on its genetic variability for deeper roots or thicker leaves to survive.

While the Red Queen hypothesis might explain the relationship between adaptation and ecological niche, it fails to account for the spectacular divergence and radiation when organisms move into a new zone, as when birds took to the air. "Clearly," writes Richard C. Lewontin, "there have been in the past ways of making a living that were unexploited and were then 'discovered' or 'created' by existing organisms." Niches, in other words, seem to exist before there are organisms to fill them. But science seems unable to identify unfilled niches if it has never encountered organisms that *can* fill them. The relationship between environment and adaptation continues to pose paradoxical puzzles for evolutionary research.

See also ADAPTATION; DIVERGENCE, PRINCIPLE OF; NEUTRAL TRAITS.

For further information:
Van Valen, L. "A New Evolutionary Law." *Evolutionary Theory* I (1974): 1–30.

RESISTANT STRAINS
Evolving Super Bugs

Some evolutionary change is a direct threat to human well-being. Over the past century, science has developed insecticides to control insect pests, toxins to kill disease-bearing rats and antibiotics to knock out infectious microbes. Yet, in many cases, these organisms develop "resistant strains," evolved populations immune to our poisons.

So commonplace is this phenomenon that public health officials must routinely take it into account. For example, penicillin was widely used for a wide variety of infections, including the sexually transmitted gonorrhea organism. After 20 years, many of these microorganisms had evolved resistant strains, and doctors found they had best rotate a variety of cures or penicillin might soon be useless.

Just as creatures in nature respond to predators with new defenses, so the targets of human medicine and pest control show an amazing resilience. It is the most common example of evolution "in action." Those who claim we never "see" evolution at work need only look at resistant strains, whose generations are usually very short compared to humans.

Two recently studied examples are the Indian meal moth, which consumes tons of stored grain, and the tropical blood parasite that causes malaria.

A microbial insecticide known as BT (*Bacillus thuringiensis*) had been used successfully for more than 20 years to control the Indian meal moth, which infests grain bins. It was thought to be the knockout punch for these destructive insects, which rapidly developed resistance to chemicals such as DDT. But the moths proved to be a classic case of selection for specific immunity.

As the moths began to make a comeback after two decades of control, laboratory experiments were done to see how quickly they developed resistant strains. In the first generation fed BT-laced food, only 19% survived; by the fourth generation, 82% and by the 15th generation, one hundred times the initial concentration of BT was needed to kill the moths.

A U.S. Department of Agriculture scientist, William H. McGaughey, had noted that there already existed "extensive diversity in the susceptibility of populations." His experiments, reported in 1985, point to a single major (recessive) genetic factor, which rapidly increased in frequency in the population.

Malaria, which was beaten down almost a century ago, is on the increase again. There are now about 100 million cases a year, of which a million and a half are fatal. Most occur in tropical Africa, Asia and South America, home of the *Anopheles* mosquito, which transmits the blood parasite by biting humans.

Chloroquine had been the cheapest, most effective drug for combating malaria; it kills an early, asexual stage of the parasite. No one knows exactly how chloroquine acts or how the parasites have developed a resistance; but the fact is, they have. Resistance began independently in southeast Asia and Latin America, then spread to cover most of Asia and South America. During the 1970s, an independent resistant strain of the parasite evolved in East Africa; they now occur globally.

New strategies are being developed to outfox the newfound immunity of malarial parasites, but these, too, will be effective only for a limited time. Public health officials now conduct "holding" actions rather than all-out warfare, for they have accepted the impossibility of absolute victory. The old idea that there can be a "cure" or medicine for every disease or a single substance that will exterminate pests has yielded to an evolutionary perspective.

Populations that contain sufficient variability in their gene pools are extremely resilient. While it may be frustrating and costly to contemplate never-ending shifts of strategy, the phenomenon of resistant strains is also part of the wonder of evolution and a telling example of the power of selection in shaping the adaptive capabilities of organisms.

RITUALIZATION
Evolution of Animal "Displays"

In Charles Darwin's *Expression of Emotions in Man and Animals* (1873), he argued that actions that are originally voluntary may somehow become involuntary and "fixed" in a species' behavior. Half a century later, Konrad Lorenz and Niko Tinbergen based ethology—their attempt to understand animal behavior—on Darwin's premise.

Focusing on "displays," they observed many kinds of animals repeatedly perform intense and (to humans) bizarre behaviors. Ducks, geese and other waterfowl perform elaborate sequences: seemingly aggressive moves, such as mock biting, and partial nesting behavior, like snatching clumps of vegetation. Some bob their heads alternately, extend wings, ruffle feathers, entwine necks, dive down quickly and up again or vibrate their tails. There is a broad repertoire of behavior, but each species elaborates only a few.

Biologist Julian Huxley's field observations of courtship in crested grebes (1906) first called attention to the phenomena. Their formalized, repetitive actions, which produced excitement and sometimes "ecstacy" in the birds, reminded Huxley of human rituals. He coined the term "ritualisation" for the evolution of ordinary behaviors (feeding, preening, attacking, parenting) into social signals. Ritualization could involve abbreviating movements, speeding up, slowing down or incorporating them into a special patterned sequence to enhance their communicative function.

Ethologists constructed elaborate theories about the presumed "original drives" behind certain behaviors and how they evolved into "diverted" or "redirected" ritualizations. It was a bold attempt at coming to grips with the origin of behavior we find especially baffling: why some birds suddenly peck at the ground in the middle of a fight, for instance. Calling it "ritualized feeding" or "redirected aggression" ultimately explains little.

Lorenz and others even attempted to reconstruct the evolution of "ritual" displays through comparison of the behavior of related species. Such conjectural "just-so" stories enjoyed a brief vogue among evolutionists, then quietly went out of fashion.

During the 1960s, Lorenz's *On Aggression* (1968), Desmond Morris's *The Naked Ape* (1967), and other "pop ethology" books promoted lingering confusion between animal displays and human rituals by seeming to equate them, naively disregarding cultural and historical dimensions. Such facile and superficial leaps showed the ritualization concept, once so stimulating, had become a hindrance to clear thought.

See also ETHOLOGY; HUXLEY, SIR JULIAN S.; LORENZ, KONRAD; TINBERGEN, NICKOLAAS.

ROMANES, GEORGE JOHN (1848–1894)
Comparative Psychologist

Canadian-born George J. Romanes, a graduate of Cambridge University (1870), was pursuing advanced studies in neurology and psychology in Europe when he read several books that would permanently alter his life: the works of Charles Darwin, including the just published *Descent of Man* (1871), in which Darwin asserted "there is no fundamental difference between man and the higher animals in their mental functions." Romanes determined to become "Darwin's disciple in animal behavior," applying evolutionary theory to comparative psychology.

Just as Darwin soon after attempted to reconstruct the development of emotional expression (*Expression of the Emotions in Man and Animals*, 1872), Romanes would trace "mental evolution" by comparing the intelligence and abilities of various organisms, including man. Like the zoologists who compared bones and muscles to infer past history of related organisms, he would compare their "mental structures." [See COMPARATIVE METHOD.]

Although not the first author to compile accounts of animal behavior, Romanes certainly founded the subject within the framework of Darwinian evolution and was also the first to publish all the available information in English. His ambitious books *Animal Intelligence* (1881), *Mental Evolution in Animals* (1883) and *Mental Evolution in Man* (1888) established him as the "father of comparative psychology."

With the brashness of youth, the 25-year-old Romanes wrote Darwin and was thrilled when the reclusive naturalist (then in his mid-sixties) took time to reply. Soon after, the ailing Darwin extended the hand of friendship and, when Romanes proposed to tackle the evolution of mind and behavior, offered enthusiastic encouragement.

On Romanes's first visit to Down House in 1874, Darwin greeted him with "How glad I am that you are so young, Mr. Romanes!" for he foresaw a full lifetime's work ahead. When Romanes began compiling material on animal abilities, Darwin gave him his 40 years' collection of notes, files and clippings on the subject. Darwin even offered his own interesting essay on "Instinct," which was published posthumously as the final chapter in Romanes's *Mental Evolution in Animals* (1883).

Oddly, for an evolutionist, Romanes focused a good deal of attention on the common domestic cat, whose mechanical abilities, he thought, indicated "a higher level of intelligence than any other animal except for monkeys or elephants." He had been very impressed when his own cat had mastered a series of complicated latches to let herself in and out of buildings.

In one of his more quirky experiments, Romanes attempted to test the common belief that cats have an uncanny sense of direction. Leave a cat out in the country, it was said, and it would always find its way back home. In 1881, Romanes wrote Darwin, "I have got a lot of cats waiting for me at different houses . . . and shall surprise our coachman by making a round of calls upon the cats, drive them several miles into the country, and then let them out of their respective bags."

Romanes stopped the coach in the middle of a country road, had an assistant release the cats and mounted the roof of the cab in order to get a good view as they scurried away in all directions. But, as he wrote Darwin, "all the cats I have hitherto let out of their respective bags have shown themselves exceedingly stupid, not one having found her way back."

Although enthusiastic about contributing to science, in taking on the "evolution of mind," Romanes chose a problem insoluble by the methods of his time, since no one then (or now) can do fruitful work based on such a concept. He was also tormented by continual flip-flops in his beliefs about science, religion and even Spiritualism.

He had tried to broach the topic of Spiritualism with Darwin in 1877. Romanes thought he had experienced "thought communication" with a supernatural intelligence and was impressed by the floating hands and luminous "disconnected faces" he had witnessed at seances with a paid professional medium (later revealed as a notorious faker). Darwin assured Romanes his interest in Spiritualism would be their secret ("never mentioned to a human being"), but that he was unsympathetic, for "I fear I am a wretched bigot on the subject."

Frustrated by his mentor's attitude, Romanes called on evolutionist Alfred Russel Wallace, Darwin's friendly rival and a publicly avowed Spiritualist, and confided his secret belief in paranormal phenomena. (A man of science risked his reputation by publicly admitting such views.) Wallace received him graciously, promised to keep his confidence, and "gave him the best advice I could" about "the usual perplexities which beset the beginner" in Spiritualism.

A few years later, Wallace and Romanes disagreed on a scientific question, and Romanes published a scathing review of Wallace's book *Darwinism* (1889). (Its demonstration of natural selection in the animal world stops at human evolution, which Wallace thought was guided "by the unseen world of Spirit.") Here, wrote Romanes, "We encounter the Wallace of spiritualism and astrology, the Wallace of vaccination and the land question, the Wallace of incapacity and absurdity."

Wounded, the usually gentle-mannered Wallace wrote Romanes, "It seems there is also a Romanes 'of incapacity and absurdity!!' But he keeps it secret. He thinks no one knows it. He is ashamed to confess it to his fellow-naturalists; but he is not ashamed to make use of the ignorant prejudice against belief in such phenomena, in a scientific discussion with one who has the courage of his opinions, which he himself has not."

The split between Romanes and Wallace widened in the debate over Weismannism. Romanes was an enthusiastic champion of the new genetics, and he helped popularize the work of August Weismann (*An Examination of Weismannism*, 1893). Wallace thought a triumph of the "germ-plasm theory" would mean the end of Darwinism and the theory of natural selection.

See also EXPRESSION OF THE EMOTIONS; SPIRITUALISM; WALLACE, ALFRED RUSSEL.

For further information:

Boakes, Robert. *From Darwin to Behaviorism: Psychology and the Minds of Animals.* Cambridge and New York: Columbia University Press, 1984.

Romanes, E., ed. *The Life and Letters of George John Romanes.* London: Longmans, Green, 1896.

Romanes, George J. *An Examination of Weismannism.* London: Longmans, 1893.

———. *Mental Evolution in Animals, with a Posthumous Essay on Instinct by Charles Darwin.* London: Kegan Paul, Trench, 1883.

ROOSEVELT, THEODORE (1858–1919)
The Naturalist President

"I sat at the feet of Darwin and Huxley," wrote President Theodore Roosevelt in 1918, recounting his lifelong interest in evolution and natural history. As a young boy, he recalled, he had loved to read about animals, birds and paleontology and—though handicapped by poor eyesight—tried to make field observations in the New York countryside near his family home.

When he was 14, Roosevelt's father, then a trustee of the American Museum of Natural History, encouraged his son's interests by giving him a shotgun

EVOLUTIONIST—NATURALIST President Theodore Roosevelt founded America's National Parks, changed from avid sportsman to visionary conservationist. When he refused to shoot a cornered bear during a hunting trip, Roosevelt became the inspiration for the Teddy Bear.

and professional lessons in mounting birds. His taxidermist-tutor was John G. Bell, who had accompanied the famed John James Audubon on his collecting trips to the far West. When his wealthy father sponsored him to field trips to Egypt, young Teddy learned he could make first-hand discoveries that were not to be found in his nature guidebooks.

As an adult, he continued traveling to wilderness areas throughout the world, including Africa and South America. His observations on behavior and ecology went into his books *African Game Trails* (1910) and (with Edmund Heller) *Life Histories of African Game Animals* (1914). He prided himself on his original observations:

> which were really obvious, but to which observers hitherto had been blind, or which they had misinterpreted partly because sportsmen seemed incapable of seeing anything except as a trophy, partly because stay-at-home systematists never saw anything at all except skins and skulls which enabled them to give Latin names to new "species" . . . partly because collectors had collected birds and beasts in precisely the spirit in which other collectors assemble postage stamps.

When he became the 26th president of the United States, Roosevelt was an active, far-sighted conservationist, the first to reserve millions of acres of lands for wilderness parks, national forests and wildlife sanctuaries. Even while president, he continued to make field trips all over the world to collect specimens for museums and to observe nature. His exploration of an unknown tributary of the Amazon, then called the Rio Mysterioso, resulted in its being renamed the Rio Teadoro in his honor. On one trip to Africa, he shot a huge bull elephant for the American Museum of Natural History; it still dominates the mounted herd in the Akeley African Hall.

Rossevelt could name dozens of genera and varieties of American mammals from their skulls alone and frequently astounded professional naturalists with his skill. His passion for natural history, he said, "has added immeasurably to my sum of enjoyment in life."

In 1902, on a hunt in Mississippi, his servants and hounds treed a young bear. When called upon to kill it, Roosevelt indignantly declared the situation "unsporting" for a gentleman and refused to shoot. (Out of sight of reporters, however, he instructed one of the servants to do so.) The incident of the conservationist president's apparent mercy for his quarry made headlines all over the country and Teddy's appealing young bear became a national mascot: the original Teddy Bear. Toy bears marketed by enterprising doll manufacturers became a fad, as a "teddy bear" craze swept the country.

Somewhat tongue in cheek, Roosevelt always credited his passion for natural history with having saved his life from an assassin's bullet. When he was a boy stalking birds in the New York countryside, a twig caught his spectacles and flung them into the snow, leaving the nearsighted youth helpless and miserable. Ever after that painful incident, he recalled,

> I never again in my life went out shooting, whether after sparrows or elephants, without a spare pair of spectacles in my pocket. After some ranch experiences I had my spectacle cases made of steel; and it was one of these steel spectacle cases which saved my life in after years when a man shot into me in Milwaukee.

See also AKELEY, CARL; JEFFERSON, THOMAS; REAGAN, RONALD.

For further information:

Cutright, Paul Russell. *Theodore Roosevelt the Naturalist.* New York: Harper & Row, 1956.
Roosevelt, Theodore. *Life Histories of African Game Animals.* New York: Scribner's, 1914.

ROTHSCHILD, LORD WALTER (1868–1937)
World's Greatest Bird Collector

Lord Walter Rothschild, scion of one of Europe's wealthiest banking families, spent a fortune amassing the world's largest and most complete collection of preserved birds. His private museum at the family's estate in Tring, England, which he built in 1889, eventually housed a quarter-million specimens. Evolutionary biologists flocked to Tring, to work on problems of species distribution, varieties, transitional forms, population variability and classification.

Rothschild's birds became legendary, along with their eccentric owner's attachment to his precious collection. But on March 10, 1932, it was suddenly announced that most (including his especially beloved Birds of Paradise) would be sold to the American Museum of Natural History. Rothschild was disconsolate, for circumstances forced him, as he put it, "to tear one's being out by the roots." But he had no choice. His generous brother had died, the family cut off further funds and he was pressed by creditors and tax debts.

He offered the collection first to the British Museum, but it could not take them. Few institutions could afford it, even at a bargain price. Finally, he accepted the American Museum of Natural History's offer of $225,000 (about a dollar a bird) for a collection most scientists agreed was well worth two million.

His only consolation was that his feathered treasure trove, which had taken 30 years to gather from all over the world, would be kept intact and remain accessible for the advancement of scientific studies.

Among his other eccentricities, the bird man of Tring had a strange way of filing his personal mail. For years, he had separated the letters into two piles, which he deposited, unopened, into large wicker laundry baskets, particularly those which bore "a well-known dreaded handwriting on the envelope." As the hampers filled up, he sealed them closed, and placed them in a storeroom.

After his death in 1937, when they opened the hampers Lord Rothschild's relatives made an astounding discovery. "It fell to the lot of his horrified sister-in-law," wrote Miriam Rothschild, "to discover the existence of a charming, witty, aristocratic, ruthless blackmailer who at one time had been Walter's mistress, and, aided and abetted by her husband, had ruined him financially, destroyed his mind for forty years and eventually forced him to sell his bird collection . . . [He] seemed to shrink visibly in the period following the sale . . . It was winter—his birds had flown."

See also BEETLES; MAYR, ERNST; PEALE'S MUSEUM.

ROUSSEAU, JEAN JACQUES

See NOBLE SAVAGE.

SABER-TOOTHED CATS

See LA BREA TAR PITS.

SACRED THEORY OF THE EARTH (1681)

See BURNET, REVEREND THOMAS.

SAINT HILAIRE

See GEOFFROY ST. HILAIRE, ETIENNE.

SALTATION
Evolutionary Leaps

Saltation, derived from the Latin, means jumping or leaping from place to place. It can be used to describe the peculiar locomotion of grasshoppers, kangaroo rats, tarsiers and other jumpers.

But in evolutionary studies, saltation means rapid change, where species seem to evolve by rapid jumps or macromutations, rather than through a slow series of intermediate forms.

When Charles Darwin first expressed his theory of evolution, he adopted this time-worn cliche as part of the evolutionary process: *Natura non facit saltum* (Nature makes no leaps). His friend Thomas Huxley thought that was an unnecessary burden for the theory to carry. Although he was a staunch defender of the general truth of evolution, Huxley's reading of the fossil record presented some puzzles about evolutionary rates. Many species appeared to be stable, showing little change over long periods, while certain groups seemed to change and diverge fairly rapidly. Recent "punctuational" theorists incline more to Huxley's view.

Of course, whether evolution is gradual or saltational still is a very relative matter. From the vantage point of a human life span, evolution is excruciatingly slow—whether change takes place over millions of years or in mere thousands.

See also HOPEFUL MONSTERS; PUNCTUATED EQUILIBRIUM.

SANDWALK
Darwin's "Thinking Path"

One of the first things Charles Darwin did when he and his wife settled in Down village, in the Kentish countryside, was to construct a circular path through the fields and woods on his property. He called it the "Sandwalk," his "thinking path," and had the gardener sprinkle its length with sand.

This was to be no idle bit of landscaping, but an essential tool for his work. Each morning and each afternoon for over 40 years, he took his turns on the Sandwalk sometimes accompanied by his little dog. Scientific friends such as Thomas Huxley or Sir Joseph Hooker, when they visited, would join him for theoretical discussions on his walks.

Darwin was in the habit of placing a small pile of flints at the crossroad of the Sandwalk, depending on how difficult a problem he was pondering. If it was "a three flint problem," he would knock a flint off with his walking stick each time he made a circuit; when the flints were gone, it was time to return home. (His method was strikingly similar to the "three pipe problems" of Sherlock Holmes.)

The Sandwalk has a dark and curvy side through the wood, and a straight and sunlit stretch adjacent to the field. The alternation of straight and curved,

"MY THINKING PATH" was how Charles Darwin described his Sandwalk, which he strolled several times a day, pondering his scientific problems. When botanist Joseph Hooker visited, the two walked the path together, discussing "old friends, old books, and things distant to eye and mind."

shaded and bright, gives a stimulating rhythm to the walk.

One can go to Down House today (it is a Darwin Museum, or, perhaps, shrine) and take the walk as Darwin did so many thousands of times. Some visitors report feeling a strong sense of the great naturalist's presence on the Sandwalk, as if he were just about to round the bend, swinging his walking stick, deep in thought.

See also DARWIN, CHARLES; DOWN HOUSE.

SARAH

See APE LANGUAGE CONTROVERSY.

SARAWAK LAW (1855)
Wallace's Evolutionary "Bombshell"

In 1854, naturalist Alfred Russel Wallace went to Sarawak, in northern Borneo, a tropical island of the Malay Archipelago (today the Islamic Federation of Malaysia). Stricken with malaria during the wet season, he was compelled to halt active field work and "look over my books and ponder over the problem which was rarely absent from my thoughts," the question of the origin of species. The result was the Sarawak Law.

One of the most important short papers in the history of biology, Wallace's essay was published in the September 1855 issue of the *Annals and Magazine of Natural History* while he was still in Borneo. Its formal title was "On the Law which has Regulated the Introduction of New Species"; its conclusion was that "Every species has come into existence coincident both in space and time with a pre-existing closely allied species."

It was a closely reasoned argument pointing to the reality of evolution based on overwhelming evidence from the geographical distribution of living plants and animals, as well as the fossil record. Where Darwin had begun with an interest in varieties and the results of domestic breeding, Wallace was led to identical conclusions by the facts of zoogeography (animal distribution), a science he pioneered.

Though he was only 32 at the time and self-educated, Wallace displayed a sure grasp of botany, geology, geography, zoology and entomology. Although he listed nine "well-known facts" to support his conclusion, several were his own ingenious generalizations, which were then scarcely known at all. He also sketched a detailed metaphor of life as a branching tree, progressing "only in a general way" and diverging into sidebranches that were neither higher nor lower than their collaterals.

"The Sarawak Law shook Darwin, who had yet to publish a line on evolutionary theory," wrote journalist-historian Arnold Brackman, "and stunned his close friend [geologist] Sir Charles Lyell, forcing Lyell into opening his species notebook."

Wallace's paper was getting awfully close to elaborating the theory on which Darwin had been laboring for 30 years. Lyell visited Darwin at Down House in 1856 and apparently advised him to hurry up and publish something on his long-held views or Wallace would surely beat him to it. Darwin wrote Lyell on May 3, 1856, that "I rather hate the idea of writing for priority, yet I certainly should be vexed if anyone were to publish my doctrine before me." Two years later, when Wallace published his Ternate Paper, Darwin wrote Lyell, "Your words have come true with a vengeance . . . that I should be forestalled."

Wallace was entirely unaware of the anxiety and panic he was causing Darwin and his friends back in England by his apparent perversity in formulating and publishing "Darwin's theory" before Darwin.

In a letter to Darwin written September 27, 1857, Wallace thanks him for praising the Sarawak Law and is delighted that "my views on the order of succession of species [are] in accordance with your own; for I had begun to be a little disappointed that my paper had neither excited discussion nor even elicited opposition."

See also "DELICATE ARRANGEMENT" (THE); LINNEAN SOCIETY; "TERNATE PAPER"; WALLACE, ALFRED RUSSEL.

For further information:
McKinney, H. L. *Wallace and Natural Selection.* New Haven: Yale University Press, 1972.
Wallace, Alfred R. "On the Law Which Has Regulated the Introduction of New Species." (London, 1855). Reprinted in Brackman, A. C. *A Delicate Arrangement: The Strange Case of Charles Darwin and Alfred Russel Wallace.* New York: Time Books, 1980.

SARICH, VINCENT

See RAMAPITHECUS.

SCALA NATURA

See GREAT CHAIN OF BEING.

SCHAAFFHAUSEN, D.

See NEANDERTAL MAN.

SCIENCE
An Evolving System

Generations of students have been taught that scientific theories or facts may change, but science itself is rock-solid. It is a particular method for arriving at an objective understanding of nature, a self-correcting system that, given more facts, produces an ever-clearer picture of reality. Unlike religion, it is not blinkered by obedience to traditional dogma or faith, nor is it "subjective" in explaining the universe. Until recently, such was science's view of itself, repeated in countless textbooks.

However, beginning in the 1960s, philosopher-historians of science began to discern a radically different picture. History shows that science is not at all a frozen, definable system of thought, but is itself evolving. Oddly enough, though "scientific" discoveries have been made over thousands of years, the word "scientist" was not coined until 1809, by English philosopher William Whewell.

Change within science (previously called by various names, such as "natural philosophy") has not simply been a matter of correcting or rejecting previous theories and conclusions. "What is remarkable about what has occurred over the past few decades," writes physicist-philosopher William B. Jones, "is the

general recognition that the methodology and rules of science themselves change." It now appears there was never any such thing as a single, objective "scientific method," independent of its time and social context. And philosophers of science have struggled mightily (so far without success) to understand whether a discernable progress or pattern exists in the history of the scientific process itself.

In this new view of science, there can be no such thing as a "fact" uncontaminated by interpretation, expectation or some kind of human bias. Embedded in a society's culture, science is heavily influenced by that larger context. Some historians (Thomas Kuhn, among others) view science as a series of paradigms or models of reality—governed more by explanatory metaphors and a unifying "fabric of thought" than by logic and experiment.

"Operationalism"—an optimistic attempt to define "facts" in terms of standardized operations by the scientist—has also failed to hold up as a method of distinguishing "meaning" from "pure facts." It has proven fruitless, for instance, to apply such measurements to very diverse phenomena, which is why physicists usually fail when they address the "messier" problems of biology.

Science is also limited to what is "do-able" within its narrow research strategies. Positivists once argued that whatever can be "scientifically proven" is necessarily meaningful or important. But it has become clear that since science can take on only certain kinds of problems—those that offer some hope of solution with known techniques—it often ignores possibly more significant questions that do not lend themselves to testing. Thus, published scientific work can (and does) "prove" an immense number of trivial, obvious conclusions. "Difficult problems rarely succumb to a frontal assault," notes Jones, "they are much more likely to yield to someone who appears on the scene with just the right tools, which were acquired for other reasons altogether.":

> Science or something very much like it has been around for a long time . . . [but] its origins, its ultimate form, the exact character of the changes it is undergoing, and even its very nature are far from clear . . . Stories are told about its beginnings, about the discovery of the "scientific method," but one doesn't know quite what to make of them. How could Bacon have discovered (invented?) the scientific method around 1600 A.D. when Archimedes, for example, did such brilliant scientific work eighteen hundred years earlier?

While we know that ancient scientists managed to reach some solid analytical conclusions, we are only beginning to realize how much of today's "advanced" knowledge is all mixed up with myths, fads, social biases and religious values. Certainly our current scientific notions must be mistaken in ways beyond our present ability to imagine. Yet science changes in novel, unpredictable ways; like human evolution, it remains open-ended. The traditional conflict between science and religion appears increasingly irrelevant, for neither appears to be what we had supposed it to be. As physicist Werner Heisenberg wrote, in articulating his famous uncertainty principle, "natural science does not simply describe and explain nature; it is part of the interplay between nature and ourselves; it describes nature as exposed to our method of questioning." And just as the history of an individual *is* that individual, so "the history of science is science itself."

See also MILITARY METAPHOR; POSITIVISM; THEORY, SCIENTIFIC.

For further information:

Coley, Noel, and Hall, Vance, eds. *Darwin to Einstein: Primary Sources on Science and Belief.* London: Longman, 1980.

Kuhn, Thomas S. *The Structure of Scientific Revolutions.* Chicago: University of Chicago Press, 1970.

Nagael, E. *The Structure of Science.* New York: Hackett, 1961.

"SCIENTIFIC CREATIONISM"
Elusive Research Program

During heated court battles over the teaching of evolutionary biology in the schools during the early 1980s, proponents of "balanced" presentations argued that "creation science" should be taught as an equally valid interpretation of the evidence.

They were not seeking to introduce sectarian religion into the public schools, argued Henry Morris, director of the Institute for Creation Research. Scientific creationism, he wrote, has "no reliance upon biblical revelation, utilizing *only scientific data* to support and expound the creation model." Morris and his colleagues referred often to a growing body of scientific research supporting creationist interpretations in biology and geology.

At the height of the debate, two university biologists, Eugenie Scott and Henry Cole, conducted a three-year scan of 1,000 scientific and technical journals to survey the kind of creationist research that might be taught in classrooms. "Nothing resembling empirical or experimental evidence for scientific creationism was discovered," they reported in 1984. A few prominent creationists who were professional scientists had published, but their topics were food processing, aircraft stress and other unrelated areas, since most were not biologists.

Scott and Cole next sought to find if there were many manuscripts submitted by creationists that had been rejected for publication by "establishment" science publications. After checking 68 journals, which had received more than 135,000 submissions during the three-year period, they found only 18 were written by scientific creationists and 12 of those were polemics on science education. The remaining six articles, mainly "refutations" of evolution, were sent to biology and zoology journals but were rejected as incompetent and unprofessional.

There is, of course, the possibility that evidence for creationism would be "censored" by exclusion from orthodox, evolutionary-biased publications no matter how professional or well presented. But six articles submitted to 68 journals over three years hardly indicates a thriving research enterprise productive of new insights and discoveries.

Virtually the entire "creation science" literature consists of the books and tracts published by the Creation Science Institute. Most are arguments against evolution, based on the logic that if evolutionary theory has flaws, weaknesses or cannot account for data, then creationism is proved correct. Their arguments assume that there are only two alternatives: creationism or Darwinian evolutionism.

Still, science teachers were pressured by public campaigns, some led by professors with science degrees (usually not in biology) who asserted that there were abundant scientific studies based on creationist concepts. Based on their claims, many parents, politicians, and educators have assumed there is a published body of scientific "creationist" research, which does not, in fact, exist.

See also FUNDAMENTALISM; INHERIT THE WIND; SCOPES TRIAL.

For further information:

Creation Science Institute. *Scientific Creationism.* Public school ed. San Diego, Calif: CLP Publishers, 1974.

Kitcher, Philip. *Abusing Science.* Cambridge: MIT Press, 1982.

SCIOLISM
An Extinct Insult

During the 19*th* century, "Sciolism" referred to theories about natural or supernatural phenomena that the scientific establishment rejected as "false" or "pretended" knowledge. Spiritualism, phrenology and astrology were among the psuedosciences attributed to "sciolists," which literally means "those who know only a little." (The word "science" is from Latin *scientia*, to know; *sciolus*, a smatterer, is the diminutive form.)

Hugh Miller, the most popular Victorian author on geology and the Earth's past, insisted the idea of evolution (or "development") belonged to sciolism.

Reacting to early evolutionist Robert Chambers's anonymously published *Vestiges of Creation* (1844), Miller expressed outrage in his *Footprints of the Creator* (1856). Chambers had dared to cite fossils and other geological evidence to support evolution, claiming professional geologists had misread the record of the rocks.

"Can this mean," asked Miller, "that he appeals from the only class of persons qualified to judge of his facts, to a class ignorant of these . . . that he appeals from astronomers and geologists to low-minded materialists and shallow phrenologers—from phytologists and zoologists to mesmerists and phreno-mesmerists? . . . No true geologist holds by the development hypothesis;—it has been resigned to scioloists and smatterers;—and there is but one other alternative. [Species] began to be, *through the miracle of creation.*"

See also CHAMBERS, ROBERT; MILLER, HUGH; PHRENOLOGY; SPIRITUALISM; VESTIGES OF CREATION.

SCOPES, JOHN T.

See SCOPES TRIAL.

SCOPES TRIAL (1925)
Evolutionists *vs.* Creationists

Beginning on July 10, 1925, a small, sweltering courtroom in the American South became the focus of worldwide attention. High school biology teacher John T. Scopes of Dayton, Tennessee, was charged with violating the state's new law against teaching human evolution or "Any theory that denies the . . . Divine Creation of man," in the public schools. The famous "Monkey Trial" pitted fundamentalist politician William Jennings Bryan against the liberal lawyer Clarence Darrow in a classic courtroom drama, which filled the newspapers of the day and inspired books, plays and movies for years afterward.

In the standard version, the Scopes trial has come down as a triumph of science over religion, in which creationism was refuted by the evidence for evolution in open court. Scopes is depicted as an unwilling participant, arrested while going about his normal duties as a teacher. Bryan is said to have died soon after his loss to Darrow of a stroke or a broken heart. And the town was said to have suffered the indignities of swarms of outsiders coming in to profiteer, turning the proceedings into a garish circus.

These are major distortions. In fact, the 24-year-old Scopes was not a regular biology teacher, but a part-time substitute. His strongest belief was in freedom of expression, not evolution. At the urging of the American Civil Liberties Union he agreed to deliberately provoke a test case to challenge the Butler Act and notified authorities of his intentions. Local merchants were delighted with the prospect of some excitement in the town and decided it would be great for business. Evidence or expert testimony by scientists about evolution vs. creationism was never permitted in the courtroom; the judge limited the case to whether or not Scopes had broken the law. Darrow, not Bryan, lost. But Bryan died soon after of diabetic complications following a binge of overeating.

Nevertheless, the public perception was that Darrow had won—and he had, in the larger sense. He had outfoxed his opponent, backed him into an intellectual corner, and made mincemeat of his most basic argument: that the Bible had one and only one clear meaning, which is not subject to human interpretation. Bryan had declared that "the one beauty about the Word of God is, it does not take an expert to understand it."

Darrow then put Bryan on the stand and asked whether he believed that the sun stood still for Joshua so the day of battle would be lengthened. "I accept the Bible absolutely" was the reply. "Do you believe at that time the entire sun went around the Earth?" Darrow pressed. "No," Bryan admitted, "I believe that the Earth goes around the sun." Noting the biblical passage was written before it was known that the Earth rotates and orbits the sun, Darrow's point was made. Even the description of a miracle depends on prevailing human assumptions about how nature works.

Bryan had a right to his own interpretation, Darrow stipulated; still, he demonstrated that any individual's understanding of the ancient text *had* to be an interpretation. At any given time, there are hundreds of Christian and Jewish sects, each claiming to offer the one and only correct reading of divine revelation.

A seasoned orator who had thrice run for president of the United States, Bryan pulled out all the stops. He told Darrow he was not interested in the age of rocks but in the Rock of Ages. "There is no place for the miracle in this train of evolution," he intoned, "and the Old Testament and the New are filled with miracles . . . [Evolutionists] eliminate the virgin birth . . . the resurrection of the body . . . the doctrine of atonement. [Scientists] believe man has been rising all the time, that man never fell; that when the Savior came there was not any reason for His coming . . . [Outsiders] force upon the children of the taxpayers of this state a doctrine that refutes . . . their belief in a Savior and . . . heaven, and takes from them every moral standard that the Bible gives us . . ."

Darrow summed up his case with equal passion.

"MONKEY TRIAL" of science teacher John T. Scopes captured worldwide attention in 1925. Accused of breaking Tennessee law against teaching "evolution theories" in public high school, Scopes's defense pitted famed attorney Clarence Darrow against fundamentalist politician William Jennings Bryan.

Freedom cannot be preserved in written constitutions, he argued:

> when the spirit of freedom has fled from the hearts of the people . . . Bigotry and ignorance are ever active . . . Always it is feeding. Today it is the public school teachers, tomorrow the private. The next day the preachers . . . the magazines, the books, the newspapers. After a while, your Honor, it is the setting of man against man and creed against creed, until with flying banners and beating drums we are marching backward to the glorious ages of the sixteenth century, when bigots lighted fagots to burn the men who dared to bring any intelligence and enlightenment and culture to the human mind.

About 100 reporters had crowded into the tiny Dayton courtroom from all over the world to catch every word uttered by the two old war-horses, both in their late sixties. A particularly nasty caricature of Bryan as a stupid, pompous old fool emerged from the dispatches of H. L. Mencken, an acerbic young reporter for the *Baltimore Evening Sun*.

Of course, the drama of the battle for ideas was what held the world's attention. Everyone knew in advance that Scopes was guilty of breaking the law; he had admitted that from the beginning. He was convicted and fined $100, but the verdict was later overturned on a technicality. Tennessee did not attempt to retry the case.

After their hollow victory in Tennessee, fundamentalist activists shifted their focus to other states. In 1926 and 1927, they successfully lobbied for new legislation prohibiting the teaching of evolution in the public schools of Mississippi and Arkansas. These laws remained on the books—as did the Butler Act—for many years, though largely unenforced. Tennessee's antievolution law was finally repealed in 1967, and the following year its Arkansas counterpart was declared unconstitutional by the United States Supreme Court.

Fundamentalist activists, however, would soon launch a new strategy, the so-called Balanced Treatment Acts. In 1981, more than a half-century after the Scopes "Monkey Trial," an Arkansas courtroom would become the setting for still another such battle—nicknamed Scopes II.

See also BRYAN, WILLIAM JENNINGS; BUTLER ACT; (THE) CHRYSALIS; "EQUAL TIME" DOCTRINE; FLAT-

For further information:

Scopes, John. *Center of the Storm*. New York: Holt, Rinehart and Winston, 1967.

SCOPES II (1981)
"Creation Science" Trial

In 1981, scientists and educators experienced historical *deja vu*, a feeling the past was repeating itself. Fifty-six years after Tennessee's famous Scopes "Monkey Trial" had faded into history, another Southern courtroom was about to rehash the old controversy. Fundamentalist Christians still did not want "Darwin's theory" taught in their public schools. But if they had to put up with it, they wanted equal time for "Moses' theory."

After several attempts to have evolution banned from high school science courses had failed, fundamentalist Christians in Arkansas had lobbied their legislators to pass Act 590, requiring "equal time" or "balanced treatment" for the theories of Creation Science. Scriptural literalists described the trial, to be held at Little Rock, Arkansas, December 7–16, 1981, as "a test of fairness for the two scientific models." Scientists called it "Scopes II."

The challenge to the "equal time" law came from the Reverend Bill McLean and the American Civil Liberties Union (ACLU), who regarded it as incompatible with America's constitutional separation of church and state. They were not suing any creationist group, but the Arkansas Board of Education, to enjoin it from enforcing an unconstitutional law.

A team of volunteer lawyers searched the country for expert witnesses on the subjects of science and religion. Among those who appeared for the ACLU were paleontologists Niles Eldredge and Stephen Jay Gould, philosopher of science Michael Ruse, biochemist Harold Morowitz and theologians Langdon Gilkey, Father Bruce Vawter and others. Though science teaching was at issue, most of the plaintiffs and one-half of the witnesses in the case were priests, ministers, theologians and historians of religion.

Theologian Langdon Gilkey, of the University of Chicago Divinity School, later summed up his motives for challenging the law: "It would represent a disaster for religion in our society . . . There can be, I believe, no healthy, creative, or significant religious faith in a modern society unless [all] the forms of that faith are free. A politically enforced or supported religious faith becomes corrupt, dead, and oppressive."

Paradoxically, many of the witnesses supporting the creationist cause were professors with doctorates in science; all maintained that their religious advocacy was, in fact, scientific. At the same time, the theologians and church historians, some of whom were well known for their critiques of scientific arrogance toward religion, had come to defend the teaching of evolutionary biology!

Creationists came to argue that there was indeed a "scientific basis" for the Scriptural version of human origins and that to exclude this "alternate model" from the classroom was a violation of academic freedom. At the same time, they labeled evolutionary biology "a humanist religion," which required more faith than Scripture. As one creationist pamphlet put it, "A frog turning instantaneously into a prince is called a fairy tale, but if you add a few million years, it's called evolutionary science."

The experts on evolution gave the court their own view of science, revealing a complex tapestry of astronomy, geophysics, paleontology, biochemistry, genetics, anthropology and more. It was a picture of interlocking disciplines, built up and tested, each feeding into the other. Theologian Gilkey, a committed Christian, was struck by the realization that "without this thesis of a universe in process over eons of time . . . there simply *is* no modern science":

> "Creation science" rejects the scientific content of evolutionary biology—if triumphant, it would discard the entire fabric of natural science. Since creationists deny the validity of science's premises and methods, and reject its tested and unified theoretical structure, it would effectively *end* science altogether.

The court agreed. On January 5, 1985, Judge William Overton released his 38-page ruling in favor of the plaintiffs, the ACLU. Overton ruled that "creation science" could not qualify as an alternative *scientific* explanation or theory. In the court's judgment, it is a religious doctrine concerning the biblical Christian God and the account of creation given in Genesis. Act 590, Overton concluded, was therefore an attempt to establish religion in a state-supported school in violation of the First Amendment of the federal Constitution.

A similar test case of a "balanced treatment" law (Act 685) occurred in Louisiana in 1981–1982; it dragged on through several years of twists and turns, countersuits and appeals. Finally, a Fifth Circuit U.S. Court of Appeals ruled on July 8, 1985: "The act's intended effect is to discredit evolution by counterbalancing its teaching at every turn with the teaching of creationism, a religious belief."

See also BRYAN, WILLIAM JENNINGS; BUTLER ACT; FUNDAMENTALISM; "SCIENTIFIC CREATIONISM"; SCOPES TRIAL.

For further information:

Futuyma, Douglas. *Science on Trial: The Case for Evolution.* New York: Pantheon, 1983.

Gilkey, Langdon. *Creationism on Trial: Evolution and God at Little Rock.* Minneapolis: Winston, 1985.

Godfrey, Laurie, ed. *Scientists Confront Creationism.* New York: Norton, 1983.

SCRIPTURAL GEOLOGY

See BURNET, REV. T.; FOUR THOUSAND AND FOUR B.C.; MILLER, HUGH; NOAH'S FLOOD; "TWO BOOKS."

SECULAR HUMANISM
Goodness Without God

According to one authority on American English, William Safire, *"Secular humanism* may be defined as: (1) a philosophy of ethical behavior unrelated to a concept of God; (2) a characterization of an emphasis on individual moral choices as having the common denominator of atheism; (3) an attempt to besmear political opponents by impugning their faith in God."

Many secular humanists would define themselves as ethical atheists who try to behave well towards their fellow men because it is right, not because they fear punishment in an afterlife. Some who accept the label are churchgoers, and some do not consider themselves atheists, but agnostics. Safire, writing in the *New York Times,* conceded that "trying to define secular humanism is like trying to nail Jell-O to a tree." Many who accept parts of the philosophy refuse the label and, therefore, it has no widely agreed-upon meaning.

Roy R. Torasco, plaintiff in a 1961 case in which the secular "bias" in public education first came to national attention, prefers the definition given in the book *The Philosophy of Humanism* (1957) by Corliss Lamont: "joyous service for the greater good of all humanity in this natural world and advocating the methods of reason, science, and democracy."

By that definition, a great many pre-Darwinian thinkers in history would qualify, including Francois Voltaire, John Stuart Mill, Jeremy Bentham and Thomas Jefferson, many of whom never used the phrase. Jefferson, for instance, looked on any form of supernaturalism as an offense against reason and the laws of nature. He believed that churchmen had so distorted the story of Jesus that "were he to return to earth, he would not recognize one feature." Accordingly, America's third president took scissors and paste and edited the New Testament, retitling it *The Life and Morals of Jesus.* He believed truths about the historical Jesus were as easily distinguishable from added supernaturalism "as diamonds in a dunghill," and that he had "restored the book to its original teachings." Everything contrary to reason or nature was edited out of the "Jefferson Bible," which he always kept at his bedside.

Clerics feared "rightly" that he would dash their hopes of establishing official national churches, Jefferson wrote, "for I have sworn on the altar of God, eternal hostility against every form of tyranny over the mind of man." The irony of swearing to a God he doubted would not escape him, but Jefferson put his basic attitude in a letter written to a nephew in 1787: "Fix reason firmly in her seat . . . Question with boldness even the existence of a God, because, if there is one, he must more approve the homage of reason, than that of blind-folded fear."

Almost a century later, Karl Marx would try to wed reason, atheism and humanity to a communist-socialist political system. Surely it was Marx and not Jefferson that modern evangelist James Kelley had in mind when he defined secular humanism as a "Godless, atheistic, evolutionary, amoral, collectivist, socialistic, communistic religion." According to Safire, that kind of rhetoric caused one editor to redefine secular humanism as "the new label employed to indict anyone who opposes school prayer, believes in evolution, or disagrees with the religious right's views on abortion."

See also AGNOSTICISM; CREATIONISM; HUXLEY, THOMAS HENRY; JEFFERSON, THOMAS; SCOPES II.

SEDGWICK, REVEREND ADAM (1785–1873)
Geologist

Like many of the foremost geologists of his day, Adam Sedgwick started out by being ordained for the ministry. He had a reputation for being forthright and sincere and was called "the First of Men" by contemporaries—a play on his biblical first name.

Despite his essentially generous and placid nature, Sedgwick became embroiled in one of the most acrimonious scientific controversies of Victorian England. From his position as Woodwardian Professor of Geology at Cambridge (a position he held from 1818 until his death) and as president of the Geological Society from 1829–1831, the Reverend Sedgwick was able to command the attention and respect of his peers both in England and Europe.

It was on a ramble through the Mendip Hills in the Southwest of England that Sedgwick first came into contact with the formations that were to be forever associated with him. He wrote that the area "affords fine specimens of the contorsions exhibited by that rock to which geologists have given the name of Greywacke. What a delightfully sounding word! It must needs make you in love with my subject."

Sedgwick had become friends with one of the

young rising stars of British geology: Roderick Impy Murchison. Together they explored the complex formations of Devonshire and Wales. Sedgwick was the first to describe the fossils of the lower Graywacke Strata, which he named the Cambrian system, after an ancient name for Wales. Eventually their studies lead them to different levels of the Graywacke, where the mercurial and territorial Murchison claimed much of Sedgwick's domain for his newly founded Silurian system.

Inevitably, almost all of the members of the Geological Society were drawn into the fray, and, when another geologist of the time, Sir Henry Thomas de la Beche, claimed part of the Graywacke for his Devonian period, the battle lines were drawn. For nearly a decade the Great Devonian Controversy, as it was called, raged on in the scientific journals. The political maneuvering behind the scenes was almost as convoluted as the Graywacke itself, and the fact that a number of important fossils were incorrectly labeled didn't help matters. Although the dispute was eventually successfully resolved with the Cambrian, Silurian and Devonian each carefully delineated, Sedgwick and Murchison were to remain implacable foes for the rest of their careers.

As one of the foremost scientific figures of his time, Sedgwick of course, was, drawn into the controversy that revolved around the publication of the *Origin of Species* in 1859. A young Charles Darwin had accompanied him on a geologizing trip through Wales just before Darwin's historic voyage on the *Beagle,* and Sedgwick had written of him, "It is the best thing in the world for him that he went out on the voyage of discovery. There was some risk of his turning out an idle man, but his character will now be fixed and if God spares his life he will have a great name among the Naturalists of Europe." Darwin sent Sedgwick a copy of his book when it first appeared, expecting the worst kind of criticism. He wasn't disappointed. The Reverend Sedgwick wrote:

> I have read your book with more pain than pleasure. Parts I laughed at till my sides were sore; others I read with absolute sorrow, because I think them utterly false and grievously mischievous . . . You have deserted the true method of induction and started off in machinery as wild as Bishop Wilkin's locomotive that was to sail with us to the Moon.

He later wrote to David Livingstone, of African fame, that he liked the theory of evolution "far better in the poetry of the Grandfather than in the prose of the Grandson." He was referring to Erasmus Darwin's work *Zoonomia* (1794–1796). Although they remained cordial, Sedgwick never did embrace Darwin's theory.

ONCE DARWIN'S MENTOR, the Reverend Adam Sedgwick turned on him after reading the *Origin of Species.* "I have read your book with more pain than pleasure," he wrote, "parts I read with absolute sorrow, because I think them utterly false and grievously mischievous. You have deserted . . . the true method of induction."

Adam Sedgwick died in Cambridge in 1873, much beloved by many generations of geology students. No more fitting tribute could be made than when, in 1903, the collections that he had done so much to assemble were put on display in the newly opened Sedgwick Museum of Geology.

See also CAMBRIAN–SILURIAN CONTROVERSY; DARWIN, CHARLES; MURCHISON, SIR RODERICK IMPY.

For further information:

Clark, John, and Hughes, Thomas. *Life and Letters of Reverend Adam Sedgwick.* 2 vols. Cambridge: Cambridge University Press, 1890.

Rudwick, M. J. S. *The Great Devonian Controversy.* Chicago: University of Chicago Press, 1985.

Woodward, H. B. *History of Geology.* New York: Putnam, 1911.

SEED BANK (FORT COLLINS, COLORADO)

See GENETIC VARIABILITY.

"SELFISH GENE"
Unit of Selection?

English evolutionist Richard Dawkins's popular book, *The Selfish Gene* (1976), emphasizes the genetic aspects of evolution. If reproductive success is the measure of "fitness," he argues, then evolution boils down to which genes continue to gain frequency in a population's gene pool.

Genes "swarm in huge colonies," writes Dawkins, "safe inside gigantic lumbering robots, sealed off from the outside world, manipulating it by remote control. They are in you and me; they created us body and mind; and their preservation is the ultimate rationale for our existence . . . we [the phenotypes or individual organisms] are their survival machines."

Selection favors those phenotypes that pass on the genes that produced them. Any behavior, any color or structure of a creature that increases the chance it will leave more of its genes will tend to be preserved and exaggerated. Within certain physical constraints, the chicken is really the egg's way of making another egg.

Historian Peter J. Bowler reports some naturalists felt Dawkins's approach "obscures the fact that it is the organism which confronts the environment and engages in the struggle for existence and reproduction." As Ernst Mayr put it, "The potentially reproducing individual, not the gene, is the target of selection."

Some years later (1982), Dawkins wrote a response in which he admitted the emphasis was lopsided. But he thought it important to encourage a genetic perspective on fitness, "so that genetic, biometric, and biochemical factors were brought into sharper focus."

He was also promoting the sociobiological view that there is no real altruism in nature, where animals sacrifice themselves "for the good of the species." Adaptive behaviors, including apparent altruism and cooperation, are the result of an individual's "selfish" genes, whose only concern is to perpetuate themselves. Sometimes saving the lives of several close relatives can keep most of an individual's genes in circulation, even if he or she dies trying to rescue family members. [See KIN SELECTION.]

Like many current controversies and arguments over emphasis, "the selfish gene" is not new at all. Before the word "gene" had even been invented, one of the pioneers of genetics, August Weismann (1834–1914), took a similar view. An accomplished scientist who helped distinguish sex cells from body cells, Weismann is best remembered for his classic demonstration (by cutting the tails off generations of mice) that acquired modifications are not inherited. Around 1900, he wrote:

> From the point of view of reproduction, the germ cells [later called sex cells, or genetic material] appear the most important part of the individual, for they alone maintain the species, and the body sinks down almost to the level of a mere cradle for the germ cells, a place in which they are formed, nourished, and multiply.

Weismann employed such strong imagery as "the body sinks down to a mere cradle" in the context of challenging those who still clung strongly to "use-inheritance" [Lamarckian inheritance], the idea that changes acquired by an individual during its lifetime could be transmitted to offspring.

Many scientists then believed human progress might be impossible if individuals could not directly pass on acquired achievements to the next generation. In Russia, where Lysenkoism became the scientific orthodoxy, some geneticists who fought for the ideas of Mendel and Weismann were punished with professional disgrace, imprisonment and even death.

Metaphors in science writing are always revealing. To Weismann's readers it was shocking enough to depict the human body as a "mere cradle," a passive, neutral incubator. Genes grow inside it unsuspected, like benign tumors. But Dawkins's "lumbering robot" not only harbors the genes; it lurches mindlessly through life, its every behavior mechanically controlled by them.

Critics thought Dawkins's imagery was reductionism carried to absurdity, a caricature of the biological study of behavior. Some biologists objected that "the selfish gene" is an extreme expression of mechanistic determinism, which brushes aside other approaches to understanding animal behavior. In the 1980s, these critiques helped trigger a backlash against the newly established discipline of sociobiology.

See also KIN SELECTION; LYSENKOISM; SOCIOBIOLOGY; WEISMANN, AUGUST.

For further information:
Dawkins, Richard. *The Selfish Gene.* New York: Oxford University Press, 1976.

SEX, ORIGIN OF
Necessity or Historical Accident?

Some biologists consider the most profound puzzle about evolution to be: Why are there sexes among most kinds of plants and animals? Reproduction without sex is by no means impossible. There are plants that propagate from cuttings, one-celled creatures that keep on dividing and even some all-female species of lizards; so evolutionists have some reason to wonder why sex is widespread in nature.

Sex carries high levels of genetic risk as a method of reproducing individuals. At every place (locus) on a chromosome where a male and female parent differ, only one individual's gene will prevail—making it only a 50% possibility that any gene will be passed on. In addition, sexual reproduction can produce harmful mutations, birth defects, sexually transmitted diseases and parasites, and sometimes even the search for a mate can be hazardous. Such heavy costs make the whole rigmarole seem downright maladaptive. Yet it is common, while asexual reproduction is rare.

Among the cliches used by population biologists is the "major advantage" of increased genetic diversity. Sexual species do indeed create a bank of flexibility within genetic populations, which can function as insurance against future changes in environment.

A few individuals can carry mutations or rare genes the whole population may one day need, and these will become more prevalent in response to selection pressures. (Cloned seed crops, for instance, can be entirely wiped out by an unexpected blight if they are genetically uniform; such a disaster befell much of the American corn crop in the 1970s.)

But even if sex provides a bank of genetic variability as a real advantage, it is a secondary effect. It neither explains how sex may have originated, nor why it is so widespread despite such high "costs."

Some argue that sexual reproduction is necessary because it is conservative, acting as a buffer against too-rapid change of species in response to environment. For instance, it would not be an advantage for temperate populations to evolve tropical adaptations in response to a couple of scorching summers.

Other biologists have taken a new look at theories developed during the 1940s by Lemuel Roscoe Cleveland of Harvard University, who studied the microcosmic world in the guts of termites and other wood-eating insects. Among the strange microbes he observed, some streamed right through each other, and some "mated" in threes. Some appeared to "eat" other kinds, but not digest them, instead living comfortably inside one another as composite creatures.

In this tiny world, which may be similar to life on Earth 2,000 million years ago, scientists seek clues to how the cycle of meiosis (halving of chromosomes) and fertilization (doubling) may have originated. Biochemist Lynne Margulis, building on Cleveland's work, believes that doubling could have started with ingestion of one cell by another.

If their habitat starts to dry out and they are packed more densely together, present-day amoebae or paramecia cannibalize each other, creating a larger individual with twice the ordinary (diploid) number of chromosomes. "Does this recall," Margulis wonders, "an earlier time when food was scarce, and our ancestral protists resorted to cannibalism? Did fertilization first evolve as an antidote to hunger or drying out?"

One possibility is that cyclic environmental changes—such as seashores that periodically dried out—could have sometimes favored survival in the doubled state. During wetter times, there may have been another advantage in splitting up again. Recent studies of reduction division (meiosis) has disclosed a sort of "roll call" ensuring that sets of genes are in order before multicellular unfolding, i.e., the development of an embryo, begins. The process lines up genes and restores those that are damaged. "Repair and rearrangement of genes," Margulis believes, "is behind the importance of sex."

Amazing as it seems, however, there are animals—rotifers, some lizards, and many species of fish—that reproduce without sex. Generation after generation, mothers produce fatherless daughters. Yet even these creatures undergo a form of "meiotic sex" within the formative cells of the offspring, which is presently of great interest to microbiologists. (Margulis and her son Dorian Sagan reviewed such studies in *The Origins of Sex*.)

The origin of sex remains one of the most challenging questions in biology. Despite these exciting research directions, it is far from solved. Gaps in our knowledge should not be an embarassment to science, but a challenge to explore the many blank spots in our understanding.

See also EUKARYOTES; LIFE, ORIGIN OF; WEISMANN, AUGUST.

For further information:

Ghiselin, M. *The Economy of Nature and the Evolution of Sex.* Berkeley: University of California Press, 1974.

Margulis, Lynne, and Sagan, Dorian. *The Origins of Sex.* New Haven: Yale University Press, 1986.

Smith, J. Maynard. *Evolution of Sex.* Cambridge: Cambridge University Press, 1978.

SEXUAL SELECTION
Survival of the Flamboyant

Even Charles Darwin thought natural selection could not account for peacocks' tails or similar fantastic structures so prominent in courtship displays. On the contrary, elaborate appendages or tailfeathers could easily get in the way when animals had to escape enemies. Secondary male sex characteristics, he thought, were actually selected by females.

Darwin first put forward this idea in *The Descent of Man and Selection in Relation to Sex* (1871), really two books in one. Almost a century earlier, his grandfather Erasmus Darwin had written that by consistently choosing stronger, better-formed males as their mates, women had gradually improved the human race. Although Charles read his grandfather's books, he seems to have reinvented the idea and doesn't mention Erasmus as a source. Rather, he came to it through the observation that among animal populations there is competition for mates, and often one male gets a larger share of females. If only a few males impregnate many females, the next generation contains a disproportionately large number of their offspring, carrying their traits.

With more emphasis on population thinking after Darwin, evolutionists realized the issue wasn't really "survival of the fittest"—it was reproductive success. The proportion of genes passed on to succeeding generations was the real battleground, not how much food or territory an individual could grab in a lifetime. Whatever characteristics made a male more successful in leaving descendants would be intensified over the generations.

Strength or size might be decisive if males competed directly, but in many species they use courtship displays, which (in Darwin's words) "stimulate, attract, and charm" the opposite sex. Female peacocks or birds of paradise select males with the most intense displays of colorful plumage.

Can the reality of sexual selection be demonstrated experimentally? And how might these displays function in evolution? During the 1980s, Swedish ethologist Malte Andersson showed that female birds do indeed prefer males with exaggerated plumage.

Males of the East African widowbird have 18-inch long black tailfeathers, which they whirl in spectacular displays visible for a half-mile. Andersson cut some of the males' ornate feathers short, and stuck the cut sections onto other males' tails. Some were glued right back on their rightful owners, to control for variables of capture and handling. (This experiment was made possible by the modern miracle of quick-setting superglue.) Results: Males with artificially extended tails attracted four times as many females as those with normal tails or those whose tails were cut short. Within their territories, the number of nests with eggs actually quadrupled.

Still, if elaborate plumage makes the birds more vulnerable to predators, why should evolution favor them, even if females do? Back in 1889, Alfred Russel Wallace suggested that birds with more robust plumage were generally stronger individuals, and perhaps they were really being selected for their vigor, not their beauty. Sexual selection, he thought, was simply a special kind of natural selection, not a different process. Recent studies have followed his lead, exploring the relationship between exaggerated plumage, displays and good health.

Some years ago, English researchers proposed that there might be a correlation between species that were prone to blood parasites and males within that species displaying bright colors. In the constant evolutionary war between the birds and infectious protozoa, the healthiest males would have to identify themselves to females in each generation to keep one step ahead of the "bugs" that weakened the population.

In 1987, Andrew Read of the University of Oxford, after an exhaustive study of thousands of species of North American and European songbirds, found such a correlation. Species plagued by much higher levels of parasites also tended to be just those species with bright, displaying males—and the males with the brightest plumage tended to have the most resistance to parasites. Female birds may exercise a primitive sense of beauty, as Darwin thought, but selection seems to be working to produce vigor, as Wallace thought. If these new studies are correct, females pick mates who are literally "glowing with health."

See also DESCENT OF MAN; WALLACE, ALFRED RUSSEL.

For further information:
Bateson, P. P. G., ed. *Mate Choice.* Cambridge: Cambridge University Press, 1983.
Darwin, Charles. *The Descent of Man and Selection in Relation to Sex.* London: John Murray, 1871.

SHAW, GEORGE BERNARD (1856–1950)
Playwright of "Creative Evolution"

George Bernard Shaw was an evolutionist, but not a Darwinian. Appalled at the idea of a mechanistic universe devoid of purpose or design, the renowned Irish playwright argued that mankind could—indeed, should—control its own evolution into a superior species of "Supermen."

A thoroughgoing Lamarckian in his view of inheritance, Shaw was convinced that conscious choice and aspirations to moral progress could be inherited to improve future generations. His play *Back to Methuselah* (1920) is based on the premise that the human lifespan is naturally hundreds of years, though it is usually cut short by unhealthy thoughts and habits.

A self-proclaimed foe of Darwinism, Shaw was fond of quoting Samuel Butler's remark that Darwin "had banished mind from the Universe." Both literary men agreed, too, that Darwin had stolen all the best parts of his theory from his grandfather Erasmus. What was new in it, the idea of natural selection, was not only wrong but morally offensive, leading to despair and abdication of responsibility.

"When its whole significance dawns on you," Shaw wrote in his "Preface" to *Back to Methuselah*, "your heart turns to sand within your breast . . . There is a hideous fatalism about it . . . [the forces of nature] modify all things by blindly starving and murdering everything that is not lucky enough to survive in the universal struggle for hogwash."

Yet Shaw did acknowledge Darwin's genius and seems to have drawn on the character of Darwin's friend Professor Thomas Henry Huxley in the char-

BERNARD SHAW FOCUSED some of his plays on the subject of evolution, particularly *Back to Methuselah* and *Don Juan in Hell*. Though better known for Fabian Socialism and ethical vegetarianism, Shaw's major vision was that the human species could progressively evolve itself into an ethically higher and biologically more perfect creature.

acter (and name) of Shaw's fictitious Professor Henry Higgins. In Shaw's *Pygmalion* (1916) (later made into the musical *My Fair Lady*) Higgins bemoans the rarity of clear English, which was also a frequent lament of Huxley's. (The plot, however, may have been inspired by Shaw's journalistic colleague Henry T. Stead, who "bought" a girl named Eliza in the London slums as an "experiment" to expose the "White Slave" trade in his newspaper.)

Although disavowing natural selection, Shaw embraced the notion that evolution makes blood relatives of man, beast and bird. Rather than viewing man's kinship with other creatures as degrading to humans, he viewed it as an upgrading of animals. Consequently, he proclaimed himself an antivivisectionist and vegetarian and remained so to the end of his life.

Shaw saw his mixture of socialism and evolutionism as a new religion; he thought artists and writers should develop new symbols and myths for it, to deliberately reshape world culture. This idea (or ideal) of "Creative Evolution," also developed by French philosopher Henri Bergson, is remarkably similar to the ancient Asian religions, though Shaw thought it Western and modern—"the genuinely scientific religion for which all wise men are now anxiously looking."

According to Creative Evolution, change is driven by an *Elan Vital*—a Life Force—that pushed matter towards complexity and progress, upward toward the Infinite. It is the "ghost in the machine," with which humans had better cooperate, or they will be swept away with "the mammoth and other mistakes."

Shaw's parables of Creative Evolution include *Don Giovanni in Hell* (1903) and *Man and Superman* (1903), through which he hoped to give playgoers a sugar-coated indoctrination while they were being entertained. But, he complained, he had made the confection too good and "nobody noticed the new religion in the centre."

A century later, audiences continue to enjoy Shaw's plays, almost entirely unaware they were intended as propaganda for the evolutionary religion he believed could save the world. *Back to Methuselah*, he had realized, would be his final attempt to get his point across; after having twice failed, he did not expect to succeed. "My powers are waning," he lamented, "but so much the better for those who found me unbearably brilliant when I was in my prime."

See also BUTLER, SAMUEL; ELAN VITAL; NEO-LAMARCKIAN; WEISMANN, AUGUST.

SHOCKLEY, WILLIAM B.

See SPERM BANK, GRAHAM'S.

SIMONS, ELWYN

See RAMAPITHECUS.

SIMPSON, GEORGE GAYLORD (1902–1984)
Paleontologist, Evolutionist

During the 1930s, George Gaylord Simpson of the American Museum of Natural History exerted a major influence on bringing paleontology into the modern Synthetic Theory of evolution, which had already become the theoretical umbrella for genetics, zoology, taxonomy and studies of plant and animal populations. While most paleontologists had become accustomed to think of a series of types "leading to" the modern forms, Simpson's attempt to approach the fossil record as a sampling of ancient breeding populations led to a revitalization of the field.

Simpson had joined the American Museum of Natural History as assistant curator of vertebrate paleontology in 1927, at the invitation of director Henry Fairfield Osborn, who had a knack for picking future "stars" in natural history. Simpson's productive field trips to the American West and to Patagonia added much new information on the evolution and distribution of extinct mammals of the New World.

Traditionally, fossil-hunters had sought magnificent specimens for their museums and exhibited them as a series of individuals, like O. C. Marsh's famous linear "progression" of individual horse skeletons. Simpson made the evolution of the horse one of his specialties; his detailed quantitative studies, published in his classic book *Horses* (1951), exploded Marsh's "single-line" evolution of the horse from a fox-sized hoofless ancestor.

Instead, Simpson showed the complex and diverse branching of the horse's ancient relatives, not only through time, but over geographical area, as early populations pushed into various habitats, adapting first to forests, then to open grasslands. Horses represented a complex, branching bush of diverging species—nothing like a line leading straight from *Eohippus* to old Dobbin.

A pioneer in tackling the problem of rates of evolution, Simpson was impressed with the pattern of long periods of stability in species, interspersed with relatively rapid change. Creationists had seen these "discontinuities" as evidence that no evolution had occurred, while Darwin considered them gaps in an imperfect fossil record. Employing Sewell Wright's idea of genetic drift, Simpson argued that important changes might occur fairly rapidly in very small populations, leaving little fossil evidence before they spread and stabilized in large numbers. In his book *Tempo and Mode in Evolution* (1944), he introduced the term "quantum evolution" for the phenomena, a precursor of the theory of punctuated equilibrium.

He was also an engaging and popular writer: His journals of his travels and explorations attracted a wide readership (*Attending Marvels: A Patagonian Journal*, 1934). Simpson's lively and often definitive discussions of evolutionary theory and its history can be found in *The Major Features of Evolution* (1965), *Evolution and Geography* (1953) and his delightful *Book of Darwin* (1982), a personal guide to the life and works of the patriarch of evolutionary biology.

See also HORSE, EVOLUTION OF; SYNTHETIC THEORY.

For further information:

Simpson, George G. *Attending Marvels: A Patagonian Journal.* New York: Macmillan, 1934.

———. *The Major Features of Evolution.* New York: Columbia University Press, 1965.

———. *Tempo and Mode in Evolution.* New York: Columbia University Press, 1944.

———. *This View of Life.* New York: Harcourt Brace Jovanovitch, 1964.

SINANTHROPUS PEKINENSIS

See BEIJING MAN; HOMO ERECTUS; PEKING MAN.

SINCLAIR DINOSAUR
Saurian Corporate Image

For 40 years, Harry Sinclair, Chief Executive of the Sinclair Oil and Refining Corporation, shared equal billing with a diplodocus. His thousands of gas stations sported the dinosaur sign, while ubiquitous billboards made it an inescapable part of the American landscape. But while Sinclair's "Dino" sold gasoline, paleontologist Barnum Brown was selling Harry Sinclair real dinosaurs.

Brown spent months courting the oil man, until he agreed to bankroll a decade of expensive bone-hunting expeditions in the American West. Throughout the 1920s and 1930s, Sinclair picked up the tab for Brown's field crew from the American Museum of National History, and announced each season's discoveries as part of a massive promotion campaign. As a tie-in Brown created popular dinosaur cards, stamp books and maps to be given out as premiums at Sinclair gas stations. (Today, the Brown/Sinclair material from the mid-1930s is considered prime "dinosaurabilia" by collectors.)

After a while, Sinclair decided the much-publicized expeditions could earn their keep in an even more direct way, by scouting for new oil fields. Barnum Brown, an acute field geologist as well as dinosaur hunter, pioneered the aerial survey. In Sinclair's four-passenger cabin plane *Diplodocus*, Brown could search for dinosaurs and oil simultaneously. Rarely had there been a happier marriage of science and commerce.

On one of the expeditions to the Big Bend region of Texas, Brown and his chief assistant R. T. Bird unearthed an enormous fossil crocodile of previously unknown species. In honor of their benefactor, it has since borne the name *Phobosuchus sinclairi*.

As recently as 1964, Sinclair Oil sponsored creation of lifesize fiberglas dinosaur models at the World's Fair held in Flushing Meadow, New York. After the Fair's run, they were donated to various parks and exhibits around the country. Several have found a permanent home in a small park near Glen Rose, Texas, the region where R. T. Bird and Barnum Brown found remains and trackways of the real thing.

See also BIRD, ROLAND T.; BROWN, BARNUM; DINOSAURABILIA; FANTASIA.

SINCLAIR, HARRY

See SINCLAIR DINOSAUR.

SINGLE SPECIES HYPOTHESIS
Human Lineage as a Ladder

Classifiers stuck so many hominid fossils "close to but not on" the human line, anthropologists Loring Brace and Milford Wolpoff complained, they were throwing out all our ancestors. Brace and Wolpoff held that all the "near-men"—*Australopithecus, Homo erectus, Homo habilis,* Neandertal—represented "grades" along a single lineage leading to modern man.

This "single species" hypothesis assumed different hominid species could not exist together without competing for the same foods and resources. Under the ecological principal of "competetive exclusion," only one species could prevail during a given time period.

Human evolution was like a progressive ladder, with each step or "grade" leading towards *Homo sapiens.* Differences in the fossil's cranial size or stature were dismissed as populational variations along the way. The notion enjoyed a scientific vogue during the 1960s, when hominid fossils were still fragmentary and scarce. It simplified a messy picture.

But by 1975, Richard Leakey's discoveries at Koobi Fora, Lake Turkana, indicated a more complex evolutionary history. A single site yielded robust *Australopithecus boisei,* slight *Homo habilis, Homo erectus,* and unidentified others, all estimated at 1.8 million years old. Evidently, at least three hominid species had coexisted at the same time and place.

"Single-species" theory collapsed; *H. sapiens* could no longer be considered a case apart from the rest of nature. Like most other groups of animals, the human species evolved as a divergent twig on a branching bush; the idea of an inevitable one-track "trend" leading toward ourselves had been another false conceit.

See also ANAGENESIS; BRANCHING BUSH.

SKULLDUGGERY (film)

See VERCORS (JEAN BRULLER).

SLADE, HENRY

See SLADE TRIAL.

SLADE TRIAL (1876)
Darwin *vs.* Wallace on Spiritualism

Charles Darwin and Alfred Russel Wallace—the two greatest naturalists of the 19th century—took opposing sides when the supernatural went on trial. The fascinating confrontation between the cofounders of evolutionary theory took place in November, 1876, but was downplayed by the Darwin family and almost lost to history. At issue was whether the famous American psychic Henry Slade was sincere about communicating with "departed spirits" or was merely a clever con man. The case marked the first time a scientist had ever brought criminal charges against a professional "medium" for conducting fraudulent experiments.

Known as the "slate-writing medium," Slade's specialty was posing questions to the spirit of his dead wife and receiving mysteriously written answers on slates. With the cream of society among his clients, his trial before a packed London courtroom was front-page news for weeks.

Professor Edwin Ray Lankester, a young evolutionary biologist who had been Thomas Henry Huxley's student, paid to attend a seance at Slade's with hope of catching him in trickery. Boldly, he snatched the slate from the medium's hand in the darkened room and found an answer written before the question had been asked. Lankester hauled Slade into police court as a "common rogue."

Alfred Russel Wallace, a staunch believer in Spiritualism, gladly appeared as star witness for the defense. Known for his honesty and as Darwin's co-discoverer of evolution by natural selection, Wallace testified he detected no fraud. Slade, he thought, was "as sincere as any investigator in a university department of natural science."

Darwin, on the contrary, was convinced that all "spirit-mediums" were "clever rogues," preying upon the credulous and bereaved. He wrote Lankester that he considered it a "public benefit" to put Slade out of business and quietly contributed funds to the cost of prosecution.

Young Lankester, an evolutionary biologist, was well aware that his entrapment of Slade would please the two men he revered most: Darwin and Huxley. They had previously attempted to expose other fraudulent mediums. Professor Huxley had even mastered the trick of snapping his big toe inside his boot to confound Spiritualist adversaries with "mysterious" rapping sounds—an achievement that has been overlooked by his many biographers. In 1874 Huxley had attended a seance incognito at Darwin's request, then sent him a written account of the chicanery he observed.

Slade claimed at his trial that he didn't know how the writing was produced on the slate. Neville Maskeleyne, the well-known stage magician, got up and performed a few tricks with slates to show how such effects might be produced, but the judge disallowed

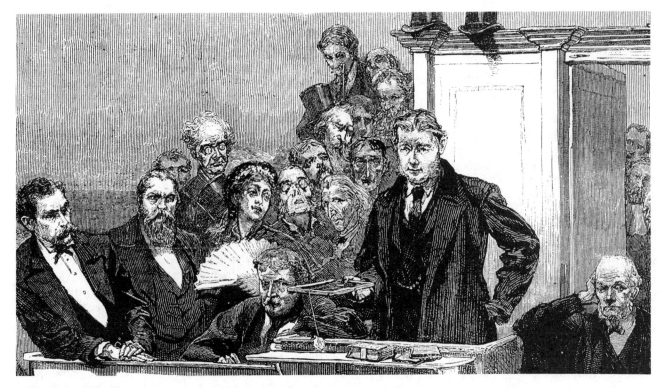

DARWIN'S GHOST-BUSTER, Dr. E. Ray Lankester holds up slate on which professional "medium" Henry Slade (left) caused "spirit-writing" to appear. Backed by Charles Darwin, the prosecution of Slade (1876) was the first time scientists had charged a psychic with criminal fraud.

the performances in a vain attempt to keep the trial from becoming a circus. Finally, the judge said that he had to rule "according to the ordinary course of nature" and convicted Slade under an old law against palmists and fortune-tellers.

But Slade never served a day in jail; a few months later his conviction was overturned on a technicality, and he fled England for Europe. He continued his spirit-writing seances in Germany, where he impressed the renowned physicist Zollner with a demonstration of "odic force," and even convinced the local police chief of the authenticity of his powers.

However, after repeated exposures, his reputation declined, and the faded celebrity ended up living in a run-down New York boarding house. On slow news days, city editors would break in cub reporters by giving them the price of a seance and sending them over to expose old Slade one more time.

One of Lankester's motives in pursuing Slade was to punish Wallace, whom he thought had "degraded" scientific meetings by permitting a paper on "thought transference" to be read. Only a few days before Lankester's exposure of Slade, Wallace had chaired the anthropology section at the British Association, and his vote had broken a bitter deadlock to allow a paper by physicist William Barrett, about a woman who claimed to "see" events in London without leaving her small Irish town.

The incident caused deep acrimony in the scientific community for which some never forgave Wallace. Years later, when Darwin wrote his friend Sir Joseph Hooker, asking help in securing a government pension for Wallace, Hooker refused. "The candidate is a leading and public Spiritualist," Hooker replied, recalling that Wallace managed to force the Spiritualist paper into the British Association "despite the opinion of the whole committee." But, although they were at opposite poles on such questions as spirits and miracles, Darwin prevailed on his colleagues to approve the pension for Wallace, in belated recognition of his extraordinary contributions to natural science.

See also LANKESTER, E. RAY; SPIRITUALISM; WALLACE, ALFRED RUSSEL.

For further information:

Milner, Richard. "Darwin for the Prosecution, Wallace for the Defense: How Two Great Naturalists put the Supernatural on Trial." *North Country Naturalist* 2 (1990).

———"Darwin for the Prosecution, Part II: Spirit of a Dead Controversy." *North Country Naturalist* 2 (1990).

SMITH, SYDNEY (1911–1988)
Zoologist, Historian of Science

Cambridge zoologist Sydney Smith made important studies of trout embryos, led a chamber music group and collected fine old wines. Combining his interests, he planned to serve von Schubert wines "during pauses in the rehearsal of Franz Schubert's Trout Quintet." Historians of science may forget these passions of Smith's, but not his great labor of love: organizing Charles Darwin's books, letters and manuscripts, which had become a scattered and inaccessible jumble.

Darwin often cut up his notes, transferred them, sometimes destroyed or gave away pages of manuscripts or made undated additions. Smith spent years sorting out the confusion, during which he acquired a deep understanding of Darwin's mind as he transcribed thousands of pages of his nearly illegible scrawl.

Among the results are the definitive edition of *Darwin's Notebooks 1836–1844*, published in 1987. (Smith's coeditors were Paul Barrett, Peter Gautrey, Sandra Herbert and David Kohn.) In 1985, he also edited (with Frederick Burkhardt) *A Calendar of the Correspondence of Charles Darwin, 1821–1882*, and helped launch publication of the *Complete Correspondence of Charles Darwin* (1980–1990s); the multivolume set will include 13,000 Darwin letters.

Smith collected the material from members of the Darwin family and other sources. Often he had to "badger, negotiate, charm and coerce" owners of precious documents to part with them. The fruits of these labors now constitute the Darwin Archive in the Cambridge University Library—one of the world's outstanding resources for historians of science.

Charles Darwin's immense impact on history had transformed him into a mythic figure—remote, inaccessible and no longer human. Sydney Smith rescued the great naturalist from his own fame. Thanks to his dedicated restoration of thousands of documents, we have direct access to Darwin's thought and development, his achievement and humanity and his obsession with discovering our origins.

SMITH, WILLIAM (1769–1839)
Discoverer of Strata

One of the most basic geological truths is that the Earth contains layers, or strata, that tell a consistent story of the past. William Smith, an English surveyor and self-taught engineer, was the first to develop and apply that idea. Orphaned at an early age and raised on a farm by an uncle, the rugged Smith was drawn to the outdoor life and made himself an expert on English terrain. After an apprenticeship to a surveyer, he found he could earn a decent living by becoming a practical expert in geology at a time when many canals were being dug for coal barges.

Canal companies hired him to survey, drain swamps, report on coal deposits and recommend the best courses for canal routes. His livelihood depended on accurate knowledge and predictions of the composition of the land, and he was very good at his job.

While making his own surveys for maps, he found that each stratum contains its own characteristic minerals and fossils—a record of creatures and events. By understanding how to identify and use key fossils, "Strata" Smith, as he became known, taught himself to trace a given layer over long distances, even where it was eroded, discontinuous or inverted. He could also work out its relationship to older and younger strata, discerning the sequence in which the deposits were laid down.

Before Smith, several geologists had noticed that there seemed to be a sequence in the strata. But they had looked only at the minerals or rocks, ignoring fossils as markers. Smith realized that each layer had a distinctive "signature," made up of the organisms it contained. Also, he noted that the oldest must be on the bottom, younger on top. Simple as that seems now, it was not at all obvious until Smith pointed out his "doctrine of superposition."

Although he hated to write, he managed to publish *Delineation of the Strata of England* (1815), *Strata Identified by Organized Fossils* (1816) and *Stratigraphical System of Organized Fossils* (1817). But his real life's work was making the first geological maps of England. A dogged field man who covered enormous distances over every kind of ground, he regretted that "the theory of geology was in the possession of one class of men, the practice in another."

He declined to theorize, but bowed to the scientific climate of his times, which accounted for strange fossil creatures by unknown supernatural events. Nevertheless, he had opened the book of the Earth, and its story would soon be read.

Smith's efforts did not go unnoticed by men of science. In 1831, the polished gentlemen of the Geological Society of London awarded their first Wollaston Medal to the weatherbeaten "Strata" Smith.

See also PROGRESSIONISM; STRATIGRAPHIC DATING; UNIFORMITARIANISM.

For further information:
Eiseley, Loren. *Darwin's Century: Evolution and the Men Who Discovered It.* New York: Doubleday, 1958.
Smith, William. *Stratigraphical System of Organized Fossils.* London: E. Williams, 1817.

SOCIAL BEHAVIOR, EVOLUTION OF
A Widespread Adaptation

Social life is neither an invention by man nor the special province of his primate relatives, but has

evolved countless times in the natural world. Disproportionate attention paid to monkeys and apes because of their evolutionary kinship to ourselves has
obscured an important fact: many kinds of creatures
have evolved remarkable societies.

Anthropologists, with the exception of Darwin's
protege Sir John Lubbock, have never bothered to
study such social insects as bees, ants or wasps,
summarily dismissing them as brainless automations
whose social behavior is instinctively programmed.
In their view, mere "herds" and "aggregations" of
elephants or social birds did not really hold a candle
to primates. But over the last few decades, even as
the world's wildlife is disappearing, dedicated field
researchers have made astounding discoveries, which
are helping to place human social behavior in evolutionary perspective.

In 1968, a field team of anthropologists, Doctors
Michael and Barbara MacRoberts, who had previously studied free-ranging Gibralter monkeys, focused their attention on a small bird—the California
acorn woodpecker. This remarkable woodpecker lives
in complex colonies that communally store food in
tree trunks, which are packed with thousands of
acorns. Each is hammered into a custom-carved com-
partment, carefully shaped by the birds for a snug
fit. These "granaries" are worked on by scores of
individuals and are passed on for the use of succeeding generations. After the MacRobertses' pioneering
monograph appeared, scores of other researchers
(zoologists rather than anthropologists) have continued to unravel the complexity of the woodpecker's
world.

With a brain no bigger than a thimble, these California acorn woodpeckers lead a social life as complex
as any monkey. They stake out and defend territories, including nesting sites as well as food resources
and storage granaries, exhibit elaborate communicative behaviors and have a kin-based network of social
relations. Also, they are "cooperative" breeders, with
unmated "helpers" assisting parents in hatching eggs
and rearing young.

Michael and Barbara MacRoberts could find no
dichotomy between the societies of monkeys and
other animals. It was heresy for anthropologists,
though not an unusual point of view for some zoologists. George Schaller, for instance, studied gorillas,
giant pandas and tigers with an even-handed approach, which never penetrated anthropology.

In the decades since, other scientists have begun
to study social behavior in animals other than primates, with a view to better understanding how
human society may have evolved.

Stephan Emlen (whose father, John Emlen, had
trained George Schaller in zoology), made interesting

STORAGE GRANARIES pecked out of trees hold thousands of
acorns for use of California woodpeckers, who store enough for
the use of future generations. With a brain the size of a thimble,
these remarkable birds have social behavior as complex as those
of monkeys. (Photo by MacRoberts.)

discoveries about the societies of an African bird
known as the bee-eater. Social bee-eaters live in sandy
cliff "apartment houses," with each burrow scooped
out by three to five "team" members who sleep
together, cooperate in food-gathering, rearing and
provisioning of young, and defense against predators. Teams sometimes allow other birds related to
them (clan members) to sleep in their burrows, but
will drive out individuals who belong to another clan.

Among mammals, recent studies of African hunting dogs, meerkats and mongooses show hitherto
unsuspected social complexity, learning, division of
labor, teamwork and manipulation of environment.
Meerkats and mongooses, for instance, take turns
standing upright, performing "sentry duty" while
the others feed, play or mate. Good sentinels learn
their roles from older individuals, who take on "ap-
prentices." Usually they divide the task, with some
searching the skies for hawks and eagles, while others scan ground and horizon in all directions.

Like social birds, meerkats and mongooses help
feed and rear their kinsmen's youngsters. Adults
routinely risk danger to retrieve straying juveniles
(not their own) to safety. One adult "baby-sitter" will

INTO THE SALT MINES, elephant herds made difficult expeditions deep into Kitum Cave at Mt. Elgon in Kenya. Worn trails and gouged walls show they had been mining minerals there for thousands of years, a social tradition that ended when groups were slaughtered by ivory poachers in 1986. (Photo by Ian Redmond.)

patiently guard, nurture and play with several of the colonies' youngsters for hours while their parents are out foraging.

Field research on African elephants continues to reveal unsuspected dimensions of social complexity, intelligence and communication. Recent studies have proved their familiar trunk-raising and ear-spreading behavior enable herds to keep in touch with one another on the open plains, even though miles apart. Electronic equipment has picked up the low-frequency calls, inaudable to human ears, with which elephants can stay aware of the movements of distant groups in all directions.

Leadership by qualified, experienced matriarchs has become a well-known feature of elephant societies. Within the past decade, elephant "traditions" have been documented, including the remarkable trek of Kenyan herds to underground caverns where they "mine" salts and minerals with their tusks to supplement their diets. Trails worn in the cave floors evidence the hundreds of generations that have made the difficult, dangerous expedition. (A misstep in the cave can mean death, which occasionally happens.)

If chimpanzees, not elephants, were filmed on a traditional social expedition to gather minerals from distant caves; if gorillas, not acorn woodpeckers, scooped out thousands of bins in tree trunks to store food to be used by their offspring; if monkeys spent as much time as meerkats in upright postures as dedicated sentinels or worked in cooperative "teams" like bee-eaters—primatologists would become ecstatic about our very remarkable relatives.

But monkeys and apes, despite their undeniable genetic closeness to us, are not at all unique in the animal world for their powers of social organization, manipulation of environment, traditions or communication. Really, they are quite ordinary—especially considering the size of their brains compared to meerkats and acorn woodpeckers. (Not to mention their human-like hands, which fashion nothing more remarkable than leafy sleeping nests or stripped twigs.)

The more we know of monkeys and apes in the wild, the less similarities we find to human social organization. Increasingly, it appears that evolution into the hominid niche was a unique historical event, not an inevitable development from primate societies.

Scrutiny of primate behavior, from lemurs to monkeys and apes, has been going on now for three or four decades. It was (and is) a worthwhile venture. But it has not yielded the sort of profound insights into the evolution of human behavior that latter-day Darwinians had expected. Instead, it has led us to reevaluate human behavior against a broad spectrum of creatures not closely related and left us to puzzle over the ecological challenges that have produced such remarkably diverse social species.

See also APE LANGUAGE CONTROVERSY; BABOONS; CHIMPANZEES; DINOSAUR HERESIES.

For further information:
Bonner, John Tyler. The Evolution of Culture in Animals. Princeton: Princeton University Press, 1980.

Chance, Michael, and Jolly, Clifford. Social Groups of Monkeys, Apes and Men. London: Dutton, 1970.

Douglas-Hamilton, Iain and Oria. Life Among the Elephants. New York: Viking, 1975.

MacRoberts, Michael, and MacRoberts, Barbara. "Social

Organization and Behavior of the Acorn Woodpecker in Central Coastal California." *Ornithological Monographs 21*, American Ornithologists' Union, 1976.

Trivers, R. L. *Social Evolution*. Menlo Park, Calif.: Benjamin-Cummings, 1985.

SOCIAL DARWINISM
Political-Economic Ideology

What is usually called "Social Darwinism" was the wedding of evolutionary ideas to a conservative political program in the 1870s. It had a particular vogue among American businessmen, elevating traditional virtues of self-reliance, thrift and industry to the level of "natural law." Based more on the writings of Herbert Spencer than of Charles Darwin, its proponents urged *laissez-faire* economic policies to weed out the unfit, inefficient and incompetent.

One of Social Darwinism's leading spokesmen, William Graham Sumner of Princeton, thought millionaires were the "fittest" individuals in society and deserved their privileges. They were "naturally selected in the crucible of competition." Andrew Carnegie and John D. Rockefeller agreed and espoused similar philosophies they thought gave a "scientific" justification for the excesses of industrial capitalism.

Like other great truths, Darwinism seemed to lend itself to the most wildly conflicting political programs, depending on who was doing the interpreting. Edward Bellamy, the Utopian social critic, thought the complete *elimination* of competition would hasten evolutionary perfection. Cooperation and socialism could come about in slow degrees; after all, Darwin had taught "nature makes no leaps." Sumner's reply was that socialism was "a plan for nourishing the unfittest and yet advancing in civilization"—an evolutionary impossibility.

Karl Marx wrote his friend Friedrich Engels that Darwin's theory was "the basis in natural history that we need" for the philosophy he called "Scientific Socialism." In Darwin's "materialism" he found ammunition against the "divine right" of kings and a social hierarchy supported by religion. And the idea that evolution is a history of competitive strife fit well with his ideology of "class struggle."

Marx sent Darwin a copy of his major work *Das Kapital* (1867), but the naturalist never read it (the pages remained uncut). Communists as well as capitalists claimed to be "Social Darwinists," though their reasons were very different. Engels eulogized Marx by claiming he had found the laws of human society, just as Darwin had found those of nature.

When Mendelian genetics came into vogue about 1900, the idea of discontinuous evolutionary jumps in nature suggested a basis for revolution in the social sphere. Some ideologists seized upon it as the antidote to Darwin's "slow, steady changes." Yet, after the Russian Revolution, "Mendelists" were reviled among doctrinaire Soviet scientists. Now a new society was be achieved through "improvement" of the peasantry, producing cumulative genetic "progress." Under the tyranny of Lysenko's falsified "proof" of "use inheritance," they refused to believe each generation must be educated anew. [See LYNSENKOISM.]

Another Darwinian social philosophy was claimed by the anarchists, whose spokesman was Peter Kropotkin, a Russian prince who despised the excesses of the nobility. Kropotkin took his cue from the cooperative social behavior of animals and certain passages in Darwin's *Descent of Man* (1871). He thought natural social cooperation was the true form of Social Darwinism.

In his book *Mutual Aid* (1902), Kropotkin argued that evolution had produced a great deal of social behavior in the natural world; survival often depended on individuals combining for their mutual benefit. His anarchist philosophy was not simply the absence of all rules and order, with everybody running wild. He deeply believed that if mankind was freed from oppressive and corrupt institutions, a natural, harmonious order would reassert itself. Cooperation for the common good in a classless society, he thought, was basic human nature in its natural state.

Theological liberals tied Darwinism to social progress as part of God's plan. Many Christians found in evolution an inevitable "ascent" of humanity. Man was not a fallen angel but a risen ape, still progressing upward.

Reverend Henry Ward Beecher, the most popular Protestant preacher in America, taught that God's plan is to constantly improve man. Moral progress toward a higher type of being lay ahead; sins were simply slips back to a more animalistic behavior. While Christian theologians shed guilt and Original Sin, Social Darwinists like William Graham Sumner seemed just as driven by their grim duty to evolutionary "competition" as any Calvinist ever was by his duty to God.

Thomas Henry Huxley viewed evolution in nature as bloody and ruthless, but thought man is obliged to leave it behind and seek a better way. He taught that humans have the choice *not* to accept the "law of the jungle." Instead, we must struggle towards a compassionate and humane society.

Germany's leading evolutionist Ernst Haeckel, on the other hand, thought man must "conform" to nature's processes, no matter how ruthless. The "fittest" must never stand in the way of the laws of evolutionary progress. In its extreme form, that social view was used in Nazi Germany to justify sterilization and mass murder of the "unfit," "incompetent" and "inferior races."

Darwin's politics were liberal (sometimes radical) for his day; he had too much compassion for the underdog to be a Social Darwinist in the Anglo-American sense. Once he laughed at a newspaper squib claiming "that I have proved 'might is right' and therefore that Napoleon is right, and every cheating tradesman is also right." He was passionately opposed to slavery, was known as a very lenient magistrate, campaigned against abusive child labor practices and was locally admired for his philanthropies.

Yet, he also was resigned to the subjugation of tribal peoples most Englishmen considered "inferior." He had seen first-hand the extermination of South American Indians by the Argentine army and thought the slaughter of indigenous Australians and Tasmanians an inevitable outcome of the clash between "advanced" and "savage" races.

Sometimes empire-building Englishmen spoke of their "white man's burden": the duty of "civilized" nations to bring material and moral progress to the "backward" races. In a private letter to a friend, Darwin cynically remarked that many native populations were being "improved right off the face of the earth."

See also BEECHER, HENRY WARD; CARNEGIE, ANDREW; HAECKEL, ERNST; HUXLEY, THOMAS HENRY; KROPOTKIN, PRINCE PETER; SUMNER, WILLIAM GRAHAM.

For further information:

Bannister, R. C. Social Darwinism: Science and Myth in Anglo-American Social Thought. Philadelphia: Temple University Press, 1979.

Hofstadter, R. Social Darwinism in American Thought. Philadelphia: University of Pennsylvania Press, 1944.

Jones, Greta. Social Darwinism and English Thought. Atlantic Highlands, N.J.: Humanities Press, 1980.

SOCIOBIOLOGY

See COUNT, EARL W.; ETHOLOGY; KIN SELECTION; "SELFISH GENE."

SPECIES, CONCEPT OF
From "Type" to Population

Few questions in evolutionary biology seem so simple, yet are really so complex as: What is a species? Surely any fool can tell a wolf from a tiger, or a bluebird from an ostrich. Creationists cite the biblical statement that God created each animal "after its kind," once and forever. To fundamentalists, a species is a fixed, unchanging "unit of creation."

Early naturalists, educated by the church, based their science on the congenial Platonic concept of "ideal types." [See ESSENTIALISM.] Animals and plants were thought to be imperfect expressions of abstract forms or ideas in the mind of God. Any particular zebra partakes of the essence of Zebra.

In Carl Linnaeus's (1707–1778) influential system of classification, the naturalist chooses a "type specimen" when naming a new species. Identifying variable (polytypic) species in the field often proved difficult, however, when captured individuals looked very different from the so-called "standard" type. Imagine the perplexity of a naturalist from Mars, for instance, trying to identify a bulldog, chihuahua, dachshund, or Great Dane as members of the same species by comparing them to an arbitrarily chosen "type specimen" of Canis familiaris.

Typological notions of species crumbled as serious field naturalists confronted two pervasive facts. The first was the tremendous variability within species. Even those that seemed at first comprised of almost identical creatures always revealed, on closer examination, an unexpectedly wide range of differences between individuals. Secondly, unaccountable mutations and hybrids blurred the boundaries of species, since they did not resemble their parents. It was this last phenomenon that caused even the great classifier Linnaeus, late in life, to question the typological species system he made famous.

In the late 18th century, some naturalists who saw the difficulties of defining species as an "essence" or "fixed type" declared that all we really see in nature are individuals. Grouping them in a class, and naming that class "zebra" is an exercise of the human mind: It may be convenient, but it has no reality as a unit in nature. (This is known as the "nominalist" definition, after a medieval school of thought that taught that reality was created by giving names to things.)

Charles Darwin began wrestling with the species concept during his voyage on H.M.S. Beagle and by 1830 had jotted some important insights in his journal. During his sojourn in the Galapagos Islands, he noticed similar but different birds and reptiles and was puzzled about whether they were species or "varieties" of the same species.

He finally decided to let the animals themselves decide the species question. Those that were attracted to one another and mated belonged to the same species. Darwin had observed courting displays and other behavioral modes of species recognition, which are known today as "isolating mechanisms."

Despite their present "repugnance" for individuals of other species, Darwin realized that at some future time such populations might mingle and produce hybrids. But he wrote that "until their instinctive impulse to keep separate" is overcome, "these animals are distinct species." He also noted that "species may be good ones and differ scarcely in any external character."

But when Darwin discussed species in the *Origin* (1859) 20 years later, as Ernst Mayr has put it, "One cannot help but feel that one is dealing with an altogether different author." In 1859, he seems to regard species as a purely arbitrary label for the convenience of taxonomists and thinks only a naturalist of "sound judgment and wide experience" should attempt to identify one.

In a private letter, he commented on the "laughable" confusion over the definition of species. What goes on, he wondered, in naturalists' minds when they think of species? In some, "resemblance is everything and descent of little weight—in some, resemblance seems to go for nothing, and creation the reigning idea—in some descent is the key—in some, fertility an unfailing test, with others it is not worth a farthing. It all comes, I believe, from trying to define the undefinable." His "solution" in the *Origin* is that species are fluid and change through time. Since they have no fixed essence, one need not worry too much about defining them.

Today, biologists define species in populational terms. Darwin often skipped back and forth between individuals, types and population. In current definitions, species are never "kinds" or "types," but breeding populations occupying a specific niche in nature and reproductively isolated from one another by geography, ecology or behavior.

This modern concept, glimpsed by young Darwin in the Galapagos, was lost for a century when he retreated into a view of species as "undefinable." Ironically, one of Darwin's last acts was to provide funds for the compilation of a comprehensive botanical species list, the *Index Kewensis* (1892–1895), cataloging thousands of "type specimens."

See also ESSENTIALISM; INDEX KEWENSIS; MAYR, ERNST; TRANSITIONAL FORMS.

For further information:

Eldredge, Niles. *Unfinished Synthesis: Biological Hierarchies and Modern Evolutionary Thought.* New York: Oxford University Press, 1985.

Mayr, Ernst. *Populations, Species, and Evolution.* Cambridge: Harvard University Press, 1970.

Salthe, S. N. *Evolving Hierarchical Systems: Their Structure and Representation.* New York: Columbia University Press, 1985.

SPECIESISM
Human Supremacy Ethic

Years before Charles Darwin published the *Origin of Species* (1859), he had come to the conclusion—contrary to the established beliefs of his day—that the human species is not separate and distinct from the rest of nature. Our kinship with other species was then a startling idea, noted in Darwin's private notebook (1837):

If we choose to let conjecture run wild, then animals, our fellow brethren in pain, disease, death, suffering, and famine—our slaves in the most labourious works, our companions in our amusements—they may partake of our origin in one common ancestor—we may all be netted together.

Darwin devoted much of his life to giving his poetic "conjecture" a scientific basis. More than a century later, the well-established kinship of all life influenced Australian philosopher Peter Singer to promote the term "speciesism" in his book *Animal Liberation* (1975)—the founding document of the current animal rights movement.

Although many outspoken advocates of humane treatment of animals, including the 19th century's influential anti-vivisectionist movement, had preceded him, Singer's reformulation of the issue struck a responsive chord in the 1970s.

On the heels of such successful political causes as human rights, women's rights, and the campaign against racism and sexism, Singer redefined the treatment of animals in terms of animal rights.

Racism is the belief that other human groups are inferior to one's own and can therefore be denied equal treatment. Analogously, "speciesism" is the belief that all living creatures other than humans are "lower" in value and that man has the right to kill them, eat them, destroy their homes or experiment on them without restriction. ("Speciesism" was coined by British psychologist Richard Ryder, a former experimenter who now campaigns against needless and cruel experiments.)

"When I say animals should have rights, that does not mean the same rights as people," says Joyce Tischler, director of the Animal Legal Defense Fund. As one journalist put it, "Pigs will never be interested in freedom of religion," but they might like to be free of being chained, butchered and eaten. Tischler was reminded that killing animals pervades our culture during a speech at a grade school. Her young audience was sympathetic to her plea for animal rights, until a sixth-grader stood up and reminded his classmates: "You realize she's talking about our hamburgers here."

See also ANIMAL RIGHTS; DARWIN, CHARLES; GREAT CHAIN OF BEING; ORIGIN OF SPECIES.

For further information:

Rachels, James. *Created from Animals: The Moral Implications of Darwinism.* New York: Oxford University Press, 1990.

SPENCER, HERBERT (1820–1903)
Philosopher of Evolution

For his evolutionary theories of technology, ethics, nature and society, English author Herbert Spencer was acclaimed by many as the greatest of Western philosophers. His works made him world famous by 1870 and, in America, his star rose higher than that of his countryman Charles Darwin.

A very successful American magazine, the *Popular Science Monthly*, was founded by Edward Youmans as a forum for Spencer's ideas. Industrialist Andrew Carnegie gave a dinner in his honor, attended by everybody who was anybody during the Gilded Age. Yet today, Spencer's works are unread, his name greeted by yawns and he is no hero even to philosophers or evolutionists—a victim of changing fashions and ideologies.

Spencer decided to dedicate his life to philosophy after a brief career as an engineer and inventor. Applying his mechanical and mathematical training to the search for general laws, he set out to be the ultimate reductionist, "interpreting all concrete phenomena in terms of the redistribution of matter and motion."

Instead, he became best known for providing an ethical rationale for *laissez-faire* industrial capitalism. Although the idea became known as Social Darwinism, it was really Social Spencerism.

Spencer's concept of social evolution had the poor classes "eliminated" by their "unfitness," while rational men abstained from interfering with the inexorable "laws" of evolution. The result, he believed, would be an evolved society that functioned smoothly and for the general good of its (future) members.

Perpetual progress was the rule of evolution, with individual and social happiness its eventual goal. However, when Europe turned toward militarism and the welfare state after 1870, Spencer became disillusioned with his theory of inevitable social progress. Complex political and economic systems, he began to see, refused to obey the evolutionary "laws" he had thought ruled the universe.

Among Spencer's major works were his *Principles of Biology* (1864), *Principles of Psychology* (1855), *Principles of Sociology* (1876), *Principles of Ethics* (1879) and essays on education, religion and political institutions. His scope was remarkable, ranging over organic nature, chemistry, technology, ethics and social science. To all these fields he brought a grand vision of unifying evolutionary synthesis, which did not survive the 19th century.

Spencer had even devised a sketch of an "organic" theory of natural selection, published it before Darwin and always insisted on his priority to Darwin and Wallace, although biologists never considered him one of their own. His science writing was speculative and airily devoid of familiarity with nature or the systematic study of biological problems.

Darwin often admired Spencer's cleverness at definitions and logic, but wished he would trade in a bit of it for greater powers of observation. Any one of the ideas he is constantly throwing out, Darwin said, would be "a fine subject for half a dozen year's work." To Darwin, Spencer's generalizations did not seem "to be of any strictly scientific use . . . They do not aid one in predicting what will happen in any particular case . . . [and have] not been of any use to me."

In his personal life, Spencer's youthful rebellion against authority evolved into crankiness as he grew older. He devoted himself to thinking and writing, never married or romanced women. In later life, he was subject to nervous disorders, disliked social company, regularly used opium and became a recluse.

Even in middle age, his nervous condition caused him to stay awake all night if he became the least bit excited or upset. Therefore, he carried with him at all times a pair of ear pads connected by a spring that passed around the back of his head.

Sir Ray Lankester, the Darwinian zoologist who became director of the British Museum of Natural History, told of being asked by Spencer to provide him with some information on a biological question. When he answered the summons and appeared at Spencer's club, the philosopher immediately expounded his own theories on the matter to the youthful Lankester.

Based on the facts he had gathered, Lankester began to point out some difficulties and objections. "Spencer hastily closed the conversation by fitting on his ear-pads," Sir Ray later recalled, "saying that his medical advisers would not allow him to enter into discussions."

Spencer's scientific friends sometimes lost patience with his propensity for spinning theories about nature with cavalier disregard for observation and experimentation. Once, when Thomas Huxley and other scientists were dining with him, Spencer announced he had just written a tragedy. "Yes," replied Huxley, "I know the catastrophe." Spencer protested it had never been shown to anyone, but Huxley insisted he could reveal the plot. Spencer's idea of a tragedy, said Huxley, is "a beautiful theory, killed by a nasty, ugly little fact."

See also CARNEGIE, ANDREW; SOCIAL DARWINISM; "SURVIVAL OF THE FITTEST."

SPERM BANK, GRAHAM'S
"Improving" the Human Stock?

Ever since Charles Darwin's cousin Sir Francis Galton suggested the idea of eugenics in 1869, sporadic attempts have been made to "improve" the human species by selective breeding. Galton's idea was that if "superior" people were mated (on the analogue of domesticated animals), the human race could deliberately evolve into a "better" species.

One major problem has always been who decides what are "desirable" traits in the parents; another is whether those chosen traits can be passed on to the offspring. In Hitler's Germany during the late 1930s, experimental breeding of a "master race" was attempted on a large scale. Carefully screened high-ranking Nazi officers were chosen as fathers, while mothers were selected for supposed "Aryan" qualities. The specially bred "Hitler youths" were cared for by the state and raised in official boarding schools. This misguided social policy resulted in producing thousands of confused orphans who proved to be genetically unremarkable.

Nevertheless, the idea persists that genetic superiority in humans can be identified and bred. In 1980, a self-made millionaire from Escondido, California, Dr. Robert Graham, began a new attempt to create an American "super breed" based on achievment and intelligence.

Dr. Graham, who pioneered the use of plastic eyeglass lenses, intends to spawn genius children by implanting the sperm of brilliant male scientists in superintelligent women. At the start of his project, Dr. Graham recruited three women through Mensa, an organization of people with high IQ scores. These women were married, with infertile husbands, and had agreed to be artificially inseminated.

Three Nobel Prize-winning scientists volunteered to donate repeatedly to the project's sperm bank. The only one publicly identified was Dr. William B. Shockley of Stanford, the septagenarian winner of the 1956 Nobel Prize in physics, who publicly endorsed the project's goal of creating "a hundred superbabies."

No mention was made of using women scientists in the program, nor did Dr. Graham attempt to demonstrate what a high score on an IQ test might really represent about "superior" genes. In fact, the qualities for which Graham tried to select are not scientifically identifiable or measurable, and even if they were, their mode of transmission is still unknown. Many studies have shown that scores on IQ tests have a great deal to do with social and cultural factors, and that the scores change at different times in an individual's life. Also, intelligence is an elusive characteristic—every genius is an idiot at something. Many years ago, George Bernard Shaw punctured the foundation of such ill-conceived eugenic programs with a classic remark. A lovely dancer had suggested to the playwright that "We two ought to have a child, so that it could inherit my beauty and your brains." "But Madame," Shaw replied, "what if it turned out to have *my* beauty and *your* brains?"

See also EUGENICS; GALTON, SIR FRANCIS.

SPERM COMPETITION
Natural Selection of Gametes

Gametes are sex cells: ova (eggs) in the female and sperm in the male. When they combine to form an embryo, natural selection may operate on the developing fetus, on the infant, the child, or the adult. But in 1970, a study of insect reproduction by Geoffrey Parker introduced an idea Darwin never considered: competition at the level of gametes.

Female insects sometimes mate with several males and store their sperm for days or, in some cases, years. "What determines which male's sperm will succeed in fertilizing her eggs?" Parker wondered.

He found in some cases a "rival's" sperm was diluted or simply washed away; in others males would insert plugs in the female after mating—the insect version of a chastity belt. Some mammalian species are thought to have evolved large testicles to increase the quantity of sperm production, hence the probability of fathering more offspring than rivals.

In recent years, the phenomenon has been studied in reptiles, amphibians, birds, and various mammals. Robert L. Smith, in a 1984 book *Sperm Competition and the Evolution of Animal Mating Systems* explores current research in the field.

One aspect under study is the role of female physiology and behavior in determining which of her several mates ultimately becomes father of her children—a research area that has rekindled interest in Darwin's idea of "sexual selection."

See also SEXUAL SELECTION.

For further information:
Smith, Robert L. *Sperm Competition and the Evolution of Animal Mating Systems*. Orlando, Florida: Academic Press, 1984.

SPIRITUALISM
Cosmic Evolution Controversy

The spectacular rise of the Spiritualist movement in 19th century England and America posed a dramatic challenge to evolutionary science.

Psychics claimed to offer "objective proof" of spiritual "phenomena," including communication with the dead—directly opposing scientific "materialism." And while they agreed humans evolved, they extended the concept into another dimension, from material body to cosmic spirit. Presented as a new philosophy, Spiritualism was really a reinterpretation of ancient Eastern beliefs, mixed with the Western desire to "secularize" the supernatural, and establish it "scientifically."

Although a tradition of ghosts and spirits has always persisted on the fringes of Western culture, the modern movement began in 1848 with Margaret and

Kate Fox, sisters born into a poor family in upstate New York. At their mysterious seances, the Fox sisters put questions to "spirits" and received loud rapping noises in response—resulting in a flurry of worldwide attention.

Their fame spread, and soon many other "mediums" arose in Europe and America. Thirty years later, after a fabulous and controversial career, one of the sisters admitted that the taps had always been produced by snapping her big toe inside her shoe.

As public interest grew, many charismatic "spirit-mediums" became wealthy celebrities. Among the more famous were Daniel Dunglass Home, Dr. Henry Slade, Mrs. Agnes Guppy, Frank Herne, and Charles Williams. All invited prominent scientists to come to their seances and observe for themselves such wonders as "automatic writing," accordians that played by themselves, glowing spirit hands and the visitations of spirit beings.

Scientists were bitterly divided on the question. A handful of respected savants, notably the chemist William Crookes, discoverer of thallium, physicist Sir Oliver Lodge, and Alfred Russel Wallace, the naturalist and evolutionist, believed they had contacted the spirit world. But most leading scientists, including Charles Darwin, physiologist William Carpenter and botanist Sir Joseph Hooker were implacable foes of Spiritualism and resented Wallace's attempt to allow papers about "psychic phenomena" to be read at a scientific meeting.

Alfred Russel Wallace, codiscoverer of the theory of natural selection, published a series of controversial essays supporting the existence of psychic phenomena. His book *Miracles and Modern Spiritualism* (1874) was endlessly quoted by mediums, who sought credibility through his scientific reputation. At the sensational trial of medium Henry Slade, Wallace appeared as star witness for the defense.

Crookes published books of his experiments, supposedly under "controlled conditions," to determine the existence of psychic phenomena. His work was ridiculed by most scientists, particularly when he was seen strolling and holding hands with the attractive "ghost" Katy King, who looked suspiciously like his favorite medium in a dark wig. In one of his naive "tests," he placed a cage around a self-playing accordion so no one could touch the keys in the dark.

Darwin expressed the majority opinion that "it was all trickery and deceit." Eventually, the skeptics were vindicated, for within the decade every one of the top dozen mediums—with the sole exception of Daniel Dunglass Home—was caught in fakery and fraud. These exposures were usually made not by scientists, but by professional stage magicians and conjurers, who were experts in detecting how the uncanny "phenomena" were produced. Darwin detested mediums for preying on the grief-stricken, and carried on a secret campaign to expose them, which has not been known to historians until recently.

After years of earnest attempts by the Association for Psychic Research to establish the truth of Spiritualism were inconclusive, the organization dissolved in disarray. Eventually, the wonders of a "materialist" science—telephone, radio, television—eclipsed the marvels of mysterious raps and self-playing accordians in darkened rooms. Although a few Spiritualist churches still exist, the movement declined after 1930, and mediums were reduced to the status of carnival tricksters in the popular mind.

However, in the late 1970s and 1980s the movement reemerged in a different form. Mediums are now called "channelers." Rather than calling up ghosts of the departed, they claim to be "channels" for ancient "entities" who wish to disseminate their wisdom to the modern world. They continue the tradition of what skeptical Victorians satirized as what "foxes say and geese believe."

Darwin was amazed when his own brother-in-law Hensleigh Wedgwood clung to his belief in the medium Charles Williams, who had been caught in the act "with false beard and dirty ghost clothes." Such tenacious belief against all evidence, Darwin said, was "a psychological curiosity." Physicist Michael Faraday compared the mind of a believer in Spiritualism to the iris of the eye. The more light that was thrown on it, the more tightly it closed.

Evolutionist Thomas Henry Huxley said he wouldn't be caught dead at a seance, in this life or the next. He thought it was "better to live a crossing sweeper [cleaning horse droppings from traffic intersections] than die and be forced to talk twaddle by a medium" who could summon you when a customer paid for a seance. His colleague Darwin, and student Ray Lankester, actively attacked the Spiritualist movement, particularly in the unprecedented criminal prosecution of medium Henry Slade.

See also MATERIALISM; ROMANES, GEORGE J.; SLADE TRIAL; WALLACE, ALFRED RUSSEL.

For further information:

Kottler, Malcolm J. "Alfred Russel Wallace, the Origin of Man, and Spiritualism." *Isis* 65: (1974) 145–192.

Podmore, Frank. *Modern Spiritualism,* 2 vols. London: Methuen & Co., 1902.

SPLIT PERSONALITY
The Primitive Beast Within

During the 1870s, a German doctor named Franz Hoffman treated a patient whom he believed could regress to an earlier stage of evolution. To describe the condition Dr. Hoffman coined the term "Doppelt-Ganger," or split personality. Soon he encountered others who lived ordinary, respectable lives—but had episodes of antisocial behavior, taking on different names, voices and mannerisms. Each personality seemed totally unware of the other's existence. Since evolutionary theories of "vestiges," "survivals" and "degenerations" were then in vogue, he believed the affliction might be a regression to an earlier stage of mental evolution.

Today psychologists view extreme fragmentations in personality as a behavioral disorder, not a "throwback" to an earlier type of human. As in the famous case dramatized in *Three Faces of Eve* (1957), the individual may even be "split" into more than two distinct personalities. The disorder seems to have been more pronounced during the last century, perhaps because Victorian society fostered an exaggerated division between acceptable behavior and what we today consider "normal" sexuality.

In the late 19th century, an openly sexual or antisocial individual was viewed as closer to the "savage" or "brutish" stage of human evolution. Under Darwin's influence, some psychologists redefined the violent or sexual side of human nature as a primitive vestige of our animal inheritance. What had formerly been explained as "evil" was now seen as an echo of our bestial origin. Sigmund Freud, who was preoccupied with the childhood history of the individual, believed that Darwin was studying "the childhood of the human race."

In taking this view of criminals and dual personalities, European theorists seem to have been influenced by ancient folktales about men who "revert" to savage beasts and are not responsible for their behavior during these episodes. One of Freud's famous cases was written up as "The Wolfman."

Fascination with this "Throwback" theory led Robert Louis Stevenson to write his haunting *Doctor Jekyll and Mr. Hyde* (1886)—a nightmarish vision that immortalized the "split-personality" in literature.

When the kindly Dr. Jekyll metamorphoses into Hyde, he undergoes instant, degenerative evolution. His nails and hair grow, face and hands become hairy and mind "regresses" to that of a savage hunter or killer-ape. Stevenson's classic tale has continued to inspire horror stories and movies a century later; but, since we no longer hold the Victorian view of criminals as "throwbacks" to more primitive beings, in most recent versions the original evolutionary slant of Jekyll-Hyde has been dropped. (An exception is *Altered States*, 1980, in which John Hurt experiments with cellular evolution and turns into a hairy, killer man-ape.)

See also DEGENERATION THEORY; FREUD, SIGMUND.

SPORT

In Darwin's day, the word "sport" was used to describe an individual born with different features than most. Early naturalists couldn't explain "sports," but breeders of domestic animals used them as the basis for new varieties. Creatures so different as to appear grotesque or deformed were called "monsters." Usually, "sports" and "monsters" are known today as "mutants," referring more to changes in their genetic material than to their phenotypic appearance.

SPRENGEL, CHRISTIAN KONRAD (1750–1816)
Botanist

In 1841, Charles Darwin read a book that amazed him—a German book that had gone almost completely unknown and unread since its publication a half-century earlier. Its title, roughly translated, was *The Secret of Nature Revealed—How Trees and Flowers Bring Themselves Forth* (1793). His friend Robert Brown had sent it to him. "It may be doubted," wrote Francis Darwin, "whether Robert Brown ever planted a more fruitful seed than in putting such a book into such hands."

Christian Sprengel had been rector of Spandau, but was fired for neglecting his official duties for his beloved plants. The "secret" revealed in his book, of course, was pollination by insects—not then generally known or appreciated. For years a "pollen controversy" had gone on: Was pollen really the "sperm" of plants? And if so, were plants at the mercy of visiting insects for their survival and reproduction? Only 200 years ago, this was still an open question.

Sprengel's book helped launch Darwin on a research program that would last the rest of his life, drawing him into thousands of experiments and observations in his greenhouse and garden. Perhaps its greatest culmination was his botanical classic *On the Various Contrivances by which Orchids are Fertilised by Insects* (1862).

See also INSECTIVOROUS PLANTS; ORCHIDS, DARWIN'S STUDY OF.

SPURTZHEIM, J.

See PHRENOLOGY.

STABILIZING SELECTION
Eliminating Directional Evolution

Variation is constantly occurring in a genetic population. If it is well adapted and its "average" (most frequent) traits are time-tested and proven, selection will eliminate such less common variations as they arise. Imagine the variation of traits as a bell-shaped curve—the narrow "tails" on both sides will be sliced off, while the curving center will be enhanced. That is stabilizing selection.

Directional selection is what we usually think of as evolution—a change in adaptation over time. If the environment changes so that rare variations suddenly became adaptive, selection must now favor the bell-curve's "tails." These traits increase in frequency until they become the new mean for the adapted population, which then returns to stabilizing its adaptive traits.

STEADMAN, DAVID (b. 1951)
Darwin's Islands Restudied

In 1835 when Charles Darwin explored the Galapagos Islands, isolated volcanic cones in the Pacific Ocean, 600 miles west of Ecuador, he found a unique little world inhabited by strange animals and plants—"creations found nowhere else." Many closely related species had diverged from a few ancestors that had arrived from the mainland long ago. Each island had its own species of giant tortoises, finches and mockingbirds, which, Darwin later realized, could yield clues to understanding the evolutionary process.

One hundred and fifty years after Darwin's voyage, American zoologist David Steadman decided to continue unraveling the story of evolution in these remarkable Pacific islands. But when Steadman tried to assess present knowledge of how many different species had evolved there, he found many problems and gaps in the inventory.

Young Darwin's time and knowledge had been limited, and his collections incomplete. He had overlooked some of the species still present today, while some others he saw and collected have since disappeared. And what about the situation before Darwin's famous visit: Was there any way to make a fuller inventory of island species as they were before any humans had intruded?

Steadman began, not in the Pacific, but at the British Museum's research collection in London. Here Darwin's original specimens—the preserved mammals, birds, fish, and reptiles collected by the great naturalist himself—are stored. Detailed examination filled out the present picture somewhat, for some species in the collection had become extinct since Darwin's day, only 150 years ago. Now Steadman began to wonder about possible extinctions a few thousand years back, before any human intrusion on the islands.

Geological experts advised him he was wasting his time. Fossils form only in sedimentary rock, when creatures become buried under layers of hardening silt. Since the Galapagos are formed of volcanic rock (cooled magma), bones could not fossilize; and if they were deposited before a lava flow, they would burn up. No specialist had ever found any.

But Steadman had the advantage of being more of an old-fashioned naturalist than a specialist and had a good working knowledge of zoology, geology, spelunking (cave exploration) and anthropology.

Seeking a way off the beaten path, Dave Steadman and his brother Lee became intrigued by the islands' systems of caves (lava tubes), which were largely unexplored. From their experience in other kinds of caves, they realized there was a possibility of finding undisturbed material that had been brought in by various creatures.

Steadman learned, to his delight, that a species of small owl nests in these caves; these owls habitually cough up compacted remains of rats, birds and other small prey. The caves turned out to be littered with thousands of owl pellets; when the tiny bones they contained were carefully identified and pieced together, they yielded important information about species that had lived there thousands of years ago.

Steadman was able to show that many extinctions had occurred in the Galapagos before Darwin or any other scientist had surveyed the wildlife. Then, too, he found many remains of small animals Darwin had described, but no one has seen since.

During the 1970s, Steadman spent months sifting and screening these delicate cave remains, adding 400,000 tiny fossils to Darwin's collection. These included numerous bones of an extinct giant rat, two other rat species and several previously unknown birds. One was a mockingbird that forms an addition to Darwin's famous series of several closely related species—all descended from mainland ancestors that somehow reached these isolated islands.

Extending his investigations to the Cook Islands, Steadman found very few living species compared to the original fauna. Polynesians had eaten their way through the wildlife, exterminating four kinds of pigeons that had formerly inhabited one island.

Steadman was fascinated to find that Polynesian people living there today have names and descriptions of various birds in their legends, which they have never seen. His fossil finds prove these birds actually did exist there a few centuries back.

Little by little, he patiently continues to gather the pieces of this immense jigsaw puzzle—the story of evolution in the Pacific Islands. "Any scientist who looks," says Steadman, "can find questions enough to outweigh the answers by a thousand to one. Darwin gave us something interesting to do here—enough for several lifetimes."

See also GALAPAGOS ARCHIPELAGO; LONESOME GEORGE; VOYAGE OF THE (H.M.S.) BEAGLE.

For further information:

Steadman, David W. *Holocene Vertebrate Fossils from Isla Floreana, Galapagos.* Washington, D.C.: Smithsonian Institution Press, 1986.

Steadman, David W., and Steven Zousmer. *Galapagos: Discovery of Darwin's Island.* Washington, D.C.: Smithsonian Press, 1988.

STEADY–STATE EARTH
Change Without Direction

For thousands of years, common observation showed the Earth was constantly in a state of change. Floods eroded hillsides, weather cracked and crumbled rocks, avalanches changed the shape of mountainsides. But such changes were not thought to modify the world in any particular direction, from one point in time to another. As the geological theorist James Hutton (1726–1797) had written it was a steady, ongoing process with "no vestige of a beginning, no prospect of an end."

The idea was common in Elizabethan times, as expressed in Shakespeare's 54th sonnet:

> When I have seen the hungry ocean gain
> Advantage on the kingdom of the shore,
> And the firm soil win of the watery main,
> Increasing store with loss and loss with store;

This notion, which the Bard called an "interchange of state," has been likened to the Hindu "cosmic dance." Everything changes, yet the total remains the same.

Since Sir Charles Lyell's (1797–1895) "uniformitarian" geology strongly influenced Charles Darwin, it is commonly—and mistakenly—assumed that Lyell believed the Earth itself "evolved," undergoing progressive or directional change. Oddly enough, Lyell never questioned the very old idea of a "steady-state" Earth and wove it into his "uniformitarian" doctrine.

His famous *Principles of Geology* (1830–1833) appears in hindsight as a mixed bag of seemingly irreconcilable ideas. Lyell actually believed time moved in great cycles, resisted Darwin's demonstrations that evolution had taken place and thought that perhaps one day the flying reptiles and dinosaurs would return.

See also ACTUALISM; CATASTROPHISM; GRADUALISM; LYELL, SIR CHARLES; UNIFORMITARIANISM.

STOPES, MARIE C. (1880–1958)
Paleobotanist and Sexual Reformer

Around 1900, when Marie Stopes explored Japan searching for 70-million-year-old flowers, women in academic work were still a novelty. In geology, they were "almost an impropriety." Nevertheless, the Scotch–English paleobotanist didn't mind being the first foreigner, first woman and first Western scientist most Japanese had ever seen.

In addition to her work on petrified plant remains from the Cretaceous rocks of Hokkaido, she wrote a charming *Journal From Japan* (1910) describing her adventures. Before 1920, her treatise on *Cretaceous Flora* was published by the British Museum, and her nontechnical book, *Ancient Plants*, successfully popularized fossil botany. Stopes's *Monograph on the Constitution of Coal* (1918), a classic in its field, solved the puzzle of the formation of so-called "coal balls." Stopes demonstrated the structures are fossils of densely matted swamp vegetation, often containing small marine animals, which lived on the plants.

Stopes was interested in the past life on Earth *and* in its future. An early champion of women's rights, she authored many books and tracts on birth control and sex education. Like Margaret Sanger and Victoria Woodhull, she became notorious for her outspoken views at a time when such matters were simply not discussed in public, especially by women.

Her manual *Married Love* (1918) was reprinted in many languages, and in 1921 she and her husband established the first birth control clinic in Britain. These pioneering efforts to change public attitudes earned her a harvest of hatred. "In those times," wrote one historian, "you might socially have got away even with spitting in church or defiling the Union Jack—but not with uttering the name Marie Stopes in respectable company. She was reviled as immoral, monstrous, obscene . . . and her life was threatened."

In order to withstand the pressures of being a feminist, social reformer, scientist and sexual guru, Marie Stopes developed a unique personality. As one of her acquaintants put it:

> Her vanity was so colossal, so uninhibited, and so unashamed, as to be positively endearing . . . Having done so much for other people's personal relationships, she mismanaged all her own. She was compassionate, headstrong, tactless, public-spirited, humourless, intellectually distinguished and wholly lacking in aesthetic taste. I count it a privilege to have known her.

Fossil plant specialist W. T. Gordon of King's College London, once enlisted Stopes's aid in raising funds for a university charity. As part of his series of "science week" lectures, he announced a rare speaking appearence by the notorious Dr. Marie Stopes. It was during the height of her infamy as an author who challenged accepted sexual practices and

**COAL BEDS AND MARRIAGE BEDS were among Marie Stopes'
favorite topics. While her geological studies won scientific applause, her advocacy of legal and social equality for women made
her a social pariah. Stopes founded the first birth control clinics
in England.**

the lecture hall was packed with paying curiosity
seekers. This expectant audience was treated to an
hour's talk by the irrepressible Marie Stopes—on the
formation of coal.

For further information:
Briant, Keith. *Passionate Paradox: The Life of Marie Stopes.*
New York: Norton, 1962.

STRATIGRAPHIC DATING
Establishing Relative Chronology

The basic chronology of Earth history was established
by identifying different strata or layers in geologic
formations and relating them to other layers. It is
based on the assumpton that lower beds were laid
down first and are therefore older, while higher
(later) beds are younger. In some areas, the sequence
may be reversed because of folding and uplifts, but
such distortions can be detected.

By relating partial sequences from various localities, a more inclusive picture of wider areas can be
built up. For example, if one sequence contains A–
B–C–D and another has C–D–G, and still another
has D–E–F–G–H, the three can be correlated into
one sequence. Stratigraphy is a relative method of
dating, which cannot give "absolute" dates unless it
is used in conjunction with geophysical "clocks."

See also CHRONOMETRY; RADIOCARBON DATING;
SMITH, WILLIAM.

SUBDUCTION ZONE
Crustal Plate Overlap

In the "new" geology of plate tectonics, the 16 or so
segments that make up the Earth's crust are believed
to be in constant motion, "drifting" across the globe.
They also act as conveyer belts to the deeper mantle,
bringing up new layers of molten rock (magma) and
returning old rock to the mantle.

Subduction is the redepositing of a portion of the
crust into the Earth's mantle. It frequently occurs
where two plates meet, one slides under the other,
and vast amounts of rock are forced downward. This
area of overlapping is known as a subduction zone.

See also CONTINENTAL DRIFT; PLATE TECTONICS.

SUMNER, WILLIAM GRAHAM (1840–1910)
American Social Darwinist

A student once asked Professor William Graham
Sumner of Yale—then the most influential spokesman for economic "survival of the fittest"—how he
would feel if he was the loser. "Any other professor
is welcome to try for my job," Sumner replied, "if
he succeeds, it is my fault. My business is to teach
the subject so well that no one can replace me."

Sumner became famous as the philosopher and
spokesman of Social Darwinism in America; he was
an energetic advocate of letting the strong survive
and the weak perish in the social and economic
spheres. Although he was a philosopher and sociologist, his social views were most influenced by the
work of a paleontologist, Othniel C. Marsh. It was
Marsh's fossil studies of the evolution of the horse
that won him over to the idea of evolution, although
he borrowed more from Herbert Spencer than from
Darwin.

Sumner became America's most strident and influential Social Darwinist, who joined his conception of
organic evolution to a conservative political program
based on free competition. He used his teaching post
to develop and promote Social-Darwinist ideas, which
brought together the Protestant ethic, doctrines of
classical economics and Darwinian natural selection.

Harsh though this might be, Sumner believed the
Protestant ideal of the hardworking, thrifty, temperate man was "the fittest." Millionaires deserve their

special privileges, he taught, because they have been tested and have survived in the only correct system—unregulated competition. Once Sumner was asked if he believed in government bailing out private industry. "No! It's root, hog, or die!" he said.

From the 1870s through the 1890s, he waged a "holy war" against reformism, protectionism, socialism and government intervention. His books include *What Social Classes Owe to Each Other* (1883), *The Forgotten Man* (1883) and *The Absurd Effort to Make the World Over* (1894). One of his chapters surveying human customs throughout the world had gotten much too long, so he brought it out as a separate book. It became the classic *Folkways* (1906), which is still a standard source.

Sumner argued it was not cruel to pit the strong against the weak in the economy, because the strong are really the "industrious and frugal," while the weak are the "idle and extravagant":

> If we do not like the survival of the fittest, we have only one possible alternative, and that is the survival of the unfittest. The former is the law of civilization; the latter is the law of anticivilization.

As for socialism, his contempt was unbounded. "It is a plan for nourishing the unfittest and yet advancing in civilization," he wrote, a combination "no man will ever find."

See also CARNEGIE, ANDREW; HORSE, EVOLUTION OF; SOCIAL DARWINISM.

SUNDAY LEAGUE
Victorian "Sabbatarian" Controversy

Professor Thomas Henry Huxley, England's chief exponent of evolutionary biology, was once charged by police with "keeping a disorderly house." The scandalous incident, not nearly so racy as it sounds, occurred when Huxley openly defied London's Sunday Laws by lecturing on Darwinism. It fanned the flames of one of the most bitter, drawn-out controversies in Victorian England—sabbatarianism.

Nineteenth-century London was a dreary place on its sacred Sundays. Shops were padlocked, recreations and concerts forbidden, and the great public cultural institutions—zoos, art galleries, museums, and libraries—were closed. In 1853, a jeweller named R. Morrell formed the National Sunday League to promote "elevating recreation" on that day. He founded a journal, *The Sunday Review*, sponsored by many liberal thinkers, literary men and scientists, including such notables as Charles Dickens, Thomas Henry Huxley and Charles Darwin.

In his novel *Little Dorrit* (1856), Dickens had described Sunday in sooty London as "gloomy, close,

and stale," punctuated by "maddening church bells of all degrees of dissonance":

> Everything was bolted and barred that could . . . furnish relief to an overworked people. No pictures, no unfamiliar animals, no rare plants or flowers, no natural or artificial wonders of the ancient world . . . nothing to see but streets, streets, streets . . . Nothing for the spent toiler to do, but to compare the monotony of this seventh day with the monotony of his six days, and think what a weary life he led.

Twenty-two years earlier, conservative "sabbatarians" had formed the Lord's Day of Observance Society (LDOS), dedicated to preserving the "sanctification of the Christian Sabbath" as "one of the few remaining threads on which [England's] destiny is awfully suspended."

After a long and bitter fight between the two factions, the Sunday League won its first victory in 1856, when Lord Palmerston allowed the Guards' bands to play in the public parks. Outraged, the archbishop of Canterbury proclaimed that "unless the Sunday band concerts cease, I can be no longer responsible for the religion of the country."

Palmerston maintained that "innocent intellectual recreations, combined with fresh air and healthy exercise [are] not at variance with the soundest and purest sentiments of religion." But he had to yield to conservative pressure, and the bands were withdrawn.

Some time later, Thomas Henry Huxley, who was unabashedly evangelical about Darwinism, decided to present the first scientific lecture ever given on a Sunday. To conform with traditional sabbath practices, the program included some singing before and after the talk. Presented on the evening of Sunday, January 7, 1866, it was attended by an enthusiastic audience of more than 2,000 at the St. Martin's Hall in Long Acre.

The event was a great success, but was followed by demands from sabbatarians for Huxley's arrest. Puzzled police could not find a law banning a "secular service," and in desperation charged him under an old law for "keeping a disorderly house." To no one's great surprise, the case was thrown out of court. Later, Huxley entitled a book of his collected lectures *Lay Sermons* (1879).

See also FUNDAMENTALISM; MILITARY METAPHOR; SECULAR HUMANISM.

SUNDEW PLANT

See INSECTIVOROUS PLANTS.

SUPERORGANIC

See TEILHARD DE CHARDIN, FATHER PIERRE.

"A DREAM OF THE FUTURE" was the title of this 1885 cartoon in an English humor magazine, lampooning the bitter fight to open the libraries and museums on weekends. A drunken, loutish working class mob is shown entering the British Museum on Sunday, and emerging refined and gentrified, studying books and carrying the banner of "Sweetness and Light."

"SURVIVAL OF THE FITTEST"
Evolutionary Slogan

Nowhere in the first edition of Charles Darwin's masterpiece *On the Origin of the Species* (1859) does he use the phrase "survival of the fittest." It was actually coined by the English philosopher Herbert Spencer, in his *Principles of Biology* (1864).

Although Spencer wrote a volume on biological evolution, he was no naturalist; evolution interested him as a "universal principle." His vague notion of "the fittest" meant those individuals best able to foster general progress and improvement for their society or species.

Alfred Russel Wallace, the cofounder of evolutionary theory, was struck by "the utter inability of many intelligent persons" to understand what he and Darwin meant by natural selection and suggested they substitute Spencer's phrase. When Darwin obliged him by using "survival of the fittest" in later editions of the *Origin*, readers were still confused; everyone seemed to have his own interpretation of what was meant by "the fittest."

However, the phrase caught the public's imagination and became completely associated with Darwin. Critics said it was a meaningless tautology—a proposition that simply repeats itself. Since the fit are the individuals who survive, they argued, wasn't it another way of saying "survival of the survivors"?

Evolutionary biologists have long been aware of that pitfall, and many have contributed to making the concept useful. Fitness, it turns out, is a relative term. Organisms that are the "fittest" in one environment, may be completely unsuccessful in another. Or they may be extremely sucessful for millions of years—as the dinosaurs were—only to be suddenly wiped out when conditions change.

In populational terms, fitness simply means reproductive success. The race does not always go to the strong or swift, but to those who manage, by whatever means, to produce the largest number of offspring. Sometimes the "fittest" may be those who attain high social status (and more matings) with bold bluffs or subterfuges rather than prowess or strength. Other methods of outcompeting rivals in production of offspring include persistent sexiness, extraordinary tail plumage or throwing your neighbor's eggs over a cliff.

Social Darwinists seized the phrase as a slogan to advocate a totally unregulated economy. Robber barons of the Gilded Age—James J. Hall, John D. Rockefeller, Andrew Carnegie—often told journalists their cutthroat business practices were, in the long run, helping society evolve. Elimination of weak and inefficient competitors was the road to progress, with future benefits for all. During the late 19th and early 20th centuries, "Survival of the Fittest" became the oft-repeated mantra of industrial capitalism.

But Thomas Henry Huxley knew full well that scoundrels were appropriating biology to exalt themselves. The problem, he pointed out in his essay "On Providence" (1892), lies "in the unfortunate ambiguity of the term 'fittest' in the formula, 'survival of the fittest.' We commonly use 'fittest' in a good sense, with an understood connotation of 'best' . . . [which we are] apt to take in its ethical sense. But the 'fittest' which survives in the struggle for existence may be, and often is, the ethically worst."

See also CARNEGIE, ANDREW; FITNESS; SOCIAL DARWINISM; SPENCER, HERBERT.

SYNTHETIC THEORY
Reformulation of Evolution

Around 1900, Darwinism was in a scientific limbo. Hugo De Vries and T. H. Morgan had brought macromutation into prominence, relegating natural selection to a minor sorting role. Other geneticists had shown that most mutations were harmful, and that the random shuffling of genes makes little change in the population. Paleontologists were talking about "straight line" evolution, and philosophers were looking for "vital forces" that guided evolution to a predetermined goal.

But during the 1920s and 1930s the rise of population genetics brought evolutionary studies full circle back to Darwin. Population concepts, shaped by new mathematical tools, dealt with changing gene frequencies in populations. And the new technology reaffirmed the importance of natural selection as a major force in the origin of species. It also gave new weight to some other forces, notably "genetic drift"—the random sampling error associated with small breeding populations.

One of the leaders in reintegrating evolutionary science was Theodosius Dobzhansky, a Russian emigre who continued his work with the fruit fly *Drosophila* in America. His controlled laboratory experiments with the flies, which reproduce every ten days, enabled him to observe evolution directly, and to study adaptation as an experimental science. In his book *Genetics and the Origin of Species* (1937), one of the founding documents of the modern synthesis, Dobzhansky showed how minor changes among a few flies in his small populations could greatly change a large number of descendants.

Another seminal book in shaping what came to be known as the modern synthesis gave the movement its name. Biologist Julian Huxley, the grandson of "Darwin's bulldog" Thomas Henry Huxley, wrote

Evolution; the Modern Synthesis in 1942. At about the same time, paleontologist George Gaylord Simpson applied populational thinking and genetics to the study of fossils in his *Tempo and Mode in Evolution* (1944). Simpson also dealt with varying rates of evolution and discredited attempts to view the fossils in terms of "straight-line" predetermined evolutionary paths or goals.

Other major contributors to the Synthetic Theory included Ernst Mayr, J. B. S. Haldane, G. Ledyard Stebbins, Sewall Wright and R. A. Fisher. Through their efforts, a new understanding of Darwinism was reached that integrated the results of genetics, mathematics, paleontology and especially populational thinking with Darwin's original principle of natural selection. The Synthetic Theory is the basis of 20th-century biology and prompted new experimental research on the nature of change in populations.

However, in the 1970s, Stephen Jay Gould, Niles Eldredge, Steven Stanley and others criticized the Synthetic Theory for promoting an unfounded "gradualist" view of evolutionary change. According to these critics, Darwin's idea that most widespread species undergo slow changes at a steady rate had become orthodoxy in the Synthetic Theory. They believed, instead, that fossil and genetic evidence points to widely varying rates of change. In 1972, Gould and Eldredge proposed the "punctuational" model of evolution: long periods of stability interrupted by episodes of rapid change. Defenders of Synthetic Theory, in turn, accuse them of setting up "gradualism" as a straw man. The Synthetic Theory, they argue, has always included the idea that evolution can proceed at different rates.

See also ESSENTIALISM; GENETIC DRIFT; PUNCTUATED EQUILIBRIUM.

For further information:
Dobzhansky, Theodosius. *Genetics and the Origin of Species.* New York: Columbia University Press, 1937.
Huxley, Julian. *Evolution: The Modern Synthesis.* London: George Allen and Unwin, 1942.
Mayr, Ernst. *Systematics and the Origin of Species.* New York: Columbia University Press, 1942.
Simpson, George G. *Tempo and Mode in Evolution.* New York: Columbia University Press, 1944.

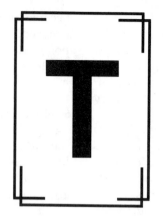

TAIEB, MAURICE

See AFAR HOMINIDS.

TANGLED BANK
Darwinian Metaphor

Darwin's *Origin of Species* (1859) is not only a scientific argument, but a work of literature as well—full of rich imagery, metaphors and appeals to the reader's imagination. The tangled bank is one of his famous metaphors for the ecological web, the complexity and interrelatedness of all life.

The final paragraph of *Origin of Species* contains Darwin's lyrical musings on the Tangled Bank. (He once said that a good book, like a fine day, should end "with a glorious sunset"):

> It is interesting to contemplate a tangled bank, clothed with many plants of many kinds, with birds singing on the bushes, with various insects flitting about, and with worms crawling through the damp earth, and to reflect that these elaborately constructed forms, so different from each other, and dependent upon each other in so complex a manner, have all been produced by laws acting around us.

In his book *The Tangled Bank: Darwin, Marx, Frazer and Freud as Imaginative Writers* (1962), Stanley E. Hyman shows how several influential thinkers clothed their abstract theories in poetic language and imagery. The Tangled Bank expresses Darwin's special view of nature: a web of interrelationships binding various plants and animals into a community. Not only is the visible vegetation entangled; the lives of different species are also intertwined "in so complex a manner."

Ecological interdependence was one hallmark of Darwin's contribution to modern thought, though the word "ecology" had yet to be invented. "The metaphor of entanglement," Gillian Beer has written, "enacts what often remains latent in his argument: the extent to which evolution is a lateral rather than simply an onward movement, whose power lies in multiple relationships as much as in selecting out."

Why was this Tangled Bank so "interesting to contemplate?" Because beneath its illusory surface confusion, science could discover underlying regularities and relationships. Darwin's later work established the painstaking research methods that could eventually "untangle" his beloved Tangled Bank.

This Tangled Bank is clearly on an English stream, familiar and commonplace—like the banks of the River Severn he played upon as a child. Evolution's mysteries could be observed not only in the exotic rain forests of South America, but in one's own country garden—which, after returning from his adventures in the tropics, is exactly what he did.

See also EARTHWORMS AND THE FORMATION OF VEGETABLE MOULD; ECOLOGY; METAPHORS IN EVOLUTIONARY WRITING; ORCHIDS, DARWIN'S STUDY OF; ORIGIN OF SPECIES.

For further information:
Hyman, Stanley E. *The Tangled Bank: Darwin, Marx, Frazer and Freud as Imaginative Writers.* New York: Atheneum, 1962.

TARZAN OF THE APES
Fictional Lord of the Jungle

Tarzan, the jungle hero raised by apes, became one of the most popular fictional creations of the 20th

JUNGLE MAN raised by apes, Edgar Rice Burroughs's popular fantasy hero was the post-Darwinian version of Rousseau's "noble savage." Johnny Weissmuller's movie Tarzan of the 1940s (left) was widely popular. The pongid madonna scene (right) is from Hollywood's more recent *Greystoke* (1985).

century. According to zoologist Jane Goodall, who has lived with wild chimpanzees for 30 years, her early exposure to Tarzan stories helped inspire a strong interest in African animals. Prior to Charles Darwin, there were fantastic tales of abandoned infants raised by she-wolves or other animals, but never by apes. Part of Tarzan's appeal stems from Rousseau's "noble savage" and part from the post-Darwinian appreciation of man's kinship with chimpanzees and gorillas. Even after the cinema-version Tarzan establishes a treetop household with Jane and Boy, Cheetah the ape remains an inseparable member of the family.

This remarkable fantasy adventure series by Edgar Rice Burroughs first appeared in the pulp magazine *All-Story* in 1912. In the original episode, Tarzan's parents Lord and Lady Greystoke are shipwrecked on the African coast, but survive for a year in a jungle hut.

He is orphaned when his mother dies of fever, and his father is killed by Kerchak, a fierce giant ape who rules the neighborhood. (Some writers have seen in Kerchak a pop culture forerunner of King Kong.) Adopted by Kala, a gentle female ape who is mourning her own dead infant, Tarzan grows up at home in the trees, thinking himself an ape until he meets other humans.

Burroughs posed the question of whether the child of English aristocrats could survive in a "state of nature." His answer was that his hero not only survived, but became benevolent ruler over all other creatures. Despite the crudity of his speech and manners, Tarzan has an innate nobility of character that leads him to perform heroic deeds on the side of right and justice.

When a visiting American woman from Baltimore, Jane Porter, comes through the jungle, she is abducted and brutalized by Kerchak. Tarzan enters into a life-or-death struggle to rescue her and avenges his father by killing the murderous ape. Jane falls in love with Tarzan, and they begin their life together in a treetop retreat.

More than 25 Tarzan books sold hundreds of millions of copies worldwide. In the 1930s, journalist Alva Johnston observed, "Millions of boys took to leaping from limb to limb and from tree to tree. The nation resounded with the Tarzan yell and the snapping of collarbones."

For the better part of this century, Tarzan's adventures have been popular on radio and television and in movies, cartoons, comic books. The first of the many movie Tarzans, in 1918, was Elmo Lincoln, who appeared in three silent films. But Olympic champion swimmer Johnny Weissmuller became the most enduring film Tarzan, although he was the ninth actor to play the role.

Now pop classics, his films were shot in California and Florida—never in Africa—during the 1930s and 1940s. Asiatic elephants were used because they are easier to handle than their African cousins. Sometimes the films showed jungle monkeys hanging from their tails, although no African species can do that. (Monkeys with grasping or prehensile tails are found only in Central and South America.)

Tarzan's famous movie yell was a combination of several different sounds, including Weissmuller yelling, a soprano singing, a note played on a violin, and hyena growls played backward. Eventually, Weissmuller himself was able to give a piercing rendition of the sound. During his terminal illness in 1979, newspapers reported that Weissmuller became sadly confused about his identity. He had to be moved from the Actor's Hospital because he frightened other patients with blood-curdling Tarzan calls in the middle of the night.

Edgar Rice Burroughs claimed he took the germ of his story from the classical myth of Romulus and Remus, the twin brothers who were abandoned, nursed by a she-wolf and grew up to found the city of Rome. He was also strongly influenced by the romantic idea of a "noble savage" first popularized by French philosopher Jean Jacques Rousseau and subsequently embedded in American popular art and literature.

Burroughs insisted the sound of the name "Tarzan" was an important factor in his character's success, and had tried hundreds of sound combinations before settling on it. A lover of the "music" of words, he also invented an entire "ape language" for his fictional anthropoids, which were never specified as chimpanzees or gorillas in the original stories.

As he grew wealthy from book and movie royalties, Burroughs bought a ranch in the San Fernando Valley, near Hollywood, where he lived happily for many years. There he founded a town and named it after his imaginary ape-man of the African jungle: Tarzana, California.

See also "APE WOMEN," LEAKEY'S; NOBLE SAVAGE.

TASADAY TRIBE (HOAX)*
"Stoneage Cavemen" Who Weren't

During the late 1960s, reports began to appear in the world press about a sensational anthropological find: a tribe of cave-dwelling people isolated in the rainforests of the Philippines since the dawn of time. Subjects of a popular book by John Nance, *The Gentle Tasaday* (1970), they were said to be food gatherers living on tadpoles, crabs, wild yams and palm pith. Aided only by simple stone axes and digging sticks, the zenith of their technology was a simple hand-rolled friction drill for making fire. Above all, the Tasaday were characterized as gentle and loving and were reported to have no words in their language for "enemy," "weapon," or "war."

According to the widely circulated story, a local trapper named Defal ran across the Tasaday in the mid-1960s and became their benefactor. When he met them, the cave people wore only a few leaves or g-strings and had no previous contact with the outside world. Defal taught them to hunt, brought them metal knives and reported their existence to Manuel Elizalde Jr., an official of the Philippine government. Elizalde in turn arranged for scientists and journalists to visit the tribe, which was under his protection as head of the Office of the Presidential Assistant for National Minorities (PANAMIN).

Author John Nance, and the rest of the world's media scribes, fell in love with the Tasaday. As the vogue of the hunting hypothesis began to fade [see HUNTING HYPOTHESIS] the Tasadays offered a peaceful alternative model of stone-age life many were eager to embrace. "We saw the Tasaday as a people who were very touching, caring, affectionate . . . [I felt] a whole new perception about who *we* were in that rain forest." John Nance said. "If our ancestors were like the Tasaday," wrote a *National Geographic* editor, "we came from far better stock than I had believed."

A *Geographic* team visited the Tasadays' jungle caves and published a lyrical cover story in 1972: "We, skyborne creatures of man's most advanced and tormented society, and they, perhaps the last of the world's innocents, watched each other across the full span of cultural evolution. And felt love for each other." Picture captions described naked innocents, living just like "Our ancestors . . . thousands of years ago . . . (in) a primeval Eden."

But in a strange burst of prescience, the *Geographic* editor also mentioned a disturbing feeling that "in the morning all this will have vanished like Brigadoon. It's not real. Or maybe we're not." He couldn't have known how prophetic his words would become.

Elizalde built a treetop helipad and strictly controlled the comings and goings of reporters and anthropologists—and such celebrated visitors as Charles Lindbergh and Gina Lollobrigida. In 1974, the Philippine government sealed off the entire area under martial law and declared it a tribal reservation. The purpose of these heavy-handed controls on investigators, Elizalde explained, was for the protection of the Tasaday from exploitation by outsiders. The

CONFESSING "CAVEMAN" HOAX, member of the "former, alleged Tasaday tribe" snicker at nude pictures of themselves as "stone age primitives" in magazines. When this previously unpublished photo was taken by Kristina Loz, they were admitting to an ABC-TV crew that their "lost tribe" was the brainchild of corrupt Philippine officials.

tribesmen seemed to venerate him, calling him "Momo Dakel de Wata Tasaday"—"great man, god of the Tasaday."

For 12 years the Tasaday were "protected" by Elizalde's soldiers, while their story became a staple in anthropology classes and television documentaries. But during that period, various researchers noticed disturbing contradictions. For example, a linguist had found in their language words for "planting," "mortar," "roof of a house," and others that didn't fit with a cave-dwelling forest people. When another anthropologist noted that cooked rice was being sneaked into the caves, his study was soon disrupted and he was forced to leave. And locals claimed to know the "cave people" as ordinary peasants who came to the village market, fully clothed and smoking cigarettes!

Within a month of the departure of the corrupt Marcos government, Swiss journalist Oswald Iten teamed up with a local writer named Lozano and slipped into the forest. They found the "Tasaday" families dressed in shirts and jeans, living in the traditional frame huts common to the area. Teams from various television networks and magazines made other forays, with ambiguous results. In 1986, a persistent group from ABC's "20/20" program finally cleared up the mystery.

Eight of the original group, fully clothed, came to meet the television crew. They poked fun at the tribal name by which foreigners had known them, chanting "Tasaday, Tasaday." After giggling at nude photos of themselves in *National Geographic* and other magazines, they admitted on-camera that the whole thing had been a fraud.

They were not born in the caves, they said, but in the area surrounding them. Their languages were Manubo and T'boli, those of the two well-known country tribes from which they had come. The caves, they said, were a place of reverence where they worship their ancestors; the general region of these holy caves was called Tasaday. No one had actually occupied them in living memory.

Why did they pretend to be primitive cave people? "Elizalde said if we went naked we'd get aid because we'd look poor." Were they rewarded with help? "None. We're poorer now than we were before . . . He lied."

In 1983, according to world news media, Elizalde fled the country, taking with him $35 million of National Minority Protection Bureau money and ". . . at least 25 tribal maidens." According to Judith Moses, who produced the television expose, "An unquestioning, receptive Western press has learned a different kind of lesson from the former, alleged Tasaday tribe."

See also CARDIFF GIANT; PILTDOWN MAN (HOAX).

TAUNG CHILD
Dart's Fabulous Fossil

In November of 1924, as he was dressing in tie and tails for a wedding, Dr. Raymond Dart received three crates of rocks from a friend. With his collar still unfastened, and the groom and guests waiting, Dart opened the boxes and pulled out a small fossil brain cast and the back of the skull into which it fitted. Immediately he knew he had something extraordinary, and the groom had to yank him away to be best man. The young anatomist at the University of Witwatersrand in South Africa had received the first known fossils of the ancient African man-apes.

Intrigued by the baboon skulls a student had shown

"DART'S BABY," from a South African limestone quarry, was the first known australopithecine skull. Anatomist Raymond Dart's discovery of this small-brained creature with large, human-like teeth opened a new chapter in the search for human origins.

him from a limeworks at Taung, Botswana, 200 miles from his Johannesburg home, Dart had requested to see further fossils from the site. A geologist friend, Dr. R. B. Young, who was visiting the quarry, packed up some odd chunks from the day's blasting and sent them along. Dart's full-dress reception of them turned out to be strangely appropriate.

Over the next three months, Dart worked constantly on separating the face from the limestone matrix in which it was embedded, using improvised tools, including his wife's knitting needles. What he uncovered was a six-year-old child's skull of a hominid creature no one had ever seen before, along with the mineral impression of its braincase (endocranial cast). Today, scientists believe the Taung child lived between one and two million years ago.

It was rare that such a cast would be formed, rarer still that it was not destroyed at the quarry—as many fossils had been—and even more remarkable that it had fallen into the hands of one of the only men in the world who was able to understand what it was.

He named the creature *Australopithecus africanus*, meaning southern ape of Africa, since he thought it must be some kind of "missing link" between apes and men. Its large brain case, small teeth (compared to apes), rounded palate, and position of the hole at the base of the skull (foramen magnum) indicated an upright hominid with many similarities to humans, but it would take decades for the scientific world to agree with him.

When Dart announced his discovery in 1925, four of Europe's leading anthropologists scoffed and ho-hummed "another" missing link. Sir Arthur Keith thought it was merely an aberrant chimpanzee and had no hesitation in placing the fossil among the true apes. Other anatomists thought it was an ancient gorilla, of no importance to the quest for human origins. The jury of experts gave their verdict, and it proved to be totally wrong.

While anthropologists turned their attentions to Peking man in the 1930s, Dart stuck to his guns, and reminded them of Darwin's half-forgotten prediction, made in 1871, that the most ancient human ancestors would probably be found in Africa, home of the chimp and gorilla. Some years later, after his colleague Robert Broom began to find adult skulls of *Australopithecus* in the southern Transvaal, the tide began to shift in Dart's favor. But it took another 40 years before he was fully vindicated; many subsequent discoveries have finally brought scientific opinion into agreement with his first assessment of the Taung child's importance. "It's no good being in front if you're going to be lonely," he once remarked—but in science that often comes with the territory.

See also AUSTRALOPITHECINES; DART, RAYMOND ARTHUR; DARTIANS.

TEILHARD DE CHARDIN, FATHER PIERRE (1881–1955)
Philosopher, Paleontologist

From his early boyhood among nine brothers and sisters in the rural French province of Auvergne, Pierre Teilhard de Chardin was remarkable for his mystical devotion and scientific curiosity. If he had developed only one of these talents, the lanky, brilliant young man might have enjoyed a comfortable, uncomplicated life. Instead, his persistent attempts to straddle both science and religion brought years of anguished struggle. As an outstanding paleontologist turned evolutionary philosopher, Father Teilhard was rebuffed, silenced and finally exiled by the church he sought only to serve.

After spending his childhood at Jesuit schools, at 18 Teilhard entered the order, then was sent to teach physics and chemistry at a Jesuit College in Cairo for several years. On returning to France, his lifelong curiosity about nature led him to the Museum of Natural History in Paris, where he studied human paleontology with the great Marcellin Boule. There, at the Institute of Human Paleontology, he also met another mentor and lifelong friend, the Abbe Henri Breuil, a churchman-archeologist who pioneered the study of Ice Age cave paintings and stone tools.

Teilhard took a doctorate in paleontology at the Sorbonne, where he found acceptance and recognition as a scientist. During his long career, Teilhard was to make important contributions to the study of early fossil mammals, serve as a missionary in China and visit early-man sites throughout Europe, Africa and Asia, earning a respected place among paleoanthropologists.

His first experience with the excitement of a fossil man dig, oddly enough, was with Charles Dawson in Sussex, England at the Piltdown site, where he found a celebrated tooth. (Decades later, it transpired that the gravels had been "salted" with doctored bones and artifacts by still-unknown hoaxers.) Between visits to England, he served as a paramedic in World War I, then returned to Paris, where he hoped to pursue a career of writing, teaching and research.

But it was not to be. While Teilhard's science had led him to the fact of evolution, his devotion fastened on an impossible dream, upon which he refused to compromise—that the Catholic church embrace the reality of evolution as part of the divine plan.

This growing preoccupation with the evolution of Catholicism itself was greeted by his superiors with hostile, intolerant suppression. He tried to explain that his views were no threat to religious values and that an evolutionist's view of man could add a new basis to the church's highest moral teachings. A scientific quest for the truth about human origins, he believed, would surely lead back to the same Creator worshipped by the church. Refusing him any kind of hearing, the hierarchy branded him a dangerous, heretical rebel who needed to be muzzled.

For the rest of his life, church authorities made sure Teilhard did not teach or write for publication

CATHOLIC EVOLUTIONIST, Father Pierre Teilhard de Chardin pursued a dual career as paleontologist and unorthodox theologian. Although his books later won world fame, he was censured by the church for attempting to synthesize science and religion.

and sought to strictly limit his influence. They sent him to China as a missionary, encouraged him to attend excavations at remote sites—anything to keep him from speaking and writing in France.

Paradoxically, his long stay in China resulted in his participation in one of the great discoveries in human evolution: the excavation of the limestone caves where Peking man (*Homo erectus*) had left an impressive number of fossil skulls.

The "internal tearing" between his passion for knowledge and obedience to church tradition gave him a new vision of Christ's agony on the cross. Mindful that the "struggle for existence" in nature could be harsh and painful, he envisioned his personal Savior "carrying the weight of an evolving world" instead of its guilty sins. Teilhard had little patience with the doctrinal Fall of Man, but instead focused on man's rise and ascent. [See CHRYSALIS.]

Holding him to a vow of strict obedience he had taken 30 years earlier, church officials forbade Teilhard to print any of his philosophical books or articles during his lifetime. However, after his death in 1955, friends published the manuscripts, beginning with his grandest work *The Phenomenon of Man (Le Phenomene Humaine)*, which he had completed in 1941. An instant sensation, for several years it enjoyed a tremendous vogue, widely discussed by philosophers, psychologists, social scientists and theologians (although some biologists dismissed it as an "an incoherent rhapsody").

Father Teilhard viewed man as a phenomenon—a species not static, but a process of being and becoming, for which he coined the term "hominisation." In the future, mankind would continue to evolve toward a distant "Omega point," drawn onward to his higher destiny.

He also coined the term *noosphere* to denote the realm of developing mind or culture, called by others "the superorganic" or "psychocultural evolution." Whereas organic evolution previously took place only in the biosphere, or world of life, the future of earthly life will be decided in this noosphere, the arena of human perceptions, thoughts and values.

One of Teilhard's favorite notions was that special creativity results when an evolutionary "splitting-off" (divergence) is followed by a "coming together" of lines that have begun to splay out on different paths. He cited such "convergent integration" in his discussion of world "races" and cultures—pursuing separate histories for a time, but not for so long they could not then "melt" back together, enriching all. His great hope was that the divergent streams of scientific and religious thought, too, would one day merge, producing an enlarged understanding of the "phenomenon of man."

Though Teilhard won the personal admiration and friendship of distinguished philosophers, scientists and theologians, they never integrated his grand synthesis into the mainstream of Western thought. His work remained, as it had during his lifetime, too mystical for scientists, and too scientific for religionists.

Despite their initial impact, his works ultimately failed to establish the new view of man and God intended by its author, except among a small, dedicated band of admirers who continue to cherish his memory and publish several journals in which to discuss his ideas. To these Friends of Teilhard Societies, he is a martyred visionary, an uncanonized saint whose persecution shames the Catholic church.

Exiled from France, denied the satisfaction of seeing his philosophical works in print, Father Teilhard spent his final years in a small New York City apartment. He grew ill and alienated, living on the philanthropy of the Wenner-Gren Fund, whose president, Paul Fejos, was a long-time admirer.

"How is it possible," he wrote during his last days, "after 'descending from the mountain' and despite the glory that I carry in my eyes, I am so little changed for the better, so lacking in peace, so incapable of passing on to others through my conduct, the vision of the marvelous unity in which I feel myself immersed? . . . As I look about me, how is it I find myself entirely alone of my kind? . . . Why am I the only one who sees?"

See also GOPI KRISHNA; "OMEGA MAN"; PEKING MAN; PILTDOWN MAN (HOAX).

For further information:

Lukas, Mary, and Lukas, Ellen. *Teilhard: The Man, the Priest, the Scientist.* New York: Doubleday, 1977.

Teilhard de Chardin, Pierre. *The Phenomenon of Man.* New York: Harper Bros., 1959.

TELEOLOGY
Search for Design in Nature

Before the quest to understand evolution became their central focus, naturalists—many of whom were also churchmen—tried to find divine design and purpose in plants and animals. Since they believed the Creator had devised a plan for every creature, they named the focus of their study from the Greek *telos*, which means "purpose."

Early zoologists marveled at the supposedly "perfect design" of organisms for their environments, such as the structure of eagle's wings or the optical mechanisms of the human eye. Ignoring abundant evidence of imperfect or quirky structures in nature, they focused on the seeming marvels of adaptation. As Bergen Evans wrote in his *Natural History of Non-*

sense (1946), teleologists are "like optimists in earthquakes who pick their way through ruins and over corpses to squeal with rapture over a statue or a kitten that has survived the general destruction. They do not perceive that all living things are survivors, and that the adaptation that has made survival possible is usually the barest minimum."

Teleologists argue backwards. Sir Thomas Browne praised God's wisdom for putting living things that need heat and sunshine in the tropics and those that thrive on cold in the Arctic. Since Darwin's day, we take it for granted that the vegetation has evolved to suit the climate, and not the 'other way round—a view that would have been considered shockingly "mechanistic" a few hundred years ago.

In Voltaire's *Candide* (1758), teleology is satirized through the character of Dr. Pangloss, who continually asserts that God made everything for the best in "this best of all possible worlds." (Pangloss could be translated "Dr. Everything-is-shiny.") At one point, the optimistic doctor remarks how wonderful it is that the nose is midway between the eyes, for God knew it would be the perfect position to support eyeglasses. Voltaire was joking, but zoologist Richard Owen was not when he suggested the gap in horse's teeth was "designed" to allow room for the bit, so men could ride them.

To some, Darwin at first seemed to "rescue" teleology by giving it a new basis, for his experiments revealed the unsuspected "purpose" of bizarre structures and colors. Now oddly shaped nectaries of orchids, for instance, could be viewed as "designed" to accommodate the heads or tongues of pollinating insects. But Darwin also demonstrated how the structures could result from the coevolution of insects and orchids—parts of a system whose components have evolved together, influencing changes in each other and creating a "fit" between them.

He also showed that most of the expected "perfection" in nature did not hold up under systematic examination. In fact, most life forms are more like odd contraptions, evolutionary makeovers that resemble the work of a make-do tinkerer rather than a "Master Planner."

See also ADAPTATION; GRAY, ASA; ORCHIDS, DARWIN'S STUDY OF; PALEY'S WATCHMAKER; PANDA'S THUMB.

For further information:

Dawkins, Richard. *The Blind Watchmaker*. New York: Norton, 1986.

Gould, Stephen Jay. *The Panda's Thumb*. New York: Norton, 1980.

TELLIAMED

See DE MAILLET, BENOIT.

TEMPLE OF NATURE
Erasmus Darwin's Evolutionary Epic

After a lifetime of pondering the phenomena of nature and man's place in it, 18th-century philosopher Erasmus Darwin expressed his mature views in an epic poem rather than a scientific treatise. His masterpiece *The Temple of Nature* (1803) was published a year after his death.

Heroic and grandoise in scope, musical in its language, *The Temple of Nature* is a remarkable presentation of the pioneer evolutionist's view of the Earth's creation, the rise of life from microscopic "filaments" in the seas, through the radiation of plants and vertebrates, including man. Combining science and art, Erasmus footnoted the text with references to the raw material on which his mind fed—horticulture, medicine, anatomy, botany, chemistry. Indeed, he parades before the reader surprisingly more "facts" than one might suppose from his reputation as a speculative visionary.

It is an extraordinary work, little read or studied today, but widely popular at the turn of the 19th century, when it was praised by philosophers and literary men alike. Its excellence and rarity justify reprinting the following excerpt:

Then, whilst the sea, at their coeval birth
Surge over surge, involv'd the shoreless earth;
Nurs'd by warm sun-beams in primeval caves,
Organic Life began beneath the waves.

First forms minute, unseen by spheric glass,
Move on the mud, or pierce the watery mass;
These, as successive generations bloom,
New powers acquire, and larger limbs assume;
Whence countless groups of vegetation spring,
And breathing realms of fin, and feet, and wing.

Next, when imprison'd fires in central caves
Burst the firm earth, and drank the headlong waves;
And, as new airs with dread explosion swell,
Form'd lava-isles, and continents of shell;
Pil'd rocks on rocks, on mountains mountains rais'd,
And high in heaven the first volcanoes blaz'd.

In countless swarms an insect-myriad moves
From sea-fan gardens, and from coral groves;
Leaves the cold caverns of the deep, and creeps
On shelving shores, or climbs on rocky steeps
Cold gills aquatic, for respiring lungs,
And sound aerial flow from slimy tongues.

Thus the tall Oak, the giant of the wood,
Which bears Britannia's thunders on the flood;
The Whale, unmeasured monster of the main,
The lordly Lion, monarch of the plain,
The Eagle, soaring in the realms of air,
Whose eye undazzled drinks the solar glare;—
Imperious man, who rules the bestial crowd,

Of language, reason, and reflection proud
With brow erect, who scorns this earthly sod,
And styles himself the image of his God;
Arose from rudiments of form and sense,
An embryon point, or microscopic ens!

Erasmus summarizes his major thesis in one of the poem's footnotes: "The Great Creator of all things has infinitely diversified the works of his hands, but has at the same time stamped a certain similitude on the features of nature, that demonstrates to us that the *whole is one family of one parent*." Half a century later, Charles Darwin's words in *Origin of Species* (1859) unmistakably echo his grandfather: "There is grandeur in this view of life . . . having been originally breathed into a few forms or into one, and that whilst this planet has gone cycling on . . . from so simple a beginning endless forms most beautiful and most wonderful have been, and are being evolved."

See also DARWIN, ERASMUS; ORIGIN OF SPECIES; ORTHOGENESIS.

For further information:

Darwin, Erasmus. *The Temple of Nature, or The Origin of Society.* London: J. Johnson, 1803.

King-Hele, D. G. *Doctor of Revolution: The Life and Genius of Erasmus Darwin.* London: Faber and Faber, 1977.

TENNYSON, ALFRED, LORD

See TENNYSON'S IN MEMORIAM.

TENNYSON'S *IN MEMORIAM* (1850)
Evolutionary Requiem

When her beloved husband Prince Albert died, Queen Victoria told the poet Alfred, Lord Tennyson (1809–1892) that "Next to the Bible, *In Memoriam* is my comfort." So popular was the 131-stanza poem that within a few months of its publication in June 1850, Tennyson was named poet laureate of England. This poem, to which so many Victorians turned for comfort during dark days, was partially inspired by Robert Chambers's *Vestiges of Creation* (1844)—an important work on evolution, which appeared before Darwin's *Origin of Species* (1859).

In Memoriam was Tennyson's tribute to the memory of his beloved friend Arthur Hallam, who died at the age of 22 in 1833. Tennyson, who had been tutored by the great William Whewell, had an intense interest in science and wove his reactions to scientific developments into his work. Near the middle of the long poem, Tennyson despairs, faced with what seems to be the meaninglessness of the natural world, according to his reading of Charles Lyell and Chambers:

Are God and Nature then at strife,
  That Nature lends such evil dreams?

So careful of the type she seems,
So careless of the single life;

That I, considering everywhere
  Her secret meaning in her deeds
  And finding that of fifty seeds
She often brings but one to bear

"So careful of the type?" but no,
  From scarped cliff and quarried stone
  She cries "A thousand types are gone:
I care for nothing, all shall go."

Toward the end of his poem, Tennyson regains his optimism through a faith in an evolutionism that seeks progress. His friend Hallam, he believes, was a premature example of a higher type of man.

A soul shall draw from out the vast
And strike his being into bounds,

And moved thro' life of lower phase
  Result in man, be born and think,
  And act and love, a closer link
Betwixt us and the crowning race . . .

Among the phrases that became common in the language are the description of "Nature, red in tooth and claw," which "shrieke'd against" the creed of divine love. And, of course, the poem's core message, which is often wrongly attributed to Shakespeare: " 'Tis better to have loved and lost/Than never to have loved at all.' " One beautiful stanza could serve as an epitaph for all naturalists, past and future:

My love has talked with rocks and trees
He finds on misty mountain ground
His own vast shadow glory-crown'd
He sees himself in all he sees.

The Victorians loved the poem and quoted it endlessly. But paradoxically, they failed to notice that its philosophy ran counter to their most cherished religious teachings. Since "its hope comes from evolutionism and its prospect of progress to a race of supermen like Hallam," writes historian Michael Ruse, "one suspects that deep down many Victorians did not much care whether organic evolutionism or the doctrinal niceties of conventional Christianity" were really true. What they did care about, according to Ruse, was "the frightening rapidity of change in their lifetimes and the essential lack of security of their society—a society supported and surrounded by the vast underprivileged, frequently starving masses. When Tennyson held out the hand of hope and progress to a better state, they grasped it thankfully, not bothering about details."

For further information:
Roppen, G. *Evolution and Poetic Belief: A Study in Some Victorian and Modern Writers.* Oslo: Oslo University Press, 1956.

TERMITES

See GAIA HYPOTHESIS; SEX, ORIGIN OF.

"TERNATE PAPER" (1858)
Wallace's Evolution Theory

One of the most important—and most obscure—founding documents of evolutionary biology is "The Ternate Paper" (1858), written in a tropical jungle by Alfred Russel Wallace. Entitled "On the Tendency of Varieties to Depart Indefinitely from the Original Type," it is a clear, concise statement of Wallace's theory of evolution by natural selection, worked out independently of Charles Darwin.

Wallace was collecting specimens on the island of Ternate in what is now Indonesia, then called the Moluccas. Like Darwin, he conceived his theory after reading Thomas Malthus's *Essay On Population* (1798), a striking example of historical parallelism. Although he and Charles Darwin had corresponded, he did not know the senior naturalist had been working secretly for 20 years on the same theory, although he had not yet published anything on the subject.

Innocently, Wallace sent the paper off to Darwin on March 9, 1858 from Ternate, asking him if he thought it was worthy of publication. It carried such topic headings as "The Struggle for Existence," "Adaptation to the Conditions of Existence," "Useful Variation will tend to Increase; Useless or Hurtful Variations to Diminish." Wallace concluded the paper with a statement that "There is a tendency in nature to the continued progression of certain classes of varieties further and further from the original type." This explanation of evolutionary process, he concluded, will "agree with all the phenomena presented by organized beings, their extinction and succession in past ages, and all the extraordinary modifications of form, instinct, and habits which they exhibit."

When Darwin received the Ternate Paper, he was thrown into a panic. He feared he had been "forestalled . . . so that all my originality . . . has gone for naught." He was astounded that Wallace's topics "could stand as my chapter headings" and appealed to his friends Charles Lyell and Joseph Hooker for help.

There followed a series of events that are still debated by historians, culminating with the joint publication of the "Darwin–Wallace" theory in the journal of the Linnean society in 1858. Darwin then frantically set to work on his long-delayed book and produced *The Origin of Species* (1859) in 13 months. When Wallace returned from the Moluccas three years later, he was gracious about not claiming priority for the discovery. If he had, evolutionary theory might be known as "Wallaceism."

See also DARWIN, CHARLES; "DELICATE ARRANGEMENT"; NATURAL SELECTION; WALLACE, ALFRED RUSSEL.

For further information:
Brooks, John L. *Just Before the Origin: Alfred Russel Wallace's Theory of Evolution.* New York: Columbia University Press, 1984.
Wallace, Alfred R. "On the Tendency of Varieties to Depart Indefinitely from the Original Type." (London, 1858). Reprinted in Brackman, A. C. *A Delicate Arrangement: The Strange Case of Charles Darwin and Alfred Russel Wallace.* New York: Times Books, 1980.

TERRACE, HERBERT

See APE LANGUAGE CONTROVERSY; NIM CHIMPSKY.

THALES (c. 600 B.C.)
Ancient Greek Scientist

Western scientific traditions—including the earliest attempts at a rational explanation of human origins—can be traced to ancient Greece, even though some of these ideas were not followed up for thousands of years.

Thales was the founder of the earliest school of scientific speculation known to us. He was prominent among a group of teachers known as the "Seven Sages" of Miletus, a city of Ionia, though his exact dates are not known.

He taught that "Nothing comes into being out of nothing, and that nothing passes away into nothing." Everything in nature is constantly changing, Thales taught, "all in motion like streams." He visited Egypt, observed how the annual flooding of the Nile brought the parched country to life and concluded that the primary substance from which life originates is water.

Thales and the Ionian philosophers did not come close to solving the questions they posed, because their method was confined to speculation. They wondered, they thought and they discussed various ideas, but they did not yet put questions to the test of systematic observation or experiment. Nevertheless, they were the first recorded askers of "Why?" and "Whence?" in the Western tradition who looked for natural, rather than supernatural, explanations.

See also ANAXIMANDER.

THEORY, SCIENTIFIC
"Truth" and Uncertainty

Critics often charge that evolution by natural selection is "only a theory" and "cannot be proved." Charles Darwin would have agreed with them—not because he didn't believe in evolution, but because he was a subtle philosopher of science. He well understood that "induction" (going from facts to general principles) could not guarantee a final, absolute truth. But he was among the first to realize the enormous power and value of "provisional" truth in advancing human understanding.

Darwin caused a great uproar, not only by the substance of his theory, but also by his redefinition of science itself. Some critics were more upset at his insistence that there is no absolute truth in science than with his belief in man's kinship with the apes. A good scientific explanation, he thought, is simply one that accounts for the most facts at a given time. It does not need to be proven beyond all doubt as "true" forever. If a more productive and comprehensive explanation is devised, the theory is superseded or becomes a "special case."

His stated procedure was to try a probable explanation on a group of facts, and—if it seemed to solve problems—attempt a broader application. It was a modest, realistic and revolutionary concept of the limitations of science as well as its strengths. Darwin rarely used the word "law" as in physics, preferring the more probablistic terms "doctrine," "principle," "theory" or "explanation."

His frankness about the uncertainty of his most deeply held beliefs was not welcome news to a shaky society looking to science or religion for unambiguous answers. But Darwin, and his advocate Thomas Henry Huxley, knew that truth-seekers would have to learn to live with a new acceptance of uncertainty—in both religion *and* science.

Darwin explained in a private letter (1838) that his theory was based on four "general considerations": (1) the "struggle for existence" or competition in nature; (2) "the certain geological fact [from fossils] that species do somehow change"; and (3) "the analogy of change under domestication by man's selection." And, most important of all: (4) the theory connects and makes intelligible "a host of facts."

He matter-of-factly admitted that he could not "prove" any living species had changed or that supposed changes are beneficial. Nor could he explain "why some species have changed and others have not," and was baffled by how the variability he observed in species might be produced.

If he could not prove with certainty that evolution was true, how could Darwin expect any thinking person to adopt his theory? The answer is simply that it is productive; it works. Applying it produced a torrent of discoveries, insights, new information. Connections arose to bridge formerly diverse disciplines: comparative psychology, geology, botany, paleontology. Working scientists found it solved many puzzles; and formerly inexplicable phenomena made sense within a coherent larger picture.

Darwin wrote a friend in 1861, "the change of species cannot be directly proved, and . . . the doctrine must sink or swim according as it *groups and explains* [disparate] *phenomena*. It is really curious how few judge it in this way, which is clearly the right way." A few years later he wrote that he was "weary of trying to explain" the point; most people could not grasp it.

See also MATERIALISM; MECHANISM; MILITARY METAPHOR; SCIENCE.

For further information:
Ghiselin, M. *The Triumph of the Darwinian Method.* Berkeley: University of California Press, 1969.
Giere, R. *Understanding Scientific Reasoning.* New York: Holt, Rinehart and Winston, 1979.

TINBERGEN, NIKOLAAS (1907–1988)
Pioneer Ethologist

Although Charles Darwin had been a great field naturalist, acceptance of evolution had the strange effect of sending several generations of scientists indoors to laboratories of psychology and anatomy. Dutch ethologist Niko Tinbergen—along with Konrad Lorenz and Karl von Frisch—took their studies back to the woodlands and rivers. After watching animals in natural settings, they devised simple experiments out in the open to clarify what the creatures were doing.

As a student in The Netherlands, Tinbergen had distinguished himself as a hockey player and pole-vaulter, but not on his zoology exams. Whenever possible, he fled the classrooms for long, solitary hikes across fields and beaches, where he became increasingly curious about the behavior of common birds and animals. Later, he taught his own students never to substitute book learning for active, direct observation of nature.

He never had the desire to trek through exotic lands to seek out rare or spectacular creatures. Common species, ignored by many naturalists, seemed to offer enough mysteries for a lifetime. He wondered about little things, like the function of bright red splotches on the beaks of gulls, or how digger wasps locate their narrow burrows after flights of several miles.

While still a student, his studies of the predatory wasps showed a talent for what he called "natural experiments." First, he caught several individuals and marked them with colored dots of paint, for identification. Next, he removed pebbles, grass clumps or pine cones from around the burrows. As he had suspected, the airborne wasps were unable to locate their homes without such "landmarks."

Moving these objects to another place—but keeping them in the same configuration—fooled returning wasps into heading for where the entrance "ought" to be. Using such simple methods, Tinbergen was able to discover how wasps can recognize their own burrows among many; more importantly, he discovered their remarkable ability to quickly learn and memorize new patterns.

Tinbergen's classic study of herring gulls revealed, among other things, the function of their bright red bill markings. Hungry chicks instinctively peck at anything red. When they target the mother's beak, they receive regurgitated food. But Tinbergen showed they will also peck at a crude cardboard cutout if it has the red splotch, and even prefer it to a real beak on which the red has been covered.

If one knows what to look for, Tinbergen pointed out, there is not much difference between such experiments and accidental observations under naturally varying circumstances. In his book *Curious Naturalists* (1958), he recalled receiving a 10-year-old girl's letter about her unexpected confrontation with a young herring gull at the shore.

It had boldly walked over and suddenly gave a vicious peck at a "very red scab" on her little sister's knee. Most children might have reacted with fear or incomprehension. But this "curious naturalist" had heard Professor Tinbergen discuss gull behavior on the radio, and thought he might like to know about the incident "because she realized that this was a kind of experiment."

Tinbergen's investigations of bird behavior culminated in such classics as *The Study of Instinct* (1950) and *The Herring Gull's World* (1953). He left Holland in 1949 to join the faculty at Oxford, where a special position had been created for him. His lectures attracted enthusiastic students, who spread his influence all over the world.

Like Darwin, Tinbergen was a man of "enlarged curiosity," who was drawn to science for the sheer fun of it. But he never could quite shake the feeling that watching animals was a very pleasurable waste of time, not a "serious" career. Certainly, he never expected that one day (when he was 74) he would share the Nobel Prize for Biology and Medicine with his friends von Frisch and Lorenz for establishing a new science of animal behavior.

Yet even after receiving this highest scientific accolade, he continued to be plagued by what he called "my little devil"—the nagging, depressing doubt that his life's work was neither worthwhile nor enduring. Frequently he felt guilty—and not quite sane—because "even the smallest new discovery about animal behavior" produced in him "such intense delight." It was almost as if he thought someday he would get caught and made to return the money and honors he had accepted for doing what he loved.

See also ETHOLOGY; LORENZ, KONRAD.

TOOL-USING, EVOLUTION OF
Former "Human" Hallmark

For centuries philosophers believed that using tools was one trait that separated mankind from animals. From the simple stone implements found in prehistoric sites, the human species has developed a tool kit that includes computers, scuba gear and rocket ships. Because these tools allow humans to enter new environments, some anthropologists speculate that once man became a tool-using creature he may have liberated himself from any future biological evolution.

If man wants to fly, he builds an airplane rather than growing wings; to dive deep in the ocean, he carries oxygen instead of evolving gills. (Despite such logic, mankind, like all living things, continues to evolve.)

Studies by field naturalists during the past 30 years have produced a growing list of a wide variety of other creatures that use tools to manipulate or modify their environment. When Egyptian vultures want to crack open large eggs, they pick up stones and fling them at the eggs until the shells break. On Darwin's beloved Galapagos Islands, the woodpecker finch uses small twigs and cactus spines held in the beak to pick insects out of small holes in trees.

It takes the finch but a moment to select a suitable spine or twig, sometimes with a bend near the end, for leverage. The bird then patiently manipulates it to poke and pry grubs from underneath bark. Between probes, it may transfer the tool to a foot, hold it while getting a better stance, then seize it again in its beak. Sometimes finches carry the tools with them for another foray. Not all members of the species have mastered the trick, indicating that it may well be a learned tradition. (Darwin himself never witnessed this remarkable behavior during his stay on the Galapagos.)

Another simple tool is used by sea otters that swim in the kelp beds off the coast of Calfornia. They often float on their backs and crack open abalone shells with a stone, using their chests as anvils. One was observed using the thick part of a Coca-Cola bottle as a hammer, in place of the usual stone.

Some anthropologists supposed that it was tool-*making*, not merely using, that was unique to humans. That objection was shattered forever when Jane Goodall discovered how chimpanzees in the African forest not only use but *make* simple tools. Chimps select twigs and sticks, bite them to just the right size, strip them of leaves, and chew them into shape for simple jobs, such as gathering termites, prying up ant nests or stealing honey from beehives. During dry seasons, the chimps go to large clay termite mounds, wet their "termiting" sticks with saliva and shove them carefully into the insect's passageways. The thirsty termites grab onto the wet stick, and the apes then eat them off without getting bitten themselves. In addition to "termiting," chimps collect and chew wads of dry leaves to use as "sponges" for gathering water from hollows in tree trunks.

When anthropologist Louis Leakey first heard of his protege Jane Goodall's discoveries, his reaction was, "either we're going to have to change the definition of tools and tool-making, or we'll have to change the definition of apes—and man."

But some naturalists think the Galapagos woodpecker finchs' accomplishments—with a brain a mere fraction the size of a chimp's—show up the apes as underachievers.

"Given their hands and huge brains, it's amazing apes and monkeys don't do a lot more tool-using. They're incredibly stupid," says Dr. Michael MacRoberts, who has conducted field studies of both free-ranging macaque monkeys and acorn-storing woodpeckers. MacRoberts argues:

> If Leakey had seen the Galapagos finch prying and stabbing hidden grubs with cactus spines, or watched California woodpeckers chisel trees into collective "granaries" for storing acorns, would he say we would have to change the definition of man—or birds? No, because primatologists are like doting parents. Anything "their" monkeys or apes do is remarkably clever, because they expect them to be bright. And anything other animals do is "just instinct," because they're supposed to be far removed from man.

After interviewing Jane Goodall on late-night television, comedian Jay Leno remarked, "I don't believe chimps are similar to humans just because they use tools. What would really make them like humans would be if they *borrowed* the tools and never gave them back."

See also ABANG; APES, TOOL USE OF.

TORRALBA, SPAIN (PREHISTORIC "KILL SITE")

See HOWELL, F. CLARK.

TRACKWAYS

See BIRD, ROLAND T.; FOSSIL FOOTPRINTS, PALUXY RIVER; "NOAH'S RAVENS."

TRANSITIONAL FORMS
Links Between Species

Creationists often charge that fossil creatures intermediate between ancient and modern species don't exist. Yes, there may be a few transitional forms between ancient elephants (mastodons) and modern elephants, they argue, but an elephant is still one "kind" of animal. Where is the fish on its way to becoming a lizard or the lizard turning into a cat?

While it is true that there are "gaps" in the fossil record between various groups, we do have many transitional forms—some of them quite spectacular. For instance, remains of a winged, feathered creature with lizard-like teeth and bones (*Archeopteryx*) was discovered in German quarries in the 1890s. "That's no half-way bird," writes creationist Wayne Gish, "it *was* a bird."

But archeopteryx also has claws on its wings, long tail vertebrae, unfused back vertebrae, rudimentary breastbones and other skeletal characteristics of reptiles. As evolutionary biologist Douglas Futuyma writes in *Science on Trial* (1983), "It has exactly the characteristics that ancestors of birds must have had if they descended from reptiles . . . it occurs at the same geological time as the small therapsid dinosaurs . . . and is almost identical to those dinosaurs in virtually every characteristic except its feathers."

And therein lies one basic problem. Creationists have a notion that fundamental "kinds" of animals don't change. But zoologists who have extensive experience examining living organisms find "kind" or "species" a slippery concept. Thousands of supposedly "established" species have turned out be variants within the same breeding populations.

Many living species, writes Futuyma, "are in various intermediate stages and cannot be clearly categorized, any more than we can divide a person's life discretely into adolescence and adulthood":

> There is no gap between thrushes and wrens, between lizards and snakes, or between sharks and skates. A complete gamut of intermediate species runs from the great white shark to the butterfly ray, and each step in the series is a small one, corresponding to the slight differences that separate similar species.

When we turn to fossils, numerous examples of transitional forms are available. Paleontologists agree that the distinction between the early mammal-like reptiles and reptile-like mammals is totally arbitrary. If the lower jaw has certain bones in a particular sequence, they call it a reptile. If the lower jaw consists of only the dentary bone, it's a mammal. These early transitional groups (synapsids and their offshoot, the therapsids) show a very peculiar transition: two bones that formed the articulation for the upper and lower jaws in reptiles became modified to the inner ear bones (the malleus and incus or hammer and anvil) of mammals.

As Stephen Gould notes, "in an excellent series of temporally ordered structural intermediates, the reptilian dentary gets larger and larger, pushing back as the other bones of a reptile's lower jaw decrease in size. We've even found a transitional form with an elegant solution to the problem of remaking jaw bones into ear bones: a double articulation which would allow one set to be 'remade' while the other set preserves the jaw structure."

It is true that we usually find sequences of structural intermediates rather than straight-line sequences of ancestors and descendants. Evolution is not a ladder, but a branching bush, sending off dozens of shoots in many directions. We may only find fossilized a twig here, a bough there—usually from periods of stability, since most major changes occur in short bursts in small populations. In William H. Calvin's analogy, "If you were an archeologist digging up a parking garage, nearly all the cars you'd uncover would be parked on one level or another; few would actually be on a ramp between levels."

Even when we turn to man-like creatures, we find a temporal sequence displaying a threefold increase of brain size, and corresponding reduction of jaws and teeth over four million years. As Gould asks, "What more could you ask from a record of rare creatures living in terrestrial environments that provide poor opportunity for fossilization?"

See also CREATIONISM; ESSENTIALISM; "MISSING LINKS."

TURKANA BOY
Most Complete Erectus

In the summer of 1984, Richard Leakey's "hominid gang" unearthed the 1.6-million-year-old remains of a gangly, adolescent boy—the most complete skeleton of a *Homo erectus* ever found.

Discovered by Kimoya Kimeu, the foreman of a team led by anatomist Alan Walker and Richard Leakey, the fossil boy came from a site called Na-riokotome, near the western shore of Lake Turkana in Kenya. Perhaps the most complete fossil hominid skeleton ever found (aside from members of our own species, *Homo sapiens*), only tiny fragments are missing from the skull and jaws. Every tooth is present, the spine and rib cage are complete, there are hand bones, arm bones, a complete pelvis and both legs. Only a few arm bones, the feet and some neck vertebrae are missing.

In contrast with the modern human cranial capacity of approximately 1300–1400 cubic centimeters, the boy's brain case held about 840cc. He stood about 5'4", considered tall for an early hominid, and had human-like body proportions. Canine milk teeth had not yet been shed and the long bones weren't fully grown, indicating that he was less mature than a *sapiens* his age would be. As an adult, his brain would have reached about 882cc—65% of what adults have today, and his body would have reached a stature of six feet. Also, his spinal cord is much narrower than that of modern humans.

Back in 1891, Eugene Dubois found the first remains of *Homo erectus*, which he called the "Java ape man" and thought was the missing link between apes and men. But, spectacular as his finds were at the time, they consisted of only a fragmentary skullcap and some molar teeth. Thirty-five years later, more complete skulls and long bones were discovered at Choukoutien in China and nicknamed "Peking man." Since then, *Homo erectus* skulls and fragments have been found in various parts of Africa, Europe and Asia—including several skulls excavated at Olduvai by the Leakeys.

But never before the Turkana boy was a specimen of such completeness, with so few gaps left to the imagination, found. Along with "Lucy," a 40% complete skeleton of a female Australopithecine from Ethiopia, the Turkana boy is one of the most exciting fossil hominid discoveries in the history of paleoanthropology.

See also HOMINID GANG; HOMO ERECTUS; "LUCY".

TURKANA MOLLUSKS

During the mid-1970s, Harvard paleontologist Peter G. Williamson worked up the best-documented fossil sequence supporting the idea of punctuated equilibrium. But his series of mollusk layers taken from Lake Turkana, in northern Kenya, also demonstrates that "suddenness" in geology may really be a very long time.

Using the rate of reproduction of living relatives of the snails as a measure, Williamson calculated the rate of change in his fossil mollusk series. The "abrupt" discontinuities or "revolutionary changes" turn out to have taken 5,000 to 50,000 years—which would be about 20,000 generations of the creatures.

In an article in *Nature*, J. S. Jones, of University College, London, translated this fossil record into the terms of an experimental geneticist. The time in which the major changes occurred in these species, he figured, would be the equivalent of 1,000 years of breeding fruit flies, 6,000 years of mouse generations, or 40,000 years of breeding dogs. (Most dog breeds we have today were developed over only a few thousand years from wolf ancestors.) Williamson's snails are therefore much more slow-changing than they look to a geologist who reads thick strata punctuated by "rapid" evolution. Ordinary Darwinian selection can account for the seemingly fast changes; they seem fast only to a geologist, not to a geneticist.

See also GOULD, STEPHEN JAY; PUNCTUATED EQUILIBRIUM.

TURNOVER PULSE HYPOTHESIS
Clues from Antelope Evolution

African antelopes may provide another clue to the puzzle of the human past. Paleontologist and evolutionist Elisabeth Vrba, formerly of South Africa's Transvaal Museum and now at Yale, has studied living and fossil antelopes in a search for patterns of evolution during the past 10 million years, or Neogene, spanning the Miocene through Pleistocene periods. The fossil record seems to indicate periods of rapid "turnover" when many new species diverged at the same time and replaced a suite of old ones that became extinct.

Vrba began by studying the distribution and food habits of such modern African antelopes as hartebeest, gazelle, wildebeest, gerenuk and springbok. All are herbivores, though most specialize in a different niche or "layer" of the ecosystem—some stick to tough grasses, others eat shrubs, still others prefer tree leaves. By finding out which antelope species are abundant in a given area, one can pretty well tell what kind of habitat and vegetation prevails there.

Applying that knowledge to fossil antelopes, Vrba noticed two dramatic shifts in their evolution—periods of rapid radiation that she suspected might be correlated with major changes in the types of available food plants. These changes are reflected in their teeth, for instance, where heavy-duty grinders are needed for the tough, dry grasses that are the only available food during periods of arid climate.

Satisfied that the pattern of antelope evolution reflected responses to severe climatic changes, Vrba suggested a hypothesis with broader implications: climatic changes are the chief motor or driving force

DRASTIC CLIMATIC changes sent "pulses" throughout the natural world, affecting creatures of forest and plain. Yale's Elisabeth Vrba has attempted to correlate data on ancient climate with fossil evidence for rates of evolutionary change.

of speciation and extinction. If that is true, then many different lineages of animals and plants ought to show a similar turnover of divergent species during the same time periods; preliminary studies seem to indicate that they do. A picture began to emerge of a "pulse" moving throughout the food chain of living things, synchronizing a divergence of species across many groups.

In the case of the African antelopes, the corroboration with a wide range of geophysical evidence was striking. In 1987, deep-sea sediment cores were gathered by an international project to map ancient climates (CLIMAP). Isotopic curves of their distribution revealed major change in temperature at 2.5 million years ago, and another near 5.6 million (the end of the Miocene), reflecting surges in the amount of polar ice. Pollen specialists tell the same story; their studies of Neogene pollen layers show that forests gave way to drier grasslands and plains in Europe, South America and Africa at the same time.

"All the continents are saying the same thing about the pulse at 2.5 million," concludes Vrba, "the climate was changing—with dramatic force, in some places. The cause of the change is not clear . . . but that it *did* happen is almost beyond doubt."

Vrba's antelopes show two such radiations—one at 5 million years ago and another around 2.5 million years ago. The first is about the same time hominid fossils first appear, perhaps having recently diverged from the pongid (ape) lineage. The second pulse, at

2.5 million years ago, roughly correlates with the estimated time *Homo habilis* appeared.

Vrba raises the possiblity that hominid and antelope evolution alike were responding to the same "pulses" in the environment—widespread, dramatic changes in climate and vegetation—and cautions further testing of the idea against other lineages in the fossil record. Donald Johanson has noted that the first known occurrence of simple stone tools near Hadar "matches the date exactly" and there are no tools before 2.5 million years. "If Elisabeth Vrba is right, and I think she is," he writes in *Lucy's Child* (1989), "then *Homo* emerged in East Africa some 2.5 million years ago, one among many species struggling to adapt to radically altered conditions."

See also "LUCY"; PUNCTUATED EQUILIBRIUM.

TWAIN, MARK (1835–1910)
On Human Evolution

Samuel Clemens (Mark Twain) took a particular delight in deflating pomposity, whether of churchmen or scientists. His satirical essays on evolutionists, written around 1900, were later collected and published in *Letters from the Earth* (1938). Still fresh almost a century later, they offer insight into the public perception of major issues in 19th-century evolutionary theory.

In 1903, Alfred Russel Wallace, codiscoverer of natural selection, marshaled evidence for what he believed was the "guiding force" or "purpose" behind human evolution. His book *Man's Place in the Universe* concluded that Earth occupied a central position in the universe and that special conditions were necessary to make the planet habitable for man.

Despite Wallace's fresh approach and compilation of current data, scientists saw in it the old church doctrine of a preordained plan and purpose for everything in nature, with man at the center of things, like a spoiled child. Twain found it an irresistible target.

In "Was the World Made for Man?" (c. 1904) he gave his own version of Wallace's anthropocentric ("man-centered") theory. "It was foreseen," he began, "that man would have to have the oyster . . . which you cannot make out of whole cloth. You must make the oyster's ancestor first. This is not done in a day."

Before the oyster, Twain wrote, there had to be an array of invertebrates, and most would have to be failures that become extinct. When the oyster was finally evolved, it started to think about how it got there:

> An oyster has hardly any more reasoning power than a scientist has; and so it is reasonably certain that

EVOLUTIONARY SPOOFS were a staple in Mark Twain's satiric arsenal. Here a monkey wanders into his hotel room and becomes very upset and destructive upon looking into one of his books, whereupon Twain hails him as "An Honest Critic."

> this one jumped to the conclusion that the nineteen million years was a preparation for *him*; but that would be just like an oyster, the most conceited animal there is, except man.

Next, there had to be fish, so man could come along later and catch them. And great fern forests had to rise up, then die and turn to coal so man could fry the fish! Mastodons, giant sloths and Irish elk wander around ducking moving ice sheets, getting soaked when the continents submerge and burned by exploding volcanoes.

"At last came the monkey, and anybody could see that man wasn't far off now. The monkey went on developing for close upon five million years, and then turned into a man—to all appearances."

In another satirical essay, "The Lowest Animal," also written around 1900, Twain argues that man is descended from the "Higher Animals," like snakes— a takeoff on the then-popular Degeneration Theory. Twain says he made the experiment of putting several calves in with an anaconda; it ate one and spared the rest. On the other hand, an English earl went out on the plains, shot 71 buffalo, and left them to rot. "Which proves," Twain wrote, "that an Earl is cruel and an anaconda isn't":

> We have descended and degenerated, from some far ancestor—some microscopic atom wandering [in] a drop of water—insect by insect, animal by animal,

reptile by reptile, down the long highway of smirchless innocence, till we reached the bottom stage of development—namable as the Human Being. Below us—nothing. Nothing but the Frenchman.

Twain was an internationally popular platform speaker during the 1870s, and he did play engagements in London. Whether he used some of his evolutionary spoofs in that program is not known. We do know, however, that in 1879, the greatest American humorist of his day made a personal appearance in the little village of Downe to call on one of his devoted readers, Charles Darwin.

See also DEGENERATION THEORY; HAPPY FAMILY; SPIRITUALISM; WALLACE'S PROBLEM.

For further information:
Ficklen, Anne, ed. *The Hidden Mark Twain: A Collection of Little-known Mark Twain.* New York: Crown, 1984.
Twain, Mark. *Letters from the Earth.* Edited by B. DeVoto. New York: Harper & Row, 1938.

"TWO BOOKS," DOCTRINE OF THE
God's Word vs. God's Works

Sir Francis Bacon (1561–1626), who took "All knowledge to be my province" helped create the modern scientific attitude though he was neither an experimenter nor a systematic observer. As philosopher as well as counsel to King James I, Bacon used his nimble mind as a great arranger and compromiser. Approving chroniclers call him "a master of the balance of power among the different faculties" of the human mind, while critics have described his complexity as a mixture of "the soaring angel and the creeping snake."

One of the tasks he set himself was to make the pursuit of natural science acceptable in an age that took Scripture as revealed authority on all matters. In 1605, he set forth his "Doctrine of the Two Books," which set the tone for the next two hundred years of natural philosophy:

> Let no man upon a weak conceit of sobriety or an ill-applied moderation think . . . that a man can search too far, or be too well studied in the book of God's word, or in the book of God's works, divinity or philosophy; but rather let men endeavor an endless progress or proficience in both; only let men beware . . . that they do not unwisely mingle or confound these learnings together.
> Our saviour saith, "You err, not knowing the scriptures, nor the power of God"; laying before us two books or volumes to study . . . first, the scriptures, revealing the will of God, and then the creatures expressing his power . . . The latter is a key unto the former . . . opening [both] our understanding [and] our belief, in drawing us into a due meditation

of the omnipotency of God, which is chiefly signed and engraven upon his works.

"Previously philosophers seldom drew the sharp distinction" between the lessons of God's work and God's words, writes philosophical historian James Moore, but "for Bacon the book of God's works is 'a key' to the book of God's word; students of nature may therefore instruct interpreters of the Bible." To attempt to make nature conform to Scripture would be to "unwisely mingle or confound these learnings together." This Baconian compromise, according to Moore, formed the basis of congenial relations between naturalists and churchmen, encouraging the growth of scientific research and observation.

The metaphor of the "Two Books" became an almost unconscious convention, embedded in the language of naturalists. Geologists still speak of "reading the record in the rocks" today as they did a century ago. Darwin himself spoke of the fossil record containing "missing pages, and even whole chapters." Nevertheless, on living plants and animals, he could see their evolutionary history "stamped in plain letters on almost every line of their structure." Bacon's definition of the Two Books was chosen by Darwin to face the title page of his own greatest book, the *Origin of Species* (1859).

For further information:
Bacon, Francis. *The Advancement of Learning,* London: Henrie Tomes, 1605. Edited by W. A. Wright. London: Oxford University Press, 1930.
Eisely, Loren. *The Man Who Saw Through Time.* (Originally issued as *Francis Bacon and the Modern Dilemma.*) New York: Scribner's, 1973.

2001: A SPACE ODYSSEY (1968)
Evolutionary Epic

Dawn on the ancient African plains a million years ago. A group of man-apes awakens and begins the daily search for food. Within a few minutes, they act out an image of our ancestral nature and evolving behavior as it was understood by anthropologists in the mid-20th century. These are killer apes brought to life on the Super Panavision screen, based upon the writings of paleoanthropologist Raymond Dart, animal behaviorist Konrad Lorenz and best-selling popularizer Robert Ardrey, author of *African Genesis* (1961) and *The Territorial Imperative* (1966).

The film is the enormously influential *2001: A Space Odyssey,* directed and produced by Stanley Kubrick and costing $10.5 million. Based on a novella by science fiction author Arthur C. Clarke, it begins with the man-apes (played by costumed French mimes) and depicts an evolutionary journey that takes man to the far planets aboard space vehicles, burrows inside the psychedelic vortex of his own brain and locks him into a "struggle for existence" with the conscious machines he has created. ("Close the pod door, Hal!")

Eventually, man is reborn as a "starchild": an embryo floating in space, about to evolve into a higher, more advanced being. Each step of his evolutionary odyssey is heralded (and perhaps accelerated) by a mysterious "monolith," which first appears among the African man-apes, then reappears on the moon and again in outer space.

First announced by MGM with the title *From the Ocean to the Stars*, the original version was to start with unicellular life and trace in detail the evolutionary history of life on Earth before arriving at the man-apes. When Stanley Kubrick was brought in, he changed the concept and wrote the screenplay with Clarke, based on the latter's story "The Sentinel," published in 1947. (The original tale did not include the apes, though it prominently featured the mysterious monoliths.)

The ape sequence shows the territoriality of two groups fighting over which is to have access to a waterhole, the killing of a tapir with a rock and the murder of one man-ape by another, using a heavy bone as weapon. In a classic juxtaposition, which has become part of the language of cinema, the bone tool is tossed in the air, turning end over end in slow motion—then dissolves into a shot of an elongated space station, rotating slowly amidst the star-studded blackness. One tool implies the unfolding of a technology without limits.

Kubrick's film received five Academy Award nominations and won the Oscar for Special Effects. A sequel, *2010* (1984), written by Arthur C. Clarke and Peter Hyams, was not particularly innovative and failed to generate excitement comparable to the original.

See also AFRICAN GENESIS; KILLER APE THEORY; KING KONG; O'BRIEN, WILLIS; QUEST FOR FIRE.

TYNDALL, JOHN

See BELFAST ADDRESS.

TYRELL, JOSEPH BURR

See TYRELL MUSEUM OF PALEONTOLOGY.

TYRELL MUSEUM OF PALEONTOLOGY
Canada's Dinosaur Treasures

Opened in 1985, the Tyrell (TEER-ell) Museum of Paleontology in Drumheller, Alberta (near Calgary) is setting new world standards for public exhibition of dinosaurs. The museum features 200 specimens of mounted dinosaur skeletons, as well as dramatic dioramas with full-sized models of the great reptiles in their natural habitats. Innovative exhibits vie with spectacular murals and paintings created by Vladimir Krb, the Tyrell's resident artistic director. (Krb is the protege of the legendary Zdenek Burian, the master painter of prehistory whose works are "National Treasures" in his native Czechoslovakia.)

Situated within the sprawling badlands of the Red Deer River Valley, the Tyrell Museum sits in Midland Provincial Park, the heart of one of the world's richest sources of dinosaur remains. Joseph Burr Tyrell, the Canadian geologist for whom the museum is named, discovered the first *Albertosaurus* skull near Drumheller in 1884, attracting international scientific attention to the region's late Cretaceous fossil beds. Although it is now rough, arid terrain, during the dinosaurs' day the area was a series of deltas and river floodplains, stretching to a warm, shallow inland sea.

The Field Station, a satellite facility the museum opened in 1987, is located in Dinosaur Provincial Park, several hours' drive (about 120 miles) into the badlands. Visitors can watch paleontologists at work and take guided tours of the wilderness areas to see the naturally occurring outcrops of dinosaur skeletons.

UNIFORMITARIANISM
Slow, Steady Change

In the early 19th century, the top geologists of England and France, among them the great Georges Cuvier, were convinced catastrophists. They believed the geology of the Earth could be explained by such biblical catastrophes as the great flood, or "Noachian Deluge" as they called it. Some even attempted to calculate the dimensions of Noah's ark; Captain Robert FitzRoy of the *Beagle*, for instance, held a pet theory that mammoths became extinct because the door of the ark was too small to admit them!

Charles Lyell (1797–1895) published a revolutionary book, *Principles of Geology* (three volumes, 1830–1833), in which he carefully developed the theory that the great features of the Earth—sedimentary strata, river deltas, eroded plains—were all caused by slow and steady natural processes. Great features had been produced by small causes working at a uniform rate over immense periods of time. These processes were common ones and could still be observed at work today, such as water carrying sediments or wearing down rocks.

When Charles Darwin left on his voyage aboard H.M.S. *Beagle* he took the newly published first volume of Lyell's *Principles* with him. It had a profound effect on his thought as he made geological observations on his travels. In 1832, when the ship stopped over at Montevideo, on the Rio de la Plata, he received mail from England containing the second volume. "I am become a zealous disciple of Mr. Lyell's views, as known in his admirable book." Darwin wrote a friend, "Geologising in South America, I am tempted to carry parts to a greater extent even than he does."

Lyell's uniformitarianism was actually a rag-tag package of ideas—vast geological timeframe, steady-state earth, actualism, gradualism, and more—woven together by a master writer.

Actualism, the concept that ordinary present processes operated in the past, is the keystone of what we usually call uniformitarian thinking; it was not original with Lyell, though he made it widely popular. (Cuvier, for instance, had written that "in order for a geologist to judge what happened in the past, as well as what may happen in the future, he must know what is happening now.")

But few noticed that Lyell spliced actualism with other ideas that seemed to be logical extensions but, in fact, were not. Gradualism and other theories were tied onto actualism like tin cans to a dog's tail. They had no necessary connection or unity, but Lyell's skillful presentation made them an accepted part of the package later called uniformitarianism. (A master of argument, Lyell had trained as a barrister before switching to geology.)

Modern geology is uniformitarian in accepting the actualist notion that the study of processes observable today can tell us what happened in the past, in postulating an immense age for the Earth, and in concluding that many great geologic features are the products of slow, steady forces causing gradual change over very long periods.

However, geology is also catastrophic in deducing radical changes in atmospheric gases since life first evolved, in attributing global mass extinctions to fairly rapid shifts inn climate and in tracing these in turn to meteoric impacts and "Nemesis" stars. There has also been a shift towards the discontinuous, or jumpy, view of evolutionary events known as punctuationalism.

Today's earth sciences claim Lyell's *Principles of Geology* as their founding document. But in reality, the modern view is a mixed deck of catastrophic and uniformitarian elements.

See also ACTUALISM; GRADUALISM; LYELL, SIR CHARLES; STEADY-STATE EARTH.

For further information:

Gould, Stephen Jay. *Time's Arrow, Time's Cycle: Myth and Metaphor in the Discovery of Geological Time.* Cambridge: Harvard University Press, 1987.

Lyell, Sir Charles. *Principles of Geology, being an attempt to explain the former changes of the earth's surface, by reference to causes now in operation.* 3 vols. London: John Murray, 1828–1832.

Rudwick, M. J. S. *The Meaning of Fossils: Episodes in the History of Paleontology.* London: Macdonald, 1972.

UNIT CHARACTERS

See "BEANBAG GENETICS."

USSHER/LIGHTFOOT CHRONOLOGY
Biblical Timekeepers

In 1650, Archbishop James Ussher of Armagh, Ireland published the result of his calculations based on the numerology of the Bible. Counting backward through all the "begats" in the Old Testament, he estimated the number of generations since Adam. Creation of man and all other creatures, he con-cluded, took place in 4004 B.C., giving the planet an age of about 6,000 years. His figures became the standard against which inquiries about the Earth's early history were measured.

Ussher's calculations were refined by another 17th century divine, Dr. John Lightfoot, vice-chancellor of Cambridge University, who computed that "Man was created by the Trinity on 23rd October, 4004 B.C., at nine o'clock in the morning." An eminent Hebrew scholar, Dr. Lightfoot had also concluded that "heaven and earth, centre and circumference, and clouds full of water, were created all together, in the same instant." A convenient feature of Lightfoot's chronology was that the time of the month of all creation coincided exactly with the annual beginning of the academic year at Cambridge.

The Ussher/Lightfoot chronology was widely accepted for many years and was challenged only when evidence began to accumulate from geological strata and fossils in the mid-18th century. But it continued to stand as dogma until the 19th century, when the rise of geology and evolutionary biology stretched the vistas of geologic time back hundreds of millions of years. Most people, including Charles Darwin, had thought the Ussher/Lightfoot chronology was part of the original scripture itself, since it was usually printed in the margins of most Bibles for 150 years.

See also CHRONOMETRY; CREATIONISM; FOUR THOUSAND AND FOUR B.C.; SMITH, WILLIAM.

VAN VALEN, LEIGH

See RED QUEEN HYPOTHESIS.

VEBLEN, THORSTEIN (1857–1929)
Sociologist, Economist

By the late 19th century, "survival of the fittest" had became the slogan of American free enterprise. Andrew Carnegie and John D. Rockefeller justified ruthless competition as the price of evolutionary progress. As big winners, they liked to believe they had been "naturally selected" as the best of men. Sociologist Thorstein Veblen, however, arrived at a quite different conclusion: they were clearly atavistic "throwbacks" to an earlier stage of human evolution.

Veblen's classic *The Theory of the Leisure Class* (1899) remains famous for its many original insights about economic behavior. (It introduced the concept of "conspicuous consumption," for instance, which explains seemingly "wasteful" extravagance as an attempt to purchase social status.) When first published, it outraged Social Darwinists by turning their theories about capitalism upside down. Veblen concluded that successful entrepreneurs were far from "the fittest" members of the human species.

Drawing on the bizarre evolutionary theories of Italian criminologist Cesare Lombroso, he lumped them with lower-class thieves and murderers as "atavistic" throwbacks. In Veblen's words, the powerful industrialists shared with criminals "predatory aptitudes and propensities carried over . . . from the barbarian past of the race . . . with the substitution of fraud and . . . administrative ability" for naked violence.

Capitalist businessmen were certainly not, in Veblen's iconoclastic view, the evolutionarily most advanced members of civilized society. Though respected and respectable, they were genetic "throwbacks" to early plundering barbarian tribes—a lingering hindrance to social progress.

See also CARNEGIE, ANDREW; SOCIAL DARWINISM; "SURVIVAL OF THE FITTEST."

VELIKOVSKY, IMMANUEL (1895–1979)
"Worlds in Collision"

A bold and imaginative amateur theorist, Immanuel Velikovsky thought he had found a new key to understanding the past in a vision of planetary cataclysms shaping geologic and human history. In his controversial books (*Worlds in Collision*, 1950, and *Earth in Upheaval*, 1955) he ignored the experts and appealed to the general public. Almost immediately, he attracted a large and receptive readership. Velikovsky's popularity sent professional scientists into upheaval and collision. (One critic nicknamed his theory "cosmic pinball.")

According to Velikovsky, Venus broke off from Jupiter within historical times, assuming the form of a comet during the Jewish exodus from Egypt. Its tail swept over Earth creating the various "miraculous" events related in Scripture, such as the Nile turning blood-red. Later, it knocked Mars out of orbit, causing a near-collision with Earth, which prompted wars and social traumas still embedded in the human "collective memory."

Although his work seemed heretically novel, it was, in essence, warmed-up leftovers from the previous century. Catastrophists William Buckland (1820) and Hugh Miller (1841) had similarly sought farfetched "natural" explanations for miraculous occurrences in the Bible using the same method. If the

Book of Joshua said that the sun stood still in the sky, they assumed it was fact, not myth, and the problem became finding a naturalistic explanation for how the sun could have halted. Despite decades of trying, scriptural geologists could not make such a system work; there were just too many facts that wouldn't fit. A few talented naturalists and geologists, such as Miller and Philip Gosse, became hopelessly depressed in the attempt.

Rejecting a uniformitarian theory of how geologic features are formed. Velikovsky invoked great floods to explain fossil beds containing thousands of animal remains (though the layers show they were laid down during the course of millennia) and great fiery catastrophes to explain deep lava formations, which also evidence gradual build-up over long periods.

Besides, a different new theory—plate tectonics—was gaining momentum in the 1950s as an explanation that could account for a much greater range of facts. Thrusting up of mountains or intense volcanic activity could be explained by the pressures of shifting continental plates whose relentless movements can be observed and measured today. Within a decade science found the theory of continental drift accounted for thousands of geological features without calling in comet tails or cosmic collisions (although there's still room for an occasional asteroidal shower!).

Most damaging, Velikovsky's arguments were riddled with half truths, errors and misreadings of the scientific literature. Yet, instead of giving reasoned refutations, many professional scientists responded venomously to the unauthorized interloper; to their lasting discredit, some even tried to suppress publication of his books. But blatant persecution by the scientific establishment won him more fans among rooters for the underdog, and he became a cult hero among the "counterculture."

See also CATASTROPHISM; MILLER, HUGH; NEMESIS STAR.

For further information:
Velikovsky, Immanuel. *Earth in Upheaval.* Garden City, N.Y.: Doubleday, 1955.
———. *Worlds in Collision.* Garden City, N.Y.: Doubleday, 1950.

VENUS FLYTRAP

See INSECTIVORUS PLANTS.

"VERCORS" (JEAN BRULLER) (b. 1902)
French Author, Illustrator

Talented writer-artist Jean Bruller ("Vercors") wondered what might happen if a tribe of primitive hominids had survived to the present day. Evolu-

MINDLESS MOVIE *Skullduggery,* featuring these hairy Hollywood hominids, was made from thought-provoking novel by French author Vercors. The film managed to avoid the central question of the book: How closely must another hominid species resemble mankind before we grant it fully human rights?

tionary kinship of species, he realized, raises ethical questions that have never been asked, let alone answered.

If "bigfoot" or "yeti" were actually found, would it be murder to shoot one—or merely cruelty to animals? How "close" to human does an ape-like creature have to be to deserve "human rights"? Vercors tackled these themes in *Les Animaux Denatures* (1952), published in English as *You Shall Know Them* (1953) and later retitled *The Murder of the Missing Link.*

Vercors set his story in Australia, where an elusive tribe of hominids is discovered. Scientists recognize them as australopithecines, a man-like species from a million years ago. When the news gets out, a factory owner sets out to capture them for exploitation as unpaid, semiskilled laborers. If they are not men, he could legally treat them as domestic animals; but if they're human, it would be slavery.

When attempts to protect the tribe fall on deaf ears, Vercors's hero takes a desperate course. He gets a female pregnant, then announces he has killed her infant with a lethal injection. Now the legal system must determine his punishment. If he is to be convicted of murder, the jury must first define the man-apes as humans, which would rescue them from enslavement. Vercors's hero puts his own life on the line to establish the tribe's humanity and win their freedom.

With elegant Gallic logic, Vercors manages to rescue both his hero and the near-men. After prolonged deliberations, the jury declares the creatures are indeed human and any future killing of them will be considered murder. Since they were not legally human at the time of the killing, the jury decides the hero has committed no crime. Like the great Swedish classifier Carl Linnaeus, Vercors concludes that man himself defines who is human and who is an animal. (Instead of describing the human species zoologically, Linnaeus had written "Man, know thyself.")

Unfortunately, Vercors's unusual, thoughtful novel was made into an uncommonly junky Hollywood film. *Skullduggery* (1970), starring Burt Reynolds, trivialized the story, and left viewers with a grotesque image of full-figured women in hairy body suits.

See also ANIMAL RIGHTS; HOMO SAPIENS, CLASSIFICATION OF.

VESTIGES OF CREATION (1844)
Controversial Exposition of Evolution

In the early 19th century, "The treasures of the whole world of nature" were pouring into Europe from the colonized tropics, evolutionist Alfred Russel Wallace later recalled, "and there was a general impression that we must spend at least another century in collecting, describing, and classifying" before science could hope to tackle that "mystery of mysteries," the origin of species. "The need of any general theory of how species came into existence was hardly felt."

Robert Chambers (1802–1883) of Edinburgh was neither scientist nor philosopher, but a writer-publisher of popular encyclopedias, biographies and reference books. In 1844, he created a furor with the first widely read book championing evolution, *Vestiges of Creation*, published 15 years before Charles Darwin's *Origin of Species* (1859). Wallace credits Chambers with being first to set down "The vague ideas of those who favored evolution . . . with much literary skill and scientific knowledge." By 1860, the *Vestiges* had gone through 11 editions.

Beginning with the astronomer Pierre-Simon Laplace's Nebular Hypothesis, which had recently gained acceptance, Chambers summarized evidence for the evolution of solar and planetary systems. Given a primitive Earth and inconceivably long periods of time, he wrote, "an impulse . . . was imparted to the forms of life, advancing them in definite lines, by generation, through grades of organization terminating in the highest plants and animals."

Though he argued for the "reasonableness" of modification through ordinary reproduction, rather than special creation, Chambers never tackled the *how* and *why* of evolution. "The book," Wallace said

in 1898, "was what we would now call mild in the extreme." Its tone was reverential, its language respectful and even religious. Yet, even as the public snapped up one printing after another, most scientific, religious and literary critics greeted it "with just the same storm of opposition and indignant abuse which assailed Darwin's work fifteen years later."

Wallace claimed it was Chambers's *Vestiges of Creation* that convinced him evolution of species took place by means of the ordinary process of reproduction, inspiring him to gather more evidence in support of the idea. (The initial results of that effort, known as Wallace's "Sarawak Law," was published in 1855.) Thomas Huxley, before he had become an ardent evolutionist, had a quite different reaction—to his later regret, he published a nasty review of the *Vestiges*.

Chambers's main objective was to extend the conception of the province of law in the universe and to establish the Theory of Development (evolution). All living organisms were connected, he concluded, obeying a system of law ordained by God. Change came about gradually and continuously, proceeding in the direction of increasing progress for all life—now spearheaded by humans, whose continuing progress was certain.

Despite its conciliatory, reverent language, Chambers correctly anticipated the scorn and abuse it would provoke, and took elaborate precautions to disguise his authorship, which was not revealed until after his death.

See also CHAMBERS, ROBERT; "MR. VESTIGES"; SARAWAK LAW; WALLACE, ALFRED RUSSEL.

VOLIVA, WILBUR GLENN

See FLAT-EARTHERS.

VON BAER, KARL ERNST (1792–1876)
German Embryologist

Evolutionary biologists consider Karl Ernst von Baer to be the "father of modern embryology." Von Baer's classic studies of how generalized embryos develop into specific organisms were an important influence on Charles Darwin, Thomas Huxley and Ernst Haeckel. Huxley was fond of pointing out that evolution over millions of years was not more remarkable than changes in form that take place during nine months in a mother's womb.

Among von Baer's many contributions: He clarified the germ layers and how they develop, first described the primitive backbone (notocord) and the mammalian egg. He documented how embryos of different species are similar in their early stages, then develop their special characteristics later. (This at a time when many believed embryos were completely "preformed" in all details from the tiniest "seed.") His probing writings, still highly readable today, include essays on medicine, comparative anatomy, geography, anthropology and education.

Von Baer's search for laws and regularities in nature—then known as "philosophical" anatomy—made him, said Thomas Huxley, "a man of the same stamp" as Darwin.

For further information:
Gould, Stephen Jay. *Ontogeny and Phylogeny.* Cambridge: Harvard University Press, 1977.

VON SHELLING, F. W. J.

See NATURPHILOSOPHIE.

VOYAGE OF H.M.S. *BEAGLE*
Darwin's Journey of Discovery

"My first real birthday" was how Charles Darwin remembered December 27, 1831, when, at the age of 23, he boarded H.M.S. *Beagle* for a voyage round the world as ship's naturalist. Five years later, after shipping back tons of natural history specimens, he returned to England with his notes, diaries and journals, the foundation for his epochal work *On the Origin of Species* (1859).

During the surveying vessel's mission to make accurate coastal maps and charts for the Admiralty, Darwin collected exotic animals, plants, fossils and rocks from all over the world—sometimes amid storms, earthquakes, wars and volcanic eruptions. He was filled with wonder in the Brazilian rain forest, rode the backs of giant tortoises in the Galapagos and confronted painted "savages" at the tip of South America.

Autocratic Captain Robert FitzRoy (1805–1865), only four years older than young Darwin, was already a seasoned commander, who had once before (1826–1830) steered this ship through the treacherous waters around Cape Horn. He had sought a naturalist who could also be an educated companion for himself, to share his meals and cabin. Darwin was not the first to apply, and his father hated the idea. Thanks to a recommendation from his Cambridge professor, Rev. John Henslow, he got the post.

FitzRoy's most intensive survey was of coastal South America, though Darwin managed some overland trips through Brazil, Chile and Argentina. After two unsuccessful attempts to navigate around the Horn, during which the vessel almost capsized, FitzRoy steered through the Strait of Magellan. Then the *Beagle*, after visiting the Galapagos Islands west of Ecuador, struck out across the Pacific Ocean.

She visited Tahiti and the Society Islands, contin-

ued on to New Zealand, Australia, Tasmania, Keeling Island and Mauritius in the Indian Ocean, and the Cape Colony in South Africa. Then, instead of heading north to England (to the chronically seasick Darwin's dismay), the *Beagle* traced the eastern coast of South America once more. Finally, on October 2, 1836, she returned to Falmouth, England.

For a good part of the voyage, Darwin was cramped into a tiny chart room in the forward poop deck, where he had to be perfectly organized about his work—there was no room for even the slightest disarray. Although he grumbled about the lack of space, he later admitted it forced him to concentrate, complete one task at a time, and arrange his materials in impeccable order, a habit that stayed with him for life. (Friends suggested his little study at his home in Downe, jammed with books, papers and specimens, was an attempt to recreate the poop cabin.)

Young Darwin of the *Beagle* was quite different than the older semi-invalid philosopher of Down House, who was easily tired and had daily bouts of headaches, abdominal pain and vomiting. As a young man, he thought nothing of riding with the "sinister" gauchos on the pampas, trekking 400 miles through wilderness, excavating fossils by hand with a geologist's hammer, and climbing unexplored mountains.

He met his share of danger—more from people than from the wild animals he sought. In Argentina, he found himself in the midst of a bloody and horrible war of extermination against the Pampas Indians led by the ruthless General Juan Manuel Rosas, who at first mistook him for a spy. Among Brazilians, he observed the slavery of Africans first-hand: It sickened and infuriated him. (Years later, he still had nightmares about a screaming serving girl, whose fingertips were crushed in screw-vises by her mistress for a small infraction.) Everywhere—among people as among birds and beasts—he saw competition, waste of life, struggle for survival.

He was impressed, also, by the fossil evidence of the "former inhabitants" of South America, giant sloths and armadillos, which bore obviously close resemblances to the modern small animals living there. It struck him that they might be ancestors and descendants. And some of the nearly naked hunters and gatherers he encountered made him think of what men must have been like before the dawn of history.

Another striking fact was that chains of volcanic islands had so many differently adapted species that appeared very closely related, as though they had arisen from a common ancestor. Their general structure, Darwin observed, was neither unique or novel, but closely resembled creatures from the nearest mainland. (At first, he nearly missed this important point. But bird expert John Gould pointed it out back in London, after studying his collection of Galapagos mockingbirds and finches.)

On the last leg of the voyage, Darwin lavished attention on the formation of coral reefs, for which he had a theory—still accepted today and confirmed with modern equipment. The results of his studies became his first scientific book, *On The Nature of Coral Reefs* (1842).

But before it was published, Darwin won fame with his travel narrative of discovery and adventure, generally known as *Voyage of the Beagle*. Actually, it was first issued as the last of a three-volume report, with the full title *Journal of Researches into the Geology and Natural History of the Various Countries Visited by H.M.S. Beagle Under the Command of Captain FitzRoy, R.N. From 1832 to 1836.*

The first volume, written by Capt. Philip King, described the ship's first voyage. Volume two was written by FitzRoy, who was somewhat irritated when Darwin's volume on the natural history gained instant public attention and popularity. By Victorian standards, it was a best seller, while the other volumes were ignored.

Sir Arthur Conan Doyle, himself no stranger to adventure books, called Darwin's "*Voyage of the Beagle*" one of the two best books of "the romance of travel and the frequent heroism of modern life." (His other choice was Alfred Russel Wallace's *Malay Archipelago*, 1872.) Doyle admired Darwin's "gentle and noble firmness of mind" and devotion to his naturalist's quest:

Nothing was too small and nothing too great for . . . alert observation. One page is occupied in the analysis of some peculiarity in the web of a minute spider, while the next deals with the evidence for the subsidence of a continent, and the extinction of a myriad animals . . . [Darwin] rode the four hundred miles between Bahia and Buenos Ayres, when even the hardy Gauchos refused to accompany him. Personal danger and a hideous death were small things to him compared to a new beetle or an undescribed fly.

But the excitement and stimulation were literally enough to last Darwin for a lifetime. Although he always maintained that he owed all later accomplishments to his strenuous voyage, he wrote his sister Carolyn in 1836 that his seafaring days were over: "I am convinced that it is a most ridiculous thing to go round the world, when by staying quietly in one place, the world will go round with you."

See also (H.M.S.) BEAGLE; CORAL REEFS; DARWIN, CHARLES; FITZROY, CAPTAIN ROBERT; ORIGIN OF SPECIES.

For further information:

Barlow, Lady Nora, ed. *Charles Darwin's Diary of the Voyage of H.M.S. Beagle.* Cambridge: Cambridge University Press, 1932.

————. *Charles Darwin and the Voyage of the Beagle.* New York: Philosophical Library, 1945.

Brosse, Jacques. *Great Voyages of Discovery.* New York: Facts On File, 1983.

Darwin, Charles. *Journal of Researches into the Geology and Natural History of the Various Countries visited by H.M.S. Beagle 1832–1836.* London: H. Colburn, 1839.

Keynes, R. D. *The Beagle Record.* Cambridge: Cambridge University Press, 1979.

Moorehead, A. *Darwin and the Beagle.* New York: Harper & Row, 1969.

WAGNER, MORITZ (1813–1887)
Founder of Isolation Theory

Famous German explorer, geographer and naturalist Moritz Wagner was troubled by a common objection to Darwinian selection in his day: domesticated varieties, if set free from man's control, would be absorbed in a few generations back into the wild populations. Besides, while collecting beetles in Algeria in 1837 (more than 20 years before the Darwin-Wallace theory), it struck Wagner that different but closely related species appeared whenever he crossed a river.

Wagner created quite a stir with his "law of migration": only when small populations are geographically isolated and prevented from crossing with the ancestral population can new species arise. Isolation, he wrote, is "the necessary condition for natural selection . . . Organisms which never leave their ancient area of distribution will never change."

Most naturalists realized isolation was a common condition of speciation—particularly on remote islands—but were not willing to admit its general importance in the formation of new species. August Weismann, for instance, noted that in fossil beds at Steinheim Lake, new species were found in the same beds as older, closely related fossil species. Some believed Wagner's view was "deprived of all foundation" by Weismann's argument that sexual species must have evolved from original hermaphrodites "on the same territory." Darwin himself objected that Wagner's theory could not explain widespread adaptation among the majority of species spread out over vast areas.

In applying his "law" to the origin of man, Wagner speculated that one or several pairs of proto-humans were driven away from their tropical forest homelands, where survival had been easy. While in less hospitable northern latitudes, their return was cut off when the great Old World mountain chains were thrust up. Trapped in the harsher environment, they had to work and develop the arts of men to survive. (Wagner and his colleagues failed to see how closely this "just-so" story retells the familiar biblical account of the expulsion from the Garden of Eden.)

Despite such flights of fancy, Wagner's notion was rooted in sound observations of the distribution of natural populations relative to geographic barriers. Revived by Ernst Mayr 75 years later, the idea that small, isolated populations fostered speciation became part of the Synthetic Theory. This time it was anchored in a mechanism, "genetic drift" or "founder's effect," which could be demonstrated in experimental populations.

Although Mayr had reestablished it on the basis of his studies of bird species in the 1930s, isolation was still not considered of great general importance by biologists until the 1970s, when paleontologists, Stephen Jay Gould and Niles Eldredge tied it to punctuated equilibrium. Today, reproductive isolation of small populations is considered a major "cradle" for the origin of new species.

See also DIVERGENCE, PRINCIPLE OF; GENETIC DRIFT; HAWAIIAN RADIATION: MAYR, ERNST; PUNCTUATED EQUILIBRIUM.

For further information:
Wagner, Moritz. *The Darwinian Theory and the Law of the Migration of Organisms.* Translated by J. L. Laird. London: E. Stanford, 1873.

WALLACE, ALFRED RUSSEL (1823–1913)
Codiscoverer of Natural Selection

After publication of the *Origin of Species* in 1859, evolution by natural selection, biology's great unifying concept, became famous as "Darwin's theory." First announced jointly the previous year, it is actually the Darwin-*Wallace* Theory. Nevertheless, Charles Darwin often called it "my theory," while Alfred Russel Wallace, his partner and coauthor, graciously insisted "It is yours and yours only." Wallace carried modesty to extremes, even calling his own book on evolution *Darwinism* (1889). Had he been more ambitious and less generous, evolutionary science might have become known as "Wallaceism."

An explorer, zoologist, botanist, geologist and anthropologist, Wallace was a brilliant man in an age of brilliant men. Famous not only as cocreator of the natural selection theory, he was the discoverer of thousands of new tropical species, the first European to study apes in the wild, a pioneer in ethnography and zoogeography (distribution of animals) and author of some of the best books on travel and natural history ever written, including *Travels on the Amazons* (1869) and *The Malay Archipelago* (1872). Among his remarkable discoveries is "Wallace's Line," a natural faunal boundary in Malaysia (now known to coincide with a junction of tectonic plates) separating Asian-derived animals from those evolved in Australia.

Born in 1823, in Usk, England, a small town near the Welsh border, Wallace was raised in genteel poverty. His first employment was helping his brother John survey land parcels for a railroad. While still in his twenties, he served a stint as a schoolmaster in Leicester, where he met young Henry Walter Bates, who shared his passion for natural history. On weekend bug-collecting jaunts, the would-be adventurers discussed such favorite books as the *Voyage of the H.M.S. Beagle* (1845) and dreamed of exploring the lush Amazon rain forests of Charles Darwin's ecstatic descriptions.

Another book also inspired them: Robert Chambers's anonymously published *Vestiges of Creation* (1844), a controversial, literary treatise on evolution. Scorned by scientists, *Vestiges* championed the idea that new species originate though ordinary sexual reproduction rather than by spontaneous creation. Wallace and Bates decided they would comb the exotic jungles to collect evidence that might prove or disprove this exciting "development hypothesis" (only later known as evolution). When Darwin had embarked on his own voyage of discovery some 20 years earlier, he had had no such clear purpose in mind.

Science was not yet a well-established profession, and naturalists were often dedicated amateurs from wealthy families. When Darwin went on his circum-global voyage, his father paid all expenses, even providing a servant to assist with his work. Wallace's achievements are all the more remarkable, for he had to finance his expeditions by selling thousands of natural history specimens, mainly insects, for a few cents apiece. When his exploring and collecting days were over, Wallace struggled to support his family on author's royalties and by grading examination papers. (He often said intelligence had nothing to do with getting rich, that "success at money-getting requires mainly *impudence*.")

Bates and Wallace reached Para, at the mouth of the Amazon, in May 1848; they collected and explored the surrounding regions for several months, then decided to split up. Wallace went up the unknown Rio Negro, leaving Bates to explore the upper Amazon regions. From 1848 until 1852, Wallace collected, explored and made numerous discoveries despite malaria, fatigue and the most meager supplies.

When he finally returned to rejoin Bates downriver, he found that his beloved younger brother had traveled across the world to join the adventure and had just died of yellow fever in Bates's camp. Grief-stricken, exhausted and suffering from malaria himself, Wallace boarded the next ship for England. With him went his precious notebooks and sketches, an immense collection of preserved insects, birds and reptiles, and a menagerie of live parrots, monkeys and other jungle creatures.

In the midst of the North Atlantic, as Wallace suffered a new attack of malaria, the ship suddenly burst into flames. He was able to rescue only a few notebooks as he dragged himself into a lifeboat. Everything else burned or sank beneath the waves, and he later recalled:

> I began to think that almost all the reward of my four years of privation and danger was lost . . . How many times, when almost overcome by ague, had I crawled into the forest and been rewarded by some unknown and beautiful species! How many places, which no European foot but my own had trodden, would have been recalled to my memory by the rare birds and insects they had furnished to my collection! . . . And now everything was gone, and I had not one specimen to illustrate the wild scenes I had beheld!

The measure of Wallace's enormous courage and resilience showed itself shortly after returning to England. With the insurance money he received for part of his lost collections, he immediately set out on a new expedition—this time to the Malay Archipelago (1854–1862).

Wallace mastered Malay and several tribal languages, for he was intensely interested (as Darwin never was) in "becoming familiar with manners, customs and modes of thought of people so far removed from the European races and European civilization." A self-taught field anthropologist, he made pioneering contributions to ethnology and linguistics and developed "a high opinion of the morality of uncivilized races." He later recalled with satisfaction that while he lived among them he never carried a gun nor locked his cabin door at night.

In Malaysia (then known as the Moluccas) he tracked orang-utans through the deep forest, shot several for the British Museum's collection and raised an orphaned infant orang in his field camp. Since local tribesmen regarded the red-haired apes as "men of the woods," they were horrified when he shot and skinned them, convinced he would next want to add their own skulls to his collection.

Wallace collected natural history specimens with an extraordinary passion:

> I found . . . a perfectly new and most magnificent species [of butterfly] . . . The beauty and brilliancy of this insect are indescribable, and none but a naturalist can understand the intense excitement I experienced . . . On taking it out of my net and opening the glorious wings, my heart began to beat violently, the blood rushed to my head, and I felt . . . like fainting . . . so great was the excitement produced by what will appear to most people a very inadequate cause.

Wallace came to the idea of evolution not through artificial selection of domestic animals, as Darwin had done, but through his observations of the natural distribution of plants, animals and human tribal groups and their competition for resources. Like Darwin, he had also been influenced by Thomas Malthus's essays *On Population* (1803), which he had read some years before.

In 1855, while in Sarawak, he composed "My first contribution to the great question of the origin of species." Combining his knowledge of plant and animal distribution with Sir Charles Lyell's account of "the succession of species in time," he came up with a conclusion about when and where species originate. ("The how," he wrote, "was still a secret only to be penetrated some years later.") His paper "On the Law which has Regulated the Introduction of New Species," stated that: "Every species has come into existence coincident both in space and time with a pre-existing, closely-allied species." This preliminary conclusion, he knew, "clearly pointed to some kind of evolution."

Published in an English natural history journal in September, 1855, Wallace's "Sarawak Law" was generally ignored by the scientific world. When he expressed his disappointment in a letter to Darwin,

BRILLIANT, ECCENTRIC and utterly his own man, Alfred Russel Wallace independently developed the theory of evolution by natural selection. He was also a pioneering anthropologist, a founder of zoogeography, explorer, travel writer, naturalist and the first European to observe orang-utans in the forest.

"He replied that both Sir Charles Lyell and Mr. Edward Blyth, two very good men, specially called his attention to it." Writing years later, Thomas Huxley said, "On reading it afresh I have been astonished to recollect how small was the impression it made."

In February 1858, Wallace was living at Ternate, one of the Moluccan Islands and was suffering from a sharp attack of intermittent malarial fever, which forced him to lie down for several hours every afternoon:

> It was during one of these fits, while I was thinking over the possible mode of origin of new species that somehow my thoughts turned to the "positive checks" to increase among savages and others described . . . in the celebrated *Essay on Population* by Malthus . . . I had read a dozen years before. These checks—disease, famine, accidents, wars, etc.—are what keep down the population . . . [Then] there suddenly flashed upon me the idea of the survival of the fittest . . . that in every generation the inferior would inevitably be killed off and the superior would remain . . . and considering the amount of individual variation that my experience as a collector had shown me to exist . . . I became convinced that I had at length found the long-sought-for-law of nature that solved the problem of the origin of species . . . On the two succeeding evenings [I] wrote it carefully in order to send it to Darwin by the next post . . .

It was this article, "On the Tendency of Varieties to Depart Indefinitely from the Original Type" (1858), that sent Darwin into a panic, convinced his friend Charles Lyell's warning, that he would be "forestalled" by Wallace, "had come true with a vengeance."

Lyell and Sir Joseph Hooker, attempting to rescue their friend's threatened prior claim, arranged to have Wallace's paper published along with some of Darwin's early drafts. The announcement of the Darwin-Wallace theory of evolution by means of natural selection was published in the Linnean Society's journal in 1858; the following year Darwin wrote the *Origin of Species* and rushed it into print.

Wallace was informed of these developments after the fact and received a copy of Darwin's book while still in Malaysia. When he returned to England in 1862, Darwin was anxious about Wallace's reaction, and was relieved to discover his "noble and generous disposition." Later Wallace maintained that even if his only contribution was getting Darwin to write his book, he would be content. But the fact remains that Wallace was not given an opportunity to exercise his nobility or generosity, since the joint publication was decided without anyone consulting him. [See "DELICATE ARRANGEMENT."]

In addition to the chronicles of his travels, Wallace turned out a remarkable series of books, all landmark contributions to evolutionary biology: *Contributions to the Theory of Natural Selection* (1870), *Geographical Distribution of Animals* (1876), *Island Life* (1882) and *Darwinism* (1889). His idea of the living Earth as a single, complex system seems, in some sense, to have foreshadowed the Gaia hypothesis:

> . . . there are now in the universe infinite grades of power, infinite grades of knowledge and wisdom, infinite grades of influence of higher beings upon lower . . . This vast and wonderful universe, with its almost infinite variety of forms, motions, and reactions of part upon part, from suns and systems up to plant life, animal life, and the human living soul, has ever required and still requires the continuous co-ordinated agency of myriads of such intelligences.

Unlike the cloistered, tactful Darwin, Wallace was imprudently outspoken about his religious and political beliefs. Outraged colleagues wanted to dismiss him as a "senile crank" for his strong advocacies of utopian socialism, pacifism, wilderness conservation, women's rights, psychic research, phrenology, Spiritualism and his campaign against vaccination. Wallace replied he was not "brain-softening" with age, but had held many of these beliefs for 30 years.

Spiritualism strongly influenced his ideas on human evolution, causing him to differ with Darwin in

1869 on whether natural selection could explain "higher intelligence" in man. Wallace thought the human mind was supernaturally injected into an evolved ape from "the unseen world of Spirit." He also rejected Darwin's concept of "sexual selection," which he dismissed as merely a special case of natural selection. Although they remained friendly and mutually respectful, they never really understood each other's perspective. [See SPIRITUALISM; WALLACE'S PROBLEM.] Nevertheless, Wallace was called upon to be an honored pallbearer at Darwin's funeral at Westminster Abbey.

In 1876, Wallace helped introduce a Spiritualist paper at the British Association's scientific meetings, which apparently touched off the notorious Slade affair. [See SLADE TRIAL.] He testified for the defense at the trial of Henry Slade and often defended other professional "spirit-mediums" who were accused of conducting fraudulent "psychic experiments." In 1881, Wallace joined the Society for Psychic Research; he headed the Land Nationalisation Society in 1882 and openly declared himself a Socialist in 1890.

Some of his admirers had recommended he be appointed director of the proposed new park at Epping Forest, but Wallace immediately lost the position by stating that he would keep the woodland exactly as it was for future generations, allowing no restaurants, hotels or other concessions. Darwin started a petition among scientists to get Wallace a civil pension, but botanist Sir Joseph Hooker and others objected to appealing for government funds on behalf of "a public and leading Spiritualist." However, Darwin and Huxley prevailed and Wallace got his pension. (Huxley, though differing with Wallace on many issues, assured him that he would never seek a Certificate of Lunacy against him!)

In his last book, *Social Environment and Moral Progress* (1913), Wallace catalogued the horrors of the urban poor, colonial exploitation and "reckless destruction of the stored-up products of nature, which is even more deplorable because more irretrievable.":

> It is not too much to say that our whole system of society is rotten from top to bottom, and the Social Environment as a whole, in relation to our possibilities and our claims, is the worst that the world has ever seen [because of] our system of Individualist Competition.

He was furious when apologists for the *status quo* told him society needed no safety net for its poor or infirm, since, according to the "law" of natural selection, they ought to be eliminated. "Having discovered the theory," he fumed, "it is rather amusing to be told . . . that I do not know what natural selection is, nor what it implies." Eugenists who sought to regulate human breeding for selective improvement he considered "dangerous and detestable . . . sure to bungle disastrously."

Influenced by the socialist Henry George, Wallace urged a policy of land nationalization and an economy in which *"all* shall contribute their share either of physical or mental labor, and . . . every one shall obtain the full and equal reward for their work. [Then] the future progress of the race will be rendered certain by the fuller development of its higher nature acted on by a special form of selection which will then come into play."

What "special form of selection" might be the salvation of humanity? Wallace argued that human populations produce many more males than females, but in his day young men were dying by the millions. Alcoholism, dangerous occupations and particularly the frequent wars left Europe with a huge proportion of unattached women. But under a just and nonmilitaristic social system, Wallace predicted, the number of males would rise dramatically, until they greatly outnumbered women:

> This will lead to a greater rivalry for wives, and will give to women the power of rejecting all the lower types of character among their suitors . . . [The well-educated, enfranchised, responsible] women of the future [will be] the regenerators of the entire human race . . . in accordance with natural laws . . .

Wallace's special hope for the salvation of mankind, then, was none other than "sexual selection," one of Darwin's favorite mechanisms for explaining the evolution of man—which Wallace had always insisted did not exist! However, Wallace added a twist to Darwinian sexual selection: an explicit acknowledgement of the large evolutionary effects of a slight change in sex ratio, a surprisingly modern way of thinking about populations.

During the 1970s and 1980s, Alfred Russel Wallace become a hero among disaffected academics and independent scholars. They saw in him a brilliant scientist, working outside the establishment, scrabbling for a living, snubbed by those with wealth and position, persecuted for unpopular social views—possibly even deprived of his rightful place in history. Yet Wallace was morally triumphant as a great human being and fearless truthseeker, cheerful, optimistic and productive into his ninetieth year.

In 1985, the British Entomological Society, of which Wallace was once president, launched a series of major expeditions to study the insects of the world's tropical rain forests. They called it "Project Wallace."

See also BATES, HENRY WALTER; BEETLES; GAIA HYPOTHESIS; HAMPDEN, JOHN; PHRENOLOGY; SEXUAL SELECTION; WALLACE'S LINE.

For further information:
Brooks, John. *Just Before the Origin: A. R. Wallace's Theory of Evolution.* New York: Columbia University Press, 1984.
Marchant, James. *Alfred Russel Wallace: Letters and Reminiscences.* New York: Harper, 1914.
McKinney, H. Lewis. *Wallace and Natural Selection.* New Haven: Yale University Press, 1972.
Wallace, Alfred Russel. *Darwinism: An Exposition of the Theory of Natural Selection with Some of its Applications.* London: Macmillan, 1889.
———. *Island Life: Or the Phenomena and Causes of Insular Faunas.* New York: Harper, 1881.
———. *The Malay Archipelago: The Land of the Orang-utan and the Bird of Paradise.* New York: Macmillan, 1872.
———. *Miracles and Modern Spiritualism.* London: Nichols, 1874.
———. *My Life.* 2 vols. New York: Dodd, Mead, 1905.
———. *Travels on the Amazon and Rio Negro.* London: Ward Lock, 1903.

WALLACEISM
Selectionist Evolution

After Charles Darwin's death in 1882, his "junior partner," Alfred Russel Wallace, promoted his own version of evolutionary theory. Wallace thought natural selection was sufficient to account for every form and structure in nature except the human brain.

If an animal or plant exhibited a color or behavior that seemed useless or even detrimental, Wallace argued that its "survival value" was simply not yet known or understood. In the early 20th century, selectionism became known as "neo-Darwinism," because Wallace (and George J. Romanes) insisted it was really the core of Darwinian thought.

In response to critics, Darwin himself had gradually retreated from natural selection as an all-sufficient principle. In later editions of his *Origin of Species,* he retained it as only one mechanism among many, including "use and disuse of organs," "correlation of parts," random or "neutral" developments, and "sexual selection."

As the surviving cooriginator of natural selection, Wallace attempted to undo his senior partner's waffling and reinstate their original insight. He summed up his views in an 1889 treatise ironically entitled *Darwinism,* an attempt to reestablish natural selection as the key to understanding evolution.

But Wallace's adamant selectionism stopped short at the human brain, which he insisted was miraculously endowed with language, soul and intelligence "by the unseen world of Spirit." Nothing could have been farther from Darwin's view; he had told Wallace that to make man an exception to natural processes would "completely murder" their theory.

Although some writers discussed "Wallaceism" seriously, most did not. Instead his strict selectionism

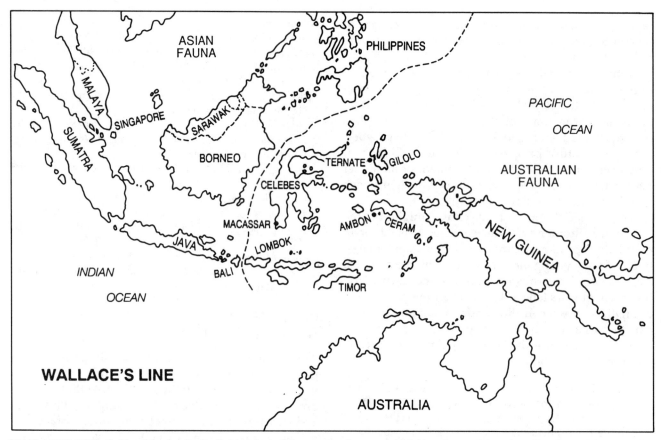

WALLACE'S LINE divides island animals derived from Asian species (western side) from those of the Australian fauna (eastern side). Alfred Russel Wallace deduced this natural boundary on the basis of animal distribution; a century later geologists confirmed that it is the perimeter of the active Indo-Australian plate.

(joined to Mendelian inheritance) become known as "neo-Darwinism," just as Wallace had intended. Science ignored his exemption of the human mind from natural selection, though Spiritualists and theologians seized upon it.

Wallace continued to be active and productive well into his ninetieth year, railing against vaccination as a "fraud," promoting socialism and defending psychic "mediums." After a lifetime of travel through exotic jungles, he cheerfully anticipated that exploring the spirit world would be his greatest adventure.

See also "DELICATE ARRANGEMENT"; WALLACE, ALFRED RUSSEL.

WALLACE'S LINE
Landmark in Zoogeography

Alfred Russel Wallace (1823–1913), the talented English naturalist who codiscovered the theory of evolution by natural selection, has often been obscured by Charles Darwin's shadow. But there is another monument to his brilliance that can still be seen today on every geologist's map of the world—Wallace's Line.

After years in the Amazon and Malaysia observing and collecting wildlife, Wallace attempted to make sense of the facts of animal geography. While he was exploring the vast 2,500 mile chain of islands known as the Malay Archipelago, he noticed similarities and differences in species that suggested a pattern and also an ever-diminishing number of species as he traveled farther from the mainland peninsula.

He wrote in *The Malay Archipelago* (1869), "I have arrived at the conclusion that we can draw a line among the islands, which shall so divide them that one-half shall truly belong to Asia, while the other shall no less certainly be allied to Australia."

Wallace's Line runs along a narrow strait between the islands of Bali and Lombok, between Borneo and the Celebes. To the east of the line are animals that came from the Australian continent, while west of it the fauna is derived from Asia.

In *The Geographical Distribution of Animals* (1876), Wallace worked out the theory more completely, and it became the founding book of zoogeography, the science of animal distribution. Islands west of the Line were once connected, he thought, but even then the eastern islands were isolated from the western by a deep channel.

Wallace had no way to observe the sea floor directly, but from his wildlife observations he deduced that the eastern islands must have been separated from the western groups for much longer than any individual islands were separated from each other. A hundred years later, geologists and oceanographers, armed with underwater mapping equipment, found the proof. Wallace's Line lies precisely on the perimeter of an area of intense crustal activity, the Indo-Australian plate.

See also PLATE TECTONICS; WALLACE, ALFRED RUSSEL.

"WALLACE'S PROBLEM"
Evolution of Human Brain

Alfred Russel Wallace, Charles Darwin's "junior partner" in discovering natural selection, had a disturbing problem: He did not believe their theory could account for the evolution of the human brain.

In the *Origin of Species* (1859), Darwin had concluded that natural selection makes an animal only as perfect as it needs to be for survival in its environment. But it struck Wallace that the human brain seemed to be a much better piece of equipment than our ancestors really needed.

After all, he reasoned, humans living as simple tribal hunter-gatherers would not need much more intelligence than gorillas. If all they had to do was gather plants and eggs and kill a few small creatures for a living, why develop a brain capable, not merely of speech, but also of composing symphonies and doing higher mathematics? Why indeed evolve a brain capable of formulating a theory of evolution?

Wallace did not share in the racist views of his time that held that tribal peoples had brains intermediate between apes and "civilized" peoples. During his travels through Malaysia and South America, he had gained great respect for the character and intelligence of nonliterate peoples. "Natural selection," he wrote, "could only have endowed the savage with a brain a little superior to that of an ape, whereas he actually possesses one but very little inferior to that of the average member of our learned societies."

But since he thought the human brain is everywhere "an instrument . . . developed in advance of the needs of its possessor," Wallace postulated a "spiritual" dimension in human evolution. While our bodies had been shaped by natural selection, he thought, at a crucial moment there had been divine intervention to expand the human brain.

There exists in man, Wallace wrote, "something which he has not derived from his animal progenitors—a spiritual essence or nature, capable of progressive development through conscious internal struggle." Although he insisted we evolved from ape-like creatures through natural selection, yet he contended that human intellectual and moral faculties "can only find adequate cause in the unseen universe of Spirit." (Not a conventionally religious man, Wallace was a staunch believer in Spiritualism; he believed he had witnessed disembodied spirits and other paranormal phenomena at seances.)

When Wallace first expressed this attempt to wed science and spirituality, Darwin wrote him: "I differ grievously from you; I can see no necessity for calling in an additional and proximate cause in regard to man . . . I hope you have not murdered too completely your own and my child." In Darwin's view, natural selection was a sufficient explanation for the emergence of the human mind, given slow, gradual evolution over millions of years.

Since Wallace's time, we have additional evidence to ponder. Both the australopithecines—bipedal hominids with brains a third the size of our own—and the larger-bodied *Homo erectus*, with intermediate-sized brains, have been discovered. Nevertheless, Wallace's problem remains unsolved; the emergence of the human mind is still a mystery. Stephen Jay Gould offers the hypothesis that many "higher" functions of the brain came along with its adaptation to more complex social behavior, including language. If you design a computer to handle business accounting, he argues, its very structure may also be capable of word processing or working out musical harmonies—side effects of its basic design for other functions.

For further information:

Wallace, Alfred Russel. *Social Environment and Moral Progress*. New York: Funk & Wagnalls, 1913.

———. *The Wonderful Century*. New York: Dodd, Mead, 1904.

WAR OF THE WORLDS

See WELLS, H. G.

WASHOE
First Signing Chimp

R. Allen and Beatrice Gardner, a married team of behavioral psychologists at the University of Nevada, were convinced chimpanzees were capable of language. Past attempts to teach them spoken words had failed, but the Gardners thought they could succeed with visual symbols. In a pioneer experiment, they taught a young female chimp to communicate in American Sign Language (Amslan) and compared her progress to that of deaf children.

Named for the Nevada county in which she was raised, Washoe had been captured as in infant in

Africa and reached the Gardners in June 1966. Her trainers took care never to speak around her, except through signs and gestures; they taught her names of things by showing the object, then immediately arranged her hands in the appropriate sign.

By the end of three years, Washoe could make 132 signs and had used three or more of them in 245 combinations, such as "hug me good" and "you tickle me." She had also become something of an ape celebrity, attracting popular interest and inspiring similar research projects into apes' capacity for language and conceptualization.

Dr. Roger Fouts, who had been the Gardners' chief assistant, took Washoe with him to the Institute for Primate Studies in Norman, Oklahoma in 1971. There, Washoe saw other chimpanzees, whom she called (in signs) "black bugs," and a swan "water bird."

After losing two infants, Washoe adopted Loulis, a one-year-old male chimp, and began signing to him. Loulis learned the gestures quickly. Fouts called it "the first case of cultural transmission of a language between generations" of chimpanzees. But other researchers were not convinced. A few years later, psychologist Herbert Terrace questioned whether the chimp's responses constituted real language behavior, or an ambiguity produced by flawed research design. The "ape language controversy" had begun.

See also APE LANGUAGE CONTROVERSY; NIM CHIMPSKY.

WATER BABIES (1863)
Classic Evolutionary Parody

Reverend Charles Kingsley, a clergyman, naturalist, poet, social reformer and novelist (*Westward Ho*, 1855) penned *The Water Babies* in 1862 as a present for his youngest son, then four years old. The tale of Tom, a child chimney sweep who is magically transformed into a "water baby" became a children's classic, reprinted for more than a century. But few modern readers (or their children) realize that it is also a spoof of some of the major issues of Victorian science and Darwinian evolution.

When Tom is transformed into a "water baby" (four inches long with external gills), he warns the reader not to decide such evolutionary "degeneration" is impossible. Scientific "experts" of his day, he complained, were supposed to cultivate open minds towards natural phenomena, yet they were always imposing their own limitations of imagination on nature.

Richard Owen, the British Museum's eminent anatomist, for example, had flatly discounted numerous sailors' reports of sea monsters, insisting such creatures could not possibly exist. But Owen also believed Charles Darwin's theories of evolution were sheer fantasies. Striking a blow for the evolutionists, Thomas Huxley had publicly announced a structure Owen insisted was unique to the human brain did not exist. And Huxley, a notorious disbeliever in churchly religion, thought Kingsley's deepest held beliefs about the liberation of the soul at death was a delusion.

All these issues of science, belief and dogma were in Kingsley's mind as he wrote his children's story. Again and again he made the point that no one was qualified to deny that "water babies"—or unknown animal species, or human souls—could exist:

> How do you know that? Have you been there to see? And if you had been there to see, and had seen none, that would not prove that there *were* none . . . And no one has a right to say that no water babies exist till they have seen no water babies existing, which is quite a different thing, mind, from not seeing water babies.

If they really exist, the young reader objects, someone would have caught one, put a story in the newspaper, "or perhaps cut it into two halves, poor little thing, and sent one to Professor Owen, and one to Professor Huxley, to see what they would each say about it." The author replies, "You must not say that this cannot be, or that is contrary to nature. You do not know what nature is, or what she can do."

> Nobody knows, not even Sir Roderick Murchison, or Professor Owen, or Professor Sedgwick, or Professor Huxley, or Mr. Darwin . . . They are very wise men; and you must listen respectfully to all they say. But even if they should say, which I am sure they never would, "That cannot exist. That is contrary to nature," you must wait a little, and see; for perhaps even they may be wrong.

The joke was that each one of them had indeed made contradictory pronouncements on what was "contrary to nature." Kingsley also advanced his favorite theory of degeneration—that evolution doesn't necessarily imply progress. He tells of a group of children who "by doing whatever they like" gradually lost the power to speak, and eventually turned into monkeys, which were shot by the African explorer Paul de Chaillu.

If one supposes that such a marvelous transformation cannot take place, that things cannot degrade or change downwards:

> does not each of us, in coming into this world, go through transformations just as wonderful as that of a sea egg or butterfly? And do not reason and analogy as well as Scripture, tell us that that transformation is not the last? . . . till you know a great deal more about nature than Professor Owen and Professor

DISGUISED ALLEGORY about science, Rev. Charles Kingsley's children's book *Water Babies* remained popular for almost a century. In one of the original illustrations, sinister zoologists resemble Professors Richard Owen and Thomas Henry Huxley (with magnifying glass).

Huxley put together, don't tell me about what cannot be, or fancy that anything is too wonderful to be true.

At the same time, Kingsley makes it clear that there are limits to his own credulity, which did not extend to the then-current fad for Spiritualism. Here he aligned himself with Darwin and Huxley, stating that a natural wood table is in itself more wonderful "than if, as foxes say, and geese believe, spirits could make it dance, or talk to you by rapping on it."

An illustration in the book shows Professors Huxley and Owen examining the bottled water baby with huge magnifying glasses. When Huxley's grandson Julian was five years old, in 1892, he saw that picture and requested an expert opinion on this matter of natural history:

DEAR GRANDPATER—Have you seen a Waterbaby? Did you put it in a bottle? Did it wonder if it could get out? Can I see it some day?—Your loving JULIAN

Huxley sent back a laboriously hand-lettered reply, legible to the precocious boy and a complete departure from his usual barely decipherable scrawl:

MY DEAR JULIAN—I never could make sure about that Water Baby.
I have seen Babies in water and Babies in bottles; the Baby in the water was not in a bottle and the Baby in the bottle was not in water.
My friend who wrote the story of the Water Baby was a very kind man and very clever. Perhaps he thought I could see as much in the water as he did—There are some people who see a great deal and some who see very little in the same things.

When you grow up I dare say you will be one of the great-deal seers, and see things more wonderful than Water Babies where other folks can see nothing.

Julian did grow up to be a distinguished evolutionary biologist, carrying on his grandfather's work in philosophy as well as science. *Evolution; the Modern Synthesis* (1942) by Sir Julian Huxley is one of the founding documents of the modern Synthetic Theory of evolution, even coining the name by which it has become known.

Strangely enough, the first biological experiments that brought fame to Julian Huxley were on the axolotl, a salamander that is a real-life counterpart to Kingsley's imaginary water baby. Like Tom, the axolotl never becomes an adult, but retains the external gills of a juvenile all its life. Huxley induced development into an adult form by feeding hormones to the creatures. They sprouted limbs and lost their gills, inspiring wild newspaper stories about the prospect of artificially speeding up human evolution.

See also HUXLEY, SIR JULIAN; HUXLEY, THOMAS HENRY; NEOTENY; OWEN, SIR RICHARD.

WATSON, JAMES D.

See GENE MAPPING.

WEGENER, ALFRED

See CONTINENTAL DRIFT; PLATE TECTONICS.

WEIDENREICH, FRANZ

See PEKING MAN.

WEISMANN, AUGUST (1834–1914)
Father of Somatic Mutation Theory

August Weismann pioneered scientific knowledge of sex, death, embryology and aging: some of the most profound questions in biology. A German evolutionist of enormous accomplishments, he was philosopher, experimentalist, naturalist, cellular physiologist and musician. Yet, like the farmer's wife in the nursery rhyme, he is chiefly remembered today for cutting off the tails of mice.

After 20 generations of amputations had no effect on mouse tails, Weismann considered his point made. If Jean-Baptiste Lamarck (and most naturalists, including Charles Darwin) had been correct about acquired characters being inherited, eventually some mice should have been born with shorter tails or none at all. Weismann's classic demonstration established once and for all that changes inflicted by environment on the bodies of individuals could not be passed on to their descendants.

These famous mutilated mice also led Weismann to the biological principle that sex (or germ) cells are separate from body (or somatic) cells. If any change (mutation) takes place, it has to occur first on the genetic level in order to be passed on.

In the turn-of-the-century confusion over mechanisms of inheritance, some biologists ("neo-Lamarckians") insisted that Weismann's theory was incompatible with Darwinism. However, Weismann saw no contradiction; in fact, independence of body cells from sex cells, he realized, gave crucial support to the idea of natural selection.

Among the critics of Weismann's experiments was the Irish playwright George Bernard Shaw, who insisted "any fool could have told him beforehand" that the mice's tails would not become reduced in their offspring. Neo-Lamarckian Shaw, with his "scientific religion" of Creative Evolution, claimed Weismann need never have taken a blade to the innocent rodents. After all:

> his experiment had been tried for many generations in China on the feet of Chinese women without producing the smallest tendency on their part to be born with abnormally small feet . . . [not to mention systematically performed] mutilations, the clipped ears and docked tails, practised by dog fanciers and horse breeders on many generations of the unfortunate animals they deal in.

Shaw argued that the proper experiment would have been to hypnotize "the mice into an urgent conviction that the fate of [their] world depended" on losing their tails. Soon, there would be a few mice born with little or no tail, and these would gain more food and mates, while "the tailed mice would be put to death as monsters by their fellows." What makes the experiment impossible, he admitted, is that "the human experimenter cannot get at the mouse's mind." Shaw had wit on his side, but history has vindicated Weismann's methods of doing science.

Having established that sex cells kept repeating and recombining themselves—and were almost immortal—Weismann wondered why the body cells (soma) had to grow old and die. His answer, which is being increasingly accepted a century later, is that death and aging were a byproduct of natural selection.

Where philosophers and most biologists had seen in death a necessary corollary of life, Weismann declared that the mortality of individual organisms was "not a primary necessity, but [it] has been secondarily acquired as an adaptation" to protect the nearly immortal genes.

Speaking to German naturalists in 1881, he suggested "that life is endowed with a fixed duration, not because it is contrary to its nature to be unlimited, but because the unlimited existence of individuals would be a luxury without a corresponding advantage" to the germ plasm.

Weismann also explored the possible origin of sex in the colonies of single-celled creatures. Unicellular life forms do not "die" when they divide into two, but continue on indefinitely, without undergoing the kind of deterioration that takes place in multicellular animals.

Early in the evolution of multicellular life, he believed, some cells in the colony began to specialize in providing nutrition, while others took over reproduction—the original separation of the body cells from the sex cells.

Weismann's advanced ideas about cooperative cell colonies evolving specialized functions, his correlation of individual longevity with the species reproductive cycle, and his sophisticated hypotheses on the origin of sex, aging and death were based on a lifetime of careful experiment and observation. (He peered through microscopes until he went blind, but continued his investigations with the help of his wife and sighted assistants.)

Once, in pondering the question of how life arose out of nonliving matter, he conceded the problem appeared to be "at least for the present, insoluble." But preoccupation with the great questions of biology did not produce despair in Weismann. "In fact," he wrote, "it is the quest after perfected truth, not its possession that falls to our lot, that gladdens us, fills up the measure of our life, nay hallows it."

Despite the stale textbook tradition of caricaturing Weismann as a mere clipper of mice tails, new generations of biologists exploring evolutionary mysteries will have to confront these deeper dimensions of his work.

One dedicated (though perhaps misguided) disciple in Germany has attempted to revive interest in Weismann's achievements by impersonating him—Victorian suit, whiskers and all. During the 1980s, audiences at scientific meetings were startled and amused by this latter-day clone, delivering his memorized Weismann lectures to all who would listen.

See also LANKESTER, E. RAY; NEO-DARWINISM; "SELF-ISH GENE"; SEX, ORIGIN OF; SHAW, GEORGE BERNARD.

For further information:

Weismann, August. *Essays Upon Heredity and Kindred Biological Subjects.* 2 vols. Edited by E. B. Poulton, et al. Oxford: Oxford University Press, 1891–1892.

———. *The Evolution Theory.* 2 vols. (English translation). London: E. Arnold, 1904.

———. *The Germ Plasm: A Theory of Heredity.* (English translation). London: Scott, 1893.

WELLS, H. G. (1866–1946)
Science Fiction Novelist

More than any other writer, H. G. Wells is the father of modern science fiction. His first "science romance" *The Time Machine* (1895) established a new popular genre, while his second, *War of the Worlds* (1898), is credited with literature's first true extraterrestrials. Humanoids from other planets were commonplace, according to science fiction historian Philip Klass, but Wells's Martians were "the first intelligent creatures who were clearly the product of an alien planetary environment and an alien evolution."

Wells was trained in science as a student of Professor Thomas Henry Huxley, the combative champion of Darwinian biology. After he had won fame as an author, Wells often acknowledged Huxley's influence; he was inspired not only by his teacher's enthusiasm for scientific ideas, but also by Huxley's skill at expressing them. Once he fondly recalled a lab session with "Darwin's bulldog":

> [Professor Huxley] spoke in clear completed sentences in words so well chosen that only afterwards did you realize how much that quiet leisurely voice had said and how swiftly it had covered the ground . . . Before him was a dead rabbit. . . . We were going to see . . . how it was made . . . how it was related to other [creatures]. That little limp furry body was the key by which we were to make our way towards the understanding of the whole incessant network of life . . .

Huxley's evolutionism colored Wells's literary works for years to come; for instance, his time-travelers encountered monsters accurately described according to paleontology. But much about the idea of evolution frightened him—and his audience.

Anxiety and reservations about Victorian science's quest for the laws of evolution became *The Island of Doctor Moreau* (1896). A mad scientist invents a serum that speedily "evolves" other mammalian species into humans. Setting himself up as their God, Dr. Moreau becomes Giver of the Law to the beast-men he has created. ("Thou shalt never attack humans" . . . "Thou shalt not go on all fours.") In the end, the doctor cannot control his world, and the island of evolutionary monsters is consumed in fiery self-destruction.

In *Man of the Year Million* (1893), Wells saw future evolution producing humans with huge heads and eyes, delicate hands and much reduced bodies—an image appropriated countless times since by less imaginative writers.

Wells also produced a popular account of the rise and fall of civilizations (*An Outline of History*, 1920) to make history accessible to "the ordinary man." A few years later, he teamed with his old professor's grandson, Julian Huxley, to write a popular compendium on evolutionary biology, *The Science of Life* (1925).

In 1929, Wells joined the editorial board of an independent literary magazine, *The Realist*, which lost money but made reputations. His fellow editors were Harold Laski, Arnold Bennett, Rebecca West and the brothers Julian and Aldous Huxley.

Wells is perhaps best remembered for *War of the Worlds*, which endured as an excellent radio play (1938) and Hollywood film (1953). A terrifying fantasy of planetary conquest, it features strange Martian creatures arriving in advanced spaceships equipped with unbeatable weapons. Although it is no longer obvious, the story's power has its roots in Victorian evolutionism.

In the 19th century, European colonial powers invaded far-flung lands, subduing native peoples with their superior technology and weapons. English explorers were fond of describing the awe of "savages" towards their sailing ships and justified atrocities of colonialism with a crude Social Darwinism. Englishmen were convinced they were "the fittest" beings on Earth; by "outcompeting" native peoples, they were only obeying the Darwinian laws of nature.

Such an excuse was often given, for instance, for the brutal slaughter of the Tasmanians. They were considered a "Stone Age people," who should have had the good manners to become extinct centuries ago. But some day soon, Wells was warning, "civilized" men might be on the receiving end of the terror. And, according to the prevailing view of the "struggle for existence," the aliens would be perfectly

BEAST-MEN were part of speeded-up evolution experiment in science fiction thriller *The Island of Doctor Moreau*, written by H. G. Wells in 1896. This scene is from the 1977 Hollywood version, which starred Burt Lancaster as Moreau.

justified in using their advanced ships and weapons to colonize the Earth.

His analogy touched a deep guilt in Western culture. Even 40 years after the novel first appeared, when the Orson Welles version was broadcast in America (Halloween Eve, 1938), thousands called the station to ask if the alien invasion was real. One and a half million people fled their homes in panic, convinced the day of reckoning had finally arrived.

An idealist as well as futurist, Wells could never resolve the conflict between his utopian dreams and Darwinian reality. In one of his last books, *Mind At the End of Its Tether* (1945) he warned that man is doomed unless he can adapt to his own advancing technology, a new element in human and planetary history. "We must stop war," he said, "before it stops us."

In *All Aboard for Ararat* (1941), Wells allegorized the aloneness of modern man in a Darwinian universe. There is to be a new cataclysm, and God asks Noah to build another ark. But Noah agrees only on one condition: This time God must ride as a noninterfering passenger, while man takes charge of his own (and the planet's) destiny. Thomas Henry Huxley would have approved.

See also HUXLEY, SIR JULIAN; HUXLEY, THOMAS HENRY; (THE) LOST WORLD.

For further information:

Wells, H. G. *Experiment in Autobiography.* New York: Macmillan, 1934.

Wells, H. G., and Huxley, Julian. *The Science of Life.* Garden City: Doubleday, 1929.

WELLS, WILLIAM CHARLES (1757–1817)
Early Natural Selectionist

Charles William Wells, the son of Scots immigrants to Charleston, South Carolina, came up with the idea of natural selection while Charles Darwin was still a young child. In 1813, four years before Wells's death, he read a paper on the subject to the Royal Society of London, applying it to the origin of the human races.

After completing medical studies at Edinburgh, Wells had returned to Charleston, where he pursued a practice and developed his interests in botany and other scientific subjects. The American Revolution sent him packing to London in 1784; soon he became established as a physician-scientist and was elected to the Royal Society.

He won recognition for his classic explanations of binocular vision ("Essay on Single Vision with Two Eyes," 1818) and why leaves were wet in the mornings ("Essay on Dew," 1814). His explanation of the formation of dew was published posthumously in

1818, together with the strange paper that contains his ingenious natural selection theory.

In this "Account of a Female of the White Race of Mankind, Part of Whose Skin Resembles That of a Negro," Wells described patches of darkly pigmented skin on one of his woman patients. Black skin in Africans, he concluded, "is no proof of their forming a different species from the white race." Also, he thought this example showed that—contrary to common opinion—a hot climate was not necessary to produce black skin. (It had been thought "Negro" skin had been deeply tanned under tropical sunlight and the accumulated color passed on to succeeding generations.)

Wells's ideas seem amazingly ahead of his time. He knew that Africans were resistant to certain tropical diseases that killed Europeans easily and quickly. Perhaps, he thought, such resistance was correlated with dark skin:

> Of the accidental varieties of man, which would occur among the first few scattered inhabitants of [Central] Africa, some would be better fitted than others to bear the diseases of the country [and might also be dark.] This race would consequently multiply, while the others would decrease, not only from their inability to sustain the attacks of disease, but from their incapacity of contending with their more vigorous neighbors . . . [A] darker and darker race would in the course of time occur, and as the darkest would be the best fitted for the climate, this would at length become the most prevalent, if not the only race, in the particular country . . .

Wells also compared artificial selection, as practiced by domestic animal breeders, with selection by natural forces. He also suggested that "accidental peculiarities" or "varieties" could become established in local populations by inbreeding, isolation and geographical barriers.

Ingenious as he was, Wells continued writing on a wide variety of disconnected topics and never sought to apply his very important insights to the data of zoology, botany, or paleontology. His limited development and application of natural selection caused no great stir in scientific thought, though it remains a striking precursor of the Darwin-Wallace theory.

Wells's paper was brought to Darwin's attention some years after he published *Origin of Species* (1859), and he acknowledged it in later editions. Patrick Matthew, an expert on tree farming, had also published an early version of the theory and was constantly insisting he deserved the credit and fame. When the story of Wells's even earlier version came to light, Darwin was delighted: "So poor old Patrick Matthew is not the first after all . . ."

See also MATTHEW, PATRICK; NATURAL SELECTION.

WHALE, EVOLUTION OF*
Origin of Largest Animal

Extremes of adaptation—such as the whale—provoke wonder about how such a creature could have evolved. Sometimes larger than a herd of elephants, this intelligent mammal feeds on tons of tiny plants and animals (plankton) it extracts from seawater. Since it is air-breathing, warm-blooded and milk-giving, it must have developed from land animals in ancient times, then gone back to the sea. But 150 years ago, who could imagine how such a transformation could come about?

Charles Darwin could. He had noticed in a traveler's account that an American black bear was seen "swimming for hours with widely open mouth, thus catching, like a whale, insects in the water." If this new food-getting habit became well established, Darwin said in the *Origin of Species* (first edition, 1859): "I can see no difficulty in a race of bears being rendered, by natural selection, more and more aquatic in their structure and habits, with larger and larger mouths, till a creature was produced as monstrous as a whale."

"Preposterous!" snorted zoologists. Such an example, they thought, sounded so wild and far-fetched it would brand Darwin as a teller of tall tales. Professor Richard Owen of the British Museum prevailed on Darwin to leave out the "whale-bear story," or at least tone it down. Darwin cut it from later editions, but privately regretted giving in to his critics, as he saw "no special difficulty in a bear's mouth being enlarged to any degree useful to its changing habits." Years later he still thought the example "quite reasonable."

Transitional forms have been scarce, but a few suggestive fossils were recently discovered in India of a four-legged land mammal whose skull and teeth resemble whales. And, during the 1980s, serum protein tests were made on whales' blood, to compare it with the biochemistry of other living animal groups. The results linked them not to bears or carnivores, but to hoofed animals (ungulates). Forerunners of whales were closely related to the ancestors of cattle, deer and sheep!

Such a conclusion fits with the general behavior of the great baleen whales, who move in pods or herds and strain the sea for plankton; they are, like antelopes or cattle, social grazers. Some species, such as the killer whales, which hunt in packs, long ago diversified into meat-eaters.

But even the "grazing" whales exhibit a potential for cooperative hunting. Recent observers describe groups of whales creating an immense "bubble trap" to gather great masses of plankton. A few whales dive down deep, form a circle and exhale a column of bubbles. Their cohorts position themselves vertically, near the surface, where they can feast on the concentrated harvest as it rushes up toward them.

As new facts emerge about how amazing whales really are, Charles Darwin's "whale-bear" story no longer seems like the product of a runaway imagination.

See also ADAPTATION; EXAPTATION; TRANSITIONAL FORMS.

WHIG HISTORY
Selective Hindsight

Sir Charles Lyell, the founder of uniformitarian geology, predicted dinosaurs would someday reappear on Earth. Charles Darwin, the architect of natural selection, believed in the "Lamarckian" inheritance of acquired characteristics. And Alfred Russel Wallace saw no difficulty in fusing natural selection and Spiritualism to explain human evolution.

Science is full of dead ends, strange combinations of ideas, which later seem incompatible, and concepts that die and reemerge in totally different contexts. It is influenced by social, political and artistic environments and quirks of personality no less than other human endeavors.

"Whig history" is the fallacy of relating past ideas to those of the present without regard for context, culture or the state of knowledge at the earlier time. The phrase was coined by British historian Herbert Butterfield (1931) to describe colleagues who viewed all constitutional history as a progressive movement toward their own political stance. Every prominent figure over the centuries was seen as either a "forward-looking" liberal (Whig) or "backward" conservative on the road inevitably leading to representative government.

Butterfield himself defined it as "the tendency in many historians to write on the side of Protestants and Whigs, to praise revolutions provided they have been successful, to emphasize certain principles of progress in the past and to produce a story which is the ratification if not the glorification of the present."

"Whiggish" history of science rates each scientist according to his "contribution" to currently held concepts, ignoring all other aspects of his work. More often than not, similarity to modern thought is only superficial; it had a totally different meaning in its time and place. Ernst Mayr, Stephen Jay Gould and Niles Eldredge all complain of too much "Whig History" in science writing and insist that earlier biological theorists must be considered with reference to the context of their ideas, state of knowledge and cultural environment.

For further information:
Butterfield, Herbert. *The Origins of Modern Science*. London: G. Bell & Sons, 1957.
———. *The Whig Interpretation of History*. New York: Norton, 1931.

WIDOWBIRD

See SEXUAL SELECTION.

WIGGAM, ALBERT E. (1871–1957)
Eugenics Advocate

If Francis Galton was the prophet of eugenics, Albert Wiggam was his messenger. As journalist, lecturer and author, Wiggam enjoyed tremendous success in the 1920s by urging scientific control of future human evolution to prevent "degeneration."

His best-selling book of 1923, *The New Decalogue of Science*, trumpeted a secular religion. (The "Old" Decalogue is the biblical Ten Commandments.) Morality, human happiness, harmony with nature and the salvation of the world, Wiggam preached, were obtainable through the "scientific" management of human populations. Eugenics, he thought, was not a sinister plot to create a master race, but "simply the projection of the Golden Rule down the stream of protoplasm."

Echoing one of Charles Darwin's fears, Wiggam thought "Civilization is making the world safe for stupidity" by allowing millions of genetically "defective" individuals to survive and reproduce. If God were to return to Earth, he would surely issue a "new biological Golden Rule, the completed Golden Rule of Science. Do unto both the born and the unborn as you would have both the born and the unborn do unto you." In fact, he thought, God's revelations no longer come through visions or burning bushes, but just as surely through the "microscope, the test tube, and the statistician's curve."

Despite his breezy and inaccurate approach to biology, writes historian Daniel J. Kevles, "Wiggam was pro-science, pro-biology, pro-evolution. In the era of the Scopes trial, scientists no doubt forgave him his errors because of the banner he carried." In post-World War I America, Wiggam's "secular religion" became a popular craze with its hopeful emphasis on progress and the rational creation of a social utopia.

See also EUGENICS; GALTON, SIR FRANCIS; HALDANE, J. B. S.; LEBENSBORN MOVEMENT.

WILSON, EDWARD

See BIOPHILIA; KIN SELECTION; SOCIOBIOLOGY.

WOLTMANN, LUDWIG (1871–1907)
German "Racial" Anthropologist

Ludwig Woltmann, physician and author, was a very influential writer on the subjects of eugenics, racial anthropology and Social Darwinism around 1900. He was the most important German representative of Count de Gobineau's theory of the Nordic race and a disciple of famed biologist Ernst Haeckel's evolutionary religion. [See MONISM.]

In 1900, Woltmann submitted an essay on "Political Anthropology" to a contest of Social Darwinist writing sponsored by the industrialist Alfred Krupp. Although Professor Haeckel was one of the judges, the first prize went to another of his disciples. Woltmann was given fourth prize, but angrily withdrew from the contest, making a permanent break with Haeckel, his old mentor.

As a result of the publicity surrounding this dispute, Woltmann became famous and published the book on his own in 1903. A year earlier, in 1902, he had begun the *Political Anthropology Revue*, in which he campaigned to prevent the "deterioration of the Nordic Race" and to maintain its supremacy—by force, if necessary.

During the first few years of the 20th century, Woltmann published a series of books on race and politics that gained a large audience in Germany. Although he criticized Haeckel's bias against socialism, he accepted his teacher's basic view that a universal law of evolution operates in society as well as in nature.

In his attempt to wed the ideas of Haeckel with those of Karl Marx, he transformed the Marxist class struggle into one of worldwide racial conflict. Germans, he thought, were the highest species of mankind, whose "perfect physical proportions" expressed a heightened spirituality and inner superiority. It was their duty to avoid the "biological deterioration" that would come of "mixing" with other races. According to historian Dan Gasman, Woltmann followed Haeckel closely in teaching that "life was a constant struggle for existence and racial purity, and sought to rearm Germany against biological decay." The extent of Woltmann's influence is impressive, since he died prematurely at the age of 36 and never lived to see many of his racial ideas adopted and implemented by the Nazis years later.

See also ARYAN "RACE," MYTH OF; HAECKEL, ERNST; MONIST LEAGUE; SOCIAL DARWINISM.

For further information:

Gasman, Daniel. *The Scientific Origins of National Socialism.* London: Macdonald, 1971.
Mosse, George L. *The Crisis of German Ideology: Intellectual Origins of the Third Reich.* New York: Grosset & Dunlap, 1964.
Woltmann, Ludwig. *Political Anthropology.* Leipzig: Daner, 1903.

WOODPECKERS

See ADAPTATION; SOCIAL BEHAVIOR, EVOLUTION OF.

WORMSTONE
Measuring Worm's Work

For his last book *On the Formation of Vegetable Mould by Earthworms* (1881), Charles Darwin investigated how fast earthworms could bury things. Objects on the surface tended to sink and become buried over time, he noticed. A field covered with jagged flints near his home gradually became smooth turf over 35 years.

He put the question to his son Horace: Could he construct a device to measure the rate of the worm's work? Horace gave the matter some thought and rose to the challenge.

A large, heavy, circular stone would serve as the object to be buried. In the center, which was cut out like a doughnut, he placed a rugged steel ruler. Over the years, the descent of the stone into the ground could be measured by the protruding calibrated rod.

Horace went on to found the Cambridge Scientific Instruments Company, which still exists a century later. One of Horace's wormstones can still be seen in the garden at Down House, the Darwin home that is now a public museum of his life and works.

WORMSTONE can still be seen in Darwin's garden at Down House. Made by his son Horace, the simple device measures the rate at which stones become buried, an indicator of local earthworm activity.

See also DOWN HOUSE; EARTHWORMS AND THE FORMATION OF VEGETABLE MOULD.

WRIGHT, SEWALL (1889–1988)
Pioneering Population Theorist

After Charles Darwin, the day of the field naturalist went into decline. Now it was the turn of lab men and mathematicians (Darwin hated mathematics) to put evolutionary theory on a new foundation. Three brilliant men, working separately, were the architects of the new understanding. British biometricians J. B. S. Haldane and R. A. Fisher and the American Sewall Wright worked out the mathematical principles of population genetics, transforming Darwinian evolution into a 20th-century science.

Natural selection, as Darwin well knew, was incomplete without a theory of inheritance. But the rise of Mendelian genetics 20 years after his death was at first thought to disprove Darwinian ideas of continuously varying traits. Mendelists insisted there could only be large, discrete mutations, combining in established ratios. [See "BEANBAG GENETICS."]

Wright had studied with an experimental geneticist, William E. Castle, whose breeding experiments with hooded rats showed that inheritance in small, in-bred populations was a lot more complicated than just sorting discrete one-trait genes. For instance, the presence of "modifier genes," altered the expression of other genes. Also, there was the phenomenon of "epistasis"—genes at one position (locus) on the chromosome showing varying effects, depending on what genes are present at other loci.

All this "extra variation" posed a challenge to Wright. Was there a way, he wondered, to mathematically describe complex systems of gene interaction? In 1920, he developed and published a mathematical model for such a system; it became one of the founding documents of population genetics. Darwinism was not, it turned out, incompatible with genetics; it simply needed to be rescued from the "Beanbag" dilemma.

Wright continued to attack the problem of how genes behaved in populations and published his influential paper *Evolution in Mendelian Populations* in 1931. Individuals in large populations, he thought, are not likely to mate randomly with any other individual in the entire range. Instead, there are smaller populations ("demes") within the larger where effects of variation can be intensified.

Working with breeders of short-horned cattle, Wright found rapid evolution is more likely to occur in these small subpopulations; they provide an effective source of continuous variation for the larger population. Wright showed mathematically how selection pressure could cause adaptive variations in a small deme to spread rapidly through the larger population. He also pictured populations as "adaptive landscapes," with their gene frequencies distributed in "peaks" and "valleys."

Although Wright's name is linked in the history of science with those of J. B. S. Haldane and R. A. Fisher, the three did not view themselves as members of the same team. Each came at the problems of population genetics with a different perspective, and their bitter quarrels and fights still echo through the literature.

Wright labored for years on a definitive textbook, *Evolution and the Genetics of Populations* (1968–1978); the first volume was published when he was 79. He continued to write theoretical papers up to the time of his death in 1988, two years short of his 100th birthday.

See also HALDANE, J. B. S.; SYNTHETIC THEORY.

For further information:
Provine, William. *Sewall Wright and Evolutionary Biology.* Chicago: University of Chicago Press, 1986.
Wright, Sewall. *Evolution and the Genetics of Populations.* 4 vols. Chicago: University of Chicago Press, 1968–1978.

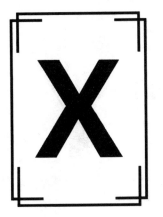

"X" CLUB
Victorian Scientific Elite

Between 1864 and 1893 in London, nine friends met regularly for dinner at six o'clock on the first Thursday of the month. Evolutionist Thomas Henry Huxley was the genial ringleader of this extraordinary dining club, comprised entirely of brilliant scientists, philosophers and mathematicians. Most had a strong interest in evolution, and were also personal friends of Charles Darwin, who rarely ventured into the city.

Members included physicist John Tyndall, director of Kew Gardens Sir Joseph Hooker, organic chemist Sir Edward Frankland, philosopher Herbert Spencer and prehistorian-naturalist-banker Sir John Lubbock. Since they had no name for the group, they called themselves simply "The 'X' Club"—the unknown quantity.

Their dinners were usually held directly before meetings of the Royal Society. "But what do they do?" asked a curious journalist. "They run British science," a professor replied, "and on the whole, they don't do it badly." Huxley insisted the group was purely social but, during its existence, the "X" Club provided three successive presidents of the Royal Society, and six presidents and several officers of the British Association. Collectively, they profoundly influenced the priorities of British science.

Once a year, they went to the countryside together on jovial excursions to which their spouses were invited. It became known as "the weekend for 'x's and their yv's."

"YERKISH" (ARTIFICIAL LANGUAGE)

See APE LANGUAGE CONTROVERSY; LANA.

YETI

See BIGFOOT.

Y-5 DENTAL PATTERN
Characteristic of Man-like Teeth

Teeth are very good diagnostic features of evolutionary relationships. They remain intact because of their protective enamel—usually longer than all other bones of the body—and their form is very distinctive. Closely related animals, such as members of the deer family or horse family, have molars that look very much alike. And the same is true for humans and apes.

Molars of monkeys are called *bilophodont*—they are shaped into two large "lophs" or cusps. A monkey running his tongue over his back teeth would feel a series of hills and valleys.

Among apes, men and other hominids, the grinding teeth exhibit what is known as the "Y-5" pattern. There are five cusps, arranged in a pattern that forms a "Y" between them. When a hominid runs his tongue over his back teeth, he feels one continuous valley between the cusps.

The Y-5 dental pattern is common to the living apes—chimps, gorillas, and orangs—as well as to man. In the fossil record, it is found in Neandertals, *Homo erectus*, and autralopithecines such as "Lucy," which go back three or four million years. Among the sketchy remains of early apes, the Y-5 makes its first known appearance among the dryopithecines, so it is also known as the "dryopithecine" dental pattern.

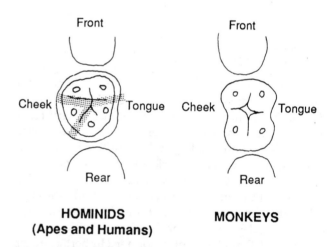

HOMINIDS (Apes and Humans)   MONKEYS

APES, HUMANS AND HOMINIDS share this distinctive structural pattern on the molar teeth; five cusps, with a Y-shaped valley between them. Monkey teeth never have more than four cusps and do not have the Y-shaped arrangement between them.

Dryopithecines or "oak-forest apes" go back 28 million years to the lower Miocene species *Aegyptopithecus*, whose fossil remains were found on the eastern edge of the Egyptian Sahara. Other dryopithecines have been dug up in East Africa, Europe, Turkey, India and southern China, indicating a successful and widespread group. The enigmatic giant ape *Gigantopithecus*, found in China, also shows the Y-5 pattern on its back teeth—which are five times the size of human molars.

See also DRYOPITHECINES; GIGANTOPITHECUS.

YOUMANS, EDWARD LIVINGSTONE (1821–1887)
American Promoter of Spencer

While essayist John Fiske and botanist Asa Gray led the campaign to make evolution respectable among intellectuals in the United States, Edward Livingstone Youmans became the self-appointed salesman of the science to America's reading public.

Youmans was an author of chemistry textbooks and a popularizer of the physical science and had good contacts in the publishing world. He talked the top men at D. Appleton and Company into publishing cheap American editions of the works of Herbert Spencer and Charles Darwin. He also persuaded them to put out a science magazine, *Appleton's Journal*, beginning in 1867, to carry popularizations of Darwin and Spencer written by himself and John Fiske.

In 1872, Youmans founded *Popular Science Monthly*, which today reports on the latest tools and do-it-yourself projects. A century ago, under Youmans, it presented a wide range of articles on such topics as hypnotism in animals, the causes of earthquakes, popularizations of evolution and essays attempting to reconcile science and religion.

Youmans also created a set of books by outstanding scientists: the red-covered International Scientific Series, which included titles by Darwin, Thomas Huxley, George Romanes, Edward B. Poulton, and other evolutionists. A century later they are sought-after collectibles.

Appleton grew to be a prosperous publisher by riding the crest of popular interest in evolutionism. Youmans was right in the middle of the excitement he helped create. He promoted Spencer's trip to America, where the British philosopher was entertained royally by steel mogul Andrew Carnegie. Spencer soon became more popular in America than he was in England. In 1871, Youmans wrote to Spencer:

> . . . things are going here furiously. I have never known anything like it. Ten thousand *Descent of Man* have been printed and I guess they are nearly all gone . . . The progress of liberal thought is remarkable. Everybody is asking for explanations. The clergy are in a flutter.

Youmans also reported to Spencer, with gleeful malice, that James McCosh, the leading Presbyterian theologian and president of Princeton, spoke before the clergy and "told them not to worry, as whatever might be discovered he would find design in it and put God behind it."

See also CARNEGIE, ANDREW; SPENCER, HERBERT.

ZDANSKY, O.

See PEKING MAN.

"ZETETIC" ASTRONOMY

See FLAT-EARTHERS.

ZINJANTHROPUS

See AUSTRALOPITHECUS BOISEI.

ZOO BREEDING PROGRAMS (OF ENDANGERED SPECIES)

See ISIS.

ZOONOMIA
Pioneering Evolutionary Treatise

Dr. Erasmus Darwin, grandfather of Charles, published a massive two-volume treatise, *Zoonomia, or the Laws of Organic Life* (1794–1796), that tapped a lifetime of medical practice and expounded his views on animal life. Its stated aim was to "unravel the theory of diseases," and it did exert wide influence on the treatment of fevers and insanity. But *Zoonomia* owes its secure place in the history of science to Erasmus Darwin's views on evolution, which in many ways anticipated those of his grandson.

As part of his attempt to establish "the laws of organic life," Darwin presented evidence to show that species do change, using examples from domestic breeds of horses, dogs, and pigeons. He discussed the function of many special organs among animals, and how they might have arisen as adaptations for feeding or defense. Most remarkable, he even proposed ideas similar to what his grandson would call sexual selection and natural selection.

For instance, Erasmus spoke of the antlers of male stags "formed for the purpose of combating other stags for the exclusive possession of the females . . . The final cause of this contest amongst the males seems to be, that the strongest and most active animal should propagate the species which should thence become improved." *Zoonomia* also proposes that in the "millions of ages before the commencement of the history of mankind . . . all warm-blooded animals have arisen from one living filament" with the "faculty of continuing to improve . . . and of delivering down these improvements by generation to its posterity, world without end!"

Desmond King-Hele, in his biography of Erasmus Darwin, claims *Zoonomia's* profound influence on Charles has been seriously undervalued. Charles abetted the neglect, for he never presented himself as building on his grandfather's work. In later life, however, he published a small memoir honoring Erasmus. "My scrutiny of the evidence," says King-Hele, "drives me to conclude that the Darwinian theory of evolution is very much a family affair, in which the shares of Erasmus and his grandson Charles are more nearly connected, and more nearly equal, than is usually supposed."

See also DARWIN, ERASMUS; TEMPLE OF NATURE.

For further information:

Darwin, Erasmus. *Zoonomia, or the Laws of Organic Life.* London: Johnson. *Vol. I,* 1794; *Vol. II,* 1796.
King-Hele, Desmond. *Erasmus Darwin.* New York: Scribners, 1963.

EVOLUTION UPDATE: RECENT DISCOVERIES

ALVAREZ THEORY
Discovery of a Candidate Crater

When Walter and Louis Alvarez proposed their "meteor impact" theory of dinosaur extinction during the 1980s, no crater huge enough—or from the appropriate time period—was known. Such a meteor would have had to strike the Earth about 65 million years ago, generating dust clouds sufficiently vast to lower global temperatures. Geologists all over the world began searching for evidence that such a meteor had ever collided with Earth. They were looking, among other signs, for significant quantities of iridium, a metal that is rare in the Earth's crust but common in meteors. Then, in 1992, a large area containing iridium, shocked quartz, and tectites (small blobs of glass, congealed from sprayed molten rock) was discovered on the north coast of Mexico's Yucatan Peninsula. The ancient crater, now entirely filled in by younger sediments, was named Chicxulub (cheek-soo-LOOB) after a nearby town. At 110 miles wide, it is the largest of Earth's 140 known meteor craters. Geologic samples tested by the argon-argon method (which measures the radioactive decay of potassium-40 into argon-40) yielded an impact date of 64.98 million years ago.

CLASSIFICATION

See DNA AMPLIFICATION.

COSQUER CAVE
Neptune's Undersea Gallery

In July 1992, archeologist Jean Clottes, of the French Ministry of Culture, announced an astonishing new find: a Mediterranean sea cave filled with rock paintings that rival those of Lascaux. Known as Cosquer Cave, after the diver who discovered it in 1985, its Paleolithic murals contain images of Pleistocene mammals as well as the first extensive drawings of sea creatures ever found. Before the cave was reported to the government, several divers died when they could not find their way out. Finally, a diving expedition headed by rock art expert Jean Courtin was sent to obtain photographs and samples for dating.

In addition to human silhouette handprints and drawings of ibex, bison, reindeer and horses, the team was greeted by painted images of seals, seabirds (auks), fish, and jellyfish or squid. The three-foot-wide underwater entrance to the cave can only be reached through a tunnel 121 feet below the Mediterranean, at the base of a cliff 7.5 miles southeast of Marseilles. Then, divers must proceed through a narrow, sloping tunnel for 400 feet until they emerge into a huge, partly air-filled chamber.

New radiometric tests of charcoal and pigments have established two periods of activity in the cave. The first, during which the handprints were made, dates back to about 27,000 years ago, making them the oldest known paintings in the world. The animal figures were incised and painted during the second period. Some of them dated to about 18,500 years ago, or about 1,000 years older than anything at Lascaux. Many of the original paintings were lost when much of the cave was submerged in salt water, but it is still spectacular. Since there is no aboveground access to the cave, however, it will never be opened to the public.

"DIMA"
New Baby Mammoth Found

A decade after the discovery of Dima, the baby mammoth found in 1977 near Magadan in Eastern Siberia, another almost intact infant was released from the ice during a spring flood. Frozen for at least 12,000 years, the new baby mammoth was first spotted in Western Siberia in September, 1988, by a freighter captain sailing off the east coast of the Yamal peninsula. Its well-preserved carcass was 3.5 feet long, covered in dark brown hair, and missing only the trunk and tail. Apparently about three months old at the time of death, the little mammoth may have fallen through ice and drowned, and then became frozen within an ice sheet.

DNA AMPLIFICATION
New Tool for Classification

A promising new tool for making species comparisons for classification is polymerase chain reaction (PCR), a technique that amplifies tiny specks of DNA. Michael Crichton and Stephen Spielberg's sci-fi film *Jurassic Park* (1993), in which dinosaurs are cloned from their amplified DNA, was inspired by PCR technology. While bringing extinct species back to life remains pure fantasy, amazing strides have been made in retrieving ancient DNA. Previously, no one had suspected that genetic material could persist for millions of years. During the past decade, however, it has been extracted from Egyptian human mummies 5 thousand years old, from mammoths and sabre-toothed cats 15 thousand years old, and from termites 30 million years old. (The mammoth's DNA seems to be equidistant between that of today's Indian and African elephants.) The very week that *Jurassic Park* was released, a team of scientists from California Polytechnic State University announced that they had amplified genetic material from a weevil preserved in amber for 130 million years. On June 10, 1993, the staid *New York Times* printed this front-page headline: DNA FROM THE AGE OF DINOSAURS IS FOUND. (Scientists insisted that the timing of their announcement with the movie's release was pure coincidence.)

Molecular biologist Svante Pääbo, of Munich's Zoological Museum, has commented: "I think it's kind of a renaissance for museums. You can obtain samples from extinct species, and you can look at populations over time, which is totally unique." Animals that were collected so that their pelts or feather patterns could be compared for classification, now have something even more valuable to offer. As Terry Yates, director of the National Science Foundation's biology program, put it: "PCR means all of a sudden that all these dead rats in museums are genetic goldmines."

GALAPAGOS ARCHIPELAGO
Older, Sunken Links

Darwin's beloved Galapagos islands, volcanic cones that rise out of the sea 600 miles west of Ecuador, posed a problem that the great evolutionist never confronted. Geologists had dated them from two to three million years old. But two biochemists at the University of California, Berkeley, thought that was far too young for evolution, operating at its usual slow pace, to have produced so many unique life forms on the islands. During the 1970s, biochemists Vincent Sarich and Jeffrey Wiles had compared the serum albumin of Galapagos land iguanas with those of marine iguanas, and concluded that they must have diverged from a common ancestor close to ten million years ago. In 1983, Sarich and Wiles published a paper entitled,

"Are the Galapagos Iguanas Older than the Galapagos?" in which they predicted that "drowned" islands would be found where the ancestral iguanas originally landed. The mystery was apparently solved in 1992, when David Christie of Oregon State University headed a team that dredged the sea floor for a month. The geophysicists found convincing evidence that the Galapagos chain once included several much older islands—about nine million years old—that have since sunk beneath the ocean. When he learned of the discovery, Sarich said he was "not at all surprised," and had been "pretty sure these sunken islands would eventually turn up."

HIMALAYAN FOSSIL HOAX
A Tangled Web

A prominent Indian geologist, Viswa Jit Gupta, published hundreds of journal articles over the last quarter of a century about his discoveries of ancient invertebrate fossils in the Himalayan region. The problem was that many of the fossils he reportedly found there are so distinctive that they could only have come from somewhere else: New York State, in one case, and Morocco in another. *Nature*, the British science journal, opined that Gupta's forgeries "will cast a longer shadow" than the famous Piltdown man hoax of 1912. It involves scores of scientific journal articles that have been frequently cited, enmeshing the phony data in the evolutionary literature on the Himalayan area. Dr. John A. Talent, an Australian paleontologist who uncovered the hoax, said the geology of the Himalayas "has been mucked up from one end to the other." When other scientists use Gupta's findings, he told the *N.Y. Times*, "they end up with highly distorted syntheses" and come to "weird conclusions." In one case, Talent charged that fossils reported by Gupta from India and Nepal—tiny jaws of segmented worms that lived 360 million years ago—actually came from Amsdell Creek in upstate New York. He also bought some fossil mollusc shells from Morocco that were virtually identical to those Gupta claimed he had found in the Himalayas. At first, Gupta denied all charges, claiming Talent was a jealous, malicious rival, and vowed to prove that his fossils were genuine. He never did, however, and in 1991 was suspended from his post as professor and director of the Institute of Paleontology at Panjab University in Chandighar.

HOMO ERECTUS
Earliest European Appearance

For many years, a major question about the dispersion of *Homo erectus* was: if the species appeared in Africa two million years ago, why did populations remain there for a million years and then migrate to Europe?

A fossil discovered in 1992 may make that question irrelevant. In that year, a German paleontologist named Antje Justus found a well-preserved *Homo erectus* fossil jaw, with teeth intact, in the medieval town of Smanisi, Georgia, in Western Asia near the Turkish border. Dated at 1.8 million years old—now the earliest known *Homo erectus* fossil found outside of Africa—it rested beneath the skull of a saber-toothed cat, in a site rich in stone tools and animal bones. If *Homo erectus* had indeed evolved in Africa about two million years ago, as many scientists believe, the Dmansi mandible indicates that it did not linger there long before wandering out across Europe. "This mandible might be a twin of the one from the Turkana Boy," says Alan Walker of Johns Hopkins Medical School, referring to the almost complete *Homo erectus* skeleton from Nariokotome, Kenya.

Some anthropologists suggest that the Turkana boy and other early *Homo erectus* fossils should be reclassified as a distinct species, *Homo ergaster*—a less specialized form known from other Turkana head fragments found in 1975. These skulls appear to be lighter, with higher vaults than later *Homo erectus*, and a cranial capacity of about 850 cubic centimeters. Ian Tattersall and Niles Eldridge of the American Museum of Natural History believe that *H. ergaster* (which means "work man," an acknowledgment that the hominid was a stone tool-maker) may have been ancestral to both the heavier boned, beetle-browed *Homo erectus*, and perhaps to a second line that led to ourselves, *Homo sapiens*.

KANZI
One Bright Bonobo

Kanzi, a pygmy chimp or bonobo, has emerged as the undeniable primate star of the 1990s—both for his language abilities and his capacity for making stone tools. Born at the Georgia State University Language Research Center, near Atlanta, Kanzi is the son of a bonobo named Matata, who was being used in communication experiments there by Sue Savage-Rumbaugh. Kanzi began to pick up use of the symbols on his own by watching the investigator's frustrating attempts to teach his mother, who never mastered them. Thus, he learned as humans do—by observation rather than conditioning. Although Kanzi cannot produce vocal speech, he points to symbols for the words on a board or punches them up on a special keyboard. By age four, his apparent understanding of 660 different English sentences compared favorably with the grammatical abilities of a two-and-a-half-year-old human child.

Kanzi's feats were performed under the most stringent controls ever attempted to eliminate the possibility of unconscious cues or bias from the experimenters (the "Clever Hans" effect). One-way mirrors prevented him from seeing who gave commands, while psychologist Rose Sevcik and others wore headphones so they could not tell whether responses came from Kanzi or from the human toddler with whom he was being compared.

In addition, Kanzi proved at least as adept as Abang, the Basel Zoo's orang-utan, in making and using stone tools. Sevcik and Nicholas Toth, a prehistorian from Indiana University, provided Kanzi with flint flakes and a tied-up box containing food or a key to a food box. Kanzi quickly learned to cut the string with a crude stone knife to get the prize. When he was shown how to hammer off sharp flakes for himself, he quickly acquired the skill. "When he formed his first flake," Sevcik reported, "he immediately vocalized excitedly. In a typical bonobo way he looked at it, picked it up, and instantly went over to cut the string." "For a Stone Age archeologist like myself," said Toth, "seeing this was almost like a religious experience." At Toth's urging, his university awarded Kanzi a prize for providing the year's greatest insight into technology.

LASCAUX CAVE

See COSQUER CAVE.

MIMICRY, BATESIAN AND MULLERIAN
Deceivers and Deceived

Nineteenth-century English naturalist Henry Walter Bates had demonstrated that a tasty butterfly species could protect itself by evolving colors that resemble a toxic species, a process known as Batesian mimicry. For 150 years, the case of the poisonous monarch butterfly and its harmless viceroy mimic has been a textbook classic. But a 1991 study by David B. Ritland and Lincoln P. Brower of the University of Florida in Gainesville revealed a surprise: when butterfly abdomens, without wings, were served up to birds, the predators found the viceroy just as unappetizing as the monarch.

Scientists, not birds, had been deceived. They assumed that the viceroy's orange warning colors were just a bluff, and that all butterflies had to acquire poisons from their food plants. Monarchs eat milkweeds to gain their toxins, while viceroys dine on harmless willows. But now it appears that viceroys manufacture noxious chemicals of their own. Fritz Muller, another nineteenth-century naturalist, first described how two or more equally distasteful butterfly species can gain greater protection from predators by evolving the same general appearance. So the viceroy may actually be a "Mullerian" mimic of the monarch. Brower suspects that the viceroy may have first evolved as a Batesian mimic, then "changed the rules of the game" and developed its own chemical defenses as well.

Another recent study has shown that the monarch butterfly gathers a second, totally different plant poison when its protective dose of milkweed cardenolide has worn thin after its annual southern migration. After the monarch's trek to Mexico, its poisons are effective against 35 out of 37 local birds and most rodents. However, one species of Mexican mouse can dine on monarch carcasses without ill effects. With this selective advantage, it has become the most successful mouse in the mountainsides of central Mexico.

MITOCHONDRIAL "EVE"
Fallen Once More

During the late 1980s, a handful of molecular biologists thought they had identified the mother of all mothers, popularly known as "Mitochondrial (MtDNA) Eve." She was an African woman who lived 200,000 years ago, from whom all living humans can trace genetic descent through the female line. Thousands of MtDNA samples and years of computer workups had pointed to her, according to geneticists at Allan Wilson's University of California laboratory. However, some stones-and-bones anthropologists were claiming that fossil evidence did not support a recent African origin for all the world's peoples. Mitochondrial Eve left many early human fossils from all continents unexplained.

Despite their uneasiness with the Eve hypothesis, most "bone men" and other anthropologists lack the technical expertise needed to process and analyze the genetic data. Then, in 1992, a journal editor asked geneticist Alan Templeton of Washington University to take a fresh look at how the Berkeley team reached their conclusions. What he saw knocked Eve out of her Garden.

First, Templeton found that the original "Eve" researchers had used their computer programs improperly. Their mitochondrial DNA had generated a parsimonious family tree all right, but there are millions of equally parsimonious family trees possible, depending on the number of computer runs and in what order the data are entered. A tree with African roots, it turned out, is no more probable than one with Asian or European roots. "The inference that the tree of humankind is rooted in Africa is not supported by the [genetic] data," Templeton wrote in the journal *Science*. There were other flaws in the research as well. The single strand of genetic information that was used can reveal only limited evolutionary information, and takes no account of the constant gene flow between local populations. Confronted with Templeton's critique, the "Eve" biochemists (except Wilson, who had

died) acknowledged the fatal flaws in their methods. If humankind had recent origins in Africa, it still remains to be proven.

RAIN FOREST CRISIS
Worse Than Feared

The world's tropical forests are vanishing at a much faster rate than scientists had expected, speeding up loss of wildlife habitat and increasing the danger of global warming. In a World Resources Institute report issued in 1990, the first new assessment of the problem in a decade, Allen Hammond reported that a total of 40 million to 50 million acres of forest disappear yearly—50 percent more than the United Nations estimated in 1980. "It's as if we cleared and burned the state of Washington in a year," he says. Local people usually believe they will reap large profits from the sale of timber and clearing for agriculture. However, they are often disappointed: only a very small proportion of the timber can be sold, and the land, once cleared, is very poor for agriculture. Leaders in destroying their forests are Brazil, India, Indonesia, Myanmar (Burma), Thailand, Vietnam, Philippines, and Costa Rica.

TASADAY TRIBE
Neither Stone Age Nor a Hoax?

Controversies continue to swirl about the "gentle Tasaday," the "primitive cave dwellers" of the Philippines. Initial studies claiming great antiquity for their language had to be revised; now linguists estimate only a few hundred years of divergence between their dialect and that spoken by nearby Manobo farmers. Millionaire "benefactor" of the tribe, Manuel Elizalde, insists he was maligned by political enemies who accused him of creating a hoax, and that his "Stone Age" tribe is genuine. When Elizalde first unveiled the Tasaday to the world in the early 1970s, fourteen scientists visited their rain forest home, and none pronounced them a fraud.

In 1988, at the World Congress of Anthropology, the original Tasaday researchers were criticized by their colleagues as romantics, in love with the idea of the noble savage. At that conference, anthropologist Gerald Berreman, of the University of California, Berkeley, said the Tasaday were "preposterous in their implausibility." They were represented to have had no hunting or fishing technology of any kind, no nets, baskets, or bags. No rituals, nor religious specialists, nor folklore. Their crude stone tools are unlike any known from past or present tribal peoples. "These are not only improbabilities," said Berreman, "but impossibilities. There's no group of people in the last 40,000 years that has been without these kinds of cultural features. . . . It sounds more like what a junior high school class assigned to invent . . . a primitive way of life, might have put together."

Nearby missionaries had originally thought the Tasaday were an invention of Elizalde's, but changed their opinion after meeting them. T'boli locals told their priests that the Tasaday do indeed speak a separate language that had been unknown to them. Meanwhile, the tribesmen themselves were inconsistent; they told ABC and British television they were not real cave people, yet insisted on Philippine television that they were genuine. Senate hearings in Manila, fostered by Elizalde, judged them authentic.

At the Summer Institute of Linguistics in Dallas, Texas, anthropologist Thomas Headland—a veteran fieldworker in the Philippines—has proposed a new answer to the riddle. The Tasaday, he believes, were part of the Manobo farming people who broke off and fled into the forest only 150 years ago. At that time, slavers were conducting raids all over the southern Philippines, and many groups hid in the forest. With no long tradition of stone-tool making or basketry, the Tasaday re-invented simple hunting techniques and crude tools. Whether the people have stories about their origins is still unclear. Still, some anthropologists remain convinced that they have an ancient history, while others consider them actors on Elizalde's payroll. And the subjects of all this acrimony still insist, more than twenty years later, "we are real Tasaday."

TYRANNOSAURUS REX
A Carnosaur Named Sue

The largest, most complete, and best preserved tyrannosaurus skeleton ever found was seized in May 1992 from the Black Hills Institute of Geological Research, a commercial fossil dealer that maintains a museum in Hill City, South Dakota. The fossil dinosaur is known as "Sue," an appropriate name when one considers the lawsuits it has inspired. Sue was discovered in 1990 by a party of Black Hills Institute collectors headed by Peter L. Larson, who spotted bones projecting from a cliff face as they prospected on a ranch. The rancher was paid $5 thousand for the right to excavate the fossil, but when its value of possibly $2 million was publicized, everyone in the area put in a claim. The rancher wanted more money, the Cheyenne River Sioux Tribe insisted the site was on their reservation land, and the Federal Government—called in to protect the Native Americans' interests—decided the tyrannosaurus was a national treasure. Sue was seized by the FBI, packed up in crates, and placed in locked storage in the name of the American people. Litigation continues.

WHALE, EVOLUTION OF
Molecular Genetics

In a new attempt to clarify the evolutionary relationships of living whales, biochemists Axel Meyer and Michel Milinkovich compared two genetic sequences from the mitochondria of sixteen species of whales. Their 1992 study found a surprisingly close genetic relationship between sperm whales and baleen whales, which had been thought to be long-separated groups. Also, they appear to have branched off comparatively recently—between 10 and 15 million years ago—from an unknown common ancestor. When the scientists compared the whale's genetic material with such other living mammals as donkeys, humans, sloths, and cows, they found that the cow was by far the whale's closest relative. If these studies are correct, cows are genetically much closer to whales than they are to such other ungulates as donkeys.

ILLUSTRATION CREDITS

AMERICAN MUSEUM OF NATURAL HISTORY: 10, left (Neg. # 311654); 10, right (#330591); 39 (#B15450); 42 (#Z99134); 50 (#ZA3077); 56, left (#34216); 82 (#31503); 103 (#325002); 193, left (#410783); 193, right (#41927); 223 (#3583); 257, top (#35827); 257, left (#B9442); 257, bottom right (#35799); 261 (#39442); 346 (#36106); 358, top (#31544722); 358, bottom (315286); 354 (#10672); 381 (#318686).

AUTHOR, PHOTOS BY: 118 (left); 144, 153, 195, 234 (both), 278, 282, 344, 394, 465.

AUTHOR'S COLLECTION: 7 (left), 13 (bottom), 24, 35, 43, 49, 57, 67, 78, 117, 118, 121, 150, 154, 161, 186, 191, 210, 213 (left), 235, 238, 239, 255, 266, 276, 284, 290, 306, 314, 331, 333, 317, 323, 325, 355, 361, 366, 378, 386, 398, 427, 447, 459, 462.

NANCY BURSON: 89, composite photo copyright by Nancy Burson.

CARNEGIE LIBRARY, PITTSBURGH: 73.

CIRCUS WORLD MUSEUM (Baraboo, Wisconsin): 197 (bottom)

MARGOT CRABTREE: 27, 28, 321 (right), 430: Copyright Margot Crabtree, from "Ancestors" exhibit.

CREATORS SYNDICATE INC.: 19: *B.C.* cartoon courtesy Johnny Hart.

STEVE AND SYVIA CZERKAS: 137 (right).

LOIS DARLING: 36: H.M.S. *Beagle* reconstruction and sketch by Lois Darling.

WALT DISNEY COMPANY: 104, 164: Copyright Walt Disney Co.; used by permission.

FACTS ON FILE: 91, 182, 194, 267, 456, 468.

STEPHEN JAY GOULD: 61, 92: Drawings by Marianne Collins; *It's a Wonderful Life.*

HAMILTON COLLEGE: 97: Hamilton Alumni Association.

INSTITUTE OF HUMAN ORIGINS, BERKELEY, Dr. Donald Johanson, Director: 5, 244, 285, 337.

RHODA KNIGHT KALT: 321 (left).

KLAGENFURT, AUSTRIA, CHAMBER OF COMMERCE: 145.

VLADIMIR KRB, TYRRELL MUSEUM, CANADA: 62.

CITY OF LONDON: 13 (left); 312 (right).

KENNETH LOVE AND NATIONAL GEOGRAPHIC TELEVISION SPECIALS: 21, 80: Copyrighted photos by Kenneth Love.

MIT PRESS: 362: Redrawn after Elliot Sober, *The Nature of Selection*

DR. MICHAEL AND BARBARA MACROBERTS: 75, 410.

JUDITH MOSES, ABC-TV: 429: Photo by Kristina Luz.

DOUG MURRAY ARCHIVES: 254, 335.

MUSEE DE L'HOMME, Paris, France: 194 (bottom), 146, 264, 431.

MUSEUM OF THE ROCKIES (Bozeman, Montana): 135 (top and bottom): Courtesy Dr. Jack Horner.

NASSAU COUNTY DEPT. OF RECREATION AND PARKS: 137 (left): Courtesy Mr. Tony Panzarella.

NATIONAL ZOOLOGICAL PARK (Washington, D.C.): 353.

NEW YORK PUBLIC LIBRARY PICTURE COLLECTION: 37, 171, 177, 184, 221, 225, 227, 243, 259, 348, 349, 405, 421, 423, 441, 453.

NEW YORK STATE HISTORICAL ASSOCIATION, Cooperstown, New York: 71

LONDON OBSERVER: 132

COLLECTION, STUART PIVAR: 2.

IAN REDMOND: 411: Copyrighted photo by Ian Redmond; used by permission.

SCIENTIFIC AMERICAN: 212: Copyrighted illustration by James F. Crow; used by permission.

FRIEDRICH SCHILLER UNIVERSITY, Jena, Germany: 54, 206.

STANFORD UNIVERSITY LIBRARY: 7 (right).

DR. SHIRLEY STRUM: 31 (left): Copyrighted photo by Marion Kaplan, Random House.

TRANSVAAL MUSEUM, Pretoria, South Africa: 56 (right).

UNIVERSAL PRESS SYNDICATE: 31 (left): Cartoon copyright by Gary Larson; used by permission.

UNIVERSITY OF THE WITWATERSRAND, Johannesburg, South Africa: 106.

ELISABETH VRBA: 83: Original phylogeny compiled by Elisabeth Vrba; used by permission. Graphic artist: Susan Hochgraf. 160: Photo by Alan C. Kemp. 440: Courtesy of Elisabeth Vrba.

DELTA WILLIS: 269: Copyrighted photo by Delta Willis; used by permission.

YALE UNIVERSITY PEABODY MUSEUM: 34, 95 (all), 387.

INDEX

Charles Darwin, from an engraving by Edward Whymper.

"You ask about my book, and all I can say is that I am ready to commit suicide; I thought it was decently written, but find so much wants rewriting, that it will not be ready to go to printers for two months, and will then make a confoundedly big book. [The publisher] will say that it is no use publishing in the middle of summer, so I do not know what will be the upshot; but I begin to think that everyone who publishes a book is a fool."

—Charles Darwin to Joseph Hooker
March 1875